New Frontiers in Integrated Solid Earth Sciences

International Year of Planet Earth

Series Editors:

The book series is dedicated to the United Nations International Year of Planet Earth. The aim of the Year is to raise worldwide public and political awareness of the vast (but often under-used) potential of Earth sciences for improving the quality of life and safeguarding the planet. Geoscientific knowledge can save lives and protect property if threatened by natural disasters. Such knowledge is also needed to sustainably satisfy the growing need for Earth's resources by more people. Earths scientists are ready to contribute to a safer, healthier and more prosperous society. IYPE aims to develop a new generation of such experts to find new resources and to develop land more sustainably.

For further volumes:
http://www.springer.com/series/8096

Sierd Cloetingh · Jörg Negendank
Editors

New Frontiers in Integrated Solid Earth Sciences

Springer

Editors
Prof. Dr. Sierd Cloetingh
VU University Amsterdam
Netherlands Research Centre
for Integrated Solid Earth Science,
Faculty of Earth and Life Sciences
De Boelelaan 1085
1081 HV Amsterdam
Netherlands
sierd.cloetingh@falw.vu.nl

Dr. Jörg Negendank
GeoForschungsZentrum
Potsdam
14473 Potsdam
Telegrafenberg
Germany
secretariat-ILP@gfz-potsdam.de

ISBN 978-94-017-7702-5 ISBN 978-90-481-2737-5 (eBook)
DOI 10.1007/978-90-481-2737-5
Springer Dordrecht Heidelberg London New York

Printed on acid-free paper

Springer is part of Springer Science+Business Media (www.springer.com)

Foreword

The International Year of Planet Earth (IYPE) was established as a means of raising worldwide public and political awareness of the vast, though frequently under-used, potential the Earth Sciences possess for improving the quality of life of the peoples of the world and safeguarding Earth's rich and diverse environments.

The International Year project was jointly initiated in 2000 by the International Union of Geological Sciences (IUGS) and the Earth Science Division of the United Nations Educational, Scientific and Cultural Organisation (UNESCO). IUGS, which is a Non-Governmental Organisation, and UNESCO, an Inter-Governmental Organisation, already shared a long record of productive cooperation in the natural sciences and their application to societal problems, including the International Geoscience Programme (IGCP) now in its fourth decade.

With its main goals of raising public awareness of, and enhancing research in the Earth sciences on a global scale in both the developed and less-developed countries of the world, two operational programmes were demanded. In 2002 and 2003, the Series Editors together with Dr. Ted Nield and Dr. Henk Schalke (all four being core members of the Management Team at that time) drew up outlines of a Science and an Outreach Programme. In 2005, following the UN proclamation of 2008 as the United Nations International Year of Planet Earth, the "Year" grew into a triennium (2007–2009).

The Outreach Programme, targeting all levels of human society from decision-makers to the general public, achieved considerable success in the hands of member states representing over 80% of the global population. The Science Programme concentrated on bringing together like-minded scientists from around the world to advance collaborative science in a number of areas of global concern. A strong emphasis on enhancing the role of the Earth sciences in building a healthier, safer and wealthier society was adopted – as declared in the Year's logo strap-line "Earth Sciences *for* Society".

The organisational approach adopted by the Science Programme involved recognition of ten global themes that embrace a broad range of problems of widespread national and international concern, as follows.

- Human health: this theme involves improving understanding of the processes by which geological materials affect human health as a means of identifying and reducing a range of pathological effects.
- Climate: particularly emphasises improved detail and understanding of the non-human factor in climate change.

- Groundwater: considers the occurrence, quantity and quality of this vital resource for all living things against a background that includes potential political tension between competing neighbour-nations.
- Ocean: aims to improve understanding of the processes and environment of the ocean floors with relevance to the history of planet Earth and the potential for improved understanding of life and resources.
- Soils: this thin "skin" on Earth's surface is the vital source of nutrients that sustain life on the world's landmasses, but this living skin is vulnerable to degradation if not used wisely. This theme emphasizes greater use of soil science information in the selection, use and ensuring sustainability of agricultural soils so as to enhance production and diminish soil loss.
- Deep Earth: in view of the fundamental importance of deep the Earth in supplying basic needs, including mitigating the impact of certain natural hazards and controlling environmental degradation, this theme concentrates on developing scientific models that assist in the reconstruction of past processes and the forecasting of future processes that take place in the solid Earth.
- Megacities: this theme is concerned with means of building safer structures and expanding urban areas, including utilization of subsurface space.
- Geohazards: aims to reduce the risks posed to human communities by both natural and human-induced hazards using current knowledge and new information derived from research.
- Resources: involves advancing our knowledge of Earth's natural resources and their sustainable extraction.
- Earth and Life: it is over two and half billion years since the first effects of life began to affect Earth's atmosphere, oceans and landmasses. Earth's biological "cloak", known as the biosphere, makes our planet unique but it needs to be better known and protected. This theme aims to advance understanding of the dynamic processes of the biosphere and to use that understanding to help keep this global life-support system in good health for the benefit of all living things.

The first task of the leading Earth scientists appointed as Theme Leaders was the production of a set of theme brochures. Some 3500 of these were published, initially in English only but later translated into Portuguese, Chinese, Hungarian, Vietnamese, Italian, Spanish, Turkish, Lithuanian, Polish, Arabic, Japanese and Greek. Most of these were published in hard copy and all are listed on the IYPE website.

It is fitting that, as the International Year's triennium terminates at the end of 2009, the more than 100 scientists who participated in the ten science themes should bring together the results of their wide ranging international deliberations in a series of state-of-the-art volumes that will stand as a legacy of the International Year of Planet Earth. The book series was a direct result of interaction between the International Year and the Springer Verlag Company, a partnership which was formalised in 2008 during the acme of the triennium.

This IYPE-Springer book series contains the latest thinking on the chosen themes by a large number of Earth science professionals from around the world. The books are written at the advanced level demanded by a potential readership consisting of Earth science professionals and students. Thus, the series is a legacy of the Science Programme, but it is also a counterweight to the Earth science information in several media formats already delivered by the numerous National Committees of the International Year in their pursuit of world-wide popularization under the Outreach Programme.

The discerning reader will recognise that the books in this series provide not only a comprehensive account of the individual themes but also share much common ground that makes the series greater than the sum of the individual volumes. It is to be hoped that the scientific perspective thus provided will enhance the reader's appreciation of the nature and scale of Earth science as well as the guidance it can offer to governments, decision-makers and others seeking solutions to national and global problems, thereby improving everyday life for present and future residents of Planet Earth.

Eduardo F.J. de Mulder
Executive Director International Secretariat
International Year of Planet Earth

Edward Derbyshire
Goodwill Ambassador
International Year of Planet Earth

Preface

This book series is one of the many important results of the International Year of Planet Earth (IYPE), a joint initiative of UNESCO and the International Union of Geological Sciences (IUGS), launched with the aim of ensuring greater and more effective use by society of the knowledge and skills provided by the Earth Sciences.

It was originally intended that the IYPE would run from the beginning of 2007 until the end of 2009, with the core year of the triennium (2008) being proclaimed as a UN Year by the United Nations General Assembly. During all three years, a series of activities included in the IYPE's science and outreach programmes had a strong mobilizing effect around the globe, not only among Earth Scientists but also within the general public and, especially, among children and young people.

The Outreach Programme has served to enhance cooperation among earth scientists, administrators, politicians and civil society and to generate public awareness of the wide ranging importance of the geosciences for human life and prosperity. It has also helped to develop a better understanding of Planet Earth and the importance of this knowledge in the building of a safer, healthier and wealthier society.

The Scientific Programme, focused upon ten themes of relevance to society, has successfully raised geoscientists' awareness of the need to develop further the international coordination of their activities. The Programme has also led to some important updating of the main challenges the geosciences are, and will be confronting within an agenda closely focused on societal benefit.

An important outcome of the work of the IYPE's scientific themes includes this thematic book as one of the volumes making up the IYPE-Springer Series, which was designed to provide an important element of the legacy of the International Year of Planet Earth. Many prestigious scientists, drawn from different disciplines and with a wide range of nationalities, are warmly thanked for their contributions to a series of books that epitomize the most advanced, up-to-date and useful information on evolution and life, water resources, soils, changing climate, deep earth, oceans, non-renewable resources, earth and health, natural hazards, megacities.

This legacy opens a bridge to the future. It is published in the hope that the core message and the concerted actions of the International Year of Planet Earth throughout the triennium will continue and, ultimately, go some way towards helping to establish an improved equilibrium between human society and its home planet. As stated by the Director General of UNESCO, Koichiro Matsuura, "Our knowledge of the Earth system is our insurance policy for the future of our planet". This book series is an important step in that direction.

R. Missotten
Chief, Global Earth Observation Section
UNESCO

Alberto C. Riccardi
President
IUGS

Introduction

In the context of the International Year of Planeth Earth (IYPE), the International Lithosphere Programme (ILP) has taken the responsibility for the scientific coordination of the IYPE theme Deep Earth.

In the preparatory phase of the IYPE, ILP has organized in June 2007 a meeting on New Frontiers in Integrated Solid Earth Sciences at the GeoForschungsZentrum Potsdam to review breakthroughs and challenges in the connection of Deep Earth and surface processes. The present volume is an outcome of this conference, providing examples of recent exciting developments in this field as well as an inventory of opportunities for future research.

The Potsdam conference was held in conjunction with the retirement of Rolf Emmermann, founding director of GFZ, one of the largest Integrated Earth Research Institutes of the world. He has also been vital in the realization of major Integrated Earth Research initiatives such as the International Continental Drilling Programme (ICDP), succeeding the first big science research project in continental geosciences in Germany drilling to 9000 m depth (KTB).

Peter Ziegler, well known for his life time activities connecting the energy industry and in-depth understanding of lithosphere evolution in space and time, is another pioneer in the domain of Integrated Solid Earth Science. His fundamental contributions to the study of the lithosphere are documented in a monumental series of atlases on the paleogeography of Europe and the North Atlantic as well as in seminal highly cited papers on sedimentary basins and lithosphere evolution. His 80th birthday in 2008 coincides with the IYPE.

ILP wishes to thank Rolf and Peter for laying the foundations both in terms of promoting scientific innovation, novel concepts and vision, on which future endeavors to move the frontiers in Integrated Solid Earth Sciences can build. This volume is dedicated to both of them.

Sierd Cloetingh
ILP President
Amsterdam

Jörg Negendank
ILP Fellow
Potsdam

Contents

Contributors

Marco Bohnhoff Department of Geophysics, Stanford University, Stanford, CA 94305-2215, USA, marcob@stanford.edu

E. Burov University Paris VI, Case 129, 4 Place Jussieu, Paris 75252, France, evgenii.burov@upmc.fr

Sierd A.P.L. Cloetingh Faculty of Earth and Life Sciences, Netherlands Research Centre for Integrated Solid Earth Science, VU University Amsterdam, Amsterdam, The Netherlands, sierd.cloetingh@falw.vu.nl

Larissa F. Dobrzhinetskaya Institute of Geophysics and Planetary Physics, Department of Earth Sciences, University of California, Riverside, CA 92521, USA larissa@ucr.edu

Doris Dransch Helmholtz-Zentrum Potsdam, Deutsches GeoforschungsZentrum (GFZ), Potsdam, Germany

Georg Dresen Helmholtz-Zentrum Potsdam, Deutsches GeoforschungsZentrum (GFZ), Potsdam, Germany, dre@gfz-potsdam.de

William L. Ellsworth United States Geological Survey; Menlo Park, CA 94025, USA, ellsworth@usgs.gov

Søren Gregersen GEUS Copenhagen

Sören Haubrock Helmholtz-Zentrum Potsdam, Deutsches GeoforschungsZentrum (GFZ), Potsdam, Germany

Hisao Ito Center for Deep Earth Exploration, Japan Agency for Marine-Earth Science and Technology, Yokohama Kanagawa 236-0001, Japan, hisaoito@jamstec.go.jp

Erik R. Ivins Jet Propulsion Laboratory

Mikhail K. Kaban Helmholtz-Zentrum Potsdam, Deutsches GeoforschungsZentrum (GFZ), Potsdam, Germany

Volker Klemann Helmholtz-Zentrum Potsdam, Deutsches GeoforschungsZentrum (GFZ), Potsdam, Germany

Elena Kozlovskaya University of Oulu

Ilmo Kukkonen Geological Survey of Finland, Finland

Björn Lund University of Uppsala

Juha-Pekka Lunkka University of Oulu

Glenn Milne University of Ottawa, Ottawa, ON K1N 6N5, Canada

Walter D. Mooney USGS, Menlo Park, CA 94025, USA, mooney@usgs.gov

Jürgen Müller University of Hannover

J.F.W. Negendank Helmholtz-Zentrum Potsdam, Deutsches GeoforschungsZentrum (GFZ), Potsdam, Germany, neg@gfz-potsdam.de

Christophe Pascal Geological Survey of Norway, N-7491 Trondheim, Norway

Federico Pasquarè Department of Chemical and Environment Sciences, University of Insubria, Como, Italy

Bjørn R. Pettersen Norwegian University of Life Science

Markku Poutanen Finnish Geodetic Institute, Masala, Finland, markku.poutanen@fgi.fi

François Roure Institut Français du Pétrole, 1-4 Avenue de Bois-Préau, 92852 Rueil-Malmaison, France; Department of Earth and Life Sciences, Vrije Universiteit, de Boelelaan 1085, 1081 HV Amsterdam, The Nederlands, Francois.ROURE@ifp.fr

Justin L. Rubinstein United States Geological Survey; Menlo Park, CA 94025, USA, jrubinstein@usgs.gov

Magdalena Scheck-Wenderoth Helmholtz-Zentrum Potsdam, Deutsches GeoforschungsZentrum (GFZ), Potsdam, Germany

Hans-Georg Scherneck Chalmers University of Technology

David R. Shelly United States Geological Survey; Menlo Park, CA 94025, USA

Holger Steffen University of Hannover; University of Calgary

Magdala Tesauro Faculty of Earth and Life Sciences, Netherlands Research Centre for Integrated Solid Earth Science, VU University Amsterdam, Amsterdam, The Netherlands; Helmholtz-Zentrum Potsdam, Deutsches GeoforschungsZentrum (GFZ), Potsdam, Germany, magdala.tesauro@falw.vu.nl

Alessandro Tibaldi Department of Geological Sciences and Geotechnologies, University of Milan-Bicocca, Italy, alessandro.tibaldi@unimib.it

Daniel Tormey ENTRIX Inc., Ventura, California, USA

Bert Vermeersen DEOS, TU Delft

Lidunka Vočadlo Department of Earth Sciences, UCL, London, WC1E 6BT, UK, l.vocadlo@ucl.ac.uk

Susan M. White USGS, Menlo Park, CA 94025, USA

Richard Wirth Helmholtz-Zentrum Potsdam, Deutsches GeoforschungsZentrum (GFZ), German Research Centre for Geosciences, Experimental Geochemistry and Mineral Physics, Potsdam, Germany

Detlef Wolf Helmholtz-Zentrum Potsdam, Deutsches GeoforschungsZentrum (GFZ), Potsdam, Germany

Peter A. Ziegler Geological-Paleontological Institute University of Basel, Bernoullistrasse 32, 4056 Basel, Switzerland, paziegler@magnet.ch

New Frontiers in Integrated Solid Earth Sciences

Group picture – ILP meeting "Frontiers in Integrated Solid Earth Science" – Potsdam 2007

Reviewers

Marco Bohnhoff (Stanford, CA, USA)
Roland Burgmann (Berkeley, CA, USA)
Evgeni Burov (Paris, France)
Cathy Busby (Berkeley, CA, USA)
Bernard Dost (De Bilt, The Netherlands)
Jeffrey T. Freymueller (Fairbanks, AK, USA)
Roy Gabrielsen (Oslo, Norway)
Georg Hoinkes (Graz, Austria)
Laurent Jolivet (Paris, France)
Fred Klein (Menlo Park, CA, USA)
Stephen R. McNutt (Fairbanks, AK, USA)
Joerg Negendank (Potsdam, Germany)
François Roure (Rueil-Malmaison, France)
Hans-Peter Schertl (Bochum, Germany)
Tetsuzo Seno (Tokyo, Japan)
Gerd Steinle-Neumann (Bayreuth, Germany)
Kazuhiko Tezuka (Chiba, Japan)
John Vidale (Seattle, WA, USA)
Marlies ter Voorde (Amsterdam, The Netherlands)
Shah Wali Faryad (Prague, Czech Republic)
Wim van Westrenen (Amsterdam, The Netherlands)
Jolante van Wijk (Los Alamos, NM, USA)
Tadashi Yamasaki (Amsterdam, The Netherlands)

Perpectives on Integrated Solid Earth Sciences

S.A.P.L. Cloetingh and J.F.W. Negendank

Abstract During the last decades the Earth sciences are rapidly changing from largely descriptive to process-oriented disciplines that aim at quantitative models for the reconstruction and forecasting of the complex processes in the solid Earth. This includes prediction in the sense of forecasting the future behaviour of geologic systems, but also the prediction of geologic patterns that exist now in the subsurface as frozen evidence of the past. Both ways of prediction are highly relevant for the basic needs of humanity: supply of water and resources, protection against natural hazards and control on the environmental degradation of the Earth.

Intensive utilization of the human habitat carries largely unknown risks of and makes us increasingly vulnerable. Human use of the outermost solid Earth intensifies at a rapid pace. There is an urgent need for scientifically advanced "geo-prediction systems" that can accurately locate subsurface resources and forecast timing and magnitude of earthquakes, volcanic eruptions and land subsidence (some of those being man induced). The design of such systems is a major multidisciplinary scientific challenge. Prediction of solid-Earth processes also provides important constraints for predictions in oceanographic and atmospheric sciences and climate variability.

The quantitative understanding of the Earth has made significant progress in the last few decades. Important ingredients in this process have been the advances made in seismological methods to obtain information on the 3D structure of the mantle and the lithosphere, in the quantitative understanding of the lithospheric scale processes as well as the recognition of the key role of quantitative sedimentary basin analysis in connecting temporal and spatial evolution of the system Earth recorded in their sedimentary fill. Similar breakthroughs have been made in the spatial resolution of the structural controls on lithosphere and (de)formation processes and its architecture by 3D seismic imaging. Earth-oriented space research is increasingly directed towards obtaining a higher resolution in monitoring vertical motions at the Earth's surface. Modelling of dynamic topography and landform evolution is reaching the phase where a full coupling can be made with studies of sediment supply and erosion in the sedimentary basins for different spatial and temporal scales.

Quantitative understanding of the transfer of mass at the surface by erosion and deposition as well as their feed back with crustal and subcrustal dynamics presents a new frontier in modern Earth sciences. This research bridges current approaches separately addressing high resolution time scales for a limited near surface record and the long term and large scale approaches characteristic so far for the lithosphere and basin-wide studies. The essential step towards a 4D approach (in space and time) is a direct response to the need for a full incorporation of geological and geophysical constraints, provided by both the quality of modern seismic imaging as well as the need to incorporate smaller scales in the modelling of solid Earth processes.

Keywords Solid earth dynamics · Earth monitoring · Reconstruction of the past · Solid earth process modelling

S.A.P.L. Cloetingh (✉)
ISES, Faculty of Earth and Life Sciences, VU University Amsterdam, Amsterdam, The Netherlands
e-mail: sierd.cloetingh@falw.vu.nl

S. Cloetingh, J. Negendank (eds.), *New Frontiers in Integrated Solid Earth Sciences*, International Year of Planet Earth,
DOI 10.1007/978-90-481-2737-5_1, © Springer Science+Business Media B.V. 2010

Introduction

The structure and processes of the deep Earth may sound remote from everyday concerns, but both have strong relevance for humanity's basic needs, such as supply of water and resources, protection against natural hazards, and control of the environmental degradation of the Earth.

In recent years geologists have come to understand the solid Earth in more measurable ("quantitative") ways. Better seismic techniques have brought us to a better understanding of the 3D structure of the Earth's mantle and lithosphere. We can describe, in numerical terms, how the deep Earth system works; at the same time, quantitative analysis of the basins in which sediments accumulate has allowed us to connect the deep Earth system with the record of those changes written in the sediments that build up over geological time.

Better ways of "seeing" through solid rock have allowed Earth scientists to understand the fine structure of the Earth's outer shell, or "lithosphere", and how it deforms under pressure from the movement of the crustal plates, in three dimensions. Recent advances in the ways geologists can give things accurate ages in years have made it possible to find out how fast tectonic and surface processes take place, with the precision necessary to distinguish between the different forces that shape the landscape.

Using space satellites to survey the Earth has allowed us to obtain ever-higher resolution when monitoring the vertical motions of Earth's surface. Modelling the way topography changes with time has now reached the stage where it is possible to couple studies of sediment supply and erosion in time and space. At a much smaller scale, we face problems of sedimentary architecture (the way different sediments are structured), and of imaging this architecture using remote sensing techniques that use seismic or electromagnetic waves to see inside them, like a "body scanner".

Despite enormous progress in the last 15 years, such remote imaging barely keeps pace with the great demands society places upon it, with urgent needs for water supplies, mineral resources, protection against natural hazards and control of the environment.

Below we highlight some key issues central in modern integrated solid Earth science.

Mass Transfer

"Mass transfer" means the way in which rocks are eroded from certain areas of the Earth's crust and redeposited in others, and the way the Earth's interior responds to those gradual changes in pressure. This presents a new frontier in modern Earth sciences – namely, trying to understand these processes quantitatively.

This needs a research strategy bridging current approaches that separately address high-resolution timescales for a limited near-surface record on the one hand, and the long-term and large-scale approaches that are more typical of studies at the scale of whole sedimentary basins. The essential step towards a four dimensional (4D) approach (i.e., involving both space and time) requires modelling solid Earth processes in a way that incorporates smaller scale data with high quality modern seismic imaging. We need to probe the deep Earth, to obtain a high-resolution image of both deep Earth structure and processes, if we are to quantify and constrain the forces that drive the Earth's crustal plates.

The deep Earth framework provides a unifying theme capable of addressing, in a process-oriented way, the full dynamics of the Earth system. Recent technical advances (including seismic tomography, Earth-oriented space observations, oceanic and continental drilling, modelling, and analytical techniques) have created fertile ground for a breakthrough by means of a global effort that integrates state-of-the-art methodology and the assembly of global databases.

Continental Topography: Interplay of Deep Earth and Surface Processes

Topography, the landscape's physical shape, is a product of the interaction between processes taking place deep in the Earth, on its surface, and in the atmosphere above it. Topography influences society, not only in terms of the slow process of landscape change, but also through climate. Topographic evolution (changes in land, water and sea levels) can seriously affect human life, as well as plants and animals. When levels of fresh water or of the sea rise, or when land subsides, the risk of flooding increases, directly affecting local ecosystems and human settlements. On the other hand, declining water levels and uplift may lead to a higher risk of erosion and even desertification.

These changes are caused both by natural processes and human activities, yet the absolute and relative contributions of each are still little understood. The present state and behaviour of the Shallow Earth System is a consequence of processes on a wide range of time scales. These include:

- long-term tectonic effects on uplift, subsidence and river systems;
- residual effects of ice ages on crustal movement (the weight of ice accumulations depresses the crust, and takes tens of thousands of years to recover following melting of the ice sheet;
- natural climate and environmental changes over the last millennia right up to the present;
- powerful anthropogenic impacts;

If we are to understand the present state of the Earth System, to predict its future and to engineer our sustainable use of it, this spectrum of processes (operating concurrently but on different time scales) needs to be better understood. The challenge to Earth science is to describe the state of the system, to monitor its changes, to forecast its evolution and, in collaboration with others, to evaluate different models for its sustainable use by human beings. Research will need to focus upon the interplay between active tectonics, topographic evolution, and related sea level changes and drainage pattern (river) development. This includes developing an integrated strategy for observation and analysis, emphasising large scale changes in vulnerable parts of the globe.

Making accurate geological predictions in complexly folded and faulted mountain belts will require collaboration between researchers from several broad fields of expertise. Among other scientific disciplines, geology, geophysics, geodesy, hydrology and climatology, as well as various fields of geotechnology, will need to be integrated.

Geoprediction: Observation, Reconstruction and Process Modelling

The increasing pressure that we are placing upon the environment makes us increasingly vulnerable. We have an urgent need for scientifically advanced "geoprediction systems" that can accurately locate subsurface resources and forecast the timing and magnitude of earthquakes, volcanic eruptions and land subsidence (some of which is caused by human activity). The design of such systems poses a major multidisciplinary scientific challenge. Prediction of solid Earth processes also imposes important constraints on predictions in oceanographic and atmospheric sciences, including climate variability.

Predicting the behaviour of geological systems requires two things: a thorough understanding of the processes, and high quality data. The biggest progress in quantitative prediction is expected to occur at the interface between modelling and observation. This is the place where scientific hypothesis is confronted with observed reality. In its most advanced version, the integrated sequence "observation, modelling, process quantification, optimization and prediction" is repeatedly carried out (in time and space) and the outcome is vital in generating fundamentally new conceptual developments.

Observing the Present

Information on the (present-day) structure of the subsurface and the deeper interior of the Earth (at various scales) is a key aspect of solid Earth science. This pertains to the study of both active processes and those that have ceased to be active but which may have contributed to present-day structures. The study of active processes plays an important role in this respect because process-related observations (concerning, for example, earthquake activity, surface deformation and the Earth's gravity field) can be made (and used) as constraints upon process models. The process-related insight gained from such exercises is very valuable in guiding our reconstruction of past processes.

Reconstructing the Past

Although the solid Earth has changed continuously through time, it still retains vestiges of its earlier evolution. Revealing the roles played in controlling rates of erosion and sedimentation by internal lithospheric processes and external forcing represents a major challenge.

The sedimentary cover of the lithosphere provides a high-resolution record of the changing environment, as well as of deformation and mass transfer at the surface and at different depths in the crust, lithosphere, and mantle system. In the last few years, pioneering contributions have helped to explain how lithosphere tectonic processes and the sedimentary record are related. These demonstrate, for example, the control exerted by stress fields in the lithospheric plates on the sequences of sediments that accumulate above them, and on the record of relative sea-level changes in sedimentary basins. Earth scientists are also becoming increasingly conscious of the way that active tectonic processes affect sedimentary basins, as well as the major implications these processes have for fluid flow and recent vertical motions in the coupled system that links the deep Earth and its surface processes.

The sedimentary cover of the lithosphere provides a record of the changing environment, involving deformation and mass transfer at the Earth's surface and at different depths within the crust, lithosphere, and mantle system. In the past few decades, sedimentary basin analysis has been in the forefront in integrating sedimentary and lithosphere components of the (previously separate) fields of geology and geophysics. Integrating active tectonics, surface processes and lithospheric dynamics in the reconstruction of the ancient topography of these basins and their surrounding areas is a key objective. A fully integrated approach (combining dynamic topography and sedimentary basin dynamics) is also important, considering the key societal role these basins play as resource locations, such as hydrocarbon reservoirs and source rocks. Moreover, given that most people alive today live either within or close to sedimentary basins (in coastal zones and deltas) both populations and their settlements remain vulnerable to geological hazards posed by Earth system activity.

Lithosphere Deformation Behaviour

The way the rocks of the mantle flow exerts controls on the thickness and strength of the lithospheric plates, the extent of coupling between plate motions and flow in the Earth's interior, and the pattern and rate of convection in the asthenosphere – as well as more local processes such as the pattern and rate of mantle flow and melt extraction at mid-ocean ridges. In order to understand the dynamic behaviour of the outer parts of the solid Earth, notably the dynamics of lithospheric extension and associated rifting and sedimentary basin development, a detailed knowledge of the way in which the different zones of the mantle flow is essential.

Process Modelling and Validation

Modelling solid Earth processes is in a transitional stage between kinematic and dynamic modelling. This development cannot take place without the interaction with (sub)disciplines addressing Earth structure and kinematics, or reconstructions of geological processes. In fact, advances in structure related research, in particular the advent of 3D seismic velocity models, have set the stage for studies of dynamic processes inside the Earth. Structural information is a prerequisite for modelling solid Earth processes. Similarly, information on present-day horizontal and vertical movements, as well as reconstructed past motion, temperatures or other process characteristics, is used in formulating and testing hypotheses concerning dynamic processes. Inversely, the results of process modelling motivate and guide research in observation of the present and reconstruction of the past.

Through the emphasis on process dynamics, it is in process modelling in particular that the full benefits of coupling spatial and temporal scales become apparent. The scale of the processes studied ranges from the planetary dimension to the small scale relevant to sedimentary processes, with the depth scale being reduced accordingly.

Challenges and New Developments

In spite of the great successes of plate tectonic theory in modern Earth science, fundamental questions still remain concerning the evolution of continents and their role in the dynamics of the Earth's lithosphere and mantle. The growth process of continents (on the scale of a differentiating planet), their thickness and the dynamic coupling with the underlying mantle are topics requiring focused attention from a series of sub-disciplines.

Equally important questions remain to be solved concerning mechanisms controlling continental tectonics and their effects on vertical movements, dynamic topography, and sedimentary basin formation. Vital in this respect are the dynamics of splitting continents, how one plate dives beneath another at a subduction zone, how mountains are built and denuded by erosion, and their effects on continental platform evolution and on the boundary processes between oceans and continents. Equally important are the rates and scales at which these processes operate.

In order to quantify the key processes involved in solid Earth science, it is essential to couple both internal and external forcing. From the extensive scale of (upper) mantle and lithospheric structure and processes, work on crustal structure and processes, the dynamics of topography and sedimentary basins and their sediment fills progresses at increasingly finer scales.

Integrated Approach to Selected Natural Laboratories and Analogues

Analogues are the key to reconstructing the past and predicting the future in the geological sciences. The natural laboratory that is the Earth allows us to observe different time slices at a range of scales, which, on their own, offer only an incomplete record of 4 billion years of its history.

Major large-scale integrated deep and surface Earth initiatives are already under development, covering Europe (TOPO-EUROPE, EUROARRAY, EPOS), the USA (EARTHSCOPE), the African Plate, and the South-American Plate. At the same time, major international research initiatives such as the Integrated Ocean Drilling Programme (IODP), the International Continental Drilling Programme (ICDP), and the International Lithosphere Programme (ILP) provide a platform from which to extend these research efforts into other areas of the globe. The complementary strength of research methods focused on selected natural laboratories will lead to a multi-method approach unparalleled so far.

Coupled Deep Earth and Surface Processes

The modern Earth system approach requires a comprehensive integration of existing databases, with the capacity to expand to allow for storage and exchange of new data collected. Unification and coupling of existing modelling techniques is needed to achieve full integration of what are currently discipline-oriented approaches and to expand fully "next generation" 3D applications. Furthermore, flexible exchange for "feedback loops" in the quantification of processes is needed at the interface between databases and modelling tools. Consequently, major investments in Information Technology are called for so as to expand existing computer hardware and software facilities.

Coupled Process Modelling and Validation

Process modelling and validation allows full implementation and optimization of the quantitative approach pursued in programmes such as TOPO-EUROPE. In this context, two interrelated aspects should be emphasised. The first is the fundamental capacity of numerical process modelling to link the geometrical/structural, mechanical properties, and kinematic and dynamic aspects (and data) of the processes studied. Second, using this capacity, quantitative modelling and reliable constraints on fluxes and timing obtained by state-of-the-art analytical tools will play a crucial role in research (and the guiding of research) aimed at deciphering complex interactions between spatial and temporal scales in Earth processes.

ILP Activities Within the International Year of Planet Earth

ILP Activities within the IYPE are centred within the following eight themes:

- *"Earth accretionary systems (in space and time)" (ERAS) (chairs: Peter Cawood, Alfred Kröner)*

The initiation and development of accretionary orogens forms the central aim of this integrated, multidisciplinary program.

- *"New tectonic causes of volcano failure and possible premonitory signals" (chairs: Alessandro Tibaldi, Alfredo F.M. Lagmay, Vera Ponomareva, Theofilos Toulkeridis)*
  More than 500 million people live in hazardous zones adjacent to active volcanoes all over the world and volcano slope instability represents one of the most extreme hazards.
- *"Lithosphere-asthenosphere Interactions" (chairs: Andrea Tommasi, Michael Kendall, Carlos J. Garrido)*
  This project focuses on the interaction between the lithosphere, the outer shell on which we live, and the asthenosphere and/or deep mantle. The dynamic processes of the Earth's interior affect our day to-day life in a profound way.
- *"Ultra-deep continental crust subduction" (UDCCS) (chairs: Larissa Dobrzhinetskaya, Patrick O'Brien, Yong-Fei Zheng)*
  Ultra-high-pressure metamorphic (UHPM) geology is a new discipline that came into being after discoveries of coesite and microdiamond within rocks of continental affinities involved in collisional orogenic belts. UHPM terranes and UHP minerals and rocks present a "special natural laboratory."
- *"Global and regional parameters of paleoseismology; implications for fault scaling and future earthquake hazard" (chair: Paolo Marco De Martini)*
  This project aims to support and promote the study of the main paleoseismological parameters at a global and regional scale to develop new ideas on fault scaling relationships and modern earthquake hazard estimates.
- *"Sedimentary basins" (chairs: François Roure, Magdalena Scheck-Wenderoth)*
  Sedimentary basins provide mankind's largest reservoir for Earth energy and natural resources. As recorder of tectonics and climate interaction they enable to reconstruct the history of the continents.
- *"Temporal and spatial change of stress and strain" (chair: Oliver Heidbach)*
  This project aims to identify, analyse and interpret variations of crustal stress and strain patterns at diverse tectonic settings characterized by return periods for strong earthquakes in the order of 50–1,000 years.
- *"Baby plumes – origin, characteristics, lithosphere-asthenosphere interaction and surface expression." (chair: Ulrich Achauer)*

The project focuses on interdisciplinary studies of baby-plumes to steer the debate on the origin and nature of plumes in general, addressing their geodynamic implications.

Research on these themes is accompanied by three coordinated efforts of a more regional nature and a global research initiative. Two of these projects (*TOPO-EUROPE*: 4D Topography Evolution in Europe: Uplift, Subsidence and Sea Level Rise *[chair: Sierd Cloetingh]* and *TOPO-CENTRAL-ASIA*: 4D Topographic Evolution in Central Asia: Lithosphere Dynamics and Environmental Changes since Mesozoic *[chairs: Qingchen Wang, Shigenori Maruyama, Boris Natal'in, Yan Chen]*) address topography evolution in space and time, adopting Europe and Central Asia as natural laboratories.

DynaQlim (Upper Mantle Dynamics and Quaternary Climate in Cratonic Areas *[chair: Markku Poutanen]*) has its focus on the study of the relations between upper mantle dynamics, its composition and physical properties, temperature, rheology, and Quaternary climate. Its regional focus lies on the cratonic areas of northern Canada and Scandinavia.

The *International Continental Scientific Drilling Program* (ICDP) *[Executive Committee: Rolf Emmermann, Ulrich Harms et al.]* coordinates continental scientific drilling efforts with research topics of high international priority. Main objectives addressed in ICDP include geodynamics and natural hazards, volcanic systems and thermal regimes, Earth's history and climate, impact structures and mass extinctions as well as deep biosphere and gas hydrates (see also http://www.scl-ilp.org for further information).

The above themes are addressed by a number of ILP Task Forces and Regional Coordinating Committees. Below we give an overview of their research activities.

Task Force 1: Earth Accretionary Systems (in Space and Time) (ERAS)

Classic models of orogens involve a Wilson cycle of ocean opening and closing with orogenesis related to continent-continent collision (Dewey, 1969; Wilson, 1966). Such models fail to explain the geological history of a significant number of orogenic belts throughout the world in which deformation, metamorphism and crustal growth took place in an environment of ongoing plate convergence. These belts are termed accretionary orogens but have also been referred to as non-collisional or exterior orogens, Cordilleran-, Pacific-, Miyashiro-, and Turkic-type orogens (Matsuda and Uyeda, 1971; Windley, 1992; Şengör, 1993; Maruyama, 1997; Ernst, 2005). Accretionary orogens form at sites of subduction of oceanic lithosphere. They consist of accretionary wedges (Fig. 1) containing material accreted from the downgoing plate and eroded from the upper plate, island arcs, ophiolites, oceanic plateaus, old continental blocks, post-accretion granitic rocks and metamorphic products up to the granulite-facies, exhumed high-pressure metamorphic rocks, and clastic sedimentary basins (Cawood et al., 2009). Accretionary orogens appear to have been active throughout much of Earth history and constitute major sites of continental growth (Cawood et al., 2006). They contain significant mineral deposits (Groves and Bierlein, 2007) and thus provide the mineralisation potential of many countries such as Australia, Canada, Zimbabwe, Saudi Arabia, Yemen, Nigeria, China, Kazakhstan and Mongolia.

Understanding of the processes for the initiation and development of accretionary orogens is moderately well established in modern orogens such as Japan, Indonesia and Alaska, the broad structure and evolution of which are constrained by seismic profiles, tomography, field mapping, palaeontology, isotope geochemistry and geochronology. However, the processes responsible for the cratonization and incorporation of accretionary orogens into continental nuclei (Cawood and Buchan, 2007) and the mechanisms of formation of pre-Mesozoic accretionary orogens are poorly understood. In a uniformitarian sense many of the features and processes of formation of modern accretionary orogens have been little applied to pre-Mesozoic orogens. Resolution and understanding of these processes form the central aim of this Task Force.

This integrated, multi-disciplinary and comprehensive program in selected accretionary orogens will provide a common framework to better understand their development. The detailed work program of Task Force *ERAS* (EaRth Accretionary Systems in space and time) will be further developed and refined through meetings and international conferences. These will encourage interested scientists to join in developing plans to implement components of the program, to

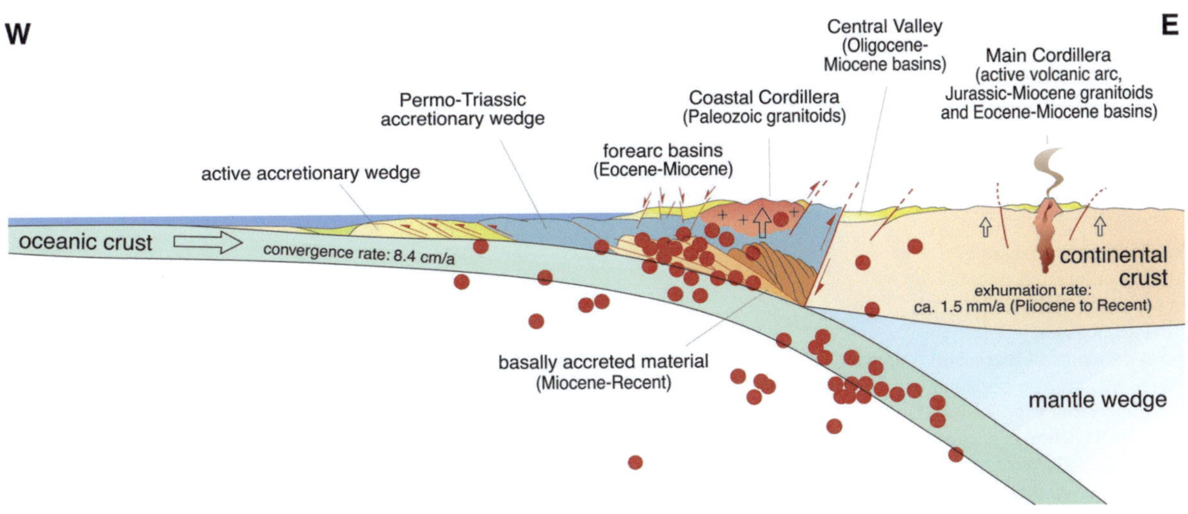

Fig. 1 Earth accretionary systems (in space and time) (ERAS): Schematic geological profile across the Chilean continental margin at ca.38° S (southern Andes). Red dots denote the distribution of recent seismicity, as registered during the ISSA 2000 experiment (Courtesy H. Echtler; GFZ-Potsdam)

advance our understanding of accretionary systems in particular and continental evolution in general. The Task Force brings together scientists from many disciplines of the Earth sciences as evidenced by a volume on "Accretionary systems in space and time" (Cawood et al., 2009) with 14 contributions covering aspects of accretionary processes from the Archaean to the Cenozoic.

Task Force 2: Tectonic Causes of Volcano Failure and Possible Premonitory Signals

"Tectonic causes of volcano failure and possible premonitory signals" is an active component of the ILP theme "Geoscience of global change". It is aimed at developing a better understanding of how volcanoes work with special emphasis on the control exerted by regional tectonics on volcano evolution. Specific objectives comprise a better understanding of the magma feeding systems, deformations in volcanoes and volcanic regions, and how these interact with hazard phenomena like lateral collapses of volcanic edifices (Tibaldi et al., 2008a). The conceptual framework is developed through an interdisciplinary study of several key sites distributed around the world that, taken together, provide insights on real volcanic systems belonging to different geodynamic settings (Tibaldi and Lagmay, 2006). These comprise the Philippines volcanic arc, the Andes of Ecuador, Chile, Bolivia and Argentina, the Mediterranean zone, Iceland and Kamchatka. During 2008–2009, several studies have also been carried out on deeply eroded volcanoes in order to identify the inner magma feeding system as analogues to modern volcanoes. These studied eroded volcanoes are located in Ireland, Great Britain, Iceland, France, Italy and USA. Applied methodologies mainly comprise structural geology, stratigraphy, petrography, geochemistry, geotechnics, engineering geology, and geophysics. These data are crosscut with numerical and analogue modelling.

Most of the involved researchers operate in countries where threats from active volcanoes and tectonic earthquakes are widespread (e.g., Eichelberger et al., 2007; Calvari et al., 2008). This means that the results of this project are useful in different nations at the same time and represent a valuable data source for scientists and administrative and government panels. It should be emphasised that several selected volcanoes, such as Mayon, Cotopaxi, Mt Etna, and Stromboli (Fig. 2), represent major hazards in terms of loss of life and economic damage.

The studies carried out in these areas allowed to gain insights into the mode of magma emplacement through the upper crust and in the interior of volcanoes, as well as to better understand how volcano slopes deform under the magma-tectonic stress field. Major work in fact has been devoted to understand the coupling of the local stresses induced by magmatic and gravity forces with the remote stresses transmitted by regional tectonics (Fig. 3). At the same time, efforts have been devoted to find out the causes of the so called "super eruptions" and their consequences (Self and Blake, 2008), whose triggering factors can comprise wide failure of volcanoes.

Physical scaled modelling has also been carried out in order to simulate deformations in volcanoes lying above reverse faults with different geometries. The

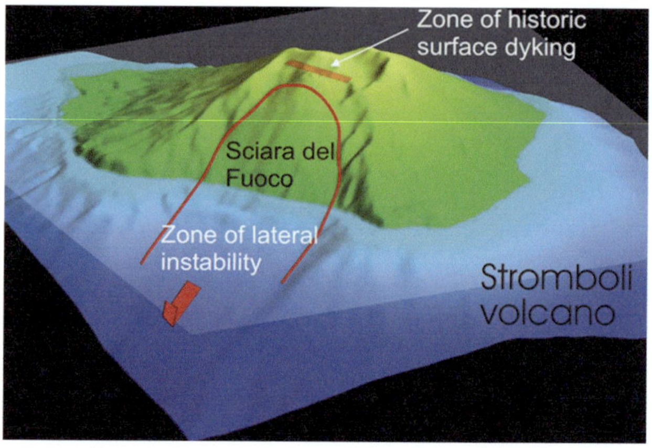

Fig. 2 Tectonic causes of volcano failure and possible premonitory signals. Three-dimensional digital elevation model of Stromboli active volcano. The horizontal pale blue colour represents the sea level. The orange strip is the preferential site of surface dyking. The *red line* encircles the zone of the major and recentmost lateral collapse of a few thousands year ago (after Tibaldi et al., 2008a)

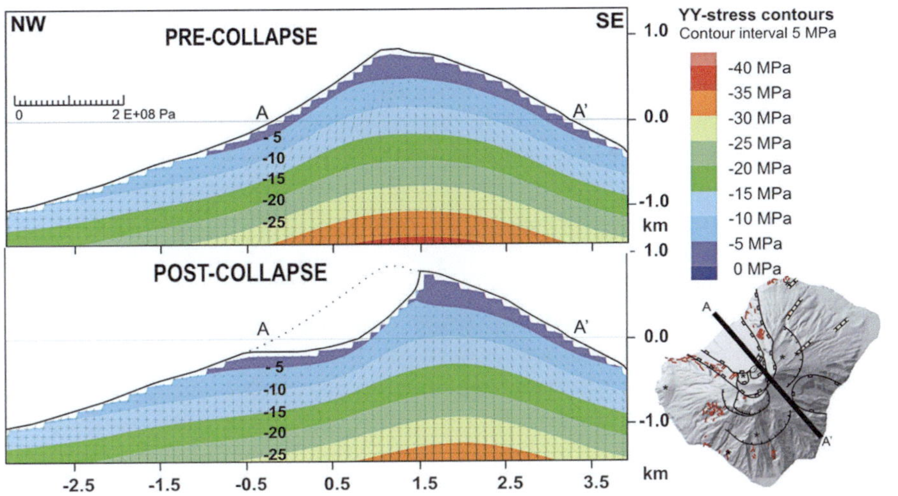

Fig. 3 State of stress computed along a NW–SE vertical section passing across the Stromboli top. Above, stress intensity before a sector collapse is portrayed by colours; below, stresses have been recalculated after the occurrence of a sector collapse: note that the collapse depression creates a debuttressing of the cone with more favourable conditions to dyke intrusions within the collapse amphitheatre (after Tibaldi et al., 2008a)

results have been compared with the key sites studied in the field, comprising volcanoes in Japan, Chile, Argentina and Ecuador. Further numerical modelling, coupled with physical and geotechnical evaluation of the rocks involved, has also been carried out. This was mainly focused on quantitative stability analysis of volcano slopes on Stromboli (Italy) and Nisyros (Greece, Tibaldi et al., 2008b).

Apart from pure scientific studies, this Task Force developed a strong interlink with industrial applications. A series of initiatives to bring closer their research to the economic productive activities have been launched: these were aimed to construct relationships with industries and private societies mostly dealing with the exploration and exploitation of geothermal energy. Collaborations started on these topics with important companies such as the ENEL in Italy and the NOCE in Philippines. Another collaboration with a major private enterprise in the USA (Entrix, CA) is now consolidated. This collaboration is aimed to improve the environment problem assessment and mitigation measures by engineering geology methods.

One of the most interesting findings both from a scientific and applied perspective, is that volcanism can occur also in compressional settings, and should not be considered as exclusive of extensional tectonic environments. In compressional settings magma migration can occur along strike-slip faults (Mazzuoli et al., 2008) as well as along reverse faults (Tibaldi, 2008). These faults can propagate from the substrate across the volcanic cone giving rise to characteristic surface fault patters. These faults crossing the volcanoes, in turn, can influence the lateral instability of the cones. Magma migration from the magma chamber to the surface can also reflect the different crustal states of stress giving rise to diverse sheet configurations, spanning from dyke swarms (Gudmundsson et al., 2008) to the newly discovered "Intrusive flower structures" (Tibaldi and Pasquarè, 2008; Tibaldi et al., 2008c). These findings, linking the regional tectonics with the more local magma-feeding paths, can contribute to elucidate the structures more prone to host fluid migration, helping, for example, the research activities linked to geothermal exploration.

Task Force 3: Lithosphere-Asthenosphere Interactions

This ILP project focuses on the interaction between the lithosphere, the outer shell on which we live, and the asthenosphere and/or deep mantle. The dynamic processes of the Earth's interior affect our day-to-day life in a profound way. Convection in the mantle shapes the Earth's surface through plate tectonics, giving rise to mountains ranges, oil-rich sedimentary basins and mineral-rich crust. It also controls catastrophic events such as earthquakes, landslides, tsunamis, and volcanic eruptions. Yet, until recently, mantle convection and its surface expression – plate tectonics – have been studied as independent systems. In convection studies, plates were generally considered as rigid rafts dragged along by the convecting mantle. On the other hand, "lithospheric" studies focused on the motion and deformation of the plates, solely represented by slab pull, ridge push, or basal drag forces. It is now understood that plate tectonics is an essential feature of mantle convection. However, the processes allowing convection in the mantle to produce plate tectonics at the Earth's surface as well as many aspects of the interaction between the convective flow and the plates remain poorly understood.

The time is ripe for collaborative research aiming to unravel the complex interactions between plates and the convective mantle as there have been multiple step-like advances in the various disciplines involved in the study of the mantle in the last 10 years. In geochemistry, advances in in-situ analysis techniques show that compositional heterogeneities may be preserved for long time periods at the crystal scale, bringing new insights on the significance of geochemical "reservoirs" in the mantle. Recent developments in seismic modelling techniques and the deployment of dense seismic networks allow high-resolution (<100 km scale) imaging of mantle structure and deformation at both global and regional scales. In mineral physics, simple shear experiments associated to new texture measurement techniques (EBSD) place constraints on the evolution of crystal preferred orientations and anisotropy in the mantle. Finally, advances in theoretical, laboratory and numerical modelling allow self-consistent generation of plates in convection models, an enhanced physical understanding of subduction and plumes dynamics, the investigation of the effects of melting, metamorphic reactions, and fluid migration on convective flow patterns, and integration of deformation effects from the grain-scale up to the convection spatial scale allowing to estimate the resulting anisotropy.

A major outcome of this research is an increased understanding of the coupling between chemical and physical processes in the lithosphere-asthenosphere boundary, in particular through partial melting, melt/fluid transport, and melt/fluid-rock reactions. From global scale seismic tomography models to regional studies on seismic anisotropy and receiver functions, recent seismic data point to a significant contribution of compositional heterogeneities, hydration, or partial melting to the seismic signal (e.g., Kendall et al., 2006; Nakajima and Hasegawa, 2007). Multi-scale numerical modelling also highlights a significant influence of chemical processes on the mantle deformation. Fluid-rock interactions are, for instance, shown to result in strong strain localization in the mantle wedge of subduction zones. Formation of a serpentine-rich, weak channel at the cold upper limit of the slab allows fast exhumation of high-pressure rocks (e.g., Gerya et al., 2002). Hydration may also directly reduce the viscosity in the olivine-rich mantle wedge, leading to thinning and possible breakup of the overriding plate above subduction zones (Arcay et al., 2005) or trigger partial melting, leading to development of instabilities and chaotic mixing in the wedge (e.g., Gorczyk et al., 2007). Melt-rock interactions probably also play a major role on the erosion of the continental lithosphere above mantle plumes (e.g., Garrido and Bodinier, 1999). Studies on peridotite massifs clearly show that magma-rock reactions play a key role on the rejuvenation (asthenospherization) of the lithospheric mantle (Lenoir et al., 2001; Le Roux et al., 2007). These observations are corroborated by studies of mantle xenoliths showing the interplay between lithospheric structure, plume channeling in the base of the lithosphere, and erosion of the continental lithospheric mantle (Lenoir et al., 2000). Finally, the presence of a low viscosity zone at the base of the lithosphere, due to partial melting or water-enrichment is also required to localize strain, favouring a plate-like behaviour in convection models (Tackley, 2000; Stein et al., 2004).

Recent work by the various teams participating to the Task Force 3 also highlighted the links between small-scale processes (crystal to outcrop, i.e., μm to cm) and the large-scale dynamics. Crystal- and rock-scale processes, such as viscous and elastic deformation, melt transport and reactions, modify the composition, microstructure, and physical properties of mantle rocks (e.g., Tommasi et al., 2004; Le Roux et al., 2008; see also Fig. 4). A full understanding of these processes is thus essential for interpreting geophysical observations in terms of temperature, composition, and deformation in the mantle. Development of a preferred orientation of olivine crystals in response to the deformation in the mantle results in a large-scale anisotropy of elastic, electrical, thermal, and mechanical properties of the mantle. The resulting anisotropic propagation of seismic waves has been largely used to map the deformation in the mantle in the last 25 years (e.g., Kendall et al., 2005). In contrast, the studies on the consequences and possible uses of the anisotropy of electrical and thermal conductivities or of an anisotropic viscosity in the upper mantle are still in their infancy, but a large number of promising results

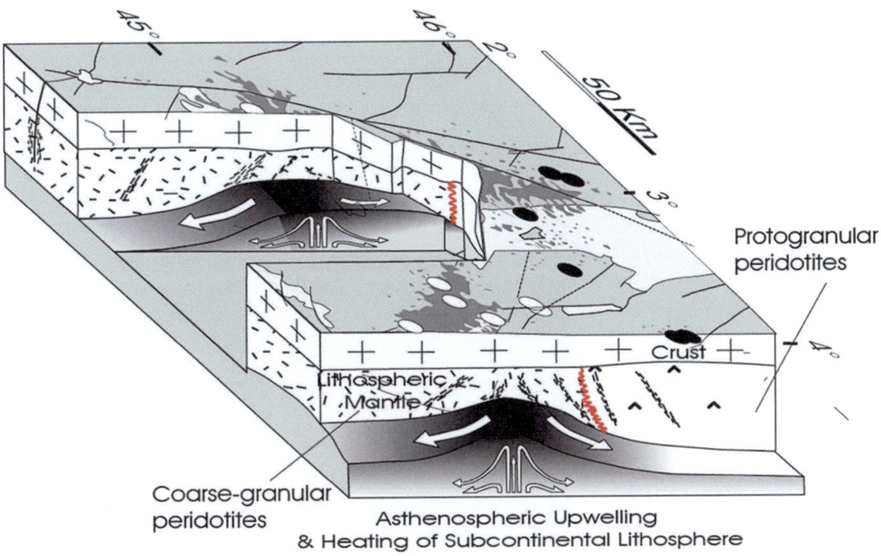

Fig. 4 French massif central: Lithospheric control on plume- lithosphere interaction (after Garrido et al., 2009)

have been published recently (e.g., Tommasi et al., 2001; Gatzemeier and Tommasi, 2006; Baba et al., 2006; Petitjean et al., 2006; Lev and Hager, 2008; Tommasi et al., 2009).

Task Force 4: Ultra-Deep Continental Crust Subduction (UDCCS)

This Task Force coordinates four major directions related to understanding of ultradeep subduction of the continental crust into the Earth deep interior: (I) Tectono-metamorphic development of ultrahigh pressure metamorphic terranes and building of the orogenic belts related to continental collisions; (II) Nanometric and atomic scale mineralogy, fluid inclusions (Fig. 5), and exsolution products in ultrahigh pressure metamorphic rocks; (III) Fluid-rock interaction due to the course of deep subduction and exhumation; (IV) Numerical modeling of the fate of continental crust subducted to the deep mantle and mantle transition zones and its geophysical implications for the plate tectonic concept. The endeavour to integrate such knowledge in the frame of deep subduction process is essential for erecting realistic large-scale models of the solid Earth's thermal and tectonic evolution.

Traditionally geologists have "known" for many years that continental crust is buoyant and cannot be subducted very deep. Microdiamonds (Fig. 6) 10–80 μm in size discovered in the 1980s within metamorphic rocks of continental affinities related to continental collisions clearly refute this statement. Their occurrence within marbles, acid gneisses and quartzite suggest that material of continental crust has been subducted to a minimum depth of >150 km and incorporated into mountain chains during tectonic exhumation. Over the past decade, the rapidly moving technological advancement has made it possible to examine these microdiamonds in detail, and to learn that they contain nanometric multiphase inclusions of crystalline and fluid phases and are characterized by a "crustal" signature of carbon stable isotopes. This discovery spurred unprecedented multidisciplinary investigations of continental collisions, mountain building, mantle enrichment in H_2O, and rare earth and lithophile elements, including ^{40}K, which has strong influence on the earth's thermal evolution. The discovery of these microdiamonds, as well as coesite, triggered a major revision in understanding of deep subduction processes, leading to the realization that continental materials can be recycled into the earth's interior.

Recently, this Task Force has produced exciting results (e.g., Dobrzhinetskaya and Gilotti, 2007; Gerya et al., 2008; Dobrzhinetskaya and Brueckner, 2009) that illustrate the fundamental importance of nanoscale minerals and their effects on geological processes on Earth and other planets. A new branch of ultrahigh-pressure metamorphism (UHPM) research – nanoscale mineralogy for studying mountain building and collisions – has become a fast-developing sub-discipline in this field. The new mineralogical/petrological data have led to the conclusion that UHPM is not limited to continent-continent

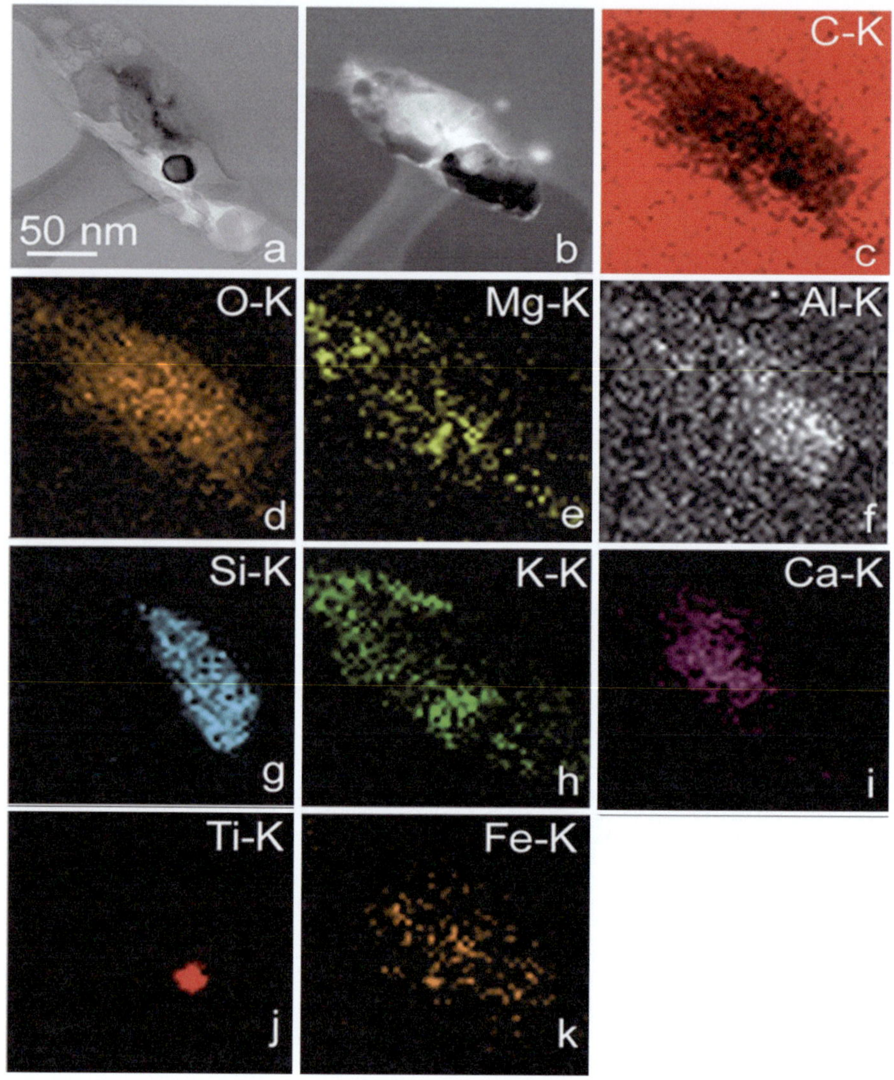

Fig. 5 Elemental mapping of multiphase (fluid + solid) inclusion using TEM-EDAX technique bright-field high resolution TEM image (**a**) of multiphase inclusion A taken early in analysis; fluid phase movement caused by beam heating as observed in upper left and lower right parts of this composite inclusion; STEM image (**b**) of he same inclusion taken late in analysis after lower right corner of the inclusion burst (*black*); *c–k* panels show individual maps of the *K-lines* of the following elements: *O*, *Mg*, *Al*, *Si*, *K*, *Ca*, *Ti* and *Fe* (after Dobrzhinetskaya et al., 2007)

collision settings, as was previously thought. Mineralogical assemblages of some lawsonite-bearing eclogite, with the oceanic basalt as protolith, are now interpreted as being the result of subduction of oceanic lithosphere and/or continental margin beneath oceanic lithosphere. Most of these studies contribute to knowledge of various aspects of the processes in subduction channels, which are the locations where the lithospheric plate plunges from the Earth's surface to deep into the mantle, forming mountain belts and volcanoes and causing earthquakes. The subduction zone processes are a top priority of the modern geosciences because without subduction, return flows, and mantle convections the plate tectonics could not exist and Earth would not be a planet suitable for life.

Task Force 5: Global and Regional Parameters of Paleoseismology; Implications for Fault Scaling and Future Earthquake Hazard

The projects of this ILP Task Force focus on three principal directions:

(1) Support and promote the study of the main paleoseismological parameters at a global and regional scale in order to develop new ideas on fault scaling relationships and modern earthquake hazard estimates. In this context, the Task Force is

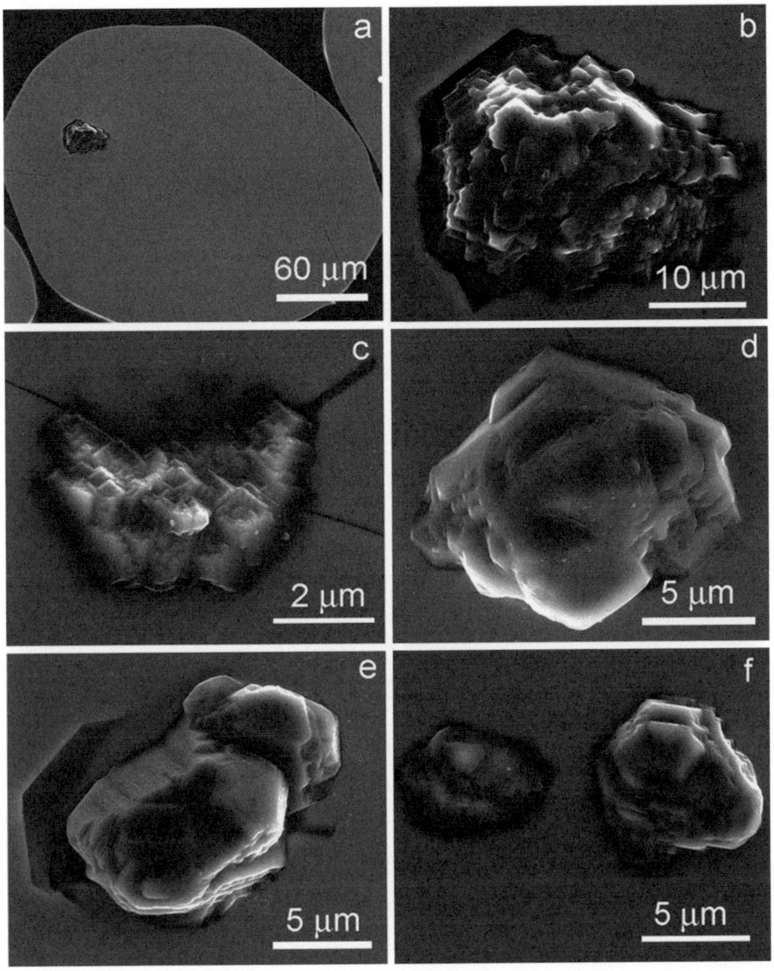

Fig. 6 Scanning electron microscope images of microdiamonds included in zircons from quartz-feldspathic gneisses of the Erzgebirge massif, Germany. Such diversity in diamond morphologies suggest that they were formed in a medium oversaturated with impurities, and that the rate of deposition of carbon atoms at the corners and on the faces of diamond nuclei was different, providing faster growth of the crystal edges (Chernov, 1974; Dobrzhinetskaya et al., 2001). The nanometric inclusions in diamonds are pristine witnesses of the medium from which diamond was crystallized. Their direct study by TEM was not possible until recently when the FIB technique became available for high quality foil preparation (Heaney et al., 2001; Dobrzhinetskaya et al., 2003; Wirth, 2004). Their results revealed that the microdiamonds contain dozens of nanometric crystalline and fluid inclusions which contain Ti, P, K, Si, Fe, Cl, S and O in non-stoichiometric combinations and proportions (Heaney et al., 2001). The thin foil (10 μm × 5 μm × 100 Å) was prepared from diamond using FIB technique (Wirth, 2004). (Courtesy: L. Dobrzhinetskaya, see also Dobrzhinetskaya and Wirth, this volume)

recognizing the importance of regional variations in surface fault rupture characteristics (Begin et al., 2005; Gràcia et al., 2006; Marco, 2007; Masana et al., 2005; McNeill et al., 2005; Michetti et al., 2005; Palyvos et al., 2005; Pantosti et al., 2008; Pucci et al., 2008; Vanneste et al., 2006).

The research deals mainly with the integration of earthquake and fault rupture parameters across the seismological/geological/geophysical boundaries, exploring the difference in scaling between large and small earthquakes. Another aspect concerns the evidence for linear or nonlinear relationship between average displacement and fault length for large dip-slip and strike-slip earthquakes. As these have been the topic of considerable debate in the last years, the Task Force selected the following main research topics:

- relationships between paleoseismological parameters (surface rupture length and average surface displacement) and earthquake magnitude estimates;
- comparison of paleoseismological parameters obtained from the pre-instrumental epoch as against modern earthquakes studies;
- comparison of source parameters obtained by different methodologies (geodesy, geology, seismology and geophysics).

Valuable results have been obtained thanks to extensive fieldwork, high-resolution geophysical measurements, remote sensing studies and analogue models. At the same time, this project highlighted some specific methodological problems concerning the under- or over-estimation of earthquake magnitude resulting from spatial variability of slip and scarp degradation processes at the surface.

(2) Maintenance of the worldwide database of independently-dated paleoearthquakes. INGV researchers provide the database maintenance and implementation. An extract of the database entries will be available at http://www.ingv.it.

(3) Develop paleoseismic research capability, especially in developing regions with high earthquake hazard. In this context the Task Force 5 participated in the organization of schools/field training courses, including the "Hokudan Symposium and School, Japan 2005" (http://home.hiroshima-u.ac.jp/kojiok/hokudan.html) and the "50th Anniversary Earthquake Conference Commemorating the 1957 Gobi-Altay Earthquake" held in Ulaanbaatar, Mongolia, 25 July–8 August 2007 (http://www.rcag.url.mn/seis/index.html; Fig. 7).

The Hokudan meeting took place on the occasion of the tenth anniversary of the 1995 Hyogo-ken Nanbu (Kobe) earthquake, which killed nearly 6,000 people and caused almost $115 billion in structural damage. The town of Hokudan has built an impressive earthquake memorial park and seminar house dedicated to earthquake education. Centerpiece of the park is a 140-m-long structure erected over the surface trace of the Nojima fault. This structure protects the surface

Fig. 7 ILP Task Force 5 participants in the field along the 1957 Gobi-Altay Earthquake surface rupture (the Toromkhon thrust fault) (Photo by David Schwartz)

rupture and offset cultural features from erosion and human modification. The Seminar House has served to educate almost four million visitors since it first opened in April, 1998. It was in this spirit that the meeting participants convened to present and discuss developments in active fault research to fellow scientists and the Japanese public. The principal aims of the 2005 symposium were to review the development of the studies on active faulting after the Kobe earthquake, and to promote advanced research in active tectonics and seismic hazard assessment in order to mitigate seismic hazards. This was an extraordinary occasion to review and to discuss the hazard assessment as well as the knowledge and techniques of active fault studies in a broad context. The presentations and discussions demonstrated that active tectonics research has progressed well in recent decades. Another important theme of the Symposium was the knowledge transfer to the general public of the scientific achievements. This culminated with several public lectures and an open house where the public was invited to view the poster presentations of the symposium participants. As part of the symposium and school, two field excursions were made: to the 1995 Nojima fault rupture and at the end of the meeting, to the Median Tectonic Line (MTL). The interactions in the field between specialists with different perspectives demonstrated the necessity for making field observations and discussing interpretations in order to establish a comprehensive international knowledge base. In addition, many concepts are more completely understood and fully discussed in the field where the original observations are made.

Task Force 6: Sedimentary Basins

The origin of sedimentary basins is a key element in the geological evolution of the continental lithosphere. During the last decades, substantial progress was made in the understanding of thermomechanical processes controlling the evolution of sedimentary basins and the isostatic response of the lithosphere to surface loads such as sedimentary basins. Much of this progress stems from improved insights into the mechanical properties of the lithosphere, from the development of new modeling techniques, and from the evaluation of new, high-quality datasets from previously inaccessible areas of the globe. The focus of this Task Force is on tectonic models of the processes controlling the evolution of sedimentary basins, and their validation by an array of different geological and geophysical data sets. After the realization that subsidence patterns of Atlantic-type margins, corrected for effects of sediment loading and palaeo-bathymetry, displayed the typical time-dependent decay characteristic of ocean-floor cooling (Sleep, 1971), a large number of studies were undertaken aimed at restoring the quantitative subsidence history of basins on the basis of well data and outcropping sedimentary sections. With the introduction of backstripping analysis algorithms (Steckler and Watts, 1982), the late 1970s and early 1980s marked a phase during which basin analysis essentially stood for backward modeling, namely reconstructing the tectonic subsidence from sedimentary sequences. These quantitative subsidence histories provided constraints for the development of conceptually driven forward basin models. For extensional basins this commenced in the late 1970s, with the appreciation of the importance of the lithospheric thinning and stretching concepts in basin subsidence (Salveson, 1976). After initiation of mathematical formulations of stretching concepts in forward extensional basin modeling (McKenzie, 1978), a large number of basin fill simulations focused on the interplay between thermal subsidence, sediment loading, and eustatic sealevel changes. To arrive at commonly observed more episodic and irregular subsidence curves, the smooth postrift subsidence behaviour was modulated by changes in sediment supply and eustatic sea-level fluctuations. Another approach frequently followed was to input a subsidence curve, thus rendering the basin modeling package essentially a tool to fill in an adopted accommodation space (Lawrence et al., 1990). For the evolution of extensional basins this approach made a clear distinction between their syn-rift and postrift stage, relating exponentially decreasing postrift tectonic subsidence rates to a combination of thermal equilibration of the lithosphere-asthenosphere system and lithospheric flexure (Watts et al., 1982). A similar set of assumptions were made to describe the syn-rift phase. In the simplest version of the stretching model (McKenzie, 1978), lithospheric thinning was described as resulting from more or less instantaneous extension. In these models a component of lithosphere mechanics was obviously lacking. On a smaller scale, tilted fault block models were introduced for modeling of the basin fill at the scale of half-graben models. Such models essentially decouple the response of the brittle upper crust from deeper lithospheric levels during rifting phases (see, e.g., Kusznir et al., 1991). A noteworthy feature of most modeling approaches was their emphasis on the basin subsidence record and their very limited capability to handle differential subsidence and uplift patterns in a process-oriented, internally consistent manner (see, e.g., Kusznir and Ziegler, 1992). To a large extent the same was true for most of compressional basin modeling. The importance of the lithospheric flexure concept, relating topographic loading of the crust by an overriding mountain chain to the development of accommodation space, was recognized as early as 1973 by Price in his paper on the foreland of the Canadian Rocky Mountains thrust belt (Price, 1973). Also, here it took several years before quantitative approaches started to develop, investigating the effects of lithospheric flexure on foreland basin stratigraphy (Beaumont, 1981). The success of flexural basin stratigraphy modeling, capable of incorporating subsurface loads related to plate tectonic forces operating on the lithosphere (e.g., Van der Beek and Cloetingh, 1992; Peper et al., 1994) led to the need to incorporate more structural complexity in these models, also in view of implications for the simulation of thermal maturation and fluid migration (e.g., Parnell, 1994). The necessary understanding of lithospheric mechanics and basin deformation was developed after a bridge was established between researchers studying deeper lithospheric processes and those who analyzed the record of vertical motions, sedimentation, and erosion in basins. This permitted the development of basin analysis models that integrate structural geology and lithosphere tectonics. The focus of modeling activities in 1990s was on the quantification of mechanical coupling of lithosphere processes to the near-surface expression of tectonic controls on basin fill (Cloetingh et al., 1995a, b, 1996, 1997). This invoked a process-oriented approach, linking different spatial and temporal scales in the basin record. Crucial in this was the testing and validation of modeling predictions in natural laboratories for which high-quality databases were available at deeper crustal levels (deep reflection- and refraction-seismic) and the basin fill (reflection-seismic, wells, outcrops), demanding a close cooperation between academic and industrial research groups (see Watts et al., 1993; Sassi et al., 1993; Cloetingh et al., 1994; Roure et al., 1996; Lacombe et al., 2007, Scheck-Wenderoth et al., 2009a,b).

These compressional systems are progressively shaped by the coupled influence of deep (flexure, plate rheology and kinematics) and surficial (erosion, sedimentation) geological processes, at different time scales, and constitute important targets for scientists interested in both fundamental and applied (fluids, hydrocarbons) aspects (e.g., Fig. 8). This provides the opportunity for Earth scientists from somehow disjoint domains, i.e., geologists, as well as geophysicists or geochemists, to share their complementary expertise on the processes governing the evolution of orogenic belts and adjacent forelands. In this context, a special emphasis must be given to make a "bridge" between the most recent advances in surface processes, geochemistry, provenance studies, field studies, analogue/numerical modelling, high resolution seismicity, and hydrocarbon prospect in forelands basins. The Task Force has been successful in bringing together many scientists from academia and industry from many countries over the Globe. As discussed in Lacombe et al. (2007), thrust belts and foreland basins provide key constraints on the orogenic evolution of adjacent mountain belts, their stratigraphic records resulting from the coupled influence of deep and surficial geological processes. This has important implications for exploration strategies for hydrocarbons in foothills areas with recent methodological and technical advances that have renewed our view on these important targets in both their fundamental and applied aspects.

Another challenge relates to Circum-Polar basins, i.e., the Arctic and Antarctic margins (Kirkwood et al., 2009). There is an obvious need for a comprehensive Atlas of Arctic regions, which would be also most welcome in industry. Some effort needs to be made to bridge the difference from the university focus on processes to the industrial interest in these regions or areas. Much existing data could be obtained if collaborative research themes were developed, as well as opportunities for new data collection on a major collaborative scale. Properly presented, with reasonable goals and time frames, Industry would probably also support such initiatives. The petroleum resources of the Polar regions are perceived to be great, and the Task Force provides an innovative and potentially effective mechanism to efficiently and effectively understand, not just the Arctic margins, but the sedimentary basins on continental margins in general, including their petroleum resources.

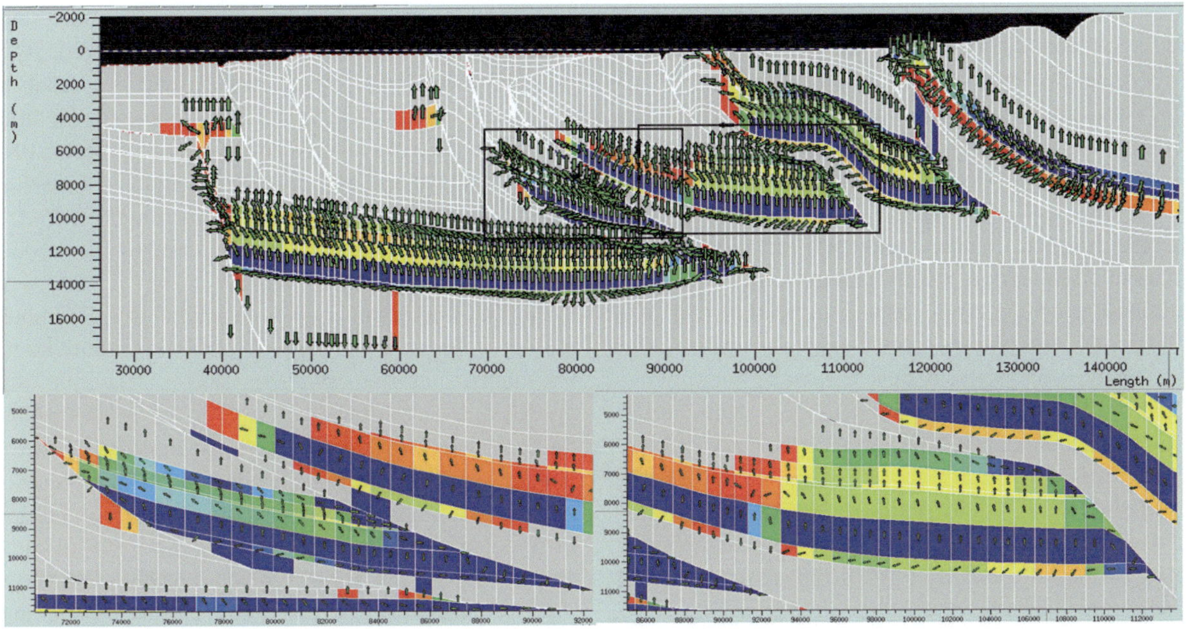

Fig. 8 Coupled kinematic and fluid flow modelling across the Outer Albanides Water saturation and hydrocarbon migration at the present day with two detailed zones: the Cika unit on the left and the Kurveleshi unit on the right inside. Note the accumulations of hydrocarbon in the Upper Cretaceous to Eocene carbonates located beneath the thrusts, sealed by Triassic evaporite and Oligocene flysch (Albania; Kendall et al., 2006)

The Task Force is also addressing the vertical movements in African basins and margins (Ghorbal et al., 2008; Bertotti et al., 2009). For example, the subsidence and inversion in the basins of the Atlas can hardly be explained by classical concepts. These basins are located between a passive non-volcanic continental margin to the west and a transform and/or convergent plate boundary to the north. In addition, the high topography of the Atlas is not supported by a lithospheric root but appears to be underlain by an unusually thin lithosphere (Missenard et al., 2006). Both long-term subsidence histories as well as recent processes can be evaluated in the context of forward thermo-mechanical models (e.g., Gouiza et al., 2009), ranging in scale from the sediment-fill to lithospheric and finally to whole mantle convection models and their surface expression. This integrated approach provides new insights on processes controlling present and past vertical movements and related stress/temperature conditions.

More lately, the Task Force initiated a research initiative in the Gulf of California rift system and young rifted margins in general (Delgado and Ortuño, 2008). The Gulf of California natural laboratory constitutes an unique site in the World where (1) deep lithospheric structure can be linked to incipient versus aborted pull-apart basins and ongoing/versus recent oceanization, and (2) only limited active faults and volcanism have been identified to date. Still, there is a need for deep geophysical controls in the area, the study of which being definitively of interest for international collaboration and more integrated studies. These could be also of major interest for the energy industry, because the unique exposures of the Gulf of California basin could provide first class analogues for the study of petroleum and geothermal systems associated with rifts and passive margins, and because in many other margins, syn-rift series are no longer exposed at the surface.

Further initiatives of the Task Force are targeted on the formation and deformation of Middle-East basins and related crustal/lithospheric processes.

Task Force 7: Temporal and Spatial Change of Stress and Strain

Temporal and Spatial Changes of Stress and Strain is a component of the second ILP theme Continental Dynamics and Deep Processes. Stress and strain are fundamental quantities which control and describe the geodynamic processes in the Earth's crust. However, their relationship on different temporal and spatial scales is a challenging research field. This Task Force aims to identify, analyse and interpret the variations of crustal stress and strain patterns at different spatial scales on time spans that range from 0.1 to 10,000 years. Major results of Task Force 7 are that (1) elastic strain accumulation can not be directly mapped to brittle strain release, indicating that other plastic, but aseismic processes such as creeping, slow earthquakes, and stress diffusion must be quantified on their characteristic spatial and temporal scales and (2) that the accumulation of strain can be very different depending on the temporal and spatial scale looked at (Friedrich et al., 2003; Heidbach et al., 2007, 2008a). Therefore, the interpolation as well as the extrapolation backwards and forward in time is probably often non-linear and depends on the aforementioned processes and the physical and structural conditions in the area of interest.

One key target in this context is to contribute to the time-dependent seismic hazard assessment by means of numerical models that describe stress changes due to inter-, co-, and postseismic stress transfer (Steacy et al., 2005; Hergert and Heidbach, 2006). These processes could not have been studied in great detail without the major progress in data acquisition. In the last decade the advancement of modern satellite techniques such as GPS and InSAR have provided excellent time-series of the surface deformation pattern that revealed new processes that were previously not anticipated (Hergert and Heidbach, 2006; Fielding et al., 2009). Also the crustal stress field observations have increased significantly. The former ILP project World Stress Map (WSM) released in 2008 a new global WSM database that almost tripled the amount of stress data records in comparison to the WSM database release 1992 (Heidbach et al., 2008b). At the same time the knowledge of the 3D structure and its physical properties increased significantly due to modern tomography studies and data compilations (e.g., Tesauro et al., 2008). This massive increase of observational data now enables to set up complex 3D model geometries with high resolution in order to investigate the geodynamic processes that control the stress state and the deformation of the Earth's crust. The vast amount of model-independent constraints enables to narrow the model parameter space.

So far, numerical models simulate stress changes mainly in terms of changes of Coulomb failure stress (Heidbach and Ben-Avraham, 2007). However, this approach is not capable to answer the question how far faults are from failure, but only describe the stress changes with respect to an arbitrarily chosen reference stress state. Thus, the demand of the next generation of geomechanical models is to simulate the absolute stress state of the Earth's crust. Furthermore, these models should also be consistent with kinematic and dynamic observations at the same time.

A project that attempts to succeed in the task of modeling the absolute stress state was performed as part of the CEDIM project Megacity Istanbul. Aim of the geomechanical, numerical model was to describe the 3D absolute stress state and the 3D velocity field and their temporal evolution for the Sea of Marmara region (Hergert et al., 2007; Hergert, 2009). Here, the North Anatolian fault system (NAF) cuts through the Marmara Sea (Fig. 9). Its seismically quiescence lasts now since mid eighteenth century imposing a major seismic risk for the city of Istanbul, a city with app. 15 Million inhabitants located 20 km north of the fault segment that is expected to fail next after the 1999 Izmit earthquake. The numerical model geometry contains the complex fault geometry as well as the structure of the topography, bathymetry, basement, and Moho (Fig. 10). Boundary conditions are the push and pull of the Anatolian block and Hellenic Arc (see also Reilinger et al., 2006), respectively as well as gravity that imposes a significant contribution to the

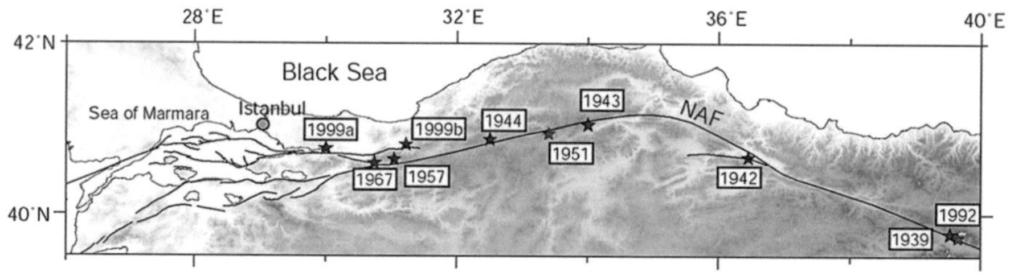

Fig. 9 Epicentres of major earthquakes (Mw > 6.5) along the North Anatolian Fault between 1939 and 1999. Box indicates the model area displayed in Fig. 10 (from Hergert et al., 2007)

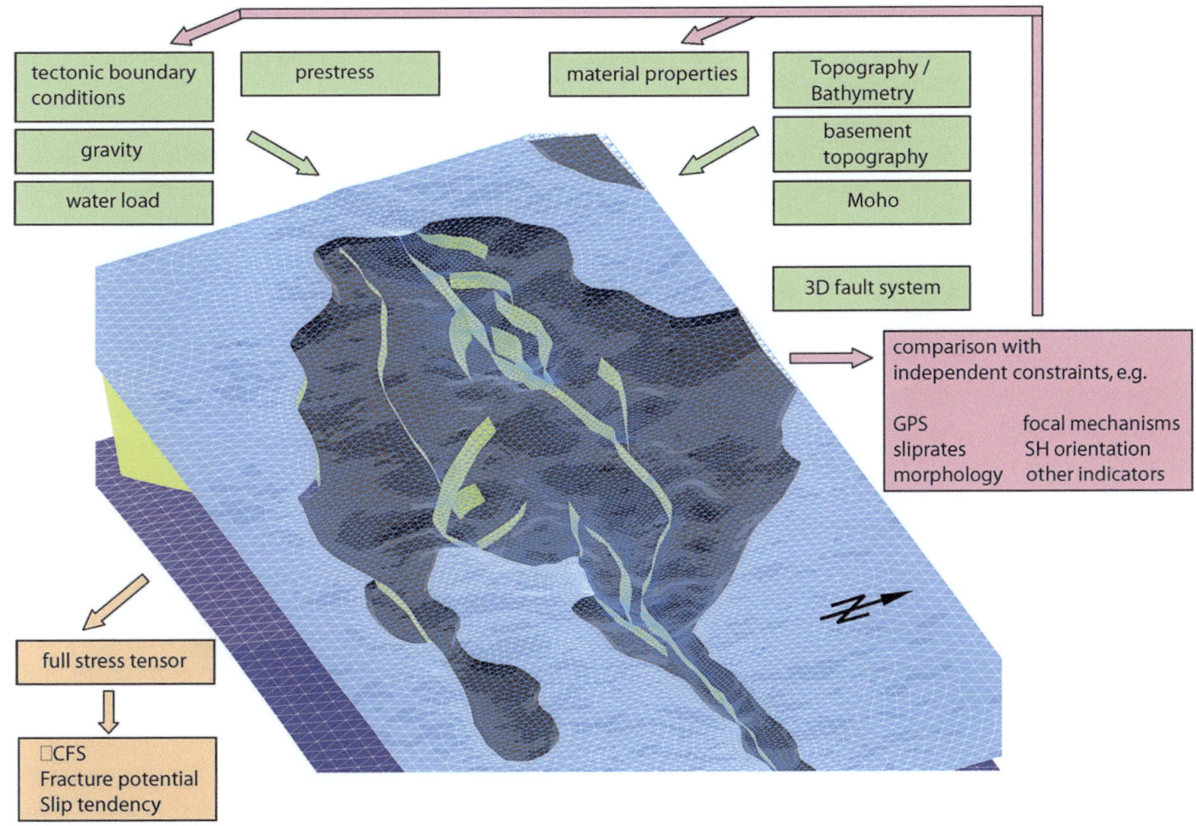

Fig. 10 Mesh of the finite element model and modeling concept. The model geometry involves the active fault system (*yellow*), topography and bathymetry (*solid* and *transparent light blue*), basement topography (*dark grey*) and the Moho (*blue*). *Green boxes* indicate the model input (geometry, boundary conditions, material properties), *pink boxes* the calibration and validation procedure and *light brown* boxes the model output and post processing (from Hergert et al., 2007)

stress field due to the lateral density inhomogeneities. This model fits the GPS data, subsidence rates, tectonic regime, and stress orientation to a high degree (Hergert et al., 2007; Hergert, 2009). In particular the model results reveal that (1) the strain accumulation is distributed amongst the three major branches of the NAF, (2) that the right-lateral slip-rates along the major fault are smaller than previously assumed, and (3) that they vary significantly along strike. Even though the right-lateral slip rates are about 30% smaller on the main fault in comparison with previous publications that were based on simpler models, the absolute stress state of the model indicates that the central segment is mature in terms of failure (Hergert, 2009). These findings contribute significantly to the geodynamic understanding of the Marmara Sea region that has so far been a topic of major debate. Furthermore the new model concept has to be incorporated into the time-dependent seismic hazard assessment for that region.

Recent results of Task Force 7 will appear in a special issue of Tectonophysics with approximately 25 manuscripts (Heidbach et al., 2009), with a focus on stress research.

Task Force 8: Baby-Plumes – Origin, Characteristics, Lithosphere-Asthenosphere Interaction and Surface Expression

In recent years a number of high-resolution integrated seismic projects across areas with Tertiary to recent volcanism in central Europe (e.g., Granet et al., 1995; Ritter et al., 2001) have been stimulated by the project TRACK (tracking a mantle plume by seismological means) in combination with detailed geochemical studies. These have demonstrated the existence of a number of small-scale, almost cylindrical, upwellings of low-velocity mantle material (~ 100–150 km in diameter) within the upper mantle, the so-called "baby-plumes". These "baby-plumes" have some very similar characteristics to classical plumes (as proposed by Schilling and others), but two distinct differences:

- They are much smaller in size than classical plumes;
- They do not seem to "have" a plume head.

These baby-plumes suggest that there might exist a number of different classes of plumes originating from different depths (i.e., different interfaces) within our planet. So far these baby plumes have mainly been identified within the European mantle, but some new results from China (Zhao et al., 2004, 2007; Zhao, 2007; Huang and Zhao, 2006; Lei et al., 2009) suggest that similar regimes exist there. This highlights the necessity to search for these kind of features on other continents as well (see also Nolet et al., 2007; Burov and Cloetingh 2009; Smith et al., 2009; Waverzinek et al., 2008).

The following features seem to be characteristic of these baby-plumes in Europe:

- Small-scale convective instabilities within the upper mantle beneath Europe appear to originate in the mantle Transition Zone (410–660 km depth);
- There is a strong correlation between the location of "upwellings" and lithospheric architecture;
- The upwellings appear to be concentrated around the edge of a region of subducted slabs at the base of the upper mantle;
- Basaltic magmas derived by decompression partial melting of the upwelling mantle "diapirs" have the distinctive geochemical signature of a common mantle source component – the European Asthenospheric Reservoir (EAR) (Fig. 11).

The PLUME task force has been involved in an International Polar Year seismic project, (LAPNET) in the arctic region. In this project it is not only the detailed knowledge of upper mantle structure that is of major concern but whether in an environment like an old craton, which has not experienced any major tectonic events for over one billion years, it is still possible to trace subduction and plume related features across the Archean-Proterozoic boundary.

The LAPNET array is part of the POLENET project and most likely the biggest array ever installed in the arctic region. The LAPNET consortium includes several seismological institutions from the Baltic states as well as other European countries and covers essentially the northernmost part of Finland and adjacent countries. LAPNET research will result in a detailed 3D seismic model of the crust and upper mantle down to the mantle Transition Zone. The new seismic experiment will provide unique, more precise information on lithospheric structure and thickness beneath the Karelian craton, with its high diamond potential, as well as the area of transition from Archean to Proterozoic

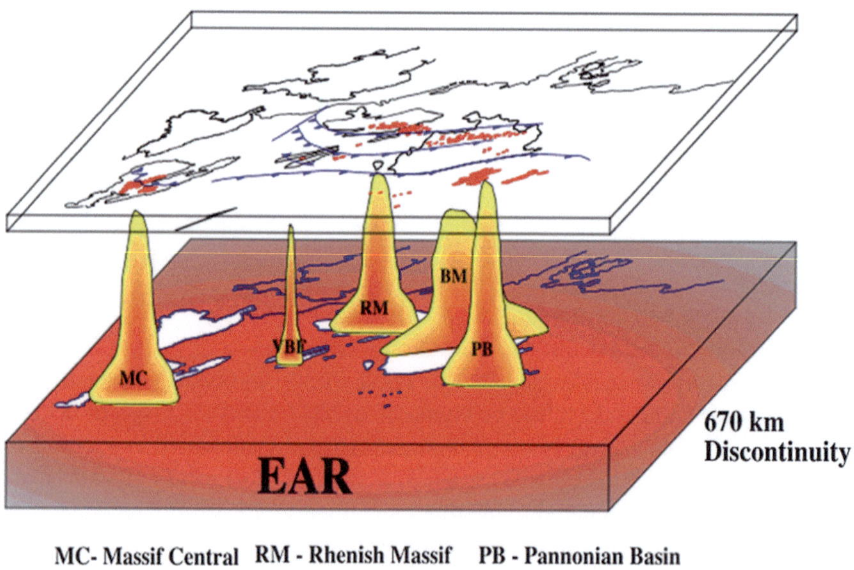

Fig. 11 EAR: European asthenospheric reservoir (after Granet et al., 1995)

lithosphere. More information can be obtained from Dr. Elena Koslovskaya, Oulu University, Finland.

A detailed geodynamic study of the Pannonian Basin in Hungary, involving a consortium of Austrian, British, French and Hungarian scientists, aims to study the geodynamic evolution of the Pannonian Basin by using broadband seismology, lithospheric tomography and numerical modelling of lithosphere dynamics. Preliminary results were presented at a workshop in Siofok, Hungary (2007) and at EGU meetings in 2007 and 2008. More details on this project can be obtained from Greg Houseman and Graham Stuart of Leeds University, UK (greg@earth.leeds.ac.uk).

Considerable progress has been made in the geochemical characterisation of the upper mantle beneath Europe and the relationship of mantle heterogeneities to mantle dynamics (e.g., Landes et al., 2007). A European Mantle workshop (EMAW) was held in August 2007 in Ferrara, Italy, followed by a symposium at the 33rd IGC in Oslo, Norway, in August 2008. Papers based on the EMAW workshop will be published in a thematic issue of the Journal of Petrology in 2009.

A new paradigm for the origin of "baby-plumes" in the upper mantle has been proposed which attributes these features to supercritical fluid streaming from the 410 km seismic discontinuity at the top of the Transition Zone (Wilson, 2008).

There are a number of issues which need further research efforts to shed light on the origin and nature of baby-plumes:

1. It still needs to be established if the small plume-like instabilities which are observed beneath the European continent are interconnected at a certain depth.
2. There is still debate on the source region of mantle plumes in general and baby-plumes in particular. Only much larger and denser seismic antennas, such as e.g., foreseen with the EPOS initiative, will eventually give the necessary data to tackle this particular question.
3. Detailed studies of the Transition Zone region are of particular interest in identifying the source of the baby-plumes. Integrated seismic studies, using all available seismic techniques, will be necessary to resolve this in combination with other geophysical and geochemical data sets.

Regional Coordinating Committee Europe: TOPO-EUROPE

TOPO-EUROPE addresses the 4-D topographic evolution of the orogens and intra-plate regions of Europe through a multidisciplinary approach linking geology, geophysics, geodesy and geotechnology (Cloetingh, 2007; Cloetingh, TOPO-EUROPE Working Group, 2007, Cloetingh et al., 2009; see also www.topo-europe.eu). TOPO-EUROPE integrates monitoring, imaging, reconstruction and modelling of the interplay between processes controlling continental topography and related natural hazards. Until now, research on neotectonics and related topography development of orogens and intra-plate regions has received little attention. TOPO-EUROPE initiates a number of novel studies on the quantification of rates of vertical motions, related tectonically controlled river evolution and land subsidence in carefully selected natural laboratories in Europe (Fig. 12). From orogen through platform to continental margin, these natural laboratories include the Alps/Carpathians-Pannonian Basin System, the West and Central European Platform, the Apennines-Aegean-Anatolian region, the Iberian Peninsula, the Scandinavian Continental Margin, the East-European Platform, and the Caucasus-Levant area. TOPO-EUROPE integrates European research facilities and know-how essential to advance the understanding of the role of topography in Environmental Earth System Dynamics. The principal objective of the network is twofold. Namely, to integrate national research programs into a common European network and, furthermore, to integrate activities among TOPO-EUROPE institutes and participants. Key objectives are to provide an interdisciplinary forum to share knowledge and information in the field of the neotectonic and topographic evolution of Europe, to promote and encourage multidisciplinary research on a truly European scale, to increase mobility of scientists and to train young scientists.

Continental topography is at the interface of deep Earth, surface and atmospheric processes (Fig. 13). In the recent past, catastrophic landslides and rock falls have caused heavy damage and numerous fatalities in Europe. Rapid population growth in river basins, coastal lowlands and mountainous regions and global warming, associated with increasingly frequent exceptional weather events, are likely to exacerbate the

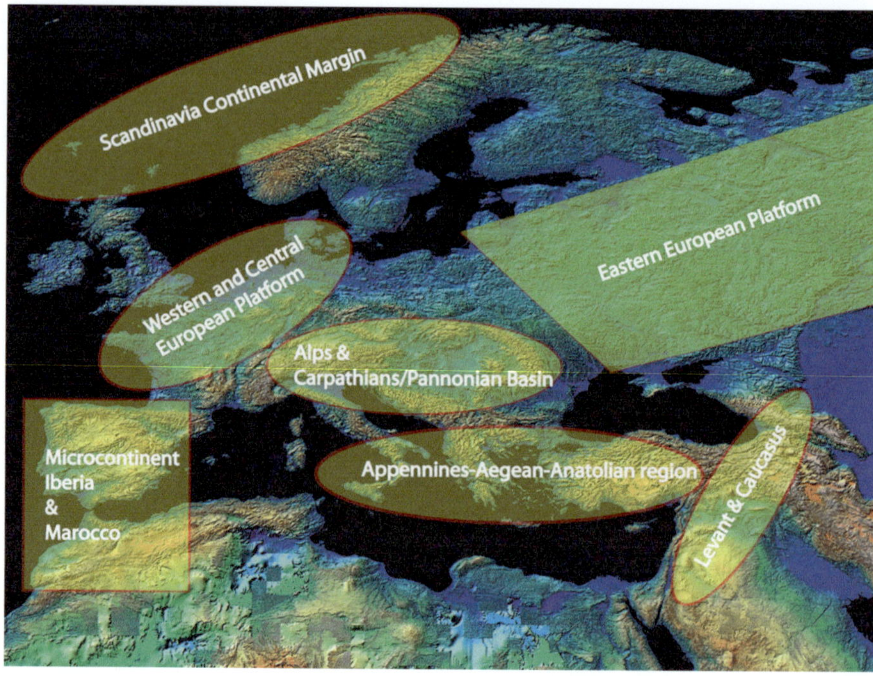

Fig. 12 Examples of natural laboratories selected for TOPO-EUROPE (Cloetingh, TOPO-EUROPE Working Group, 2007)

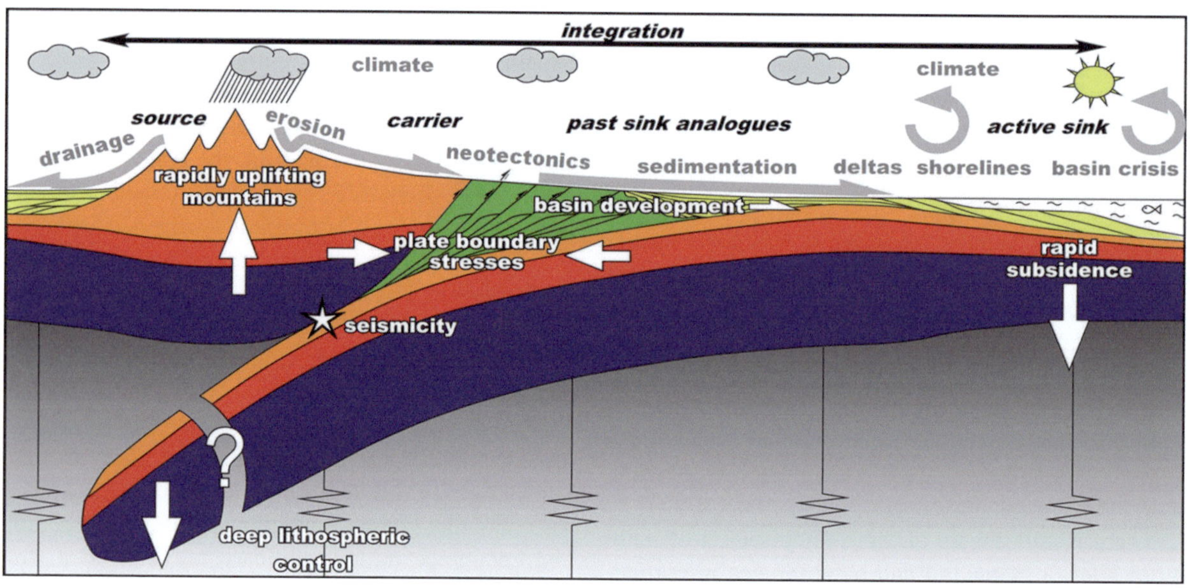

Fig. 13 Schematic source-to-sink systematics and coupled orogen-basin evolution in the aftermath of continental collision, taking the Pannonian Basin – Carpathians – Black Sea System as an example (from Cloetingh, TOPO-EUROPE Working Group, 2007)

risk of flooding and devastating rock failures. Along active deformation zones, earthquakes and volcanic eruptions cause short-term and localized topography changes. These changes may present additional hazards, but at the same time permit, to quantify stress and strain accumulation, a key control for seismic and volcanic hazard assessment. Although natural processes and human activities cause geohazards and environmental changes, the relative contribution of the respective components is still poorly understood.

That topography influences climate is known since the beginning of civilization, but it is only recently that we are able to model its effects in regions where good (paleo-) topographic and climatologic data are available.

An important step has been the selection in early 2008 by the European Science Foundation (ESF) of TOPO-EUROPE as one of its large scale European collaborative research initiatives (EUROCORES). The EUROCORES (ESF Collaborative Research) Scheme is an innovative ESF instrument to stimulate collaboration between researchers based in Europe to maintain European research at an international competitive level. The principle behind the EUROCORES Scheme is to provide a framework for national research funding organisations to fund collaborative research, in and across all scientific areas. Participating funding agencies (national research councils and academies and other funding organisations) jointly define a research programme, specify the type of proposals to be requested and agree on the peer review procedure to be followed. The ESF provides support for the networking of funded scientists while the funding of the research stays with national research funding organisations. Further background information on EUROCORES Scheme may be found on the ESF web site (http://www.esf.org/activities/eurocores/programmes/topo-europe.html). In response to the ESF call for proposals, 42 outline proposals were submitted, resulting in 22 full proposals submitted for international peer-review. Out of these, ten collaborative research projects (CRP's) were selected for funding with a total amount of 13.5 M€ for the ESF EUROCORES TOPO-EUROPE, opening up research positions for more than 60 PhD students and postdocs.

Regional Coordinating Committee Asia: TOPO-CENTRAL-ASIA: 4D Topographic Evolution in Central Asia

"Lithosphere Dynamics and Environmental Changes since Mesozoic" is a sister project of TOPO-EUROPE. The topography of the Earth has changed dramatically since Mesozoic, as a result of amongst others the dismembering of Gondwana, the opening of the Atlantic Ocean, closing of the Tethys Ocean, collision between the Indian and Eurasian Plates. These changes have shaped Central Asia into a unique region in the world: the North China Craton lost its thick Archean lithospheric root; the Tethys Ocean closed and many oil-gas fields formed along its previous seaway; the Tibetan Plateau rose and changed atmospheric circulation as well as the global climate; and intensive intraplate deformation developed in areas far from the India-Eurasia collision zone. These important events and their natural records have made Central Asia an international natural laboratory of Earth science. The project TOPO-CENTRAL-ASIA (Fig. 14) studies the Earth's mantle-lithosphere interaction process, its coupling to the shallow Earth System and feedback mechanisms between Solid-Earth processes and topography.

The main themes of the TOPO-CENTRAL-ASIA include:

- Deep structure of Central Asia;
- Tectonic evolution of Central Asia;
- Tibetan Plateau Uplift: deep processes and environmental effects;
- Dynamic topography and processes of removal of the subcontinental lithosphere below the North China Craton;
- Modelling the response of the lithosphere to surface processes.

The central Tibetan Plateau was uplifted by 40 Ma ago (Wang et al., 2008), whereas regions south and north of the central plateau gained elevation significantly later. To the south, the Himalaya rose during the Neogene, and to the north, the Qilian Shan was rapidly uplifted in the Late Cenozoic. These findings differ from inferences from previous studies that the Tibetan Plateau was constructed at the same time in the early Paleocene (50 Ma) in the Tethyan Himalaya to the south and in the Nan Shan and Qilian Shan 1,400 km to the north (Yin and Harrison, 2000), or that the Tibetan Plateau grew from the south, in Eocene, to the north, in the Late Miocene -Quaternary (Tapponnier et al., 2001). These different uplift models put forward a new challenge for future studies of the links between the Tibetan Plateau uplift and global climate changes. Different uplift processes, whether the Tibetan Plateau was uplifted totally or partially, have different effects on global and Asian climate scenarios.

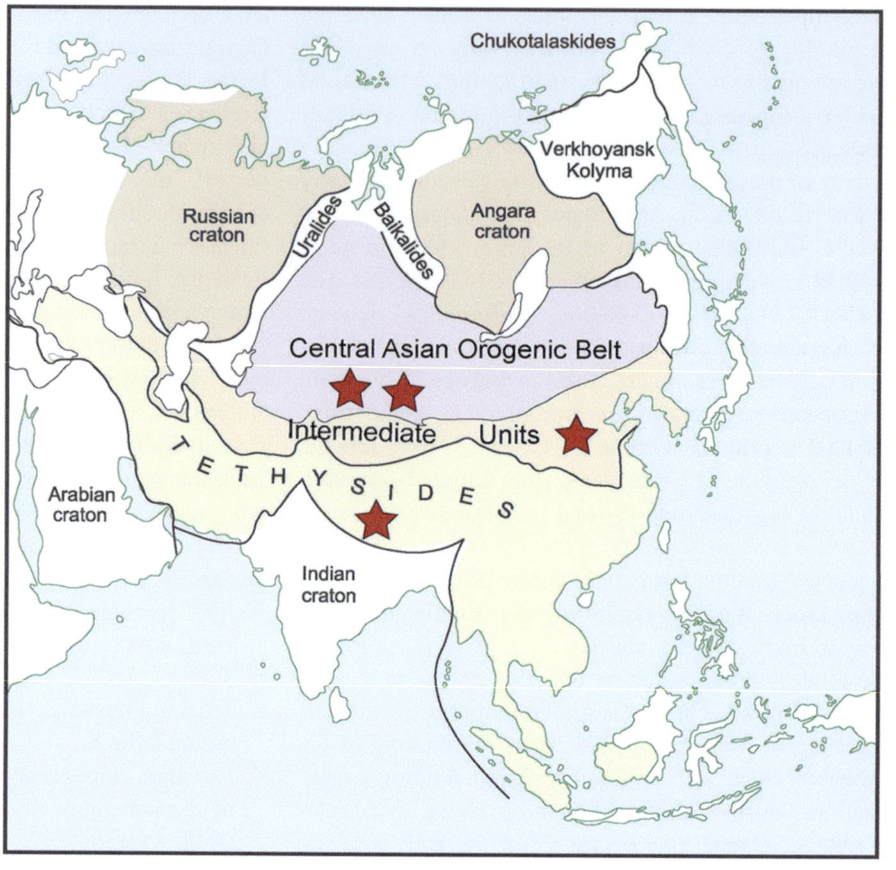

Fig. 14 The project TOPO-CENTRAL-ASIA studies the Tibetan Plateau, the Central Asian Orogenic Belt, Cenozoic orogenesis-climate-erosion processes in Northwest China, as well as dynamic topography and processes of removal of the subcontinental lithosphere in North China. Stars mark the recent study areas. The Asian tectonic map is modified from Xiao et al. (2009a)

Central Asia has had a long, complex geological history. The Central Asian Orogenic Belt is an important tectonic unit formed by accretion of island arcs, ophiolites, oceanic islands, seamounts, accretionary wedges, oceanic plateaux and micro-continents (Windley et al., 2007). Therefore, both TOPO-CENTRAL-ASIA and ERAS (Earth Accretionary Systems in space and time) have studied this system. Some new results of the joint research of the two projects show evidence for Paleozoic multiple subduction-accretion processes in Northwest China (Xiao et al., 2009a, b).

An important Mesozoic topographical change in Central Asian has been the formation and destruction of a Mesozoic peneplain-plateau pair. Recent studies revealed a large Mesozoic peneplanation surface, which has been preserved from middle Jurassic to Neogene, in Central Asia (e.g., Jolivet et al., 2007). At the same time, in the east of China a major Mesozoic plateau existed, known as the East China Plateau. The existence of the Mesozoic East China Plateau is evidenced by paleontology (Chen, 1979) and petrology (e.g., Zhang et al., 2001). However, the Mesozoic peneplanation surface has been uplifted to an altitude of 4,000 m, in Gobi Altay and Altay, Mongolia (Jolivet et al., 2007), while the Mesozoic East China Plateau has changed into the North China Plain and the Bohai Bay basin. A geophysical survey (Chen et al., 2008) carried out in the framework of the project displayed that the thickness of the lithosphere beneath the North China Craton has been thinned to < 80 km (Fig. 15). This dramatic topographical change reflects Mesozoic lithosphere dynamics in Central Asia. A number of questions remain on, for example, the spatial extension of the peneplain, which processes and conditions have lead to the formation and the preservation of a peneplain, when and how the Mesozoic East China Plateau was formed, and when and how it was destructed. Obviously, more comprehensive and multidisciplinary studies are required, including paleotopography, paleo-altitude, as well as research on magmatic activities and deformation.

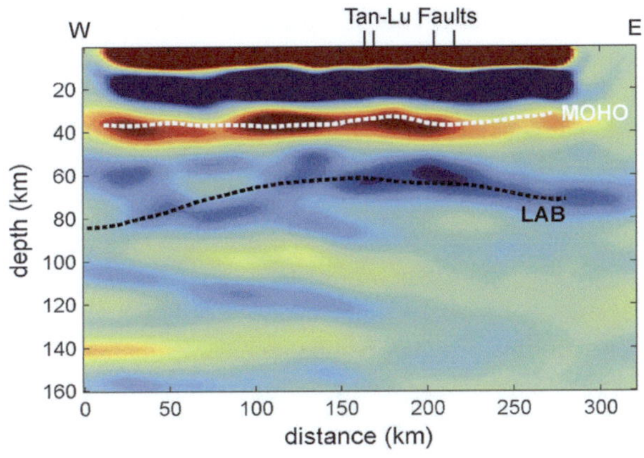

Fig. 15 Geophysical survey displaying that the thickness of the lithosphere beneath the North China Craton has been thinned to < 80 km (after Chen et al., 2008). LAB: Lithospere-Asthenosphere Boundary

Regional Coordinating Committee – DynaQlim: Upper Mantle Dynamics and Quaternary Climate in Cratonic Areas

DynaQlim aims to integrate existing data and models on glacial isostatic adjustment (GIA) processes, including both geological and geodetic observations. The themes of DynaQlim include Quaternary climate and glaciation history, postglacial uplift and contemporary movements, ice-sheets dynamics and glaciology, postglacial faulting, rock rheology, mantle xenoliths, past and present thermal regime of the lithosphere, seismic structure of the lithosphere, and gravity field modelling. Combining geodetic observations with seismological investigations, studies of the postglacial faults and continuum mechanical modelling of GIA, DynaQlim offers new insights into properties of the lithosphere. Another step toward a better understanding of GIA has been the joint inversion of different types of observational data – preferentially connected with geological relative sea-level evidence of the Earth's rebound during the last ten thousand years.

The main objectives of this project are:

- to collect all existing relevant geo-data for improving geophysical models;
- to create more accurate kinematic 3-D models in the Fennoscandian (Fig. 16) and other previously or currently glaciated areas, and in general, understand better the dynamics of the Earth's crust;

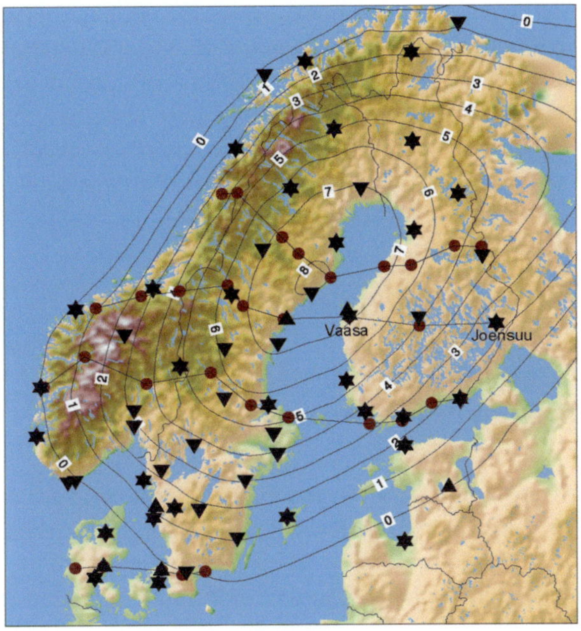

Fig. 16 DynaQlim -upper mantle dynamics and Quaternary climate in Cratonic areas: Observing the contemporary post-glacial rebound in Fennoscandia. The upside-down triangles are permanent GNSS stations, triangles are stations where regular absolute gravity is regularly measured, and dots with joining lines are the land uplift gravity lines, measured since the mid-1960's. *Contour lines* show the apparent land uplift relative to the Baltic mean sea level 1892–1991, based on Nordic uplift model NKG2005LU (Vestøl, 2006; Ågren and Svensson, 2007)

- to understand the relations between the upper mantle dynamics, mantle composition, physical properties, temperature and rheology;
- to understand the effect of Quaternary climate, glaciation cycles and ice thickness on contemporary rebound rate;
- to study the Quaternary climate variations and Weichselian (Laurentian and other) glaciations;
- to study the response of the gravity field with different ice loads and to understand the secular change of gravity in the rebound area and especially in its margin areas,
- to apply the results in the polar regions, like the glaciation history in Antarctica and the Holocene rebound history in the Fennoscandian area;
- to offer reliable data and stable reference frames for other disciplines in geoscience and for studies of global change, sea level rise and climate variation, and therefore, to maintain national, regional and global reference frames also in the future.

The DynaQlim project is expected to lead to a more comprehensive understanding of the Earth's response to glaciations, improved modelling of crustal and upper mantle dynamics and rheology structure. An important goal is to construct and improve coupled models of glaciation and land-uplift history and their connection to climate evolution on the time scales of glacial cycles.

International Continental Scientific Drilling Programme (ICDP)

A major hurdle in modern Earth science research is the very limited approach to the subsurface, where active processes happen and where structures are preserved unaffected by surface influence. Although it is costly and time consuming, drilling is the only means to directly access the underground for sampling, measuring, and to validate models based on geophysical exploration from surface. The International Continental Scientific Drilling Program, ICDP, is since 1996 paving the road to ease scientific drilling. It has been founded with support of the International Lithosphere Program to co-finance and coordinate continental scientific drilling efforts with research topics of high international priority. More than twenty major drilling projects have been executed to date within the framework of ICDP. The program focuses on challenging themes of geoscientific and socio-economic relevance such as Climate Dynamics and Global Environments, Impact Craters and Impact Processes, the Geobiosphere, Active Faults and Earthquake Processes, Hotspot Volcanoes and Large Igneous Provinces, Convergent Plate Boundaries and Collision Zones, Volcanic Systems and Thermal Regimes and Natural Resources (Fig. 17). A key to the success of the program is that it provides the necessary start-up financing for cost-intensive projects at locations of global significance. As of 2009, 17 member countries provide membership fees which are allocated for projects in a concept of commingled funding and international cost sharing.

Outstanding examples of the scientific progress accomplished in the framework of ICDP (Harms et al., 2007) include coring through and instrumenting of the active traces of the San Andreas Fault Zone at 3 km depth to elucidate fundamental processes of earthquake cycling. A major improvement in this project includes coring of sidetrack wells into a constantly deforming (~4 cm/year) fault at more than 3 km depth (Fig. 18). Furthermore, the orientation of horizontal stress could be measured proving the weakness of the fault at depth leading to slipping at low levels of shear stress (Boness and Zoback, 2006). Long-term observations are conducted at levels where minor earthquakes occur repeatedly and regularly to register physical and chemical properties that control deformation and earthquake generation in a transform plate boundary.

Other in-situ studies in ICDP aim at active volcanic processes such as the hazardous Unzen Volcano in Japan where the magma conduit of a recent eruption has been cored at degassing depth of about 2 km below

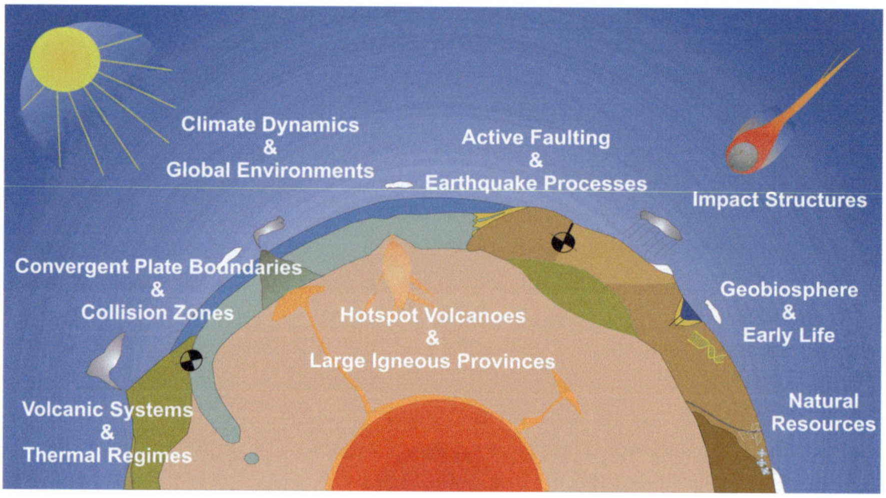

Fig. 17 Scientific topics of the ICDP depicted on a global scale Earth section (after Harms and Emmermann, 2007)

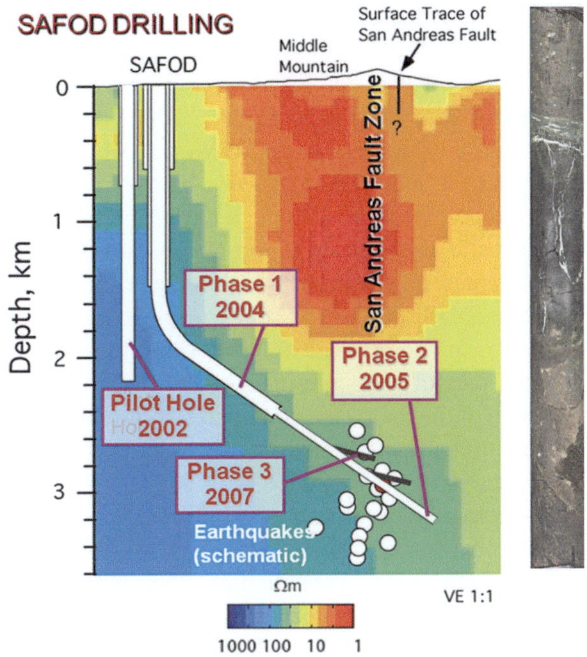

Fig. 18 Schematic section of San Andreas fault zone observatory at depth with different phases of drilling. Background color shows electric resistivity measured along a profile perpendicular to the faults strike. The *bold black lines* at the bottom show sidetrack coreholes drilled through the active trace of the San Andreas Fault. The core photo shows a large black serpentine clast cut by calcite veins embedded in foliated fault gouge (courtesy ICDP, see also http://www.icdp-online.org)

the dome. The detected dacitic dyke system which feeds both effusive and explosive eruptions was unexpectedly cool due to enhanced groundwater circulation (Sakuma et al., 2008). Even the structure of oceanic hot spots, a highly debated topic in Earth sciences, has been tested by scientific drilling. At the most prominent volcanic edifice on the globe, Hawaii, one Million years of volcanic built-up is documented in the pancake-like pile of lava flows of Mauna Kea. This layered structure allows charting of the complex channeled buoyancy of lowermost mantle materials entraining as plume upper mantle on its passage to surface (Stolper et al., 2009).

Almost 180 craters on Earth are known currently that have been formed by astrophysical chance when celestial bodies such as asteroids collided with our planet. Drilling to study cratering processes provides data not only for modeling the impactor size but also for modeling the energy release through melting, evaporation, ejection and, most importantly, for modeling of the environmental consequences of such dramatic events. ICDP drilled the 200-km-wide Chicxulub Crater in Mexico (Hecht et al., 2004; Dressler et al., 2003) and the 60 km Chesapeake Bay Crater in the Eastern U.S. The latter underwent a complex microbiological evolution initiated by an impact-related thermal sterilization and subsequent post-impact repopulation (Gohn et al., 2008). For such large craters fluidization of target rocks leads to the formation of a central uplift, whereas the peak of the small, 10 km Bosumtwi Crater in Ghana (Ferrière et al., 2008) was formed by brittle deformation processes.

With smart, cost-effective drilling, paleo-climate and paleo-environmental evolution is being studied on continuous lake sediments from Lakes Titicaca, Malawi, Bosumtwi, Qinghai, and Peten Itza. The latter for example provided new insights into the changes of precipitation patterns due to variations of the Intertropical Convergence Zone over Central America (Hodell et al., 2008). Sediments in the African tropical Lakes Malawi and Bosumtwi shed new light on a megadrought at about 100 K years before present with implications for migration of early humans out of Africa (Scholz et al., 2007).

Several other ICDP-funded projects provided novel awareness about active processes and geological resources (Harms et al., 2007), while ongoing and future exploration can be monitored on the programs web resource (http://www.icdp-online.org).

Perspectives on Integrated Solid Earth Sciences

The papers in this IYPE volume provide a review of recent developments in aspects of integrated solid Earth sciences that can be considered as frontier research.

Tesauro et al. (2009a) (this volume) present EuCRUST-07, a new 3D model of the crust for Western and Central Europe that offers a starting point in any kind of numerical modelling to remove the crustal effect beforehand. The digital model (35°N, 71°N; 25°W, 35°E) consists of three layers: sediments and two layers of the crystalline crust. The latter are characterized by average P-wave velocities (Vp), which were defined based on seismic data. The model was obtained by assembling together at a uniform $15' \times 15'$ grid available results of deep seismic reflection, refraction and receiver function studies. The Moho depth variations were reconstructed by these authors by merging the most robust and recent maps existing for the European region and compiled using published interpretations of seismic profiles. EuCRUST-07 demonstrates large differences in Moho depth with previous compilations: over ± 10 km in some specific areas (e.g., the Baltic Shield). The basement is outcropping in some parts of Eastern Europe, while in Western Europe is up to $\sim$16 km deep, with an average value of 3–4 km, reflecting the presence of relatively shallow basins. The velocity structure of the crystalline crust turns to be much more heterogeneous than demonstrated in previous compilations, average Vp varying from 6.0 to 6.9 km/s. In comparison to existing models, the new model shows average crustal velocity values distributed over a larger and continuous range. Furthermore, the results of EuCRUST-07 are used by Tesauro et al. (this volume) to make inferences on the lithology, which is typical for different parts of Europe. The new lithology map shows the Eastern European tectonic provinces represented by a granite-felsic granulite upper crust and a mafic granulite lower crust. Differently, the younger Western European tectonic provinces are mostly characterized by an upper and lower crust of granite-gneiss and dioritic composition, respectively.

In the companion paper by Tesauro et al. (2009b) (this volume), a new thermal and rheological model of the European lithosphere (10°W–35E; 35 N–60 N) is implemented based on a combination of recently obtained geophysical models. To determine temperature distribution they use a new tomography model, which is improved by correcting a-priori for the crustal effect using the digital model of the European crust (EuCRUST-07). The uppermost mantle under Western Europe is mostly characterized by temperatures in a range of 900–1,100°C with the hottest areas corresponding to the basins, which have experienced recent extension (e.g., Tyrrhenian Sea and Pannonian Basin). By contrast, the mantle temperatures under Eastern Europe are about 550–750°C at the same depth and the minimum values are found in the north-eastern part of the study area. EuCRUST-07 and the new thermal model are used to calculate strength distributions within the European lithosphere. Lateral variations of lithology and density derived from EuCRUST-07 are used to construct the new model. Following the approach of Burov and Diament (1995), the lithospheric rheology is employed to calculate variations of the effective elastic thickness of the lithosphere. According to these estimates, in Western Europe the lithosphere is more heterogeneous than that in Eastern Europe. Western Europe with its predominant crust-mantle decoupling is mostly characterized by lower values of strength and elastic thickness. The crustal strength provides a large contribution ($\sim$50% of the integrated strength for the whole lithosphere) in most part of the study area ($\sim$60%). The results reviewed in this paper shed light on the current debate on the strength partition between crust and mantle lithosphere.

As pointed out by Burov (2009) (this volume), simple mechanical considerations show that many tectonic-scale surface constructions, such as mountain ranges or rift flanks that exceed certain critical height (about 3 km in altitude, depending on rheology and width) should flatten and collapse within a few My as a result of gravitational spreading that may be enhanced by flow in the ductile part of the crust. The elevated topography is also attacked by surface erosion that, in case of static topography, would lead to its exponential decay on a time scale of less than 2.5 My. However, in nature, mountains or rift flanks grow and stay as localized tectonic features over geologically important periods of time (> 10 My). To explain the long-term persistence and localized growth of, in particular, mountain belts, a number of workers have emphasized the importance of dynamic feedbacks between surface processes and tectonic evolution. Surface processes modify the topography and redistribute tectonically significant volumes of sedimentary material, which acts as vertical loading over large horizontal distances. This results in dynamic loading and unloading of the underlying crust and mantle lithosphere, whereas topographic contrasts are required to set up erosion and sedimentation processes. As demonstrated by Burov (2009), tectonics therefore could be a forcing factor of surface processes and vice versa. He suggests that the feedbacks between tectonic and surface processes are realised via 2 interdependent mechanisms: (1) slope, curvature and height dependence of the erosion/deposition rates; (2) surface load-dependent subsurface processes such as isostatic rebound and lateral ductile flow in the lower or intermediate crustal channel. Loading/unloading of the surface due to surface processes results in lateral pressure gradients that, together with low viscosity of the ductile crust, may permit rapid relocation of the matter both in horizontal and vertical direction (upward/downward flow in the ductile crust). In his paper, Burov (2009) provides an overview of a number of coupled models of surface and tectonic processes, with a particular focus on 3 representative cases: (1) slow convergence and erosion rates (Western Alps), (2) intermediate rates (Tien Shan, Central Asia), and (3) fast convergence and erosion rates (Himalaya, Central Asia).

Roure et al. (2009) (this volume) point out that thanks to a continuous effort for unravelling geological records since the early days of coal and petroleum exploration and water management, the architecture and chrono-litho-stratigraphy of most sedimentary basins has been accurately described by means of conventional geological and geophysical studies, using both surface (outcrops) and subsurface (exploration wells and industry seismic reflection profiles) data. However, the understanding of the early development and long term evolution of sedimentary basins usually requires the integration of additional data on the deep Earth, as well as kinematic-sedimentological and thermo-mechanical modelling approaches coupling both surface and deep processes.

In the last twenty years, major national and international efforts, frequently linking academy and industry, have been devoted to the recording of deep seismic profiles in many intracratonic sedimentary basins and offshore passive margins, thus providing a direct control on the structural configuration of the basement and the architecture of the crust. At the same time, needs for documenting also the current thickness of the mantle lithosphere and the fate of subducted lithospheric slabs have led to the development of more academic and new tomographic techniques. When put together, these various techniques now provide a direct access to the bulk 3D architecture of sedimentary basins, crystalline basement and Moho, as well as underlying mantle lithosphere.

Inherited structures, anisotropies in the composition of the sediments, crust and underlying mantle as well as thermicity and phase transitions are now taken into account when predicting the localization of deformation in the lithosphere during compression and extension episodes, and reconstructing the geodynamic evolution of rift basins, passive margins and foreland fold-and-thrust belts.

Source to sink studies also provide accurate estimates of sedimentary budget at basin-scale. Extensive use of low temperature chrono-thermometers and coupled kinematic, sedimentological and thermal models allow a precise control on the amount and timing of erosion and unroofing of source areas, but also the reconstruction of the sedimentary burial, strata architecture and litho-facies distribution in the sink areas.

Coupling deep mantle processes with erosion and climate constitutes a new challenge for understanding the present topography, morphology and long term evolution of continents, especially in such sensitive areas as the near shore coastal plains, low lands and intra-mountain valleys which may be subject to devastating flooding and landslides.

In addition to the search for hydrocarbon resources and geothermal energy, other societal needs such as CO_2 storage and underground water management will benefit from upgraded basin modelling techniques. New 2D and 3D basin modelling tools are progressively developed, coupling in different ways deep thermo-mechanical processes of the mantle (asthenosphere and lithosphere), geomechanics of the upper crust and sediments (compaction, pressure-solution and fracturing of seals and reservoirs), basin-scale fluid and sediment transfers (development of overpressures, hydrocarbon generation and migration). As pointed out by Roure et al. (2009), further challenges related to CO_2 storage will soon require the integration of fluid-rock interactions (reactive transport) in basin and reservoir models, in order to cope with the changes induced by diagenesis in the overall mechanical properties, and the continuous changes in fluid flow induced by compaction, fracturing and cementation.

As pointed out by Mooney and White (2009) (this volume), seismology has greatly advanced in the past century. Starting with the invention of the pen-and-paper seismograph in the 1880s and the advent of plate tectonics theory in the 1960s, scientists have been made progress in understanding, forecasting and preparing for earthquakes and their effects. Tectonic plate theory explains the occurrence of earthquakes as two or more plates meeting one another at plate boundaries where they may collide, rift apart, or drag against each other. These authors point out that diffuse plate boundaries, unlike convergent, divergent and lateral boundaries, are not completely defined and spread over a large area thereby spreading seismic hazards over a broad region. Intraplate earthquakes occur far away from any plate boundary, cause a great loss of life and cannot be explained by classical plate tectonics. However, classical plate tectonics is evolving, and now there are more theories behind earthquake generation dealing not only with the Earth's crust but also the hot, viscous lower lithosphere. These authors draw attention to the notion that in addition to damaging buildings and infrastructure and taking lives, earthquakes may also trigger other earthquakes due to stress changes once seismic energy is released.

Bohnhoff et al. (2009) (this volume) draw attention to an important discovery in crustal mechanics that the Earth's crust is commonly stressed close to failure, even in tectonically quiet areas. As a result, small natural or man-made perturbations to the local stress field may trigger earthquakes. To understand these processes, Passive Seismic Monitoring (PSM) with seismometer arrays is a widely used technique that has been successfully applied to study seismicity at different magnitude levels ranging from acoustic emissions generated in the laboratory under controlled conditions, to seismicity induced by hydraulic stimulations in geological reservoirs, and up to great earthquakes occurring along plate boundaries. In all these environments the appropriate deployment of seismic sensors, i.e., directly on the rock sample, at the Earth's surface or in boreholes close to the seismic sources allows for the detection and location of brittle failure processes at sufficiently low magnitude-detection threshold and with adequate spatial resolution for further analysis. One principal aim is to develop an improved understanding of the physical processes occurring at the seismic source and their relationship to the host geologic environment. In their paper, Bohnhoff et al. (2009) (this volume) review selected case studies and future directions of PSM efforts across a wide range of scales and environments. These include induced failure within small rock samples, hydrocarbon reservoirs, and natural seismicity at convergent and transform plate boundaries. They demonstrate that each example represents a milestone with regard to bridging the gap between laboratory-scale experiments under controlled boundary conditions and large-scale field studies. The common motivation for all studies is to refine the understanding of how earthquakes nucleate, how they proceed and how they interact in space and time. This is of special relevance at the larger end of the magnitude scale, i.e., for large devastating earthquakes due to their severe socio-economic impact.

As pointed out by Rubinstein et al. (2009) (this volume), the recent discovery of non-volcanic tremor in Japan and the coincidence of tremor with slow-slip in Cascadia have made Earth scientists re-evaluate models for the physical processes in subduction zones and on faults in general. Subduction zones have been studied very closely since the discovery of slow-slip and tremor. This has led to the discovery of a number of related phenomena including very low frequency earthquakes. All of these events fall into what some have called a new class of events that are governed under a different frictional regime than simple brittle failure. While this model is appealing to many, consensus as to exactly what process generates tremor has yet to be reached. As demonstrated by Rubinstein et al., tremor and related events also provide a window into the deep roots of subduction zones, a poorly understood region that is largely devoid of seismicity. Given that such fundamental questions remain about non-volcanic tremor, slow-slip, and the region in which they occur, these authors expect that this will be a fruitful field for a long time to come.

The paper by Tibaldi et al. (2009) (this volume) examines recent data demonstrating that volcanism also occurs in compressional tectonic settings (reverse and strike-slip faulting), rather than the traditional view that volcanism requires an extensional state of stress in the crust. Data describing the tectonic setting, structural analysis, analogue modelling, petrology, and geochemistry are integrated to provide a comprehensive presentation. An increasing amount of field data describes stratovolcanoes in areas of coeval reverse faulting, and stratovolcanoes, shield volcanoes and monogenic edifices along strike-slip faults, whereas calderas are associated with pull-apart structures in transcurrent regimes. Physically-scaled analogue experiments simulate the propagation of magma in these settings and taken together with data from subvolcanic magma bodies provide insight into the magma paths followed from the crust to the surface. In several transcurrent tectonic plate boundary regions, volcanoes are aligned along both the strike-slip faults and along fractures normal to the local least principal stress. As pointed out by these authors, at subduction zones, intra-arc tectonics is frequently characterised by contraction or transpression. In intra-plate tectonic settings, volcanism can develop in conjunction with reverse faults or strike slip faults. In most of these cases, magma appears to reach the surface along fractures striking perpendicularly to the local least principle stress, although in some cases there is a direct geometric control by the substrate strike-slip or reverse fault. Magma is transported beneath the volcano to the surface along the main faults, irrespective of the orientation of the least principle stress. The petrology and geochemistry of lavas erupted in compressive stress regimes indicate longer crustal residence times, and higher degrees of lower crustal and upper crustal melts contributing to the evolving magmas. Small volumes of magma tend to rise to shallow crustal levels, and magma mixing is common. In detailed studies from the Andes and Anatolia, with geographic and temporal coverage to compare contractional, transcurrent and extensional episodes, there do not appear to be changes to the mantle or crustal source materials that constitute the magmas. These authors demonstrate that, as the stress regime becomes more compressional, the magma transport pathways become more diffuse, and the crustal residence time and crustal interaction increases.

The isostatic adjustment of the solid Earth to glacial loading (GIA, Glacial Isostatic Adjustment) with its temporal signature offers a great opportunity to retrieve information on the Earth's upper mantle. As described by Poutanen et al. (2009) (this volume) the programme DynaQlim (Upper Mantle Dynamics and Quaternary Climate in Cratonic Areas) studies the relations between upper mantle dynamics, its composition and physical properties, temperature, rheology, and Quaternary climate. Its regional focus lies on the cratonic areas of northern Canada and Scandinavia. Geodetic methods like repeated precise levelling, tide gauges, high-resolution observations of recent movements, gravity change and monitoring of postglacial faults have given information on the GIA process for more than 100 years. They are accompanied by more recent techniques like GPS observations and the GRACE and GOCE satellite missions which provide additional global and regional constraints on the gravity field. Combining geodetic observations with seismological investigations, studies of the postglacial faults and continuum mechanical modelling of GIA, DynaQlim offers new insights into properties of the lithosphere. Another step toward a better understanding of GIA has been the joint inversion of different types of observational data – preferentially connected with geological relative sea-level evidence of the Earth's rebound during the last ten thousand years. Due to changes in the lithospheric stress state large faults ruptured violently at the end of the last glaciation resulting in large earthquakes. Whether the rebound stress is still able to trigger a significant fraction of intraplate seismic events in these regions is not completely understood due to the complexity and spatial heterogeneity of the regional stress field. Glacial ice sheet dynamics are constrained by the coupled process of the deformation of the viscoelastic solid Earth, the ocean and climate variability. How exactly the climate and oceans reorganize to sustain growth of ice sheets that ground to continents and shallow continental shelves is poorly understood. Incorporation of non-linear feedback in modelling both ocean heat transport systems and atmospheric CO_2 is a major challenge.

The paper by Dobrzhinetskaya and Wirth (2009) (this volume) summarizes recent achievements in studies of superdeep mantle rocks and diamonds from kimberlite and ultrahigh-pressure metamorphic (UHPM) terranes using advanced analytical techniques and instrumentations such as focused ion beam (FIB)-assisted transmission electron microscopy (TEM) and synchrotron-assisted infrared spectroscopy. As pointed out by these authors, mineralogical characterisations of the ultradeep earth materials using novel techniques with high spatial and energy resolution are resulting in unexpected discoveries of new phases, thereby providing better constraints on deep mantle processes. One of the unexpected results is that the nanometric fluid inclusions in diamonds from kimberlite and UHPM terranes contain similar elements such as Cl, K, P, and S. Such similarity reflects probably the high solubility of these elements in a diamond-forming C–O–H supercritical fluid at high pressures and temperatures. The paper by Dobrzhinetskaya and Wirth emphasizes the necessity of further studies of diamonds occurring within geological settings (oceanic islands, fore-arcs and mantle sections of ophiolites) previously unrecognized as suitable places for diamond formation.

As pointed out by Vočadlo (2009) (this volume), there are many unresolved problems concerning our understanding of the Earth's inner core; even fundamental properties, such as its internal structure and composition, are poorly known. Although it is well established that the inner core is made of iron with some alloying element(s), the structural state of the iron and the nature of the light element(s) involved remain controversial. Furthermore, seismically observed P-waves show the inner core to be anisotropic and layered, but the origins of this are not understood; seismically observed S-waves add to the complexity as they have unexpectedly low velocities. Seismic interpretation is hampered by the lack of knowledge of the physical properties of core phases at core conditions. Moreover, the resolution of seismic data are hampered by the need to de-convolve inner core observations from seismic structure elsewhere in the Earth; this is particularly relevant in the case of shear waves where detection is far from straightforward. If sufficiently well constrained seismological data were available, together with accurate high-pressure, high-temperature elastic properties of the candidate materials, it would be, in principle, possible to fully determine the structure and composition of the inner core – an essential prerequisite to understanding its evolution. As pointed out by Vočadlo, unfortunately, the extreme conditions of pressure and temperature required make results from laboratory experiments unavoidably inconclusive. However, computer simulations of materials at inner core conditions are now achievable. Ab initio molecular dynamics simulations have been used to determine the stable phase(s) of iron in the Earth's core and to calculate the elasticity of iron and iron alloys at core conditions. The calculated shear wave velocities are significantly higher than those inferred from seismology. Vočadlo argues that if the seismological observations are robust, then a possible explanation for this discrepancy is if the inner core contains a significant amount of melt. The observed anisotropy can only be explained by almost total alignment of crystals present.

Acknowledgements François Roure is acknowledged for a constructive review of this paper. We thank all the reviewers for their rigorous and constructive criticism of the chapters presented in this book. Financial support and scientific input from ILP, GeoForschungsZentrum Potsdam and the Netherlands Research Centre for Integrated Solid Earth Science is greatly acknowledged. Mrs. Christine Gerschke is thanked for dedicated support to ILP reports and for her effort in organising the Potsdam conference. All ILP Task Force and Regional Committee chairs are thanked for contributing to this review paper. We thank Thomas Kruijer for his valuable editorial assistance.

References

Ågren J. and Svensson R., 2007, Postglacial Land Uplift Model and System Definition for the New Swedish Height System RH 2000. Reports in Geodesy and Geographical Information Systems Rapportserie, LMV-Rapport 2007, v. 4, Lantmäteriet,Gävle.

Arcay D., Tric E. and Doin M.P., 2005, Numerical simulations of subduction zones: Effect of slab dehydration on the mantle wedge dynamics, Physics of The Earth and Planetary Interiors, v. 149, pp. 133–153.

Baba K., Chave A.D., Evans R.L., Hirth G. and Mackie R.L., 2006, Mantle dynamics beneath the East Pacific Rise at 17° S: Insights from the Mantle Electromagnetic and Tomography (MELT) experiment, Journal of Geophysical Research-Solid Earth, v. 111(B2). doi:10.1029/2004JB003598.

Begin B.Z., Steinberg D.M., Ichinose G.A. and Marco S., 2005, A 40,000 years unchanging of the seismic regime in the Dead Sea rift, Geology, v. 33, pp. 257–260.

Bertotti G., Frizon de Lamotte D., Teixell A. and Charoud M. (Eds), 2009, The geology of Vertical Movements: Proceedings of ILP 2007 workshop in Marrakech, Special issue of Tectonophysics, in press.

Beaumont C., 1981, Foreland basins, Geophysical Journal of the Royal Astronomical Society, v. 65, pp. 291–329.

Bohnhoff M. et al., 2009, Passive Seismic Monitoring of Natural and Induced Earthquakes: Case Studies, Future Directions and Socio-Economic Relevance, in Cloetingh S. and Negendank J. (Eds), New Frontiers in Integrated Solid Earth Sciences, Springer, New York.

Boness N.L. and Zoback M.D., 2006, A multiscale study of the mechanisms controlling shear velocity anisotropy in the San Andreas Fault Observatory at Depth, Geophysics, v. 71, pp. F131–F146.

Burov E.B. and Diament M., 1995, The effective elastic thickness of continental lithosphere: What does it really mean? (constraints from rheology, topography and gravity), Journal of Geophysical Research, v. 100, pp. 3905–3927.

Burov E., 2009, Thermo-mechanical models for coupled lithosphere-surface processes, in Cloetingh S. and Negendank J. (Eds), New Frontiers in Integrated Solid Earth Sciences, Springer, New York.

Burov E. and Cloetingh S., 2009, Controls of mantle plumes and lithospheric folding on modes of intraplate continental tectonics: differences and similarities: Geophysical Journal International, v. 178, p. 1691–1722.

Calvari S., Inguaggiato S., Puglisi G., Ripepe M. and Rosi M. (Eds), 2008, "The Stromboli Volcano: an integrated study of the 2002–2003 eruption": Geophysical Monograph Series. v 182, AGU, ISBN 978-0-87590-447-4.

Cawood P.A., Kröner A. and Pisarevsky S., 2006, Precambrian plate tectonics: Criteria and evidence, GSA Today, v. 16, pp. 4–11.

Cawood P.A. and Buchan C., 2007, Linking accretionary orogenesis with supercontinent assembly. Earth-Science Reviews, v. 82, pp. 217–256.

Cawood P.A., Kröner A., Collins W.J., Kusky T.M., Mooney W.D. and Windley B.F., 2009, Accretionary orogens through Earth history, in Cawood, P.A. and Kröner, A. (Eds), Earth accretionary systems in space and time. Geological Society, London, Special Publications, v. 318, pp. 1–36.

Chen P., 1979, Jurassic-Cretaceous paleogeography of China, Journal of Peking Universtiy (Natural Science Series), v. 3, pp. 90–108 (in Chinese).

Chen L., Wang T., Zhao L. and Zheng T., 2008, Distinct lateral variation of lithospheric thickness in the Northeastern North China Craton, Earth and Planetary Science Letters, v. 267, pp. 56–68.

Chernov A.A., 1974, Stability of faceted shapes, Journal of Crystal Growth, v. 24/25, pp. 11–31.

Cloetingh S., Sassi W. and Task Force Team, 1994, The origin of sedimentary basins: a status report from the task force of the International Lithosphere Program, Marine and Petroleum Geology, v. 11, pp. 659–683.

Cloetingh S., d'Argenio B., Catalano R., Horvath F. and Sassi W. (Eds), 1995a, Interplay of extension and compression in basin formation, Tectonophysics, v. 252, pp. 1–484.

Cloetingh S., Van Wees J.D., Van der Beek P.A. and Spadini G., 1995b, Role of pre-rift rheology in kinematics of basin formation: constraints from thermo-mechanical modelling of Mediterranean basins and intracratonic rifts, Marine and Petroleum Geology, v. 12, pp. 793–808.

Cloetingh S., Ben-Avraham Z., Sassi W. and Horváth F. (Eds), 1996, Dynamics of strike slip tectonics and basin formation, Tectonophysics, v. 266, pp 1–523.

Cloetingh S, Van Balen R.T., Ter Voorde M., Zoetemeijer B.P. and Den Bezemer T., 1997, Mechanical aspects of sedimentary basin formation: development of integrated models for lithospheric and surface processes, International Journal of Earth Sciences, v. 86, pp. 226–240.

Cloetingh S. (Ed), 2007. TOPO-EUROPE: the geoscience of coupled deep earth-surface processes, special issue, Global and Planetary Change, v. 58, 454 pp.

Cloetingh S.A.P.L. and TOPO-EUROPE Working Group, 2007, TOPO-EUROPE: The geoscience of coupled deep Earth-surface processes, Global and Planetary Change, v.58, pp. 1–118.

Cloetingh S., Thybo H. and Facenna C. (Eds), 2009, TOPO-EUROPE: continental topography, tectonics and surface processes, Tectonophysics, v. 474, p. 1–416.

Delgado L. and Ortuño F., 2008. ILP workshop in Ensenada, Abstracts and Programme, GEOS, 28, 1.

Dewey J.F., 1969, Evolution of the Appalachian-Caledonian orogen, Nature, v. 222, pp. 124–129.

Dobrzhinetskaya L., Green H.W., Mitchell T.E. and Dickerson R.M., 2001, Metamorphic diamonds: Mechanism of growth and inclusion of oxides, Geology, v. 29, pp. 263–266.

Dobrzhinetskaya L.F., Green H.W., Weschler M., Darus M., Wang Y.-C., Massonne H.-J. and Stöckhert B., 2003, Focused ion beam technique and transmission electron microscope studies of microdiamonds from the Saxonian Erzgebirge, Germany, Earth and Planetary Science Letters, v. 210, pp. 399–410.

Dobrzhinetskaya L.F. Liu Z, Cartigny P., Zhang J., Tchkhetia N.N., Green II H.W., and Hemley R.J., 2006, Synchrotron infrared and Raman spectroscopy of microdiamonds from Erzgebirge, Germany, Earth and Planetary Science Letters, v. 248, pp. 325–334.

Dobrzhinetskaya L. and Gilotti J. (Eds), 2007, Special Issue on Multidisciplinary approaches to ultrahigh-pressure metamorphism: a celebration of the career contribution of Juhn G. Liou, Journal of Metamorphic Petrology, v. 25.

Dobrzhinetskaya L.F., Wirth R. and Green H.W., 2007, A look inside of diamond-forming media in deep subduction zones, Proceedings of National Academy Sciences of the United States of America, v. 104, pp. 9128–9132.

Dobrzhinetskaya L. and Brueckner H. (Eds), 2009, Ultra-high pressure metamorphism: A window into the earth's interior (in memory of T. Carswell), Lithos, special volume in press.

Dobrzhinetskaya L.F. and Wirth R., 2009, Integrated geosciences: from atomic scale to mountain buildings, in Cloetingh S. and Negendank J. (Eds), New Frontiers in Integrated Solid Earth Sciences, Springer.

Dressler, B.O., Sharpton, V. L., Morgan, J., Buffler, R., Moran, D., Smit, J., and Urrutia, J., 2003, Investigating a 65-Ma-old smoking gun: Deep drilling of the Chicxulub impact structure. Eos, Transactions American Geophysical Union, v. 84, 14, pp.125–130, doi:10.1029/2003EO140001.

Eichelberger J., Gordeev E., Kasahara M., Izbekov P. and Lees J. (Eds), 2007, Volcanism and Tectonics of the Kamchatka Peninsula and Adjacent Arcs, Geophysical Monograph Series, AGU, v. 172.

Ernst W.G., 2005, Alpine and Pacific styles of Phanerozoic mountain building: Subduction-zone petrogenesis of continental crust, Terra Nova, v. 17, pp. 165–188.

Ferrière L., Koeberl C., Ivanov B.A. and Reimold W.U., 2008, Shock metamorphism of Bosumtwi impact crater rocks, shock attenuation, and uplift formation, Science, v. 322, pp. 1678–1681, doi: 10.1126/science.1166283.

Fielding E.J., Lundgren P.R., Bürgmann R. and Funning G.J., 2009, Shallow fault-zone dilatancy recovery after the 2003 Bam earthquake in Iran, Nature, v. 458, doi:10.1038/nature07817.

Friedrich A.M., Wernicke B., Niemi N.A., Bennett R.A. and Davis J.L., 2003, Comparison of geodetic and geologic data from the Wasatch region, Utha, and implications for the spectral character of Earth deformation at periods of 10 to 10 million years, Journal of Geophysical Research, v. 108, p. 2199, doi.2110.1029/2001JB000682.

Garrido C.J. and Bodinier J.L., 1999, Diversity of mafic rocks in the Ronda peridotite: Evidence for pervasive melt-rock reaction during heating of subcontinental lithosphere by upwelling asthenosphere, Journal of Petrology, v. 40, pp. 729–754.

Garrido C.J., Tommasi A., Lenoir X., Marchesi C. and Gibert B., 2009. Correlating geothermometry and texture of French Massif Central peridotite xenoliths with geophysical observations on the continental lithosphere structure and asthenospheric upwelling, in preparation.

Gatzemeier A. and Tommasi A., 2006, Flow and electrical anisotropy in the upper mantle: Finite-element models constraints on the effects of olivine crystal preferred orientation and microstructure, Physics of the Earth and Planetary Interiors, v. 158, pp. 92–106.

Gerya T.V., Stoeckhert B. and Perchuk A.L., 2002, Exhumation of high-pressure metamorphic rocks in a subduction channel – a numerical simulation, Tectonics, v. 21, no. 6, pp. 6–19.

Gerya T. Connolly J. and Perchuk L. (Eds), 2008, Rocks generated under extreme pressure and temperature conditions: Mechanisms, concepts, models (special volume), Lithos, v. 103.

Ghorbal B., Bertotti G., Foeken J. and Andriessen P.A.M., 2008, Unexpected Jurassic to Neogene vertical movements in "stable" parts of NW Africa revealed by low temperature geochronology, Terra Nova, v. 20, pp. 355–363.

Gohn G.S., Koeberl C., Miller K.G., Reimold W.U., Browning J.V., Cockell C.S., Horton J.W., Kenkmann T., Kulpecz A.A., Powars D.S., Sanford W.E., Voytek M.A., 2008,

Deep drilling into the Chesapeake bay impact structure, Science, v. 320, pp. 1740–1745, doi:10.1126/science.1158708.

Gorczyk W., Gerya T.V., Connolly J.A.D. and Yuen D.A., 2007, Growth and mixing dynamics of mantle wedge plumes, Geology, v. 35, pp. 587–590.

Gouiza M., Bertotti G., Hafid M., Cloetingh S., 2009, The tectonic evolution of the passive margin of Morocco along a transect from the Atlantic Ocean to the anti-Atlas, Tectonophysics, submitted.

Gràcia E., Pallàs R., Soto J.I., Comas M., Moreno X., Masana E., Santanach P., Díez S., García M., Dañobeitia J.J. and HITS Team (incl. G. Lastras) 2006, Active faulting offshore SE Spain (Alboran Sea): Implications for earthquake hazard assessment in the South Iberian Margin, Earth and Planetary Science Letters doi: 10.1016/j.epsl.2005.11.009, v. 241 (3–4), pp. 734–749.

Granet M., Wilson M. and Achauer U., 1995, Imaging a mantle plume beneath the French Massif Central, Earth and Planetary Science Letters, v. 136, pp. 281–296.

Groves D.I. and Bierlein F.P., 2007, Geodynamic settings of mineral deposit systems, Journal of the Geological Society, London, v. 164, p. 19–30.

Gudmundsson A., Friese N. and Galindo I. et al., 2008, Dike-induced reverse faulting in a graben, Geology, v.36, pp. 123–126.

Harms U. and Emmermann R., 2007, History and Status of the International Continental Scientific Drilling Program. In Harms U., Koeberl C., Zoback M.D. (Eds), Continental Scientific Drilling. A decade of progress, and challenges for the future, Springer, New York, pp. 1–53.

Harms U., Koeberl C. and Zoback M.D. (Eds), 2007, Continental Scientific Drilling. A decade of progress, and challenges for the future. Springer, New York, 355 pp.

Heaney P. J., Vicenzi E.P., Giannuzzi L.A. and Livi, K.J.T., 2001, Focused ion beam milling: A method of site-specific sample extraction for microanalysis of Earth and planetary materials, American Mineralogist, v.86, pp. 1094–1099.

Hecht L., Wittmann A., Schmitt R.T. and Stöffler D., 2004, Composition of impact melt particles and the effects of post-impact alteration in suevitic rocks at the Yaxcopoil-1 drill core, Chicxulub crater, Mexico, Meteoritics and Planetary Science, v. 39, pp.1169–1186.

Heidbach O. and Ben-Avraham Z., 2007. Stress evolution and seismic hazard of the Dead Sea fault system, Earth and Planetary Science Letters, v. 257, pp. 299–312.

Heidbach O., Reinecker J., Tingay M., Müller B., Sperner B., Fuchs K. and Wenzel F., 2007, Plate boundary forces are not enough: Second- and third-order stress patterns highlighted in the World Stress Map database, Tectonics, v. 26, TC6014, doi:10.1029/2007TC002133.

Heidbach O., Iaffaldano G. and Bunge H.-P., 2008a, Topography growth drives stress rotations in the Central Andes – observations and models, Geophysical Research Letters, doi:10.1029/2007GL032782.

Heidbach O., Tingay M., Barth A., Reinecker J., Kurfeß D. and Müller B., 2008b. The World Stress Map database release 2008 doi:10.1594/GFZ.WSM.Rel2008.

Heidbac O., Tingay M. and Wenzel F., 2009, Frontiers in stress research – Observation, integration, and application, Tectonophysics, Special Issue, in press.

Hergert T. and Heidbach O., 2006. New insights in the mechanism of postseismic stress relaxation exemplified by the June 23rd 2001 Mw = 8.4 earthquake in southern Peru: Geophysical Research Letters, v. 33, doi:1029/2005GL024585.

Hergert T., Heidbach O., Bécel A., Hirn A. and Wenzel F., 2007, The seismic hazard of Istanbul: an approach with numerical stress field modelling. in 8. Forum DKKV/CEDIM: Disaster Reduction in Climate Change, pp. 4, Karlsruhe.

Hergert T., 2009, Numerical modelling of the absolute stress state in the Marmara Sea region – a contribution to seismic hazard assessment, Ph. D. thesis, 152 pp., Karlsruhe Universität, Germany.

Hodell D.A., Anselmetti F.S., Ariztegui D., Brenner M., Curtis J.H., Escobar J., Gilli A., Grzesik D.A., Guilderson T.J., Kutterolf S. and Müller A.D., 2008, An 85-Ka record of climate change in lowland Central America, Quaternary Science Reviews, v. 27, pp.1152–1165.

Huang J. and Zhao D., 2006, High-resolution mantle tomography of China and surrounding regions, Journal of Geophysical Research, v. 111, B09305. doi:10.1029/2005JB004066.

Jolivet M., Ritz J.F., Vassallo R., Larroque C., Braucher R., Todbileg M., Chauvet A., Sue C., Arnaud N., De Vicente R., Arzhanikova A., Arzhanikov S., 2007, The Mongolian summits: An uplifted, flat, old but still preserved erosion surface, Geology, v. 35, pp. 871–874.

Kendall J.M., Pilidou S., Keir D., Bastow I.D., Stuart G.W. and Ayele A., 2006, Mantle upwellings, melt migration and the rifting of Africa: insights form seismic anisotropy: Geological Society, London, Special Publications, v. 259, p. 55–72.

Kirkwood D., Lavoie D., Malo M. and Osadetz K. (Eds), 2009. The North American Arctic Margins (from the Beaufort Sea to Nares Strait). Proceedings of ILP 2006 workshop in Québec, Special issue of the Bulletin of Canadian Petroleum Geology.

Kusznir N.J., Marsden G. and Egan S.S., 1991, A flexural-cantilever simple-shear/pure-shear model of continental lithosphere extension: application to the Jeanne d'Arc Basin, Grand Banks and Viking Graben, North Sea. In: Roberts A M, Yielding G, Freeman B (Eds), The Geometry of Normal Faults: Geological Society of London, London, Special Publications, v. 56, pp. 41–60.

Kusznir N.J. and Ziegler P.A., 1992, Mechanics of continental extension and sedimentary basin formation: a simple-shear/pure-shear flexural cantilever model, Tectonophysics, v. 215, pp. 117–131.

Lacombe O., Lavé O., Roure F. and Vergés J. (Eds), 2007, Thrust belts and foreland basins: From fold kinematics to hydrocarbon systems. Proceedings of ILP 2005 workshop in Rueil-Malmaison, Frontiers in Earth Sciences, Springer, New York, 492 pp.

Landes M., Ritter J.R.R. and Readman P.W., 2007, Proto-Iceland plume caused thinning of Irish lithosphere, Earth and Planetary Science Letters, v. 255, pp. 32–40, doi:10.1016/j.epsl.2006.12.003.

Lawrence D.T., Doyle M. and Aigner T., 1990, Stratigraphic simulation of sedimentary basins: Concepts and calibration, AAPG Bulletin, v.74, pp. 273–295.

Lei J., Zhao D., Steinberger B., Wu B., Shen F. and Li Z., 2009, New seismic constraints on the upper mantle structure of the Hainan plume, Physics of the Earth and Planetary Interiors, v. 173, pp. 33–50.

Lenoir X., Garrido C.J., Bodinier J.L. and Dautria J.M., 2000, Contrasting lithospheric mantle domains beneath the Massif Central (France) revealed by geochemistry of peridotite xenoliths, Earth and Planetary Science Letters, v.181, pp. 359–375.

Lenoir X., Garrido C., Bodinier J.-L., Dautria J.-M. and Gervilla F., 2001, The recrystallization front of the Ronda peridotite: Evidence for melting and thermal erosion of lithospheric mantle beneath the Alboran basin, Journal of Petrology, v. 42, pp. 141–158.

Le Roux V., Bodinier J.L., Tommasi A., Alard O., Dautria J.M., Vauchez A. and Riches A., 2007, The Lherz spinel-lherzolite: Refertilized rather than pristine mantle, Earth and Planetary Science Letters, v. 259, pp. 599–612, doi:10.1016/j.epsl.2007.05.026.

Le Roux V., Tommasi A. and Vauchez A., 2008, Feedback between melt percolation and deformation in an exhumed lithosphere-asthenosphere boundary, Earth and Planetary Science Letters, doi: 10.1016/j.epsl.2008.07.053.

Lev E. and Hager B.H., 2008, Rayleigh-Taylor instabilities with anisotropic lithospheric viscosity, Geophysical Journal International, v. 173, pp. 806–814.

Maruyama S., 1997, Pacific-type orogeny revisited: Miyashiro-type orogeny proposed. The Island Arc, v. 6, pp. 91–120.

Marco S., 2007, Temporal variation in the geometry of a strike-slip fault zone: Examples from the Dead Sea Transform: Tectonophysics, doi:10.1016/j.tecto.2007.08.014.

Masana E., Pallàs R., Perea H., Ortuño M., Martínez-Díaz J.J., García-Meléndez E. and Santanach P. 2005, Large Holocene morphogenic earthquakes along the Albox fault, Betic Cordillera, Spain, Journal of Geodynamics, v. 40, pp. 119–133.

Matsuda T. and Uyeda S., 1971, On the Pacific-type orogeny and its model: Extension of the paired belts concept and possible origin of marginal seas, Tectonophysics, v. 11, pp. 5–27.

Mazzuoli R., Vezzoli L., Omarini R., Acocella V., Gioncada A., Matteini M., Dini A., Guillou H., Hauser N., Uttini A. and ScailletS., 2008, Miocene magmatic and tectonic evolution of the easternmost sector of a transverse structure in Central Andes at 24°S, Geological Society of America Bulletin, v. 120, pp. 1493–1517.

McKenzie D.P., 1978, Some remarks on the development of sedimentary basins, Earth and Planetary Science Letters, v. 40, pp. 25–32.

McNeill L.C., Collier R.E.L., De Martini P.M., Pantosti D. and D'Addezio G.2005, Recent history of the Eastern Eliki Fault, Gulf of Corinth: Geomorphology, paleoseismology and impact on palaeoenvironments, Geophysical Journal International, v. 161, pp. 154–166, doi: 10.1111/j.1365-246X.2005.02559.

Michetti A.M., Audemard F. and Marco S., 2005, Future trends in Paleoseismology: Integrated study of the Seismic Landscape as a vital tool in Seismic Hazard Analyses, Tectonophysics, v. 408, 1–4, pp. 3–21.

Missenard Y., Zeyen H., Frizon de Lamotte D., Leturmy P., Petit C., Sébrier M. and Saddiqi O., 2006, Crustal versus asthenospheric origin of relief of the Atlas Mountains of Morocco, Journal of Geophysical Research, v. 111, B03401, doi:10.1029/2005JB003708.

Mooney W.D. and White S.M., 2009, Recent Developments in Earthquake Hazards Studies, in Cloetingh S. and Negendank J., (Eds), New Frontiers in Integrated Solid Earth Sciences, Springer, New York.

Nakajima J. and Hasegawa A., 2007, Tomographic evidence for the mantle upwelling beneath southwestern Japan and its implications for arc magmatism, Earth and Planetary Science Letters, v. 254, pp. 90–105.

Nolet G., Allen R. and Zhao D., 2007, Mantle plume tomography, Chemical Geology, v. 241, pp. 248–263.

Palyvos N., Pantosti D., De Martini P.M., Lemeille F., Sorel D. and Pavlopoulos K., 2005, The Aigion-Neos Erineos coastal normal fault system (western Corinth Gulf Rift, Greece): Geomorphological signature, recent earthquake history and evolution, Journal of Geophysical Research- Solid Earth, v. 110, B09302, doi: 10.1029/2004JB003165

Pantosti D., Pucci S., Palyvos N., De Martini P.M., D'Addezio G., Collins P.E.F. and Zabci C., 2008. Paleoearthquakes of the Düzce fault (North Anatolian Fault Zone): insights for large surface faulting earthquake recurrence, Journal of Geophysical Research – Solid Earth, v. 113, B01309, doi:10.1029/2006JB004679

Parnell J. Ed., 1994, Geofluids: origin migration and evolution of fluids in sedimentary basins: Geological Society of London, London, Special Publications, v. 78, pp. 1–372.

Peper T., Van Balen R.T. and Cloetingh S., 1994, Implications of orogenic wedge growth intraplate stress variations and sea level change for foreland basin stratigraphy: inferences from numerical modeling. In: Dorobek S, Ross G (Eds), Stratigraphic development in foreland basins. SEPM Special Publication, v. 52, pp. 25–35.

Petitjean S., Rabinowicz M., Grégoire M. and Chevrot S., 2006, Differences between Archean and Proterozoic lithospheres: Assessment of the possible major role of thermal conductivity, Geochemistry Geophysics Geosystems, v. 7: Q03021, doi:10.1029/2005GC001053.

Poutanen M., Dransch D., Gregersen S., Haubrock S., Ivins E.R., Klemann V., Kozlovskaya E., Kukkonen I., Lund B., Lunkka J.-P., Milne G., Müller J., Pascal C., Pettersen B.R., Scherneck H.G., Steffen H., Vermeersen B., Wolf D., 2009, DynaQlim – Upper Mantle Dynamics and Quaternary Climate in Cratonic Areas, in Cloetingh S. and Negendank J. (Eds), New Frontiers in Integrated Solid Earth Sciences. Springer Verlag, New York.

Price R.A., 1973, Large scale gravitational flow of supracrustal rocks, southern Canadian Rockies. In: de Jong K. and Scholten R.A. (Eds), Gravity and tectonics: Wiley, New York, pp. 491–502.

Pucci S., De Martini P.M. and Pantosti D., 2008, Preliminary slip rate estimates for the Düzce segment of the North Anatolian Fault Zone from offset geomorphic markers: Geomorphology, v.97, 538–554, doi: 10.1016/j.geomorph. 2007.09.002

Reilinger R., McClusky S., Vernant P., Lawrence S., Ergintav S., Cakmak R., Ozener H., Kadirov F., Guliev I., Stepanyan R., Nadariya M., Hahubia G., Mahmoud S., Sakr K., ArRajehi A., Paradissis D., Al-Aydrus A., Prilepin M., Guseva T., Evren E., Dmitrosta A., Filikov S.V., Gomez F., Al-Ghazzi R. and Karam G., 2006, GPS constraints on continental deformation in the Africa-Arabia-Eurasia continental collision zone and implications for the dynamics of plate interaction, Journal of Geophysical Research, v. 111, doi:10.1029/2005JB004051.

Ritter J.R.R., Jordan M., Christensen U.R. and Achauer U., 2001, A mantle plume below the Eifel volcanic fields, Germany: Earth and Planetary Science Letters, v. 186, p. 7–14.

Roure F., Shein V.S., Ellouz N. and Skvortsov L. (Eds), 1996, Geodynamic evolution of sedimentary basins: Editions Technip, Paris, pp. 1–453.

Roure F., Cloetingh S., Scheck-Wenderoth M. and Ziegler P.A., 2009, Achievements and Challenges in Sedimentary Basins Dynamics, in Cloetingh S. and Negendank J. (Eds), New Frontiers in Integrated Solid Earth Sciences, Springer.

Rubinstein J.L., Shelly D.R. and Ellsworth W.L., 2009, Non-Volcanic Tremor and Slow Slip, in Cloetingh S. and Negendank J. (Eds), New Frontiers in Integrated Solid Earth Sciences, Springer.

Sakuma S., Kajiwara T., Nakada S., Uto K. and Shimizu H., 2008, Drilling and logging results of USDP-4 – Penetration into the volcanic conduit of Unzen Volcano, Japan, Journal of Volcanology and Geothermal Research, v. 175, pp. 1–12, doi:10.1016/j.jvolgeores.2008.03.039.

Salveson J.O., 1976, Variations in the oil and gas geology of rift basins: Egyptian General Petroleum Corp, 5th Explor Sem, Cairo, Egypt, 15–17 November, 1976.

Sassi W., Colletta B., Bale P. and Paquereau T., 1993, Modeling of structural complexity in sedimentary basins: the role of pre-existing faults in thrust tectonics, Tectonophysics, v. 226, pp. 97–112.

Scheck-Wenderoth M., Bayer U. and Roure F. (Eds), 2009a. Progress in understanding sedimentary basins. ILP Task Force, Special issue of Teconophysics, in press.

Scheck-Wenderoth M., Bayer U. and Roure F. (Eds), 2009b. Progress in understanding sedimentary basins. ILP Task Force, Special issue of Marine and Petroleum Geology, in press.

Scholz C.A., Johnson T.C., Cohen A.S., King J.W., Peck J., Overpeck J.T., Talbot M.R., Brown E.T., Kalindekafe L., Amoako P.Y.O., Lyons R.P., Shanahan T.M., Castaneda I.S., Heil C.W., Forman S.L., McHargue L.R., Beuning K.R., Gomez J. and Pierson J., 2007, East African megadroughts between 135–75 kyr ago and bearing on early-modern human origins, Proceedings of the National Academy of Sciences, v. 104, pp. 16416–16421.

Self S. and Blake S., 2008, Consequences of explosive super eruptions, Elements, v. 4, 1, pp. 41–46.

Şengör A.M.C., 1993, Turkic-type orogeny in the Altaids: Implications for the evolution of continental crust and methodology of regional tectonic analysis (34th Bennett Lecture), Transactions of the Leicester Literature and Philosophical Society, v. 87, pp. 37–54.

Sleep N.H., 1971, Thermal effects of the formation of Atlantic continental margins by continental break up, Geophysical Journal of the Royal Astronomical Society, v. 24, pp. 325–350.

Smith R. et al., 2009, Geodynamics of the Yellowstone Hotspot and Mantle Plume: Seismic and GPS, Imaging, Kinematics, Mantle Flow, Journal of Volcanology and Geothermal Research, in prep.

Steacy S., Gomberg J. and Cocco M., 2005, Introduction to special section: Stress transfer, earthquake triggering, and time-dependent seismic hazard, Journal of Geophysical Research, v. 110, doi:10.1029/2005JB003692.

Steckler M.S. and Watts A.B., 1982, Subsidence history and tectonic evolution of Atlantic-type continental margins. In: Scrutton R.A. (Ed) Dynamics of Passive Margins: AGU Geodynamics Series, v. 6, pp. 184–196.

Stein C., Schmalzl J. and Hansen U., 2004, The effect of rheological parameters on plate behaviour in a selfconsistent model of mantle convection, Physics of The Earth and Planetary Interiors, v. 142, pp. 225–255.

Stolper E.M., DePaolo D.J. and Thomas D.M., 2009, Deep drilling into a Mantle Plume Volcano: The Hawaii scientific drilling project, Scientific Drilling, v. 7, pp. 4 – 14, doi:10.2204/iodp.sd.7.02.2009.

Tackley P., 2000, Self-consistent generation of tectonic plates in time-dependent, three-dimensional mantle convection simulations. 2. Strain weakening and asthenosphere, Geochemistry, Geophysics, Geosystems, v. 1, 8, doi:10.1029/2000GC000043 .

Tapponnier P., Zhiqin X., Roger F., Meyer B., Arnaud N., Wittlinger G. and Jingsui Y., 2001, Oblique stepwise rise and growth of the Tibetan Plateau, Science, v. 294: 1671–1677.

Tesauro M., Kaban M.K. and Cloetingh S.A.P.L., 2008. EuCRUST-07: A new reference model for the European crust, Geophysical Research Letters, v. 35, doi:10.1029/2007GL032244.

Tesauro M., Kaban M.K. and Cloetingh S., 2009a, 3D crustal model of Western and Central Europe as a basis for modelling mantle structure, in Cloetingh S. and Negendank J. (Eds), New Frontiers in Integrated Solid Earth Sciences, this volume, Springer.

Tesauro M., Kaban M.K. and Cloetingh S., 2009b, Thermal and rheological model of the European lithosphere, in Cloetingh S. and Negendank J. (Eds), New Frontiers in Integrated Solid Earth Sciences, this volume, Springer.

Tibaldi A. and Lagmay A.F.M. (Eds), 2006. Interaction between Volcanoes and their basement, Journal Volcanology and Geothermal Research, Special Issue, v. 158, 220 pp.

Tibaldi A., 2008, Contractional tectonics and magma paths in volcanoes, Journal of Volcanology and Geothermal Research, v. 176, pp. 291–301.

Tibaldi A. and Pasquarè F., 2008, A new mode of inner volcano growth: The "flower intrusive structure": Earth Planetary Science Letters, v.271, pp. 202–208.

Tibaldi A., Corazzato C., Kozhurin A., Lagmay A.F.M., Pasquaré F.A., Ponomareva V., Rust D., Tormey D. and Vezzoli L., 2008a, Influence of substrate tectonic heritage on the evolution of composite volcanoes: Predicting sites of flank eruptions, lateral collapse, and erosion, Global and Planetary Change, v. 61 (3), pp. 151–174.

Tibaldi A., Pasquarè F.A., Papanikolaou D. and Nomikou P., 2008b, Discovery of a huge sector collapse at the resurgent caldera of Nisyros, Greece, by onshore and offshore geological-structural data, Journal of Volcanology Geothermal Research, v.177, pp. 485–499.

Tibaldi A., Vezzoli L., Pasquarè F.A. and Rust D., 2008c, Strike-slip fault tectonics and the emplacement of sheet-laccolith systems: The Thverfell case study (SW Iceland), Journal of Structural Geology, v.30, pp. 274–290.

Tibaldi A., Pasquarè F. and Tormey D., 2009, Relationship between compressional fault tectonics and volcanism, in Cloetingh S. and Negendank J. (Eds), New Frontiers in Integrated Solid Earth Sciences, Springer, New York.

Tommasi A., Gibert B., Seipold U. and Mainprice D., 2001, Anisotropy of thermal diffusivity in the upper mantle. Nature, v. 411, pp. 783–787.

Tommasi A., Godard M., Coromina G., Dautria J.-M. and Barsczus H., 2004, Seismic anisotropy and compositionally induced velocity anomalies in the lithosphere above mantle plumes: A petrological and microstructural study of mantle xenoliths from French Polynesia, Earth and Planetary Science Letters, v. 227, no. 3–4, pp. 539–556.

Tommasi A., Knoll M., Vauchez A., Signorelli J.W., Thoraval C. and Loge R., 2009, Structural reactivation in plate tectonics controlled by olivine crystal anisotropy: Nature Geosciences, v. 2, p. 423–427.

Van der Beek P.A. and Cloetingh S., 1992, Lithospheric flexure and the tectonic evolution of the Betic Cordillers, Tectonophysics, v. 203, pp. 325–344.

Vanneste K., Radulov A., De Martini P.M., Nikolov G., Petermans T., Verbeeck K., Camelbeeck T., Pantosti D., Dimitrov D. and Shanov S. 2006, Paleoseismologic investigation of the fault rupture of the 14 April 1928 Chirpan earthquake (M 6.8), southern Bulgaria, Journal of Geophysical Research – Solid Earth, v. 111, B01303, doi:10.1029/2005JB003814

Vestøl O., 2006, Determination of postglacial land uplift in Fennoscandia from leveling, tide-gauges and continuous GPS stations using least squares collocation, Journal of Geodesy, v. 80, pp. 248–258. doi 10.1007/s00190-006-0063-7.

Vilasi N., Malandain J., Barrier L., Callot J.-P., Amrouch K., Guilhaumou N., Lacombe O., Muska K., Roure F. and Swennen R., 2009, From outcrop and petrographic studies to basin-scale fluid flow modelling: The use of the Albanian natural laboratory for carbonate reservoir characterisation: Tectonophysics, v. 474, p. 367–392.

Vočadlo L., 2009, Geomaterials Research – ab initio simulation of the Earth's core, in Cloetingh S. and Negendank J. (Eds), New Frontiers in Integrated Solid Earth Sciences, Springer.

Wang C., Zhao X., Liu Z., Lippert P.C., Graham S.A., Coe R.S., Yi H., Zhu L., Liu S. and Li Y., 2008, Constraints on the early uplift history of the Tibetan Plateau. PNAS, v. 105, pp. 4987–4992.

Watts A.B., Karner G.D. and Steckler M.S., 1982, Lithospheric flexure and the evolution of sedimentary basins. In: Kent P, Bott M H P, McKenzie D P, Williams C A (Eds) The Evolution of Sedimentary Basins: Philosophical Transactions of the Royal Society of London, Ser. A., v. 305, pp. 249–281.

Watts A.B., Platt J. and Buhl P., 1993, Tectonic evolution of the Alboran Sea basin, Basin Research, v. 5, pp. 153–177.

Wawerzinek B., Ritter J.R.R., Jordan M. and Landes M., 2008, An upper-mantle upwelling underneath Ireland revealed from non linear tomography, Geophysical Journal International, v. 175, pp. 253–268, doi:10.1111/j.1365-246X.2008.03908.

Wilson J.T., 1966, Did the Atlantic close and then re-open?, Nature, v. 211, pp. 676–681.

Wilson M. 2008. Fluid streaming from the Transition Zone as a trigger for within-plate magmatism, Geophysical Research Abstracts, v. 10, EGU2008-A-05636.

Windley B.F., 1992, Proterozoic collisional and accretionary orogens, in Condie K.C. (Ed), Proterozoic crustal evolution. Developments in Precambrian Geology. Elsevier, Amsterdam, pp. 419–446.

Windley B.F., Alexeiev D., Xiao W., Kroner A. and Badarch G., 2007, Tectonic models for accretion of the Central Asian Orogenic Belt, Journal of the Geological Society, London, v. 164, pp. 31–47.

Wirth R., 2004, Focused ion beam (FIB): A novel technology for advanced application of micro- and nanoanalysis in geosciences and applied mineralogy, European Journal of Mineralogy, v. 16, pp. 863–876.

Xiao W., Windley B.F., Yuan C., Sun M., Han C., Lin S.F., Chen H., Yan Q., Liu D., Qin K., Li J. and Sun S., 2009a, Paleozoic multiple subduction-accretion processes of the Southern Altaids: American Journal of Science, v.309, in press.

Xiao W., Windley B.F., Yong Y., Yan Z., Yuan C., Liu C. and Li J., 2009b, Early Paleozoic to Devonian multiple-accretionary model for the Qilian Shan, NW China. Journal of Asian Earth Sciences, in press.

Yin A. and Harrison T.M., 2000, Geologic evolution of the Himalayan-Tibetan orogen. Annual Review of Earth and Planetary Sciences, v. 28, pp. 211–280.

Zhang Q., Qian Q., Wang E., Wang Y., Zhao T., Hao J. and Guo G., 2001, An East China Plateau in mid-late Yanshanian period: Implication from adakites, Chinese Journal of Geology, v. 36, pp. 248–255 (in Chinese).

Zhao D., Lei J. and Tang Y., 2004, Origin of the Changbai volcano in northeast China: Evidence from seismic tomography: Chinese Science Bulletin, v. 49, pp. 1401–1408.

Zhao D., 2007, Seismic images under 60 hotspots: Search for mantle plumes. Gondwana Research, v. 12, pp. 335–355.

Zhao D., Maruyama S. and Omori S., 2007, Mantle dynamics of western Pacific to East Asia: New insight from seismic tomography and mineral physics, Gondwana Research, v. 11, pp. 120–131.

3D Crustal Model of Western and Central Europe as a Basis for Modelling Mantle Structure

Magdala Tesauro, Mikhail K. Kaban, and Sierd A.P.L. Cloetingh

Abstract EuCRUST-07 is a new 3D model of the crust for western and central Europe. It offers a starting point in any kind of numerical modelling, which requires an a priori removal of the crustal effect. The digital model (35°N, 71°N; 25°W, 35°E) consists of three layers: sediments and two layers of the crystalline crust. The latter are characterized by average *P-wave* velocities (V_p), which were defined based on seismic data. The model was obtained by assembling together at uniform $15' \times 15'$ grid available results of deep seismic reflection, refraction and receiver function studies. The Moho depth variations were reconstructed by merging the most robust and recent Moho depth maps existing for the European region and compiled using published interpretations of seismic profiles. EuCRUST-07 demonstrates large differences in Moho depth with previous compilations: over ± 10 km in some specific areas (e.g., the Baltic Shield). The basement is outcropping in some part of eastern Europe, while in western Europe it is up to $\sim$16 km deep, with an average value of 3–4 km, reflecting the presence of relatively shallow basins. The velocity structure of the crystalline crust turns out to be much more heterogeneous than demonstrated in previous compilations, having an average V_p varying from 6.0 to 6.9 km/s. In comparison to existing models, the new model shows average crustal velocity values distributed over a larger and continuous range. The sedimentary thickness appears underestimated by CRUST2.0 by $\sim$10 km in several basins (e.g., the Porcupine basin), while it is overestimated by $\sim$3–6 km along part of the coastline (e.g., the Norwegian coast). EuCRUST-07 shows a Moho 5–10 km deeper than previous models in the orogens (e.g., the Cantabrian Mountains) and in the areas where the presence of magmatic underplating increases anomalously the crustal thickness. EuCRUST-07 predicts a Moho shallower 10–20 km along parts of the Atlantic margin, and in the basin (e.g., the Tyrrhenian Sea), where previous models overestimate the average crustal velocity. Furthermore, the results of EuCRUST-07 are used to make inferences on the lithology for various parts of Europe. The new lithology map shows the eastern European tectonic provinces represented by a granite-felsic granulite upper crust and a mafic granulite lower crust. By contrast, the younger western European tectonic provinces are mostly characterized by an upper and lower crust of granite-gneiss and dioritic composition, respectively.

Keywords 3D crustal model · EuCRUST-07

Introduction

The crust is the most heterogeneous layer in the Earth and its impact on the interpretation of deep structures can mask the effect of deep seated heterogeneities. It is, for instance, nearly impossible to separate the crustal and mantle effects in potential field and geothermal modeling without additional data on the crustal structure (e.g., Kaban et al., 2004). It is still very

M. Tesauro (✉)
Faculty of Earth and Life Sciences, Netherlands Research Centre for Integrated Solid Earth Science, VU University, Amsterdam, The Netherlands GeoForschungsZentrum (GFZ), Potsdam, Germany
e-mail: magdala.tesauro@falw.vu.nl

S. Cloetingh, J. Negendank (eds.), *New Frontiers in Integrated Solid Earth Sciences*, International Year of Planet Earth,
DOI 10.1007/978-90-481-2737-5_2, © Springer Science+Business Media B.V. 2010

difficult to minimize the trade-off between the crustal and upper mantle heterogeneities in seismic tomography, which remains the main tool to investigate the structure of the mantle (e.g., Piromallo and Morelli, 2003). This is clearly demonstrated by two tomography models for North America: NA95 (Van der Lee and Nolet, 1997) and NA00 (Goes and Van der Lee, 2002), produced from nearly the same datasets. The only difference is that in NA00 trade-offs between Moho depth, crustal and subcrustal velocities are reduced by including independent constraints on Moho depth from the database compiled by Chulick and Mooney (1998). Although NA00 does not significantly alter the waveform fits compared to model NA95, it does provide a better estimate of subcrustal velocities. The differences between the models reach ± 0.3 km/s in the uppermost mantle (*S-wave*), which is more than a half of the total amplitude (± 0.55 km/s). In Europe only a few local travel time tomography models are obtained including an a priori crustal model (e.g., Lippitsch et al., 2003; Sandoval et al., 2003; Martin et al., 2005 and Martin et al., 2006). Also these cases demonstrate the importance of a careful a priori correction of the teleseismic traveltime residuals caused by 3-D crustal structure. In particular, Martin and Ritter (2005) have reported that teleseismic tomography for SE Romania without applying a sophisticated 3-D crustal correction gives results, which are strongly contaminated in the uppermost 100–150 km depth by a significant effect from an incorrect crustal model. Waldhauser et al. (2002) have demonstrated that the non-linear inversion of the synthetic residuals without correcting for the 3-D crustal structure erroneously maps the crustal anomalies into the upper mantle. In some cases neglecting a priori crustal correction in the travel time tomography might even lead to an error in the anomaly sign detected. For instance, Piromallo and Morelli (2003) have found a strong negative anomaly in the uppermost mantle under the southeastern Carpathians and Focşani foredeep (up to −7%). By contrast, the seismic tomography inversion of Martin and Ritter (2006) corrected for the custal effect leads to high velocities (+3.5%) in the upper mantle in this area.

Crustal models primarily based on existing reflection and refraction seismic profiles have been used for these purposes during the last decade. The first global model CRUST5.1 (Mooney et al., 1998) has clearly demonstrated that even coarse data on the crustal structure could remarkably improve modelling results of other methods. The global model S20 (Ekström and Dziewonski, 1998), which is constructed including these data, up to now remains one of the most used tomography models. CRUST5.1 was successfully used in global gravity and geothermal modelling (Kaban et al., 1999; Kaban and Schwintzer, 2001; Artemieva and Mooney, 2001), respectively. CRUST2.0 (Bassin et al., 2000), a successor of CRUST5.1, already offers a resolution of $2^\circ \times 2^\circ$, which is sufficient to employ this model not only in global but also in large-scale regional modelling. However, this resolution is not supported in many cases by experimental data (e.g., Koulakov and Sobolev, 2006). Furthermore, different models of the European crust are still inconsistent in many respects. In particularly, differences of the existing Moho maps often reach and even exceed ± 15 km (e.g., CRUST2.0, Bassin et al., 2000; Ziegler and Dèzes, 2006; Kozlovskaya et al., 2004). Consequently, the obtained results after corrections for crustal structure are different in many cases. For example, the mantle gravity anomalies obtained by various authors may differ up to about 100 mGal (e.g., Yegorova and Starostenko, 2002 1999; Kaban et al., 2004). The same problem exists in other applications, e.g., in seismic tomography as in the above example for the Carpathians where the negative velocity anomaly in the upper mantle appears due to a trade-off with the crustal structure (Piromallo and Morelli, 2003).

In this chapter a new digital crustal model (EuCRUST-07) is presented, which can be used as a starting point in a wide range of lithosphere and upper mantle studies. EuCRUST-07 is constructed for Western and Central Europe (35°N–71°N, 25°W–35°E), Fig. 1 and is available from open sources (ftp://ftp.agu.org/apend/gl/2007gl032244, Tesauro et al., 2008). The new crustal model (average velocities and depth of each layer) together with surface heat flow values are used to characterize the lithology of different tectonic provinces in Europe. The new lithology map should be considered as an attempt to estimate possible predominant lithotypes of the upper and lower crust. This is principal for various types of geophysical modelling. For instance, the lithotypes can be used to calculate the strength distribution in the European lithosphere.

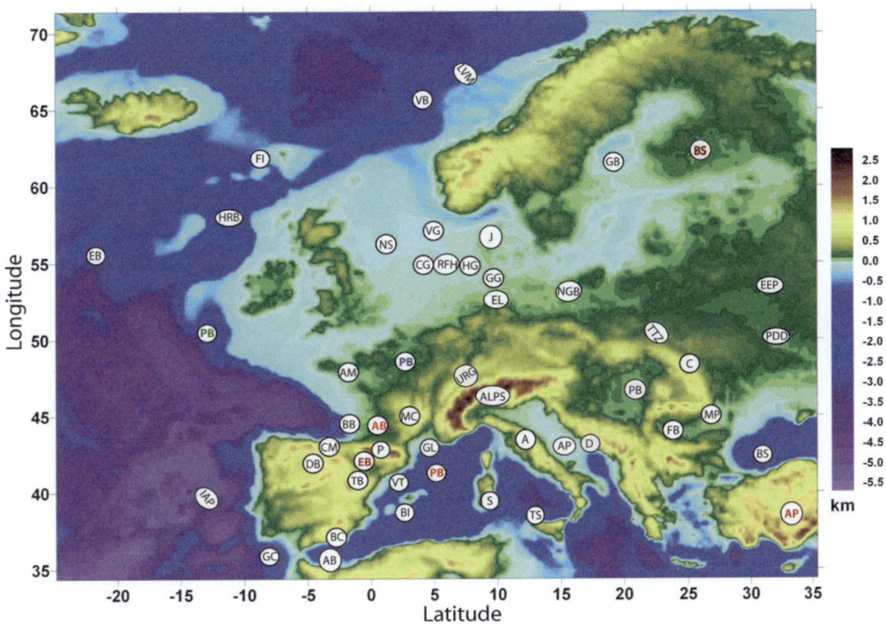

Fig. 1 ETOPO2 European topography (km) averaged to 15′×15′ resolution. Abbreviations are as follows: *A*, Apennines; *AB*, Alboran Basin; *AB*, Aquitane Basin; *AM*, Armorican Massif; *AP*, Adriatic Promontory; *AP*, Anatolian Plateau; *BB*, Bay of Biscay; *BC*, Betic Cordillera; *BI*, Balearic Islands; *BS*, Black Sea; *BS*, Baltic Shield; *C*, Carpathians; *CG*, Central Graben; *CM*, Cantabrian Mountains, *D*, Dinarides; *DB*, Duero Basin; *EB*, Edoras Bank; *EB*, Ebro Basin; *EL*, Elbe Lineament; *EEP*, East European Platform; *FB*, Focşani Basin, *FI*, Faeroe Islands; *GB*, Gulf of Bothnia; *GC*, Gulf of Cadiz; *GG*, Glueckstadt Graben; *GL*, Gulf of Lyon; *HG*, Horn Graben; *HRB*, Hatton-Rockall Basin; *IAP*, Iberian Abyssal Plain; *J*, Jutland; *LVM*, Lofoten–Vesterålen margin; *MC*, Massif Central; *MP*, Moesian Platform; *NGB*, North German Basin; *NS*, North Sea; *P*, Pyrenees; *PB*, Pannonian Basin; *PB*, Provençal Basin; *PB*, Paris Basin; *PB*, Porcupine Basin; *PDD*, Pripyat-Dniepr-Donets rift; *RFH*, Ringkøbing-Fyn High, *S*, Sardinia; *TS*, Tyrrhenian Sea; *TESZ*, Trans European Suture Zone; *TB*, Tajo Basin; *URG*, Upper Rhine Graben; *VB*, Vøring Basin; *VG*, Viking Graben; *VT*, Valencia Trough

Basic Model Assumptions

EuCRUST-07 is largely compiled from the results of seismic refraction, reflection and receiver function studies, most of them carried out within recent international projects, such as CELEBRATION2000 (Guterch et al., 2003), SUDETES 2003 (Grad et al., 2003), ALP 2002 (Brück et al., 2003), ESCI-N (Fernández-Viejo, 2005), CROP (Finetti, 2005a). Available local models based on seismic data (e.g., SVEKALAPKO, Kozlovskaja et al., 2004) were also incorporated. The study area is limited to 35°N–71°N and 25°W–35°E. The model consists of three layers: sediments and 2 layers of the crystalline crust, the latter characterized by an average *P*-wave velocity determined from seismic data.

Depth to the crystalline basement and Moho are the parameters most reliably determined in all kinds of seismic data. The situation with the inner crustal boundaries is more complicated. As at least two layers within the crystalline crust are detected in most seismic sections, it has been decided to maintain this division in the generalized model. In the areas, where the crystalline crust consists of only one layer, having a constant velocity (e.g., in the Tyrrhenian Sea) or characterized by a gradual change (e.g., in the western part of the Black Sea), the crust is arbitrarily divided in two layers of equal thickness having average velocities consistent with the seismic data. In the opposite case, several layers are joint to form one equivalent layer, e.g., in the EEP, where the velocity in the upper layer is calculated as a weighted average between the upper and the middle crust velocity. The mean velocities in the crystalline crust were evaluated as a weighted average between the upper and lower crust velocities. The final model is presented on a uniform 15′×15′ grid.

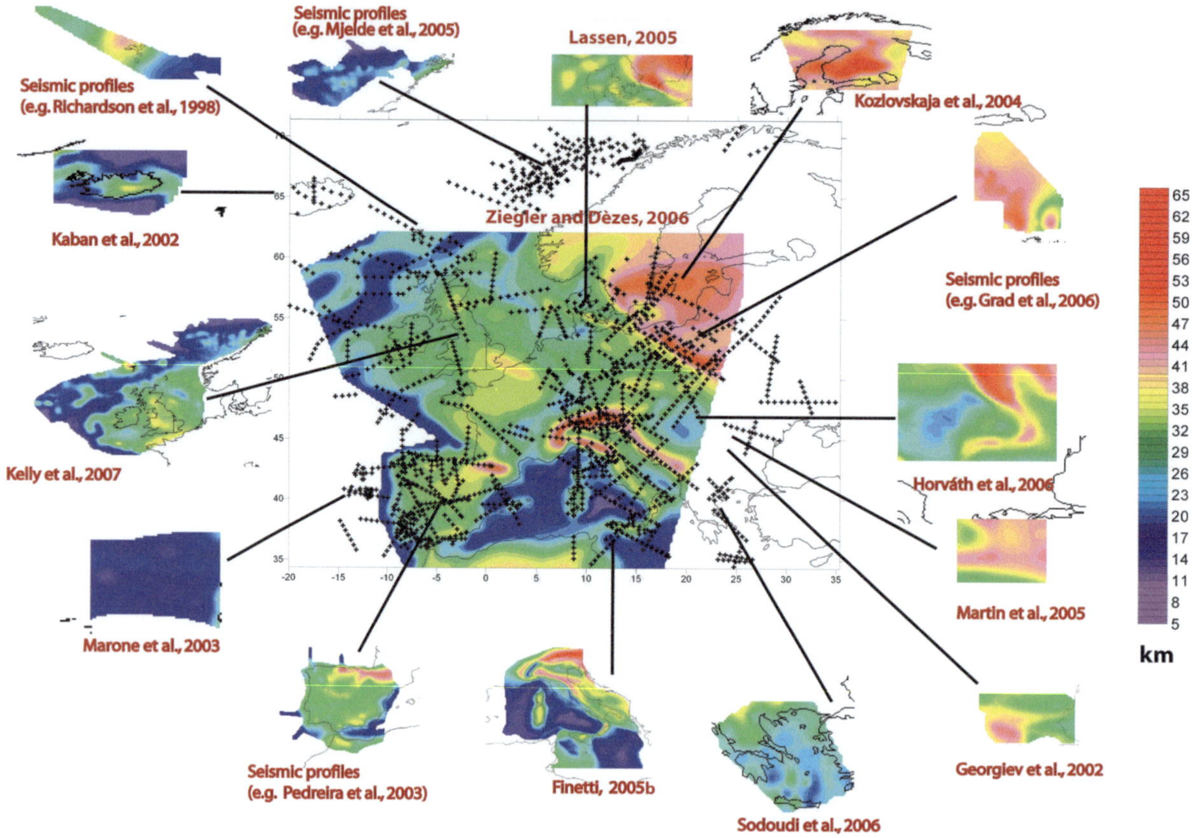

Fig. 2 Moho depth (km) updated from Ziegler and Dèzes (2006) (34°N–62°N, 18°W–25°E) and extended (35°N–71°N, 25°W–35°E) including an array of new datasets. Dashed lines show the location of the seismic profiles incorporated

Locations of the original seismic data and existing local Moho maps are shown in Fig. 2. The velocity distributions within the crystalline crust layers are mostly based on the interpolated determinations (wide-angle seismic data), whereas for significant part of Fennoscandia the model of Kozlovskaja et al. (2004) is used. In several cases when the non-uniform data coverage is not sufficient for a robust interpolation, the number of data is increased by adding extra points in accordance with the position of local tectonic units with reliable determinations in other parts (e.g., in Norway). In the oceanic domain without seismic data (the same as for the Moho map) *P*-wave velocities of 5.5 km/s and of 6.75 km/s to the upper and lower crust are assigned, respectively.

In the first stage average values for each of the $15' \times 15'$ grid cells was determined, which contain at least one determination of the crustal parameters (average velocity and thickness of each layer). In the second stage the remaining gaps were filled using a standard kriging technique (SURFER, Golden Software package). Most of the gaps do not exceed significantly the grid resolution and, therefore, a choice of the interpolation method is not principal for the final result. In addition to kriging, the "minimum curvature" and "inverse distance" techniques have been tested. In all cases the existing grid cells were not modified while the differences of the Moho depth in the interpolated points are less than 1 km with using of the different techniques, which is less than the accuracy of the seismic determinations. The kriging scheme provides slightly less artefacts for several relatively wide gaps and for this reason it was chosen for the final map.

For the major part of the area the most recent local Moho compilations are employed (e.g., Ziegler and Dèzes, 2006; Kozlovskaja et al., 2004). When possible, these compilations have been verified and modified in some details using available seismic data. All the maps have been converted to the same resolution $15' \times 15'$.

In the regions densely covered by recent seismic data, not included in existing Moho maps (e.g., in the Iberian Peninsula), seismic profiles were interpolated using the kriging technique to trace the crust/mantle boundary. All compilations were merged in a unified model giving a preference to the most robust. For instance, in Iceland and surrounding areas we used the Moho map of Kaban et al. (2002), constrained by various seismic data including reflection/refraction profiles and receiver function data, instead of employing the most recent compilation of Kumar et al. (2007), which is based solely on receiver function studies, which might have some problems with accurate detection of the Moho in the areas characterized by a broad crust-mantle transition zone. For the part of the oceans not covered by seismic profiles, the Moho depth is assigned from the global model CRUST2.0 (Bassin et al., 2000), for a part of Norway from the Geothermal Atlas of Europe (Hurtig et al., 1992) and for a part of the EEP and the Black Sea from the compilation of Kaban (2001).

The basement depth was determined using available maps (e.g., EXXON, 1985) and sedimentary thickness compilations (e.g., Scheck-Wenderoth and Lamarsche, 2005). All the compilations employed were verified and some of them were modified in several areas (e.g., Scheck-Wenderoth and Lamarche, 2005 in the North German Basin), according to the seismic data employed in order to trace the top of the crystalline crust. Concerning the continental domain, sediments and soft crust (e.g., in the Apennines) having an average *P*-wave velocity lower than 6 km/s, are included in the sedimentary layer. Few exceptions are represented by the Polish Trough, where the metamorphic sediments/volcanic strata layer [referred in Grad et al. (2005) as a "transition zone"], having $V_p < 6.0$ km/s, is included in the upper crust and the Adriatic Sea, where only the first 4 km of sediments, having an average velocity less than 5 km/s, are included in the sedimentary layer. Concerning the oceanic domain, fractured basaltic lavas having average *P*-wave velocity lower than 5 km/s are included in the sedimentary layer. The integration of the separate compilations in a unique map ($15' \times 15'$) was done using the same method employed to trace the Moho boundary. The velocity structure of the sediments is not specified a priori. Due to extremely strong heterogeneity (both lateral and vertical) of this layer it is difficult to integrate relatively sparse published data into a uniform model. On the other hand, the material properties of sediments (e.g., density) are much less related to velocity variations, while seismic tomography results are mostly biased by crystalline crust heterogeneity.

Seismic experiments detect the internal crustal layering usually with more uncertainties than the Moho and basement depth. Furthermore, very few, old and limited to restricted areas, 3D compilations of these boundaries depth exist (e.g., Gajewski et al., 1987; Scarascia et al., 1998). On the other hand, since at least two layers within the crystalline crust are observed in most of the seismic sections, this division was kept in the model (Fig. 3a and b). This makes EuCRUST-07 more consistent, since the upper and lower parts of the crust are usually characterized by strong differences of the velocities.

Uncertainties of the new crustal model presented cannot be estimated systematically, as they result from the merging and averaging of various compilations (e.g., of Moho depth) and seismic data having different error margins. Furthermore, as mentioned before, the seismic data used are unevenly distributed in the study area (Fig. 2). However, since the uncertainties of EuCRUST-07 mostly depend on those of the input data used, it is worth considering that the most recent seismic experiments (e.g., CELEBRATION 2000) estimate the Moho depth and the crustal velocity with an accuracy of $\pm$ 1 km and $\pm$ 0.1 km/s, respectively.

Crustal Model of Western and Central Europe

The main crustal boundaries and average *P*-wave velocity values are displayed in Figs. 1, 2, 3, 4, 5, 6. The crustal structure is rather heterogeneous, even without considering the regional differences among western Europe, eastern Europe and the Baltic Shield, which are basically shown in previous models (e.g., CRUST 2.0). The eastern part of the study area is mostly characterized by large crustal velocities ($V_p \sim 6.6$ km/s) and deep Moho (~40–45 km). By contrast, west from the TESZ the crust is more heterogeneous, composed of Variscan crust with reduced average velocities ($V_p \sim 6.2$–6.4 km/s) and thickness

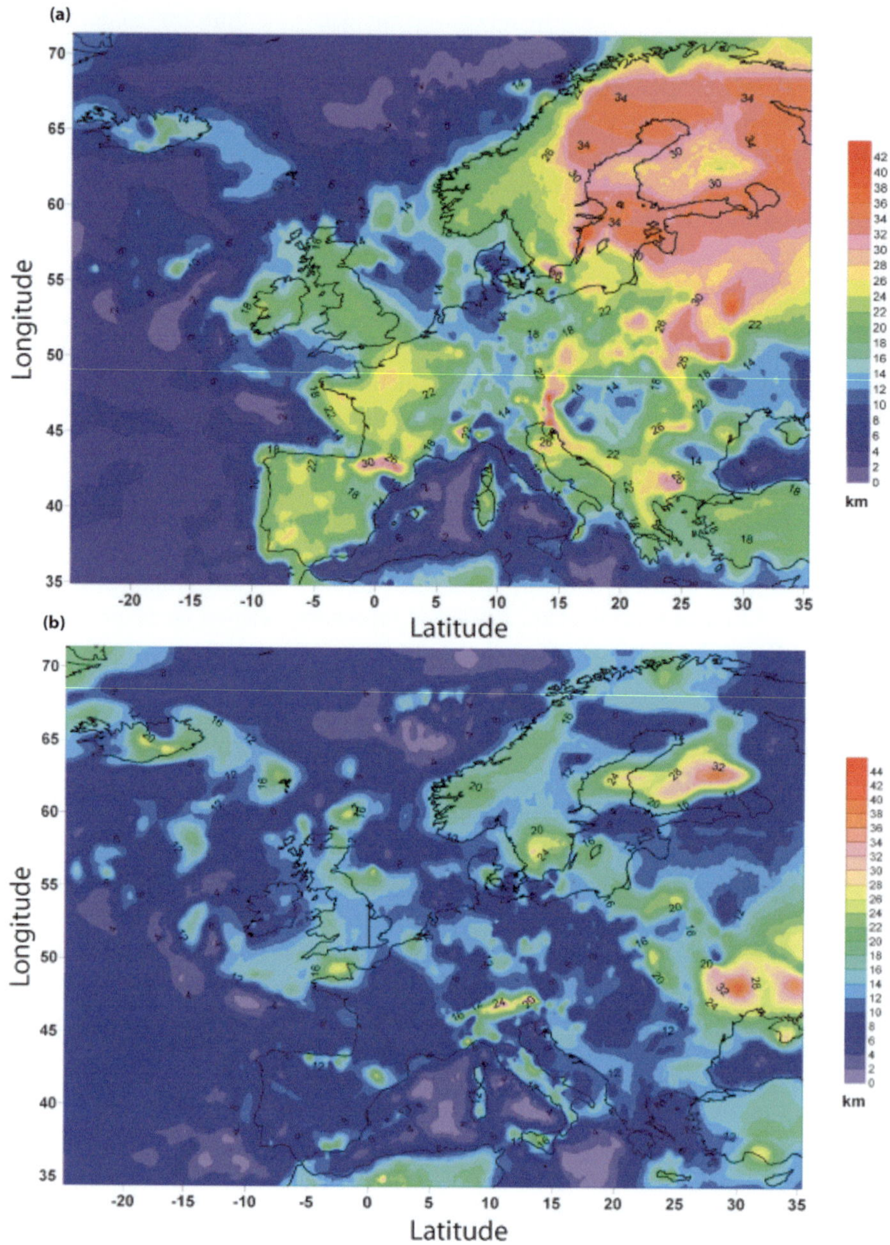

Fig. 3 Crystalline crustal thickness (km). (**a**) Upper crust. (**b**) Lower crust

(30–35 km), orogens with crustal thickness increasing up to 45–50 km, and areas that experienced strong extension having thin crust and low velocities (e.g., the Pannonian Basin and the Tyrrhenian Sea).

Below the main features of EuCRUST-07 for the principal tectonic units of Europe are discussed in detail.

Southeastern Europe

The new Moho map evidences crustal thicknesses of 32–38 km beneath the Hellenides mountain range in western Greece, and 25–28 km beneath the Peleponnesus. Moving toward the west and north coasts of the Aegean Sea the Moho shallows to

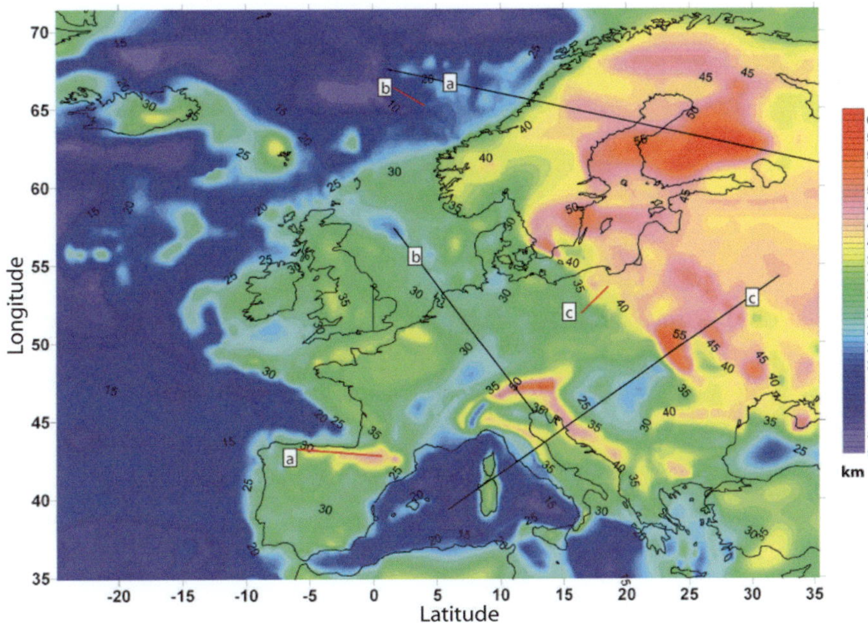

Fig. 4 Moho depth (km). *Red lines* depict location of three seismic profiles used in EuCRUST-07 displayed in Fig. 7a–c. *Black lines* depict location of EuCRUST-07 cross-sections displayed in Fig. 8a–c

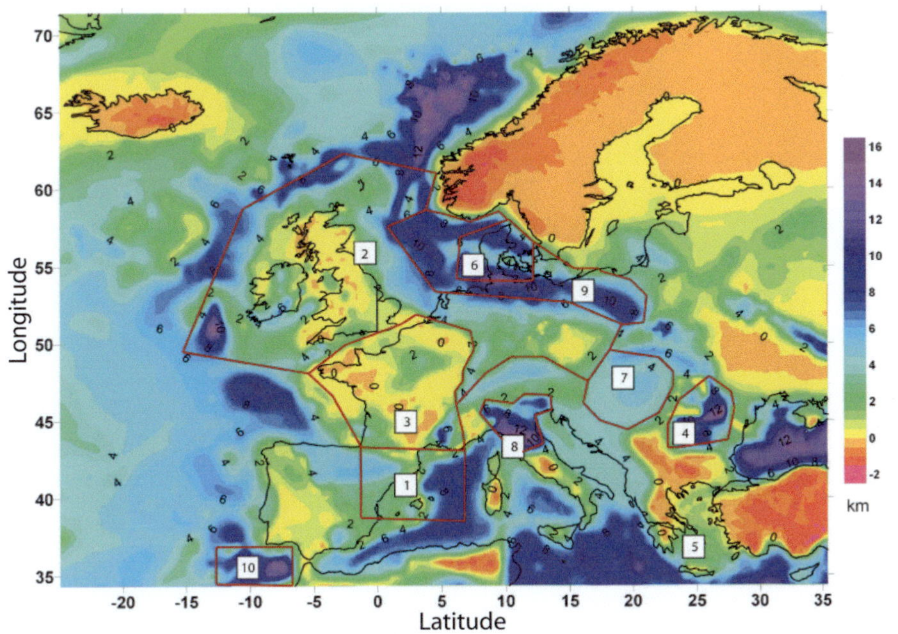

Fig. 5 Depth to basement (km) determined using the following compilations: *1*, Ayala et al. (2003); *2*, Bourgeois et al. (2007); *3*, BRGM (2006); *4*, (Diehl and Ritter (2005); 5, EXXON (1985); *6*, Lassen (2005); *7*, Lenkey (1999); *8*, Pieri and Groppi (1981); *9*, Scheck-Wenderoth and Lamarche (2005); *10*, Thiebot and Gutscher (2006). The red lines depict the limits of each compilation used. In the area outside the red boundaries the sedimentary thickness map of EXXON (1985) is employed. All the compilations are verified and some of them (*1*, *2*, *5*, *7*, *8*, *9*) are modified in several areas according to the seismic data used in order to trace the top of the crystalline crust

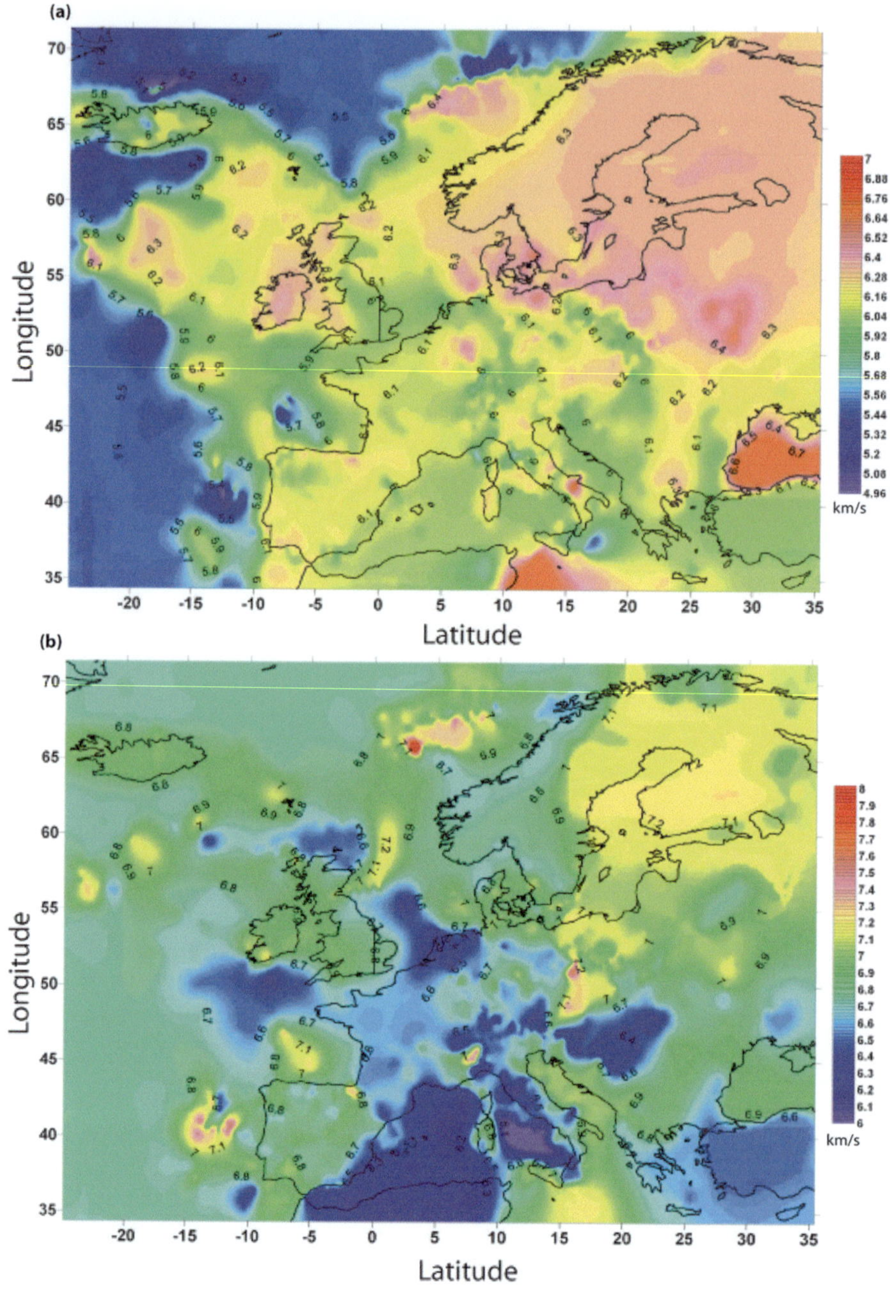

Fig. 6 Average *P-wave* velocity in the crystalline crust (km/s). (**a**) Upper crust. (**b**) Lower crust

28–30 km. This crustal thickness reduction is related to the Cenozoic extensional tectonics that affected the whole Aegean and adjacent parts of Greece (Sodoudi et al., 2006).

A low average crustal velocity (~6.30 km/s) is found in the Hellenic arc from tomography data (e.g., Papazachos et al., 1995), possibly due to the presence of thick unconsolidated material (Alessandrini et al., 1997). In the Aegean Sea the values of these parameters decrease to 6.18 km/s due to the crustal thinning, mostly at the expense of the lower crust. In the Aegean Basin the Moho depth changes from 25 to 28 km in the northern part to 26–30 km in the central part across the Cyclades region and to 20–22 km in the southern part.

The average crustal thickness beneath western and central Crete is 30 km and decreases to 21–25 km under the eastern part of this island (Sodoudi et al., 2006). In western Turkey the Moho depth increases from about 28 km along the coastline to 35–37 km beneath the unextended Anatolian Plateau, while low crustal velocities between 6.0 and 6.5 km/s (Akyol et al., 2006) are observed. In these areas the sedimentary thickness reaches 2–3 km only in the Aegean Basin. The crustal structure of the western part of the Black Sea is characterized by 8–13 km thick sediments overlying a crystalline layer having P-wave velocity similar to that of basalt (~6.8 km/s) and a thickness of 5–10 km. This layer is interpreted as a relict ocean crust or as probably formed in an extensional stress regime (Cloetingh et al., 2003; Çakir and Erduran, 2004). The crustal thickness increases from the centre of the western Black Sea, where the Moho depth is only 19 km, toward its margins with the growing thickness of materials having a velocity similar to that of granite (6.0–6.4 km/s, Neprochnov et al., 1970). The deepest Moho in this area (~48–50 km) is observed beneath southern Crimea (Starostenko et al., 2004). West of the Black Sea, the Moho depth is 30–32 km in the east Srednogorie–Balkan zone and in the Moesian Platform, while it deepens up to ~45 km beneath the southwestern part of Bulgaria (Georgiev et al., 2002). The average velocity in these areas is about 6.45 km/s. Beneath the Carpathians the Moho depth varies from over 50 km in the eastern part to about 35 km in the western part (Horváth et al., 2006). These recent determinations exceed the values from the Ziegler and Dèzes map (ZDm) for more than 10 km. In the Southern Carpathians the Moho is between 37–42 km and reaches a maximum of about 44 km in the Focşani Basin (Martin and Ritter, 2005), which is also characterized by very thick sediments, up to ~16 km (Diehl et al., 2005). The crystalline crust displays in these areas relatively high velocities of ~6.45 km/s, on account of a rather relatively thick (~45%), high velocity lower crust (~6.85 km/s). The Carpathians are bordered to the west by the Pannonian Basin, an area subjected to strong Miocene extension (e.g., Horváth et al., 2006), characterized by a very shallow Moho (~25 km, Horváth et al., 2006) and low crustal velocities (~6.15 km/s, Środa et al., 2006) possibly related to its high heat flow (120 mW/m^2, Lenkey, 1999). The thickness of the Tertiary and Quaternary sediments in this area is about 2–3 km (Lenkey, 1999), while the basement is deeper up to 5–6 km (Środa et al., 2006; EXXON, 1985). West from the Pannonian Basin, beneath the Dinarides, the Moho deepens to 45 km, while the average crustal velocities increase to ~6.35 km/s, presumably due to the presence of a high velocity lower crust (~6.80 km/s).

Italian Peninsula

Strong variations of the Moho depth in the Alpine belt are reflected in the new map. Beneath the western and eastern Alps the European Moho plunges southward to ~40 and ~55 km, respectively. On the Adriatic side the Moho depth reaches ~45 km beneath the eastern Alps, while beneath the western Alps a fragment of mantle-like material is observed, which is imbricated into the Alpine crust at the Insubric line (Finetti, 2005b). Therefore, the Moho depth is significantly reduced here [~20 and ~7 km, compared to the ZDm and the model of Kaban (2001)] and the lower crust velocity is increased ($V_p \sim 7.3$ km/s). In the other parts of the Alpine chain the mean crustal velocities (~6.34 km/s) are relatively low due to reduced lower crust velocities (6.4–6.6 km/s, Aichrot et al., 1992; Scarascia and Cassinis, 1997) representing 50% of the crystalline crust. Similar values are observed in the Molasse Basin, where the Moho shallows to a depth of ~30 km and the crystalline crust thins to ~25 km. The Adriatic Moho has a depth beneath the Adriatic Sea of about 35 km reaching a minimum of 30 km in its southern part, while it deepens westward beneath the Apennines up to over 40 km in the Po Plain. This area represents the foredeep of the Apennines and, differently from the Molasse Basin, is characterized by a very high thickness of sediments (>8 km, Pieri and Groppi, 1981), high tectonic subsidence rates (~1 mm/yr) and a high dip ranging between 10° and 20° (Mariotti and Doglioni, 2000). These peculiarities are possibly related to the Adriatic plate subducting in a westward direction (e.g., Carminati et al., 2004). The crust of the Apennines and surrounding areas is quite complex and clearly stratified, being characterized by horizontally and vertically alternating velocity values (e.g., Morelli, 1998). In particular, a velocity inversion is evidenced in the Gargano promontory, where the velocity decreases from 6.75 km/s in the upper crust to 6.30 km/s in the lower crust

(Scarascia et al., 1994). A general increase of the average crustal velocities from the western ($V_p \sim 6.20$ km/s) to the eastern part ($V_p \sim 6.40$ km/s) of the Apennines at the transition zone to the Adriatic plate is observed, due to the increase of the lower crust velocity from ~6.4 to ~6.7 km/s. Beneath western Tuscany the Moho shallows to a depth of about 20 km, which is probably related to the late Cenozoic opening of the Tyrrhenian back-arc basin (Doglioni, 1991; Carminati et al., 2004). The mean crustal velocities are low there (~6.0 km/s), on account of the velocity decrease in the lower part of the crust, which is also evidenced by tomography data (e.g., Amato et al., 1998). The low velocities and high heat flow (~200 W/m^2, Zito et al., 2003) are probably related to partial melting of the lower crust, as a consequence of the asthenosphere uprising (e.g., Di Stefano et al., 1999). The low P_n velocities (~7.7 km/s, Morelli, 1998) provide some evidence for a delamination of the continental crust in this area. Halfway between Liguria and Corsica the crust is likely suboceanic and its thickness is reduced to less than 20 km. The crust is thickened again up to 33–35 km beneath Corsica and Sardinia having a typical continental structure (Finetti, 2005b), being characterized by average velocities of ~6.40 km/s. Southward, in the Sardinian Channel the Moho rises to 20–15 km, while velocities are typical for a transitional type of crust. The Moho deepens again to ~28 km beneath the Tunisian coast, and to ~35 km beneath Sicily showing a continental type crustal structure ($V_p \sim 6.60$ km/s). Westward of Sardinia the crystalline crust thins to ~3 km in the Tyrrhenian Sea and the Moho rises up to a minimum depth of ~10 km beneath the Vavilov and Marsili volcanoes, where a gradual transition from continental to oceanic crust is observed. The average crustal velocities in the southeastern part of the Tyrrhenian Sea are about 6.0 km/s. The origin of such low values is attributed to lithosphere thinning, as a consequence of the back-arc extension (e.g., Morelli, 1998). This hypothesis is confirmed by the high heat flow (>200 W/m^2) and low Pn velocity (~7.5–7.7 km/s) observed in this area (e.g., Contrucci et al., 2001). Approaching the coast of Italy, the Moho depth gradually increases to 20–25 km together with the crustal velocity. Beneath the Ionian Sea the crystalline crust is of oceanic type and very thin (up to 3 km). The Moho plunges sharply from 12–20 km in this area to ~40 km beneath the Calabrian arc over a horizontal distance of ~100 km (Finetti, 2005b).

Iberian Peninsula and Central Atlantic Margin

In the central part of the Iberian Peninsula the Moho depth is between 30 and 34 km and the average velocities of the crust are relatively low (6.20–6.30 km/s), on account of a low velocity layer ($V_p \sim 5.6$ km/s) located in the upper part of the crust (Banda et al., 1981; Suriñach and Vegas, 1988; Paulssen and Visser, 1991). In addition, the lower crust is thin (< 10 km) representing less than 35% of the crystalline crust. The basement depth is relatively shallow (2–3 km), reaching a maximum of ~5 km in the basins, in the Cantabrian Mountains and the External Betics. The maximum depth of the Moho found beneath the Pyrenees and Cantabrian Mountains (~45 km) is 8–10 km deeper than in the previous maps (e.g., ZDm), on account of the subduction of the Iberian plate beneath the European one. The average velocity in the crust is also increased to 6.6 km/s due to high-velocity bodies in the mid-crust (e.g., Pedreira et al., 2003, Fig. 7a). These features are interpreted as portions of the European lower crust embedded in the Iberian

Fig. 7 Example of three seismic profiles used in EuCRUST-07. Profiles location is depicted as red lines in Fig. 4. (**a**) Two-dimensional *P-wave* velocity (km/s) model from Pedreira et al. (2003). Numbers in white boxes indicate representative velocities (km/s). *Dashed lines* indicate surfaces with no velocity contrast. *Bold lines* indicate layer segments directly sampled by seismic reflections. Notice that Moho deepens up to 45 km under the Cantabrian Mountains, on account of the subduction of the Iberian plate beneath the European plate, and the high-velocity bodies embedded in the upper part of the crystalline crust. (**b**) Two-dimensional *P-wave* velocity (km/s) model from Raum et al. (2006). P-wave velocities (km/s) are shown as small numbers and the bold numbers represent *Vp* /*Vs* ratios. Notice the high velocity body (~8.4 km/s) in the deepest part of the lower crust interpreted as a deep crustal root of partially eclogitized rocks. (**c**) Two-dimensional *P-wave* velocity (km/s) models for the LT-2 profile from Grad et al. (2005). Notice the large thickness (~10 km) of metamorphic sediments and volcanic strata with low velocity (< 6.0 km/s) present in the upper part of the crystalline crust of the TESZ

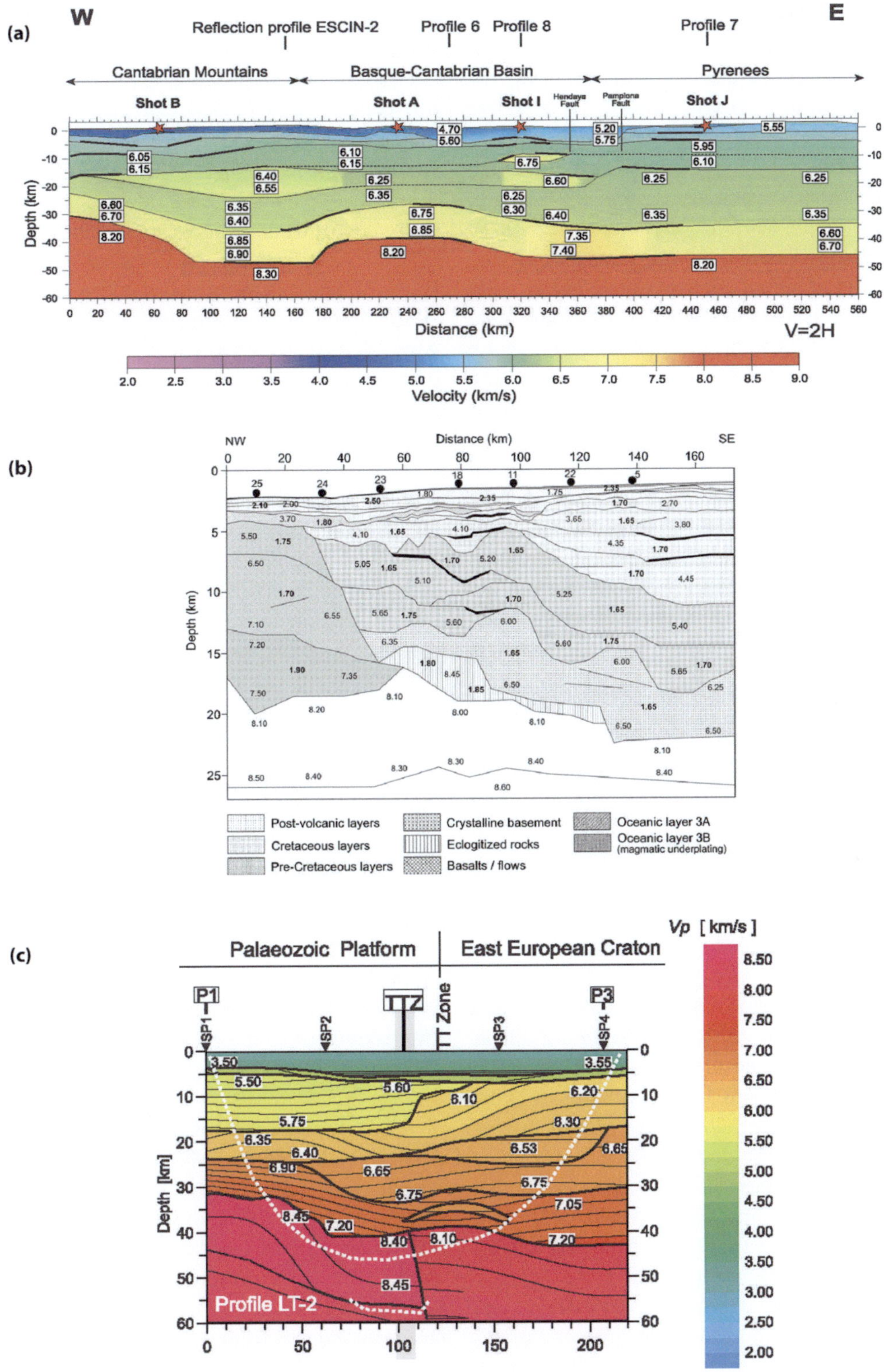

Fig. 7 (continued)

mid-crust (Pedreira et al., 2003). South of the Cantabrian shoreline, the Moho raises to ~30 km in the Ebro and Duero basins, while north of it, the Moho shallows from ~30 km beneath the shoreline to ~20 km beneath the Bay of Biscay over a distance of ~120 km. The substantial thinning of the crust, related to the Atlantic opening, has occurred mostly at the expense of the lower crust (Pinet et al., 1987). Consequently, a decrease of the mean crustal velocity (~6.30 km/s) is observed. Beneath the Catalan coast and the Balearic islands the Moho is ~25 km (e.g., Banda et al., 1980; Gallart et al., 2001) and rises up to 20 km beneath the Gulf of Lyon and up to 15 and 12 km beneath the Valencia Trough and the centre of the Provençal Basin, respectively. In these areas the sedimentary thickness increases up to 8 km, while the crystalline crust thins to less than 5 km. The average crustal velocities decrease to 6.2 km/s due to relatively low velocities at deep levels (~6.45 km/s) attributed to the removal of part of the lower crust during the rifting process (Gallart et al., 2001). On the other hand, an increase of these values to ~6.45 km/s is observed westward, beneath the Corsica and Provençal margins on account of the serpentinized peridotite ($V_p \sim 7.25$ km/s) in the lower part of the crust (Contrucci et al., 2001). In the southeastern part of the Iberian Peninsula a sharp Moho depth transition from a maximum of 38 km in the Betics to 25 km beneath the shoreline and to 15 km in the eastern part of the Alboran Basin is observed (e.g., Fullea et al., 2007). The crystalline crust thins to 7–8 km marking the transition to oceanic type in the south Balearic Basin. Furthermore, the average velocities decrease from 6.35 to 6.15 km/s, probably on account of the lack of a high velocity lower crust (e.g., Galindo-Zaldivar et al., 1997) and the high heat flow observed (up to 120 mW/m^2, Polyak et al., 1996). In the western part of the Alboran Sea the Moho is ~20 km, deepening sharply towards the Gibraltar Arc down to 30–32 km, while the mean velocities increase to ~6.5 km/s. Continental type crust underlies the entire Gulf of Cadiz with a progressive thinning (from 30 to 20 km) from east to west and from north to south, on account of the Atlantic oceanic spreading (e.g., González-Fernández et al., 2001). The lower crust thins to about 4 km and represents less than 20% of the crystalline crust. Southwestward, the crust becomes of oceanic type and the Moho shallows up to 11–14 km beneath the Tagus and Horseshoe abyssal Plains, over a distance of ~120 km. The mean depth of the basement is ~9 km with a gradual eastward increase up to 13 km (Thiebot and Gutscher, 2006). The northwestern part of the Iberian Peninsula is characterized by a Moho depth of ~30 km in central Galicia, which decreases to 27–25 km beneath the coastline (e.g., Córdoba et al., 1987). Beneath the southwestern part (South Portuguese Zone) the Moho deepens to ~34 km (e.g., Matias, 1996; Simancas et al., 2006), while the crustal velocities increase from 6.35 to 6.50 km/s, on account of basic intrusions ($V_p \sim 7.0$ km/s) in the lower crust (e.g., Simancas et al., 2006). From the western Iberian coast to the Iberian Abyssal Plain the Moho shallows from 30 to 13 km over a distance of ~300 km, showing a transition from continental to oceanic crust. The Galicia Bank and the Iberia Abyssal Plain represent the ocean-continent transition and are characterized by an average crustal velocity of about 6.75 km/s, which is much higher than for standard oceanic crust. This effect is presumably caused by a presence of thick (up to 5 km) serpentinized lower crust ($V_p \sim 7.4$ km/s, Whitmarsh et al., 1993, 1996; Chian et al., 1999).

England and North Atlantic Margin

Beneath England and Ireland the Moho reaches a depth of 30–35 km (Kelly et al., 2007), and the velocity in the upper crust is between 6.0 and 6.4 km/s, while it increases in the lower crust from 6.7 to 7.0 km/s. The basement in this area is close to the surface (1–3 km deep) and deepens then to 3–4 km in the vicinity of the Shannon Estuary in western Ireland and to 6–8 km along the southern coastline. The Moho shallows southwest of the Irish coast from 27 to less than 20 km over a horizontal distance of 100 km, while the average crustal velocities decrease to 6.25 km/s. West of the Irish coast the continental crust shows a variable thickness, with a Moho depth between 12 and 25 km (Kelly et al., 2007), on account of the different amount of stretching to which it was subjected during the opening of the Atlantic ocean (e.g., O'Reilly et al., 1995). The difference with CRUST2.0 is up to 15 km in this area. In the basins the depth of the basement ranges from 7 to 8 km (e.g., in the Hatton and Rockall Basin) to 15 km (e.g., in the Porcupine Basin, Kimbell et al., 2004), while the crystalline crust thins up to 3 km. High average crustal velocities (~6.90 km/s) are observed beneath the Edoras and the Hatton Bank, possibly related to mantle underplating, which results in high lower crustal velocities (~7.15 km/s).

The Moho deepens to 37–38 km beneath the Faeroe Islands and to 40–42 km approximately 100 km west of them (Richardson et al., 1998), while the average crustal velocity is about 6.75 km/s, due to the thick, ultramafic (V_p>7.0 km/s) bodies present in lower crust. The thick crust [~10 km thicker than in CRUST2.0 and in the model of Kaban (2001)] is a consequence of the Icelandic mantle plume, whose hottest core was located under Faeroe Islands at the time of continental breakup (Richardson et al., 1998). East to the Faeroe Islands, beneath the Faeroe-Shetland Basin, the Moho shallows to 20 km and the velocity decreases to 6.6 km/s, while west of them, beneath the Faeroe-Iceland Ridge, the Moho depth reaches a maximum of 28 km, while the velocity in the lower crust is relatively high (~6.9 km/s, Kelly et al., 2007). In Iceland the depth of the Moho ranges from less than 20 km along its northern and western coastline to 38–40 km in its southern part (Kaban et al., 2002). This thickening of the crust is caused by abnormally thick lower crust presumably related to increased melt generation occurring in the plume core (Darbyshire et al., 1998), while the upper crust remains very thin (~4 km, Menke et al., 1998). The average crustal velocity (~6.55 km/s) is quite high due to a thick and high velocity lower crustal layer (~6.9 km/s).

Central Europe

The cross-section in Fig. 8b shows general tendencies of the crustal structure in central Europe. The crystalline crust thickness tends to increase from ~15 km under the North Sea to 25–28 km under the German Plain with a medium lower crust velocity of 6.5–6.6 km/s. To the southeast the crust becomes thick (up to ~50 km) under the eastern Alps and thins again to 25–20 km under the northern Adriatic coast, while the lower crust velocity increases up to 6.8 km/s. In France the Moho depth ranges from ~28 km in the

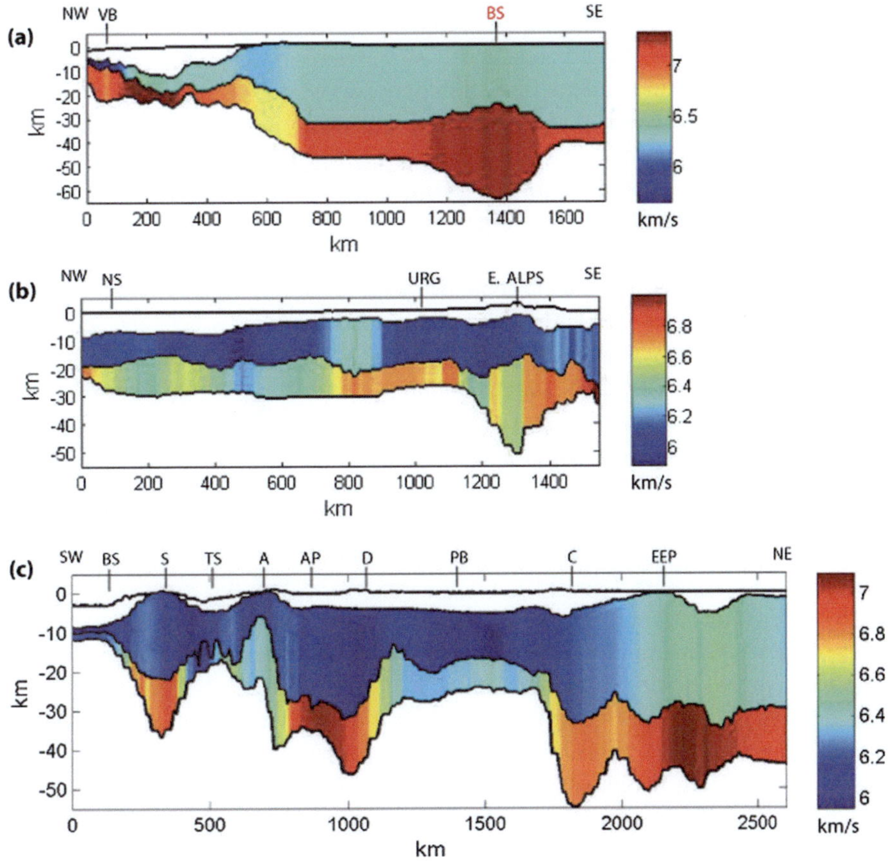

Fig. 8 Average *P*-wave velocity distribution (km/s) along three cross-sections. Black lines mark the crustal layers division. Sections location is depicted as black lines in Fig. 4. Abbreviations are as follows: *A*, Apennines; *AP*, Adriatic Promontory; *BS*, Balearic Sea, *BS*, Baltic Shield; *C*, Carpathians; *D*, Dinarides; *E. Alps*, Eastern Alps; *EEP*, East European Platform; *NS*, North Sea; *PB*, Pannonian Basin; *S*, Sardinia; *TS*, Tyrrhenian Sea; *URG*, Upper Rhine Graben; *VB*, Vøring Basin

Aquitane Basin to ~35 km in the Paris Basin and in the Armorican Massif (Ziegler and Dèzes, 2006), while the sedimentary thickness is increased up to 3–4 km in the basins. The average velocities are relatively low, on account of the large thickness of the upper crust (~75% of the crystalline crust). Across the north Armorican margin, in the transition zone to the oceanic domain the Moho shallows to ~18 km and the crystalline crust thins to less than 10 km. The mean crustal velocity increases to ~6.75 km/s due to a thick (~5 km) high velocity body (~7.45 km/s) in the lowest part of the crust (Thinon et al., 2003). The rifting, occurred during the Cenozoic, has affected the crust in the Massif Central by forming several basins (e.g., the Limagne graben) and thinning of the crust up to 4 km, mainly at the expense of the lower crust (Zeyen et al., 1997). As a consequence, the mean crustal velocity is low (~6.25 km/s), while the Moho rises from a depth of 30–32 km in the western part to ~28 km in the eastern part and up to a minimum of 24 km in the Limagne graben. The shallow position of the Moho is probably related to a thermal uplift of the crust as a result of a rising plume (Granet et al., 1995).

In southern Germany the Moho depth is about 25–27 km beneath the Black Forest and Vosges, while in the URG it varies from 24 km in the southern part to 30 km in the northern part (Ziegler and Dèzes, 2006). In the URG the mean crustal velocity is relatively low (~6.30 km/s), on account of the large thickness of the upper crust (70% of the crystalline crust). Underneath the central part of the graben the lower crust thins to ~5 km, since the mantle tends to bulge into it (Zucca, 1984). The average velocity in the lower crust is ~6.75 km/s. This layer is characterized by a very strong vertical gradient suggesting an increase of the mafic material content, probably derived from the mantle during the formation of the graben (e.g., Edel et al., 1975; Zucca, 1984). In the Black Forest a decrease of the mean crustal velocity (~6.20 km/s) is due to a low velocity zone (~5.4 km/s) present in the upper crust between 7 and 14 km (Gajewski et al., 1987). Although the low velocity layer tends to vanish toward southeast, similar average crustal velocities (~6.15 km/s) are observed due to a lateral decrease of the velocity in the lower crust from 6.7 to 6.4 km/s (Gajewski et al., 1987). In these areas the depth of the basement reaches a maximum of ~4 km in the central part of the URG. In the Rhenish Massif the Moho deepens to 30–32 km and the crustal structure is quite heterogeneous being characterized by alternation of high and low velocity zones with average crustal velocity ~6.35 km/s (Franke et al., 1990; Dekorp Research Group, 1990). Beneath the Bohemian Massif the Moho has an average depth of 35 km and the crust thickens gradually from northwest to southeast (Wilde-Piórko et al., 2002; Majdanski et al., 2006). The average velocity increases sharply from the western (~6.35 km/s) to the eastern (6.50 km/s) part of the unit, where mafic garnet granulite ($V_p \sim 7.2$ km/s) is probably present in the lower crust (Enderle et al., 1998).

In the North German Basin the Elbe Lineament represents, like the TESZ, another important geological boundary, where significant changes in the crustal thickness and composition are observed. The Moho beneath the North German Basin shallows from 35–34 km to 28–25 km from north to south of the Elbe Lineament, while the average crustal velocity decreases from ~6.70 to ~6.27 km/s. The strong change, occurring at this boundary, is related to a sharp decrease in the lower crust velocity from ~7.0 to ~6.3 km/s and possibly reflects a compositional transition from mafic to meta-sediments and granitic rocks (e.g., Scheck et al., 2002). In the southeastern part of the North Sea similar changes in the velocity and crustal thickness are present suggesting that the WNW–ESE-striking Elbe Lineament continues into the southern North Sea (e.g., Scheck et al., 2002). More in detail, the Moho is at 32–36 km depth in the Ringkøbing-Fyn High (RFH), while it becomes shallower (up to 25 km) westward beneath the Central Graben. Furthermore, the average crustal velocity decreases from RFH to the graben from 6.60 to 6.50 km/s. The Moho depth varies between 25 and 38 km beneath the Norwegian Danish Basin (Lassen, 2005), while from the western to the eastern part of Denmark the Moho depth increases from 27 to 34 km together with the average crustal velocities (from 6.55 to 6.90 km/s). The increase is related to the presence of a magmatic body ($V_p \sim 7.0$ km/s) of Carboniferous-Permian age in the lower part of the crust (Thybo et al., 2006). The top of the crystalline crust in the North German Basin and the Norwegian Danish Basin is quite deep, on account of a thick Paleozoic sedimentary sequence (e.g., Nielsen et al., 2005; Lyngsie and Thybo, 2007). It reaches 7–8 km beneath Jutland (Lassen, 2005), 12 km beneath the Glueckstadt Graben and the Viking Graben (Christiansson et al., 2000), 6–10 km beneath the Horn and Central Graben (Scheck-Wenderoth and Lamarche, 2005) and up to 4–6 km beneath the RFH (Nielsen et al., 2005, Lyngsie and Thybo, 2007).

Northern and Eastern Europe

Recent seismic experiments (SVEKALAPKO, Kozlovskaja et al., 2004) have demonstrated that the Moho depression beneath central and southern Finland and the Gulf of Bothnia is much more pronounced (>60 km) than in previous compilations (e.g., Hurtig et al., 1992), while the average crustal velocity tends to increase towards the southeastern part of Finland (up to 6.90 km/s) together with the thickness of the mafic lowermost crust (up to 30 km) characterized by extremely high velocities ($V_p \sim 7.2$ km/s). This is demonstrated in the cross-section in Fig. 8a. The Moho rises to the west reaching a value of ~40 km in Sweden and about 30–32 km beneath the Oslo rift and the Norwegian coast. Two local maxima of 40–43 km have been recently detected in southern Norway by a receiver functions study (Svenningsen et al., 2007). Westward to the Norwegian coast the Moho gradually shallows to ~20 km, while it deepens again up to 25 km at the transition zone in the Vøring Basin and Lofoten–Vesterålen margin (e.g., Mjelde et al., 2005). The Vøring Basin is characterized by high crustal velocities (6.8–7.0 km/s) reflecting the presence of thick mafic intrusions in the lower crust having velocities above 7.0 km/s. The maximal lower crust velocity (~8.0 km/s) observed in this area is interpreted as a deep crustal root of partially eclogitized rocks that were formed during the Caledonian orogeny (Raum et al., 2006, Fig. 7b). Northward, in the Lofoten–Vesterålen margin, a decrease of the lower crustal thickness is accompanied by a decreasing amount of breakup intrusives and extrusives. The crust in this area experienced only moderate extension, in contrast to the occurrence of major crustal extension in the southern Vøring margin (e.g., Tsikalas et al., 2005). As a result, the depth of the basement ranges in this area from ~5 km (in Lofoten– Vesterålen margin) to ~15 km (in the Vøring Basin).

The depth to the Moho discontinuity ranges from 35 to 50 km within the part of EEP considered in this study. The upper crust is rather heterogeneous and is made of a set of high and low velocity layers with an average value of ~6.3 km/s. The mafic lower crust is thick and characterized by high velocities (~7.0 km/s). Consequently the average crustal velocities are also high (~6.7 km/s). These values become lower in the southwestern edge of the EEP, on account of a decrease of the lower crustal velocities (~6.75 km/s). The cross-section in Fig. 8c demonstrates the difference in the crustal structure between the EEP with high velocities and crustal thickness and the heterogeneous areas west to the TESZ comprising stretched thin crust with slow upper and lower crust velocities (e.g., the Pannonian Basin and the Tyrrhenian Sea) and the structures with deep Moho and medium-high lower crust velocities (e.g., the Carpathians, the Dinarides and the Apennines). The Moho depth beneath the TESZ in central Poland is intermediate between that of the EEP and western Europe ranging from ~45 km in the eastern part to 30–35 km in the Palaeozoic Platform in the southwest. The central part of the TESZ is characterized by high thicknesses of sediments (8–10 km) (e.g., Scheck-Wenderoth and Lamarche, 2005; Grad et al., 2006). The average crustal velocity in this area is relatively low (~6.30 km/s), reflecting the large thickness of metamorphic sediments and volcanic strata (<6.0 km/s) present in the upper part of the crystalline crust. Although the lower crust is characterized by high velocities (~7.0 km/s), which is very similar to that of the EEP (Guterch and Grad, 2006), its contribution to the average crustal values is relatively small since it represents only 30% of the crystalline crust, Fig. 7c.

EuCrust-07: Statistical Analysis and Difference with Previous Models

Bar plots of the crustal velocities, boundaries and thickness of the main layers of EuCRUST-07 are shown in Fig. 9a–g. It can be observed that 93% of the average crustal velocity values fit to a range of 6.20 to 6.80 km/s (Fig. 9a), with two pronounced peaks, the first one at ~6.30 km/s (~28% of samples), mostly corresponding to western Europe, and the second one at ~6.50 km/s (~23%), a typical value characterizing the east European crust and some structures of west Europe, e.g., the Atlantic Margin (Fig. 10). Outside from this range, the highest values (V_p >6.8 km/s), constituting 1% of the entire distribution, reflect a mafic composition of the crust and are mainly observed in the areas affected by magmatic underplating (e.g., the Vøring Basin and the Baltic Shield). The lowest determinations (5.90–6.10 km/s) representing ~5% of the

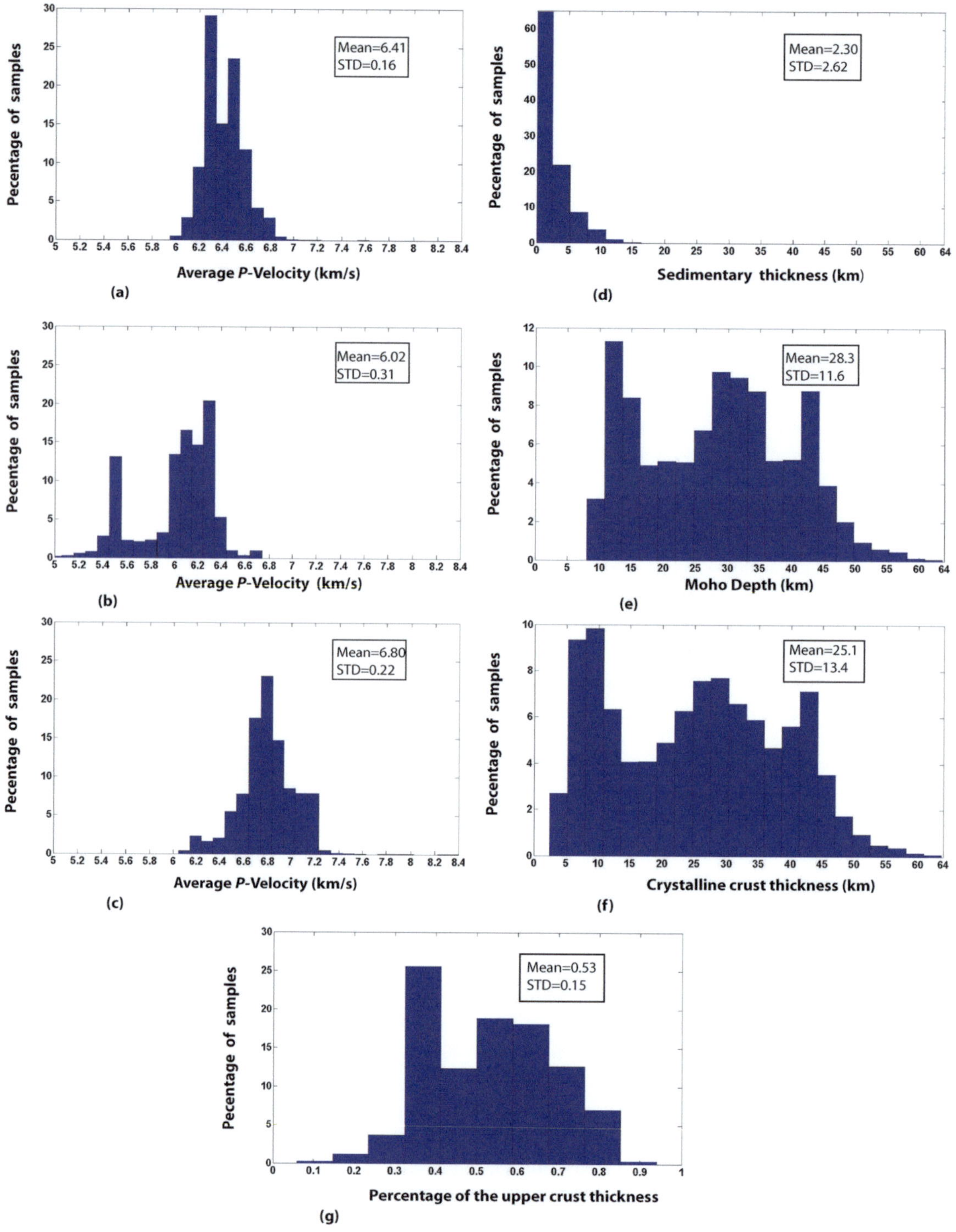

Fig. 9 Bar plot of the values distribution of EuCRUST-07 model. **(a)** Average *P-wave* velocity distribution (km/s) in the crystalline crust, **(b)** in the upper crust, **(c)** in the lower crust. **(d)** Sedimentary thickness distribution (km), **(e)** Moho depth distribution (km), **(f)** crystalline crust thickness (km), **(g)** percentage of upper crust thickness

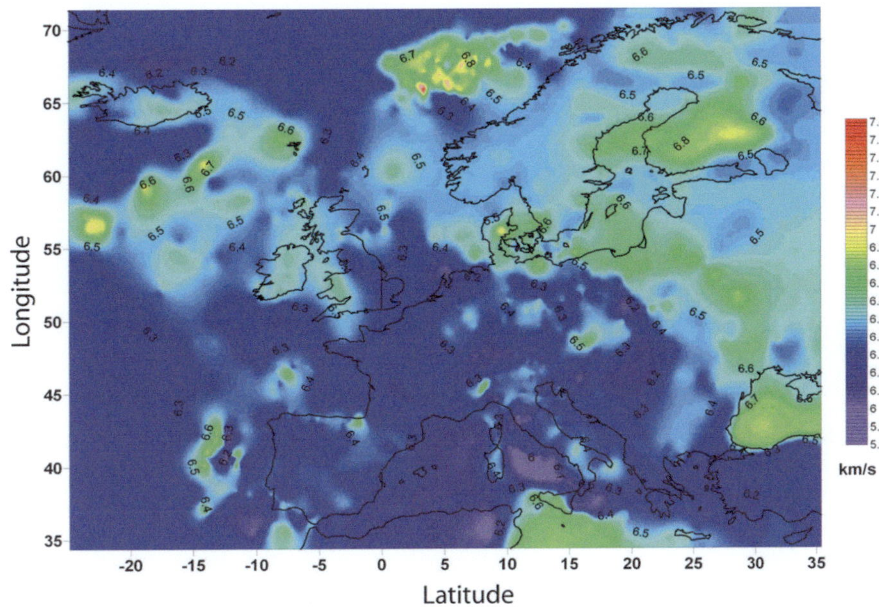

Fig. 10 Average *P-wave* velocity in the crust (km/s)

distribution are, instead, associated with areas subjected to strong extension (e.g., the Tyrrhenian Sea).

In the upper crust ~97% of the velocity values fall between 5.4 and 6.6 km/s (Fig. 9b), showing the highest peaks at 6.0–6.3 km/s (~50%) and the smallest one at ~5.5 km/s (~14%). These peaks are primarily related to the continental and oceanic areas, respectively. By contrast, the lower crust is more homogeneous: ~94% of the velocity values are distributed in a relatively narrow range from 6.3 to 7.2 km/s (Fig. 9c), around one strong peak at 6.70–6.90 km/s (~48%). The highest values (V_p >7.2 km/s), representing less than 3% of the entire distribution, are mostly found in the Baltic Shield, where an ultramafic composition of the lower crust is observed (e.g., Kozlovskaja et al., 2004). By contrast, the lowest determinations ($V_p <$ 6.3 km/s), constituting ~3.5% of the entire distribution, represent thin, undifferentiated crust, observed, for instance in the Tyrrhenian Sea.

In order to provide a consistent comparison with previous crustal models, the new model values were averaged within $1^\circ \times 1^\circ$ and $2^\circ \times 2^\circ$ compartments, which correspond to the resolution of the global models of Kaban (2001) and CRUST2.0 (Bassin et al., 2000), respectively. The main parameters of the models and most principal differences are shown in the bar plots of Figs. 11a–h and 12a–d. The difference between EuCRUST-07 and CRUST2.0 is especially visible when comparing average crustal velocities. The values of EuCRUST-07 are distributed in a larger and continuous range than CRUST2.0 (6.0–6.94 km/s compared to 6.28–6.72 km/s) and their mean value is substantially lower (~0.09 km/s), Fig. 11a and b. The velocity distribution of CRUST2.0 shows a gap between two very narrow peaks. In fact, this model attributes nearly constant velocities to two domains, representing the oceanic crust together with the continental crust of western Europe (~6.35 km/s, ~35%) and the EEP with the Atlantic continental margins (6.60–6.65 km/s, ~42%). The differences between the new and the old model, within a range of ±0.6 km/s, are negligible (±0.1 km/s) in ~44% of the area, principally represented by the EEP and the oceanic domain (Fig. 13a). Positive systematic differences (0.2–0.4 km/s) are observed in several areas (e.g., Baltic Shield, Lofoten–Vesterålen margin, Vøring Basin, Hatton and Edoras bank, Denmark), where EuCRUST-07 evidences a presence of an ultramafic lower crust with higher velocities. Smaller positive differences are found in England-Ireland, in Iceland and in the western Black Sea, where the new model shows a more mafic crust (e.g., Neprochnov et al., 1970). Negative systematic differences (0.15–0.4 km/s) are observed primarily in the TESZ and

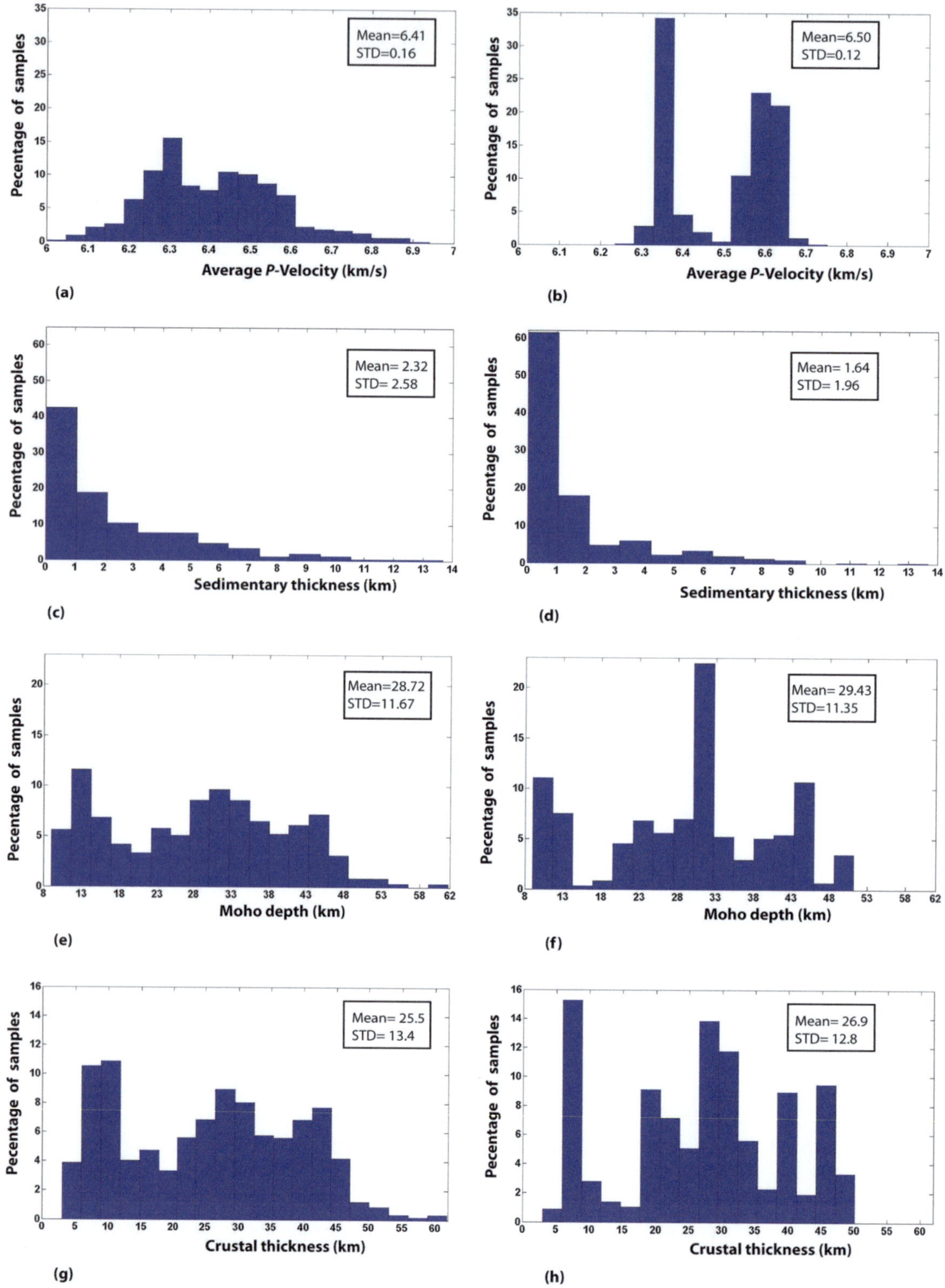

Fig. 11 Bar plot of the values distribution of EuCRUST-07 (averaged to $2° \times 2°$ resolution) and of CRUST2.0 model. **(a–b)** Average *P-wave* velocity distribution (km/s) in the crystalline crust of EuCRUST-07 and CRUST2.0 model, **(c–d)** sedimentary thickness distribution (km) of EuCRUST-07 and CRUST2.0 model, **(e–f)** Moho depth distribution (km) of EuCRUST-07 and CRUST2.0 model, **(g–h)** crystalline crust thickness distribution of EuCRUST-07 and CRUST2.0 model

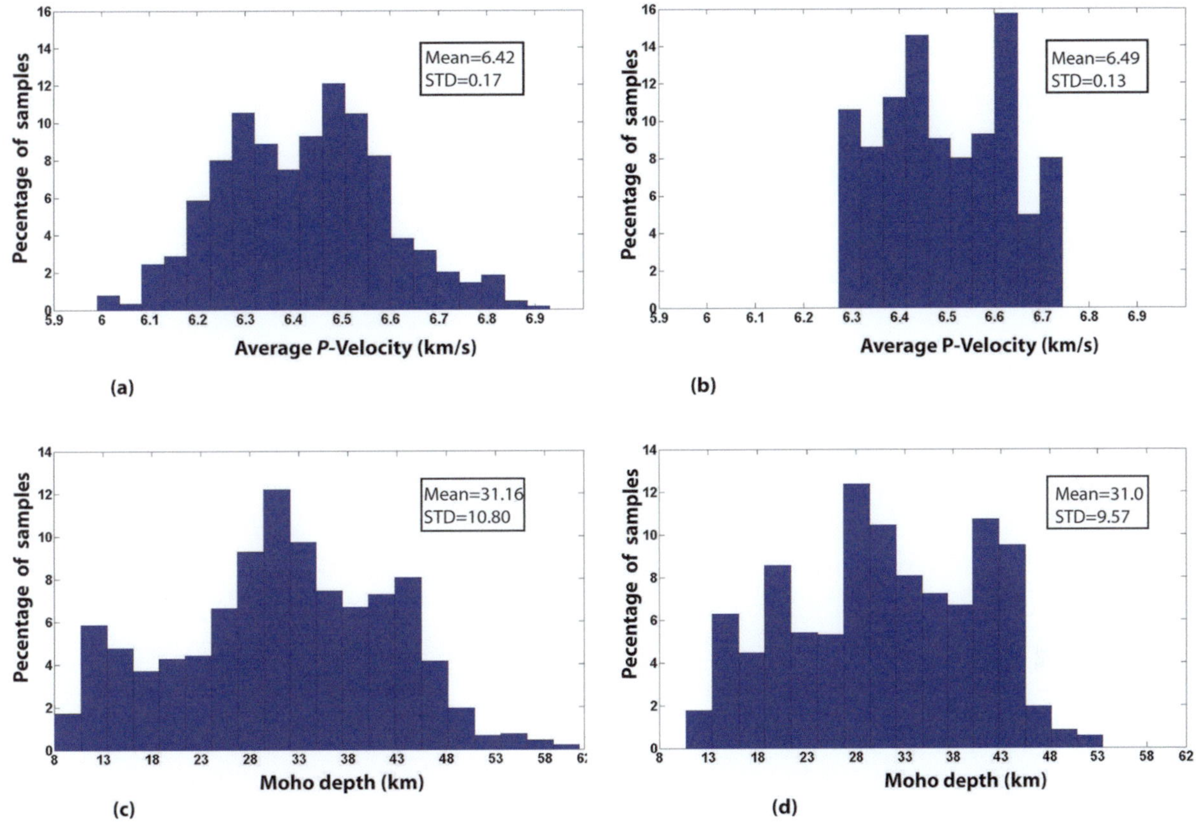

Fig. 12 Bar plot of the values distribution of EuCRUST-07 (averaged to 1°×1° resolution) and of Kaban (2001) model. **(a–b)** Average *P-wave* velocity distribution (km/s) in the crystalline crust of EuCRUST-07 and Kaban (2001) model, **(c–d)** Moho depth distribution (km) of EuCRUST-07 and Kaban (2001) model

west to the TESZ (Fig. 13a). The first difference is presumably due to a thick layer of metasediments observed in recent seismic experiments (e.g., Grad et al., 2006). West to the TESZ, EuCRUST-07 shows smaller V_p values than CRUST2.0, primarily in the areas subjected to extension, such as the Atlantic margin and some back-arc basins (Pannonian, Aegean, Tyrrhenian and Balearic Basin). A comparison of EuCRUST-07 and the model of Kaban (2001) shows similar mean values of the average velocity distribution (6.42 km/s and 6.49 km/s, respectively), Figs. 12a and b. On the other hand, the values of the old model are distributed within a more restricted range than EuCRUST-07 (6.28–6.74 km/s compared to 6.0–6.94 km/s). Two main peaks of velocities, observed in both models, mainly represent crustal velocities in eastern and western Europe, but the peaks of the new model are visibly shifted relatively to the old one: ~6.3 (~11%) and ~6.5 (~13%) km/s versus ~6.45 (~15%) and ~6.65 (~16%) km/sec, respectively. For about half part of the area (52%) the velocity differences exceed ±0.1 km/s with a maximum of ±0.55. These differences have the same sign and are mostly located in the same areas where the differences between EuCRUST-07 and CRUST2.0 are observed.

The bar plots of the sedimentary thickness (Fig. 9d) show that relatively deep basins (>2 km) occupy about ~35% of the area (Fig. 5). Most of them (about 30% of the area) are represented by the basins, which are 2–6 km deep, like the Paris Basin, while the basins with sedimentary thickness larger than 9 km (e.g., the western Black Sea and the North German Basin) represent less than 3% of the area of study. The highest values (14–16 km) are observed in the Atlantic basins (e.g., Porcupine Basin, Fig. 5) The values of the sedimentary thickness span for EuCRUST-07 and CRUST2.0 from 0 to ~13 km, showing both a main peak (~43 and ~61%, respectively) at 0–1 km and similar mean value (2.32 km and 1.64 km, respectively, Fig. 11c–d). The differences between EuCRUST-07 and CRUST2.0 are in a range of –7 to +11 km, altough they are moderate (±2 km) in 71% of the area (Fig. 13b).

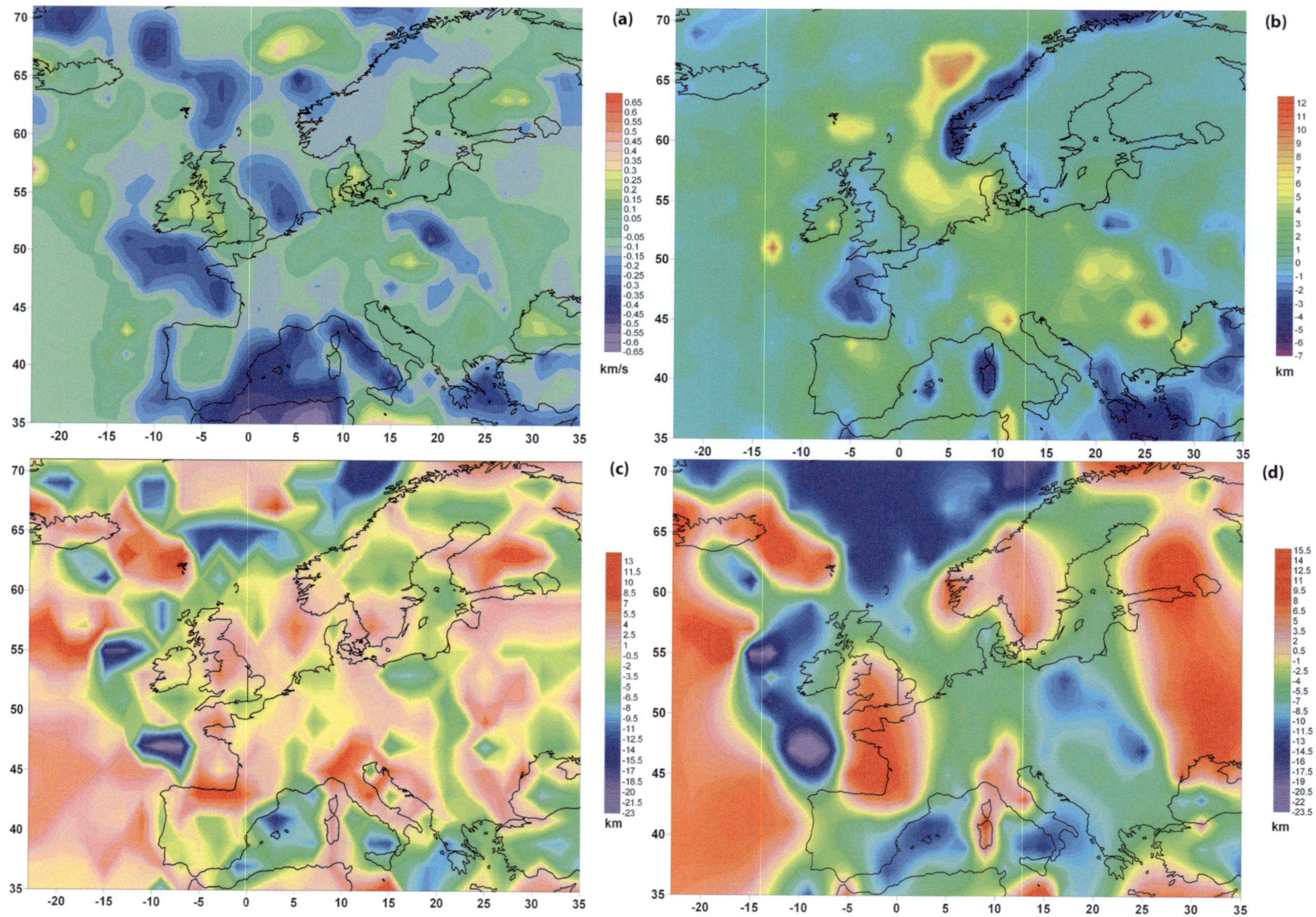

Fig. 13 Difference between crustal parameters of EuCRUST-07 averaged to $2^\circ \times 2^\circ$ and CRUST2.0 model. **(a)** Difference of the average *P-wave* velocity (km/s) in the crystalline crust between EuCRUST-07 and CRUST2.0 model. **(b)** Difference of the sedimentary thickness (km). **(c)** Difference of the Moho depth (km). **(d)** Difference of the crystalline crust thickness (km)

The sedimentary thickness appears underestimated by CRUST2.0 in the Atlantic basins, like the Porcupine Basin (~10 km) or the Vøring Basin (8–10 km), in the Central European Basin System (4–6 km), in the Po plain (~10 km), in the Focşani Basin (~10 km) and in the Cantabrian basins (4–6 km). These differences are the result of the availability of new seismic data (e.g., Mjelde et al., 2005) and sedimentary compilations (e.g., Diehl and Ritter, 2005; Scheck-Wenderoth and Lamarsche, 2005) used in EuCRUST-07. On the other hand, the sedimentary thickness is overestimated by CRUST2.0 along most part of the Norwegian coast (3–5 km), the western margin of Sardinia and Corsica (4–6 km) and along the southern Aegean coast (4–6 km). In this case, the differences can be a result of the coarser resolution of CRUST2.0, which can hardly detect small scale features, like the abrupt sedimentary thickness variation along the coastline observed in the seismic profiles.

Concerning the Moho depth values, it can be observed that ~95% of them are between 8 km and 46 km, distributed around several peaks (Fig. 9e). The lowest determinations (8–14 km), representing 15% of the distribution, with the highest peak at ~12 km are associated to the oceanic areas and the basins subjected to major extension (e.g., the Tyrrhenian Basin), Fig. 5. The highest values (>46 km) correspond to orogens, such as the Alps and the Carpathians, and the Baltic Shield. The highest peak of the Moho depth distribution (~27%) is at 28–36 km and corresponds to the average Moho depth in the study area. Another significant peak (~9%) is at ~42–44 km, typical values of the Moho depth in the EEP and in some orogens (e.g., the Dinarides and Pyrenees-Cantabrian mountains). The Moho of EuCRUST-07 evidences a maximum value in southern Finland (61.7 km), 10 km deeper than CRUST2.0 (50.1 km), although the mean depth value of the former (28.7 km) is slightly lower than the latter (29.4 km, Fig. 11e and f). Somewhat similarly to the crustal velocity distribution, the CRUST2.0 values appear to be distributed primarily around several peaks: 9–14 km (~18%), corresponding to the oceanic domain; 30–33 km (~24%), typical values for stable areas in western Europe; 43–46 km (~11%) mostly located in the EEP and Fennoscandia. By contrast, EuCRUST-07 does not show any significant gap. This means that the CRUST2.0 model rather represents general peculiarities of the main tectonic domains than providing homogeneous crustal structure even on the coarser grid. The differences are within a range of –23.0 to 13.6 km (Fig. 13c). EuCRUST-07 shows a shallower Moho of 10–20 km along a part of the Atlantic margin, and 5–10 km in western Anatolia. The same situation also exists for those basins, where CRUST2.0 overestimates the average crustal velocity (e.g., Tyrrhenian Sea, Fig. 13b and c). On the other hand, EuCRUST-07 shows a Moho 5–10 km deeper than CRUST2.0 in the orogens (e.g., the Alps and the Cantabrian Mountains) reflecting a flexure of the subducting plate (e.g., Pedreira et al., 2003). The highest positive differences (10–13 km) are observed in the Baltic Shield, in the Faeroe-Iceland margin and the Hatton and Edoaras bank, where the presence of magmatic underplating (e.g., Richardson et al., 1998; Kozlovskaja et al., 2004) increases anomalously the crustal thickness (Fig. 13c).

EuCRUST-07 spans over a larger range even compared to the Moho depth of Kaban (2001) (8–61.5 km, compared to 11.9–53 km), although the average Moho depth is similar in both models (~31 km), Fig. 12c and d. The new and the old model are distributed around similar main peaks: 11–16 km (10%) compared to 13–16 km (~7%), 27–35 (~32%) compared to 27–32 (~32%) and 40–45 km (~16 and ~21%, respectively). For both models the values of the first two peaks are typically for the Moho depth of the oceanic and continental domain, respectively, while the last one is characteristic of the Moho observed in the orogens and in the areas east to the TESZ. In addition, the model of Kaban (2001) shows another significant peak (~9%) at 20 km, mostly represented by the Moho depth in the Atlantic margin and in the Mediterranean Sea. The differences between the new and the old model are within a range of –14.9 to 17.2 km, with moderate values (within ±5 km) observed in ~79% of the area. In contrast with CRUST2.0, the model of Kaban (2001) overestimates the Moho depth (7–14 km) only in some basins (e.g., Tyrrhenian and Ionian Sea), in western Anatolia and in the Ivrea zone, and the differences are not strongly correlated with the average crustal velocity. The strongest positive differences (10–13 km) are observed in the Baltic Shield.

The thickness of the crystalline crust distribution is similar to that one of the Moho depth except for a slightly different depth range (3–64 km, Fig. 9f). The peaks of the distributions are also shifted 2–4 km towards low values. The first peak at 6–10 km (~15%) represents not only thin crystalline crust in the oceanic domain as in the Moho depth distribution, but also some basins in the thinned continental area (e.g., the Focşani Basin). One can observe that the peak at 42–44 km presented in the Moho depth distribution is also significant in the crystalline crust plot (~7%), being a typical value for the orogens and shields within the EEP. Compared to CRUST2.0 (Fig. 11g and h), EuCRUST-07 demonstrates a more continuous distribution similarly to the Moho depth. Differences between the new and the old model span from –23.5 to 15.5 km and are moderate (within ±5 km) in the ~67% of the area (Fig. 13d). However, for approximately one third of the area the differences are very large. The most pronounced differences are found in the Baltic Shield where the EuCRUST-07 thickness is ~10 km higher and along the Atlantic margin, where it is 10–23 km lower.

The relative thickness of the upper crust, corresponding to the entire crystalline crust, falls mainly (>93%) in a range from 0.3 to 0.8 (Fig. 11g). Maximal values (>0.5) correspond to continental areas (~42%), while in the oceanic domain the ratio decreases to 0.35–0.4 (~22%). The highest values (>0.7, ~15% of samples) are observed in areas where extensional deformations occurred mostly at the expense of the lower crust (e.g., in the Alboran Sea) and in areas where a middle crust, which is considered in EuCRUST-07 as a part of the upper crust, was identified in the seismic profiles (e.g., in the EEP).

Lithology of the European Crust

The parameters of EuCRUST-07 were used to make inferences on lithology of different European tectonic provinces (Fig. 14a–d), (Table 1). The average velocity values in each crustal layer have been related to the values reported by Christensen and Mooney (1995) for different lithotypes, which are determined in laboratory experiments under various temperature and pressure conditions. The pressure was estimated in the middle of each layer. The depths to the mid points of the upper and lower crust were divided into 5 and 9 different groups, respectively, with a range of 5 km (Fig. 14a and b). The group 1 of the upper and lower crust contains the values related to part of the oceanic crust. Concerning the upper crust, the values of the group 5 are located where a thick upper crust and sedimentary layer, characterizing the deepest basins in Europe (the Focşani Basin, the Po Plain and the Polish Through), is observed. The assumption, adopted in our model for the middle crustal layer (mostly detected in eastern Europe) as a part of the upper crust, increases the difference in the middle point depth of this layer (as well as in the average upper crustal velocities), existing between the areas east and west to the TESZ. Concerning the lower crust, the deepest values belonging to the group 8 and 9 correspond mostly to the areas of very thick crystalline crust without a significant sedimentary layer (e.g., the eastern Alps and the central part of the Baltic Shield). Thermal conditions of the crystalline crust were estimated using the surface heat flow (HF) values provided in the Geothermal Atlas of Europe (Hurtig et al., 1992) for the continental part and by local studies in other parts of the area of study (e.g., Zito et al., 2003 in the Tyrrhenian Sea). These values were distributed in three groups with high (HF> 90 mWm^{-2}), average (60 <HF> 90 mWm^{-2}) and low (HF< 60 Wm^{-2}) crustal thermal regime. This approach is probably a reasonable first order approximation, since the contribution of the crust to the surface HF values is quite strong, varying from 25 to 75%, as suggested by previous studies (e.g., Morgan, 1985; Artemieva and Mooney, 2001).

The estimated pressure and thermal conditions together with the velocity distributions within the crustal layers were compared with experimental rock data (Christensen and Mooney, 1995) to select the representative lithotypes of the upper and lower crust (Fig. 14c and d). An obvious limitation of this approach is due to the fact that the model provides average *P*-wave velocities for the crystalline crustal layers, whereas more robust inferences on lithology could be obtained using direct seismic data for a specific depth possibly for both *P*- and *S*-wave. Furthermore, it should be taken into account that the interpretation of *P*-wave velocities in terms of rock composition is non-unique, as a given velocity value may characterize a number of different rock types (Christensen and Mooney, 1995). Therefore, among various lithotypes the most common upper and lower crust components were adopted, while rocks having strong anisotropy (e.g., phyllite) were excluded, due to a lack of information about variations of this parameter in the European crust. In addition, the analysis was limited to the continental part of the study area, since the oceanic domain is not covered by a

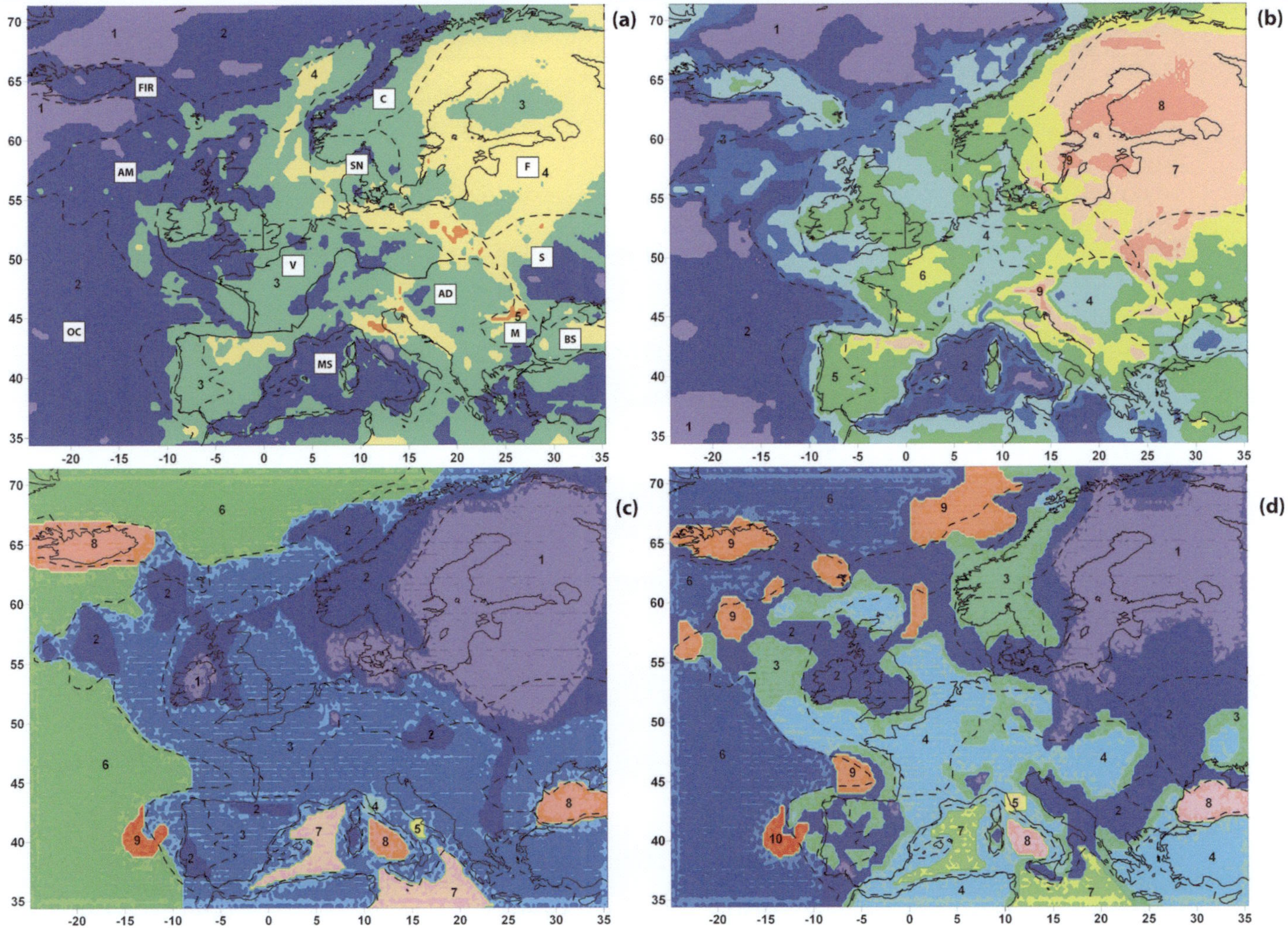

Fig. 14 Depths to the mid-points and lithotypes of the crustal layers. Dashed lines show the boundaries between tectonic provices. **(a)** Depths to the mid-points of the upper crust (km). Numbers show the depth range as follows: *1*, 0–5 km; *2*, 5–10 km, *3*, 10–15 km, *4*, 15–20 km, *5*, 20–25 km. Abbreviations stand for tectonic provinces name as follows: *F*, Fennoscandia; *S*, Sarmatia, *SN*, SvecoNorvegian; *BS*, Black Sea, *M*, Moesian Platform; *C*, Caledonides; *V*, Variscides; *AD*, Alpine Domain; *AM*, Atlantic Margin; *MS*, Mediterranean Sea, *OC*, Oceanic Crust; *FIR*, Faeroe Iceland Ridge. **(b)** Depths to the mid-points of the lower crust (km). Numbers show the depth range as follows: *1*, 5–10 km; *2*, 10–15 km; *3*, 15–20 km; *4*, 20–25 km; *5*, 25–30 km; *6*, 30–35 km; *7*, 35–40 km; *8*, 40–45 km; *9*, 45–50 km. **(c)** Lithology of the upper crust. Numbers represent the following lithotypes: *1*, Felsic Granulite; *2*, Granite-Granodiorite; *3*, Granite-Gneiss; *4*, Molten Granite; *5*, Anomalous High velocity upper crust; *6*, Oceanic Crust; *7*, Transition to Oceanic Crust; *8*, Basalts; *9*, Serpentinized peridotite. **(d)** Lithology of the lower crust. Numbers represent the following lithotypes: *1*, Mafic Garnet Granulite; *2*, Mafic Granulite; *3*, Amphibolites; *4*, Diorite; *5* Molten Granite; *6*, Oceanic Crust; *7*, Transition to Oceanic crust; *8*, Basalt; *9*, Ultrafemic crust; *10* Serpentinized Peridotite

Table 1 Main crustal parameters values, thermal regime and lithotypes of the European tectonic provinces

Region	Age	Average crystalline crust thickness/std (km)	Average upper crust thickness/std (km)	Average lower crust thickness/std (km)	Average crystalline crust velocity/std (km/s)	Average upper crust velocity/std (km/s)	Average lower crust velocity/std (km/s)	Average depth middle point upper crust (groups)	Average depth middle point lower crust (groups)	Thermal regime	Predominant lithotype upper crust	Predominant lithotype lower crust
Fennoscandia	3.0–1.45 Ga	43.9/5.2	30.3/3.7	13.6/6.0	6.56/0.096	6.33/0.03	7.09/0.11	2–5	5–9	Low	Felsic granulite	Mafic garnet granulite
Sarmatia	3.7–3.0 Ga	41.4/4.5	19.3/7.2	22.1/6.5	6.57/0.069	6.27/0.12	6.84/0.12	1–5	4–8	Low	Felsic granulite/granodiorite/gneiss	Mafic granulite
SvecoNorvegian	1.2–0.9 Ga	30.8/7.8	17.6/4.6	13.2/4.5	6.54/0.093	6.32/0.058	6.83/0.093	2–4	3–8	Low-average	Felsic granulite/granodiorite	Mafic granulite/amphibolite
Caledonides	520–400 Ma	29.7/7.6	17.1/5.7	12.6/4.0	6.44/0.095	6.17/0.13	6.82/0.17	1–5	2–9	Low-average	Felsic granulite/granodiorite/gneiss	Mafic granulite/amphibolite
Variscides	370–310 Ma	30.2/3.6	19.3/4.8	10.8/3.7	6.32/0.098	6.11/0.094	6.70/0.19	2–4	3–6	Average-high	Gneiss granite	Diorite
Alpine domain	130–0 Ma	29.7/6.6	17.5/5.3	12.1/4.4	6.29/0.12	6.08/0.11	6.62/0.22	1–5	2–3	Average-high	Gneiss granite	Mafic granulite/diorite
Western black Sea	90–71 Ma	17.7/6.6	8.4/3.4	9.4/3.4	6.77/0.06	6.65/0.10	6.89/0.03	2–4	3–7	Low	Basalt	Basalt
Moesian platform	550 Ma	30.0/4.3	17.0/3.63	13.0/2.4	6.42/0.061	6.11/0.09	6.81/0.083	2–5	4–7	Low	Gneiss granite	Mafic granulite
Mediterranean	5–290 Ma	14.1/7.8	7.4/4.5	6.7/4.1	6.31/0.21	6.10/0.18	6.54/0.32	1–4	1–4	Average-high	Transitional crust/basalt gneiss	Transitional crust/diorite
Iceland-faeroe ridge	0–58 Ma	27.6/4.6	12.0/2.1	15.6/3.7	6.50/0.08	5.97/0.15	6.91/0.05	1–3	2–5	High	Basaltic lavas	Ultramafic crust
Atlantic margin	58–105 Ma	17.2/6.4	8.6/4.5	8.5/3.5	6.44/0.14	6.06/2.01	6.81/0.16	1–4	1–6	Average	Gneiss granite	Mafic garnet granulite/mafic granulite/
Oceanic crust	0–105 Ma	10.0/4.7	3.7/1.9	6.3/3.0	6.34/0.13	5.57/0.22	6.78/0.12	1–4	1–5	Average	Basaltic lavas	Gabbro

sufficiently dense net of seismic profiles and for most oceanic areas a "standard" oceanic crust composed of fractured basaltic lavas and gabbro was assumed. On account of the above mentioned limitations, this lithology classification cannot represent a reference for petrologists, but rather a first attempt to evaluate possible predominant lithotypes of the upper and lower crust, which should improve further estimates of the crustal parameters (e.g., rheology). The new lithology map can be employed in geophysical modelling, e.g., as input for the calculation of strength distribution in the crust. On the other hand, as very few direct determinations exist for the composition of the lower crust, interpretation of geophysical data appears to remain the only option to obtain such (preliminary) estimates. It should be noted that petrological studies of the upper crust have also restrictions due to the limited number of deep boreholes. Therefore, the crustal model provides an opportunity for a consistent comparison of lithology of the crust for the main tectonic units of Europe.

As the main representative lithotypes of the upper crust were selected felsic granulite (covering 17% of the study area), granite-granodiorite (covering 13% of the study area) and granite-gneiss (covering 40% of the study area). Among these three lithologies, only felsic granulite is not observed in many outcropping rocks. However, this is the most suitable lithotype which can be associated to a thick (~30 km) high velocity (~6.30 km/s) upper crust including also the middle crustal layer. The granite-gneiss lithotype could be attributed to most part of the upper crust of western Europe. These areas are, characterized by various pressure and thermal crustal conditions, but similar velocities (6.0–6.1 km/s), which correspond to the values obtained for this type of rocks in laboratory experiments (Christensen and Mooney, 1995). Following the same approach, several lithotypes for the lower crust were defined: mafic garnet granulite (covering 12% of the study area) mafic granulite (covering 25% of the study area), amphibolites (covering 14% of the study area) and diorite (covering 19% of the study area). At corresponding depths in the lower crust all the lithotypes should be of metamorphic type. Although, diorite is an igneous rock, it was selected as a lithotype, possibly representative for the lower crust, because of its average composition and physical characteristics. Other lithotypes are defined for smaller areas characterized by particular upper and lower crust conditions, as discussed in more detail below.

Among large tectonic provinces, we can identify the Archean-Proterozoic crust (SvecoNorvegian, Fennoscandia and Sarmatia provinces), which is characterized by a thick (~40 km) and high velocity (~6.55 km/s) crystalline crust mostly at a low thermal regime (<60 mW/m^2). In these regions also the average depth of the middle points of both crustal layers is within a similar range (between 10–20 km and 25–35 km, respectively). Some differences are found for the Sarmatia province, which is characterized by approximately equal thickness (~20 km) of both crustal layers. The SvecoNorvegian province shows a thinner crystalline crust (~30 km), with the average velocity values in both layers, similar to the other two provinces (Table 1).

On account of these crustal parameters, the Fennoscandia province can mostly be represented by felsic granulite in the upper crust (covering 91% of the area) and mafic garnet granulite in the lower crust (covering 76% of the area). By contrast, Sarmatia shows a more heterogeneous upper crust being characterized by felsic granulite, granodiorite and gneiss granite (covering 28, 23, 45% of the area, respectively), while the lower crust is prevalently represented by mafic granulite (75% of the area). The SvecoNorvegian province, instead, may be characterized by two prevalent lithotypes in the upper (felsic granulite and granite-granodiorite) and in the lower crust (amphibolites and mafic granulite). Petrological studies in Fennoscandia (e.g., Korja et al., 2006), evidence that the uppermost part of the crust is generally made of granite-granodiorite and migmatite and only in smaller percentage of felsic granulite. However, since the lithotypes selected are attributed for the entire upper crust, felsic granulite, which has the physical parameters corresponding to the crustal conditions of this area, was chosen.

The oldest provinces in western and central Europe, the Moesian Platform and the Caledonides, demonstrate similar characteristics in terms of the average crustal thickness (30 km), upper/lower crust ratio (0.6/0.4) and average velocity (6.43 km/s) (Figs. 3a, b and 10). However, the last province is much more extended than the first one and is characterized by a more heterogeneous crust taking also into account the pressure and thermal conditions (Table 1). Therefore, the upper crust of the Caledonides is prevalently represented by granodiorite and granite gneiss (~41 and ~45%, respectively). The characteristics of the lower crust, instead correspond to mafic granulite (~47% of the area) and amphibolites (~30% of the area). By contrast, the upper and the lower crust of the Moesian Platform is mostly represented by granite-gneiss (~80% of the area) and mafic granulite (~70% of the area).

The Alpine domain and the Variscides are characterized by similar average crustal thicknesses (~30 km) and velocities (~6.30 km/s), noticeably lower than in the other European provinces. The Variscides compared to the Alpine domain are characterized by a slightly lower upper/lower crust ratio (0.36, compared to 0.41) and higher crustal velocity (Table 1). However, the crust in the Alpine province is more heterogeneous considering also the pressure conditions (Table 1), since it comprises different tectonic types: orogens (e.g., the Alps) and basins subjected to extension (e.g., the Pannonian Basin). The thermal regime varies in both provinces from the average to the high level (e.g., the Massif Central and the Pannonian Basin). On account of these crustal characteristics, the upper crust of both provinces is mainly represented by granite-gneiss (~90% of the area), and for the lower crust by diorite (~53–57% of the area), mafic granulite (~16–22% of the area) and amphibolites (~18–25% of the area). The mafic granulite in the Alpine domain mostly corresponds to the Adriatic plate, which has a different origin, being interpreted by some authors as an African promontory (e.g., Channel et al., 1979). Furthermore, a molten granite lithotype to the upper and lower crust of western Tuscany was associated, on account of the specific crustal conditions observed here (hot, thin and low velocity), while the Gargano promontory is characterized by an anomalous high velocity upper crust.

Although the Atlantic margin, the Mediterranean Sea, the Iceland-Faeroe Ridge and the western Black Sea demonstrate crustal characteristics (velocity and thickness), which are generally similar to some other continental areas (Table 1), we cannot adopt for these provinces the same lithology as for the other parts of the study area subjected to the same conditions, on account of the different origin of the crust. Therefore, in these areas or in part of them a correspondence with the laboratory data on the "standard" rocks was not searched. Instead, the representative lithotypes were chosen using the interpretations of the seismic profiles employed in EuCRUST-07. The crust of the western Black Sea has been interpreted, as before mentioned, as a relict of ocean crust (Çakir and Erduran, 2004). Therefore, the basalt is the lithotype representative for the upper and lower crust of this province. Furthermore, the same lithotype is also associated to the upper crust of Iceland, since in this area basaltic lavas outcrop and compose the entire layer. The anomalously thick high velocity lower crust observed below part of the Atlantic Margin and the Iceland-Faeroe ridge is generally interpreted as a result of underplating and is characterized by an ultramafic composition. On the other hand, most of the upper crust of the Atlantic Margin, being mainly of continental origin, was associated with the gneiss-granite lithotype. An exception is represented by a part of the Iberian Abyssal Plain where the entire crust is interpreted as peridotite affected by a different degree of serpentinization (Chian et al., 1999). The opening of the Balearic and Tyrrhenian Sea as back-arc extensional basins resulted in an extremely thin low velocity oceanic crust under high thermal conditions (e.g., Zito et al., 2003). We attribute the basalt lithotype to the central part of the Tyrrhenian Sea. The most eastern part of the Mediterranean (the Aegean Sea) is of continental origin and the granite-gneiss and diorite lithotypes were defined as representative for the upper and lower crust in accordance with the physical crustal conditions.

Conclusions

Data on the crustal structure are used in various models of the underlying mantle structure to remove a priori the very strong effect produced by this very heterogeneous layer in almost all fields observed at the Earth surface. It is nearly impossible without this prerequisite step to resolve a trade-off between the crustal and upper mantle effects in gravity and geothermal modelling, seismic tomography and other geophysical fields. Many seismic studies performed during the last decade have clearly demonstrated that existing crustal models contain significant errors, which could remarkably affect the results based on these models. Significant differences are even observed between the most recent Moho map of Ziegler and Dèzes (2006) and results of new seismic studies, reaching –25 km in the Ivrea zone where the new data show the updoming of a 10–20 km shallower Adriatic Moho, and +17 km in the eastern Alps (Finetti, 2005b).

In this paper EuCRUST-07, a new digital model of the crust of western and central Europe, has been constructed by assembling available results of deep seismic reflection, refraction and receiver functions studies including most recent studies. All the data collected were analysed and cross-checked to select the most robust. The new model covers the area within 35°N, 71°N; 25°W35°E and is represented at a uniform $15' \times 15'$ grid. It consists of three layers: sediments and two layers of the crystalline crust. Each layer of the crystalline crust is characterized by average *P*-wave velocities. The new model demonstrates large differences with existing regional/global compilations, mostly resulting from inclusion of recently acquired seismic data and detailed local compilations of the principal crustal boundaries. The Moho depth values are mostly distributed between 8 and 46 km (95%). The lowest determinations (8–14 km) are associated with the oceanic areas and the basins subjected to major extension (e.g., the Tyrrhenian Basin), while the highest ones (> 46 km) correspond to orogens and the Baltic Shield. The comparison between the new model and regional Moho compilations used in the gravity modelling of the mantle structure reveal also significant discrepancies. EuCRUST-07 shows a Moho 5–10 km deeper than that one used in previous studies in the orogens (e.g., the Cantabrian Mountains), reflecting a flexure of the subducting plate (e.g., Pedreira et al., 2003) and even up to 10–13 km in the areas where a presence of magmatic underplating (e.g., Baltic Shield and Faeroe-Iceland margin) strongly increases crustal thickness. On the other hand, EuCRUST-07 entails a Moho depth 10–20 km shallower along a part of the Atlantic margin and in some basins (e.g., Tyrrhenian Sea), characterized by thin crust due to the back- arc extension experienced.

Although, the basement depth is more consistent with previous regional and global maps, several remarkable differences are also found. CRUST2.0 compared to the new model underestimates thickness of sediments up to ~10 km in many basins (e.g., the Porcupine Basin), while sedimentary thickness is overestimated up to ~3–6 km along some coastlines (e.g., the Norwegian Coast).

The velocity structure of the crust is even more different from the one of CRUST2.0. While the latter shows only general differences among few tectonic units of the region, EuCRUST-07 demonstrates a very complex velocity structure of the crystalline crust with sharp horizontal changes. The velocity structure of the crust turns out to be very heterogeneous, with values distributed in a larger and continuous range than CRUST2.0 (6.0–6.94 km/s, compared to 6.28–6.72 km/s) and the model of Kaban (2001) (6.0–6.94 km/s compared to 6.28–6.74 km/s). A ninety three percentage of the values of EuCRUST-07 velocity fit to a range from 6.20 to 6.80 km/s, with two pronounced peaks, the first one at ~6.30 km/s (~28%) mostly corresponds to western Europe, while the second one at ~6.50 (~23%) to the east European crust and restricted structures of western Europe, e.g., the Atlantic Margin. The differences between EuCRUST-07 and the previous models are particularly strong (up to 0.6 km/s) in areas where EuCRUST-07 reveals very high average crustal velocity (V_P> 6.6 km/s), on account of an ultramafic lower crust (e.g., in the Vøring Basin), and low average crustal velocity ($V_P \sim 6.0$–6.2 km/s) observed in the basins subjected to strong extension, mostly at the expense of the lowest crustal layer (e.g., in the Alboran Basin).

The parameters of EuCRUST-07 and crustal thermal regime information, reflected in the surface heat flow, were used to define possible lithotypes of the European crust, whose data are required for further geophysical modelling. The new lithology map for the continental domain demonstrates that the older eastern European tectonic provinces, characterized by a high velocity, thick upper and lower crust and a low thermal regime are mostly represented by felsic and mafic granulite, respectively. On the other hand, the younger and more heterogeneous upper and lower crust of western Europe might have predominantly a granite-gneiss and dioritic composition, respectively. In the oceanic domain not covered by seismic profiles a "standard" oceanic crust composed of fractured basaltic lavas and gabbro was assumed. The opening of the Atlantic margin has produced in some areas an anomalous thick high velocity lower crust due to mantle underplating. By contrast, the opening of the Balearic and Tyrrhenian Sea has resulted in an extreme thin low velocity transitional crust characterized by a high thermal regime. The different origin of the crust in these areas and the anomalous crustal characteristics in some other parts of the study area (e.g., in the Black Sea) prevented us from following the same approach in the lithotypes determination, which is used in the continental part. Due to the heterogeneous composition of the Earth's crust, the *P*-wave average velocity and thickness of the upper and lower crust are not sufficient to discern among all possible lithotypes. On the other hand, up till now a consistent study of the crustal lithology did not exist for the whole area. Therefore, this study represents a first attempt to define representative lithologies of the upper and lower crust using the new crustal model. The determined lithotypes can be useful to construct a new rheology model of the European crust.

Acknowledgments We would like to thank Pierre Dèzes and Peter Ziegler (University of Basel), Richard W. England (University of Leichester), Elena Kozlovskaja (University of Oulu), Federica Marone (University of California), Michael Martin and Joachim Ritter (University of Karlsruhe), Luis Matias (University of Lisboa), Forough Sodoudi and Rainer Kind (GFZ, Potsdam), for providing Moho data. We are grateful to Conxi Ayala (IGME), Olivier Bourgeois (Nantes University), Patrick Ledru (BRGM), Thomas Diehl (ETH), Laszlo Lenkey (Eötvös University) and Magdalena Scheck-Wenderoth (GFZ, Potsdam), for providing sedimentary thickness and basement depth compilations. We are grateful to Charlotte Krawczyk, Yuriy Maystrenko, Magdalena Scheck-Wenderoth (GFZ, Potsdam), Adele Manzella and Giovanni Ruggeri (Istituto di Geoscienze e Georisorse), Roland Oberhänsli (University of Potsdam) and Peter Ziegler (University of Basel) for fruitful discussions. Funds were kindly provided by NWO (Netherlands Organization for Scientific Research) and SRON (Space Research Organization Netherlands) DFG (German Research Foundation) Ro-2330/4-1.

References

Aichrot, B., C. Prodehl, and H. Thybo: Crustal structure along the Central Segment of the EGT from seismic-refraction studies. Tectonophysics **207**, 43–64 (1992)

Akyol, N., L. Zhu, B. J. Mitchell, H. S. Ozbilir, and K. Kekovalı: Crustal structure and local seismicity in western Anatolia. Geophys. J. Int. **166**, 1259–1269 (2006)

Alessandrini, B., L. Beranzoli, G. Drakatos, C. Falcone, G. Karantonis, F. M. Mele, and G. Stavrakakis: Back arcs basins and P-wave crustal velocity in the Ionian and Aegean regions. Geophys. Res. Lett. **24**, 5, 527–530 (1997)

Amato, A., R. Azzara, A. Basili, C. Chiarabba, M. G. Ciaccio, G. B. Cimini, M. Di Bona, A. Frepoli, I. Hunstad, F. P. Lucente, L. Margheriti, M. T. Mariucci, P. Montone, C. Nostro, and G. Selvaggi: Geodynamic evolution of the northern Appenines from recent seismological studies. Mem. Soc. Geol. It. **52**, 337–343 (1998)

Artemieva, I. M. and W. D. Mooney: Thermal thickness and evolution of Precambrian lithosphere: A global study. J. Geophys. Res. **106B**, 16387–16414 (2001)

Ayala, C., M. Torne, and J. Pous: The lithosphere-asthenosphere boundary in the western Mediterranean from 3D joint gravity and geoid modeling: Tectonic implications. Earth Planet. Sci. Lett. **209**, 275–290 (2003)

Banda, E., J. Ansorge, M. Boloix, and D. Cordoba: Structure of the crust and upper mantle beneath the Balearic islands (western Mediterranean). Earth Planet. Sci. Lett. **49**, 219–230 (1980)

Banda, E., E. Surinach, A. Aparicio, and E. Ruiz de la Parte: Crust and upper mantle structure of the central Iberian Meseta (Spain). Geophys. J.R. Atr. Soc. **67**, 779–789 (1981)

Bassin, C., G. Laske, and G. Masters: The current limits of resolution for surface wave tomography in North America. EOS Trans AGU **81**, F897 (2000)

Bourgeois, O., M. Ford, M. Diraison, C. Le Carlier de Veslud, M. Gerbault, R. Pik, N. Ruby, and S. Bonnet: Separation of rifting and lithospheric folding signatures in the NW-Alpine foreland. Int. J. Earth Sci. (Geol Rundsch) DOI 10.1007/s00531-007-0202-2 (2007)

Brück, E., T. Bodoky, E. Hegedüs, P. Hrubcová, A. Gosar, M. Grad, A. Guterch, Z. Hajnal, G. R. Keller, A. Pičák, F. Sumanovac, H. Thybo, and F. Weber and Alp 2002 Working Group: ALP 2002 SEISMIC EXPERIMENT. Stud. Geophys. Geod. **47**, 671–679 (2003)

Carminati, E., C. Doglioni, and D. Scrocca: Alps Vs Apennines. Special Volume of the Italian Geological Society for the IGC 32 Florence-2004 (2004)

Channel, J. E. T., B. D'Argenio, and F. Horwath: Adria, the African promontory, Mesozoic Mediterranean paleogeography. Earth Sci. Rev. **15**, 213–292 (1979)

Chian, D., K. F. Louden, T. Minshull, and R. Whitmarsh: Deep structure of the ocean-continent transition in the southern Iberian Abyssal Plain from seismic refraction profiles: Ocean Drilling Program (Legs 149 and 173) transect. J. Geophys. Res. **104**, B4, 7443–7462 (1999)

Christensen, N. I. and W. D. Mooney: Seismic velocity structure and composition of the continental crust: A global view. J Geophys Res. **100**, 9761–9788 (1995)

Christiansson, P., J. I. Faleide, and A. M. Berge: Crustal structure in the northern North Sea: an integrated geophysical study, in: Nottvedt, A. (ed.), Dynamics of the Norwegian Margin. Geological Society of London, Special Publication, **167**, 15–40 (2000)

Cloetingh, S., G. Spadini, J. D. Van Wees, and F. Beekman: Thermo-mechanical modelling of Black Sea Basin (De)formation. Sedimentary Geology **156**, 169–184 (2003)

Córdoba, D., E. Banda, and J. Ansorge: The Hercynian crust in northwestern Spain: A seismic survey. Tectonophysics **132**, 321–333 (1987)

Contrucci, I., A. Nercessian, N. Béthoux, A. Mauffret, and G. Pascal: A Ligurian (Western Mediterranean Sea) geophysical transect revisited. Geophys. J. Int. **146**, 74–97 (2001)

Chulick, G. S. and W. D. Mooney: New maps of North American crustal structure. Seismol. Res. Lett. **69**, 160 (1998)

Çakir, O. and M. Erduran: Constraining crustal and uppermost mantle structure beneath station TBZ (Trabzon, Turkey) by receiver function and dispersion analysis.. Geophys. J. Int. **158**, 955–971 (2004)

Darbyshire, F. A., I. Th. Bjarnason, R. S. White, and O. G. Flóvenz: Crustal structure above the Iceland mantle plume imaged by the ICEMELT refraction profile. . Geophys. J. Int. **135**, 1131–1149 (1998)

DEKORP Research Group: Wide-angle Vibrosesis data from the western Rhenish Massif. Tectonophysics **173**, 83–93 (1990)

Diehl, T. and J. R. R. Ritter and the CALIXTO Group: The crustal structure beneath SE Romania from teleseismic receiver functions. Geophys. J. Int. **163**, 238–251 (2005)

Di Stefano, R., C. Chiarabba, F. Lucente, and A. Amato: Crustal and uppermost mantle structure in Italy from the inversion of P-wave arrival times: geodynamic implications. Geophys. J. Int. **139**, 483–498 (1999)

Doglioni, C.: A proposal for the kinematic modelling of W-dipping subductions-possible applications to the Tyrrhenian-Apennines system. Terra Nova **3**, 423–434 (1991)

Edel, G. B., K. Fuchs, C. Gelbke, and C. Prodehl: Deep structure of the southern Rhinegraben area from seismic refraction investigations. J. Geophys. **41**, 333–356 (1975)

Ekström, G. and A. M. Dziewonski: The unique anisotropy of the Pacific upper mantle. Nature **394**, 168–172 (1998)

Enderle, U., K. Schuster, C. Prodhel, A. Schulze, and J. Bribach: The refraction seismic experiment GRANU 95 in the Saxothuringin Belt, southeastern Germany. . Geophys. J. Int. **133**, 245–259 (1998)

Exxon: Tectonic map of the World, Scale 1:10,000,000. Exxon Prod. Research Co., AAPGF, Tulsa, OK, USA (1985)

Fernández-Viejo, G.: The ESCI-N Project after a decade: a synthesis of te results and open questions. Trabajos de Geologia Univ. de Oviedo **25**, 9–25 (2005)

Finetti, I. R.: CROP Project, Deep Seismic Exploration of the Central Mediterranean and Italy. edited by Finetti, I.R., Elsevier, pp. 606 (2005a)

Finetti, I. R.: Depth contour Map of the Moho discontinuity in the Central Mediterranean region from new CROP seismic data, in CROP Project, Deep Seismic Exploration of the Central Mediterranean and Italy, edited by I. R. Finetti, pp. 597–606, Elsevier (2005b)

Franke, W., R. K. Bortfeld, M. Brix, G. Drozdewski, H. J. Duerbaum, P. Giese, W. Janoth, H. Jödicke, Ch. Reichert, A. Scherp, J. Schmoll, R. Thomas, M. Thuenker, K. Weber, M. G. Wiesner, and H. K. Wong: Crustal structure of the Rhenish Massif: results of the deep seismic reflection lines DEKORP 2-North and 2North-Q. Geologische Rundschau **79/3**, 523–566 (1990)

Fullea, J., M. Fernàndez, H. Zeyen, and J. Vergés: A rapid method to map the crustal and lithospheric thickness using elevation, geoid anomaly and thermal analysis. Application to the Gibraltar Arc System, Atlas Mountains and adjacent zones. Tectonophysics **430**, 97–117 (2007)

Galindo-Zaldivar, J., A. Jabaloy, F. Gonzáles-Lodeiro, and F. Aldaya: Crustal structure of the central sector of the Betic Cordillera (SE Spain). Tectonics **16, (1),** 18–37 (1997)

Gallart, J., J. Diaz, A. Nercessian, A. Mauffret, and T. Dos Reis: The eastern end of the Pyrenees: Seismic features at the transition to the NW Meidterrenean. Geophs. Res. Lett. **28**, 11, 2277–2280 (2001)

Gajewski, D., W. S. Holbroock, and C. Prodehl: A three-dimensional crustal model of southwest Germany derived from seismic refraction data. Tectonophysics **142**, 49–70 (1987)

Georgiev, G., C. Dabovski, and G. Stanisheva-Vassileva: East Srednogorie-Balkan Rift Zone, in Peri-Tethys Memoir 6: Peri-Tethyan Rift/Wrench Basins and Passive Margins, vol. 186, edited by P. A. Ziegler, et al., pp. 259– 293. Memoires Museum National d'Histoire Naturelle, Paris (2002)

Goes, S. and S. van der Lee: Thermal structure of the North American uppermost mantle inferred from seismic tomography. J. Geophys. Res. **107**, 10:1029/2000JB000049 (2002)

González-Fernández, A., D. Cordoba, L. Matias, and M. Torné: Seismic crustal structure in the Gulf of Cadiz (SW Iberian Peninsula. Marine Geophys. Res. **22**, 207–223 (2001)

Grad, M., A. Guterch, and A. Polkowska-Purys: Crustal structure of the Trans-European Suture Zone in Central Poland reinterpretation of the LT-2, LT-4 and LT-5 deep seismic sounding profiles. Geol. Quart. **49**, **(3)**, 243–252 (2005)

Grad, M., T. Janik, A. Guterch, P. Oeroda, and W. Czuba: EUROBRIDGE'94–97, POLONAISE'97 , CELEBRATION 2000 Seismic Working Groups,. Lithospheric structure of the western part of the East European Craton investigated by deep seismic profiles. Geol. Quart. **50**, **1**, 9–22 (2006)

Grad, M., A. Pičák, G. R. Keller, A. Guterch, M. Bro., and E. Hegedüs, and SUDETES 2003 Working Group: SUDETES 2003 seismic experiment. Stud. Geophys. Geod. **47**, 681–689 (2003)

Granet, M., M. Wilson, and U. Achauer: Imaging a mantle plume beneath the Massif Central (France). Earth Planet. Sci. Lett. **136**, 281–296 (1995)

Guterch, A. and M. Grad: Lithospheric structure of the TESZ in Poland based on modern seismic experiments. Geol. Quart. **50, (1)**, 23–32 (2006)

Guterch, A., M. Grad, A. Pičák, E. Brückl, E. Hegedüs, G. R. Keller, and H. Thybo and Celebration 2000, Alp 2002, Sudetes 2003 Working Groups: An overview of recent seismic refraction experiments in central Europe. Stud. Geophys. Geod. **47**, 651–657 (2003)

Horváth, F., G. Bada, P. Szafián, G. Tari, A. Adám, and S. Cloetingh: Formation and deformation of the Pannonian Basin: constraints from observational data, in European Lithosphere Dynamics. vol. 32, edited by D. Gee and R. Stephenson, pp. 191–206, Geological Society, London, Memoirs (2006)

Hurtig, E., Cermak, V., Haenel, R., Zui, V. (Eds.): *Geothermal* Atlas of Europe, International Association for *Seismology* and *Physics* of the *Earth's Interior.* 156 pp, Hermann Haack Verlagsgesellschaft mbH-Geographisch-Kartographische Anstalt Gotha(1992)

Kaban, M. K., P. Schwintzer, and P. S. A. Tikhotsky: Global isostatic gravity model of the Earth. Geophys. J. Int. **136**, 519–536 (1999)

Kaban, M. K.: A gravity model of the North Eurasia crust and upper mantle: 1. Mantle and Isostatic Residual Gravity Anomalies.. Russian Journal of Earth Sciences, (maintained and distributed by AGU, http://www.agu.org/wps/rjes/), 3, **(2)**, 143–163 (2001)

Kaban, M. K. and P. Schwintzer: Oceanic upper mantle structure from experimental scaling of Vs and density at different depths. Geophys. J. Int. **147**, 199–214 (2001)

Kaban, M. K., O. G. Flovenz, and G. Palmason: Nature of the crust-mantle transition zone and the thermal state of the upper mantle beneath Iceland from gravity modelling. Geophys. J. Int. **149**, 281–299 (2002)

Kaban, M. K., P. Schwintzer, and Ch. Reigber: A new isostatic model of the lithosphere and gravity field. J. Geod. **78**, 368–385 (2004)

Kelly, A., R. W. England, and P. K. H. Maguire: A three-dimensional seismic velocity model for northwestern Europe. Geophys. J. Int. **171**, 1172–1184 (2007)

Kimbell, G. S., R. W. Gatliff, J. D. Ritchie, A. S. D. Walker, and J. P. Williamson: Regional three-dimensional gravity modelling of the NE Atlantic margin.. Basin Research **16**, 259–278, doi: 10.1111/j.1365-2117.2004.00232 (2004)

Korja, A., R. Lahtinen, and M. Nironen: The Svecofennian orogen: a collage of microcontinents and island arcs, in European Lithosphere Dynamics. vol. 32, edited by D. Gee and R. Stephenson, pp. 11–41, Geological Society, London, Memoirs (2006)

Koulakov, I. and S. V. Sobolev: Moho depth and three-dimensional P and S structure of the crust and uppermost mantle in the Eastern Mediterranean and Middle East derived from tomographic inversion of local ISC data. Geophys. J. Int. **164**, 218–235 (2006)

Kozlovskaja, E. G., E. Elo, S. -E. Hjelt, J. Yliniemi, and M. Pittijärvi: SVEKALAPKO Seismic Tomography Working Group: 3-D density model of the crust of southern and central Finland obtained from joint interpretation of the SVEKALAPKO crustal P-wave velocity models and gravity data. Geophys. J. Int. **158**, 827–848 (2004)

Kumar, P., R. Kind, K. Priestley, and T. Dahl-Jensen: Crustal structure of Iceland and Greenland rom receiver function studies. J. Geophys. Res. **112**, B03301, doi:10.1029/2005JB003991 (2007)

Lassen, A.: Structure and evolution of the pre-Permian basement in the Danish Area. PhD-thesis, Copenhagen University, pp. 130 (2005)

Lenkey, L.: Geothermics of the Pannonian basin and its bearing on the tectonics of Basin evolution. PhD Thesis,Vrije Universiteit, Amsterdam, pp. 215 (1999)

Lippitsch, R., E. Kissling, and J. Ansorge: Upper mantle structure beneath the Alpine orogen from high-resolution teleseismic tomography. J. Geophys. Res. doi:10.1029/2002JB002016, **108**, B8, 2376 (2003)

Lyngsie, S. B. and H. Thybo: A new tectonic model for the Laurentia Avalonia Baltica sutures in the North Sea: A case study along MONA LISA profile 3. Tectonophysics **429**, 201–227 (2007)

Majdanski, M., M. Grad, and A. Guterch and SUDETES 2003 Working Group: 2-D seismic tomographic and ray tracing modelling of the crustal structure across the Sudetes Mountains basing on SUDETES 2003 experiment data. Tectonophysics **413**, 249–269 (2006)

Mariotti, G. and C. Doglioni: The dip of the foreland monocline in the Alps and Apennines. Earth Planet. Sci. Lett. **181**, 191–202 (2000)

Martin, M. and J. R. R. Ritter, and the CALIXTO working group: High-resolution teleseismic body-wave tomography beneath SE Romania – I. Implications for three- dimensional versus one-dimensional crustal correction strategies with a new crustal velocity model. Geophys. J. Int.**162**, 448–460 (2005)

Martin, M. and J. R. R. Ritter, and the CALIXTO working group: High-resolution teleseismic body wave tomography beneath SE-Romania – II. Imaging of a slab detachment scenario. Geophys. J. Int. **164**, 579–595 (2006)

Matias, L. M.: A Sismologia Experimental na Modelação da Estrutura da Crosta em Portugal Continental. PhD thesis, Faculdade de Ciências, Universidade de Lisboa, 390 pp (in Portuguese) (1996)

Menke, W., M. West, B. Brandsdóttir, and D. Sparks: Compressional and shear velocity structure of the lithosphere in Northern Iceland. Bull. Seism. Soc. Am. **88**, 1561–1571 (1998)

Mjelde, R., T. Raum, B. Myhren, H. Shimamura, Y. Murai, T. Takanami, R. Karpuz, and U. Næss: Continent-ocean transition on the Vøring Plateau, NE Atlantic, derived from densely sampled ocean bottom seismometer data. J. Geophys. Res. **110**, B4, doi:10.1029/2004JB003026 (2005)

Mooney, W. D., G. Laske, and T. G. Masters: CRUST 5.1: A global crustal model at 5°X5°. J. Geophys. Res **103B**, 727–747 (1998)

Morelli, C.: Lithospheric structure and geodynamics of the Italian Peninsula derived from geophysical data: A review. Mem. Soc. Geol. It. **52**, 113–122 (1998)

Morgan, P.: Crustal radiogenic heat production and the selective survival of ancient continental crust, Proc. Lunar Planet. Sci. Conf. 15th, Part 2. J. Geophys. Res. suppl. **90**, C561–C570 (1985)

Neprochnov, Y. P., I. P. Kosminskaya, and Y. P. Malovitsky: Structure of the crust and upper mantle of the Black and Caspian seas. Tectonophysics **10**, 517–538 (1970)

Nielsen, L., H. Thybo, and M. Glendrup: Seismic tomography interpretation of Paleozoic sedimentary sequences in the southeastern North Sea. Geophysics **70**, **(4)**, 45–56 (2005)

O'Reilly, B. M., F. Hauser, A. W. B. Jacob, P. M. Shannon, J. Makris, and U. Vogt: The transition between the Erris and the Rockall basins: new evidence from wide-angle seismic data. Tectonophysics **241**, 143–163 (1995)

Paulssen, H. and J. Visser: The crustal structure in Iberia inferred from P-wave coda. Tectonophysics **221**, 111–123 (1991)

Papazachos, C. B., P. M. Hatzidimitriou, D. G. Panagiotopoulos, and G. N. Tsokas: Tomography of the crust and upper mantle in southeast Europe. J. Geophys. Res. **100**, 12405–12422 (1995)

Pedreira, D., J. A. Pulsar, J. Gallart, and J. Dıaz: Seismic evidence of Alpine crustal thickening and wedging from the western Pyrenees to the Cantabrian Mountains (north Iberia).. J. Geophys. Res. **108**, B4, doi:10.1029/2001JB001667 (2003)

Pieri, M. and G. Groppi Subsurface geological structure of the Po plain, Italy. Consiglio Nazionale delle Ricerche – Progetto finalizzato geodinamica, sottoprogetto "modello strutturale", pubblicazione 414, 13 pp (1981)

Pinet, B., L. Montadert, R. Curnelle, M. Cazes, F. Marillier, J. Rolet, A. Tomssino, A. Galdeno, Ph. Patriat, F. Brunet, J. L. Olivet, M. Schaming, J. P. Lefort, A. Arrieta, and C. Riaza: Crustal thinning on the Aquitaine shelf, Bay of Biscay, from deep seismic data. Nature **325**, 513–516 (1987)

Piromallo, C. and A. Morelli: P wave tomography of the mantle under the Alpine-Mediterranean area.. J. Geophys. Res. **108**, B2 2065, doi:10.1029/2002jb001757 (2003)

Polyak, B. G., M. Fernàndez, M. D. Khutorskoy, J. I. Soto, I. A. Basov, M. C. Comas, V. Ye, B. Khain Alonso, G. V. Agapova, I. S. Mazurova, A. Negredo, V. O. Tochitsky, J. de la Linde, N. A. Bogdanov, and E. Banda: Heat flow in the Alboran Sea, western Mediterranean. Tectonophysics **263**, 191–218 (1996)

Raum, T., R. Mjelde, H. Shimamura, Y. Murai, E. Bråstein, R. M. Karpuz, K. Kravik, and H. J. Kolstø: Crustal structure and evolution of the southern Vøring Basin and Vøring Transform Margin, NE Atlantic. Tectonophysics **415**, 167–202 (2006)

Richardson, K. R., J. R. Smallwood, R. S. White, D. B. Snyder, and P. K. H. Maguire: Crustal structure beneath the Faroe Islands and the Faroe–Iceland Ridge. Tectonophysics **300**, 159–180 (1998)

Sandoval, S., E. Kissling, and J. Ansorge and the SVEKALAPKO Seismic TomographyWorking Group: High-resolution body wave tomography beneath the SVEKALAPKO array: I. A priori three-dimensional crustal model and associated traveltime effects on teleseismic wave fronts. Geophys. J. Int. **153**, 75–87 (2003)

Scarascia, S. and R. Cassinis: Crustal structures in the central-eastern Alpine sector: a revision of the available DSS data. Tectonophysics **271**, 157–188 (1997)

Scarascia, S., R. Cassinis, and F. Federici: Gravity modelling of deep structures in the northern-central Apennines. Mem. Soc Geol. It. **52**, 231–246 (1998)

Scarascia, S., A. Lozej, and R. Cassinis: Crustal structures of the Ligurian, Tyrrhenian and Ionian seas and adjacent onshore areas interpreted from wide-angle seismic profiles. Boll. Geofis. Teor. **XXXVI, 141–144, 5–19 (1994)**

Scheck, M., U. Bayer, V. Otto, J. Lamarche, D. Banka, and T. Pharaoh: The Elbe Fault System in North Central Europe – a basement controlled zone of crustal weakness. Tectonophysics 360, 281– 299 (2002)

Scheck-Wenderoth, M. and J. Lamarche: Crustal memory and basin evolution in the Central European Basin System—new insights from a 3D structural model. Tectonophysics **397**, 143–165 (2005)

Simancas, J. F., R. Carbonell, F. Gonzáles Lodeiro, A. Pérez Estaun, C. Juhlin, P. Ayarza, A. Kashubin, A. Azor, D. Martinez Poyatos, R. Sáez, G. R. Almódovar, E. Pascual, I. Flecha, and D. Marti: Transpressional collision tectonics and mantle plume dynamics: the Variscides of southwestern Iberia, in European Lithosphere Dynamics. vol. 32, edited by D. Gee and R. Stephenson, pp. 11–41, Geological Society, London, Memoirs (2006)

Sodoudi, F., R. Kind, D. Hatzfeld, K. Priestley, W. Hanka, K. Wylegalla, G. Stavrakakis, A. Vafidis, H. -P. Harjes, and M. Bohnhof: Lithospheric structure of the Aegean obtained from P and S receiver functions. J. Geophys. Res. **111**, (B12), doi:10.1029/2005JB003932 (2006)

Środa, P., W. Czuba, M. Grad, A. Guterch, A. K. Tokarski, T. Janik, M. Rauch, G. R. Keller, E. Hegedus, and J. Vozar, CELEBRATION 2000 Working Group: Crustal and upper mantle structure of the Western Carpathians from CELEBRATION 2000 profiles CEL01 and CEL04: seismic models and geological implications. Geohys. J. Int.**167**, 737–760 (2006)

Starostenko, V., V. Buryanov, I. Makarenko, O. Rusakov, R. Stephenson, A. Nikishin, G. Georgiev, M. Gerasimov, R. Dimitriu, O. Legostaeva, V. Pchelarov, and C. Sava: Topography of the crust–mantle boundary beneath the Black Sea Basin. Tectonophysics **381**, 211–233 (2004)

Suriñach, E. and R. Vegas: Lateral inhomogeneities of the Hercynian crust in central Spain. Phys. Earth Planet. Int. **51**, 226–334 (1988)

Svenningsen, L., N. Balling, B. H. Jacobsen, R. Kind, K. Wyllegalla, and J. Schweitzer: Crustal root beneath the highlands of Southern Norway resolved by teleseismic receiver functions.. Geophys. J. Int. **170**, **3**, 1129–1138 (2007)

Tesauro, M., M. K. Kaban, and S. A. P. L. Cloetingh: EuCRUST-07: A new reference model for the European crust. Geophys. Res. Lett. **35**, L05313, doi:10.1029/2007GL032244 (2008)

Thybo, H., A. Sandrin, L. Nielsen, H. Lykke-Andersen, and G. R. Keller: Seismic velocity structure of a large mafic intrusion in the crust of central Denmark from project ESTRID. Tectonophysics **420**, 105–122 (2006)

Thiebot, E. and M. -A. Gutscher: The Gibraltar Arc seismogenic zone (part 1): Constraints on a shallow east dipping fault plane source for the 1755 Lisbon earthquake provided by seismic data, gravity and thermal modelling. Tectonophysics **426**, 135–152 (2006)

Thinon, I., L. Matias, J. P. Réhault, A. Hirn, L. Fidalgo-Gonzales, and F. Avedick: Deep structure of the Armorican Basin (Bay of Biscay): a review of Norgasis seismic reflection and refraction data. J. Geol. Soc. London **160**, 99–116 (2003)

Tsikalas, F., T. O. Eldholm, and J. I. Faleide: Crustal structure of the Lofoten–Vesterålen continental margin, off Norway. Tectonophysics **404**, 151– 174 (2005)

van der Lee, S. and G. Nolet: Upper-mantle S-velocity structure of North America. J. Geophys. Res. **102**, 22815–22838 (1997)

Waldhauser, F., R. Lippitsch, E. Kissling, and J. Ansorge: High-resolution teleseismic tomography of upper mantle structure using an a priori 3D crustal model. Geophys. J. Int. **150**, 141–403 (2002)

Wilde-Piorko, M., M. Grad and TOR Working Group: Crustal structure variation from the Precambrian to Palaeozoic platforms in Europe imaged by the inversion of teleseismic receiver functions—project TOR. Geophys. J. Int. **150**, 261–270 (2002)

Whitmarsh, R. B., R. S. White, S. J. Horsefield, J. C. Sibuet, M. Recq, and V. Louvel: The ocean-continent boundary off the western continental margin of Iberia: Crustal structure west of Galicia Bank. J. Geophys. Res **101**, B12, 28291–28314 (1996)

Whitmarsh, R. B., L. M. Pinheiro, P. R. Miles, M. Recq, and J. -C. Sibuet: Thin crust at the Western Iberia ocean-continent transition and ophiolites. Tectonics **12**, **5**, 1230–1239 (1993)

Yegorova, T. P. and V. I. Starostenko: Lithosphere structure of Europe and Northern Atlantic from regional three-dimensional gravity modelling. Geophys. J. Int. **151**, 11–31 (2002)

Zeyen, H., O. Novak, M. Landes, C. Prodehl, L. Driad, and A. Him: Refraction seismic investigations of the northern Massif Central, France. In: K. Fuchs, R. Altherr, B. Mueller, C. Prodehl, (Eds.), Stress and Stress Release in the Lithosphere. Structure and Dynamic Processes in the Rifts of Western Europe. Tectonophysics, **275**, 99–118 (1997)

Ziegler, P. A. and P. Dèzes, 2006. Crustal evolution of Western and Central Europe. In: European Lithosphere Dynamics, vol. 32. edited by: Gee, D. and Stephenson, R., Geological Society, London, Memoirs, pp. 43–56.

Zito, G., F. Mongelli, S. de Lorenzo, and C. Doglioni: Heat flow and geodynamics in the Tyrrhenian Sea. Terra Nova **15**, 425–432 (2003)

Zucca, J. J.: The crustal structure of the southern Rhinegraben from re-interpretation of seismic refraction data. J. Geophys. 55, 13–22 (1984)

Thermal and Rheological Model of the European Lithosphere

Magdala Tesauro, Mikhail K. Kaban, and Sierd A.P.L. Cloetingh

Abstract A thermal and rheological model of the European lithosphere (10°W-35E; 35N-60N) is constructed based on a combination of new geophysical models. To determine temperature distribution a tomography model is used, which was improved by corrections for the crustal effect using a new digital model of the European crust (EuCRUST-07). The uppermost mantle under western Europe is generally characterized by temperatures ranging between 900 and 1,100°C, with the hottest areas corresponding to basins, that experienced recent extension (e.g., Tyrrhenian Sea and Pannonian Basin). By contrast, upper mantle temperatures at this depth under eastern Europe are about 550–750°C, whereby the lowest values are found in the northeastern part of the study area. EuCRUST-07 and the new thermal model are used to calculate the strength distribution within the European lithosphere. Differently from previous approaches, lateral variations of lithology and density derived from EuCRUST-07 are used to construct the new strength distribution model. Following the approach of Burov and Diament (1995), the lithospheric rheology is employed to calculate variations of the elastic thickness of the lithosphere. According to these estimates, in western Europe the lithosphere is more heterogeneous than in eastern Europe. Western Europe, with dominantly crust-mantle decoupling is mostly characterized by lower values of strength and elastic thickness. The crustal strength provides a large contribution to the integrated strength (~50% of the integrated strength for the whole lithosphere) in most part of the study area (~60%). The new results are important in view of recent disputes on the strength distribution between crust and mantle lithosphere.

Keywords Crust and mantle lithosphere

Introduction

Rheology and strength of the Earth's lithosphere have been debated since the beginning of the last century, when the concept of a strong lithosphere overlying viscous astenosphere was introduced (Barrel, 1914). The issue of strength of the lithospheric plates and their spatial and temporal variations is important for many geodynamic applications (e.g., Jackson, 2002; Burov and Watts, 2006). For rocks with given mineralogical composition and microstructure, temperature is one of the most important parameters controlling rheology. Up to now, temperature distribution can only be reliably determined using borehole measurements for the shallowest part of the crust (0–5 km) in most parts of Europe (e.g., Hurtig et al., 1992). Temperature estimates for the deeper horizons of the lithosphere, where the heat transport is mostly conductive, require a precise knowledge of many crustal parameters (mainly thermal conductivity and heat production) and are extremely uncertain, even in the case, when the steady state thermal conductivity equation (e.g., Chapman, 1986) can be used. In addition, the near-surface heat flow determinations depend on numerous local and shallow effects (e.g., hydrothermal fluids circulation,

M. Tesauro (✉)
Faculty of Earth and Life Sciences, Netherlands Research Centre for Integrated Solid Earth Science, VU University, Amsterdam, The Netherlands GeoForschungsZentrum (GFZ), Potsdam, Germany
e-mail: magdala.tesauro@falw.vu.nl

S. Cloetingh, J. Negendank (eds.), *New Frontiers in Integrated Solid Earth Sciences*, International Year of Planet Earth,
DOI 10.1007/978-90-481-2737-5_3,

sedimentation and climatic changes), which influence might be hardly removable beforehand. Furthermore, in many areas affected by transient processes (e.g., mantle upwellings) the steady state thermal conductivity equations cannot be applied, while other approaches require a precise knowledge about thermal history, which normally is not defined with sufficient accuracy.

Therefore, indirect approaches are needed to determine temperature distribution within the lithosphere. Seismic tomography is commonly used for this purpose (e.g., Sobolev et al., 1996; Goes et al., 2000). The strong effect of temperature on the seismic velocity and elastic moduli has been known for a long time from laboratory studies (e.g., Birch, 1943; Hughes and Cross, 1951). Therefore, temperature changes in the mantle lithosphere can be derived from variations of seismic velocities. However, seismic velocities also depend on many other factors (e.g., partial melt, anisotropy and composition), which strongly affect temperature estimations, as discussed in the following section.

In this paper an improved temperature model for the European lithosphere is presented. This model is obtained from inversion of a recent tomography model (Koulakov et al., 2009) supplemented by the new reference model of the crust (EuCRUST-07, Tesauro et al., this volume). EuCRUST-07 has a higher resolution (15'×15') and is more robust than previous compilations (e.g., CRUST5.1, Mooney et al., 1998; CRUST2.0 Bassin et al., 2000), primarily because of a significant number of recent seismic data assembled. Therefore, EuCRUST-07 offers a starting point for various types of numerical modelling to remove a-priori the crustal effect and to exclude a trade-off with mantle heterogeneities. Previous studies (e.g., Waldhauser et al., 2002; Martin et al., 2006) demonstrated that the use of an a-priori crustal model may significantly improve determination of the seismic velocities in the uppermost mantle. Consequently, employment of a more robust seismic tomography model will increase the reliability of the thermal model derived by its inversion.

The determined temperature variations in the upper mantle together with the data from EuCRUST-07 are used to construct a new strength model of the lithosphere for the entire region. Besides employment of the new models for the crust and upper mantle, the strength calculations are improved by incorporating lithology variations (and corresponding variations of physical parameters) in different tectonic provinces in Europe. The lithotypes have been defined based on the reference crustal model and surface heat flow distribution, as discussed in the previous chapter. This provides an opportunity to further refine existing rheology and strength determinations (e.g., Cloetingh et al., 2005). Following the approach of Burov and Diament (1995), the lithosheric rheology is employed to calculate variations of the elastic thickness of the lithosphere.

Thermal Model of the European Lithosphere

A recent seismic tomography model (Koulakov et al., 2009) provides a basis for determination of the temperature distribution within the upper mantle. However, this model, as well as other body-wave models, is only accurate in providing lateral velocity variations, which are not so sensitive to a choice of the one-dimensional reference model (Koulakov et al., 2009). Consequently the absolute velocities required to determine mantle temperatures are usually not well constrained (e.g., Cammarano et al., 2003). It is also critical that the 1D global models (e.g., ak135, Kennett et al., 1995), normally used in most tomography studies, represent an average of the laterally heterogeneous Earth structure, but on account of the non-linear relationship of seismic velocities and temperatures (e.g., Goes et al., 2000), the average seismic velocity profile does not necessarily translate into the average temperature distribution (Cammarano et al., 2003). Furthermore, the global reference models, as determined for the whole Earth, provide a better adjustment for the oceanic areas. The average depth-dependent velocity profile in the continental areas may differ by 0.15–0.2 km/s for *P*-wave velocities from the typical oceanic profile (e.g., Gudmundsson and Sambridge, 1998). This value corresponds to several hundred degrees difference in the temperature estimates. In order to solve this problem, a new reference velocity model according to the specific tectonic settings of the study area is defined. The employment of this new regional reference model resulted in consistent lateral temperature variations in the mantle, which are then extrapolated to the surface. For this purpose, typical crustal isotherms determined for different tectonic provinces on the base of characteristic values of the radiogenic heat production for each crustal layer are used (Čermák, 1993).

It has been already demonstrated that temperature is the main parameter affecting seismic velocities in a depth range of about 50–250 km (e.g., Jordan, 1979; Sobolev et al., 1996; Goes et al., 2000). On the other hand, it should be realized that other factors also affect seismic velocities, likely differently for different types of wave (*P* or *S*). Below a brief description of these factors is given:

Anharmonicity. Anharmonicity refers to behaviour of the materials in which elastic properties change because of temperature (or pressure) caused by the deviation of lattice vibration from the harmonic oscillator (e.g., Anderson, 1995). This process does not involve any energy dissipation, but produces thermal expansion. Therefore, elastic properties of materials may vary due to the change in mean atomic distances.

Anelasticity. Anelasticity is a dissipative process involving viscous deformation (e.g., Karato and Spetzler, 1990). The degree to which viscous deformation affects seismic wave velocities is measured by the attenuation parameter Q and depends on the frequency of seismic waves. Consequently, the anelasticity results in the frequency dependence of seismic wave velocities. For temperatures <900°C, rocks behave essentially elastically with very low levels of dissipation ($Q^{-1} < 10^{-2}$), while above this threshold, the dissipation is progressively increased (Karato, 1993). There are a lot of uncertainties in the anelasticity calculations. However, even coarse and approximate estimations of this effect remarkably improve reliability of the estimated temperatures (e.g., Goes et al., 2000).

Partial Melt. The effect of partial melting on seismic velocities is likely large (Sato et al., 1989; Schmeling, 1985), but not well constrained by experimental/theoretical results. The main uncertainty is due to the strong dependence on melt geometry and whether or not melt pockets are interconnected (Mavko, 1980). Modelling results (e.g., Schmeling, 1985) demonstrate a stronger decrease of shear modulus than of the bulk modulus. Furthermore, a preferred orientation of the melt pockets in the mantle may cause anisotropy of the seismic velocities and attenuation (Karato and Jung, 1998).

Water. Formation of even small amounts of free water through a dehydration of the water-bearing minerals is known to significantly decrease seismic velocities in crustal rocks (e.g., Popp and Kern, 1993). The presence of water may affect the velocities even at temperatures below the solidus (Sobolev and Babeyko, 1994). Furthermore, the presence of water results in a decrease of the melting temperature (*Tm*), which reduces mantle seismic velocities, through enhanced anelasticity (e.g., Karato and Jung, 1998).

Anisotropy. The crustal anisotropy strongly related to tectonic processes, which generate rock fabric and structural alignments such as preferred orientations of foliation, schistosity, fractures or folds. In the upper mantle, the anisotropy is usually caused by lattice-preferred orientation (LPO) of olivine and pyroxene with their axes being aligned in the direction of the old tectonic movements (in the lithosphere) and of the plate motion (in the asthenosphere). The effect of anisotropy on the velocity estimates may be strong in areas where seismic sampling is dominated by one propagation or polarization direction. Since the direction of anisotropy appears to vary throughout a large area (e.g., in Europe), its effect should not result in a systematic bias on the inversion for temperature. In addition, this might produce discrepancies of the temperature derivations based on different type of velocities (*P* or *S*) in the regions where the ray coverage is not good enough (Goes et al., 2000).

Composition. Previous studies have demonstrated that variations of the iron content in the lithosphere mantle have a large effect on seismic velocities (higher on *S*- than on *P*-waves) than any other variations in composition and mineralogy (e.g., Deschamps et al., 2002). However, the effect of the composition changes is significantly smaller than the effect of temperature variations: for example 1% of the *Vs* anomaly can be explained either by a 4% variation of iron content or by a thermal anomaly of 50–100°C (Nolet and Zielhuis, 1994; Deschamps et al., 2002).

The anharmonic and anelastic effects may be relatively easy quantified than estimating temperatures from seismic velocities. The other factors need more precise knowledge about structure and composition of the lithosphere, which may not be easily derived from the crustal and tomography models. This could result in an uncertainty in temperature determinations and discrepancy between temperatures derived from *P* and *S*-wave velocities. Assuming that the seismic model is well resolved and the composition is known, the uncertainty of the inferred temperatures is about ±100°C above 400 km (e.g., Cammarano et al., 2003).

The temperature distribution in the upper mantle is evaluated by inverting the new tomography model of Koulakov et al. (2009). The limited marginal areas not covered by the original seismic model were supplemented by data from the model of Bijwaard and Spakman (2000), which is based on nearly the same data-set. The seismic anomalies from this model have been interpolated for the same locations and depth as in the original grid of Koulakov et al. (2009) and have been corrected for the mean difference existing at each depth in the common part of the area. To produce a smooth transition between the two compilations, a buffer zone between them of about 50 km is left, which has been filled using a kriging interpolation. As was stated before, the absolute velocities should be employed for temperature estimates (e.g., Goes et al., 2000). Both seismic tomography anomalies of Koulakov et al. (2009) and Bijwaard and Spakman (2000) are referred to the same 1D global seismic reference model (ak135, Kennett et al., 1995). However, even in this case, the velocity models are shifted relative to each other in the European region (by approximately 0.043), which clearly demonstrates their limitation in deriving reliably absolute values.

The ak135 reference model was adjusted for the study area in order to better constrain the absolute velocities and resulting temperature estimates. A systematic difference between the oceanic and continental areas normally persists to a depth of about 300 km (e.g., Nolet et al., 1994), which should be also reflected in the regional reference model. One important source of information about the absolute values of the seismic velocities in upper mantle are the long-range refraction/reflection profiles (>2,500 km) reaching the transition zone (e.g., Gudmundsson and Sambridge, 1998). Unfortunately, such seismic sections are only available east from the study area, along the EEP and Siberia. However, these profiles show that the most important differences in the continental areas exist at depths down to about 150 km (e.g., Quartz profile; Pavlenkova and Pavlenkova, 2008). At greater depths both the old EEP and the younger West Siberian Basin are characterized by similar velocities from $V_p = 8.45$ km/s at a depth of 150 km to about 8.6 km/s at 350 km. These data suggest us to use a similar model also for western Europe. This approach provides a preliminary adjustment, while the final tuning should be done by comparison with the characteristic geotherms and independent determinations of the lithosphere-asthenosphere boundary (LAB) depth. Based on these considerations, the velocity values of ak135 are increased by 0.1–0.18 km/s in the uppermost part of the mantle (up to ~200 km), and less than 0.1 km/s at greater depths. The maximum offset corresponds to the depth of 135 km. At depths >200 km the velocity difference between ak135 and the new reference model becomes smaller, it disappears below 250 km. In this way, the new reference model has a velocity in the uppermost mantle, which is higher than in the oceanic areas but lower than the values observed in the EEP (Pavlenkova and Pavlenkova, 2008). This is a reasonable compromise before further studies will better constrain the European seismic reference model. However, it is shown below that already this model provides the opportunity to construct a more realistic thermal model of the European mantle. In order to eliminate small scale artefacts, the velocity field in each layer has been processed by a low-pass filter leaving the wavelengths greater than 350 km. These data were finally used to estimate the mantle temperatures.

An iterative inversion similar to the one carried out by previous authors is performed (e.g., Sobolev et al., 1997; Goes et al., 2000). From a given starting temperature the final one is obtained through iteration at a given point using the velocity (*P*-or *S-wave*) and velocity derivative calculated for anharmonicity and anelasticity effect:

$$T^{n+1} = T^n + F_{damp}\left\{ \left[V_{obs} - V_{syn}(T^n)\right] \Big/ \left[\left(\frac{\partial V}{\partial T}\right)_{syn}(T^n)\right]\right\} \qquad (1)$$

where T is temperature, n is the iteration number, F_{damp} the damping factor and V_{obs} and V_{syn} are observed (i.e., tomographic) and synthetic seismic velocity, respectively. A strong damping effect (F_{damp}) is necessary, since the velocity derivative depends very non-linearly on temperature due to the effect of anelasticity (Goes et al., 2000). V_{syn} takes into account both the anharmonic (V_{anh}) and anelastic (V_{anel}) effects and can be expressed as follows (Minster and Anderson, 1981):

$$V_{syn}(P,T,X,\omega) = V_{anh}(P,T,X)V_{anel}(P,T,\omega) \qquad (2)$$

where X stands for composition and P for pressure.

The synthetic velocity derivative is given by a sum of the derivatives related to the anharmonic and anelastic effect:

$$\left(\frac{\partial V}{\partial T}\right)_{syn} = \left(\frac{\partial V}{\partial T}\right)_{anh} + \left(\frac{\partial V}{\partial T}\right)_{anel} \qquad (3)$$

The anharmonic part of the velocities is calculated using the infinitesimal strain approximation, which is valid to a depth of ~200 km (Leven et al., 1981) and already used in previous studies (e.g., Goes et al., 2000). The estimation of density and elastic parameters of rocks of a given mineralogical composition was done using the Voigt-Reuss-Hill (VRH) averaging of the parameters for the individual minerals (Hill, 1963) (Appendix). The values of the elastic parameters and of their derivatives used in the calculation are taken from Cammarano et al. (2003). Since the area of study is mostly continental and is not extended far to the regions affected by a mantle strongly depleted in iron such as the Baltic Shield (Kaban et al., 2003), the average continental garnet lherzolite composition (Jordan, 1979), which was already adopted in the previous study of Goes et al. (2000), was used as a reference composition for the entire area (Table 1).

The anelasticity part of the velocity depends on the attenuation parameter Q, as expressed below:

$$V_{anel} = \left[1 - \frac{Q^{-1}(\omega,T)}{2\tan(\pi a/2)}\right] \quad (4)$$

Q for S-wave velocities (Q_s) is given by

$$Q_s = A\omega^a \exp\left(\frac{aH}{RT}\right) \quad (5)$$

with

$$H(P) = E + PV \quad (6)$$

where A is the normalization factor, a is the exponent describing the frequency dependence of the attenuation (between 0.1 and 0.3, consistent with the seismic observations), ω the seismic frequency (equal to 1 Hz), H is the activation enthalpy, E is the activation energy, T the temperature, R the gas constant and V the activation volume. Q for P-wave velocities (Q_P) is given by (e.g., Anderson and Given, 1982):

$$Q_P^{-1} = (1 - L)\, Q_K^{-1} + LQ_\mu^{-1} \quad (7)$$

where

$$L = \left(\frac{4}{3}\right)\left(\frac{V_s}{V_P}\right)^2 \quad (8)$$

and Q_K is the bulk attenuation and taken to be a constant equal to 1,000 (e.g., Goes et al., 2000).

According to Karato (1993), a useful homologous temperature scaling is:

$$g = \frac{H(P)}{RT_m(P)} \quad (9)$$

where the dimensionless factor g is a function of the activation enthalpy H, the melting temperature Tm and the gas constant R. From experimental results g is between 20 and 30 for olivine in the uppermost mantle (Karato, 1993). The melting temperature between 0 and 10 GPa has been calculated using the peridotite solidus KLB1 (Hirschmann, 2000). The effect of anelasticity was estimated using the model based on the homologous temperature scaling approach (model Q_4 defined in Cammarano et al., 2003), since large uncertainties exist in the estimation of the activation enthalpy (Karato, 1993). By contrast, the uncertainties in the melting temperature (<100°C) are negligible compared to other uncertainties (e.g., Cammarano et al., 2003). However, also an attenuation model based almost completely on mineral physics data (model Q_1 defined in Goes et al., 2000) was tested. The difference between the temperature distributions for two models is ~100°C at the high temperatures, at which the anelasticity produces a remarkable effect (>900°C). Furthermore, in order to estimate the uncertainties expected due to the choice of a mantle composition, additional tests were made. In particular, the temperature was estimated at 60 km depth for a piclogite and for a harzburgite mantle model. The first lithotype is

Table 1 Mantle models composition: average continental garnet lherzolite composition from Jordan (1979); piclogite from Bass and Anderson (1984); harzburgite from Irifune and Ringwood (1987)

Mineralogy (mode, vol.%)	Garnet lherzolite	Piclogite	Harzburgite
Olivine	67	40	82
CPX	4.5	22	–
OPX	23	8	14.4
Garnet	5.5	22	3.6
Ca-Garnet	–	8	–
FeO/MgO + FeO	0.11	0.11	0.11

extreme in its low olivine content, while the second one represents the lithosphere of the subducted slab (Table 1). The average difference in the temperature estimates between the two compositional models and the garnet lherzolite is significant only for the piclogite (+215°C), while it is relatively small for the harzburgite (+60°C). On the other hand, the piclogite composition might be representative only of a very small part of the European mantle. Therefore, the average uncertainties related to the compositional contribution in the study area are probably much less than the average values estimated.

The obtained temperature distributions at the top of the mantle and at the depths of 60 and 100 km are displayed in Figs. 1, 2 and 3. In addition, three vertical cross-sections through the main tectonic structures of Europe are shown in Fig. 3. Mean geotherms for the main geological domains of Europe are displayed in Fig. 4. The linear trend of the temperature distribution evidences the reliability of the new regional reference velocity model adopted in the inversion. The mantle temperature in the uppermost part varies from 550 to 800°C in the EEP and the Black Sea to 900–1,100°C in some parts of western Europe. A sharp temperature change of about 200°C occurs across the TESZ and persists also in the deeper layers of the upper mantle. The hottest area in the eastern part of the study area corresponds to the Anatolian Plateau, where also a high heat flow is observed (e.g., Hurtig et al., 1992). In western and central Europe the isotherms updome beneath the areas subjected to strong extension (e.g., the ECRIS and the Tyrrhenian Sea) and the regions of active Tertiary volcanism (e.g., Pannonian Basin and Massif Central) (Fig. 3). The mean geotherms in these areas are very similar showing temperatures, which are close to ~1,200°C at a depth of 100 km and even shallower (e.g., in the Tyrrhenian Sea). By contrast, lower temperatures are observed beneath the Pyrenees, the Alps and the Dinarides-Hellenic arc (between 750–850°C at 60 km and 900–1,050°C at 100 km), likely due to a presence of deep lithospheric roots and subducted slabs (e.g., Koulakov et al., 2009). Furthermore, the temperature in the Aegean Sea is not as high as expected for a basin that experienced recent extension. The mean geotherm here shows a lower thermal gradient compared to other areas (Fig. 4), likely on account of the cold African slab subducting under this basin. The lowest geotherms are observed between North Denmark and southern Norway and beneath the North Sea. The mantle temperature in these areas is between 550 and 800°C at the depth of 60 and 100 km, respectively. However, in the region close to the borders of the study area the thermal inversion might be affected by larger errors in the amplitudes of the seismic anomalies, on account of the poorer density ray coverage.

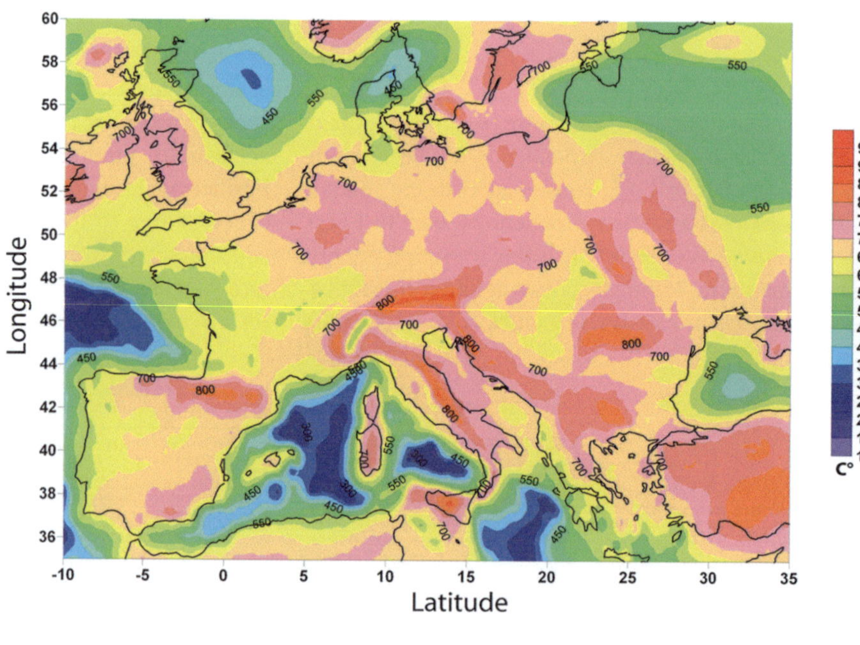

Fig. 1 Temperature variation (C°) at Moho depth. The values are extrapolated from the mantle temperature using typical crustal isotherms determined for different tectonic provinces defined in Čermák (1993)

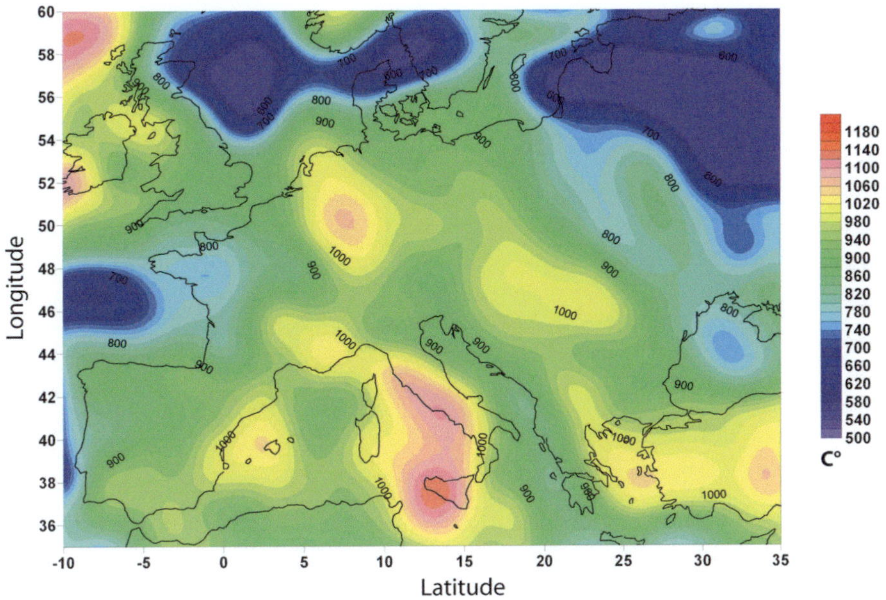

Fig. 2 Temperature variation (C°) at a depth of 60 km

Lithosphere Thickness of Europe

The term "lithosphere" comes from the Greek (*lithos* = rock) and was first used by Barrell (1914), while later it was defined by Isacks et al. (1968) as a "near surface layer of strength" of the Earth. Nowadays, there are various geophysical definitions of the Earth's lithosphere and consequently, different methods can be applied to trace it. The most common definitions identify the lithosphere as a cold outer shell of the Earth, which can support stresses elastically (Anderson, 1989), or as the layer in which density and other mechanical properties are controlled by chemical composition and temperature (Jordan, 1978). Furthermore, below the base of the lithosphere, anisotropy is controlled by convective shear stresses and should be aligned with the direction of the present mantle flow. On the other hand, within the lithosphere anisotropy probably reflects fabrics inherited from past tectonic events (e.g., Silver, 1996). Therefore, the depth, at which a transition between the fossil and flow-related anisotropy takes place, migth also be interpreted as the base of the lithosphere (e.g., Plomerová et al., 2002).

The lithosphere-asthenosphere boundary (LAB) may be detected using *P*- and *S*- receiver functions determinations (e.g., Sodoudi et al., 2006). The methods, which provide isotropic tomography images of the mantle using body waves (e.g., Arlitt, 1999) or surface waves (e.g., Cotte et al., 2002), can estimate also position of the LAB. Magnetotelluric measurements (Praus et al., 1990; Korja et al., 2002) provide another means for determination of the LAB, showing the layer with increased electrical conductivity, possibly on account of partial melting at the lithosphere base. The widely adopted thermal definition considers the lithosphere as the layer, in which heat transfer occurs prevalently by conduction, below a temperature threshold of about 1,300°C, at which starts partial melting (e.g., Anderson, 1989; Artemieva and Mooney, 2001). However, since mantle convection depends on viscosity, which is also temperature dependent, the base of the thermal lithosphere is defined sometimes as 0.85 of the solidus temperature (i.e., 1,100°C for the mantle solidus of 1,300°C) (e.g., Pollack and Chapman, 1977). On the other hand, mechanical properties of the mantle may change gradually in the vicinity of the solidus. Consequently no sharp boundary between the lithosphere and the asthenosphere possibly exists (e.g., Cammarano et al., 2003). Seismic velocities are very sensitive to temperature variations near the melting point, thus the thermal definition of the lithosphere should be coincident with the seismological definition.

On the base of the above considerations, the LAB was traced along the 1,200°C isotherm (Fig. 5). The largest values of lithospheric thickness between 150 and 230 km are observed beneath the EEP and are in a general agreement with previous estimates in this region (e.g., Babuška and Plomerová, 2006). On account of the vertical resolution of the tomograpy

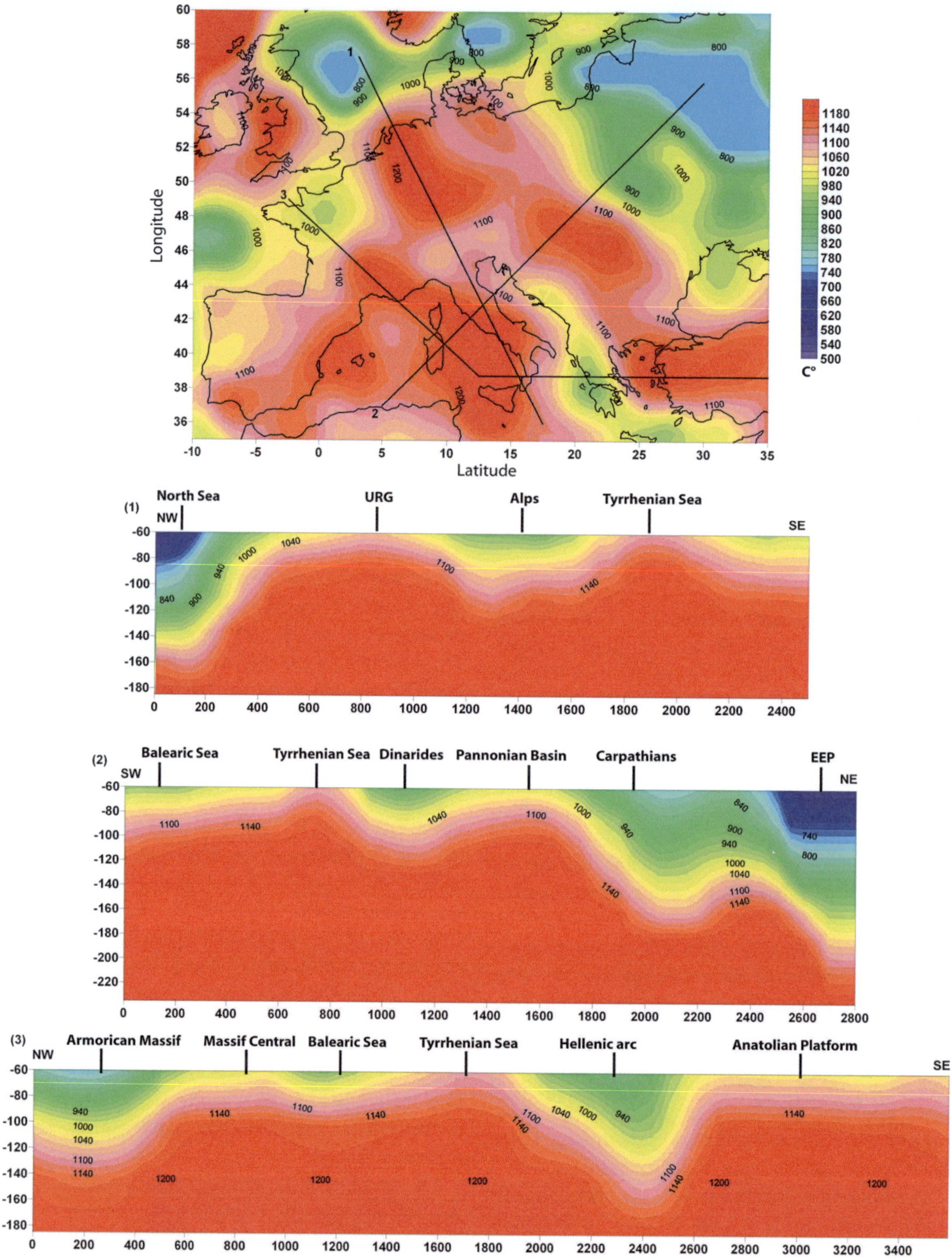

Fig. 3 Temperature variation (C°) at a depth of 100 km (top) and temperature distribution in the upper mantle along three cross-sections shown by the three *black lines* (bottom). Vertical and horizontal axes display depth and distance in km, respectively

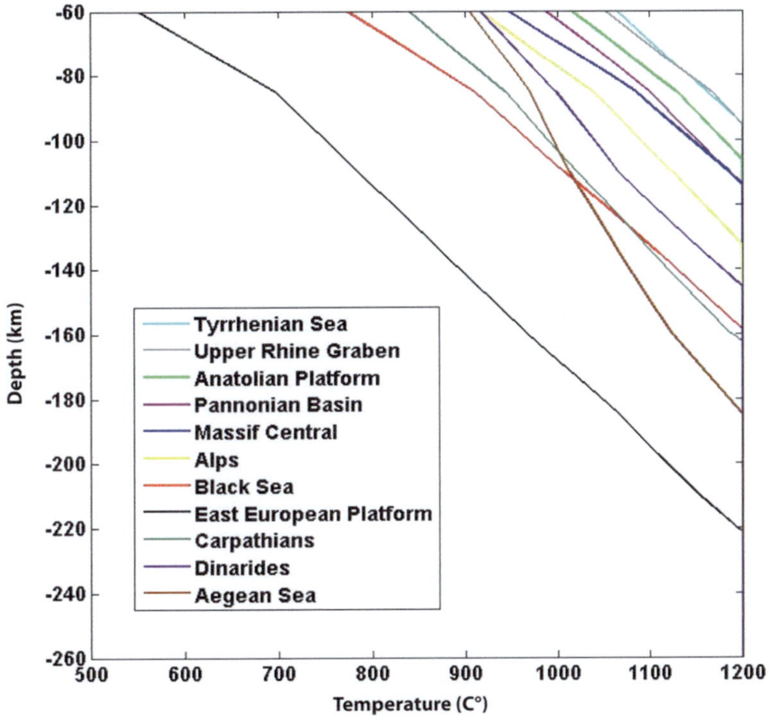

Fig. 4 Average geotherms for the main tectonic provinces of Europe

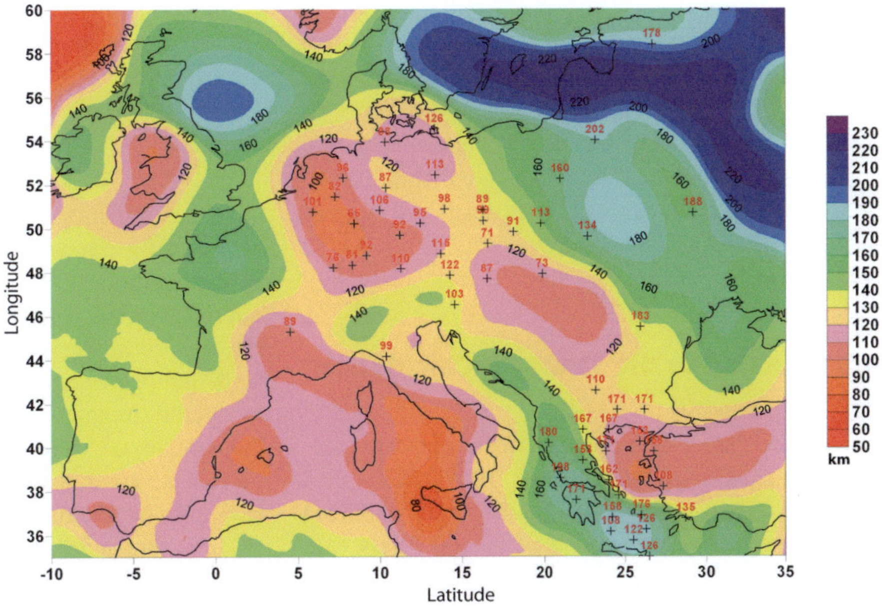

Fig. 5 Depth to the isotherm of 1,200°C, which is commonly used to mark the lithosphere-astenosphere transition (km). Black crosses and red numbers show location and values of lithospheric thickness according to receiver functions data (Sodoudi et al., 2006, 2008)

model (25 km or more) and of the filter used to smooth the velocity fields, it is not possible to determine small scale LAB variations for narrow tectonic structures. Therefore, the LAB depth is slightly underestimated in several areas (e.g., beneath the Alps) and overestimated in some others (e.g., beneath the Pannonian Basin). In particular, the lithospheric thickness beneath the Tyrrhenian Sea is significantly overestimated (~50 km) compared to previous models (e.g., Calcagnile and Panza, 1990; Panza and Raykova, 2008). The reason of such a strong discrepancy needs more detailed investigations. In general, in most part of the study area a good agreement between the lithospheric thickness variations and previous local models of European lithosphere was found (e.g., Praus et al., 1990; Babuška and Plomerová, 2006). The thinnest lithosphere (<100 km) is observed in the ECRIS and in the Tyrrhenian Sea, where also an updoming of the Moho is observed (see previous chapter). A regional thinning also appears beneath the Massif Central, possibly relating to the presence of a mantle plume (e.g., Sobolev et al., 1997), and in other regions affected by Tertiary volcanism (e.g., Pannonian Basin). The obtained results are mostly consistent with recent receiver functions determinations (Sodoudi et al., 2008), which estimate the LAB between 80 and 120 km in this area (Fig. 5). The lithosphere becomes thicker to 120–140 km toward the flanks of the Pannonian Basin, beneath the Bohemian Massif and the Alpine foredeep and to ~150 km beneath the Carpathians. The thickening of the lithosphere continues to the south beneath the Alps (~150 km), where the roots are associated with the collision of the European and the Adriatic plates. Large lithospheric thicknesses (140–160 km) are also observed along the Dinarides-Hellenic arc with a maximum of ~180 km beneath the Aegean Sea. These values slightly exceed the receiver functions determinations, which trace the LAB at ~160 km (Sodoudi et al., 2006) (Fig. 5). However, the bottom of the thermal lithosphere cannot be clearly distinguished in this area from the top of the African slab subducting beneath the Aegean plate.

Introduction to the Strength Calculation

Rheological models proposed since the late seventies (e.g., Goetze and Evans, 1979; Brace and Kohlstedt, 1980) indicate that the thermally stabilized continental lithosphere consists of several layers with a rheologically strong upper crust separated by weaker lower crust from a strong subcrustal layer, which in turn overlies the weak lower part of the lithosphere. Goetze and Evans (1979) were the first to combine data on experimental rock properties and extrapolate them onto geological time and spatial scales. They have introduced the yield strength envelope (YSE) for the oceanic lithosphere, which shows the maximal rock strength as a function of depth. In the YSE rheology models, depth dependence of rock strength integrates multiple processes such as increase of both brittle and ductile strength with pressure, decrease of the ductile strength with depth-increasing temperature, lithological structure and fluid content. The strength profiles are represented by curves of two different types. The straight lines correspond to brittle fracture and demonstrate an increase of strength with depth. The curved lines describe viscous deformation according to the Power law creep: strength decreases downwards exponentially due to the increase of temperature with the corresponding decrease of viscosity (Burov and Diament, 1995). The depth, at which the brittle and ductile strengths are equal, denotes the brittle-ductile transition (BDT). This transition can be found in the crust, as well as in the uppermost mantle, resulting in a rheological layering of the lithosphere (Ranalli and Murphy, 1987), where the brittle and ductile domains alternate throughout the lithosphere depending on depth, mineralogical composition, and thermal structure. The total lithospheric strength (σ_L), is calculated through a vertical integration of the yield envelope:

$$\sigma_L = \int_0^h (\sigma_1 - \sigma_3) \cdot dz \qquad (10)$$

where h is the thickness of the lithosphere.

One of the major experimental rheology laws used for construction of YSE's is Byerlee's law of brittle failure (Byerlee, 1978). Byerlee's law demonstrates that the brittle strength is a function of pressure and depth indipendent of rock type. On the other hand, the ductile strength strongly depends on rock type and temperature, as well as on the other specific conditions (e.g., grain size, macro and microstructure). In particular, the ductile behaviour non-linearly depends on strain rate and thus on the time scale of the deformation process. The mechanism of ductile deformation is highly versatile: diffusion creep and various mechanisms of dislocation creep. The first mechanism is predominant at a small grain size and relatively low stresses, which are specific for highly sheared material (ductile shear zones) or for very high temperatures. By contrast, at high stresses and moderate temperatures (<1,330°C), the creep rate is dominated by dislocation creep (Power law, Dorn law). Other ductile flow mechanisms can occur at low temperature conditions (e.g., pressure solution occurring at temperatures below 200°C). The rheological parameters in the brittle regime are usually assumed to be constant for all rock types. Pre-existing faults are often taken to be cohesion less, with a coefficient of friction ~0.75. The uncertainties introduced by these approximations are small compared to those generated by a lack of constraints on the pore fluid factor (ratio of hydrostatic to lithostatic pressure) (e.g., Fernàndez and Ranalli, 1997). On the other hand, the rheological parameters of the ductile regime for various rock types imply more uncertainties on account of the following main reasons: (1) Experiments usually refer to simplified conditions compared to which the real rocks are subjected (e.g., temperature-pressure (*P-T*) conditions of experiments do not represent natural *P-T* conditions of loading paths); (2) The experimental strain rates are in the order of 10^{-8} -10^{-4} s^{-1}, which is about 10^{10} times faster than the geological strain rates (10^{-18}–10^{-14} s^{-1}); (3) The experiments refer to simple monophase minerals or selective "representative"; rocks, while the extension of their results to real aggregate compositions has to be demonstrated (e.g., Kohlstedt et al., 1995). It is often assumed that the weakest of the most abundant mineral species defines the mechanical behaviour of the entire rock (e.g., quartz for granite). However, very small amounts of weak phases (e.g., micas) may result in significantly smaller strength than that of quartzite. It is also noted that poly-phase aggregates are weaker than their constituents; (4) The experiments are conducted on small rock samples of homogeneous structure, while at larger scales (>0.1–1 m), rocks may be structured; (5) Water content influences rock strength, but in nature the amount of water present in the rock is unknown; (6) Chemical and thermodynamical reactions (basically unknown factors in nature) modify the mechanical behaviour of rocks. Due to these uncertainties, Brace and Kohlstedt (1980) and Kohlstedt et al. (1995) have suggested that the real crustal rocks may be significantly "softer"; than the experimental estimates. In addition to the uncertainties of the rheology laws, even defined as "methodological uncertainties"; (Fernàndez and Ranalli, 1997) there are also "operational uncertainties"; deriving from various factors (e.g., imperfect knowledge of composition and structure of the lithosphere, errors in estimations in temperature distribution). In particular, different thermal models produce strong differences in the strength estimates (e.g., Kohlstedt et al., 1995). In fact, the geotherm not only controls the ductile strength of the lithosphere, but also indirectly, its brittle strength through the influence of temperature on the depth of the BDT.

This conventional rheology model (known as "jelly sandwich";) has been recently confuted by some authors (e.g., Jackson, 2002), who proposed for the continental lithosphere a model, which is based on the rheology envelope from Mackwell et al. (1998), in which the crust is strong, but the mantle is weak (known as "crème-brûlée"; model). This model suggests that continents are thin and hot (>800°C at 60 km) and have water-saturated mantle, which cause a concentration of the continental plate strength in the crust. The "crème-brûlée"; model has arisen because of conflicting results from rock mechanics, earthquakes and elastic thickness data (Maggi et al., 2000). Since earthquakes are mainly observed above 40 km depth (Maggi et al., 2000) both in continents and oceans, Maggi et al. (2000) and Jackson (2002) claim that all continental microseismicity originates in the crust. This theory has been recently confuted by a study of Monsalve et al. (2006), which demonstrates that continental microseismicty is bimodal, with crustal and mantle locations as deep as 100 km. However, other studies (e.g., Watts and Burov, 2003) disagree with the idea of a direct seismic depth-strength correlation, claiming the validity of the "jelly sandwich" model. They suggest that seismicity should be interpreted as a manifestation of mechanical weakness, not strength, of the seismogenic layer that fails at region specific intraplate stress level. In this approach, crust-mantle decoupling and depth-growing confining pressure that inhibits brittle failure explain the absence of deep earthquakes. It should also be noted that seismicity refers to short-time scale behaviour, which may be unrelated to long-term rheology because at this time scale the entire lithosphere should deform only in the brittle-elastic mode. Consequently, there may be no direct correlation between the seismic and long-term ductile behaviour. Indeed, the observations of plate flexure below orogens (Watts, 2001) suggest that many continental plates have strong elastic cores (*Te*) that are probably 2–2.5 times thicker than the seismogenic layer thickness (*Ts*).

Rheological Model of the European Lithosphere

The first strength distribution in the European lithosphere (Cloetingh et al., 2005) has been estimated using a simplified compositional model consisting of two homogeneous crustal layers overlain by a sedimentary cover. The thermal structure of the lithosphere was defined using the heat flow data from the global compilation (Pollack et al., 1993), and regional surface heat flow studies (e.g., Fernàndez et al., 1998; Lenkey, 1999). In this section new strength maps of the European lithosphere are presented. The new strength results are obtained employing the thermal model above described, while the composition was defined using the EuCRUST-07 model (Tesauro et al., this volume). The seismic velocities were used to estimate density variations in the upper and lower crust following the Christensen and Mooney (1995) approach. The density assigned to the sediments is a weighted average of the values estimated for each sublayer composing the sedimentary package (Kaban et al., 2009). The mantle density is based on the values obtained at different temperatures derived from the inversion of seismic tomography data (Koulakov et al., 2009). In order to estimate the crustal rheology, the lithology map presented in the previous chapter was simplified, due to the lack of rock creep parameters values defined by the lab experiments (Fig. 6). However, the relationship between the crustal rheology and lithology might be different. For instance, the southern Tyrrhenian and the western Black Sea, which are characterized by different seismic velocities and thermal regime conditions, are assigned to a different rheology although having similar lithology. The lithology of the sediments was not specified, since they are normally affected only by brittle deformations. The exceptions might correspond to the very deep basins having extremely high thermal conditions, which are not presented in the study area. A uniform strain rate of 10^{-15} s^{-1} has been adopted, as it is commonly observed for intraplate compressional and extensional settings (Carter and Tsenn, 1987). However, a lateral change of this parameter can occur due to horizontal stresses and pre-existing weak zones (e.g., faults). The friction coefficient used is equal to 0.75 and 3, for extensional and compressional conditions, respectively (e.g., Ranalli, 2000; Afonso and Ranalli, 2004). The pore fluid factor is assumed equal to 0.36, which is a typical hydrostatic value. The brittle deformation was calculated using Byerlee's law, while the Power and Dorn law has been used to estimate the ductile deformation in the crust and in the mantle, respectively, being dislocation glide (Dorn creep) the dominant creep process in mantle olivine for stresses exceeding 200 MPa. The rheology parameters values and the brittle strength and creep equations are displayed in Table 2.

It is worth noticing that the strength estimates in the mantle lithosphere are referred to a "dry olivine". A "wet" mantle model might be suitable for areas recently affected by subduction of oceanic lithosphere and tectonothermal events (e.g., Afonso and Ranalli, 2004). A previous study (Lankreijer, 1998) has demonstrated that the total integrated lithospheric strength in the Carpathians-Pannonian basin system can decrease up to 35–40% when the mantle rheology is changed from "dry" to "wet". On the other hand, since the fluid content in the upper mantle is still not well resolved, it is difficult to properly delimit the areas where a "wet" mantle can be adopted. As a consequence, only the integrated strength for a "dry" mantle is computed. Therefore, strength values obtained in this study can be considered as upper bounds of those possible for the estimated thermal and crustal rheological conditions. The integrated strength of the lithosphere under compression and extension is shown in Fig. 7a and b, while Fig. 8 displays strength profiles calculated in some selected points. Since the European stress field is mostly the result of compressional forces (e.g., Zoback, 1992; Grünthal and Stromeyer, 1992), only the lithospheric strength estimated under compressional conditions is discussed.

The European lithosphere is characterized by large spatial strength variations, with a pronounced increase in the EEP east of the TESZ compared to the relatively weak but more heterogeneous lithosphere of western and central Europe. In this part of the study area the strength distribution reflects the effect of different factors, such as crustal thickness, rheology and thermal gradient. Therefore, the high strength is localized in the regions characterized by a strong crustal rheology and average thermal regime (e.g., the Bohemian Massif), as well as in the areas having thin crust and low thermal gradient (e.g., North Sea). By contrast, the weak zones are found in areas affected by Tertiary volcanism and mantle plumes, such as the ECRIS and the Massif Central (Figs. 7a and 8, points **O** and **M**), which are

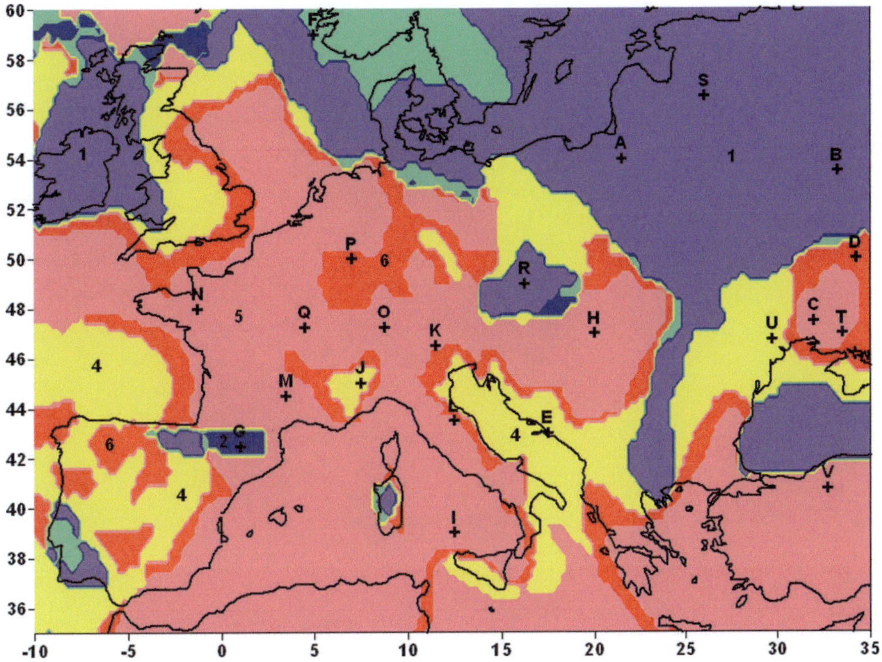

Fig. 6 Rheological model of the European crust. Numbers mark the couple of lithologies representative of the upper and lower crust as follows: *1*, Granite (dry)-Mafic granulite (not defined); *2*, Granite(dry)-Diorite (Wet); *3*, Granite(dry)-Diabase (dry) *4*, Quartzite (dry)-Mafic granulite (not defined); *5*, Quartzite (dry)-Diorite (wet); *6*, Quartzite (dry)-Diabase (dry). Crosses with capital letters depict the points location in which the strength and the elastic thickness values and/or the strength profiles are displayed (Table 4, Figs. 8 and 15)

Table 2 Rheological model parameters. Numbers in square brackets stand for the original source as it follows: [1] Carter and Tsenn (1987); [2] Wilks and Carter (1990); [3] Goetze and Evans (1979)

Parameter	Symbol	Units	Sediments	Upper crust	Lower crust	Upper mantle
Composition	–	–	–	Quarzite (dry) [1]/ Granite (dry) [1]	Mafic ganulite[2]/ Diorite (wet) [1]/ Diabase (dry) [1]	Olivine (dry) [3]
Density min-max/mean	ρ	Kg m^{-3}	1700-2749/2227	2265-2960/2715	2709-3368/2990	3232-3267/3245
Layer depth min-max/mean	z	km	0–16.5/3	4–43.5/21	9.5–57/33	85–235/154
Friction coefficient ext/com	f	–	0.75/3	0.75/3	0.75/3	0.75/3
Pore fluid factor	λ	–	0.36	0.36	0.36	0.36
Power law exponent	n	–	–	2.72/3.3	4.2/2.4/3.05	3
Power law activation energy	E_P	KJ mol^{-1}	–	134/186	445/212/276	510
Power law strain-rate	A_P	$Pa^{-n}s^{-1}$	–	$6.03e^{-24}/3.16e^{-26}$	$8.83e^{-22}/1.26e^{-16}/6.31e^{-20}$	$7.00e^{-14}$
Dorn law activation energy	E_D	KJ mol^{-1}	–	–	–	535
Dorn law strain-rate	A_D	s^{-1}	–	–	–	$5.70e^{11}$
Dorn law stress	σ_D	Pa	–	–	–	$8.50e^{9}$
Strain rate	ε	s^{-1}	–	$10e^{-15}$	$10e^{-15}$	$10e^{-15}$
Brittle strength						
	$\sigma = f\rho gz(1-\lambda)$					
Creep equations						
Power law creep	$\sigma = \left[\frac{\varepsilon'}{A_P}\right]^{\frac{1}{n}} \cdot \exp\left[\frac{E_P}{nRT}\right]$					
Dorn law creep	$\sigma = \sigma_D\left(1 - \left[-\frac{RT}{E_D} \cdot \ln\left(\frac{\varepsilon'}{A_D}\right)\right]^{\frac{1}{2}}\right)$					

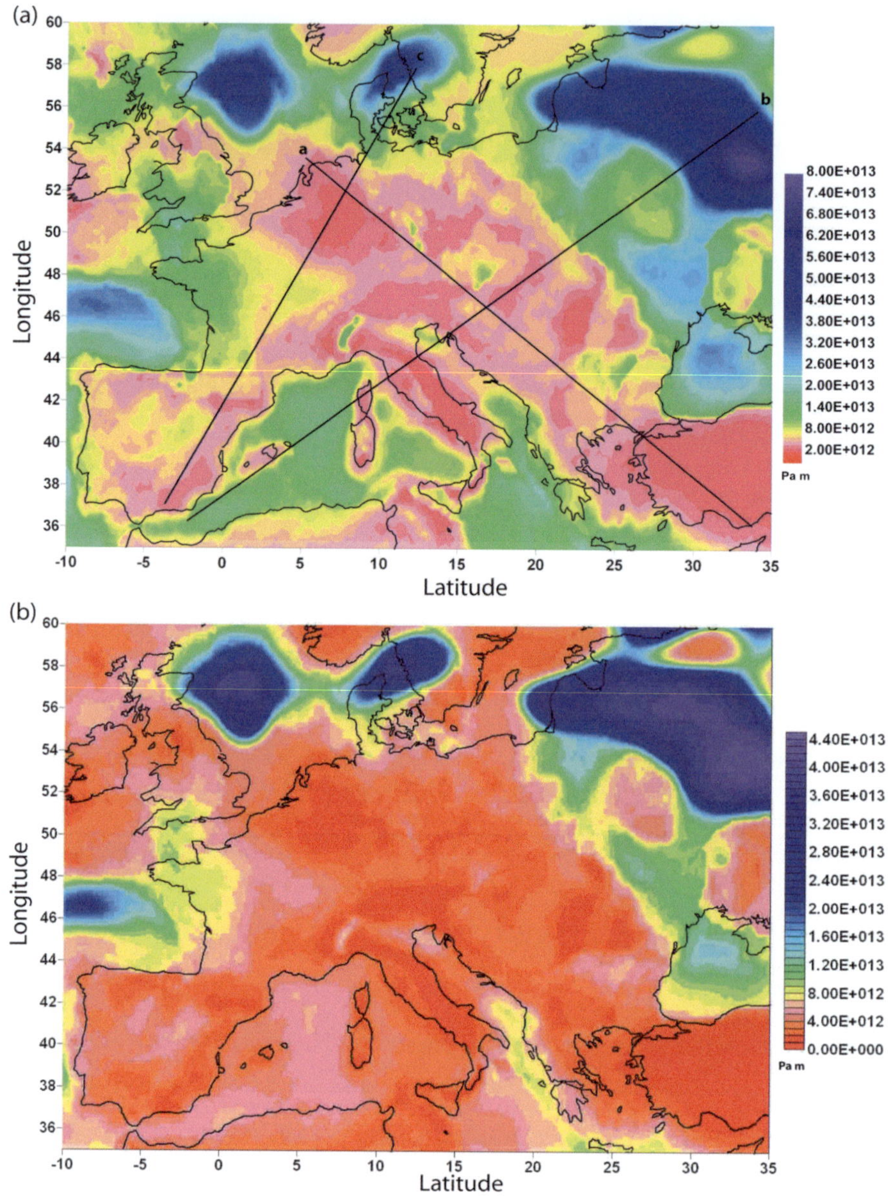

Fig. 7 Integrated strength of the European lithosphere (Pa m). (**a**) Integrated strength estimated under conditions of compression. Black lines depict locations of the cross-sections displayed in Fig. 11a–c. (**b**) Integrated strength estimated under conditions of extension

separated by high-strength regions, such as the North German Basin, the Paris Basin and the Armorican Massif (Figs. 7a and 8, points **N** and **Q**). Since the crustal rheology (being quartz dominant) is softer than the rheology of mantle olivine, a sharp increase of strength is observed in the zones, where a decrease of the crustal thickness is observed, like the zone from the Apennines to the Tyrrhenian Sea (Fig. 7a). However, the strength of the Tyrrhenian Sea (Figs. 7a and 8, point **I**) is likely overestimated due to the presence of a thicker mantle lithosphere ($\sim$80 km) not confirmed by previous studies. Furthermore, in this area affected by subduction (Koulakov et al., 2009) mantle fluids, not considered in the rheological model adopted, might cause a further strength decrease.

In order to analyse distinctly the influence of the crust on the total lithospheric strength, the integrated strength of the crust, the contribution of the crustal strength to the total lithospheric strength and the integrated crustal and total lihospheric strength variation along three cross-sections through the main tectonic structures of Europe are displayed in Figs. 9, 10 and

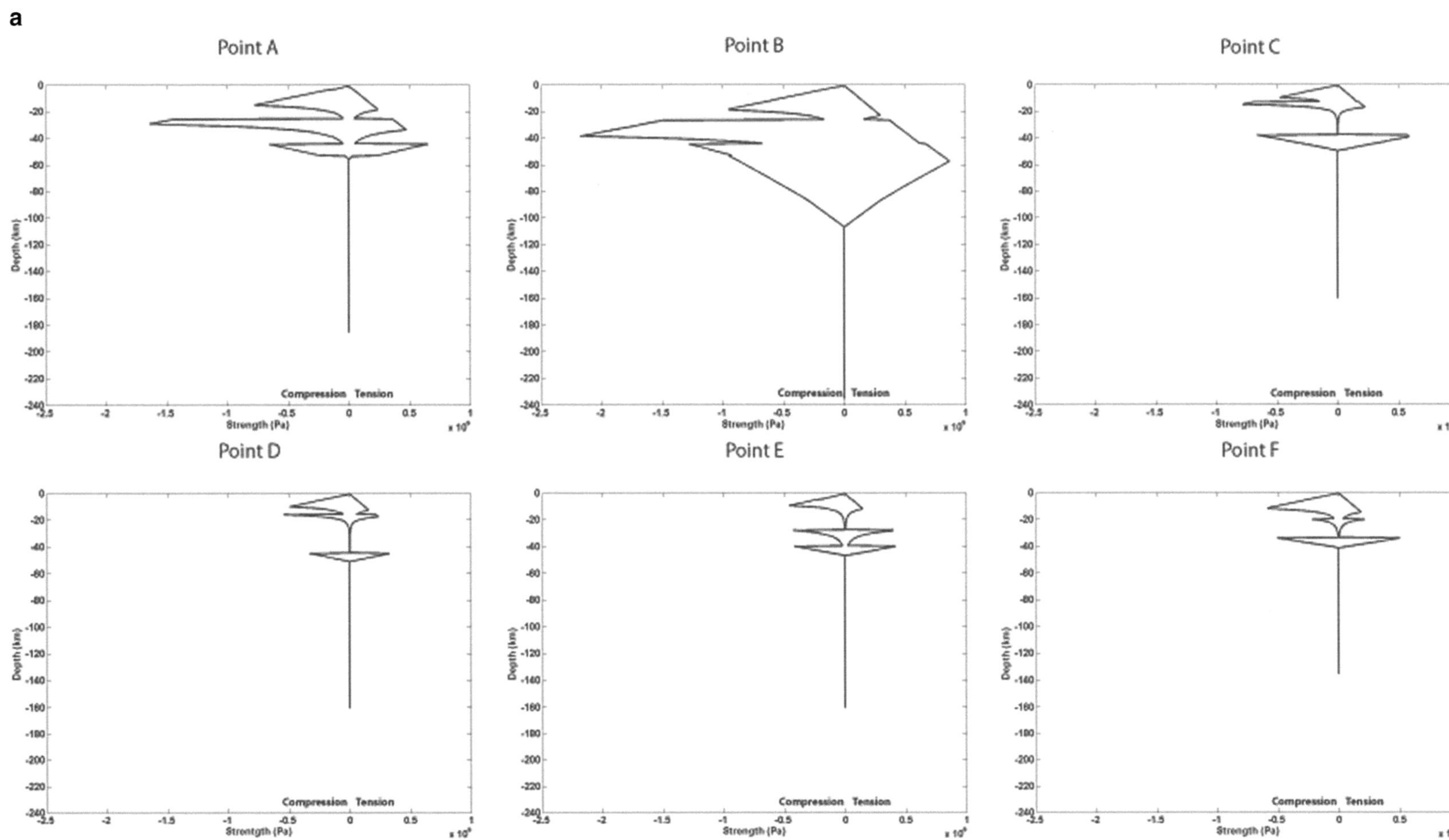

Fig. 8 Strength profiles for selected points. Points location is depicted in Fig. 6. For convention values estimated under compressional and extensional conditions are assumed negative and positive, respectively. Lithospheric layers are coupled in points **A**, **B** and **I**. Crustal layers are coupled in points **C**, **D** and **F**. Lower crust and upper mantle are coupled in points **E** and **R**. Lithospheric layers are decoupled in points **G**, **H**, **J**, **K**, **L**, **M**, **N**, **O**, **P**, and **Q**. The lithosphere structure can be represented by a "jelly sandwich" model in points **B**, **H**, **I**, **N**, **O** and **Q** and by a "crème brûlée" model on points **A**, **C**, **D**, **E**, **F**, **G**, **J**, **K**, **L**, **M**, **P** and **R**

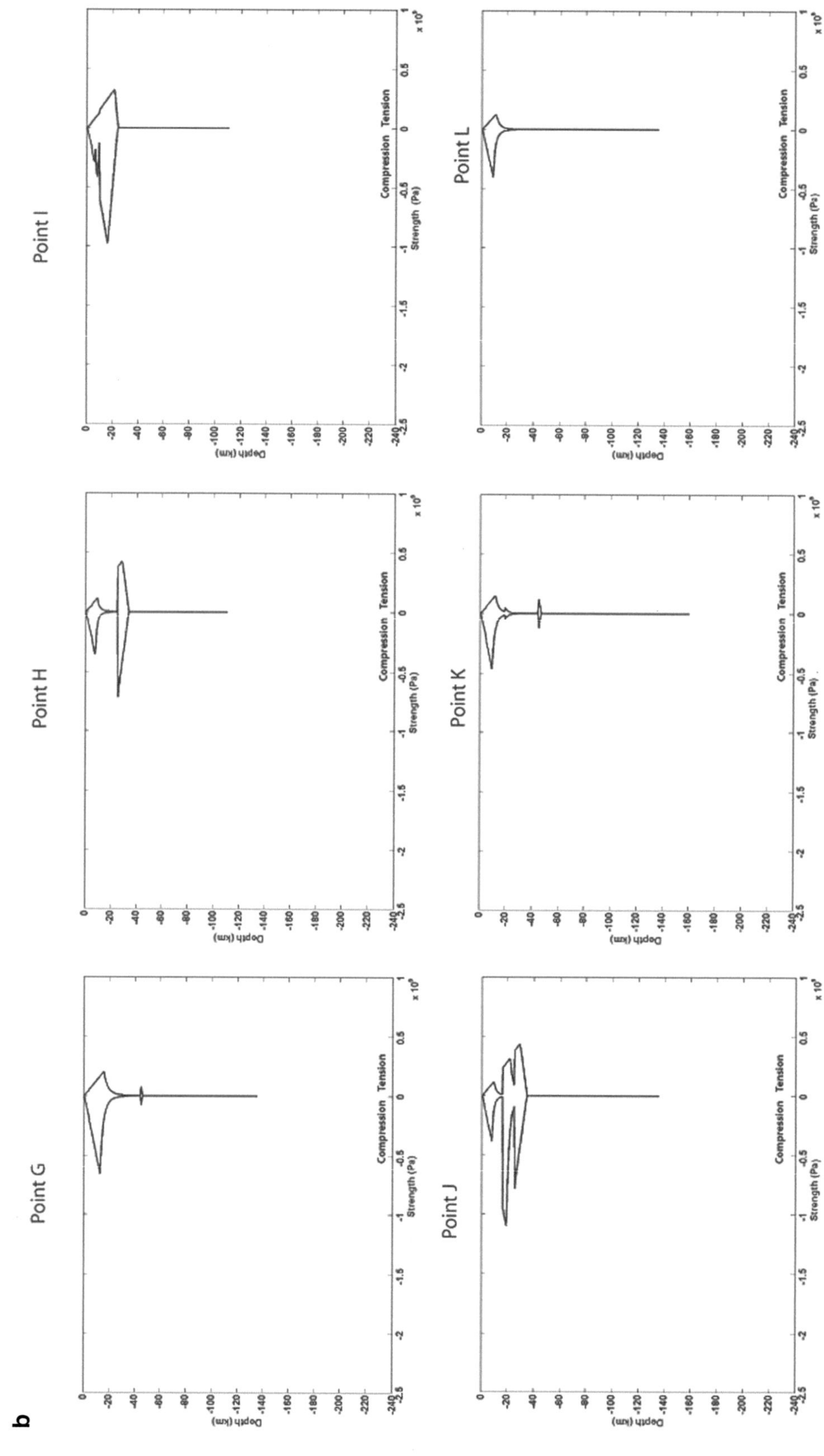

Fig. 8 (continued)

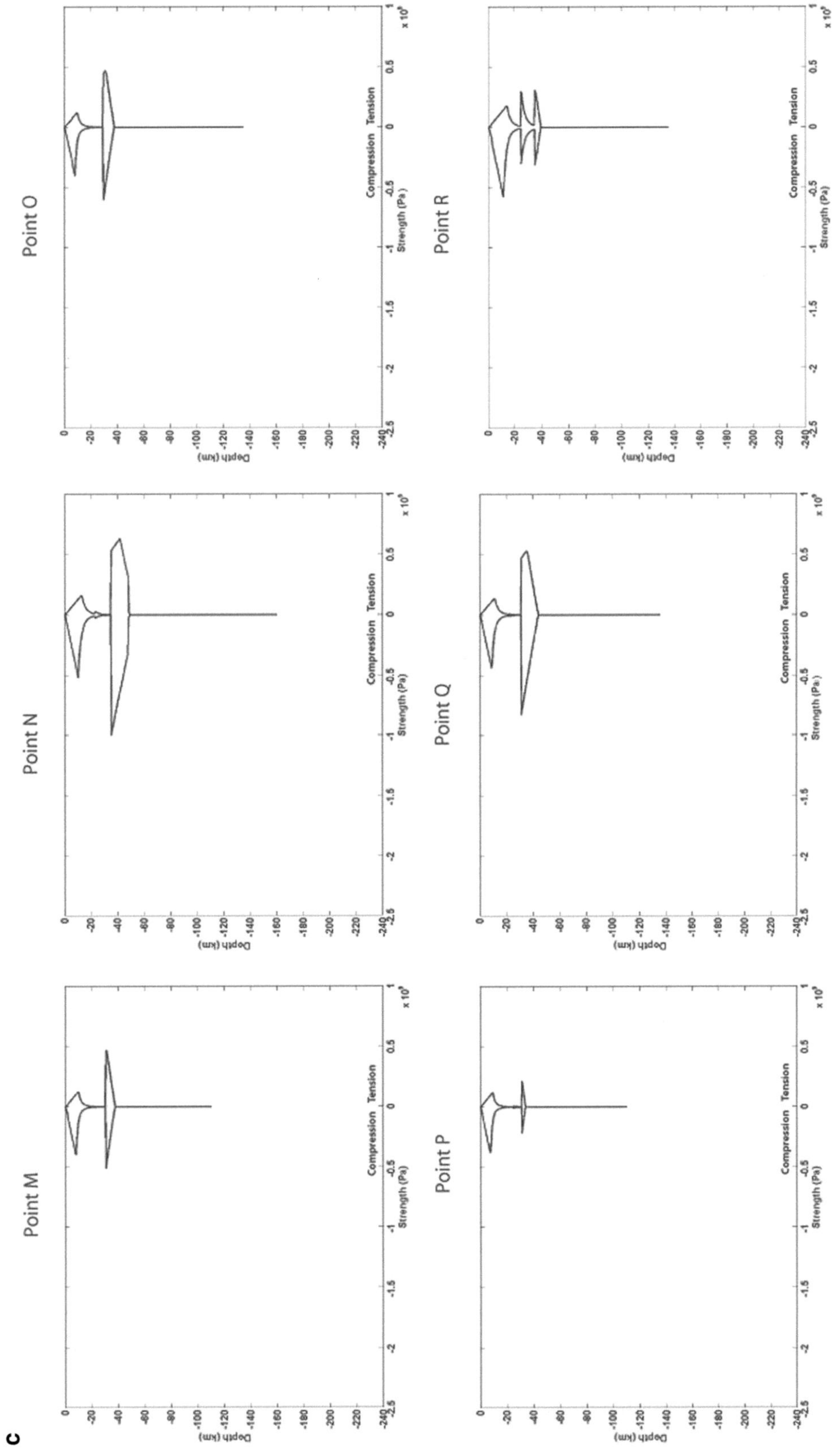

Fig. 8 (continued)

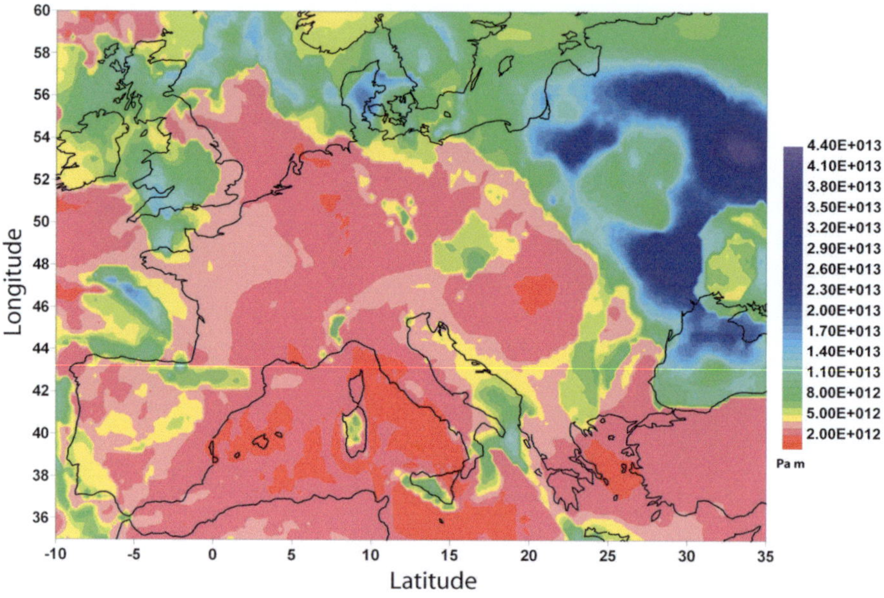

Fig. 9 Integrated strength estimated under compression of the European crust (Pa m)

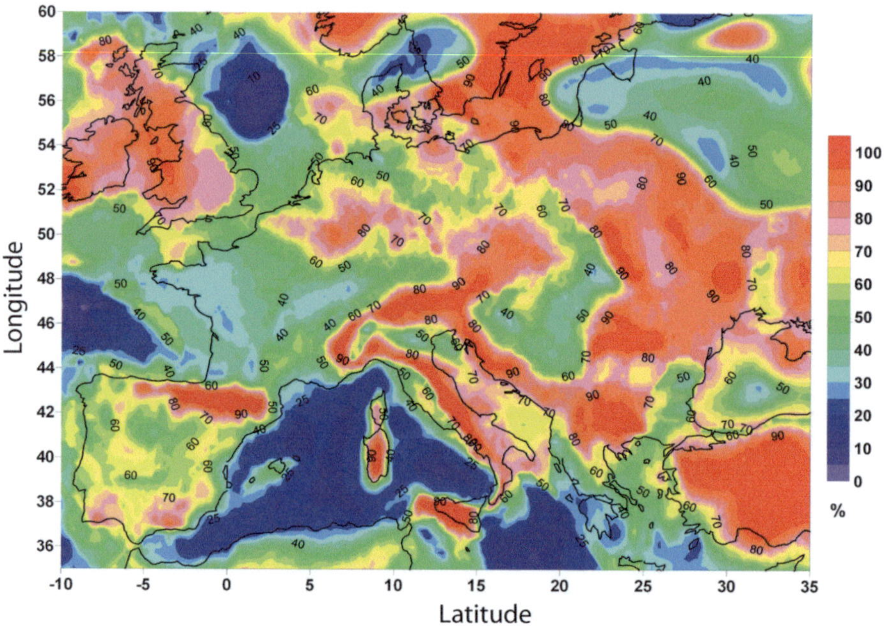

Fig. 10 Proportion of the integrated crustal strength relative to the total lithosphere value

11a–c. The crustal strength values are in a range from 9.5×10^{11} to 4.2×10^{13} Pa m and depend more on the lateral compositional variations than on the crustal thickness and thermal regime. The lowest values ($< 4 \times 10^{12}$ Pa m) are mostly found in the regions characterized by soft rheology, like the Pannonian Basin (Fig. 8, point **H**). By contrast, the variations of the mantle part of the lithosphere strength mainly depend on the thermal structure of the lithosphere and Moho depth. Therefore, in the regions that experienced recent thermal activity (e.g., the Eifel Province and Anatolian Platform) and areas characterized by large crustal thickness (e.g., the Alps and the Pyrenees) the strength of the lithospheric mantle is significantly reduced (Fig. 8, points **G** and **K**). It can be observed that the crustal contribution to the total strength dominates in the study area: about 60% of the European crust retains >50% of the total integrated strength of the lithosphere.

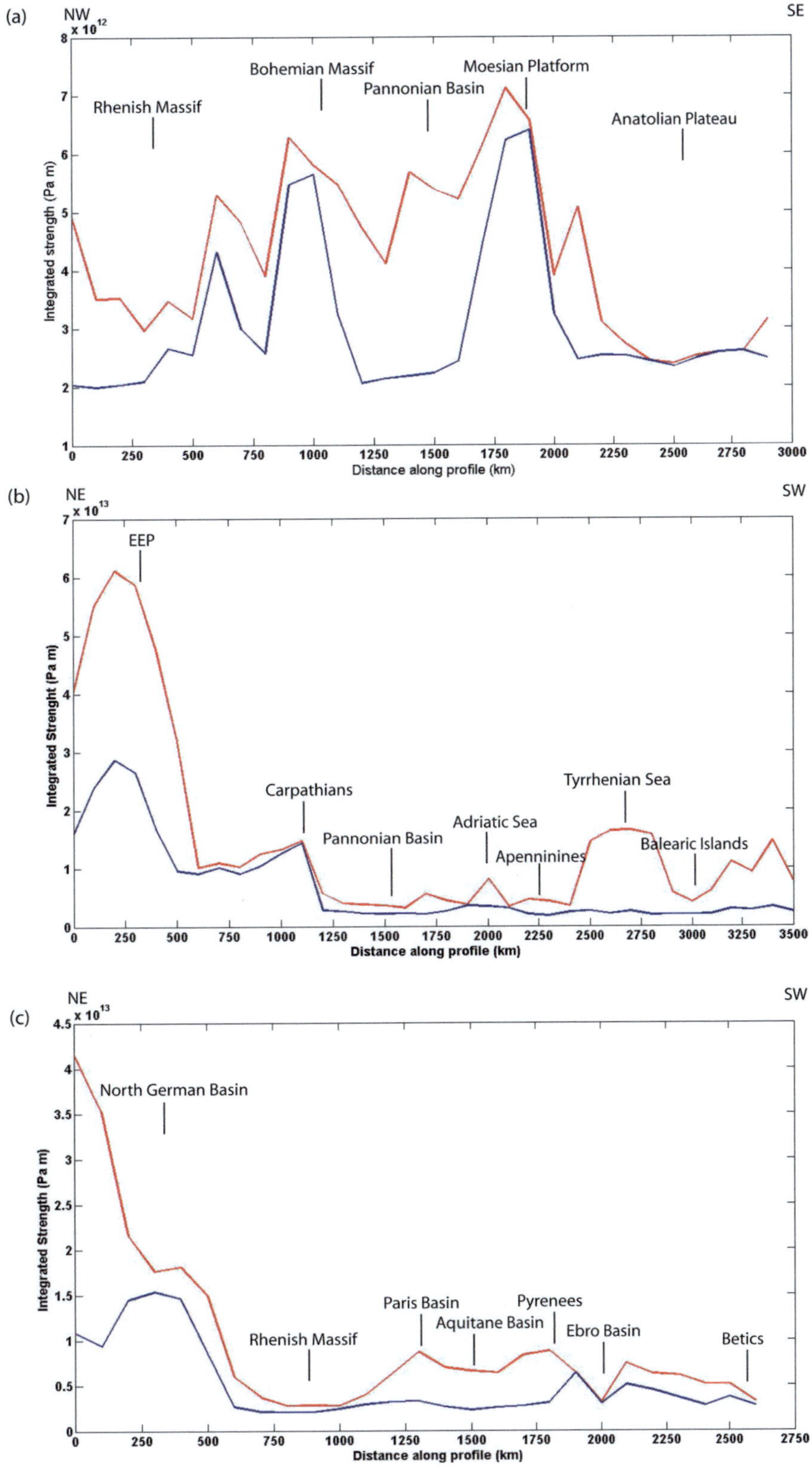

Fig. 11(a–c) Integrated strength of the lithosphere (in *red*) and of the crust (in *blue*) along the 3 sections crossing the main tectonic structures of Europe. The profiles are displayed in Fig. 7a

A low crustal strength contribution (<20%) is observed only in 7% of the area, while over 35% of the European regions are characterized by the crustal component exceeding 70% of the total lithospheric strength (Fig. 10). The highest proportion of the crustal strength (over 70%) is also found in the areas characterized by large crustal thickness (>40 km) and by medium-high thermal regime (e.g., the orogens). Also thick crust having a soft rheology (like in the Alps and the Apennines, Fig. 6) may retain over 90% of the total strength (Fig. 8, points **K** and **L** and Fig. 10). By contrast, low and moderate values of the crustal strength proportion (<50%) are observed in both hot (e.g., Tyrrhenian Sea and Pannonian Basin) and cold (e.g., North Sea) regions with a thin crust (Fig. 8, points **I** and **H** and Fig. 10). The sharp decrease of the thermal gradient in the EEP produces in this area a strong reduction of the crustal strength from ~80% beneath the TESZ to 30–40% (Fig. 10), demonstrating how the strength of the lithospheric mantle grows faster than the strength of the crust when the lithosphere becomes cold (Fig. 8, points **A** and **B**). These results confirm the hypothesis that the upper mantle of the thermally stabilized, old cratonic regions is considerably stronger than the strong part of its upper crust (e.g., Moisio et al., 2000). Furthermore, they demonstrate that both "jelly sandwich" (Fig. 8, points **B**, **H**, **I**, **N**, **Q** and **O**) and "crème brûlée" (Fig. 8, points **A**, **C**, **D**, **E**, **F**, **G**, **J**, **K**, **L**, **M**, **P** and **R**) models, are valid for the European lithosphere, depending on specific thermal and rheological conditions of the area considered, as also demonstrated in the study of Afonso and Ranalli (2004). Both the total lithospheric and the crustal integrated strength show a similar trend. The main difference is observed in the Tyrrhenian Sea, where the total integrated lithospheric strength shows a peak around 1.8×10^{13} Pa m, while the integrated crustal strength has an amplitude similar to the surrounding areas (Fig. 11b).

In comparison with the previous study of Cloetingh et al. (2005) the total integrated lithospheric strength demonstrates a more heterogeneous distribution. Nearly 60% of the area is characterized by low values ($<1\times10^{13}$ Pa m), while the largest strength values are mostly concentrated in the coldest part of the EEP. Furthermore, the new European strength maps, which are based on the improved thermal and compositional models, reveal a higher contribution of the crustal strengths to the total lithospheric strength, which is not limited to the orogens. The strongest differences with the previous results are observed in the North Sea, where the new maps show much higher strength (Fig. 7a), mostly on account of the low thermal regime. However, more investigations are required since this area is characterized by large uncertainties of the temperature estimates. Another principal difference is found in the Adriatic plate and the Bohemian Massif, where Cloetingh et al. (2005) estimate an integrated lithospheric strength as high as in the EEP, while the new results show a more gradual transition from the weaker areas surrounding these structures. The obtained strength estimates demonstrate an overall good consistency with other geophysical parameters, such as mantle gravity anomalies (Kaban et al., 2009). In particular, a correspondence is found between the low and high strength values along the ECRIS and in the North Sea, supporting the presence of a weak and strong lithosphere, respectively, and the negative and positive mantle anomalies observed in these areas.

Effective Elastic Thickness (*Te*) of the European Lithosphere

The effective elastic thickness of the lithosphere (*Te*) corresponds to the thickness of a homogeneous elastic layer, which is characterized by the same flexural rigidity as the lithosphere plate. This parameter was initially introduced in the experimental studies investigating the response of the lithosphere to the external load by means of the cross-spectral analysis of the gravity data (e.g., Banks et al., 1977). Using this method Pérez-Gussinyé and Watts (2005) have recently estimated *Te* of the European lithosphere. However, different methods used for *Te* estimates might provide essentially different results. For instance, the *Te* values obtained from foreland flexure represent rather a paleo-situation than current changes across the foreland basin. Previous studies (e.g., Watts et al., 1980) have shown that *Te* variations in the oceanic areas are mainly controlled by the thermal structure of the oceanic lithosphere related to the thermal age. The oceanic lithosphere cools, becomes stronger with time and the *Te* increases. It was demonstrated that *Te* of the oceanic plate approximately corresponds to a depth of the 450–600°C isotherm (e.g., Watts, 1978). By contrast, the continental lithosphere demonstrates a more complex rheological stratification than the oceanic plates, in particularly due to the thicker and more heterogeneous crust and due to the upper mantle, which is modified by various processes (e.g., mantle underplating). During its long tectonic history the lithosphere might experience additional warming, which can lead to its thermal rejuvenation resetting its thermomechanical age (e.g., Adriatic lithosphere, Kruse and Royden, 1994). Therefore, there is no clear *Te*-age relationship for the continental lithosphere. According to previous studies (e.g., Burov and Diament, 1995), *Te* of the continents has a wide range of values (5–110 km), which can vary within the plate and shows a bimodal distribution around two peaks at 10–30 km and 70–90 km. This clustering is probably related to influence of the plate structure: depending on the ductile strength of the lower crust, the continental crust can be mechanically coupled or decoupled with the mantle resulting in highly different *Te* (Burov and Diament, 1995). The crust-mantle decoupling occurs if the temperature of the creep activation is lower than the temperature at the Moho boundary. Therefore, to evaluate the effective elastic plate thickness of the continental lithosphere it is necessary to consider many factors describing its complicated structure and history.

The *Te* distribution within the European domain is estimated based on the integrative model of the lithosphere, which is presented above. Rheological properties of the continental upper crust are primarily controlled by content of quartz (Brace and Kohlstedt, 1980), while mechanical behaviour of the lower and middle crust may be conditioned by a variety of lithologies such as quartz, diorite, diabase or plagioclase. In general, if the crust is thick (>35 km), the lower crustal temperatures are high enough to reduce the creep strength of the rocks in the vicinity of the Moho (Burov and Diament, 1995). By contrast, when the stress is below the yield limits, the lower crust and mantle are mechanically coupled and the lithosphere behaves like a single plate, similar to the oceanic lithosphere. In this case, the *Te* value gradually depends on temperature and should be coincident with the base of the mechanical lithophere, corresponding to the depth of an isotherm of 700–750°C, below which the yielding stress is less than 10–20 Ma. On the other hand, the crust-mantle decoupling results in a drastic reduction of the total effective strength and *Te* of the lithosphere (Burov and Diament, 1995) and implies a possibility of lateral flow in the lower crust enhanced by other processes (e.g., grain-size reduction) (e.g., Burov et al., 1993). For the "normal" quartz-dominated crust decoupling should be permanent, except for the thin (e.g., rifted) crust (<20 km). For other crustal compositions (e.g., diabase, quartz-diorite, etc.) decoupling might take place in most cases, except for very old (>750 Ma), cold lithosphere. Based on the above considerations, Burov and Diament (1995) proposed a unified model of the lithosphere that relates *Te* with thermal age, crustal thickness and flexural curvature. According to these authors, *Te* of the plate consisting of *n* detached layers is equal to:

$$T_e^{(n)} = \left(\sum_{i=1}^{n} \Delta h_i^3 \right)^{1/3} \tag{11}$$

where Δh_i is the effective elastic thickness of the layer *i*. According to the Equation (11), *Te* is less than the total thickness of the competent layers in case of decoupling.

For the coupled rheology, the crust and mantle are mechanically "welded" together, and the upper limit of *Te* represents simply a sum of all competent layers:

$$T_e^{(n)} = \left(\sum_{i=1}^{n} \Delta h_i \right) \tag{12}$$

The thickness of each competent layer (Δh_i) is defined as the depth at which the yielding strength is below 1–5% of the lithostatic pressure or as the depth, at which the vertical yield stress gradient is less than 10–20 MPa/km. In the last case the values of the competent layers can be associated with a specific geotherm for each lithotype (e.g., ~750°C for olivine and ~350°C for quartzite). The two different definitions of the thickness of a competent layer provide the lower and upper bounds for the corresponding values of Δh_i (Cloetingh and Burov, 1996). Following the approach of Burov and Diament (1995), the *Te* distribution in the study area has been calculated using the second definition for the mechanically strong layers. For this purpose the pressure scaled minimum yield strength of 10 MPa/km has been adopted. Therefore, when the strength decreases below this threshold the layers are decoupled, while they are welded in the opposite case. The coupling and decoupling conditions and the elastic thickness distribution are shown in Figs. 12 and 13. In order to demonstrate different

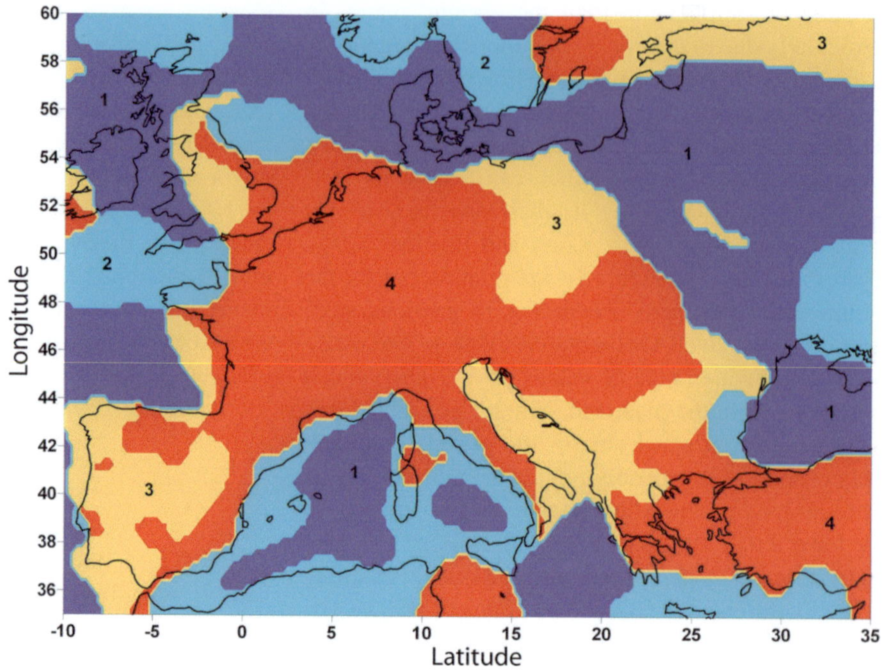

Fig. 12 Coupling and decoupling conditions of the European lithosphere. Numbers are as follows: *1*, Crustal layers and mantle lithosphere coupled; *2*, Crustal layers coupled and mantle lithosphere decoupled; *3*, Crustal layer decoupled and mantle lithosphere coupled; *4*, Crustal layers and mantle lithosphere decoupled

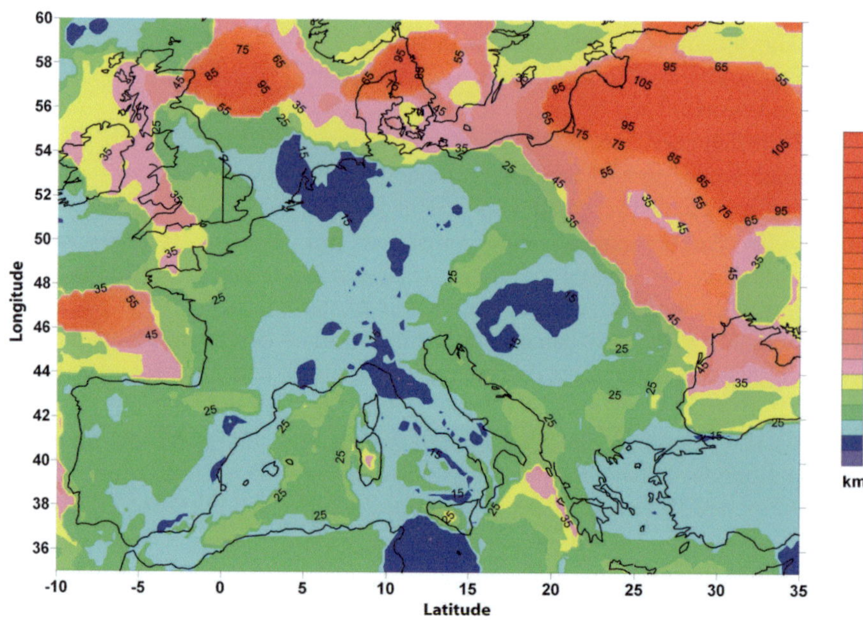

Fig. 13 Effective elastic thickness (T_e) of the European lithosphere as determined from the integrated strength of the lithosphere (km)

contributions to the total *Te* value, thicknesses of each competent layer of the lithosphere corresponding to the mechanically strong upper crust (MSUC), lower crust (MSLC) and mantle (MSL) are displayed in Fig. 14a–c.

Local studies of *Te* in Europe (e.g., Poudjom Djomani et al., 1999) have demonstrated that the largest changes of *Te* occur at the sutures that separate different provinces characterized by major changes in the lithospheric strength. *Te* is generally consistent with other physical properties of the lithosphere: high *Te* regions correspond to cold areas having large thermal thickness and fast seismic velocities and vice versa. In agreement with these considerations, the

Fig. 14 **(a)** Thickness (km) of the mechanically strong upper crust (MSUC). **(b)** Thickness (km) of the mechanically strong lower crust (MSLC). **(c)** Thickness (km) of the mechanically strong upper mantle (MSL)

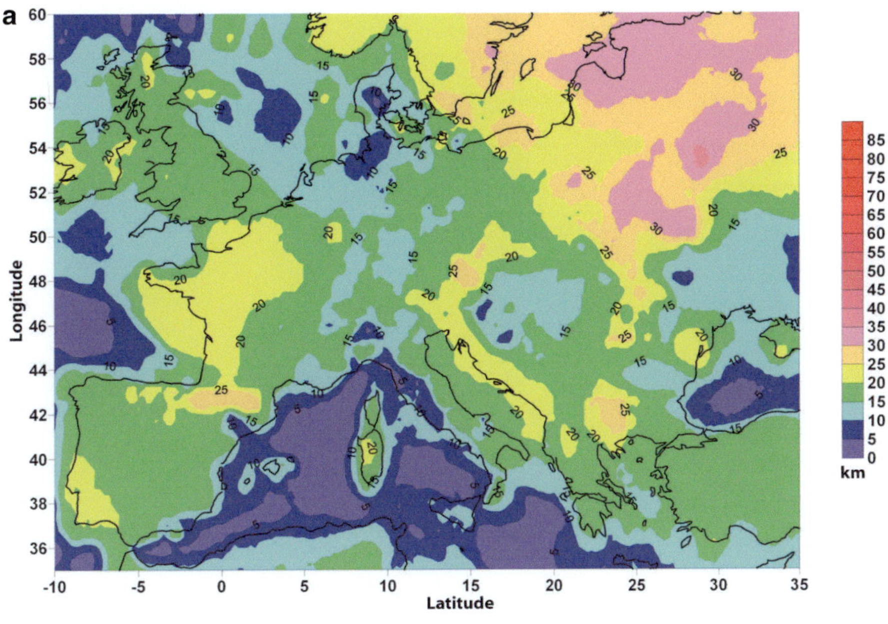

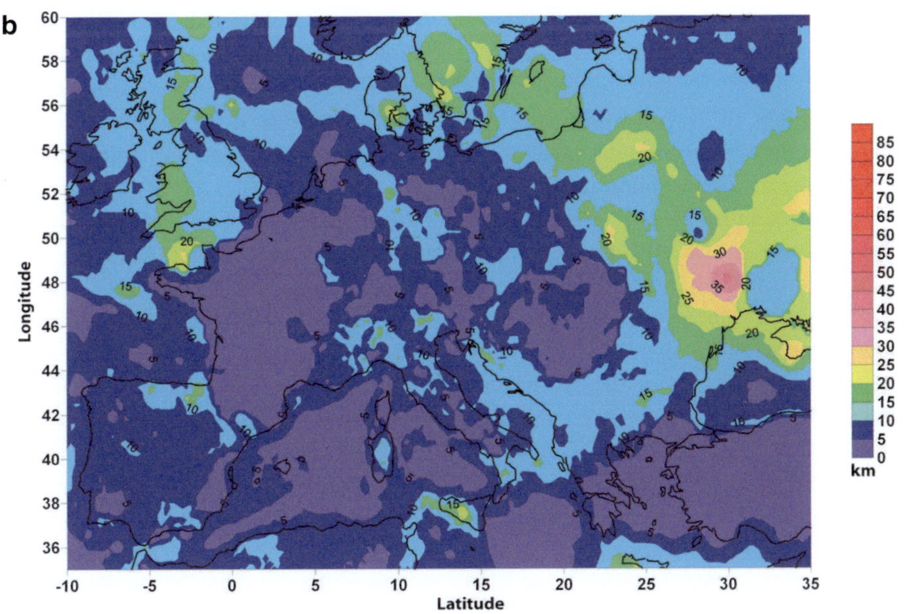

Fig. 14 (continued)

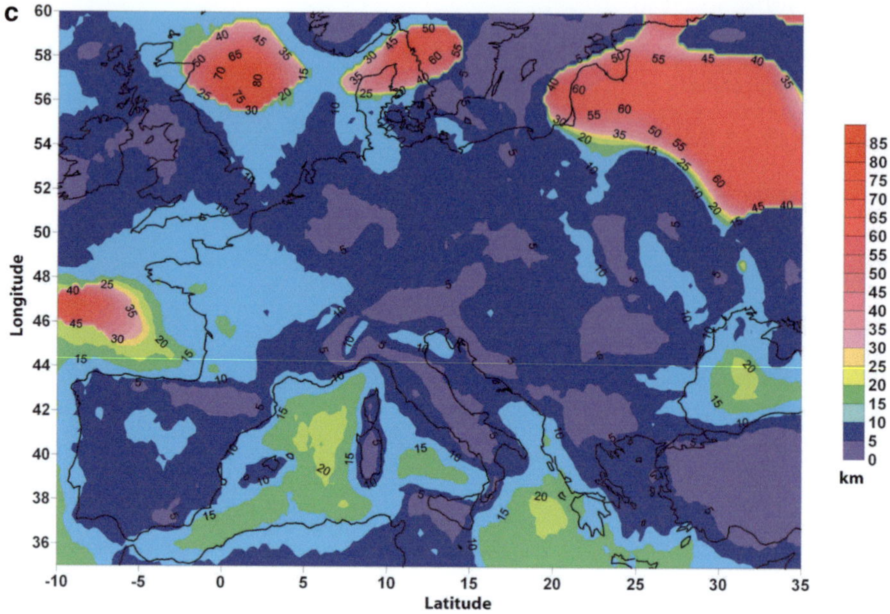

new results show a good correspondence between the distribution of *Te* values and the geological features, with the sharp decrease of *Te* west to the TESZ (<30 km). In most part of the EEP, which is characterized by high crustal and lithospheric thickness and a low thermal gradient, both crust and mantle layers are coupled (Fig. 8, points **A**, and **B** and Fig. 12), as it might be expected from its age (>750 Ma). In this area the largest values of *Te* (up to 80–100 km) are observed (Fig. 13), mainly on account of the thick MSL (>60 km) (Fig. 14c). These conditions are also responsible for significant strength at subcrustal levels (Fig. 7a). Similar values of *Te* and MSL are also found in the North Sea and North German Basin, mostly due to the low thermal regime. However this part of the tomography model is not well-resolved.

In general, lateral heterogeneity of the crust has a strong effect on the coupling conditions: in most part of the areas characterized by the "soft" crustal lithology ("dry quartzite" and "wet diorite") the mantle and the crustal layers are decoupled (Fig. 8, points **H**, **K**, **L**, **M**, **N**, **O** and **Q** and Fig. 12). The exceptions are presented by the areas with a very thin crust. In these cases, also layers of very young (<10 Ma) and hot lithosphere (like the Tyrrhenian Sea) can be coupled (Fig. 8, point **I**) and result in *Te* of about 20 km, mostly due to the contribution of MSL (15–20 km) (Figs. 13 and 14c). On the other hand, the mantle and often the crustal layers are decoupled in the young lithosphere of the Variscan and Alpine domains characterized by average thermal conditions and mean/high crustal thickness (Figs. 12 and 8, points **G**, **H**, **J**, **K**, **L**, **M**, **N**, **O** and **P**). In these areas a strong reduction of the MSL (<10 km) is observed as the result of a decrease of the upper mantle strength (Fig. 14c). MSL values are even lower (<5 km) in the areas characterized by large crustal thickness and average/high thermal conditions (e.g., the orogens and the Anatolian Platform). Therefore, the low values of *Te* (<20 km) are found in the Massif Central, the ECRIS, the German Plain, in the Pannonian Basin and the Alps (Fig. 13). In these areas *Te* mostly depends on the MSUC values, which span from 15 to 25 km, while the contribution of the MSLC is negligible (mostly <10 km) (Fig. 14a–b). The MSUC demonstrates a quite heterogeneous distribution, with the highest values (>20 km) concentrated in the areas of the large crustal thickness (e.g., the Dinarides) and low thermal regime (e.g., the Armorican Massif and the Paris Basin) (Fig. 14a). By contrast, the values of MSLC higher than 10 km are observed only in the regions characterized by strong lower crustal rheology (Figs. 6 and 14b), on account of the high thermal gradient present at the Moho boundary in the most parts of Europe. In comparison to the previous study of Tesauro et al. (2007) a general correspondence of the new and old *Te* maps can be observed (e.g., the difference between eastern and western Europe). On the other hand, the difference is very strong in details. A more gradual transition from the low values of rigidity characterizing weak areas, such as the Apennines and the Pannonian Basin, to the higher ones found in the Adriatic Sea and the Bohemian Massif (20–35 km) was found (Fig. 13). The *Te* distribution is also in agreement with the results obtained by Pérez-Gussinyé and Watts (2005), which indicate that the *Te* of old tectonic provinces (>1.5 Ga) is significantly larger (>60 km) than their mean crustal thickness (~40 km).

In order to make a comparison between different methods, *Te* has also been estimated independently from the integrated strength as a function of age and crustal thickness ($Te_{\text{age/h}}$) following the approach of Burov and Diament (1995). The values obtained for the different tectonic provinces are shown together with the average *Te* estimates discussed before (Te_{strength}) in Table 3. It can be noticed that Te_{strength} is generally smaller than $Te_{\text{age/h}}$. In the Archean provinces both parameters are similar (about 70 km in Fennoscandia). The maximum difference is observed in the Sveco-Norwegian province, where $Te_{\text{age/h}}$ is about 80 km, while Te_{strength} is reduced to 44 km, on account of the decoupling between the crust and the mantle (Fig. 12). In the Proterozoic provinces, $Te_{\text{age/h}}$ is remarkably higher than the values estimated by Te_{strength} (55–60 km versus 22–35 km). However, in these areas repeated tectonic events could have modified the lithospheric thermal regime and thickness, also reducing the strength. By contrast, the tectonic provinces younger than approximately 85 Ma (the Alpine domain, the Atlantic Margin and the western Black Sea) are characterized by a Te_{strength}, which is significantly larger (20–34 km) than $Te_{\text{age/h}}$ (13–18 km). Therefore, the strength grew fast enough in these young provinces (especially in the western Black Sea) to increase *Te* over the theoretically expected values.

The observed differences confirm that *Te* in the continental areas is influenced by numerous tectonic processes besides thermal age. $Te_{\text{age/h}}$ is referred to the mean age and crustal thickness, which are averaged over large areas, and might be not a representative value due to the non-linear relationship between the parameters. One of the main uncertainties of Te_{strength} could arise from not considering the effect of the horizontal regional stresses, which may facilitate weakening of the lower crust with subsequent crust-mantle decoupling. Previous studies (e.g., Cloetingh and Burov, 1996) have shown that this parameter might have a strong effect on *Te*. The tectonic stresses of 200–500 MPa can decrease *Te* values of the mid-age lithosphere (400 Ma) by 15–20% and of the lithosphere younger than 200 Ma by 30%.

How Reliable are the Strength Estimates?

In the previous sections new maps of strength and effective elastic thickness for the European continent were presented. To assess the reliability of these new results is a major challenge, since all the input parameters used are affected by a margin of uncertainty. For the sake of simplicity, some parameters were taken as uniform (e.g., strain rate, pore fluid factor, mantle rheology), representing the "average conditions" which likely approach the "real conditions" of the study area. However, it should be noted that an increase of 10% in the strain rates can enhance the integrated strength with about 25% and might result in a mechanical coupling

Table 3 Comparison between average values of T_e (km) obtained from the integrated strength of the lithosphere and from the age and crustal thickness (km) data for different tectonic provinces of Europe

Region	Mean age	Mean crustal thickness (km)	Mean $Te_{\text{age/h}}$ (km)	Mean Te_{strength} (km)
Fennoscandia	2.225 Ga	44.4	70	66
Sarmatia	3.35 Ga	43.1	70	55
SvecoNorvegian	1.05 Ga	34.5	80	44
Caledonides	460 Ma	32.6	57	35
Variscides	340 Ma	31.9	55	22
Alpine domain	65 Ma	32.5	13	20
Western Black Sea	80.5 Ma	28.0	15	34
Moesian platform	550 Ma	35.7	60	25
Mediterranean	147.5 Ma	20.3	25	21
Atlantic margin	81.5 Ma	22.5	18	28

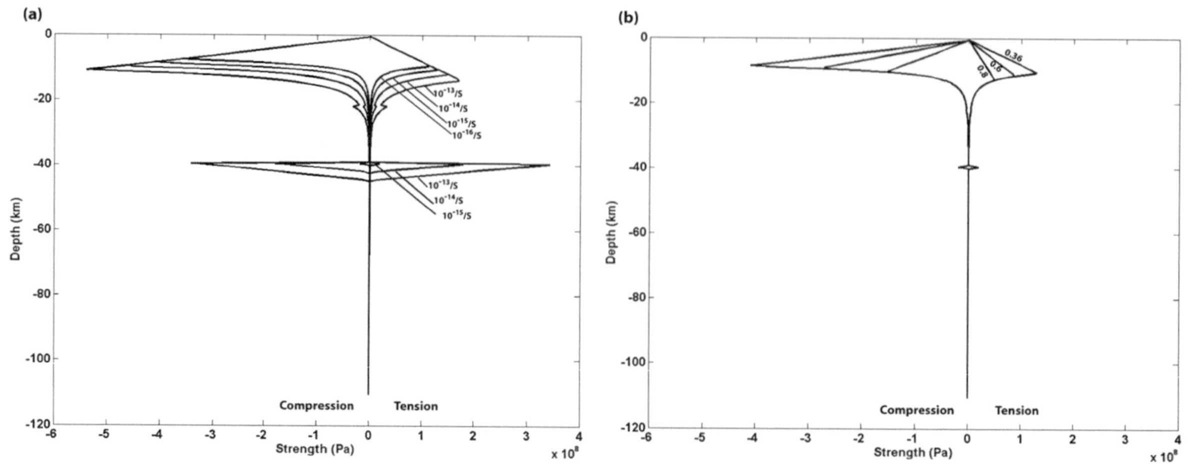

Fig. 15 Strength profiles for point V depicted in Fig. 6. For convention values estimated under compressional and extensional conditions are assumed negative and positive, respectively. (**a**) Strength profiles estimated adopting different strain rates varying from $\dot{\varepsilon} = 10^{-13}$ to 10^{-16} s^{-1}. Mantle strength is reduced to zero with the lowest value of strain rates. Coupling between the upper and the lower crust appears with the highest value of strain rate. (**b**) Strength profiles estimated adopting a pore fluid factor of 0.36, 0.6 and 0.8, respectively

of the individual lithospheric layers, as demonstrated in Fig. 15a. At the same time, an increase of the pore fluid factor (to simulate the presence of super-hydrostatic pressures) to values of 0.6 and 0.8 might lead to a decrease of the integrated strength over 25 and 50%, respectively (Fig. 15b).

On the other hand, the knowledge of possible deviations of the above mentioned parameters from the adopted standard values is limited. Therefore, this section primarily discusses the effects of the other parameters assumed to be laterally variable (e.g., temperature) on strength estimates. To this purpose, the EEC was selected as a sample area to test for some specific points the effect of differences in temperature, crustal thickness and most common rheologies adopted in the study area on the strength and elastic thickness values (Fig. 6 and Table 4). As at temperatures in excess of 750°C the vertical yield stress gradient is strongly reduced (10–20 MPa km), only the thermal variations in the shallow part of the upper mantle (<100 km) influence the results. In addition, the effects of crustal temperature and upper and lower crustal thickness variation are minor and not discussed here.

From inspection of Table 4, it can be observed that points **A** and **S**, having the same rheology and crustal thickness, show similar values of integrated crustal strength and coupling of the crustal and mantle layer. On the other hand, the integrated lithospheric strength in point **S** compared to point **A** is more than 50% higher, leading to a similar difference in the crustal strength proportion and in *Te*. These variations, which result in a different mechanical behaviour of the litho-

Table 4 Strength and elastic thickness values in selected points displayed in Fig. 6 for various crustal rheologies, lithospheric temperatures and crustal and lithospheric thickness estimations. Abbreviations are as follows: Lon, Longitude (decimal degree); Lat, Latitude (decimal degree), Str_L, Integrated strength of the lithosphere (Pa m); Str_C, Integrated strength of the crust (Pa m); % Str_C, Crustal Strength proportion; T_e, Effective elastic thickness (km); Lith. Code, Lithology Code (see Fig. 6 for explanations); Cp. Code, Coupling and decoupling conditions (see Fig. 12 for explanations); T_{60}, Temperature at 60 km (C°); T_{100}, Temperature at 100 km (C°); Moho, Moho depth (km); LAB, lithosphere-asthenosphere boundary depth (km)

Point	Lon	Lat	Str_L	Str_C	%Str_C	T_e	Lith.Code	Cp.Code	T_{60}	T_{100}	Moho	LAB
A	21.5	54	2.41E+13	2.00E+13	83%	50	1	1	763	921	43.3	171
C	32	47.5	1.01E+13	6.10E+12	61%	25.8	5	2	798	930	37.4	158.5
D	34.25	50	6.27E+12	5.44E+12	87%	33	6	2	767	942	40.9	200
S	26	56.5	5.46E+13	2.44E+13	41%	105	1	1	550	750	43.2	226
T	33.5	47	5.71E+12	4.63E+12	81%	25.6	5	2	837	967	43.6	155
U	29.75	46.75	2.30E+13	2.30E+13	89%	51	4	1	808	973	42.9	163.5

sphere ("crème brûlée" in point **A** and "jelly sandwich" in point **S**, respectively), are related to a decrease of temperature from point **A** to point **S** of about 200 and 170°C at 60 and 100 km, respectively. At the same time, points **C** and **T**, being close to each other, having the same crustal rheology and characterized by small differences in temperature (<40°C), display a variation in the integrated lithospheric strength similar to points **A** and **S**, which result in a smaller difference in the crustal strength proportion (~25%), almost equal values of *Te* (~25 km) and crust-mantle decoupling. Such a variation in lithospheric integrated strength can be mainly attributed to a difference in the crustal thickness in the two points of ~6 km. Furthermore, it can be observed how mostly the presence of a softer crustal rheology adopted in point **C** with respect to point **A** causes decoupling of the mantle from the crust and consequently a 50% decrease of *Te*. The importance of the crustal rheology for the strength calculation is evident from a comparison of points **U** and **T**, which are characterized by very similar temperatures and crustal thickness, but a different rheology of the lower crust (Fig. 6 and Table 4). The variation of the latter parameter produces in point **U** an increase close to 80% in both the integrated lithospheric and crustal strength and about 50% in *Te*. In addition, the stronger lower crust in point **U** determines coupling of all the lithospheric layers, which results in a higher *Te* (about 50%) with respect to point **T**. Finally, the differences in the strength and elastic thickness estimates in points **A** and **D** are similar to those observed in points **U** and **T**. However, the first two points considered differ for the rheological model employed in the upper and lower crust as well (Fig. 6 and Table 4). Therefore, the use of a much softer lower crust rheology in point **T** compared to point **U** has an effect similar to the replacement of a moderately softer upper and lower crust rheology in point **D** with respect to point **A** (Fig. 6, Tables 2 and 4). On the other hand, points **T** and **D**, having similar thermal conditions and crustal thickness, show comparable values of strength and elastic thickness, since the different rheological model adopted for the lower crust has only a marginal effect.

Conclusions

A new thermal model of the European lithosphere has been constructed based on the improved tomography model of Koulakov et al. (2009), which is corrected a-priori for the crustal effect using EuCRUST-07 (Tesauro et al., this volume). Although the computational approach is similar to previous studies (e.g., Goes et al., 2000), the employment of the more robust tomography model increases reliability of the results. The absolute values of the tomography model have been corrected by applying the 1D reference model, which corresponds to specific tectonic settings of the study area. The employment of the new regional reference model makes it possible to obtain a consistent temperature distribution in the mantle, which is extrapolated then to the surface.

The obtained results demonstrate a temperature range in the uppermost mantle of about 550–750°C in Eastern Europe and of 900–1,100°C in Western Europe. A sharp change of the temperature occurs beneath the TTZ and extends to the deeper layers of the upper mantle. The hottest areas correspond to the basins that experienced recent back-arc extension (e.g., Tyrrhenian Sea and Pannonian Basin). Low temperatures are found beneath the Pyrenees, the Alps and the Dinarides-Hellenic arc, likely due to a presence of deep lithospheric roots and subducted slabs. The new temperature estimates are used to trace the lithosphere-astenosphere thermal boundary, as a depth of the isotherm of 1,200°C. The lithospheric thickness is less than 100 km beneath the ECRIS and the hot basins (e.g., Tyrrhenian Sea), while the maximum values are observed beneath the East European Platform (200–230 km), the Alps and the Dinarides-Hellenic Arc (150–180 km).

The new strength maps of the European lithosphere are obtained employing the new thermal model and the compositional model derived from EuCRUST-07, which implies lateral variations of lithology and density. High strength values are found in areas with an average/low thermal regime and strong crustal rheology (the EEP, the North German Basin and the Bohemian Massif). Weak zones correspond to areas affected by Tertiary volcanism and mantle plumes, such as the ECRIS and the Massif Central. Both the integrated strength of the lithosphere and of the crust demonstrate similar trends in most parts of the study area. One of the surprising results is the high contribution of the crustal strength (50% of the integrated strength for the whole lithosphere) in a large part (~60%) of the study area. In particular, the regions with large crustal thickness (e.g., the Anatolian Plateau) are characterized by a high proportion of the crustal strength, which contribution is much larger (>80%) than of the mantle lithosphere.

The obtained strength maps are used to calculate the elastic thickness of the European lithosphere. Western Europe is mostly characterized by decoupled lithospheric layers and lower values of the calculated strength and T_e (<30 km). The contribution of the thickness of the mechanical strong lithosphere to T_e is generally low (<10 km) in western Europe. By contrast, the lithosphere of eastern Europe shows high values of T_e (80–100 km) and coupled layers. No straightforward relationship between T_e and thermal age is found in the continental part of the study area: in the tectonic provinces older than 85 Ma T_e values are significantly smaller than the theoretically predicted as a function of age and crustal thickness, while the opposite is true for the younger provinces.

Concerning the effect of the main input parameter variations on the strength estimates, it has been demonstrated that a relative high lateral variation of temperature and crustal thickness (likely above the margin of uncertainty) can produce up to 50% difference in integrated strength and elastic thickness values. The effect of the crustal rheology is more controversial: depending on the specific rheology adopted, the difference in the results obtained can be negligible or pronounced (up to 80%), especially in case the crustal strength proportion is high.

Acknowledgments We are grateful to Forough Sodoudi and Rainer Kind (GFZ, Potsdam), for providing receiver functions data. Wim Spakman (Department of Earth Science, Utrecht) is thanked for supplying seismic tomography data.

Funds were kindly provided by NWO (Netherlands Organization for Scientific Research), SRON (Space Research Organization Netherlands) and DFG (German Research Foundation) RO-2330/4–1.

Appendix: Anharmonicity Correction

For shallow depths (pressure up to about 6GPa), density ρ, compressibility K and rigidity μ can be computed at given (P,T) condition from their values at the reference state (P_0, T_0) using the infinitesimal strain approximation:

$$M(P,T) = M(P_0,T_0) + (T - T_0)\frac{\partial M}{\partial T} + (P - P_0)\frac{\partial M}{\partial T} \tag{13}$$

where M stands for either of the elastic parameters K or μ.

For density the equation is based on the definitions of α (thermal expansion) and K (compressibility):

$$\rho(P,T) = \rho(P_0,T_0)\left[1 - \alpha_0 (T - T_0) + (P - P_0)/K\right] \tag{14}$$

Voigt-Reuss-Hill (VRH) Averaging

The Voigt-Reuss-Hill (VRH) averaging scheme approximates the parameters for a combination of minerals by taking the average of the mean elastic parameters for a constant stress (Reuss) and a constant strain (Voigt) situation:

$$<\rho> = \sum \lambda_i \rho_i \tag{15}$$

$$<M> = \frac{1}{2}\left(M^{voigt} + M^{reuss}\right) \tag{16}$$

$$M^{reuss} = \left(\sum \frac{\lambda_i}{M_i}\right)^{-1} \quad M^{voigt} = \sum \lambda_i M_i \tag{17}$$

where <M> = μ (for S-wave velocities) or K+(4/3)μ (for P-wave velocities), M_i = μ or K and λ_i is the volumetric proportion of mineral i.

To calculate partial derivatives of seismic velocity V, the equations for the VRH averaging scheme is the following expression:

$$\left(\frac{\partial V}{\partial T}\right)_{anh} = \left[\frac{\partial <M>}{\partial T} - V^2\frac{\partial <\rho>}{\partial T}\right] \Big/ [2<\rho>V] \tag{18}$$

where:

$$\frac{\partial <\rho>}{\partial T} = \sum \lambda_i \frac{\partial \rho_i}{\partial T} \text{ and } \frac{\partial <M>}{\partial T} = \sum \lambda_i \frac{\partial M_i}{\partial T} + \left(M^{reuss}\right)^{-2} \sum \frac{\lambda_i \partial M_i}{M_i^2 \partial T} \tag{19}$$

References

Afonso, J.C., Ranalli, G.: Crustal and mantle strengths in continental lithosphere: is the jelly sandwich model obsolete?. Tectonophysics **394**, 221– 232 (2004)

Anderson, D.L.: Theory of the Earth. Blackwell, Boston, pp. 360 (1989)

Anderson, O.L.: Equation of State of Solids for Geophysics and Ceramic Science. Oxford University Press, New York, pp. 405 (1995)
Anderson, D.L., Given, J.W.: Absorption band Q model for the Earth. J. Geophys. Res. **87**, 3893–3904 (1982)
Arlitt, R.: Teleseismic body wave tomography across the Trans-European Suture Zone between Sweden and Denmark. Ph.D. Thesis of ETH Zurich No. 13501, 110 pp (1999)
Artemieva, I., Mooney, W.: Thermal thickness and evolution of Precambrian lithosphere: a global study. J. Geophys. Res. **106**, 16387–16416 (2001)
Babuška, V., Plomerová, J.: European mantle lithosphere assembled from rigid microplates with inherited seismic anisotropy. Phys. Earth Planet. Int. **158**, 264–280 (2006)
Banks, R.J., Parker, R.L., Huestis, S.P.: Isostatic compensation on a continental scale: Local versus regional mechanisms. Geophys. J. R. Astron. Soc. **51**, 431–452 (1977)
Barrell, J.: The strength of the Earth's crust. Part I: Geologic tests of the limits of the strength. J. Geol. **22**, 28–48 (1914)
Bass, J.D., Anderson, D.L.: Composition of the upper-mantle: geophysical test of two petrological models. Geophys. Res. Lett. **11**, 237–240 (1984)
Bassin, C., Laske, G., Masters, G.: The Current Limits of Resolution for Surface Wave Tomography in North America. EOS Trans. AGU **81**, F897 (2000)
Bijwaard, H., Spakman, W.: Non-linear global P-wave tomography by iterated linearized inversion. Geophys. J. Int. **141**, 71– 82 (2000)
Birch, F.: Elasticity of igneous rocks at temperature and pressures. Geol. Soc. Am. Bull. **54**, 263–286 (1943)
Brace, W.F., Kohlstedt, D.L.: Limits on lithospheric stress imposed by laboratory experiments. J. Geophys. Res. **94**, 3967–3990 (1980)
Burov, E.B., Diament, M.: The effective elastic thickness of (Te) continental lithosphere. What does it really means?. J. Geophys. Res. **100**, (B3), 3905–3927 (1995)
Burov, E.B., Lobkovsky, L.I., Cloetingh, S., Nikishin, S.: Continental lithosphere folding in Central Asia (part 2), Constraints from gravity and topography. Tectonophysics **226**, 73–87 (1993)
Burov, E.B., Watts, A.B.: The long-term strength of continental lithosphere: "jelly sandwich" or "crème brûlée"?. GSA Today **16**, 1, 4–10 (2006)
Byerlee, J.D.: Friction of rocks. Pure Appl. Geophys. **116**, 615–626 (1978)
Calcagnile, G., Panza, G.F.: Crustal and upper mantle structure of the Mediterranean area derived from surface-wave data. Phys. Earth Planet. Int. **60**, 163–168 (1990)
Cammarano, F., Goes, S., Vacher, P., Giardini, D.: Inferring upper-mantle temperatures from seismic velocities. Phys. Earth Planet. Int. **138**, 197–222 (2003)
Carter, N.L., Tsenn, M.C.: Flow properties of continental lithosphere. Tectonophysics **136**, 27–63 (1987)
Cermak, V.: Lithospheric thermal regimes in Europe. Phys. Earth Planet. Int. **79**, 179–193 (1993)
Chapman, D.S.: Thermal gradients in the continental crust, in: The Nature of the Lower Continental Crust. In: J.B. Dawson et al. (eds.), pp. 63–70, Geol. Soc., London (1986)
Christensen, N.I., Mooney, W.D. : Seismic velocity structure and composition of the continental crust: A global view. . J. Geophys. Res. **100**, 9761–9788 (1995)
Cloetingh, S., Burov, E.B.: Thermomechanical structure of European continental lithosphere: constraints from rheological profiles and EET estimates. Geophys. J. Int. **124**, 695–723 (1996)
Cloetingh, S., Ziegler, P.A., Beekman, F., Andriessen, P.A.M., Maţenco, L., Bada, G., Garcia-Castellanos, D., Hardebol, N., Dèzes, P., Sokoutis, D.: Lithospheric memory, state of stress and rheology: neotectonic controls on Europe's intraplate continental topography. Quat. Sci. Rev. **24**, 241–304 (2005)
Cotte, N., Pedersen, H.A. and TOR Working Group: Sharp contrast in lithospheric structure across the Sorgenfrei–Tornquist Zone as inferred by Rayleigh wave analysis of TOR1 project data. Tectonophysics **360**, 75–88 (2002)
Deschamps, F., Trampert, J., Snieder, R.: Anomalies of temperatures and iron in the uppermost mantle inferred from gravity data and tomographic models. Phys. Earth Planet. Int. **129**, 245–264 (2002)
Fernàndez, M., Marzám, I., Correia, A., Ramalho, E.: Heat flow, heat production, and lithospheric thermal regime in the Iberian Peninsula. Tectonophysics **291**, 29–53 (1998)
Fernàndez, M., Ranalli, G.: The role of rheology in extensional basin formation modeling. Tectonophysics **282**, 129–145 (1997)
Goes, S., Govers, R., Vacher, P.: Shallow mantle temperatures under Europe from P and S wave tomography.. J. Geophys. Res. **105**, (B5), 11153–11169 (2000)
Goetze, C., Evans, B.: Stress and temperature in the bending lithosphere as constrained by experimental rock mechanics. Geophys. J. R. Astron. Soc. **59**, 463–478 (1979)
Grünthal, G., Stromeyer, D.: The Recent crustal stress field in Central Europe, trajectories and finite element modelling. J. Geophys. Res. **97**, 11805–11820 (1992)
Gudmundsson, O., Sambridge, M.: A regionalized upper mantle (RUM) seismic model. J. Geophys. Res. **103**, B4, 7121–7136 (1998)
Hill, R.: Elastic properties of reinforced solids: Some theoretical principles. J. Mech. Phys. Solids **11**, 357–372 (1963)
Hirschmann, M.M.: Mantle solidus: experimental constraints and the effects of peridotite composition. Geochem. Geophys. Geosyst. **1**, paper no. 2000GC000070 (2000)
Hughes, D.S., Cross, J.H.: Elastic wave velocities at high pressures and temperatures. Geophysics **16**, 577–593 (1951)
Hurtig, E., Cermak, V., Haenel, R. and Zui, V. (Eds.), 1992. Geothermal Atlas of Europe, International Association for Seismology and Physics of the Earth's Interior, 156 pp, Hermann Haack Verlagsgesellschaft mbH -Geographisch -Kartographische Anstalt Gotha.
Irifune, T., Ringwood, A.E.: Phase transformations in a harzburgite composition to 26 GPa: implications for dynamical behaviour of the subducting slab. Earth Planet. Sci. Lett. **86**, 365–376 (1987)
Isacks, B.L., Oliver, J., Sykes, L.R.: Seismology and the new global tectonics.. J. Geophys. Res. **73**, 5855–5899 (1968)
Jackson, J.: Strength of the continental lithosphere: Time to abandon the jelly sandwich?. GSA Today **12**, 410 (2002)
Jordan, T.H.: Composition and development of the continental tectosphere. Nature **274**, 544–548 (1978)
Jordan, T.H.: Mineralogies, densities and seismic velocities of garnet lherzolites and their geophysical implications, in The Mantle Sample: Inclusions in Kimberlites and Other

Volcanos, ed. By F.R. Boyd and H.O.A. Myer, pp. 1–14, AGU Washington, D.C. (1979)
Kaban, M.K., Schwintzer, P., Artemieva, I.M., Mooney, W.D.: Density of the continental roots: compositional and thermal contributions. Earth Planet. Sci. Lett. **209**, 53–69 (2003)
Kaban, M.K., Tesauro, M., Cloetingh, S.A.P.L.: A new gravity model of the crust and upper mantle of Europe.. Earth Planet. Sci. Lett., (2009) (under review)
Karato, S.I.: Importance of anelasticity in the interpretation of seismic tomography. Geophys. Res. Lett. **20**, 1623–1626 (1993)
Karato, S., Jung, H.: Water, partial melting and the origin of the seismic low velocity and high attenuation zone in the upper-mantle. Earth Planet. Sci. Lett. **157**, 193–207 (1998)
Karato, S., Spetzler, H.A.: Defect microdynamics in minerals and solid-state mechanisms of seismic wave attenuation and velocity dispersion in the mantle. Rev. Geophys. **28**, 399–421 (1990)
Kennett, B.L.N., Engdahl, E.R., Buland, R.: Constraints on seismic velocities in the Earth from traveltimes. Geophys. J. Int. **122**, 108–124 (1995)
Kohlstedt, D.L., Evans, B., Mackwell, S.J.: Strength of the lithosphere: Constraints imposed by laboratory experiments. J. Geophys. Res. **100**, 17587–17602 (1995)
Korja, T., Hjelt, S.-E., Kaikkonen, P., Kozlovskaya, E., Lahti, I., Pajunpiaia, K., Pulkkinen, A., Viljanen, A. BEARWorking Group: Crustal and upper mantle beneath Fennoscandia as imaged by the Baltic Electromagnetic Array Research (BEAR), In: Lahtinen, R., Korja, A., Arhe, K., Eklund, O., Hjelt, S.E., Pesonen, L.J. (Eds.),, Second Symposium on the Structure, Composition and Evolution of the Lithosphere in Finland. Institute of Seismology, University of Helsinki, pp. 41–48, (Report S-42) (2002)
Koulakov, I., Kaban, M.K., Tesauro, M., Cloetingh, S.: P and S velocity anomalies in the upper mantle beneath Europe from tomographic inversion of ISC data.. Geophys. J. Int., (2009) (in press)
Kruse, S., Royden, L.: Bending and unbending of an elastic lithosphere: The Cenozoic history of the Apennine and Dinaride foredeep basins. Tectonics **13**, 278–302 (1994)
Lankreijer, A.C., 1998. Rheology and basement control on extensional basin evolution in Central and Eastern Europe: Variscan and Alpine-Carpathian-Pannonian tectonics. Ph.D. Thesis, Vrije Universiteit, Amsterdam, pp. 158.
Lenkey, L., 1999. Geothermics of the Pannonian basin and its bearing on the tectonics of Basin evolution. PhD Thesis, Vrije Universiteit, Amsterdam, pp. 215.
Leven, J.H., Jackson, I., Ringwood, A.E.: Upper mantle seismic anisotropy and lithospheric decoupling. Nature **289**, 235–239 (1981)
Mackwell, S.J., Zimmerman, M.E., Kohlstedt, D.L.: High-temperature deformation of dry diabase with applications to tectonics on Venus. J. Geophys. Res. **103**, 975–984 (1998)
Maggi, A., Jackson, J.A., Priestley, K., Backer, C.: A re-assesement of focal depth distribution in southern Iran , the Tien Shan and northern India: Do earthquakes occur in the continental mantle? Geophys. J. Int. **143**, 629–661 (2000)
Martin, M., Ritter, J.R.R. & the CALIXTO working group: High-resolution teleseismic body wave tomography beneath SE-Romania – II. Imaging of a slab detachment scenario. Geophys. J. Int. **164**, 579–595 (2006)
Mavko, G.M.: Velocity and attenuation in partially molten rocks. J. Geophys. Res. **85**, 5173–5189 (1980)
Minster, J.B., Anderson, D.L.: A model of dislocation-controlled for the mantle. Philos. Trans. R. Soc. London **299**, 319–356 (1981)
Moisio, K., Kaikkonen, P., Beekman, F.: Rheological structure and dynamical response of the DSS profile BALTIC in the SE Fennoscandian shield. Tectonophysics **320**, 175–194 (2000)
Mooney, W.D., Laske, G., Masters, T.G.: CRUST 5.1: A global crustal model at 5°X5°. J. Geophys. Res. **103**, 727–747 (1998)
Monsalve, G., Sheehan, A., Shulte-Pelkum, V., Rajaure, S., Pandey, M.R., Wu, F.: Seismicity and one-dimensional velocity structure of the Hymalayan collision zone: Earthquakes in the crust and upper mantle. J. Geophys. Res. **111**, (doi: 10.1029/2005JB004062) (2006)
Nolet, G., Grand, S.P., Kennett, B.L.N.: Seismic heterogeneity in the upper mantle. J. Geophys. Res. **99**, 23753–23766 (1994)
Nolet, G., Zielhuis, A.: Low S-velocities under the Tornquist–Teisseyre zone—evidence for water injection into the transition zone by subduction. J. Geophys. Res. **99**, 15813–15820 (1994)
Panza, G.F., Raykova, R.B.: Structure and rheology of lithosphere in Italy and surrounding. Terra Nova **20**, 194–199 (2008)
Pavlenkova, G.A., Pavlenkova, N.I.: Results of Joint Processing of Data on Nuclear and Chemical Explosions Recorded on the Long-Range Quartz Profile (Murmansk–Kyzyl). Phys. S. Earth **44**, 4, 316–326 (2008)
Pérez-Gussinyé, M., Watts, A.B.: The long-term strength of Europe and its implications for plate-forming processes. Nature **436** (2005), doi:10.1038/nature03854
Pollack, H.N., Chapman, D.S.: On the regional variation of heat flow, geotherms and lithospheric thickness. Tectonophysics **38**, 279–296 (1977)
Pollack, H.N., Hurter, S.J., Johnson, J.R.: Heat flow from the Earth's interior: Analysis of the global data set. Rev. Geophys. **31**, 267–280 (1993)
Popp, T., Kern, H.: Thermal hydration reaction characterized by combined measurements of electrical conductivity and elastic wave velocities. Earth Planet. Sci. Lett. **120**, 43–57 (1993)
Plomerová, J., Kouba, D., Babuška, V.: Mapping the lithosphere–asthenosphere boundary (LAB) through changes in surface-wave anisotropy. Tectonophysics **358**, 175–185 (2002)
Poudjom Djomani, Y.H., Fairhead, J.D., Griffin, , W.L.: The flexural rigidity of Fennoscandia: Reflection of the tectonothermal age of the lithospheric mantle. Earth Planet. Sci. Lett. **174**, 139–154 (1999)
Praus, O., Pčová, J., Petr, V., Babuška, V., Plomerová, J.: Magnetotelluric and seismological determination of the lithosphere asthenosphere transition in Central Europe. Phys. Earth Planet. Int. **60**, 212–228 (1990)
Ranalli, G.: Rheology of the crust and its role in tectonic reactivation. J. Geodyn. **30**, 3–15 (2000)
Ranalli, G., Murphy, D.C.: Rheological stratification of the lithosphere. Tectonophysics **132**, 281–295 (1987)
Sato, H., Sacks, I.S., Murase, T.: The use of laboratory data for estimating temperature and partial melt fraction in the low-velocity zone: Comparison with heat flow and electrical conductivity studies. . J. Geophys. Res. **94**, 5689–5704 (1989)

Schmeling, H.: Numerical models on the influence of partial melt on elastic, anelastic and electric properties of rocks, part 1, Elasticity and anelasticity. Phys. Earth Planet. Inter. **41**, 34– 57 (1985)

Silver, P.: Seismic anisotropy beneath the continents: Probing the depths of geology. Ann. Rev. Earth Planet. Sci. **2**, 385–432 (1996)

Sobolev, S.V., Babeyko, A.Y.: Modeling of mineralogical composition, density and elastic wave velocities in anhydrous magmatic rocks. Surv. Geophys. **15**, 515–544 (1994)

Sobolev, S.V., Zeyen, H., Granet, M., Achauer, U., Bauer, C., Werling, F., Altherr, R., Fuchs, K.: Upper mantle temperatures and lithosphere–asthenosphere system beneath the French Massif Central constrained by seismic, gravity, petrologic and thermal observations. Tectonophysics **275**, 143–164 (1997)

Sobolev, S.V., Zeyen, H., Stoll, G., Werling, F., Altherr, R., Fuchs, K.: Upper-mantle temperatures from teleseismic tomography of French massif central including effects of composition, mineral reactions, anharmonicity, anelasticity and partial melt. Earth Planet. Sci. Lett. **139**, 147–163 (1996)

Sodoudi, F., Kind, R., Hatzfeld, D., Priestley, K., Hanka, W., Wylegalla, K., Stavrakakis, G., Vafidis, A., Harjes, H.-P., Bohnhof, M.: Lithospheric structure of the Aegean obtained from P and S receiver functions. J. Geophys. Res. **111**, (B12) (2006), doi:10.1029/2005JB003932

Sodoudi, F., Geissler, W.H., Kind, R.: The thickness of the European Lithosphere as seen by S receiver functions. EGU (poster presentation) (2008)

Tesauro, M., Kaban, M.K., Cloetingh, S.A.P.L., Hardebol, N.J., Beekman, F.: 3D strength and gravity anomalies of the European lithosphere. Earth Planet. Sci. Lett. **263**, 56–73 (2007)

Waldhauser, F., Lippitsch, R., Kissling, E., Ansorge, J.: High-resolution teleseismic tomography of upper mantle structure using an a priori 3D crustal model. Geophys. J. Int. **150**, 141–403 (2002)

Watts, A.B.: An analysis of isostasy in the world's oceans: 1. Hawaiian-Emperor Seamount Chain. J. Geophys. Res. **83**, 5989–6004 (1978)

Watts, A.B., Bodine, J.H., Steckler, M.S.: Observations of flexure and the state of stress in the oceanic lithosphere. J. Geophys. Res. **85**, 6369–6376 (1980)

Watts, A.B.: Isostasy and Flexure of the Lithosphere. Cambridge University Press,Cambridge 458 pp (2001)

Watts, A.B., Burov, E.B.: Lithospheric strength and its relationship to the elastic and seismogenic layer thickness. Earth Planet. Sci. Lett. **213**, 113–131 (2003)

Wilks, K.R., Carter, N.L.: Rheology of some continental lower crustal rocks. Tectonophysics **182**, 57– 77 (1990)

Zoback, M.L.: First- and Second-Order Patterns of Stress in the Lithosphere: The World tress Map Project. J. Geophys. Res. **97**, B8, 11703–11728 (1992)

Thermo-Mechanical Models for Coupled Lithosphere-Surface Processes: Applications to Continental Convergence and Mountain Building Processes

E. Burov

Abstract Simple mechanical considerations show that many tectonic-scale surface constructions, such as mountain ranges that exceed certain critical height (about 3 km in altitude, depending on rheology and width) should flatten and collapse within few My as a result of gravitational spreading that may be enhanced by flow in the ductile part of the crust. The elevated topography is also attacked by surface erosion that, in case of static topography, would lead to its exponential decay on a time scale of less than 2.5 My. However, in nature, mountains or rift flanks grow and stay as localized tectonic features over geologically important periods of time (>10 My). To explain the long-term persistence and localized growth of, in particular, mountain belts, a number of workers have emphasized the importance of dynamic feedbacks between surface processes and tectonic evolution. Surface processes modify topography and redistribute tectonically significant volumes of sedimentary material, which acts as vertical loading over large horizontal distances. This results in dynamic loading and unloading of the underlying crust and mantle lithosphere, whereas topographic contrasts are required to set up erosion and sedimentation processes. Tectonics therefore could be a forcing factor of surface processes and vice versa. One can suggest that the feedbacks between tectonic and surface processes are realised via 2 interdependent mechanisms: (1) slope, curvature and height dependence of the erosion/deposition rates; (2) surface load-dependent subsurface processes such as isostatic rebound and lateral ductile flow in the lower or intermediate crustal channel. Loading/unloading of the surface due to surface processes results in lateral pressure gradients, that, together with low viscosity of the ductile crust, may permit rapid relocation of the matter both in horizontal and vertical direction (upward/downward flow in the ductile crust). In this paper, we overview a number of coupled models of surface and tectonic processes, with a particular focus on 3 representative cases: (1) slow convergence and erosion rates (Western Alpes), (2) intermediate rates (Tien Shan, Central Asia), and (3) fast convergence and erosion rates (Himalaya, Central Asia).

Keywords Mountain building · Surface processes · Continental collision · Lithosphere rheology · Modeling

Introduction

Continental mountain belts, such as, for example, Tien Shan (Central Asia, Fig. 1), are characterized by highly localized topography elevations persistently growing over tens of millions years. The fact that gravitational potential energy per unit surface $g\rho h^2$ scales as h^2 implies that a thrust belt should grow more easily in width than in height (Molnar and Lyon-Caen, 1988, h is the mean topography elevation above sea level, ρ is density and g is acceleration due to gravity). A portion of continental crust submitted to quasi-static horizontal shortening should tend to thicken homogeneously. This can be treated another way round by considering that a mountain range results from thrusting on faults that cut

E. Burov (✉)
University Paris VI, Case 129, 4 Place Jussieu, Paris 75252, France
e-mail: evgenii.burov@upmc.fr

S. Cloetingh, J. Negendank (eds.), *New Frontiers in Integrated Solid Earth Sciences*, International Year of Planet Earth, DOI 10.1007/978-90-481-2737-5_4, © Springer Science+Business Media B.V. 2010

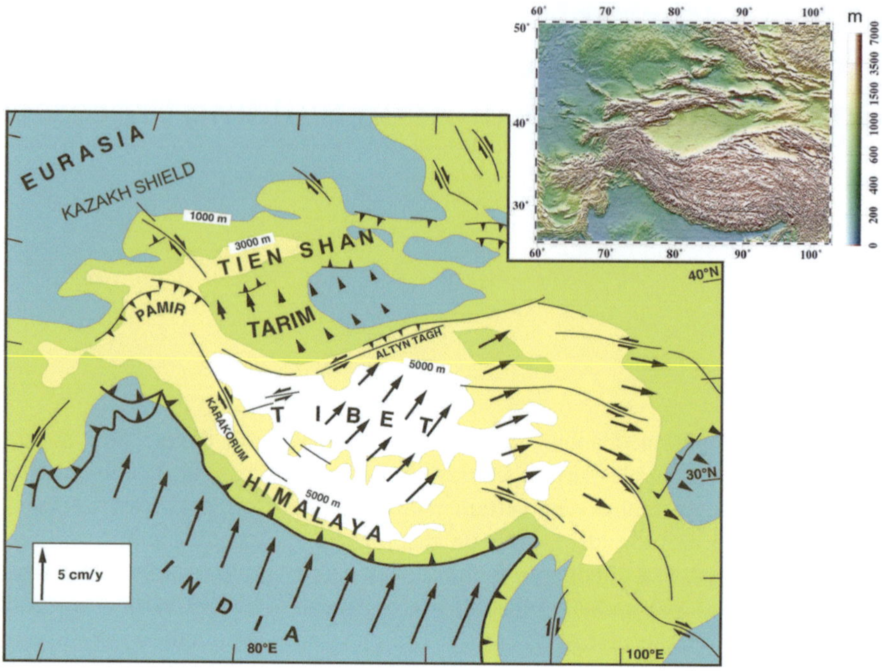

Fig. 1 Actively growing intercontinental belts and plateaux: an example showing a schematic map of India–Eurasia collision with its main features such as the Himalayan mountain belt, Tibetan plateau, Tarim basin, Pamir and Tien Shan mountain belt. Insert shows a digital elevation map of the same area. The topography peaks to 8,800 m in Hymalaya (Everest) and 7,500 m in Tien Shan (Pobeda Peak). Modified after (Avouac and Tapponier, 1993)

through the upper crust and root into the lower crust. Uplift of the range implies an increase in the vertical stress acting on the fault. This acts to oppose further frictional sliding on the fault, inhibiting further thrusting. A new fault will then form farther away from the range front leading to widening of the range. In addition, erosion and sedimentation at the surface, together with flow in the lower crust, should favor smoothing of topographic irregularities. At the pressure and temperature conditions of the lower crust, most crustal rocks are thought to flow easily at very low deviatoric stresses (e.g., Brace and Kohlstedt, 1980; Wang et al., 1994, Fig. 3). The deviatoric stresses associated with slopes of the topography and of the Moho (e.g., Fleitout and Froidevaux, 1982) should therefore be relaxed by viscoplastic flow in the ductile lower crust inducing decay of surface topography (Kusznir and Matthews, 1988; Gratton, 1989; Bird, 1991, Fig. 2).

The growth and maintenance of topographic features at the surface of continents might be taken to indicate that the strength of the crust exceeds the deviatoric stresses associated with slopes of both surface and Moho topography. Yet, as mentioned above, laboratory experiments indicate that at pressure and temperature conditions of lower crust in convergent zones, crustal rocks should be mostly ductile. Surface

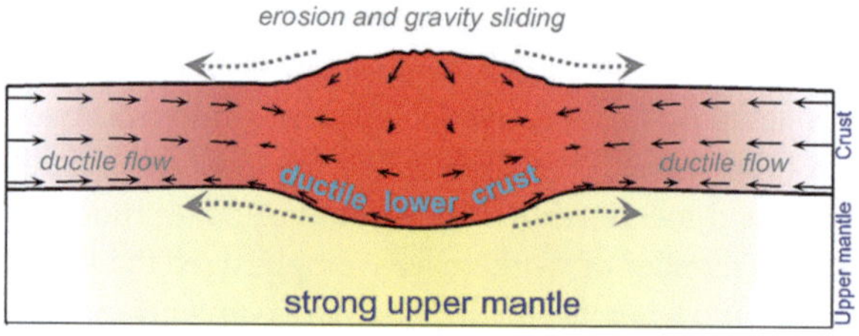

Fig. 2 Erosional and gravity collapse of a mountain range (e.g., Gratton, 1989; Bird, 1991). In the conceptual model shown here, there is no balance between surface and subsurface processes. Even if such range was created somehow, it will not persist, as its root and topography will be flattened in about 2 My in the absence of some compensating mechanisms. Tectonic convergence may not solely compensate this flattening; it may only grant an overall thickening of the crust; some additional localizing mechanisms are needed to concentrate thickening in a narrow range

and Moho topography undulations should therefore be relaxed by viscoplastic flow in the ductile lower crust and decay with time (Kusznir and Matthews, 1988; Gratton, 1989; Bird, 1991). Consider, for example, the Tien Shan range, which is, except for the Himalayas, the largest and most active intra-continental range in the world (Fig. 1). Tien Shan (traduced as "Heavenly Mountains") is 300–400 km wide in its central area, with a mean elevation of about 3,500 m and local peaks of up to 7,500 m, in a zone of relatively thick and

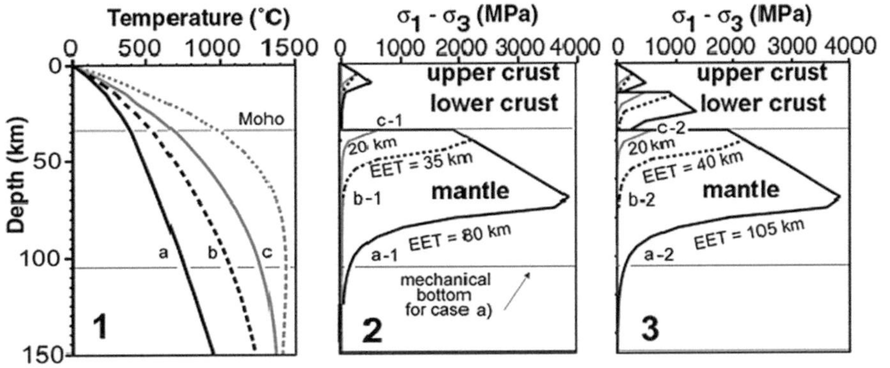

Fig. 3 Typical rheology profiles for continental lithosphere indicate the possibility for lower-crustal flow: (*1*) geotherms that yield YSEs shown in the middle and on the right; (*2*) yield stress envelope (YSE) for quartz-rich upper and lower crust and olivine mantle; (*3*) yield stress envelope (YSE) for quartz-rich upper crust, diabase lower crust, olivine mantle. EET – equivalent elastic thickness of the lithosphere computed for each of YSEs

tectonized crust (Moho depths from 50 to 70 km) (e.g., Avouac et al., 1993). The Tien Shan is a continuously growing range, that have started to rise 10–15 My ago. A simple dimensional analysis (Gratton, 1989) as well as numerical simulations (Bird, 1991; Avouac and Burov, 1996; Burov and Watts, 2006) show that topography of such range should, instead of growing, be reduced by half in a few My (Fig. 2). This estimate is based on the assumption of ductile rheology of the lower crust, which is supported for this area by multiple data starting from seismic data (Vinnik and Saipbekova, 1984; Makeyeva, 1992; Roecker et al., 1993; Vinnik et al., 2006) and ending by gravity-flexural analysis (Burov et al., 1990, 1993; Avouac and Burov, 1996). Only short topographic wavelengths, typically less than a few tens of kilometres that can be supported by the strength of the upper crust, would be maintained over geological periods of time, yet provided that they are not removed by erosion (which is faster on short wavelength topography). In addition, surface processes might be thought to contribute to an even more rapid smoothing of the topography. Similarly, in the absence of strong rheological heterogeneities or of strain localization processes, a portion of a continental crust submitted to horizontal shortening should tend to thicken homogeneously, so that no mountain should form. The growth and maintenance of an intracontinental mountain range over long periods of time must therefore involve dynamical processes allowing for long-term localisation of lithospheric strain below the mountain.

Several mechanisms have been advocated to explain localization of major thrust faults and, by proxy, stability of mountain belts. Intrinsic strain softening properties of rocks could sustain localized thrust faulting at the crustal scale. Alternatively, a range could result from shear stresses at base of the crust induced by lithospheric under-thrusting or by mantle dynamics (e.g., Beaumont et al., 1994; Ellis et al., 1995). Such a mechanism may be suggested for mountains associated with subduction zones or with hotspots (Vogt, 1991), but seems inappropriate to explain most intra-continental mountains. In case of, for example, Tien Shan mountain belt, a particular mantle dynamics has been inferred from gravity modelling (Burov et al., 1990, 1993) and seismic anisotropy (Makeyeva, 1992; Roecker et al., 1993), but we contend that it might not be the key factor because we believe that coupling between surface processes and flow in the lower crust could provide an alternative and more general explanation (Avouac and Burov, 1996; Burov and Cloetingh, 1997).

To explain the paradox of long-term mountain persistence and localized growth, a number of workers have emphasized the importance of dynamic feedbacks between surface processes and tectonic evolution (e.g., Molnar and England, 1990; Masek et al., 1994a; Avouac and Burov, 1996; Molnar, 2001). Indeed, surface processes modify the topography and redistribute tectonically significant volumes of sedimentary material (vertical, or normal loads) over large horizontal distances. This may result in dynamic loading and unloading of the underlying crust and mantle lithosphere, whereas topographic contrasts are required to set up erosion and sedimentation processes. Tectonics therefore could be a forcing factor of surface processes.

In this paper, we first review the existing models of surface processes and the thermo-mechanical properties of the lithosphere that condition its response to surface and tectonic loading-unloading. We then review our own and other previous modelling studies that show that surface and tectonic processes are not independent processes and can interact. We show in particular that advection of material at the Earth's surface and horizontal flow in the crust might be coupled so as to permit mountain growth in response to horizontal shortening. This mechanism is then validated and investigated on the basis of semi-analytical and numerical experiments in which the rheological layering of the lithosphere and surface processes are modelled. We then find that, depending of the erosion rate compared to horizontal shortening, flow in the lower crust can be "outward" (from under the high topography) or "inward" (toward the crustal root of a high topography). When inward flow occurs, a mountain range can actually grow and no other mechanism is required to explain localized uplift. Some implications about the role of climate on continental tectonics and on the geomorphology of mountain ranges are then derived.

We suggest an additional feedback mechanism by lateral crustal flow (Fig. 4). According to this mechanism, erosional removal of material from topographic heights (dynamic unloading) and its deposition in the foreland basins (dynamic loading) should result in horizontal ductile crustal flow that may oppose gravitational spreading of the crustal roots and may eventually drive a net influx of material towards the orogeny. We finally test our ideas on 3 representative and

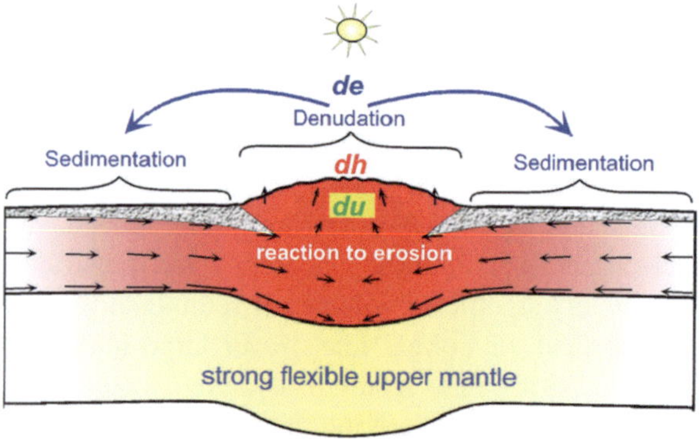

Fig. 4 Solution to the problem stated in Fig. 2: a conceptual model of continental collision in which strong feedback between surface processes, isostatic reaction and subsurface crustal flow results in accelerated growth of topography in the area of strongest subsurface uplift

well-studied cases: (1) slow convergence and erosion rates (Western Alpes), (2) intermediate rates (Tien Shan, Central Asia), and (3) fast convergence and erosion rates (Himalaya, Central Asia).

Interplays Between Surface and Tectonic Processes

Tectonic Forcing on Surface Processes

Surface topography elevations are required to set up erosion and sedimentation processes. Tectonics is therefore a forcing factor of surface processes. Following Ahnert (1970), and Pinet and Souriau (1988), Summerfield and Hulton (1994) have compiled rates of denudation at the scale of major river basins. These studies indicate that denudation is primarily influenced by basin topography so that rates of denudation appear to be systematically high in areas of active tectonic uplift. Common values of mean denudation rates in such areas would be of the order of a few 0.1 mm/y to about 1 mm/y at the scale of large drainage basin. Such rates are generally consistent with estimates derived from balancing sediment volumes over geological periods of time (Leeder, 1991; Summerfield and Hulton, 1994). Thermochronologic studies indicate however local values as great as 1 mm/y (see Leeder, 1991 and Molnar and England, 1990, for critical reviews). The discrepancy between local and basin averaged estimates is due to the fact that tectonic uplift is probably distributed in brief pulses over localized domains within a drainage basin (Copeland and Harrison, 1990). In the absence of any tectonic feedback, common values of denudation rates should lead to the disappearance of a major mountain belt like the Alpes or Tien Shan in a few million years. Pinet and Souriau (1988) demonstrated that denudation leads to an exponential decay of the topography of a range with a characteristic time constant of the order of 2.5 m.y.

Coupling Between Denudation and Tectonic Uplift Due To Isostasy

Many recent models have investigated feedbacks between surface processes and isostatic reaction (e.g.,

Kooi and Beaumont, 1994; Snyder et al., 2000; Basile and Allemand, 2002; Garcia-Castellanos, 2002; Garcia-Castellanos et al., 2002, 2003; Simpson and Schlunegger, 2003; Persson et al., 2004; Casteltort and Simpson, 2006). Redistribution of surface loads by erosion and sedimentation must induce tectonic deformation to maintain isostatic balance. Vertical isostatic uplift is expected to partly compensate unloading in the area subjected to erosion while subsidence should occur in response to loading by sedimentation. This feedback mechanism may lead to coupling between denudation and tectonic uplift (e.g., Ahnert, 1970). A first consequence is that the time needed to erode surface topography is controlled both by the erosion rate and the rate of re-equilibration of the compensating crustal root. If local isostasy is assumed and horizontal strains are neglected, removal of surface topography by erosion is instantaneously largely compensated by isostatic rebound, so that the characteristic time of decay of an average mountain range would be on the order of 10 m.y. (Leeder, 1991). Furthermore, it has been argued that in this case a strong positive feedback between surface and subsurface processes may arise (Molnar and England, 1990; Masek et al., 1994b). In particular, if the slopes of a river valley steepen during river incision, isostatic readjustment following denudation in a mountain range may result in a net uplift of the higher summits in spite of the average lowering of reliefs. Alternatively, regional compensation due to the bending stresses in elastically resistant lithosphere might lead to the uplift of the eroded edge of a plateau. Erosion might therefore induce some uplift of topographic summits leading in turn to enhanced erosion. The uplift of the Himalayan belt during the last few million years may have resulted from such a coupling rather than from thrusting at the Himalayan front (Burbank, 1992; Burbank and Verges, 1994). Note however that, while the peaks might reach higher elevations following isostatic adjustment, the net effect of erosion is crustal thinning. Thus, these models may be valid for short periods of time but cannot explain the growth of mountains over long geological time spans.

The strongest feedback between erosion and isostatic reaction must take place in the case of local isostasy. It will be mitigated in case of more regional compensation to become negligible for lithospheres, which equivalent elastic thickness exceeds 60 km. Active orogens require stiff plates to support high horizontal tectonic stresses. Consequently, isostatic compensation can solve only part of the problem and some additional feedback mechanisms should take place in collisional settings.

Coupling Between Surface Processes and Horizontal Strains

As mentioned in the introduction, small lateral variations of the crustal thickness should drive horizontal flow in the lower crust. Some studies have already pointed out to the importance of such a process in continental tectonics (e.g., Lobkovsky, 1988; Lobkovsky and Kerchman, 1991; Burov and Cloetingh, 1997). For example, Kruse et al. (1991) have shown that horizontal flow in the lower crust has regulated isostatic equilibrium during extension in the Basin and Range. The lower crust would have been extruded from under the high topography during that process. Following Westaway (1994) we dubbed this sense of flow "outward". On the other hand, Gregory and Chase (1994) inferred "inward" flow, toward the crustal root, during the Laramide orogeny of the Frontal Range, Colorado. The characteristic time associated with flow in the lower crust induced by a mountain range of few thousand m high and few hundreds of km wide, is in the order of a few m.y. The characteristic times of erosional decay of the surface topography and of lateral collapse of its crustal root are thus of the same order of magnitude. Coupling between these processes may arise because they are both driven by topographic slopes. Although it is not often pointed out, it has long been recognized that this kind of process might play a major role in elevation changes within continents (see Westaway, 1994 for an review of historical development of these ideas). Westaway (1994) made a case for such a coupling, with inward flow, in the context of extensional tectonics in western Turkey. He proposed that sediment loading in the sedimentary basins would have driven flow toward the uplifted areas. This kind of processes was first modelled by King and Ellis (1990) who modelled crustal extension using a thin elastic plate (upper crust) overlying an inviscid fluid (lower crust).

We propose that this type of coupling might also appear in compressional contexts. Let us consider a lithospheric block that is submitted to horizontal compression and is vertically loaded by isostatically balanced mountain range. Horizontal stress gradients, resulting from slopes of surface and Moho topography, must drive horizontal flow. The lithosphere in the region of the range is weakened, both because the crust is thick and hot, and because bending of inelastic lithosphere (due to the mountain loading) tends to reduce its strength (Burov and Diament, 1992, 1995; Ranalli, 1995). Higher strain rates in the area below the range should therefore be expected. A low viscosity channel in the lower crust beneath the high topography might therefore allow for lateral flow. In the absence of horizontal shortening and erosion, the lower crust below the range would be extruded laterally as discussed by Bird (1991). If erosion takes place, a special equilibrium regime may be established in which horizontal shortening would be preferentially accommodated by crustal thickening in the area below the range:

(a) Surface processes remove material from the range and feed the adjacent flexural basins inducing isostatic imbalance.
(b) This imbalance produces a temporary excess of normal stress below the foreland basins and deficit below the range favouring flow in the lower crust towards the crustal root. The range uplifts and the basins subside.

Ultimately this coupled regime might lead to some dynamic equilibrium in which the amount of material removed by erosion would balance the amount of material supplied to the range by tectonic shortening.

Apart of the direct mechanical effect of erosion/sedimentation (loading-unloading) on the lithosphere, it also has a very important thermal, and, by proxy, mechanical consequences, because the removal and accumulation of sedimentary matter modifies surface heat flux and thermal conditions in the upper crust (e.g., England and Richardson, 1977). Accumulation of sediments in the forelands leads to (1) cooling of the accretion wedge at short term, in case of rapid advection/filling (initial stages of collision when the convergence rate is highest); (2) heating of the accretion wedge at long term in case of slow advection (when collision rate slows down), due to heat screening (sediments have low thermal conductivity) and the abundance of heat producing radiogenic elements in the sedimentary matter. Furthermore, penetration of the mechanically weak sediment in the subduction channel should serve as lubrification and may enhance the conditions for subduction processes.

Coupling of Surface Processes and Tectonic Input/Reaction in Full Scale Mechanical Models: Major Stages

A number of earlier modelling studies (e.g., Beaumont, 1981; Beaumont et al., 1992, 1994; Willett, 1999) have investigated various relationships between erosion and tectonic processes. However, tectonic reaction was not fully accounted as most of these models that have exploited semi-kinematic formulations for the crust or the mantle lithosphere. One of the first full-scale parametric semi-analytical models was developed by Avouac and Burov (1996) in order to validate the coupled regime between surface and subsurface processes. This model accounted for: (1) surface processes, (2) the effect of topographic loads and variations of crustal thickness on the mechanical behaviour of the lithosphere, (3) ductile flow in the lower crust, (4) depth and strain dependent rheology of the lithosphere.

In the following sections we first discuss the components needed to build a coupled tectonic model of orogenic building: (1) the existing models of surface processes; (2) the rheology data needed to for proper account of the mechanical response of the lithosphere; (3) thermal models of the lithosphere needed for account of thermally dependent ductile rheology. We then describe the design and major results of the coupled semi-analytical model of Avouac and Burov (1996). This semi-analytical model has a number of limitations in terms of model geometry and its inability to handle some key deformation modes such as formation of major thrust faults. For this reason, in the final sections of this study, we go further by introducing an unconstrained fully coupled numerical thermo-mechanical model of continental collision/subduction similar to that used by (Burov et al., 2001; Toussaint et al., 2004a, b). This model takes into account more realistic, than in most previous studies, geometry of the convergent plates, accounts for large strains and brittle-elastic-ductile rheology including localized brittle (faulting) and ductile deformation.

Surface Processes Modelling: Principles and Numerical Implementation

Basic Models of Surface Processes

A growing amount of field and experimental studies have investigated and validated various forms of long and short range erosion and sedimentary transport laws and models (Ahnert, 1970; Beaumont, 1981; Beaumont et al., 1992, 2000; Burbank, 1992; Burbank and Verge, 1994; Ashmore, 1982; Mizutani, 1998; Lavé and Avouac, 2001; Lague et al., 2000, 2003; Davy and Crave, 2000; Lague et al., 2000; Molnar, 2001; Crave and Davy, 2001; Densmore et al., 1997, 1998; Pinet and Souriau, 1988).

Short range erosion. A simple two-dimensional law may be used to simulate erosion and sedimentation at the scale of a mountain range. The evolution of a landscape results from the combination of hillslope and stream processes such as weathering processes that prepare solid rock for erosion, and transportation processes that deliver the eroded material to sedimentary basins (see Carson and Kirkby, 1972 for extensive review). Although many factors, depending on lithology and climate (e.g., Fournier, 1960; Nash, 1980), may control this evolution, quite simple mathematical models describing the geometrical evolution of the morphology at the small scale have been proposed and tested successfully (e.g., Kirkby, 1971; Smith and Bretherton, 1972; Chorley et al., 1984, 1986; Luke, 1972, 1974; Kirkby et al., 1993). For example, the two-dimensional evolution of a scarp-like landform can be modelled assuming that the rate of downslope transport of debris, q, is proportional to the local slope, ∇h (Culling, 1960, 1965; Hanks et al., 1984; Avouac, 1993; Kooi and Beaumont, 1994, 1996; Braun and Sambridge, 1997).

$$q = -k\nabla h, \tag{1}$$

where k is the mass diffusivity coefficient, expressed in units of area per time (e.g., m^2/y). Assuming conservation of matter along a 2-D section and no tectonic deformation, h must obey:

$$dh/dt = -\nabla q. \tag{2}$$

With constant k, Equations (1) and (2) lead to the linear diffusion equation:

$$dh/dt = k\nabla^2 h. \tag{3}$$

This model of surface processes holds only for particular conditions. The regolith must form more rapidly than it is removed by surface transport and slopes must not exceed the frictional angle of the material. Even for scarps formed in loose alluvium, some complications arise when high scarps are considered. Scarps with height typically in excess of about 10 meters in arid climatic zones, tend to have systematically sharper curvatures at crest than at base (e.g., Andrews and Bucknam, 1987). Gravity-driven erosion processes such as hillslope landsliding impose strong limitations on the applicability of the diffusion equation since the processes are rather slope- then curvature dependent, which basically requires one to introduce slope and height dependent terms in the Equation (3). At larger scale, hillslope and stream processes interact and the sediment transport then depend non-linearly on the slope and on other factors such as the slope gradient, the area drained above the point, the distance from the water divide, so that the simple 2-D linear diffusion does not apply in general (e.g., Gossman, 1976). In spite of these limitations, we have chosen to stick to a linear diffusion law to model erosion in the upland. This model does not accurately mimic spatial distribution of denudation in the mountain range but it leads to a sediment yield at the mountain front that is roughly proportional to the mean elevation of the basin relative to that point (a rough approximation to the sediment yield resulting from a change of elevation h over a horizontal distance d is $k \times h/d$) and therefore accounts for the apparent correlation between elevation and denudation rates (Ahnert, 1970, Pinet and Souriau, 1988; Summerfield and Hulton, 1994). We do not apply the diffusion model to the whole system, however. We felt that we should take into account the major discontinuity in surface processes that occurs at the mountain front. As a river emerges into the adjacent basin its gradient is sharply reduced and deposition occurs. The streams shift from side to side and build up alluvial fans and tend to form a broad gently sloping pediment at the base of the mountain range. In addition, a lateral drainage often develops along the foothills of mountain ranges. The Ganges along the Himalayan foothills, the Parana along the Andes or the Tarim along the Tien Shan are good examples. Altogether the formation of the pediment and lateral drainage tend to maintain gentle slopes in the foreland. There is therefore a sharp contrast between river incision that maintains a rugged topography with steep slopes in the mountain range and widespread deposition of alluvium in the foreland. This discontinuity of processes must be considered to model the sharp break-in-slope at the mountain front that is generally observed on topographic profiles across mountain belts. In order to simulate this major change in surface processes, sedimentation in the lowland is modelled assuming flat deposition by fluvial network: we assume conservation of matter along the section and the sediment at the mountain front is distributed in order to maintain a flat horizontal topography in the foreland. We arbitrarily set the change from the diffusion erosion to sedimentation ("flat deposition") at a differential elevation of 500 m, which is, however, representative for the transition from highlands to forelands.

We consider values for k varying between 10^2 to 10^4 m^2/y that yield denudation rates on the order of few 0.01 mm/y to 1 mm/y for a 200–400 km wide mountain range with few thousand meters of relief. In order to test the sensitivity of our model for the assumed erosion law we also considered non-linear erosion laws of the form:

$$dh/dt = k^*(x, h, \nabla h)\nabla^2 h \quad (4a)$$

where $k^*(x, h, h) = k(x)(h)^n$ (e.g., Gossman, 1976; Andrews and Bucknam, 1987). We will refer to the cases with $n = 1, 2$ as first and second order diffusion, respectively. In these cases we did not introduce the change in regime at the mountain front since the non-linear effects already tend to form relatively smooth pediments. It should be noted that Equation (4) differs from one obtained assuming a non-linear diffusion coefficient in the Equation (1). In that case conservation of mass would lead to an additional term $\nabla k^* \nabla h$:

$$dh/dt = k^*(x, h, \nabla h)\nabla^2 h + \nabla k^*(x, h, \nabla h)\nabla h \quad (4b)$$

However, the Equation (4a) is phemenological one and may reflect the possibility of material loss from the system. It is also noteworthy that the existing non-linear erosion laws are not limited to the Equation (4a) (e.g., Newman, 1983; Newman and Turcotte, 1990), which only presents the simplest way to account for the dependence of erodibility on the morphology.

Long-range surface processes. The long range surface processes are associated with fluvial transport, i.e., with river incision, slope geometry, character of sediment matter and conditions for deposition (Flint, 1973, 1974; Sheperd and Schumm, 1974; Hirano, 1975; Schumm et al., 1987; Seidl and Dietrich, 1992; Govers 1992a, b; Hairsine and Rose, 1992; Sklar and Dietrich, 1998, 2001; Howard et al., 1994; Howard, 1998; Smith, 1998; Davy and Crave, 2000; Snyder et al., 2000, 2001; Hancock and Willgoose, 2001; Simpson, 2004). The characteristic laws for this range are different because these mechanisms are dependent on the incision and transport capacity of the fluvial network, local slope and type of sediment. Deep steep rivers can carry sediment to longer distances as it can be catched in turbulent flow layer. Shallow rivers would deposit sediment rapidly resulting in rapid river blockage and frequent change of the direction of the fluvial network. There is also a strong dependence of transport capacity on the grain size and climate episodicity (e.g., Davy and Crave, 2000). The long-range fluvial models were used with success by Kooi and Beamount (1994, 1996), Garcia-Castellanos (2002), Garcia-Castellanos et al. (2002, 2003), Persson et al. (2004). The cumulative material flow, q_{fe}, due to the fluvial transport can be presented, in most simple form, as:

$$q_{fe} = -K_r q_r dh/dl \quad (4c)$$

where q_r is the river discharge, K_r is non-dimensional transport coefficient and dh/dl is the slope in the direction of the river drainage with l being the distance along the transport channel. The diffusion Equation (4a), except if it is not strongly non-linear, provides symmetrical, basically over-smoothed shapes, while fluvial transport Equation (4c) may result in realistic asymmetric behaviours, because, locally the direction of each bi-furcation of the fluvial network is affected by negligibly small factors, even though the overall direction of the flow is controlled by the regional slope of topography (Fig. 5). Any important change in the regional slope of topography, such as at the transition from tectonically built steep highlands to flat sedimentary built forelands, may result, at some moment, in a drastic change of the direction of the fluvial network, which may choose a principally new stream direction orthogonal that of the highland network (as it is the case of the Ganges river, for example). This happens when the sedimentary basin is filled to a point that the inclination of its surface in the direction of tectonic convergence becomes less important than that in some other direction (basically in the direction of the boundary between the steep highlands and flat lowlands).

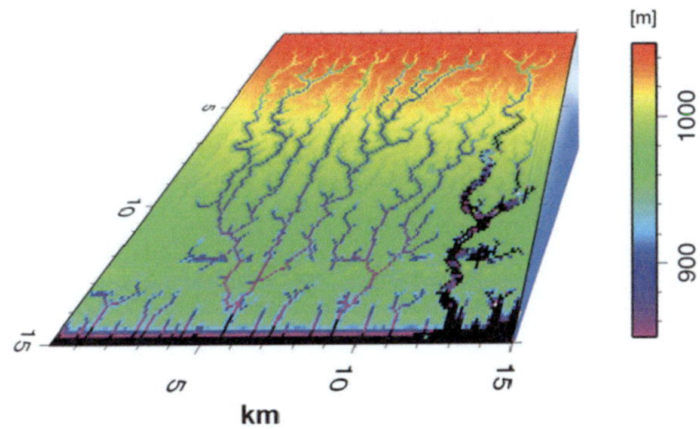

Fig. 5 Example of a typical numerical morphology model with surface erosion and sedimentation based on linear diffusion erosion equation and fluvial transport equation (B. Poisson, E. Burov, J.-P. Brun, 1996). Diffusion equation, except if it is not strongly non-linear, provides symmetrical shapes whereas fluvial transport equation may result in asymmetric behaviour because, locally, the direction of each bi-furcation of the fluvial network may be affected by negligibly small factors, even though the overall direction of the flow is controlled by the regional slope of topography

Although river networks in mountain ranges owe their existence to the competing effects of tectonic uplift and climate-controlled erosion, it was also argued that some universal geometric properties of river networks may be relatively independent of both tectonics and climate (Casteltort and Simpson, 2006). These authors have proposed that the geometry of river networks is established on the lowland margins of incipient uplifts, and is quenched into the erosion zone as the mountain belts widen with time. In that model, the geometry of river networks simply reflects the downward coalescence of alluvial rivers on undissected surfaces outside of mountain belts, and is therefore independent of erosion processes. Yet, the amount of the transported matter, incision rates and other major dynamic parameters of the network are definitely tectonic-and-climate dependent.

Alternative Models of Surface Processes

The diffusion equation reflects the integrated effect of various processes acting at micro and macro scale: chemical and physical erosion and weathering, gravity hillslope sliding etc. Some of these processes, for example, chemical erosion, are well described by the diffusion equation, since it reflects the physics of propagation of chemical interactions. On the other hand, gravity-driven processes are not diffusive. These processes are primarily slope dependent and thus do not fit well within linear diffusion model (Densmore et al., 1997, 1998; Hasbargen and Paola, 2000; Roering et al., 2001; Schorghofer and Rothman, 2002; Pelletier, 2004). Indeed, it has been noted that the diffusion equation tends to over-smooth the predicted topography and fails to reproduce the usually sharp transitions from tectonically modified uplifted landscape to typically flat deposition surfaces in the foreland basins. To remedy this problem, either enhanced split (bi-mode) erosion models that discriminate between diffusion and gravity driven processes (e.g., Simpson and Schlunegger, 2003) or alternative stochastic (based on methods of artificial intellect such as cellular automates) and analogue models were proposed (Crave et al., 2000; Davy and Crave, 2000; Crave and Davy, 2001; Bonnet and Crave, 2003). Crave et al. (2001) or Tucker and Bras (1998, 2000), for example, used stochastic methods based on cellular automates that "learn" how to reproduce erosion/sedimentation from pre-imposed logical rules that establish relations between a given grid cell and its neighbours, as function of the local slope, height, precipitation, regolith type and other conditions. If the rules and their relations are well established, they may form the "vocabulary" and "grammar" (= "language") for description of topography evolution. This approach may eventually produce more realistic landscapes than the common diffusion-fluvial transport models. However, for each new application, it requires one justify the local applicability of the previously established rules. Analogue (physical) erosion models were used to study erosional response to tectonic forcing (e.g., Lague et al., 2003). These models may produce naturally looking landscapes, yet their applicability is rather limited since it is highly difficult to control, scale and interpret their parameters.

Linear and non-linear diffusion short-range models combined with fluvial transport long-range models (Fig. 5) remain to be most widely used for tectonic scale modelling. In particular, diffusion and fluvial transport equations can be generalized (Simpson and Schlunegger, 2003) as following:

$$\begin{cases} \dfrac{dh}{dt} = \nabla \cdot \left((k + cq^n)\, \nabla h \right) \\ \nabla \cdot \left(\dfrac{\nabla h}{|\nabla h|} q \right) = -\alpha \\ De = \dfrac{c\alpha^n L^n}{k} \end{cases} \tag{5}$$

Where c is sediment discharge; α is effective rainfall, q is surface fluid discharge and $k + cq^n$ has a sense of a variable non-linear diffusion coefficient that incorporates both the effects of diffusion-driven (k-term: chemical and physical erosion, weathering) processes and gravity-driven, i.e., fluvial, processes (cq^n term: slope-dependent flow, sliding, creep etc.). The respective role of dispersive processes and hillslope creep processes is characterised by dimensionless De number (L is characteristic length scale).

Structure and Rheology of the Lithosphere

Rheology

Many studies of the interplay between erosion and tectonics have been conducted assuming either local isostasy (Ahnert, 1970; Leeder, 1991) or thin plate flexural behaviour of the lithosphere (Beaumont, 1981; Flemings and Jordan, 1989, 1990; Beaumont et al., 1992; Masek et al., 1994a, b; Garcia-Castellanos, 2002; Garcia-Castellanos et al., 2002; Garcia-Castellanos et al., 2003). Some authors have considered the possibility for ductile flow in the lower crust and treated the lower crust as an inviscid fluid overlaid by a thin elastic plate (King et al., 1988; King and Ellis, 1990; Avouac and Burov, 1996; Burov and Cloetingh, 1997; Burov et al., 2001). The effect of variations in the surface loading and in the crustal thickness on the mechanical behaviour of the lithosphere have been often neglected except several studies (e.g., Beaumont et al., 1992, 2000; Avouac and Burov, 1996; Burov and Cloetingh, 1997; Burov et al., 2001; Toussaint et al., 2004a, b). The coupled erosion-tectonics regime described in the previous sections assumes that strain localization below a tectonic load, range or basin, results from weakening of the lithosphere due to crustal thickening and bending stresses. In order to account for these processes one cannot treat the lithosphere as a simple mono-layer plate with vertically integrated properties (e.g., England and McKenzie, 1983; Vilotte et al., 1982). We have to consider lithological and mechanical rheological layering of the lithosphere. For the model demonstrated here, three lithological layers were defined: the upper crust, the lower crust and the mantle (Fig. 3). Each layer has specific properties (density, mechanical and thermal) that are given in Table 1. We assume no compositional changes due to deformation or cooling. The lithological boundary between the upper and lower is fixed at a depth of 20 km. The bottom of the mantle lithosphere (the depth of 250 km) also corresponds to the depth of 1,330°C isotherm. At small differential stresses the rocks behave elastically. In terms of principal components, the relationship between the stress tensor, σ, and the strain tensor, ε, can be written as:

$$\sigma_j = 2\mu_e\varepsilon_j + \lambda(\varepsilon_1 + \varepsilon_2 + \varepsilon_3) \text{ where } j = 1,2,3. \quad (6)$$

λ and μ_e are Lamé's constants related to Young's modulus (E) and Poissons's ratio ν as $\lambda = E\nu((1+\nu)(1-2\nu))^{-1}$; $\mu_e = E/2(1+\nu)$. Typical values for E and ν are $6.5\sim8 \times 10^{10}$ N/m^2 and 0.25 respectively (e.g., Turcotte and Schubert, 1982).

Weakening by brittle failure or ductile flow occurs when elastic stresses reach some threshold value that determines the condition for brittle failure or ductile deformation. At this threshold level, rocks no longer behave elastically, and unrecoverable strain grows without increase of stress. The conditions of brittle failure are independent of rock type and temperature, but strongly controlled by pressure (Byerlee, 1978):

$$\sigma_3 = (\sigma_1 - \sigma_3)/3.9 \ at\ \sigma_3 < 120 \text{ MPa};$$
$$\sigma_3 = (\sigma_1 - \sigma_3)/2.1 - 100 \text{ MPa} \ at\ \sigma_3 \geq 120 \text{ MPa} \quad (7)$$

where σ_1, σ_2, σ_3 are principal stresses (MPa). This law corresponds to Mohr-Coulomb plastic behaviour.

Ductile flow in the lithosphere essentially results from dislocation creep (e.g., Kusznir, 1991). This mechanism is thermally activated and results in relationship between strain rate and stress:

$$\dot{\varepsilon} = A^*\exp(-H^*/RT)\,(\sigma_1 - \sigma_3)^n. \quad (8)$$

The material constants A, n, H for different rocks are provided in Table 2. The ratio of the stress to strain rate yields the effective non-Newtonian viscosity:

$$\mu_{\text{eff}} = \mu_{\text{eff}} = \left(\frac{\partial\varepsilon}{\partial t}\right)_{\text{II}}^{d(1-n)/n} (A^*)^{-1/n} \exp(H/nRT) \quad (9)$$

where

$$\left(\frac{\partial\varepsilon}{\partial t}\right)_{\text{II}}^{d} = \left(Inv_{II}\left(\left(\frac{\partial\varepsilon}{\partial t}\right)_{\text{II}}^{d}\right)\right)^{1/2}$$

Table 1 Definition of variables

Variable	Values and units	Definition	Comments
$\tau_{xx}, \tau_{xy}, \tau_{yy},$	Pa, MPa	Viscous stress components	
$\sigma_{xx}, \sigma_{xy}, \sigma_{yy}$	Pa, MPa	Full stress components	$\boldsymbol{\sigma} = \boldsymbol{\tau} - P\mathbf{I}$, $\sigma_{xy} = \tau_{xy}$ etc.
P	Pa, MPa	Pressure	
$\mathbf{v}$	m/s, mm/y	Total velocity vector	
u	m/s, mm/y	Horizontal velocity	x component of $\mathbf{v}$
v	m/s, mm/y	Vertical velocity	y component of $\mathbf{v}$
μ	Pa s	Effective viscosity	10^{19} to 10^{25} Pa s
k	m^2/y	Coefficient of erosion	~ mass diffusivity
dh	m, km	Topographic uplift	Or subsidence
du	m, km	Tectonic uplift	Don not mix with u
de	m, km	Erosion	Or sedimentation
ψ	m^2/s	Stream function	$u = \partial\psi/\partial y$, $v = -\partial\psi/\partial y$
ξ	s^{-1}	Vorticity function	$\partial u/\partial y - \partial v/\partial x = \Delta\psi$
ε		Strain	
$\dot{\varepsilon}$	s^{-1}	Average strain rate	$\dot{\varepsilon} = \left({}^1\!/_2\dot{\varepsilon}_{ij}\dot{\varepsilon}_{ij}\right)^{1/2}$
q	m^2/s	Integrated flux	Ductile crust
q_e	(m^2/s)/m	Erosional flux	Per unit length
E	8×10^{10} N/m^2	Young's modulus	In the semi-analytical model
ν	0.25	Poisson's ratio	In the semi-analytical model
λ, μ_e	N/m^2	Lamé's constants	
A^*	Pa^{-n} s^{-1}	Material constant	Power law
n	3 to 5	Stress exponent	Power law
H^*	kJ mol^{-1}	Activation enthalpy	Power law
R	8.314 J/mol K	Gas constant	Power law
T	C°, K	Temperature	
$\gamma(y)$	Pa/m, MPa/km	Depth gradient of yield stress	$\gamma(y) \propto d\sigma(\varepsilon)/dy$,
w	m, km	Plate deflection	~ Deflection of mantle lithosphere
T_e, $\tilde{T}_e(x,w,w',w'',t)$	m, km	Effective elastic thickness	~ Instant integrated strength
T_{ec}	m, km	Effective elastic thickness of the crust	$T_e \approx \left(T_{ec}^3 + T_{em}^3\right)^{1/3}$ $T_{ec} \le h_{c1}$
T_{em}	m, km	Effective elastic thickness of mantle lithosphere	$T_e \approx \left(T_{ec}^3 + T_{em}^3\right)^{1/3}$ $T_{em} \le h_2 - T_c$
$\tilde{M}_x$	N m/m	Flexural moment	Per unit length
$\tilde{T}_x$	N	Longitudinal force	
$\tilde{Q}_x$	N/m	Shearing force	Per unit length
p^+	Pa, N/m^2	Surface load	
p^-	Pa/m	Restoring stress	Per unit length
$h(x,t)$	m, km	Surface topography	
$\tilde{h}(x,t)$	m, km	Upper boundary of ductile channel	
h_c, T_c	m, km	Moho depth	Moho boundary
h_{c2}	m, km	Lower boundary of ductile crustal cannel	$h_{c2} \le T_c$
$h_{c1}(x, t, w)$	m, km	Maximal mechanical thickness of the upper crust	Here, 10–20 km depending on stress
$\Delta h_{c2}(x, t, w, u, v)$	m, km	Thickness of crustal channel	
yij	m, km	Depths to lithological and mechanical interfaces	i is number of a detached layer, j is number of the mechanical sub-layer
ρ_{c1}	2,650 kg/m^3	Density	Of upper crust
ρ_{c2}	2,900 kg/m^3	Density	Of lower crust
ρ_m	3,330 kg/m^3	Density	Of mantle
α	Degrees	Inclination of upper boundary of channel	$\alpha \sim \partial\tilde{h}/\partial x$
β	Degrees	Inclination of lower boundary of channel	$\beta \sim \partial w/\partial x$

Table 1 (continued)

Variable	Values and units	Definition	Comments
g	9. 8 m/s^2	Acceleration due to gravity	
T	°C	Temperature	
t_a	m.y.	Thermal age	≤ Geological age
a	250 km	Thermal thickness of the lithosphere	
Tm	1,330°C	T at depth a	
χ	m^{2}s^{-1}	Thermal diffusivity	$\chi = k/\rho C_p$
χ_{c1}	8.3×10^{-7} m^{2}s^{-1}	Thermal diffusivity	Upper crust
χ_{c2}	6.7×10^{-7} m^{2}s^{-1}	Thermal diffusivity	Lower crust
χ_m	8.75×10^{-7}m^{2}s^{-1}	Thermal diffusivity	Mantle
k_{c1}	2.5 Wm^{-1}°K^{-1}	Thermal conductivity	Upper crust
k_{c2}	2 Wm^{-1}°K^{-1}	Thermal conductivity	Lower crust
k_m	3.5 Wm^{-1}°K^{-1}	Thermal conductivity	Mantle
h_r	10 km	Decay scale of radiogenic heat production	Upper crust
H_s	9.5×10^{-10} W kg^{-1}	Surface radiogenic heat production rate per unit mass	Upper crust
$H_{c2}C_{c2}^{-1}$	1.7×10^{-13} °K s^{-1}	Radiogenic heat	Lower crust
ρCp	J (m^{3}K)$^{-1}$	Density × specific heat	

Table 2 Summary of rheology parameters used in model calculations

Parameter	Value
Lamé elastic constants $\lambda = G$	30 GPa (in numerical models)
Friction angle (Mohr-Coulomb criterion)	30°
Cohesion (Mohr-Coulomb criterion)	20 Mpa
Specific upper and weak lower-crust properties	
ρ (upper crust)	2,800 kg m^{-3}
ρ (lower crust)	2,900 kg m^{-3}
N	2.4
A	6.7×10^{-6} MPa^{-n}·s^{-1}
Q	1.56×10^5 KJ·mol^{-1}
Specific strong lower-crust properties	
ρ	2,980 kg m^{-3}
N	3.4
A	2×10^{-4} MPa^{-n}·s^{-1}
Q	2.6×10^5 KJ·mol^{-1}
Specific mantle properties	
ρ (lithosphere)	3,330 kg m^{-3}
ρ (oceanic slab)	3,350 kg m^{-3}
ρ (asthenosphere)	3,310 kg m^{-3}
N	3
A	1×10^4 Mpa^{-n}·s^{-1}
Q	5.2×10^5 KJ·mol^{-1}

Note: Compilation by Burov et al. (2001). Q, n, A are parameters of the ductile flow law (activation energy, material constant, and power exponent, respectively). See also (Brace and Kohlstedt, 1980; Kirby, 1983; Kirby and Kronenberg, 1987; Kohlstedt et al., 1995; Byerlee, 1978; Carter and Tsenn, 1987; Tsenn and Carter, 1987).

is the effective strain rate and $A^* = \frac{1}{2}A \cdot 3^{(n+1)/2}$ is the material constant, H is the activation enthalpy, R is the gas constant, n is the power law exponent. A temperature of about 250–300°C is sufficient for activation of prominent ductile deformation in quartz, whereas for olivine it should be as high as 600–700°C (e.g., Brace and Kohlstedt, 1980; Carter and Tsenn, 1987; Kohlsted et al., 1995). For given background strain rate, combination of rheological laws (Equations (6), (7), (8), (9)) yields a piece-wise continuous yield-stress envelope (YSE) in $\Delta\sigma$–y space (Fig. 3), defined as a contour $\sigma^f = \sigma^f(x, y, t, \dot{\varepsilon})$ such that:

$$\sigma^f = sign(\varepsilon)\min\left(\left|\sigma^b(x, y, t, \dot{\varepsilon}, sign(\varepsilon))\right|, \left|\sigma^d(x, y, t, \dot{\varepsilon})\right|\right) \quad (10)$$

where $\sigma^b(x, y, t, \dot{\varepsilon}, sign(\varepsilon))$, $\sigma^d(x, y, t, \dot{\varepsilon})$ are the "brittle" and "ductile" yield stresses from (6) and (7). $sign(\varepsilon) = 1$ if $\varepsilon \geq 0$ and –1 if $\varepsilon \leq 0$. The differential stress $\sigma(x, y)$ at a point is taken to be equal to the minimum of σ^e and σ^f, computed as a function of the local strain $\varepsilon = \varepsilon(x, y, t, \dot{\varepsilon})$:

$$\sigma(\varepsilon) = \text{sign}(\varepsilon)\min\left(\left|\sigma^f\right|, \left|\sigma^e(\varepsilon)\right|\right) \quad (11)$$

where $\sigma^e(\varepsilon)$ is the elastic differential stress according to (5). If σ^e exceeds σ^f, the material is considered as ductile or brittle, depending on which rheology limits the yield strength. Due to asymmetry of the Byerlee's law (7), the yield stress depends on the mode of deformation, sign(ε) (for extension sign(ε) = 1; for compression sign(ε) = –1). Equation (11) implies that the lithosphere remains elastic if imposed stress does not exceed the yield stress. Most of the upper crust remains elastic or brittle-elastic (depth interval between approx. 5 and 15–20 km). The crust is mostly ductile below 15–20 km. Depending on the geotherm and strain rates, first 30–70 km of the mantle lithosphere remains elastic. This formulation reflects the fact that the lithosphere gets weaker when submitted to increasing horizontal forces or flexural stresses and when the crust gets thicker.

Thermal Model

A thermal model is required to define the rheological profile of the lithosphere and to fully account for the effect of crustal thickening on the rheology of the lithosphere. In this paper, the initial geotherm is computed according to a half-space plate cooling model (for details see Burov et al., 1993 and Burov and Diament, 1995). Heat transfer equations are solved separately for the upper crust, lower crust and mantle with conditions of temperature and heat flux continuity across the interfaces. These equations have a form:

$$\dot{T} + uT'_x + vT'_y - \chi_f \Delta T = H_d + H_r + v\Omega, \quad (12)$$

where primes mean spatial differentiation by respective coordinate. The thermal diffusivity χ_f equals χ_{c1}, χ_{c2}, or χ_m, depending on the lithological layer (see Table 1). $H_r = \chi_{c1}k_{c1}^{-1}\rho_c H_s \exp(-yh_r^{-1})$ is the radiogenic heat. H_r equals to constant heat generation $H_{c2}C_{c2}^{-1}$ in the lower crust and to zero in the mantle. H_d refers to heat generation due to frictional heating. The adiabatic temperature gradient in the asthenosphere, Ω, is 0.3°C/km (Turcotte and Schubert, 1982).

The boundary and initial conditions are: $T(x,0,t_a) = 0$°C (temperature at the upper surface = *const* at time t_a, where t_a is the thermotectonic age of the lithosphere); $T(x, a, t) = T_m = 1{,}350$°C ($a \approx 250$ km is the depth to the thermal bottom, or *thermal thickness* of the lithosphere); $T(x, y, 0) = T_m$ (homogeneous temperature distribution at the beginning).

Implementation of Coupled Models

Semi-Analytical Model for Intermediate Tectonic Convergence Rate

In this section we describe the numerical procedure used in the semi-analytical model by (Avouac and Burov, 1996; Burov and Cloetingh, 1997).

In the semi-analytical model, we distinguish between competent and weak layers (Fig. 6, top). The competent layers are those that contribute significantly to the flexural strength of the lithosphere. This encompasses not only the elastic domains but also high stress (> 5% of the lithostatic pressure) brittle and ductile domains (e.g., Burov and Diament, 1992, 1995). Only the competent layers are taken into account in the computation of the flexural response of the lithosphere. Geometry and thickness the competent layers depend on the lithological layering and on the stress field. Since both evolve during the numerical experiment, the mechanical structure has to be re-computed at each numerical step. Vertical deflections (w) of the competent portions of the crust and mantle lithosphere due

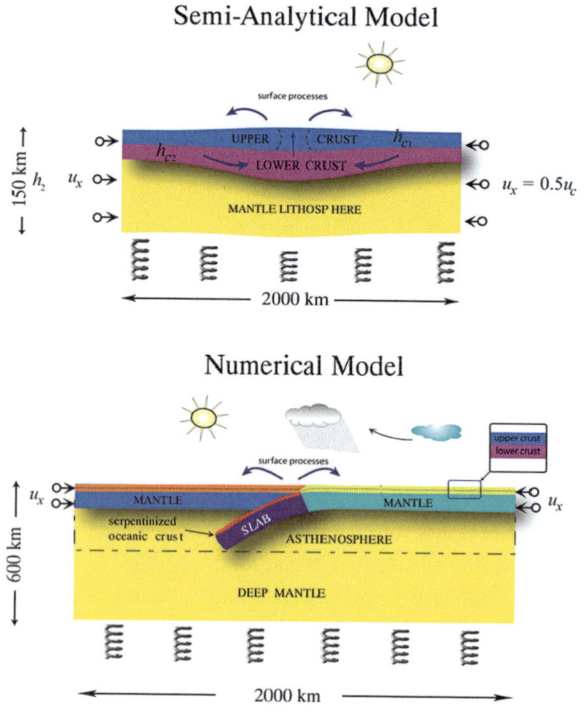

Fig. 6 Model setups. *Top*: Setup of a simplified semi-nalytical collision model with erosion-tectonic coupling (Avouac and Burov, 1996). In-eastic flexural model is used to for competent parts of crust and mantle, channel flow model is used for ductile domains. Both models are coupled via boundary conditions. The boundaries between competent and ductile domains are not predefined but are computed as function of bending stress that conrols brittle-ductile yielding in the lithosphere. Diffusion erosion and flat deposition are imposed at surface. In these experiments, initial topography and isostatic crustal root geometry correspond to that of a 3 km high and 200 km wide Gaussian mount. *Bottom*. Setup of fully coupled thermo-mechanical collision-subduction model (Burov et al., 2001; Toussaint et al., 2004b). In this model, topography is not predefined and deformation is solved from full set of equilibrium equations. The assumed rheology is brittle-elastic-ductile, with quartz-rich crust and olivine-rich mantle (Table)

to change in the stress applied at their boundaries are treated as instantaneous deflections of flexible layers (Appendix 1). Deformation of the ductile lower crust is driven by deflection of the bounding competent layers. This deformation is modelled as a viscous non-Newtonian flow in a channel of variable thickness. No horizontal flow at the axis of symmetry of the range ($x = 0$) is allowed. Away from the mountain range, where the channel has a nearly constant thickness, the flow is computed from thin channel approximation (Appendix 2). Since the conditions for this approximation are not satisfied in the thickened region, we use a semi-analytical solution for the ascending flow fed by remote channel source (Appendix 3). The distance a_l at which the channel flow approximation is replaced by the formulation for ascending flow, equals 1 to 2 thicknesses of the channel. The latter depends on the integrated strength of the upper crust (Appendixes 2 and 3). Since the common brittle-elastic-dutile rheology profiles imply mechanical decoupling between the mantle and the crust (Fig. 3), in particular in the areas where the crust is thick, deformation of the crust is expected to be relatively insensitive to what happens in the mantle. Shortening of the mantle lithosphere can be therefore neglected. Naturally, this assumption will not directly apply if partial coupling of mantle and crustal lithosphere occurs (e.g., Ter Voorde et al., 1998; Gaspar-Escribano et al., 2003). For this reason, in the next sections, we present unconstrained fully numerical model, in which there is no pre-described conditions on the crust-mantle interface.

Equations that define the mechanical structure of the lithosphere, flexure of the competent layers, ductile flow in the ductile crust, erosion and sedimentation at the surface are solved at each numerical iteration following the flow-chart:

$$
\begin{array}{llll}
\text{input} & & & \text{output} \\
\left\{\begin{array}{l}
\text{I. } u_{k-1}, v_{k-1}, T_{c(k-1)}, w_{k-1}, h_{k-1} \\
\quad +B.C.\& I.C._{\cdot k} \\
\text{II. } T,\ \dot{\varepsilon}, A, H^*, n, T_{c\ (k-1)} \\
\text{III. } \sigma^f, h_{c1}, h_{c2}, h_m, h_{k-1}, \\
\quad p^-_{k-1}, p^+_{k-1}\ +B.C._{\cdot k} \\
\text{IV. } w_k,\ \sigma(\varepsilon),\ y_{ij(k)}, \tilde{h}_{k-1},\ \sigma^f, \\
\quad \dot{\varepsilon},\ h_{k-1}, T_{ck} + B.C._{\cdot k} \\
\text{V. } h_k (\text{i.e., } I.C._{\cdot k})
\end{array}\right.
&
\begin{array}{l}
\\
\rightarrow \\
\rightarrow \\
\\
\rightarrow \\
\\
\rightarrow \\
\rightarrow
\end{array}
&
\begin{array}{l}
\\
(A1,12,14) \\
(6\text{–}11) \\
\\
(A1) \\
\\
(B5,B6,C3) \\
(3-4)
\end{array}
&
\begin{array}{l}
\\
\rightarrow T \\
\rightarrow \sigma^f, h_{c1}, h_{c2}, h_m \\
\\
\rightarrow w_k,\ T_{c(k)}, \sigma(\varepsilon),\ y_{ij(k)} \\
\\
\rightarrow u_k, v_k,\ \tilde{h}_k, h_k, T_{ck+1},\ \tau_{xy},\ \delta T_1 \\
\rightarrow h_{k+1},\ \delta T_2
\end{array}
\end{array}
\qquad (13)
$$

B.C. and *I.C.* refer to boundary and initial conditions, respectively. Notation (k) implies that the related value is used on k-th numerical step. Notation (k–1) implies that the value is taken as a predictor from the previous time step, etc. All variables are defined in Table 1. The following continuity conditions are satisfied at the interfaces between the competent layers and the ductile crustal channel:

$$\begin{array}{ll} \text{continuity of vertical velocity} & v_{c1}^{-} = v_{c2}^{+};\ v_{c2}^{-} = v_{m}^{+} \\ \text{continuity of normal stress} & \sigma_{yyc1}^{-} = \sigma_{yyc2}^{+};\ \sigma_{yyc2}^{-} = \sigma_{yym}^{+} \\ \text{continuity of horizontal velocity} & u_{c1}^{-} = u_{c2}^{+};\ u_{c2}^{-} = u_{m}^{+} \\ \text{continuity of the tangential stress} & \sigma_{xyc1}^{-} = \sigma_{xyc2}^{+};\ \sigma_{xyc2}^{-} = \sigma_{xym}^{+} \\ \text{kinematic condition} & \dfrac{\partial \tilde{h}}{\partial t} = v_{c2}^{+};\ \dfrac{\partial w}{\partial t} = v_{c2}^{-} \end{array} \qquad (14)$$

Superscripts "+" and "–" refer to the values on the upper and lower interfaces of the corresponding layers, respectively. The subscripts c_1, c_2, and m refer to the strong crust ("upper"), ductile crust ("lower") and mantle lithosphere, respectively. Power-law rheology results in the effect of self-lubrication and concentration of the flow in the narrow zones of highest temperature (and strain rate), that form near the Moho. For this reason, there is little difference between the assumption of no-slip and free slip boundary for the bottom of the ductile crust.

The spatial resolution used for calculations is $dx = 2$ km, $dy = 0.5$ km. The requirement of stability of integration of the diffusion Equations (3), (4) ($dt < 0.5dx^2/k$) implies a maximum time step of $< 2{,}000$ years for $k = 10^3$ m^2/y and of 20 years for $k = 10^5$ m^2/y. It is less than the relaxation time for the lowest viscosity value ($\sim$50 years for $\mu = 10^{19}$ Pa s). We thus have chosen a time step of 20 years in all semi-analytical computations.

Unconstrained Fully Coupled Numerical Model

To fully demonstrate the importance of interactions between the surface processes, ductile crustal flow and major thrust faults, and also to verify the earlier ideas on evolution of collision belts, we used a fully coupled (mechanical behaviour – surface processes – heat transport) numerical models that also handle brittle-elastic-ductile rheology and account for large strains, strain localization and erosion/sedimentation processes (Fig. 6, bottom).

We have extended the Paro(a)voz code (Polyakov et al., 1993, Appendix 4) based on FLAC (Fast Langrangian Analysis of Continua) algorithm (Cundall, 1989). This explicit time-marching, large-strain Lagrangian algorithm locally solves Newtonian equations of motion in continuum mechanics approximation and updates them in large-strain mode. The particular advantage of this code refers to the fact that it operates with full stress approximation, which allows for accurate computation of total pressure, P, as a trace of the full stress tensor. Solution of the governing mechanical balance equations is coupled with that of the constitutive and heat-transfer equations. Parovoz v9 handles free-surface boundary condition, which is important for implementation of surface processes (erosion and sedimentation).

We consider two end-member cases: (1) very slow convergence and moderate erosion (Alpine collision) and (2) very fast convergence and strong erosion (India–Asia collision). For the end-member cases we test continental collision assuming commonly referred initial scenario (Fig. 6, bottom), in which (1) rapidly subducting oceanic slab entrains a very small part of a cold continental "slab" (there is no continental subduction at the beginning), and (2) the initial convergence rate equals to or is smaller than the rate of the preceding oceanic subduction (two-sided initial closing rate of 2 × 6 mm/y during 50 My for Alpine collision test (Burov et al., 2001) or 2 × 3 cm/y during the first 5–10 My for the India–Asia collision test (Toussaint et al., 2004b)). The rate chosen for the India–Asia collision test is smaller than the average historical convergence rate between India and Asia (2 × 4 to 2 × 5 cm/y during the first 10 m.y. (Patriat and Achache, 1984)).

For continental collision models, we use commonly inferred crustal structure and rheology parameters derived from rock mechanics (Table 1; Burov et al., 2001). The thermo-mechanical part of the model that computes, among other parameters, the upper free surface, is coupled with surface process model based on the diffusion equation (4a). On each type step the geometry of the free surface is updated with account for erosion and deposition. The surface areas affected by sediment deposition change their material properties according to those prescribed for sedimentary matter (Table 1). In the experiments shown below, we used linear diffusion with a diffusion coefficient that has been varied from 0 m^2 y^{-1} to 2,000 m^2 y^{-1}(Burov et al., 2001). The initial geotherm was derived from the common half-space model (e.g., Parsons and Sclater, 1977) as discussed in the section "Thermal mode" and Appendix 4.

The universal controlling variable parameter of all continental experiments is the initial geotherm (Fig. 3), or thermotectonic age (Turcotte and Schubert, 1982), identified with the Moho temperature T_m. The geotherm or age define major mechanical properties of the system, e.g., the rheological strength profile (Fig. 3). By varying the geotherm, we can account for the whole possible range of lithospheres, from very old, cold, and strong plates to very young, hot, and weak ones. The second major variable parameter is the composition of the lower crust, which, together with the geo-therm, controls the degree of crust-mantle coupling. We considered both weak (quartz dominated) and strong (diabase) lower-crustal rheology and also weak (wet olivine) mantle rheology (Table 1). We mainly applied a rather high convergence rate of 2 × 3 cm/y, but we also tested smaller convergence rates (two times smaller, four times smaller, etc.).

Within the numerical models we can also trace the amount of subduction (subduction length, s_l) and compare it with the total amount of shortening on the borders, Δx. The subduction number S, which is the ratio of these two values, may be used to characterize the deformation mode (Toussaint et al., 2004a):

$$S = \delta x / s_l \tag{15}$$

When $S = 1$, shortening is likely to be entirely accommodated by subduction, which refers to full subduction mode. In case when $0.5 < S < 1$, pure shear or other deformation mechanisms participate in accommodation of shortening. When $S < 0.5$, subduction is no more leading mechanism of shortening. Finally, when $S > 1$, one deals with full subduction plus a certain degree of "unstable" subduction associated with stretching of the slab under its own weight. This refers to the cases of high s_l (>300 km) when a large portion of the subducted slab is reheated by the surrounding hot asthenosphere. As a result, the deep portion of the slab mechanically weakens and can be stretched by gravity forces (slab pull). The condition when $S > 1$ basically corresponds to the initial stages of slab breakoff. $S > 1$ often associated with the development of Rayleigh-Taylor instabilities in the weakened part of the slab.

Experiments

Semi-Analytical Model

Avouac and Burov (1996) have conducted series of experiments, in which a 2-D section of a continental lithosphere, loaded with some initial range (resembling averaged cross-section of Tien Shan), is submitted to horizontal shortening (Fig. 6, top) in pure shear mode. Our goal was to validate the idea of the coupled (erosion-tectonics) regime and to check whether it can allow for stable localized mountain growth. Here we were only addressing the problem of the growth and maintenance of a mountain range once it has reached some mature geometry.

We consider a 2,000 km long lithospheric plate initially loaded by a topographic irregularity. Here we do not pose the question how this topography was formed, but in later sections we show fully numerical experiments, in which the mountain range grows from initially flat surface. We chose a 300–400 km wide "Gaussian" mountain (a Gaussian curve with variance 100 km, that is about 200 km wide). The model range has a maximum elevation of 3,000 m and is initially regionally compensated. The thermal profile used to compute the rheological profile corresponds approximately to the age of 400 My. The initial geometry of Moho was computed from the flexural response of the competent cores of the crust and upper mantle and neglecting viscous flow in the lower crust (Burov et al., 1990). In this computation, the possibility of the internal deformation of the mountain range or of its crustal root was neglected. The model is then submitted to horizontal shortening at rates from about 1 mm/y to several cm/y. These rates largely span the range of most natural large scale examples of active intracontinental mountain range. Each experiment modelled 15–20 m.y. of evolution with time step of 20 years. The geometries of the different interfaces (topography, upper-crust-lower crust, Moho, basement-sediment in the foreland) were computed for each time step. We also computed the rate of uplift of the topography, *dh/dt*, the rate of tectonic uplift or subsidence, *du/dt*, the rate of denudation or sedimentation, *de/dt*, (Fig. 7–10), stress, strain and velocity field. The relief of the range, Δh, was defined as the difference between the elevation at the crest $h(0)$ and in the lowlands at 500 km from the range axis, $h(500)$.

In the case where there are no initial topographic or rheological irregularities, the medium has homogeneous properties and therefore thickens homogeneously (Fig. 8). There are no horizontal or vertical gradients of strain so that no mountain can form. If the medium is initially loaded with a mountain range, the flexural stresses (300–700 MPa; Fig. 7) can be 3–7 times higher than the excess pressure associated with the weight of the range itself (~100 MPa). Horizontal shortening of the lithosphere tend therefore to be absorbed preferentially by strain localized in the weak zone beneath the range. In all experiments the system evolves vary rapidly during the first 1–2 million years because the initial geometry is out of dynamic equilibrium. After the initial reorganisation, some kind of dynamic equilibrium settles, in which the viscous forces due to flow in the lower crust also participate is the support of the surface load.

Case 1: No Surface Processes: "Subsurface Collapse"

In the absence of surface processes the lower crust is extruded from under the high topography (Fig. 8). The crustal root and the topography spread out laterally. Horizontal shortening leads to general thickening of the medium but the tectonic uplift below the range is smaller than below the lowlands so that the relief of the range, Δh, decays with time. The system thus evolves towards a regime of homogeneous deformation with a uniformly thick crust. In the particular case of a 400 km wide and 3 km high range it takes about 15 m.y. for the topography to be reduced by a factor of 2. If the medium is submitted to horizontal shortening, the decay of the topography is even more rapid due to in-elastic yielding. These experiments actually show that assuming a common rheology of the crust without intrinsic strain softening and with no particular assumptions for mantle dynamics, a range should collapse in the long term, as a result of subsurface deformation, even the lithosphere undergoes intensive horizontal shortening. We dubbed "subsurface collapse" this regime in which the range decays by lateral extrusion of the lower crustal root.

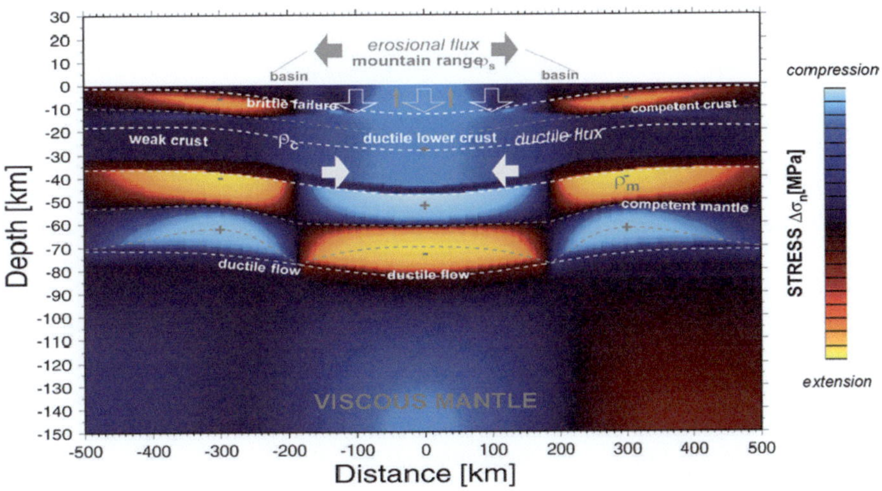

Fig. 7 Example of normalized stress distribution in a semi-analytical experiment in which stable growth of the mountain belt was achieved (total shortening rate 44 mm/y; strain rate 0.7×10^{-15} sec^{-1} erosion coefficient 7,500 m^2/y)

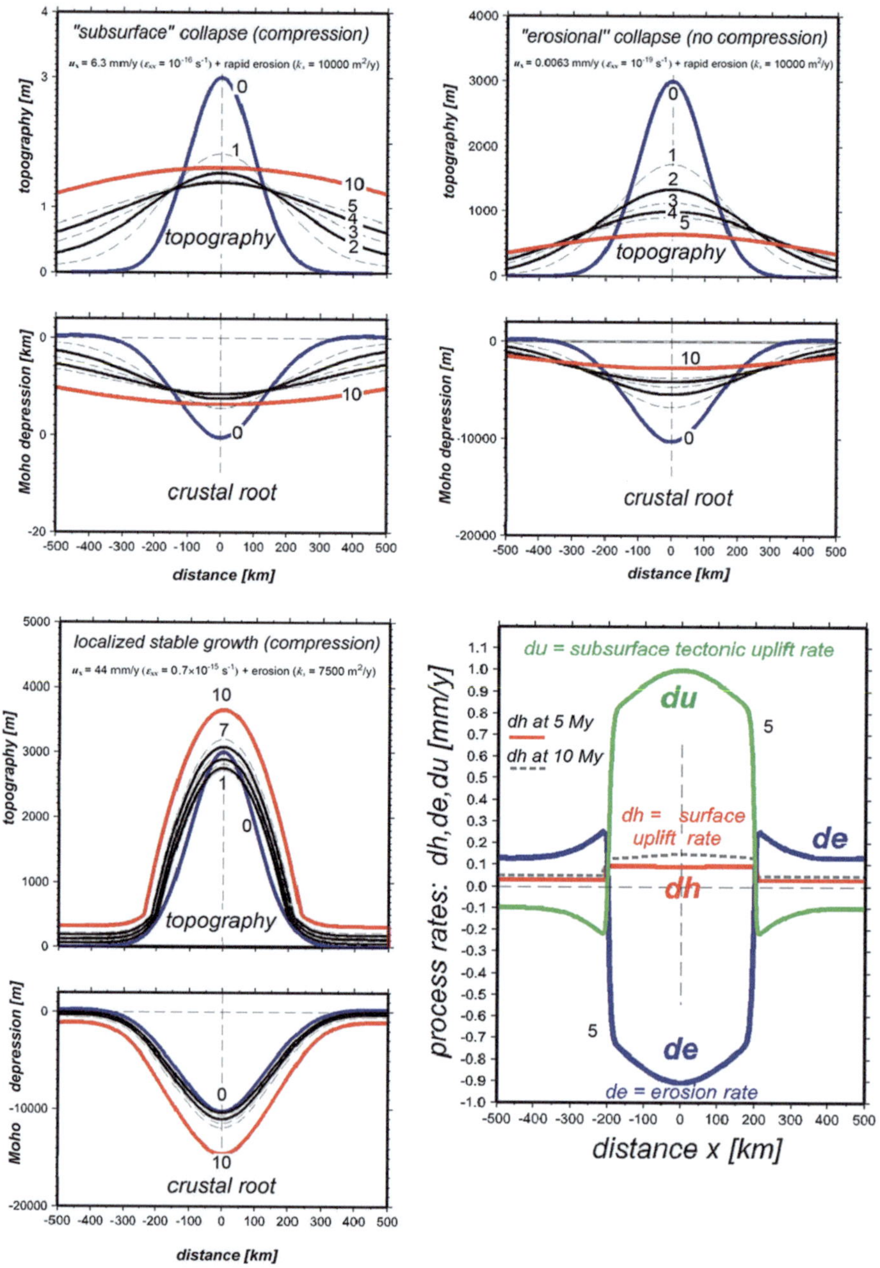

Fig. 8 Results of representative semi-analytical experiments: topography and crustal root evolution within first 10 My, shown with interval of 1 My. Top, right: Gravity, or subsurface, collapse of topography and crustal root (total shortening rate 2 × 6.3 mm/y; strain rate 10^{-16} sec^{-1} erosion coefficient 10,000 m^2/y). Top, left: erosional collapse (total shortening rate 2 × 0.006.3 mm/y; strain rate 10^{-19} sec^{-1} erosion coefficient 10,000 m^2/y). Bottom, left: Stable localised growth of the topography in case of coupling between tectonic and surface processes observed for total shortening rate 44 mm/y; strain rate 0.7 × 10^{-15} sec^{-1} erosion coefficient 7,500 m^2/y. Bottom, right: distribution of residual surface uplift rate, dh, tectonic uplift rate, du, and erosion-deposition rate de for the case of localised growth shown at bottom, left. Note that topography growth in a localized manner for at least 10 My and the perfect anti-symmetry between the uplift and erosion rate that may yield very stable steady surface uplift rate

Case 2: No Shortening: "Erosional Collapse"

If erosion is intense (with values of k of the order of 10^4 m^2/y.) while shortening is slow, the topography of the range vanishes rapidly. In this case, isostatic readjustment compensates for only a fraction of denudation and the elevation in the lowland increases as a result of overall crustal thickening (Fig. 8). Although the gravitational collapse of the crustal root also contributes to the decay of the range, we dubbed this regime "erosional", or "surface" collapse. The time constant associated with the decay of the relief in this regime depends on the mass diffusivity. For $k = 10^4$ m^2/y, denudation rates are of the order of 1 mm/y at the beginning of the experiment and the initial topography was halved in the first 5 My. For $k = 10^3$ m^2/y the range topography is halved after about 15 My. Once the crust and Moho topographies

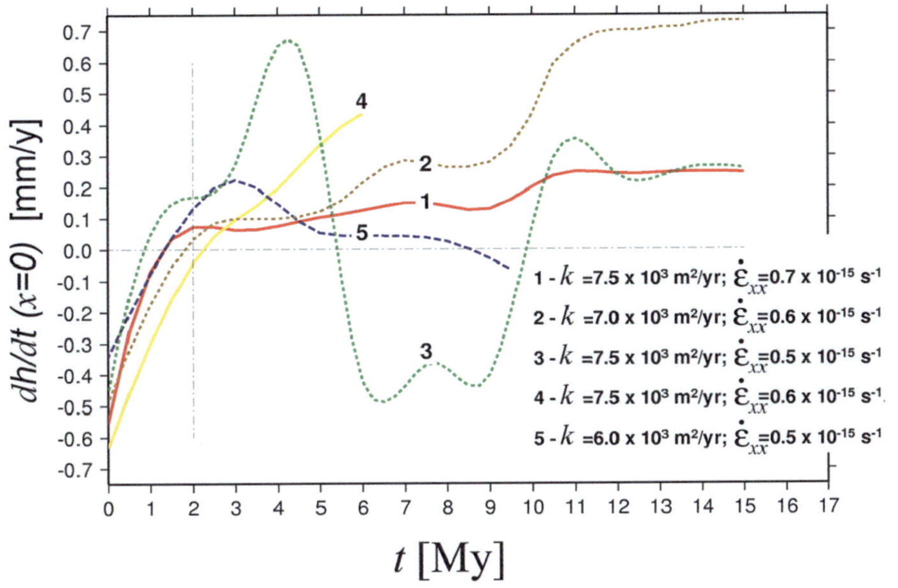

Fig. 9 Tests of stability of the coupled "mountain growth" regime. Shown are the topography uplift rate at the axis ($x = 0$) of the range, for various deviations of the coefficient of erosion, k, and of the horizontal tectonic strain rates, $\partial\varepsilon_{xx}/\partial t$, from the values of the most stable reference case "*1*", which corresponds to the mountain growth experiment from the Fig. 8 (*bottom*). Feedback between the surface and subsurface processes maintains the mountain growth regime even for large deviations of k_s and $\partial\varepsilon_{xx}/\partial t$ (curves *2*, *3*) from the equilibrium state (1). Cases *4* and *5* refer to very strong misbalance between the denudation and tectonic uplift rates, for which the system starts to collapse. These experiments suggest that the orogenic systems may be quite resistant to climatic changes or variations in tectonic rates, yet they rapidly collapse if the limits of the stability are exceeded

have been smoothed by surface processes and subsurface deformation, the system evolves towards the regime of homogeneous thickening.

Case 3: Dynamically Coupled Shortening and Erosion: "Mountain Growth"

In this set of experiments, we started from the conditions leading to the "subsurface collapse" (significant shortening rates), and then gradually increased the intensity of erosion. In the experiments where erosion was not sufficiently active, the range was unable to grow and decayed due to subsurface collapse. Yet, at some critical value of k, a regime of dynamical coupling settled, in which the relief of the range was growing in a stable and localised manner (Fig. 8, bottom). Similarly, in the other set of experiments, we started from the state of the "erosional collapse", kept the rate of erosion constant and gradually increased the rate of shortening. At low shortening rates, erosion could still erase the topography faster then it was growing, but at some critical value of the shortening rate, a coupled regime settled (Figs. 7, 8). In the coupled regime, the lower crust was flowing towards the crustal root (inward flow) and the resulting material influx exceeded the amount of material removed from the range by surface processes. Tectonic uplift below the range then could exceed denudation (Figs. 7, 8, 9, 10) so that the elevation of the crest was increasing with time. We dubbed this regime "mountain growth". The distribution of deformation in this regime remains heterogeneous in the long term. High strains in the lower and upper crust are localized below the range allowing for crustal thickening (Fig. 7). The crust in the lowland also thickens owing to sedimentation but at a smaller rate than beneath the range. Figure 8 shows that the rate of growth of the elevation at the crest, dh/dt ($x = 0$), varies as a function of time allowing for mountain growth. It can be seen that "mountain growth" is not monotonic and seems to be very sensitive, in terms of surface denudation and uplift rate, to small changes in parameters. However, it was also found that the coupled regime can be self-maintaining in a quite broad parameter range, i.e., erosion automatically accelerates or decelerates to compensate eventual variations

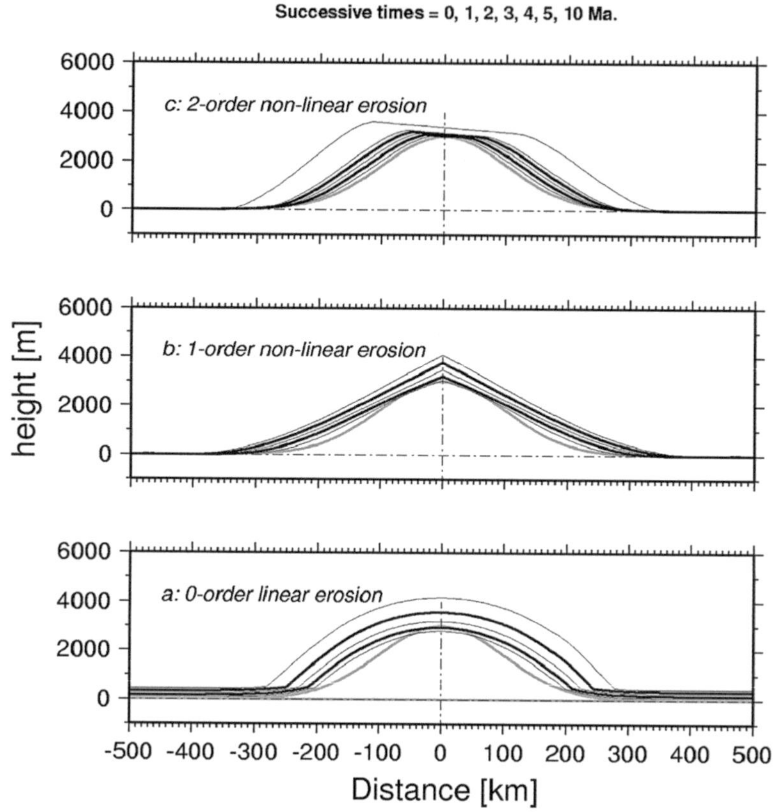

Fig. 10 Influence of erosion law on steady-state topography shapes: 0 (*a*), 1 (*b*), and 2nd (*c*) order diffusion applied for the settings of the "mountain growth" experiment of Fig. 8 (*bottom*). The asymmetry in (*c*) arrives from smallwhite noise (1%) that was introduced in the initial topography to test the robustness of the final topographies. In case of highly non-linear erosion, the symmetry of the system is extremely sensitive even to small perturbations

in the tectonic uplift rate (Fig. 9). The Fig. 9 shows that the feedback between the surface and subsurface processes can maintain the mountain growth regime even for large deviations of k_s and $\partial \varepsilon_{xx}/\partial t$ from the equilibrium state. These deviations may cause temporary oscillations in the mountain growth rate (curves 2 and 3 in Fig. 9) that are progressively damped as the system finds a new stable regime. These experiments suggest that orogenic systems may be quite resistant to climatic changes or variations in tectonic rates, yet they may very rapidly collapse if the limits of the stability range are exceeded (curves 3, 4 in Fig. 9). We did not further explore the dynamical behaviour of the system in the coupled regime but we suspect a possibility of chaotic behaviours, hinted, for example, by complex oscillations in case 3 (Fig. 9). Such chaotic behaviours are specific for feedback-controlled systems in case of delays or other changes in the feedback loop. This may refer, for example, to the delays in the reaction of the crustal flow to the changes in the surface loads; to a partial loss of the sedimentary matter from the system (long-distance fluvial network or out of plain transport); to climatic changes etc.

Figures 11 and 12 shows the range of values for the mass diffusivity and for the shortening rate that can allow for the dynamical coupling and thus for mountain growth. As a convention, a given experiment is defined to be in the "mountain growth" regime if the relief of the range increases at 5 m.y., which means that elevation at the crest ($x = 0$) increases more rapidly than the elevation in the lowland ($x = 500$ km):

$$dh/dt(x = 0\,\text{km}) > dh/dt(x = 500\,\text{km})\ \text{at}\ t = 5\text{My} \quad (16)$$

As discussed above, higher strain rates lead to reduction of the effective viscosity (μ_{eff}) of the non-Newtonian lower crust so that a more rapid erosion is needed to allow the feedback effect due to surface processes. Indeed, μ_{eff} is proportional to$\dot{\varepsilon}^{1/n-1}$. Taking into account that n varies between 3 and 4, this provides a half-order decrease of the viscosity at one-order increase of the strain rate from 10^{-15} to 10^{-14} s^{-1}. Consequently, the erosion rate must be several times higher or slower to compensate 1 order increase or decrease in the tectonic strain rate, respectively.

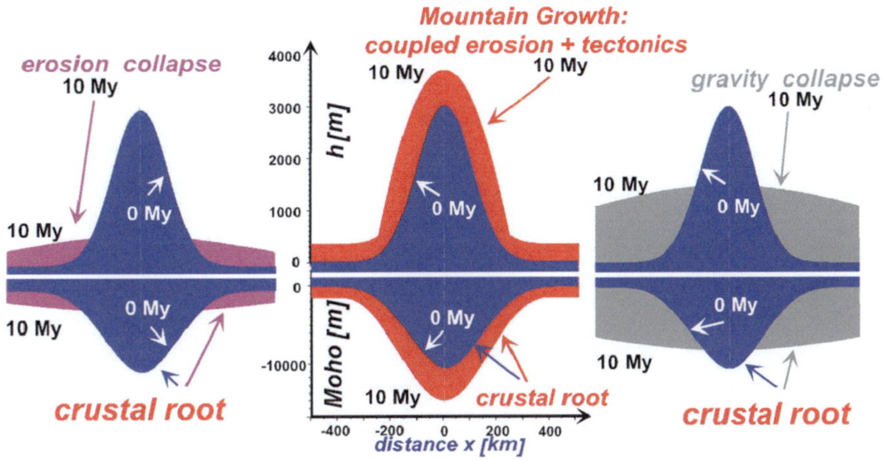

Fig. 11 Summary of semi-analytical experiments: 3 major styles of topography evolution in terms of coupling between surface and sub-surface processes

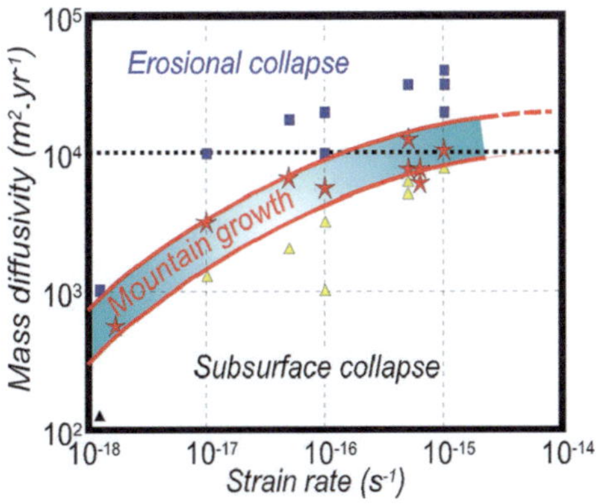

Fig. 12 Semi-analytical experiments: Modes of evolution of mountain ranges as a function of the coefficient of erosion (mass diffusivity) and tectonic strain rate, established for semi-analytical experiments with spatial resolution of 2 km × 2 km. Note that the coefficients of erosion are scale dependent, they may vary with varying resolution (or roughness) of the surface topography. Squares correspond to the experiments were erosional (surface) collapse was observed, triangles – experiments were subsurface collapse was observed, stars – experiments were localized stable growth of topography was observed

Coupled Regime and Graded Geometries

In the coupled regime the topography of the range can be seen to develop into a nearly parabolic graded geometry (Fig. 8). This graded form is attained after 2–3 My and reflects some dynamic equilibrium with the topographic rate of uplift being nearly constant over the range. Rates of denudation and of tectonic uplift can be seen to be also relatively constant over the range domain. Geometries for which the denudation rate is constant over the range are nearly parabolic since they are defined by

$$de/dt = k\, d^2 h/dx^2 = const. \qquad (17)$$

Integration of this expression yields a parabolic expression for $h = x^2(de/dt)/2k + C_1 x + C_0$, with C_1 and C_0 being constants to be defined from boundary conditions. The graded geometries obtained in the experiments slightly deviate from parabolic curves because they do not exactly correspond to uniform denudation over the range (h is also function of du/dt, etc.). This simple consideration does however suggest that the overall shape of graded geometries is primarily controlled by the erosion law. We then made computations assuming non linear diffusion laws, in order to test whether the setting of the coupled regime might depend on the erosion law. We considered non-linear erosion laws, in which the increase of transport capacity downslope is modelled by a 1st order or 2d order non linear diffusion (Equation 4). For a given shortening rate, experiments that yield similar erosion rates over the range lead to the same evolution ("erosional collapse", "subsurface collapse" or "mountain growth") whatever is the erosion law. It thus appears that the emergence of the coupled regime does not depend on a particular erosion law but rather on the intensity of erosion relative to the effective viscosity of the lower crust. By contrast, the graded geometries obtained in the mountain growth regime strongly depend on the erosion law (Fig. 10). The first order diffusion law leads to more realistic, than parabolic, "triangular" ranges whereas the 2d order diffusion leads to plateau-like geometries. It appears that the graded geometry of a range may reflect the macroscopic characteristics of erosion. It might therefore be possible to infer empirical macroscopic laws of erosion from the topographic profiles across mountain belts provided that they are in a graded form.

Sensitivity to the Rheology and Structure of the Lower Crust

The above shown experiments have been conducted assuming a quartz rheology for the entire crust (= weak lower crust), which is particularly favourable for channel flow in the lower crust. We also conducted additional experiments assuming more basic lower crustal compositions (diabase, quartz-diorite). It appears that even with a relatively strong lower crust the coupled regime allowing for mountain growth can settle (Avouac and Burov, 1996). The effect of a less viscous lower crust is that the domain of values of the shortening rates and mass diffusivity for which the coupled regime can settle is simply shifted: at a given shortening rate lower rates of erosion are required to allow for the growth of the initial mountain. The domain defining the "mountain growth" regime in Figs. 11, 12 is thus shifted towards smaller mass diffusivities when a stronger lower crust is considered. The graded shape obtained in this regime does not differ from that obtained with a quartz rheology. However, if the lower crust was strong enough to be fully coupled to the upper mantle, the dynamic equilibrium needed for mountain growth would not be established. Estimates of the yield strength of the lower crust near the Moho boundary for thermal ages from 0 to 2,000 My. and for Moho depths from 0 to 80 km, made by Burov and Diament (1995), suggest that in most cases a crust thicker than about 40–50 km implies a low viscosity channel in the lower crust. However, if the lithosphere is very old (>1,000 My) or its crust is thin, the coupled regime between erosion and horizontal flow in the lower crust will not develop.

Comparison With Observations

We compared our semi-analytical models with the Tien Shan range (Fig. 1) because in this area, the rates of deformation and erosion have been well estimated from previous studies (Avouac et al., 1993: Metivier and Gaudemer, 1997), and because this range has a relatively simple 2-D geometry. The Tien Shan is the largest and most active mountain range in central Asia. It extends for nearly 2,500 km between the Kyzil Kum and Gobi deserts, with some peaks rising to more than 7,000 m. The high level of seismicity (Molnar and Deng, 1984) and deformation of Holocene alluvial formations (Avouac et al., 1993) would indicate a rate of shortening of the order of 1 cm/y. In fact, the shortening rate is thought to increase from a few mm/y east of 90°E to about 2 cm/y west of 76°E (Avouac et al., 1993). Clockwise rotation of the Tarim Basin (at the south of Tien Shan) with respect to Dzungaria and Kazakhstan (at the north) would be responsible for this westward increase of shortening rate as well as of the increase of the width of the range (Chen et al., 1991; Avouac et al., 1993). The gravity studies by Burov (1990) and Burov et al. (1990) also suggest westward decrease of the integrated strength of the lithosphere. The westward increase of the topographic load and strain rate could be responsible for this mechanical weakening. The geological record suggests a rather smooth morphology with no great elevation differences and low elevations in the Early Tertiary and that the range was reactivated in the middle Tertiary, probably as a result of the India–Asia collision (e.g., Tapponnier and Molnar, 1979; Molnar and Tapponnier, 1981; Hendrix et al., 1992, 1994). Fission track ages from detrital appatite from the northern and southern Tien Shan would place the reactivation at about 20 m.y. (Hendrix et al., 1994; Sobel and Dumitru, 1995). Such an age is consistent with the middle Miocene influx of clastic material and more rapid subsidence in the forelands (Hendrix et al., 1992; Métivier and Gaudemer, 1997) and with a regional Oligocene unconformity (Windley et al., 1990). The present difference of elevation of about 3,000 m between the range and the lowlands would therefore indicate a mean rate of uplift of the topography, during the Cenozoic orogeny, of the order of 0.1–0.2 mm/y. The foreland basins have collected most of the material removed by erosion in the mountain. Sedimentary isopachs indicate that $1.5 +/- 0.5 \times 10^6$ km^3 of material would have been eroded during the Cenozoic orogeny (Métivier and Gaudemer, 1997), implying erosion rates of 0.2–0.5 mm/y on average. The tectonic uplift would thus have been of 0.3–0.7 mm/y on average. On the assumption that the range is approximately in local isostatic equilibrium (Burov et al., 1990; Ma, 1987), crustal thickening below the range has absorbed 1.2 to 4 10^6 km^3 (Métivier and Gaudemer, 1997). Crustal thickening would thus have accomodated 50–75% of the crustal shortening during the Cenozoic orogeny, with the remaining 25–50% having been fed back to the lowlands by surface processes. If we now place approximately the Tien Shan on the plot in Figs. 7, 8, 9, 10 the 1 to 2 cm/y shortening corresponds to a basic strain rate of $\varepsilon_{xx} = 1.5 \times - 3 \times 10^{-16}$ s^{-1} and the 0.2–0.5 mm/y denudation rate implies a mass diffusivity of a few 10^3 to 10^4 m^2/y. These values actually place the Tien Shan in the "mountain growth" regime (Figs. 8, 11, 12). We therefore conclude that the localized growth of a range like the Tien Shan indeed could result from the coupling between surface processes and horizontal strains. We do not dispute the possibility for a complex mantle dynamics beneath the Tien Shan as has been inferred by various geophysical investigations (Vinnik and Saipbekova, 1984; Vinnik et al., 2006; Makeyeva et al., 1992; Roecker et al., 1993), but we contend that this mantle dynamics has not necessarily been the major driving mechanism of the Cenozoic Tien Shan orogeny.

Numerical Experiments

Fully numerical thermo-mechanical models were used to test more realistic scenarios of continental convergence (Fig. 6 bottom), in which one of the continental plates under-thrusts the other (simple shear mode, or continental "subduction"), the raising topography undergoes internal deformations, and the major thrust faults play an active role in localisation of the deformation and in the evolution of the range. Also, in the numerical experiments, there is no pre-defined initial topography, which forms and evolves in time as a result of deformation and coupling between tectonic deformation and erosion processes. We show the tests for two contrasting cases: slow convergence and slow erosion (Western Alps, 6 mm/y, $k = 500$–1,000 m^2/y) and very fast convergence and fast erosion (India–Himalaya collision, 6 cm/y during the first stage of continent-continent subduction, up to 15 cm/y at the preceding stage of oceanic subduction, $k =$ 3,000–10,000 m^2/y). The particular interest of testing the model for the conditions of the India–Himalaya–Tibet collision refers to the fact that this zone of both intensive convergence (Patriat and Achache, 1984) and erosion (e.g., Hurtrez et al., 1999) belongs to the same geodynamic framework of India–Eurasia collision as the Tien Shan range considered in the semi-analytical experiments from the previous sections (Fig. 1).

For the Alps, characterized by slow convergence and erosion rates (maximum 6 mm/y (Schmid et al., 1997; Burov et al., 2001; Yamato et al., 2008), $k =$ 500–1,000 m^2/y according to Figs. 11, 12), we have studied a scenario in which the lower plate has already subducted to a 100 km depth below the upper plate (Burov et al., 2001). This assumption was needed to enable the continental subduction since, in the Alps, low convergence rates make model initialisation of the subduction process very difficult without perfect knowledge of the initial configuration (Toussaint et al., 2004a). The previous (Burov et al., 2001) and recent (Yamato et al., 2008) numerical experiments (Figs. 13, 14) confirm the idea that surface processes, which selectively remove the most rapidly growing topography, result in dynamic tectonically-coupled unloading of the lithosphere below the thrust belt, whereas the deposition of the eroded matter in the foreland basins results in additional subsidence. As a result, a strong feedback between tectonic and surface processes can be established and regulate the processes of mountain building during very long period of time (in the experiments, 20–50 My): the erosion-sedimentation prevent the mountain from reaching gravitationally unstable geometries. The "Alpine" experiments demonstrate that the feedback between surface and tectonic processes may allow the mountains to survive over very large time spans (> 20–50 My). This feedback favours localized crustal shortening and stabilizes topography and thrust faults in time. Indeed even though slow convergence scenario is not favourable for continental subduction, the model shows that once it is initialised, the tectonically coupled surface processes help to keep the major thrust working. Otherwise, in the absence of a strong feedback between surface and subsurface processes, the major thrust fault is soon locked, the upper plate couples with the lower plate, and the system evolution turns from simple shear subduction to pure shear collision (Toussaint et al., 2004a; Cloetingh et al., 2004). Moreover, (Yamato et al., 2008) have demonstrated that the feedback with the surface processes controls the shape of the accretion prism, so that in cases of strong misbalance with tectonic forcing, the prism would not be formed or has an unstable geometry. However, even in the case of strong balance

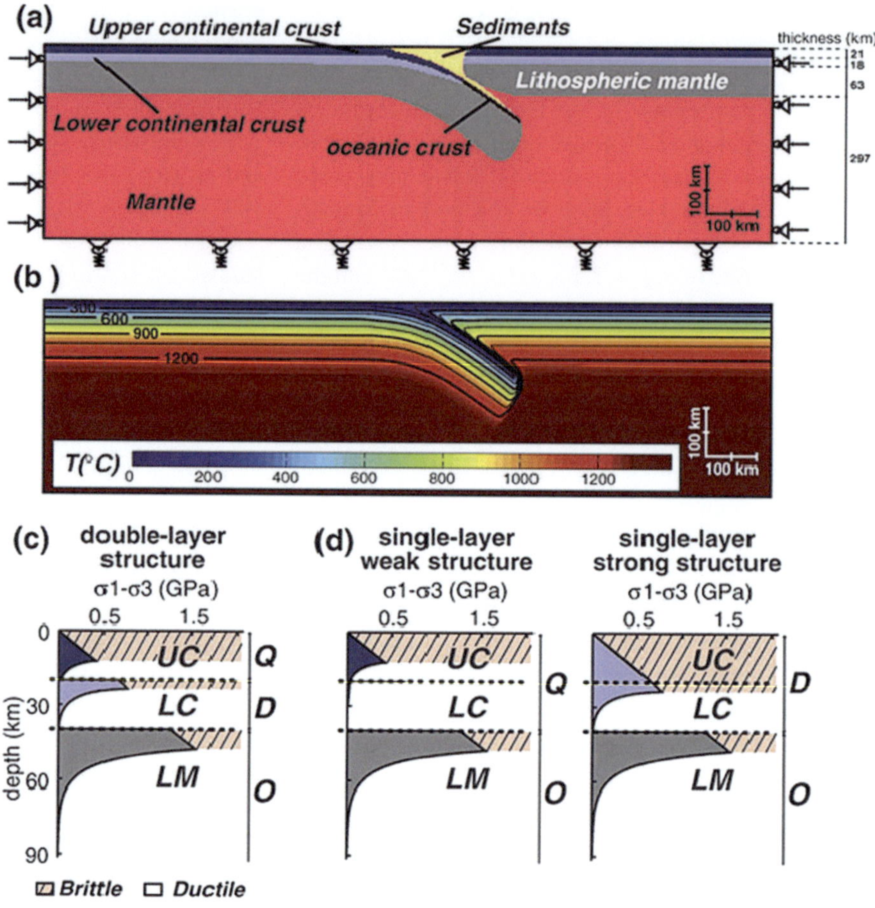

Fig. 13 Coupled numerical model of Alpine collision, with surface topography controlled by dynamic erosion. Model setup. *Top*: Initial morphology and boundaries conditions. The horizontal arrows correspond to velocity boundary conditions imposed on the sides of the model. The basement is Winkler isostatic. The top surface is free (plus erosion/sedimentation). *Middle*: Thermal structure used in the models (*bottom*) representative viscous-elastic-plastic yield strength profile for the continental lithosphere for a double-layer structure of the continental crust and the initial thermal field assuming a constant strain rate of 10^{-14} s^{-1}. In the experiments, the strain rate is highly variable both vertically and laterally. Abbreviations: UC, upper crust; LC, lower crust; LM, lithospheric mantle

between surface and subsurface processes, topography cannot infinitely grow: as soon as the range grows to some critical size, it cannot be supported anymore due to the limited strength of the constituting rocks, and ends up by gravitational collapse. The other important conclusion that can be drawn from slow-convergence Alpine experiments is that in case of slow convergence, erosion/sedimentation processes do not effect deep evolution of the subducting lithosphere. Their primary affect spreads to the first 30–40 km in depth and generally refers to the evolution of topography and of the accretion wedge.

In case of fast collision, the role of surface processes becomes very important. Our experiments on fast "Indian–Asia" collision were based on the results of Toussaint et al. (2004b). The model and the entire setup (Fig. 6, bottom) are identical to those described in detail in Toussaint et al. (2004b). For this reason, we send the interested reader to this study (see also Appendix 4 and description of the numerical model in the previous sections). Toussaint et al. (2004b) tested the possibility of subduction of the Indian plate beneath the Himalaya and Tibet at early stages of collision (first 15 My). This study used by default the "stable" values of the coefficient of erosion (3,000 ± 1,000 m^2/y) derived from the semi-analytical model of (Avouac and Burov, 1996) for shortening rate of 6 cm/y. The coefficient of erosion was only slightly varied in a way to keep the topography within reasonable limits, yet, Toussaint et al. (2004b) did not test sensitivities of the Himalayan orogeny to large variations in the erosion rate. Our new experiments fill this gap by testing the stability of the same model for large range of k, from 50 m^2/y to 11,000 m^2/y. These experiments (Figs. 15, 16, 17) demonstrate that, depending on the intensity of surface processes, horizontal compression of continental lithosphere can lead either to strain localization below a growing range and continental subduction, or to distributed thickening or buckling/folding (Fig. 16). The experiments suggest

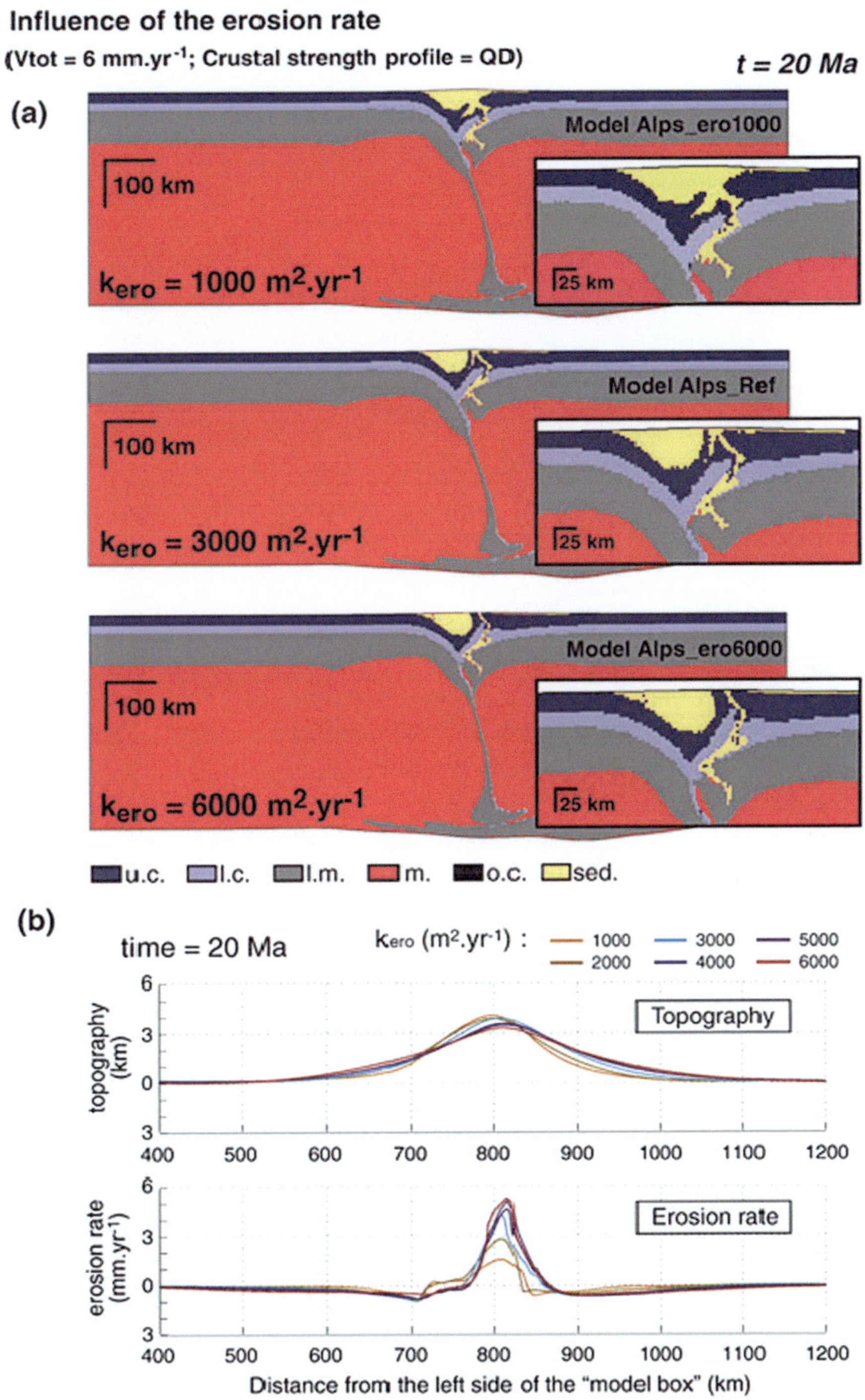

Fig. 14 Morphologies of the models after 20 Myr of experiment for different erosion coefficients. Topography and erosion rates at 20 Myr obtained in different experiments testing the influence of the erosion coefficients.This model demonstrates that erosion-tectonics feedback help the mountain belt to remain as a localized growing feature for about 20–30 My

that homogeneous thickening occurs when erosion is either too strong ($k>1{,}000\ m^2/y$), in that case any topographic irregularity is rapidly erased by surface processes (Fig. 17), or when erosion is too weak ($k<50\ m^2/y$). In case of small k, surface elevations are unrealistically high (Fig. 17), which leads to vertical overloading and failure of the lithosphere and to increase of the frictional force along the major thrust fault. As consequence, the thrust fault is locked up leading to coupling between the upper and lower plate; this

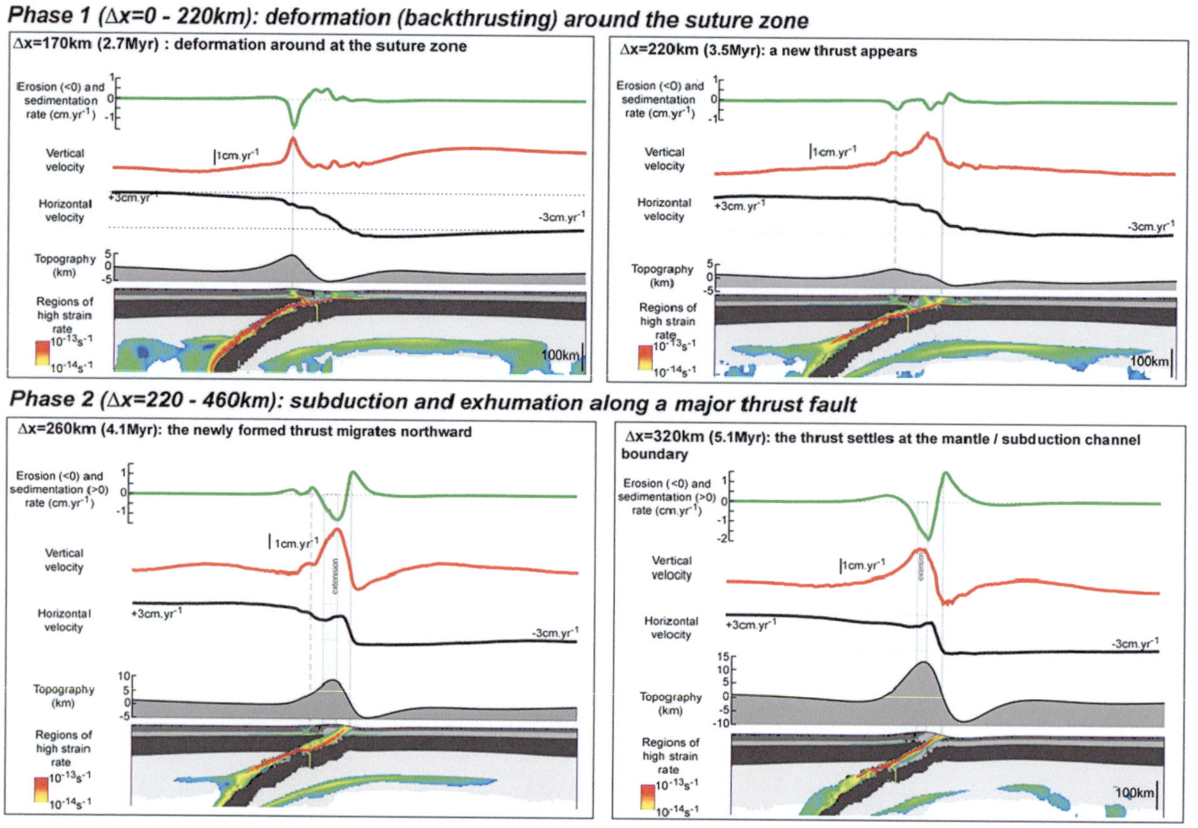

Fig. 15 Coupled numerical models of India–Eurasia type of collision as function of the coefficient of erosion. These experiments were performed in collaboration with G. Toussaint using numerical setup (Fig. 6, *bottom*) identical to (Toussaint et al., 2004b). The numerical method is identical to that of (Burov et al., 2001 and Toussaint et al., 2004a, b; see also the experiment shown in Figs. 13, 14). Shown are initial phases of rapid continental subduction that demonstrate strong correlation between the evolution of the surface erosion/sedimentation rate (k = 3,000 m^2/y), vertical and horizontal uplift rate, and the inner structure of the thrust zone and subducting plate

results in overall buckling of the region whereas the crustal root below the range starts to spread out laterally with formation of a flat "pancake-shaped" topographies. On the contrary, in case of a dynamic balance between surface and subsurface processes (k = 2,000–3,000 m^2/y, close to the predictions of the semi-analytical model, Fig. 11, 12), erosion/sedimentation resulted in long-term localization of the major thrust fault that kept working during 10 My. In the same time, in the experiments with k = 500–1,000 m^2/y (moderate feedback between surface and subsurface processes), the major thrust fault and topography were almost stationary (Fig. 16). In case of a stronger feedback (k = 2,000–5,000 m^2/y) the range and the thrust fault migrated horizontally in the direction of the lower plate ("India"). This basically happened when both the mountain range and the foreland basin reached some critical size. In this case, the "initial" range and major thrust fault were abandoned after about 500 km of subduction, and a new thrust fault, foreland basin and range were formed "to the south" (i.e., towards the subducting plate) of the initial location. The numerical experiments confirm our previous idea that intercontinental orogenies could arise from coupling between surface/climatic and tectonic processes, without specific help of other sources of strain localisation. Given the differences in the problem setting, the results of the numerical experiments are in good agreement with the semi-analytical predictions (Figs. 11, 12) that predict mountain growth for k on the order of 3,000–10,000 m^2/y for strain rates on the order of 0.5 × 10^{-16} s^{-1}–10^{-15} s^{-1}. The numerical experiments, however, predict somewhat smaller values of k than the semi-analytical experiments. This can be explained by

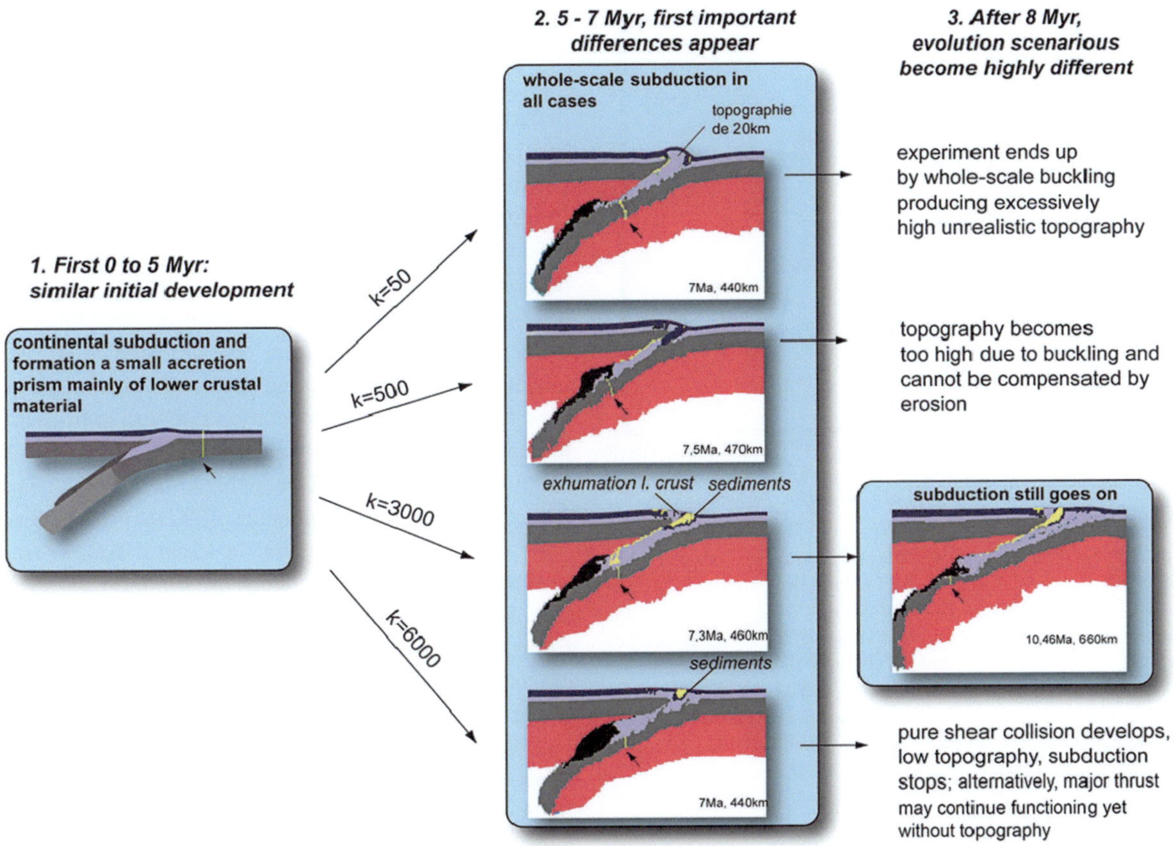

Fig. 16 Evolution of the collision as function of the coefficient of erosion. Sub-vertical stripes associated with little arrows point to the position of the passive marker initially positioned across the middle of the foreland basin. Displacement of this marker indicates the amount of subduction. Δx is amount of shortening. Different brittle-elastic-ductile rheologies are used for sediment, upper crust, lower crust, mantle lithosphere and the asthenosphere (Table 2)

the difference in the convergence mode attained in the numerical experiments (simple shear subduction) and in the analytical models (pure shear). For the same convergence rate, subduction resulted in smaller tectonic uplift rates than pure shear collision. Consequently, "stable" erosion rates and k values are smaller for subduction than for collision.

Conclusions

Tectonic evolution of continents is highly sensitive to surface processes and, consequently, to climate. Surface processes may constitute one of the dominating factors of orogenic evolution that not only largely controls the development and shapes of surface topography, major thrust faults and foreland basins, but also deep deformation and overall collision style. For example, similar dry climatic conditions to the north and south of the Tien Shan range favour the development of its highly symmetric topography despite the fact that the colliding plates have extremely contrasting, asymmetric mechanical properties (in the Tarim block, the equivalent elastic thickness, EET = 60 km whereas in the Kazakh shield, EET = 15 km (Burov et al., 1990)).

Although there is no a perfect model for surface processes, the combination of diffusion and fluvial transport models provides satisfactory results for most large-scale tectonic applications.

In this study, we investigated interactions between the surface and subsurface processes for three representative cases: (1) very fast convergence rate, such as India–Himalaya–Tibet collision; (2) intermediate rate convergence settings (Tien Shan); (3) very slow convergence settings (Wetern Alps).

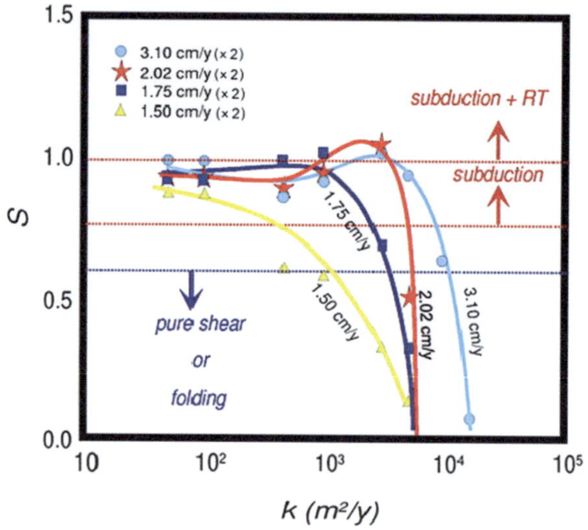

Fig. 17 Summary of the results of the numerical experiments showing the dependence of the "subduction number" S ($S =$ amount of subduction to the total amount of shortening) on the erosion coefficient, k, for different values of the convergence rate (values are given for each side of the model). Data (sampled for $k = 50$, 100, 500, 1,000, 3,000, 6,000 and 11,000 m^2/y) are fitted with cubic splines (*curves*). Note local maximum on the S-k curves for $u > 1.75$ cm/y and $k > 1{,}000$ m^2/k

The influence of erosion is different in case of very slow and very rapid convergence. In case of slow Alpine collision, the persistence of tectonically formed topography and the accretion prism may be insured by coupling between the surface and tectonic processes. Surface processes basically help to initialize and maintain continental subduction for a certain amount of time (5–7 My, maximum 10 My). They can stabilize, or "freeze" dynamic topography and the major thrust faults for as long as 50 My.

In case of rapid convergence (>5 cm/y), surface processes may affect deep evolution of the subducting lithosphere, down to the depths of 400–600 km. The way the Central Asia has absorbed rapid indentation of India may somehow reflect the sensitivity of large scale tectonic deformation to surface processes, as asymmetry in climatic conditions to the south of Himalaya with respect to Tibet to the north may explain the asymmetric development of the Himalayn-Tibetan region (Avouac and Burov, 1996). Interestingly, the mechanically asymmetric Tien Shan range situated north of Tibet, between the strong Tarim block and weak Kazakh shield, and characterised by similar climatic conditions at both sides of the range, is highly symmetric. Previous numerical models of continental indentation that were also based on continuum mechanics, but neglected surface processes, predicted a broad zone of crustal thickening, resulting from nearly homogeneous straining, that would propagate away from the indentor. In fact, crustal straining in Central Asia has been very heterogeneous and has proceeded very differently from the predictions of these models: long lived zone of localized crustal shortening has been maintained, in particular along the Himalaya, at the front of the indentor, and the Tien Shan, well north of the indentor; broad zones of thickened crust have resulted from sedimentation rather than from horizontal shortening (in particular in the Tarim basin, and to some extent in some Tibetan basins such as the Tsaïdam (Métivier. and Gaudemer, 1997)). Present kinematics of active deformation in Central Asia corroborates a highly heterogeneous distribution of strain. The 5 cm/y convergence between India and stable Eurasia is absorbed by lateral extrusion of Tibet and crustal thickening, with crustal thickening accounting for about 3 cm/y of shortening. About 2 cm/y would be absorbed in the Himalayas and 1 cm/y in the Tien Shan. The indentation of India into Eurasia has thus induced localized strain below two relatively narrow zones of active orogenic processes while minor deformation has been distributed elsewhere. Our point is that, as in our numerical experiments, surface processes might be partly responsible for this highly heterogeneous distribution of deformation that has been maintained over several millions or tens of millions years. First active thrusting along the Himalaya and in the Tien Shan may have been sustained during most of the Cenozoic time, thanks to continuous erosion. Second, the broad zone of thickened crust in Central Asia has resulted in part from the redistribution of the sediments eroded from the localized growing reliefs. Moreover, it should be observed that the Tien Shan experiences a relatively arid intracontinental climate while the Himalayas is exposed to a very erosive monsoonal climate. This disparity may explain why the Himalaya absorbs twice as much horizontal shortening as the Tien Shan. In addition, the nearly equivalent climatic conditions on the northern and southern flanks of the Tien Shan might have favoured the development of a nearly symmetrical range. By contrast the much more erosive climatic conditions on the southern than on the northern flank of the Himalaya may have favoured the development of systematically south vergent structures. While the Indian upper crust would have been delaminated and brought to the surface of erosion by north dipping thrust faults the Indian lower crust would have flowed below Tibet. Surface processes might therefore have facilitated injection of Indian lower crust below Tibet. This would explain crustal thickening of Tibet with minor horizontal shortening in the upper crust, and minor sedimentation. We thus suspect that climatic zonation in Asia has exerted some control on the spatial distribution of the intracontinental strain induced by the India–Asia collision. The interpretation of intracontinental deformation should not be thought of only in terms of boundary conditions induced by global plate kinematics but also in terms of global climate. Climate might therefore be considered as a forcing factor of continental tectonics.

To summarize, we suggest three major modes of evolution of thrust belts and adjacent forelands (Figs. 11, 12):

1. Erosional collapse (erosion rates are higher than the tectonic uplift rates. Consequently, the topography cannot grow).
2. Localized persistent growth mode. Rigid feedback between the surface processes and tectonic uplift/subsidence that may favour continental subduction at initial stages of collision.
3. Gravity collapse (or "plateau mode", when erosion rates are insufficient to compensate tectonic uplift rates. This may produce plateaux in case of high convergence rate).

It is noteworthy (Fig. 9) that while in the "localized growth regime", the system has a very important reserve of stability and may readapt to eventual changes in tectonic or climatic conditions. However, if the limits of stability are exceeded, the system will collapse in very rapid, catastrophic manner.

We conclude that surface processes must be taken into full account in the interpretation and modelling of long-term deformation of continental lithosphere. Conversely, the mechanical response of the lithosphere must be accounted for when large-scale topographic features are interpreted and modelled in terms of geomorphologic processes. The models of surface process are most realistic if treated in two dimensions in horizontal plane, while most of the current mechanical models are two dimensional in the vertical cross-section. Hence, at least for this reason, a next generation of 3D tectonically realistic thermo-mechanical models is needed to account for dynamic feedbacks between tectonic and surface processes. With that, new explanations of evolution of tectonically active systems and surface topography can be provided.

Acknowledgments I am very much thankful to T. Yamasaki, the anonymous reviewer and M. Ter Voorde for their highly constructive comments.

Appendix 1: Model of Flexural Deformation of the Competent Cores of the Brittle-Elasto-Ductile Crust and Upper Mantle

Vertical displacements of competent layers in the crust and mantle in response to redistribution of surface and subsurface loads (Fig. 6, top) can be described by plate equilibrium equations in assumption of non-linear rheology (Burov and Diament, 1995). We assume that the reaction of the competent layers is instantaneous (response time $dt \sim \mu_{min}/E < 10^3$ years, where μ_{min} is the minimum of effective viscosities of the lower crust and asthenosphere)

$$\frac{\partial}{\partial x}\left(\frac{\partial}{\partial x}\left(\frac{E}{12(1-\nu^2)}(\tilde{T}_e^3(\phi)\frac{\partial^2 w(x,t)}{\partial x^2}\right) + \tilde{T}_x(\phi)\frac{\partial w(x,t)}{\partial x}\right)$$
$$+ p_-(\phi)w(x,t) - p_+(x,t) = 0$$

$$\tilde{T}_e(\phi) = \left(\frac{\tilde{M}_x(\phi)}{L}\left(\frac{\partial^2 w(x,t)}{\partial x^2}\right)^{-1}\right)^{\frac{1}{3}} \quad (18)$$

$$\tilde{M}_x(\phi) = -\sum_{i=1}^{n}\sum_{j=1}^{m_i}\int_{y_{ij}^-(\phi)}^{y_{ij}^+(\phi)} \sigma_{xx}^{(j)}(\phi)y_i^*(\phi)dy$$

$$\tilde{T}_x(\phi) = -\sum_{i=1}^{n}\sum_{j=1}^{m_i}\int_{y_{ij}^-(\phi)}^{y_{ij}^+(\phi)} \sigma_{xx}^{(j)}(\phi)dy$$

$$\sigma_{xx}^{(j)}(\phi) = sign(\varepsilon_{xx})\mathbf{min}\left(\left|\sigma^f\right|, \sigma_{xx}^{e(j)}(\phi)\right)$$
$$\sigma_{xx}^{e(j)}(\phi) = (y_i^*(\phi)\frac{\partial^2 w(x,t)}{\partial x^2}E_i(1-\nu_i^2)^{-1}$$

where $w = w(x,t)$ is the vertical plate deflection (related to the regional isostatic contribution to tectonic uplift du_{is} as $du_{is} = w(x,t)–w(x,t–dt)$), $\phi \equiv \{x,y,w,w',w'',t\}$, y is downward positive, $y_i^* = y - y_{ni}(x)$, y_{ni} is the depth to the ith neutral (i.e., stress-free, $\sigma_{xx}|_{y_i^*=0} = 0$) plane; $y_i^-(x) = y_i^-$, $y_i^+(x) = y_i^+$ are the respective depths to the lower and upper low-strength interfaces. σ^f is defined from Equations (10) and (11). n is the number of mechanically decoupled competent layers; m_i is the number of "welded" (continuous σ_{xx}) sub-layers in the ith detached layer. P_w is a restoring stress ($p\_ \sim (\rho_m - \rho_{c1})g$) and p^+ is a sum of surface and subsurface loads. The most important contribution to p^+ is from the load of topography, that is, $p^+ \sim \rho g h(x,t)$, where the topographic height $h(x,t)$ is defined as $h(x,t) = h(x,t–dt)+dh(x,t) = h(x,t–dt)+du(x,t)–de(x,t)$, where $du(x,t)$ and $de(x,t)$ are, respectively tectonic uplift/subsidence and denudation/sedimentation at time interval $(t–dt, t)$, counted from the sea level. The thickness of the ith competent layer is $y_i^+ - y_i^- = \Delta h_i(x)$. The term w'' in (18) is inversely proportional to the radius of plate curvature $R_{xy} \approx -(w'')^{-1}$. Thus the higher is the local curvature of the plate, the lower is the local integrated strength of the lithosphere. The integrals in (18) are defined through the constitutive laws (6–9) and Equations (10) and (11) relating the stress σ_{xx} and strain $\varepsilon_{xx} = \varepsilon_{xx}(\phi)$ in a given segment $\{x,y\}$ of plate. The value of the unknown function $\tilde{T}_e(\phi)$ has a meaning of a "momentary" effective elastic thickness of the plate. It holds only for the given solution for plate deflection w. $\tilde{T}_e(\phi)$ varies with changes in plate geometry and boundary conditions. The effective integrated strength of the lithosphere (or $Te = \tilde{T}_e(\phi)$) and the state of its interiors (brittle, elastic or ductile) depends on differential stresses caused by local deformation, while stresses at each level are constrained by the YSE. The non-linear Equations (18) are solved using an iterative approach based on finite difference approximation (block matrix presentation) with linearization by Newton's method (Burov and Diament, 1992). The procedure starts from calculation of elastic prediction $w_e(x)$ for $w(x)$, that provides predicted $w_e(x)$, $w'_e(x)$, w''_e used to find subiteratively solutions for $y_{ij}^-(\phi)$, $y_{ij}^+(\phi)$, and $y_{ni}(\phi)$ that satisfy (5), (6), (7), (10). This yields corrected solutions for $\tilde{M}_x$ and $\tilde{T}_x$ which are used to obtain $\tilde{T}_e$ for the next iteration. At this stage we use gradual loading technique to avoid numerical oscillations. The accuracy is checked directly on each iteration, through back-substitution of the current solution to (18) and calculation of the discrepancy between the right and left sides of (18). For the boundary conditions on the ends of the plate we use commonly inferred combination of plate-boundary shearing force $Q_x(0)$,

$$\tilde{Q}_x(\phi) = -\sum_{i=1}^{n}\sum_{j=1}^{m_i}\int_{y_{ij}^-(\phi)}^{y_{ij}^+(\phi)} \sigma_{xy}^{(j)}(\phi)dy, \tag{19}$$

and plate boundary moment $M_x(0)$ (in the case of broken plate) and w=0, w'= 0 (and h = 0, $\partial h/\partial x = 0$) at $x \to \pm\infty$. The starting temperature distribution and yield-stress profiles (see above) are obtained from the solution of the heat transfer problem for the continental lithosphere of Paleozoic thermotectonic age, with average Moho thickness of 50 km, quartz-controlled crust and olivine-controlled upper mantle, assuming typical horizontal strain rates of $\sim 0.1 \div 10 \times 10^{-15}$ sec^{-1}. (Burov and Diament, 1995). These parameters roughly resemble Tien Shan and Tarim basin (Fig. 1).

Burov and Diament (1995) have shown that the flexure of the continental lithosphere older than 200–250 My is predominantly controlled by the mechanical portion of mantle lithosphere (depth interval between T_c and h_2). Therefore, we associate the deflection of Moho with the deflection of the entire lithosphere (analogously to Lobkovsky and Kerchman, 1991; Kaufman and Royden, 1994; Ellis et al., 1995). Indeed, the effective elastic thickness of the lithosphere (T_e) is approximately equal to $\sqrt[3]{T_{ec}^3 + T_{em}^3}$, where T_{ec} is the effective elastic thickness of the crust and T_{em} is the effective elastic thickness of the mantle lithosphere (e.g., Burov and Diament, 1995). $\lim \sqrt[3]{T_{ec}^3 + T_{em}^3} \approx \max(T_{ec}, T_{em})$. T_{ec} cannot exceed h_{c1}, that is 15–20 km (in practice, $T_{ec} \leq 5$–10 km). T_{em} cannot exceed $h_2 - T_c \sim 60$–70 km. Therefore $T_e \approx T_{em}$ which implies that total plate deflection is controlled by the mechanical portion of the mantle lithosphere.

Appendix 2: Model of Flow in the Ductile Crust

As it was already mentioned, our model of flow in the low viscosity parts of the crust is similar that formulated by Lobkovsky (1988), Lobkovsky and Kerchman (1991) (hereafter referred as L&K), or Bird (1991). However, our formulation can allow computation of different types of flow ("symmetrical", Poiseuille, Couette) in the lower crust (L&K considered Couette flow only). In the numerical experiments shown in this paper we will only consider cases with a mixed Couette/Poiseuille/symmetrical flow, but we first tested the same formulation as L&K. The other important difference with L&K's models is, naturally, the use of realistic erosion laws to simulate redistribution of surface loads, and of the realistic brittle-elastic-ductile rheology for modeling the response of the competent layers in the lithosphere.

Tectonic uplift $du(x,t)$ due to accumulation of the material transported through ductile portions of the lower and upper crust ($dh(x,t) = du(x,t)–de(x,t)$) can be modelled by equations which describe evolution of a thin subhorizontal layer of a viscous medium (of density $\rho c2$ for the lower crust) that overlies a non-extensible pliable basement supported by Winkler forces (i.e., flexural response of the mantle lithosphere which is, in-turn, supported by hydrostatic reaction of the astenosphere) (Batchelor, 1967; Kusznir and Matthews, 1988; Bird and Gratz, 1990; Lobkovsky and Kerchman, 1991; Kaufman and Royden, 1994).

The normal load, which is the weight of the topography $p+(x)$ and of the upper crustal layer (thickness h_{c1} and density ρ_{c1}) is applied to the surface of the lower crustal layer through the flexible competent upper crustal layer. This internal ductile crustal layer of variable thickness $\Delta h_{c2} = \Delta h_0(x,0)+\tilde{h}(x)+w(x)$ is regionally compensated by the strength of the underlying competent mantle lithosphere (with density ρ_m). Variation of the elevation of the upper boundary of the ductile layer ($d\tilde{h}$) with respect to the initial thickness ($\Delta h_0(x,0)$) leads to variation of the normal load applied to the mantle lithosphere. The regional isostatic response of the mantle lithosphere results in deflection (w) of the lower boundary of the lower crustal layer, that is the Moho boundary, which depth is $h_c(x,t) = T_c(x,t) = \Delta h_{c2}+y_{13}$ (see Table 1). The vertical deflection w (Equation 18) of the Moho depends also on vertical undulation of the elastic-to-ductile crust interface y_{13}.

The absolute value of $\tilde{h}$ is not equal to that of the topographic undulation h by two reasons: first, h is effected by erosion, second, $\tilde{h}$ depends not only on the uplift of the upper boundary of the channel, but also on variation of thickness of the competent crust given by value of $y_{13}(x)$. We can require $\tilde{h}(x, t)–\tilde{h}(x, t–dt) = du–dy_{13}$. Here $dy_{13} = y_{13}(\phi, t)–y_{13}(\phi, t–dt)$ is the relative variation in the position of the lower boundary of the elastic core of the upper crust due to local changes in the level of differential (or deviatoric) stress (Fig. 7). This flexure- and flow-driven differential stress can weaken material and, in this sense, "erode" the bottom of the strong upper crust. The topographic elevation $h(x,t)$ can be defined as $h(x,t) = h(x,t–dt)+d\tilde{h}–de(t)–dy_{13}$ where dy_{13} would have a meaning of "subsurface or thermomechanical erosion" of the crustal root by local stress.

The equations governing the creeping flow of an incompressible fluid, in Cartesian coordinates, are:

$$-\frac{\partial \sigma_{xx}}{\partial x}+\frac{\partial \tau_{xy}}{\partial y}+F_x=0;\quad -\frac{\partial \sigma_{yy}}{\partial y}+\frac{\partial \tau_{xy}}{\partial x}+F_y=0$$

$$\sigma_{xx}=-\tau_{xx}+p=-2\mu\frac{\partial u}{\partial x}+p$$

$$\sigma_{xy}=\tau_{xy}=\mu\left(\frac{\partial u}{\partial y}+\frac{\partial v}{\partial x}\right) \tag{20}$$

$$\sigma_{yy}=-\tau_{yy}+p=-2\mu\frac{\partial v}{\partial x}+p$$

$$\frac{\partial u}{\partial x}+\frac{\partial v}{\partial y}=0 \tag{21}$$

$$\mu=\frac{\sigma}{2\dot{\varepsilon}}$$
$$\dot{\varepsilon}=\sigma^n A^* \exp(-H^*/RT)\,. \tag{22}$$

Where μ is the effective viscosity, p is pressure, u and v are the horizontal and vertical components of the velocity $\mathbf{v}$, respectively. F is the body force. $u = \partial\psi/\partial y$ is the horizontal component of velocity of the differential movement in the ductile crust, $v = -\partial\psi/\partial x$ is its vertical component; and $\partial u/\partial y = \dot{\varepsilon}_{c20}$ is a component of shear strain rate due to the differential movement of the material in the ductile crust (the components of the strain rate tensor are consequently: $\dot{\varepsilon}_{11} = 2\partial u/\partial x$; $\dot{\varepsilon}_{12} = \partial u/\partial y + \partial v/\partial x$; $\dot{\varepsilon}_{22} = 2\partial v/\partial y$).

Within the low viscosity boundary layer of the lower crust, the dominant basic process is simple shear on horizontal planes, so the principal stress axes are dipped approximately $\pi/2$ from x and y (hence, σ_{yy} and σ_{xx} are approximately equal). Then, the horizontal component of quasi-static stress equilibrium equation $div\boldsymbol{\sigma} + \rho g = 0$, where tensor $\boldsymbol{\sigma}$ is $\boldsymbol{\sigma} = \boldsymbol{\tau} - P\mathbf{I}$ ($\mathbf{I}$ is identity matrix), can be locally simplified yielding thin layer approximation (e.g., Lobkovsky, 1988; Bird and Gratz, 1990):

$$\frac{\partial \tau_{xy}}{\partial y}=\frac{\partial p}{\partial x}-F_x=-\frac{\partial \tau_{yy}}{\partial x}\,. \tag{23}$$

A basic effective shear strain-rate can be evaluated as $\dot{\varepsilon}_{xy} = \sigma_{xy}/2\mu_{eff}$, therefore, according to the assumed constitutive relations, horizontal velocity u in the lower crust is:

$$u(\tilde{y}) = \int_0^{\tilde{y}} 2\dot{\varepsilon}_{xy}\partial\tilde{y} + C_1$$
$$= \int_0^{\tilde{y}} 2^n A^* \exp\left(-H^*/RT(y)\right)\left|\tau_{xy}\right|^{n-1}\tau_{xy}\partial\tilde{y} + C_1. \tag{24}$$

Here $\tilde{y} = y - y_{13} \cdot y_{13} = y_{13}(\phi)$ is the upper surface of the channel defined from solution of the system (18). C_1 is a constant of integration defined from the velocity boundary conditions. τ_{xy} is defined from vertical integration of (23). The remote conditions $h = 0$, $\partial h/\partial x = 0$, $w = 0$, $\partial w/\partial x = 0$ for the strong layers of the lithosphere (Appendix 1) are in accordance with the condition for ductile flow: tx $\rightarrow \infty$ $u^+_{c2} = u_c$; $u^-_{c2} = u_m$; $\partial p/\partial x = 0$, $\partial p/\partial y = \bar{\rho}_c g$; $p = P_0$ a.

In the trans-current channel flow the major perturbation to the stress (pressure) gradients is caused by slopes of crustal interfaces $\alpha \sim \partial\tilde{h}/\partial x$ and $\beta \sim \partial w/\partial x$. These slopes are controlled by flexure, isostatic re-adjustments, surface erosion and by "erosion" (weakening) of the interfaces by stress and temperature. The later especially concerns the upper crustal interface. In the assumption of small plate defections, the horizontal force associated with variation of the gravitational potential energy due to deflection of Moho (w) is $\rho_{c2}g$ tan(β)~ $\rho_{c2}g$ $sin\beta$ ~ $\rho_{c2}g\partial w/\partial x$; the vertical component of force is respectively ~$\rho_{c2}g cos$ β~ $\rho_{c2}g$ (1–$\partial w/\partial x$) ~ $\rho_{c2}g$. The horizontal and vertical force components due to slopes of the upper walls of the channel are respectively $\rho_{c2}g$tan(α)~gsin(α) ~ $\rho_{c2}gd\,\tilde{h}/dx$ and ρ_{c2} gcos(β) ~ ρ_{c2} $g(1–d\,\tilde{h}/dx)$. The equation of motion (Poiseuille/Couette flow) for a thin layer in the approximation of lubrication theory will be:

$$\begin{aligned}\frac{\partial\tau_{xy}}{\partial y} &= -\frac{\partial\tau_{yy}}{\partial x} \approx \frac{\partial p}{\partial x} - \rho_{c2}g\frac{\partial(\tilde{h}+w)}{\partial x}\\ \frac{\partial\tau_{yy}}{\partial y} + \frac{\partial\tau_{yx}}{\partial x} - \frac{\partial p}{\partial y} &\approx -\rho_{c2}g(1 - \frac{\partial(\tilde{h}+w)}{\partial x})\\ \frac{\partial u_{c2}}{\partial x} + \frac{\partial v_{c2}}{\partial y} &= 0.\end{aligned} \tag{25}$$

where pressure p is $p \approx P_0(x) + \bar{\rho}_c g(\ \tilde{y} + y_{13} + h\ )$; h is taken to be positive above sea-level; $\bar{\rho}_c$ is averaged crustal density.

In the simplest case of local isostasy, w and $\partial w/\partial x$ are approximately $\bar{\rho}_c/\Delta(\bar{\rho}_c - \rho_m) \sim 4$ times greater than $\tilde{h}$ and $d\,\tilde{h}/dx$, respectively. The pressure gradient due to Moho depression is $\rho_m g\partial(\tilde{h} + w)/\partial x$. "Correction" by the gradient of the gravitational potential energy density of crust yields $(\rho m\text{-}\bar{\rho}_c)g\partial(\tilde{h} + w)/\partial x$ for the effective pressure gradient in the crust, with w being equal to $\tilde{h}(\rho m–\bar{\rho}_c)/\rho m)$. In the case of regional compensation, when the mantle lithosphere is strong, the difference between $\tilde{h}$ and w can be 2–3 times less. To obtain w, we solve the system (A.1). Substitution of (B.3) to (B.4) gives:

$$\begin{aligned}\frac{\partial\tau_{xy}}{\partial y} &\approx \frac{\partial p}{\partial x} - \rho_{c2}g\frac{\partial(\tilde{h}+w)}{\partial x}\\ \frac{\partial p}{\partial y} &\approx \rho_{c2}g\left(1 - \frac{\partial(\tilde{h}+w)}{\partial x}\right)\\ \frac{\partial u}{\partial y} &= 2^n A^*\left(-H^*/RT\right)\left|\tau_{xy}\right|^{n-1}\tau_{xy}\\ \frac{\partial v}{\partial y} &= -\frac{\partial u}{\partial x}.\end{aligned} \tag{26}$$

The value $1 - \partial(\tilde{h} + w)/\partial x \approx 1$ due to the assumption of small deflections ($w/T_e << L/T_e$, $\tilde{h}$~0.2–0.5w, where L is the length of the plate). One has to note that strain rates of the lower crustal rocks (assuming quartz-controlled rheology) increase approximately by factor of 2 for each ~20°C of temperature increase with depth (e.g., Bird, 1991). This results in that the flow is being concentrated near the Moho, and the effective thickness of the transporting channel is much less than Δh_{c2}.

Depth integration of (26) gives us the longitudinal and vertical components of the basic material velocity in the lower crust. For example, we have:

$$\begin{aligned}u &= \int_0^{h_{c2}-y_{13}} 2^n A^* \exp\left(-H^*/RT(\tilde{y})\right)\left|\tau_{xy}\right|^{n-1}\tau_{xy}\partial\tilde{y}\\ v|_0^{h_{c2}-y_{13}} &= -\frac{\partial}{\partial x}\int_0^{h_{c2}-y_{13}} u\partial\tilde{y} \approx \frac{\partial(\tilde{h}+w)}{\partial t}.\end{aligned} \tag{27}$$

The later equation gives the variation of thickness of the ductile channel in time (equal to the difference between the vertical flow at the top and bottom boundaries). Lobkovsky (1988) (see also (Lobkovsky and

Kerchman, 1991)), Bird (1991) already gave an analytical solution for evolution of the topography *dh/dt* due to ductile flow in the crustal channel for the case of local isostatic equilibrium (zero strength of the upper crust and mantle). Kaufman and Royden (1994) provide a solution for the case of elastic mantle lithosphere but for Newtonian rheology. In our case, the irregular time-dependent load is applied on the surface, and non-linear rheology is assumed both for the ductile and competent parts of the lithosphere. Hence, no analytical solution for u and v can be found and we choose to obtain u and v through numerical integration.

The temperature which primarily controls the effective viscosity of the crust, is much lower in the uppermost and middle portions of the upper crust (first 10–15 km in depth). As a result, the effective viscosity of the middle portions of the upper crust is 2–4 orders higher than that of the lower crust (10^{22} to 10^{23} Pa sec compared to 10^{18} to 10^{20} Pa sec, Equations (7), (8)). Therefore, we can consider the reaction of the lower crust to deformation of the upper crust as rapid. The uppermost parts of the upper crust are brittle (Figs. 3, 4, 7, 8, 9, 10), but in calculation of the flow they can be replaced by some depth-averaged viscosity defined as $\bar{\mu}_{eff} = \bar{\sigma}^d/2\dot{\varepsilon}$ (Beekman, 1994). In spite of some negligence by the underlying principles, this operation does not introduce significant uncertainties to the solution because the thickness of the "brittle" crust is only 1/4 of the thickness of the competent crust. Analogously to the ductile (mostly lower) crust, we can extend the solution of the equations for the horizontal flow to the stronger upper portions of the upper crust. However, due to higher viscosity, and much lower thickness of the strong upper crustal layers, one can simply neglect by the perturbations of the flow velocity there and assume that $v = v(y \leq y_{13})$, $u = u(y \leq y_{13})$ (y is downward positive). For numerical reasons, we cut the interval of variation of the effective viscosity at 10^{19} to 10^{24} Pa sec.

Solution for the channel flow implies that the channel is infinite in both directions. In our case the channel is semi-infinite, because of the condition $u = 0$ at $x = 0$ beneath the axis of the mount. Thin flow approximation thus cannot be satisfied beneath the mount because of the possibility of sharp change of its thickness. Therefore, we need to modify the solution in the vicinity of $x = 0$. This could be done using a solution for the ascending flow for $x < al$. An analytical formulation for the symmetric flow in the crust and definition for the critical distance al are given in the Appendix 3. There we also explain how we combine the solution for the ascending symmetric flow beneath axis of the mountain range with the asymptotic solution for Poiseuille/Couette flow for domains off the axis. A similar approach can be found in literature dealing with cavity-driven problems (e.g., Hansen and Kelmanson, 1994). However, most authors (Lobkovsky and Kerchman, 1991; Bird and Gratz, 1990) ignore the condition $u = 0$ at $x = 0$ and the possibility of large thickness variations and simply considered a thin infinite channel.

Boundary conditions: We have chosen simplest boundary conditions corresponding to the flow approximations. Thus, the velocity boundary conditions are assumed on the upper and bottom interfaces of the lower crustal channel. Free flow is the lateral boundary condition. The velocity condition could be also combined with pre-defined lateral pressure gradient.

Link between the competent parts of the lithosphere and flow in the ductile parts is effectuated through the conditions of continuity of stress and velocity.

The problem of choice of boundary conditions for continental problems has no unique treatment. Most authors apply vertically homogeneous stress, force or velocity on the left and right sides of the model plate, Winkler-type restoring forces as bottom vertical condition, and free surface/normal stress as a upper boundary condition (e.g., England and McKenzie, 1983; Chery et al., 1991; Kusznir, 1991). Other authors use shear traction (velocity/stress) at the bottom of the mantle lithosphere (e.g., Ellis et al., 1995). Even choice between stress and force boundary conditions leads to significantly different results. Yet, the only observation that may provide an idea on the boundary conditions in nature comes from geodetic measurements and kinematic evaluations of surface strain rates and velocities. The presence of a weak lower crust leads to the possibility of differential velocity, strain partitioning between crust and mantle lithosphere and to possibility of loss of the material from the system due to outflow of the ductile crustal material (e.g., Lobkovsky and Kertchman, 1991; Ellis et al., 1995). Thus the relation between the velocities and strain rates observed at the surface with those on the depth is unclear. It is difficult to give preference to any of the mentioned scenarios. We thus chosen a simplest one.

Appendix 3: Analytical Formulation for Ascending Crustal Flow

In a general case of non-inertial flow (low Reynolds number), a symmetric flow problem (flow ascending beneath the mount) can be resolved from the solution of the system of classical viscous flow equations (Fletcher, 1988; Hamilton et al., 1995):

$$\begin{aligned} 0 &= \rho_{c2}F_x - \frac{dp}{dx} + \frac{\partial}{\partial y}\left(2\mu\left(\left(\frac{\partial u}{\partial y} + \frac{\partial v}{\partial x}\right)\right)\right) \\ 0 &= \rho_{c2}F_y - \frac{dp}{dy} + \frac{\partial}{\partial x}\left(2\mu\left(\left(\frac{\partial u}{\partial y} + \frac{\partial v}{\partial x}\right)\right)\right) \\ &\frac{\partial u}{\partial x} + \frac{\partial v}{\partial y} = 0 \end{aligned} \quad (28)$$

We define $\partial p/\partial x \approx \partial \tilde{p}/\partial x + g(\rho_{c2}\partial w/\partial x + \rho_{c1}\partial(du)/\partial x)$, $du \approx d\tilde{h}$ and $\partial p/\partial y = \partial \tilde{p}/\partial y - g\rho_{c2}$ where $\tilde{p}$ is dynamic, or modified pressure. The flow is naturally assumed to be Couette/Poiselle flow away from the symmetry axis (at a distance *al*). *al* is equal to 1–2 thicknesses of the channel, depending on channel thickness-to-length ratio. In practice *al* is equal to the distance at which the equivalent elastic thickness of the crust (T_{ec}) becomes less than ~5 km due to flexural weakening by elevated topography. For this case, we can neglect by the elasticity of the upper surface of the crust and use the condition of the stress-free upper surface. The remote feeding flux q at $x \to \pm a_l$ is equal to the value of flux obtained from depth integration of the channel source (Couette flow), and free flow is assumed as a lateral boundary condition. The flux q is determined as $q \sim \int u dy$ (per unit length in z direction). This flux feeds the growth of the topography and deeping of the crustal root. Combination of two flow formulations is completed using the depth integrated version of the continuity equation and global continuity equation (Huppert, 1982):

$$\int \frac{\partial v}{\partial y}dy + \frac{\partial}{\partial x}\left(\int u dy\right) = 0 = \frac{\partial(\tilde{h}+w)}{\partial t} + \frac{\partial q}{\partial x}$$

$$q|_{x=a_l, x\le a_l} = q|_{x=a_l, x\ge a_l} \quad (29)$$

$$\underbrace{\int_o^{a_l(\phi)} (\tilde{h}+w)dx}_{\text{ascending flow}} + \underbrace{\int_{a_l(\phi)}^{\infty} (\tilde{h}+w)dx}_{\text{channel flow}} = qt^{\theta},$$

where θ is some non-negative constant, θ=1 in our case. With that we can combine solutions for horizontal flow far off the mount axis (Couette/Poiseuille flow) with solutions for ascending flow below the mount (e.g., Hansen and Kelmanson, 1994). Assuming a new local coordinate system $x' = x$, $y' = -y-(h_{c2}+(h_{c2}-y_{13})/2)$, the boundary conditions for the flow ascending near the symmetry axis would be $u = v = 0$; $du/dy' = 0$ at $x' = 0$, $y' = 0$ (beneath the mount axis). Then, we assume that the viscosity (μ) in the ascending flow is constant and equal to = $\bar{\mu}(al)$ where $\bar{\mu}(a_l)$ is the depth-averaged value of the effective non-linear viscosity defined from the solution for the channel flow (Appendix 2) at distance $x = a_l$. Use of constant viscosity is, however, not a serious simplification for the problem as a whole, because *al* is small and thus this simplification applies only to a small fraction of the problem.

Introducing vorticity function $\xi = \text{rot}\mathbf{v} = \frac{\partial u}{\partial y} - \frac{\partial v}{\partial x} = \nabla^2\psi$, assuming laminar flow, we then write Stoke's equations as (Talbot and Jarvis, 1984; Fletcher, 1988; Hamilton et al., 1995):

$$\begin{aligned} \mu\frac{\partial \xi}{\partial x} &= \frac{\partial p}{\partial y} \\ \mu\frac{\partial \xi}{\partial y} &= -\frac{\partial p}{\partial x} \\ \xi &= \nabla^2\psi \end{aligned} \quad (30)$$

At the upper surface of the fluid, streamline $\psi = 0$, is taken to be stress-free (low T_{ec}, see above) which leads to following conditions: $p\cos 2\alpha = 2\mu\partial^2\psi/\partial y'\partial x$; $p\sin 2\alpha = \mu(\partial^2\psi/\partial x^2 - \partial^2\psi/y'^2)$. Here α is downward inclination of the surface to the horizontal. Finally, the symmetry of the flow requires $\psi(-x,y') = -\psi(x,y')$.

The general solution in dimensionless variables (Talbot and Jarvis, 1984): $X = h_{max}x'$; $Y = h(0)y'$; $p = (\mu q/\pi h_{max}{}^2)p'$; $\psi = (q/\pi)\psi'$, where h_{max} is the maximum height of the free surface, is:

$$\begin{aligned} \psi =\ & \tan^{-1} X/Y + XY/(X^2+Y^2) \\ & + \sum_{n=0}^{\infty}(-1)^n(n+1)Y^{2n+2}((2n+2)!)^{-1}f^{(2n)}(X) \\ & + \sum_{n=0}^{\infty}(-1)^n(n+1)Y^{2n+3}((2n+3)!)^{-1}\gamma^{(2n)}(X); \\ p =\ & K - \lambda Y + 2(Y^2-X^2)/(X^2+Y^2)^2 \\ & + \sum_{n=0}^{\infty}(-1)^n X^{2n+1}((2n+1)!)^{-1}f^{(2n+1)}(X) - G(X) \\ & + \sum_{n=0}^{\infty}(-1)^n Y^{2n+2}((2n+2)!)^{-1}\gamma^{(2n+1)}(X)G(X) \\ =\ & \int_0^x \gamma(s)ds. \end{aligned} \quad (31)$$

f and γ are arbitrary functions of expansion series and $f(j)$, $\gamma(j)$ are their jth derivatives, $\lambda = \pi\rho g h_{max}{}^3/\mu q$, K is constant. γ and f are determined numerically because the expressions for and p are non-linear and cannot be solved analytically. The calculation of ψ and p is done on the assumption of small curvature of the free surface which allows linear approximation of γ and f, i.e., as $\gamma = AX$ and $f = BX$. Then the free surface can be searched in the form of a parabolic function, e.g., $\tilde{h}\sim C-DX^2$ (Talbot and Jarvis, 1984).

The assumptions of constant viscosity and stress-free upper surface are discussible. To avoid this problem, we can solve the Equations (20), (21), (22) for the ascending flow analogously to how it was done for the channel flow (Appendix 2). The solution to (20), (21, (22) in the case of the ascending flow can be obtained assuming $\mu = \mu(y)$, $\tau_{xy}(x, 0) \neq 0$ and $U = u$ and $V = \Phi(y)$ where $\Phi(y)$ is to be determined. Here we simplify (22) by assuming that the viscosity is only depth dependent which is a better approximation to the non-linear law (8) than the assumption of constant viscosity.

The primary boundary conditions are $U(0,y) = 0$ (symmetric flow), $\tau_{xy}(x, y^*) = 0$ (assumption of the existence of a shear-free surface at some depth y^*, e.g., depth of compensation), $\tau_{xy}(x,0) = \tau_e$, $p(x,0) = \bar{\rho}_c g(h+y_{13})$ (shear stress and pressure continuity on the boundary with the overlying competent upper crustal layer of effective thickness y_{13}).

From $U(0,y) = 0$, (20), $v = \Phi(y)$, we get:

$$\begin{aligned} U &= x\partial\Phi(y)/\partial y; \sigma_{xx,yy} \\ &= \pm 2\mu\partial\Phi(y)/\partial y; \tau_{xy} = x\mu\partial^2\Phi(y)/\partial y^2. \end{aligned} \tag{32}$$

With the assumption that $y^* << al$ and from (20) this yields:

$$\begin{aligned} &\partial^2\tau_{xy}/\partial x^2 - \partial^2\tau_{xy}/\partial y^2 \\ &+2\partial^2\sigma_{xx}/\partial x\partial y - \partial F_x/\partial y + \partial F_y/\partial x = 0 \end{aligned} \tag{33}$$

which provides the expression for $\Phi(y)$ ($\partial F_x/\partial y$ and $\partial F_y/\partial x \approx 0$):

$$\partial^2(\mu\partial^2\Phi(y)/\partial y^2)\partial y^2 = 0, \text{or}\ \partial^2\tau_{yy}/\partial y^2 = 0 \tag{34}$$

With the conditions $\tau_{xy}(\tau_x, y^*) = 0$, $\tau_{xy}(x,0) = \tau_e$ and under assumption that $\partial y^*/\partial x$ is small, we can obtain: $\tau_{xy}(x,y)\approx\tau_e(1-y/y^*)$ and $p(x,y)\approx x^2\partial(\mu\partial^2\Phi(y)/\partial y^2)\partial y+C_1(y)$. $C_1(y)$ is to be found from the boundary conditions on p (Davies, 1994).

Since the expressions for stress are defined, the velocities U and V can be obtained from integration of the expressions (20), (21) relating stress components and du/y, dv/y. We have to determine the constants of integration in a way providing continuity with the solution for the channel flow at $x = a_l$. For that we define the boundary conditions at $x = \pm a_l$: $U(\pm a_l,y) = u_l$; $V(\pm a_l, y) = v_l(\pm a_l, y)$ (where u_l and v_l are provided by the solution for channel flow).

As pointed out by Davies (1994), it is impossible to provide an analytical or simplified semi-analytical solution for the case when the viscosity is defined exactly through the power law (8).

Appendix 4: Numerical Algorithm for the Full Thermo-Mechanical Model

This mixed finite-element volume/finite difference code Parovoz is based on the FLAC technique (Cundall, 1989). It solves simultaneously Newtonean dynamic equations of motion (18), in a Lagrangian formulation, coupled with visco-elasto-plastic constitutive equations (19), heat transport equations and state equation (see Appendix 1, (Burov et al., 2001; Le Pourhiet et al., 2004) for details concerning numerical implementation).

$$\left\langle \rho\frac{\partial^2\mathbf{u}}{\partial t^2} - \right\rangle div\boldsymbol{\sigma} - \rho\mathbf{g} = 0 \tag{35}$$

$$\frac{D\sigma}{Dt} = F(\sigma,\mathbf{u},\mathbf{V},\nabla\mathbf{V},...T...) \tag{36}$$

$$\begin{aligned} &\rho C_p\partial T/\partial t + \mathrm{u}\,\nabla T - k_c div(\nabla T) - H_r - frac \\ &\times\sigma_\pi\,\partial\varepsilon_\pi/\partial t = 0 \end{aligned} \tag{37}$$

assuming adiabatic temperature dependency for density and Boussinesq approximation for thermal body forces:

$$\rho = \rho_0(1 - \alpha\Delta T) \tag{38}$$

Here $\mathbf{u}$, $\boldsymbol{\sigma}$, $\boldsymbol{g}$, k_c are the respective terms for velocity, stress, acceleration due to body forces and thermal conductivity. The brackets in (18) specify conditional use of the related term: in quasi-static mode, the inertia is dumped using inertial mass scaling (Cundall, 1989). The terms t, ρ, C_p, T, H_r, α, $frac\times\sigma_{II}\partial\varepsilon_{II}/\partial t$ designate respectively time, density, specific heat, temperature, internal heat production, thermal expansion coefficient and shear heating term moderated by experimentally defined *frac* multiplier (*frac* was set to 0 in our experiments). The terms $\partial/\partial t$, $D\sigma/Dt$, F are a time derivative, an objective (Jaumann) stress time derivative and a functional, respectively. In the Lagrangian framework, the incremental displacements are added to the grid coordinates allowing the mesh to move and deform with the material. This enables solution of large-strain problems locally using small-strain formulation: on each time step the solution is obtained in local coordinates, which are then updated in the large strain mode. Volume/density changes due to phase transitions are accounted via application of equivalent stresses to affected material elements.

Solution of (18) provides velocities at mesh points used for computation of element strains and of heat advection $\mathbf{u}\nabla T$. These strains are used in (19) to calculate element stresses, and the equivalent forces are used to compute velocities for the next time step.

All rheological terms are implemented explicitly. The rheology model is serial viscous-elastic-plastic (Table 1). The plastic term is given by explicit Mohr-Coulomb plasticity (non-associative with zero dilatency) assuming linear Navier-Coulomb criterion. We imply internal friction angle ϕ of 30° and maximal cohesion S of 20 Mpa, which fit best the experimental Byerlee's law of rock failure (Byerlee, 1978):

$$\tau = S + \sigma_{\mathrm{n}} \mathrm{tg}\,\phi \tag{39}$$

where τ is the shear stress and σ_{n} is the normal stress. Linear cohesion softening is used for better localization of plastic deformation ε_p ($S(\varepsilon_p) = S_0$ min (0, 1 – $\varepsilon_p/\varepsilon_{p0}$) where ε_{p0} is 0.01).

The ductile-viscous term is represented by non-linear power law with three sets of material parameters (Table 1) that correspond to the properties of four lithological layers: upper crust (quartz), middle-lower crust (quartz-diorite), mantle and asthenosphere (olivine):

$$\mu_{\mathrm{eff}} = \mu_{\mathrm{eff}} \left(\frac{\partial\varepsilon}{\partial t}\right)_{\mathrm{II}}^{d(1-n)/n} (\mathrm{A}^*)^{-1/n} \exp\left(H/nRT\right) \tag{40}$$

where

$$\left(\frac{\partial\varepsilon}{\partial t}\right)_{\mathrm{II}}^{d} = \left(Inv_{II}\left(\left(\frac{\partial\varepsilon}{\partial t}\right)_{\mathrm{II}}^{d}\right)\right)^{1/2}$$

is the effective strain rate and $\mathrm{A}^* = \frac{1}{2}\mathrm{A}\cdot 3^{(n+1)/2}$ is the material constant, H is the activation enthalpy, R is the gas constant, n is the power law exponent (Table 2). The elastic parameters (Table 1) correspond to commonly inferred values from Turcotte and Schubert (1982).

Surface processes are taken into account by diffusing (D7) the topographic elevation h of the free surface along x using conventional Culling erosion model (Culling, 1960) with a diffusion coefficient k.

$$\frac{\partial h}{\partial t} = k\frac{\partial^2 h}{\partial x^2} \tag{41}$$

This simple model is well suited to simulate fan deltas, which can be taken as a reasonably good analogue of typical foreland basin deposits. This model is not well adapted to model slope dependent long-range sedimentation, yet, it accounts for some most important properties of surface processes such as dependency of the erosion/sedimentation rate on the roughness of the relief (surface curvature).

PARA(O)VOZ allows for large displacements and strains in particular owing to an automatic remeshing procedure, which is implemented each time the mesh becomes too destorted to produce accurate results. The remeshing criterion is imposed by a critical angle of grid elements. This angle is set to 10° to reduce frequency of remeshing and thus limit the associated numerical diffusion. The numerical diffusion was effectively constrained by implementation of the passive marker algorithm. This algorithm traces passively moving particles that are evenly distributed in the initial grid. This allows for accurate recovering of stress, phase and other parameter fields after each remeshing. PARA(O)VOZ has been already tested on a number of geodynamical problems for subduction/collision context (Burov et al., 2001; Toussaint et al., 2004a, b).

References

Ahnert, F., Functional relationships between denudation, relief and uplift in large mid-latitude drainage basins, *Am. J. Sci.*, 268, 243–263, 1970.

Ashmore, P. E., Laboratory modelling of gravel braided stream morphology, *Earth Surf. Processes Landforms*, 7, 201–225, 1982.

Andrews, D. J., R. C., Bucknam, Fitting degradation of shoreline scarps by a nonlinear diffusion model, *J. Geophys. Res.*, 92, 12857–12867, 1987.

Avouac, Analysis of scarp profiles: evaluation of errors in morphologic dating, *J. Geophys. Res.*, 98, 6745–6754, 1993.

Avouac, J. -P., P., Tapponnier, M., Bai, H., You, G., Wang, Active thrusting and folding along the northern Tien Shan and late Cenozoic rotation of the Tarim relative to Dzungaria and Kazakhstan, *J. Geophys. Res.*, 98, 6755–6804, 1993.

Avouac, J. -P., P., Tapponnier, Kinematic model of active deformation in Central Asia, *Geophys. Res. Lett.*, 20, 895–898, 1993.

Avouac, J. P., E. B., Burov, Erosion as a driving mechanism of intracontinental mountain growth, *J. Geophys. Res.*, 101 (B8), 17747–17769, 1996.

Basile, C., P., Allemand, Erosion and flexural uplift along transform faults, *Geophys. J. Int.*, 151, 646–653, 2002.

Batchelor, G. K., An introduction to fluid dynamics, Cambridge University Press, Cambridge, 615 p., 1967.

Beaumont, C., Foreland basins, *R. Astr. Soc. Geophys. J.*, 65, 389–416, 1981.

Beaumont, C., P., Fullsack, J., Hamilton, Erosional control of active compressional orogens, Thrust Tectonics, Ed. K. R. McClay, Chapman & Hall, London, 1–31, 1992.

Beaumont, C., P., Fullsack, J., Hamilton, Styles of crustal deformation in compressional orogens caused by subduction of the underlying lithosphere, submitted to: Proceedings of 5th International Symposium on Seismic Reflection Probing of the Continents and their Margins, Eds.R. Clowes and A. Green, 1994.

Beaumont, C., H., Kooi, S., Willett, Coupled tectonic-surface process models with applications to rifted margins and collisional orogens, in Geomorphology and Global Tectonics, Eds. M. A. Summerfield, pp. 29–55, John Wiley, New York, 2000.

Bonnet, S., A., Crave, Landscape response to climate change: Insights from experimental modeling and implications for tectonic versus climatic uplift of topography, *Geology*, 31, 123–126, 2003.

Braun, J., M., Sambridge, Modelling landscape evolution on geological time scales: A new method based on irregular spatial discretization, *Basin Res.*, 9, 27– 52, 1997.

Beekman, F., Tectonic modelling of thick-skinned compressional intraplate deformation, PhD thesis, Free University, Amsterdam, 1994.

Bird, P., A. J., Gratz, A theory for buckling of the mantle lithosphere and Moho during compressive detachments in continents, *Tectonophysics*, 177, 325–336, 1990.

Bird, P., Lateral extrusion of lower crust from under high topography in the isostatic limit, *J. Geophys. Res.*, 96, 10275–10286, 1991.

Brace, W. F., D. L., Kohlstedt, Limits on lithospheric stress imposed by laboratory experiments, *J. Geophys. Res.*, 85, 6248–6252, 1980.

Byerlee, J. D., Friction of rocks, *Pure Appl. Geophys.*, 116, 615–626, 1978.

Burbank, D. W., Causes of recent Himalayan uplift deduced from deposited patterns in the Ganges basin, *Nature*, 357, 680–683, 1992.

Burbank, D. W., J., Vergés, Reconstruction of topography and related depositional systems during active thrusting, *J. Geophys. Res.*, 99, 20281–20297, 1994.

Burov, E. V., M. G., Kogan, H., Lyon-Caen, P., Molnar, Gravity anomalies, the deep structure, and dynamic processes beneath the Tien Shan, *Earth Planet. Sci. Lett.*, 96, 367–383, 1990.

Burov, E. B., M., Diament, Flexure of the continental lithosphere with multilayered rheology, *Geophys. J. Int.*, 109, 449–468, 1992.

Burov, E. B., L. I., Lobkovsky, S., Cloetingh, A. M., Nikishin, Continental lithosphere folding in Central Asia (part 2), constraints from gravity and topography, *Tectonophysics*, 226, 73–87, 1993.

Burov, E. B., S., Cloetingh, Erosion and rift dynamics: new thermomechanical aspects of post-rift evolution of extensional basins, *Earth Planet Sci. Lett.*, 150, 7–26, 1997.

Burov, E. B., M., Diament, The effective elastic thickness (Te) of continental lithosphere: What does it really mean?, *J. Geophys. Res.*, 100, 3905–3927, 1995.

Burov, E. B., L., Jolivet, L., Le Pourhiet, A., Poliakov, A thermemechanical model of exhumation of HP and UHP methamorphic rocks in Alpine mountain belts, *Tectonophysics*, 113–136, 2001.

Burov, E., A. B., Watts, The long-term strength of continental lithosphere: "jelly-sandwich" or "crème-brûlé"?, *GSA Today*, 16, 1, doi: 10.1130/1052-5173(2006)016<4:TLTSOC, 2006.

Carson, M. A., M. J., Kirkby, Hillslope Form and Processes, Cambridge University Press, Cambridge, 475p., 1972.

Carter, N. L., M. C., Tsenn, Flow properties of continental lithosphere, *Tectonophysics*, 36, 27–63, 1987.

Castelltort, S., G., Simpson, Growing mountain ranges and quenched river networks, CRAS, 2006.

Chen, Y., J. P., Cogne, V., Courtillot, J. P., Avouac, P., Tapponnier, E., Buffetaut, G., Wang., M., Bai, H., You, M., Li, C., Wei, Paleomagnetic study of Mesozoic continental sediments along the northern Tien Shan (China) and heterogeneous strain in Central Asia, *J. Geophys. Res.*, 96, 4065–4082, 1991.

Chéry, J., J. P., Vilotte, M., Daignieres, Thermomechanical evolution of a thinned continental lithosphere under compression: Implications for Pyrenees, *J. Geophys. Res.*, 96, 4385–4412, 1991.

Chorley, R. J., S. A., Schumm, D. E., Sugden, Hillsopes in Geomorpholgy, 255–339, Methuen, London, 1984.

Cloetingh, S., E., Burov, L., Matenco, G., Toussaint, G., Bertotti, Thermo-mechanical constraints for the continental collision mode in the SE Carpathians (Romania), *Earth Planet Sci. Lett.*, 218 (1–2), pp. 57–76, 2004.

Copeland, P., T. M., Harrison, Episodic rapid uplift in the Himalya revealed by 40Ar/39Ar analysis of detrital K-feldspar and muscovite, Bengal fan, *Geology*, 18, 354–357, 1990.

Crave, A., P., Davy, A stochastic ''precipiton'' model for simulating erosion/sedimentation dynamics, *Comput. Geosci.*, 27, 815– 827, 2001.

Crave, A., D., Lague, P., Davy, J., Kermarrec, D., Sokoutis, L., Bodet, R., Compagnon, Analogue modelling of relief dynamics, *Phys. Chem. Earth A*, 25 (6–7)), 549–553, 2000.

Culling, W. E. H., Analytical theory of erosion, *J. Geol.*, 68, 336–344, 1960.

Culling, W. E. H., Theory of erosion on soil-covered slopes, *J. Geol.*, 73, 230–254, 1965.

Cundall, P. A., Numerical experiments on localization in frictional material: Ingenieur-Archiv, v. 59, p. 148–159, 1989.

Davies, G. F., Thermomechanical erosion of the lithosphere by mantle plumes, *J. Geophys. Res.*, 99, 15709–15722, 1994.

Davy, P., A., Crave, Upscaling local-scale transport processes in largescale relief dynamics, *Phys. Chem. Earth A*, 25 (6–7), 533–541, 2000.

Densmore, A. L., R. S., Anderson, B. G., McAdoo, M. A., Ellis, Hillslope evolution by bedrock landslides, *Science*, 275, 369–372, 1997.

Densmore, A. L., M. A., Ellis, R. S., Anderson, Landsliding and the evolution of normal fault-bounded mountain ranges, *J. Geophys. Res.*, 103 (B7), 15203–15219, 1998.

Ellis, S., P., Fullsack, C., Beaumont, Oblique convergence of the crust driven by basal forcing: implications for length-scales of deformation and strain partitioning in orogens, *Geophys. J. Int.*, 120, 24–44, 1995.

England, P. C., D. P., McKenzie, A thin viscous sheet model for continental deformation, *Geophys. J. R. Astron. Soc.*, 73, 523–532, 1983.

England, P., S. W., Richardson, The influence of erosion upon the mineral facies of rocks from different metamorphic environments, *J. Geol. Soc. Lond.*, 134, 201–213, 1977.

Flint, J. J., Experimental development of headward growth of channel networks, *Geol. Soc. Am. Bull.*, 84, 1087–1094, 1973.

Flint, J. -J., Stream gradient as a function of order magnitude, and discharge, *Water Resour. Res.*, 10 (5), 969–973, 1974.

Fleitout, L., C., Froidevaux, Tectonics and topography for a lithosphere containing density heterogeneities, *Tectonics*, 1, 21–56, 1982.

Flemings, P. B., T. E., Jordan, A synthetic stratigraphic model of foreland basin development, *J. Geophys. Res.*, 94, 3851–3866, 1989.

Flemings, P. B., T. E., Jordan, Stratigraphic modelling of foreland basins: interpreting thrust deformation and lithosphere rheology, *Geology*, 18, 430–434, 1990.

Fletcher, C. A. J., Computational techniques for fluid dynamics 2, Springer-Verlag, Berlin Heidelberg, 552 pp., 1988.

Fournier, F., Climat et Erosion: la relation entre l'érosion du sol par l'eau et les précipitations atmosphériques, Presse Universitaire de France, Paris, 201 pp., 1960.

Gaspar-Escribano, J. M., M., Ter Voorde, E., Roca, S., Cloetingh, Mechanical (de-)coupling of the lithosphere in the Valencia Through (NW Mediterranean): What does it mean? Earth and Planet Sci. Lett, 210, 291–303, 2003.

Garcia-Castellanos, D., J., Vergés, J., Gaspar-Escribano, S., Cloetingh, Interplay between tectonics, climate, and fluvial transport during the Cenozoic evolution of the Ebro Basin (NE Iberia), *J. Geophys. Res.*, VOL. 108, NO. B7, 2347, doi:10.1029/2002JB002073, 2003.

Garcia-Castellanos, D., Interplay between lithospheric flexure and river transport in foreland basins, *Basin Res.*, 14, 89–104, 2002.

Garcia-Castellanos, D., M., Fernàndez, M., Torne, Modeling the evolution of the Guadalquivir foreland basin (southern Spain), *Tectonics*, 21, (3), 1018, doi:10.1029/2001TC001339, 2002.

Gossman, H., Slope modelling with changing boundary conditions – effects of climate and lithology, *Z. Geomoph. N.F.*, Suppl. Bd. 25, 72–88, 1976.

Govers, G., Evaluation of transporting capacity formulae for overland flow, in Overland Flow: Hydraulics and Erosion Mechanics, edited by A. J. Parsons and A. D. Abrahams, pp. 243 – 273, UCL Press, London, 1992a.

Govers, G., Relationship between discharge, velocity and flow area for rills eroding loose, non-layered materials, *Earth Surf. Processes Landforms*, 17, 515–528, 1992b.

Gratton, J., Crustal shortening, root spreading, isostasy, and the growth of orogenic belts: a dimensional analysis, *J. Geophys. Res.*, 94, 15627–15634, 1989.

Gregory, K. M., C., Chase, Tectonic and climatic significance of a late Eocene low-relief, high-level geomorphic surface, Colorado, *J. Geophys. Res.*, 99, 20141–20160, 1994.

Hamilton, J. M., J., Kim, F., Waleffe, Regeneration mechanisms of near-wall turbulence structures, *J. Fluid. Mech.*, 287, 317–348, 1995.

Hanks, T. C., R. C., Buckham, K. R., LaJoie, R. E., Wallace, Modification of wave-cut and fault-controlled landforms, *J. Geophys. Res.*, 89, 5771–5790, 1984.

Hansen, E. B., M. A., Kelmanson, An integral equation justification of the boundary conditions of the driven-cavity problem, *Comput. Fluids*, 23 (1), 225–240, 1994.

Hairsine, P. B., C. W., Rose, Modeling water erosion due to overland flow using physical principles, 1, Sheet flow, *Water Resour. Res.*, 28(1), 237–243, 1992.

Hancock, G., G., Willgoose, Use of a landscape simulator in the validation of the SIBERIA catchment evolution model: Declining equilibrium landforms, *Water Resour. Res.*, 37(7), 1981– 1992, 2001.

Hasbargen, L. E., C., Paola, Landscape instability in an experimental drainage basin, *Geology*, 28(12), 1067–1070, 2000.

Hendrix, M. S., S. A., Graham, A. R., Caroll, E. R., Sobel, C. L., McKnight, B. J., Schulein, Z., Wang, Sedimentary record and climatic implications of recurrent deformation in the Tien Shan: Evidence from Mesozoic strata of the north Tarim, south Junggar, and Turpan basins, northwest China, *Geol. Soc. Am. Bull.*, 104, 53–79, 1992.

Hendrix, M. S., T. A., Dumitru, S. A., Graham, Late Oligocene-early Miocene unroofing in the Chinese Tian Shan: An early effect of the India Asia collision, *Geology*, 22, 487–490, 1994.

Hirano, Simulation of developmental process of interfluvial slopes with reference to graded form, *J. Geol.*, 83, 113-123, 1975.

Howard, A. D., Long profile development of bedrock channels: Interaction of weathering, mass wasting, bed erosion and sediment transport, in Rivers Over Rock: Fluvial Processes in Bedrock Channels, Geophys. Monogr. Ser., vol. 107, edited by K. J. Tinkler and E. E. Wohl, pp. 297– 319, AGU, Washington, D.C., 1998.

Howard, A. D., W. E., Dietrich, M. A., Seidl, Modeling fluvial erosion on regional to continental scales, *J. Geophys. Res.*, 99(B7), 13971 – 13986, 1994.

Huppert, H. E., The propagation of two dimensional and axisymmetric gravity currents over a rigid horizontal surface, *J. Fluid. Mech.*, 121, 43–58, 1982.

Hurtrez, J. -E., F., Lucazeau, J., Lave', J. -P., Avouac, , Investigation of the relationships between basin morphology, tectonic uplift, and denudation from the study of an active fold belt in the Siwalik Hills, central Nepal, *J. Geophys. Res.*, 104(B6), 12779–12796, 1999.

Kaufman, P. S., L. H., Royden, Lower crustal flow in an extensional setting: Constraints from the Halloran Hills region, eastern Mojave Desert, California, *J. Geophys. Res.*, 99, 15723–15739, 1994.

King, G. C. P., R. S., Stein, J. B., Rundle, The growth of geological structures by repeated earthquakes, 1. Conceptual framework, *J. Geophys. Res.*, 93, 13307–13318, 1988.

King, G., M., Ellis, The origin of large local uplift in extensional regions, *Nature*, 348, 689–693, 1990.

Kirby, S. H., Rheology of the lithosphere, *Rev. Geophys.*, 21, 1458–1487, 1983.

Kirby, S. H., A. K., Kronenberg, Rheology of the lithosphere: Selected topics, *Rev. Geophys.*, 25, 1219–1244, 1987.

Kirkby, M. J., Hillslope process-response models based on the continuity equation, *Spec. Publ. Inst. Br. Geogr.*, 3, 15– 30, 1971.

Kirkby, M. J., A two-dimensional model for slope and stream evolution, in Abrahams, A. D. ed., Hillslope Processes, Boston, Allen and Unwin., 203–224, 1986.

Kirkby, M., M., Leeder, N., White, The erosion of actively extending tilt-blocks: a coupled for topography and sediment budgets, application to the B&R, 13 pp., 1993.

Kohlstedt, D. L., B., Evans, S. J., Mackwell, Strength of the lithosphere: Constraints imposed by laboratory experiments, *J. Geophys. Res.*, 100, 17587–17602, 1995.

Kooi, H., C., Beaumont, Escarpment evolution on high-elevation rifted margins: Insights derived from a surface processes model that combines diffusion, advection and reaction, *J. Geophys. Res.*, 99, 12191–12209, 1994.

Kooi, H., C., Beaumont, Large-scale geomorphology: Classical concepts reconciled and integrated with contemporary ideas via a surface processes model, *J. Geophys. Res.*, 101(B2), 3361– 3386, 1996.

Kruse, S., M., McNutt, J., Phipps-Morgan, L., Royden, Lithospheric extension near lake Mead, Nevada: A model for ductile flow in the lower crust, *J. Geophys. Res.*, 96(3), 4435–4456, 1991.

Kusznir, N. J., D. H., Matthews, Deep seismic reflections and the deformational mechanics of the continental lithosphere, *J. Petrol. Spec.*, Lithosphere Issue, 63–87, 1988.

Kusznir, N. J., The distribution of stress with depth in the lithosphere: thermo-rheological and geodynamic constraints, *Phil. Trans. R. Soc. Lond. A*, 337, 95–110, 1991.

Lague, D., P., Davy, A., Crave, Estimating uplift rate and erodibility from the area– slope relationship: Examples from Brittany (France) and numerical modelling, *Phys. Chem. Earth A*, 25(6–7), 543–548, 2000.

Lague, D., A., Crave, Ph., Davy, Laboratory experiments simulating the geomorphic response to tectonic uplift, *J. Geophys. Res.*, VOL. 108, NO. B1, 2008, doi:10.1029/2002JB001785, 2003.

Lavé, J., J. P., Avouac, Fluvial incision and tectonic uplift across the Himalayas of central Nepal, *J. Geophys. Res.*, 106(B11), 26561– 26591, 2001.

Leeder, M. R., Denudation, vertical crustal movements and sedimentary basin infill, *Geologische Rundschau, Stuttgart*, 80(2), 441–458, 1991.

Le Pourhiet, L., E., Burov, I., Moretti, Rifting through a stack of inhomogeneous thrusts (the dipping pie concept), *Tectonics*, 23, 4, TC4005, doi:10.1029/2003TC001584, 2004.

Lobkovsky, L. I., Geodynamics of Spreading and Subduction zones, and the two-level plate tectonics, Nauka, Moscow, 251 pp., 1988.

Lobkovsky, L. I., V. I., Kerchman, A two-level concept of plate tectonics: application to geodynamics, *Tectonophysics*, 199, 343–374, 1991.

Luke, J. C., Mathematical models for landform evolution, *J. Geophys. Res.*, 77, 2460–2464, 1972.

Luke, J. C., Special Solutions for Nonlinear Erosion Problems, *J. Geophys. Res.*, 79, 4035–4040, 1974.

Ma, X., Lithospheric dynamic Atlas of China, China Cartographic Publishing House, Beijing, China, 1987.

Makeyeva, L. I., L. P., Vinnik, S. W., Roecker, Shear-wave splitting and small scale convection in the continental upper mantle, *Nature*, 358, 144–147, 1992.

Masek, J. G., B. L., Isacks, E. J., Fielding, Rift flank uplift in Tibet: Evidence for a viscous lower crust, *Tectonics*, 13, 659–667, 1994a.

Masek, J. G., B. L., Isacks, T. L., Gubbels, E. J., Fielding, Erosion and tectonics at the margins of continental plateaus, *J. Geophys. Res.*, 99, 13941–13956, 1994b.

Metivier, F., Y., Gaudemer, Mass transfer between eastern Tien Shan and adjacent basins (centralAsia): constraints on regionaltectonics and topography, *Geophys. J. Int.*, 128, 1–17, 1997.

Molnar, P., Climate change, flooding in arid environments, and erosion rates, *Geology*, 29(12)), 1071– 1074, 2001.

Molnar, P., Q., Deng, Faulting associated with large earthquakes and the average rate of deformation in central and eastern Asia, *J. Geophys. Res.*, 89, 6203–6228, 1984.

Molnar, P., H., Lyon-Caen, Some simple physical aspects of the support, structure, and evolution of mountain belts, in: Processes in continental lithospheric deformation, *Geol. Soc. Am. Spec. Rap.*, 218, 179–207, 1988.

Molnar, P., P., Tapponnier, A possible dependence of the tectonic strength on the age of the crust in Asia, *Earth Planet. Sci. Lett.*, 52, 107–114, 1981.

Molnar, P., P., England, Late Cenozoic uplift of mountain ranges and global climate change: chicken or egg, *Nature*, 346, 29–34, 1990.

Mizutani, T., Laboratory experiment and digital simulation of multiple fillcut terrace formation, *Geomorphology*, 24, 353–361, 1998.

Nash, D. B., Morphologic dating of degraded normal fault scarps, *J. Geol.*, 88, 353–360, 1980.

Newman, W. I., Nonlinear diffusion: Self-similarity and travelling-waves, *PAGEOPH*, 121(3), 417–441, 1983.

Newman, W. I., D. L., Turcotte, Cascade model for fluvial geomorphology, *Geophys. J. Int.*, 100, 433–439, 1990.

Parson, B., J., Sclater, An analysis of the variation of ocean floor bathymetry and heat flow with age, *J. Geophys. Res.*, 93, 8051–8063, 1977.

Patriat, P., J., Achache, India-Eurasia collision chronology has implications for crustal shortening and driving mechanism of plates, *Nature*, 311, 615–621, 1984.

Pelletier, J. D., Persistent drainage migration in a numerical landscape evolution model, *Geophys. Res. Lett.*, 31, doi:10.1029/2004GL020802, 2004.

Persson, K. S., D., Garcia-Castellanos, D., Sokoutis, River transport effects on compressional belts: First results from an integrated analogue-numerical model, *J.Geophys.*

Res., VOL. 109, B01409, doi:10.1029/2002JB002274, 2004.

Pinet, P., M., Souriau, Continental erosion and large-scale relief, *Tectonics*, 7(3)), 563–582, 1988.

Poliakov, A.N.B., P., Cundall, Y., Podladchilov, V., Laykhovsky, An explicit inertial method for the simulation of visco-elastic flow: an evaluation of elastic effects on diapiric flow in two- or three-layers models, in D. B. Stone and S. K. Runcorn (Eds), Flow and creep in the solar system: observations, modelling and theory. Dynamic Modelling and Flow in the Earth and Planet Series, 175–195, 1993.

Ranalli, G., Rheology of the Earth, Chapman & Hall, Sec. Edition, London, 413 pp., 1995.

Roecker, S. W., T. M., Sabitova, L. P., Vinnik, Y. A., Burmakov, M. I., Golvanov, R., Mamatkanova, L., Minirova, Three dimensional elastic wave velocity structure of the western and central Tien Shan, *J. Geophys. Res.*, 98, 15779–15795, 1993.

Roering, J. J., J. W., Kirchner, L. S., Sklar, W. E., Dietrich, Hillslope evolution by nonlinear creep and landsliding: An experimental study, *Geology*, 29(2), 143– 146, 2001.

Schmid, S. M., O. A., Pffifner, G., Schönborg, N., Froitzheim, E., Kissling, Integrated cross-sections and tectonic evolution of the Alps along the Eastern Traverse. In: Deep structures of the Swiss Alps, O. A. Pffifner, P. Lehner, P. Heitzmann, S. Mueller and A. Steck (Editors), Birkhäuser, Basel, pp. 289–304, 1997.

Schorghofer, N., D. H., Rothman, Acausal relations between topographic slope and drainage area, *Geophys. Res. Lett.*, 29(13), 1633, doi:10.1029/2002GL015144, 2002.

Schumm, S. A., M. P., Mosley, W. E., Weaver, Experimental Fluvial Geomorphology, John Wiley, New York, 1987.

Seidl, M. A., W. E., Dietrich, The problem of channel erosion into bedrock, *Catena Suppl.*, 23, 101– 124, 1992.

Sheperd, R. G., S. A., Schumm, Experimental study of river incision, *Geol. Soc. Am. Bull.*, 85, 257– 268, 1974.

Simpson, G., F., Schlunegger, Topographic evolution and morphology of surfaces evolving in response to coupled fluvial and hillslope sediment transport, *J. Geophys. Res.*, VOL. 108, NO. B6, 2300, doi:10.1029/2002JB002162, 2003.

Sklar, L., W. E., Dietrich, River longitudinal profiles and bedrock incision models: Stream power and the influence of sediment supply, in Rivers Over Rock: Fluvial Processes in Bedrock Channels, Geophys. Monogr. Ser., vol. 107, edited by K. J. Tinkler and E. E. Wohl, pp. 237–260, AGU, Washington, D.C., 1998.

Sklar, L. S., W. E., Dietrich, Sediment and rock strength controls on river incision into bedrock, *Geology*, 29(12), 1087–1090, 2001.

Smith, C. E., Modeling high sinuosity meanders in a small flume, *Geomorphology*, 25, 19– 30, 1998.

Smith, T. R., F. P., Bretherton, Stability and the conservation of mass in drainage basin evolution, *Water Resour. Res.*, 8fA two-dimensional model, 6, 1506–1529, 1972.

Snyder, N. P., Bedrock channel response to tectonic, climatic, and eustatic forcing, Ph.D thesis, Dep. of Earth, Atmos., and Planet. Sci., Mass. Inst. of Technol., Cambridge, Mass., 2001.

Snyder, N. P., K. X., Whipple, G. E., Tucker, D. J., Merritts, Landscape response to tectonic forcing: DEM analysis of stream profiles in the Mendocino triple junction region, northern California, *Geol. Soc. Am. Bull.*, 112, 1250– 1263, 2000.

Simpson, G., Role of river incision in enhancing deformation, *Geology*, 32, 2004, 341–344.

Sobel, E., T. A., Dumitru, Exhumation of the margins of the western Tarim basin during the Himalayan orogeny, Tectonics, in press, 1995.

Summerfield, M. A., N. J., Hulton, Natural control on fluvial denudation rates in major world drainage basins, *J. Geophys. Res.*, 99, 13871–13883, 1994.

Talbot, C. J., R. J., Jarvis, Age, budget and dynamics of an active salt extrusion in Iran, *J. Struct. Geology*, 6, 521–533, 1984.

Tapponnier, P., P., Molnar, Active faulting and Cenozoic tectonics of the Tien Shan, Mongolia and Baykal regions, *J. Geophys. Res.*, 84, 3425–3459, 1979.

Ter Voorde, M., R. T., Van Balen, G., Bertotti, S. A. P. L., Cloetingh, The influence of a stratified rheology on the flexural response of the lithosphere to (un)loading by extensional faulting, *Geophys. J. Int.*, 134, 721–735, 1998.

Toussaint, G., E., Burov, L., Jolivet, Continental plate collision: unstable versus stable slab dynamics, *Geology*, 32(No. 1), 33–36, 2004a.

Toussaint, G., E., Burov, J. -P., Avouac, Tectonic evolution of a continental collision zone: a thermo mechanical numerical model, *Tectonics*, 23, TC6003, doi:10.1029/2003TC001604, 2004b.

Tsenn, M. C., N. L., Carter, Flow properties of continental lithosphere, *Tectonophysics*, 136, 27–63, 1987.

Tucker, G. E., R. L., Bras, Hillslope processes, drainage density, and landscape morphology, *Water Resour. Res.*, 34(10), 2751– 2764, 1998.

Tucker, G. E., R. L., Bras, A stochastic approach to modeling the role of rainfall variability in drainage basin evolution, Water Resour. Res, 36(7), 1953– 1964, 2000.

Turcotte, D. L., G., Schubert, Geodynamics. Applications of continuum physics to geological problems, J. Wiley & Sons, New York, 450 p., 1982.

Vinnik, L. P., A. M., Saipbekova, Structure of the lithosphere and asthenosphere of the Tien Shan, *Ann. Geophys.*, 2, 621–626, 1984.

Vinnik, L. P., I. M., Aleshin, M. K., Kaban, S. G., Kiselev, G. L., Kosarev, S. I., Oreshin, Ch., Reigber, Crust and Mantle of the Tien Shan from Data of the Receiver Function Tomography, Izvestiya, *Phys. Solid Earth*, 42, pp. 639–651, Pleiades Publishing, Inc., 2006.

Vilotte, J. P., M., Daignières, R., Madariaga, Numerical modeling of intraplate deformation: simple mechanical models of continental collision, *J. Geophys. Res.*, 87, 10709–10728, 1982.

Vogt, P. R., Bermuda and Applachian-Labrador rises, common hotspot processes, *Geology*, 19, 41–44, 1991.

Yamato, P., E., Burov, P., Agard, L., Le Pourhiet, L., Jolivet, 2008, HP-UHP exhumation during slow continental subduction: Self-consistent thermodynamically and thermomechanically coupled model with application to the Western Alps, *Earth Planet Sci. Lett.*, 271, 63–74.

Wang, J. N., B. E., Hobbs, A., Ord, T., Shimamoto, M., Toriumi, Newtonian dislocation creep in quartzites: Implications for the rheology of the lower crust, *Science*, 265, 1204–1206, 1994.

Westaway, R., Evidence for dynamic coupling of surface processes with isostatic compensation in the lower crust during active extension of western Turkey, *J. Geophys. Res.*, 99, 20203–20223, 1994.

Willett, S. D., Orogeny and orography: The effects of erosion on the structure of mountain belts, *J. Geophys. Res.*, 104(B12), 28957–28982, 1999.

Windley, B. F., M. B., Allen, C., Zhang, Z. Y., Zhao, G. R., Wang, Paleozoic accretion and Cenozoic redeformation of the Chinese Tien Shan range, central Asia, *Geology*, 18, 128–131, 1990.

Achievements and Challenges in Sedimentary Basin Dynamics: A Review

François Roure, Sierd Cloetingh, Magdalena Scheck-Wenderoth, and Peter A. Ziegler

Abstract Thanks to a continuous effort for unravelling geological records since the early days of coal and petroleum exploration and water management, the architecture and chrono-litho-stratigraphy of most sedimentary basins has been accurately described by means of conventional geological and geophysical studies, using both surface (outcrops) and subsurface (exploration wells and industry seismic reflection profiles) data. However, the understanding of the early development and long term evolution of sedimentary basins usually requires the integration of even more data on the deep Earth, as well as quantifications by means of kinematic-sedimentological and thermo-mechanical modelling approaches coupling both surface and deep processes.

In the last twenty years, huge national and international efforts, frequently linking academy and industry, have been devoted to the recording of deep seismic profiles in many intracratonic sedimentary basins and offshore passive margins, thus providing a direct control on the structural configuration of the basement and the architecture of the crust. Seemingly, needs for documenting also the current thickness of the mantle lithosphere and the fate of subducted lithospheric slabs have led to the development of more academic and new tomographic techniques. When put together, these various techniques now provide a direct access to the bulk 3D architecture of sedimentary basins, crystalline basement and Moho, as well as underlying mantle lithosphere.

Inherited structures, anisotropies in the composition of the sediments, crust and underlying mantle as well as thermicity and phase transitions are now taken into account when predicting the localization of deformation in the lithosphere during compression and extension episodes, and reconstructing the geodynamic evolution of rift basins, passive margins or even foreland fold-and-thrust belts.

Source to sink studies also provide accurate estimates of sedimentary budget at basin-scale. Extensive use of low temperature chrono-thermometers and coupled kinematic, sedimentological and thermal models allow a precise control on the amount and timing of erosion and unroofing of source areas, but also the reconstruction of the sedimentary burial, strata architecture and litho-facies distribution in the sink areas.

Coupling deep mantle processes with erosion and climate constitutes a new challenge for understanding the present topography, morphology and long term evolution of continents, especially in such sensitive areas as the near shore coastal plains, low lands and intramountain valleys which may be subject to devastating flooding and landslides.

In addition to the search for hydrocarbon resources, other societal needs such as CO_2 storage and underground water management will benefit from upgraded basin modelling techniques. New 2D and 3D basin modelling tools are progressively developed, coupling in different ways deep thermo-mechanic processes of the mantle (asthenosphere and lithosphere), geomechanics of the upper crust and sediments (compaction, pressure-solution and fracturing of seals and reservoirs), basin-scale fluid and sediment transfers (development of overpressures, hydrocarbon generation and migration). Further challenges related to CO_2 storage will soon require the integration of fluid-rock

F. Roure (✉)
Institut Français du Pétrole, 1-4 Avenue de Bois-Préau, 92852 Rueil-Malmaison, France
e-mail: Francois. ROURE@ifp.fr

S. Cloetingh, J. Negendank (eds.), *New Frontiers in Integrated Solid Earth Sciences*, International Year of Planet Earth, DOI 10.1007/978-90-481-2737-5_5, © Springer Science+Business Media B.V. 2010

interactions (reactive transport) in basin and reservoir models, in order to cope with the changes induced by diagenesis in the overall mechanical properties, and the continuous changes in fluid flow induced by compaction, fracturing and cementation.

Keywords Sedimentary basins · Lithosphere · Basin modelling · Dynamic topography · Source to sink · ILP task force on sedimentary basins

Introduction

Sedimentary basins are the archives, which document the evolution of the Earth during Phanerozoic times, both in terms of deep mantle processes and climatically controlled surface processes. Moreover, sedimentary basins host the hydrocarbon and coal resources of the planet, as well as major water resources and huge ore deposits. In addition many large population centres are located in sedimentary basins, often in coastal lowlands and are thus vulnerable to climate-induced sea-level fluctuations and a wide range of geohazards. The geohazards are controlled by the interaction of processes inherent to the deep Earth and its surface.

Documenting the present-day architecture of sedimentary basins and understanding processes controlling their long-term evolution are a prime task of the Earth Sciences community as a whole, which comprises geologists, geophysicists and basin modellers both in academic as well as industrial organizations. Feed back from basin modelling studies is of prime importance for a wide range of end-users, such as private companies concerned with resource assessment and national agencies aiming at quantifying geohazards, including the long-term evolution of the environment, and their impact on the population. The sedimentary basin community, and Earth Sciences as a whole, face also new societal challenges owing to on-going climate changes and the needs for CO_2 sequestration. Progress in understanding natural processes that control the long-term evolution of the Planet Earth requires integrated and multidisciplinary efforts by geologists and geophysicists.

We need to better understand the interactions between the processes that generate basins, how they become filled or not by sediments, how they behave mechanically after active basin formation has stopped and how migration of fluids and transfer of heat takes place. Clearly, these objectives cannot be achieved without linking processes at great and middle depth in the Earth and those active at and near the surface.

In this context, the International Lithosphere Programme (ILP, a programme supported by IUGS and IUGG) took the initiative, almost twenty years ago, to build a task force on the "Origin of Sedimentary Basins". A first workshop of this task force was held in 1990 at Rueil-Malmaison, France (Cloetingh, 1990, 1991, 1992; Cloetingh et al., 1993a,b). Subsequent project meetings, including dedicated field trips, were held in 1991 at Matrahaza (Hungary; Cloetingh et al., 1993a, b), in 1992 at Sundvollen (Norway; Cloetingh et al., 1994), in 1993 at Benevento (Italy; Cloetingh et al., 1995a,b), in 1994 near the Dead Sea (Israel; Cloetingh et al., 1996), in 1995 at Sitges (Spain; Cloetingh et al., 1997a,b), in 1996 at Torshavn on the Faeroe Islands (Northern Atlantic; Cloetingh et al., 1998), in 1997 in the Palermo Mountains of Sicily (Cloetingh et al., 1999), and in 1998 in Oliana in the southeastern Pyrenees (Spain; Cloetingh et al., 2002a). The resulting sequence of special volumes and papers stimulated new initiatives at the European level.

Since 2005, one of the main objectives of the new ILP Task Force 6 on "Sedimentary Basins" is to promote its activities at a worldwide scale. To this end yearly workshops and scientific sessions are organized, which focus on the integration of deep and surface processes for the study of sedimentary basins, linking Earth scientists from the universities, research institutes and industry from both sides of the Atlantic as well as from the Middle-East, Africa, the Circum-Pacific region and the Southern Hemisphere. After a first meeting on foreland fold-and-thrust belts, held in 2005 again at Rueil-Malmaison (Lacombe et al., 2007), the 2006 workshop of this new task force that was dedicated to Circum-Polar sedimentary basins (Arctic and Antarctic) was held at Laval University in Québec (Kirkwood et al., 2009) and included a field trip across the Appalachian thrust front in the vicinity of Québec City. The 2007 Task Force workshop was held at Marrakech, in association with the Moroccan and American Associations of Petroleum Geologists, focusing on uplift and subsidence of African basins and margins (Bertotti et al., 2009). The 2008 workshop at Ensenada (Mexico) was dedicated to the dynamics of Mexican and other Latin American and Circum-Pacific basins. The 2009 workshop, to be held at Abu-Dhabi (United Arab Emirates), will address the geodynamics of the Middle Eastern, Asian and Australian sedimentary basins.

In this review paper we summarize the results of the past and present ILP task forces on Sedimentary Basins, related EGS and EGU sessions (Bertotti et al., 2002; Cloetingh and Ben-Avraham, 2002; Cloetingh et al., 2002; Scheck-Wenderoth et al., 2009a,b), and the proceedings of a range of networks supported by European Science Foundation (ESF), IGCP as well as other sponsors, involving combined academy-industry communities.[1] Moreover, this paper will focus on the state of the art of integrated techniques used for the study of deep and surface processes operating in sedimentary basins and underlying lithosphere, and on the challenges for new methodology development and continuous data acquisition.

Deep Controls on the Architecture of Sedimentary Basins

Information on the present-day structure of the Earth's crust and its deeper interior at various scales is of fundamental importance to the Solid-Earth sciences in general and sedimentary basin studies in particular. Close constraints on the crustal and mantle structure permit to model presently active internal processes, as well as processes that have ceased to be active, the memory of which is preserved in the present-day configuration of the sedimentary sequences, the crust and lithospheric mantle. Controls on the gravity field and internal structure of the Earth, combined with monitoring of active processes (e.g., earthquake activity, surface deformation) play an important role by providing constraints for modelling of the past and present dynamic state of the crust and mantle system and underlying processes. In turn, information on the detailed history of the crust-mantle system provides the backbone for paleo-topography reconstructions relevant for the evolution of the sedimentary record and for explaining controls on present-day surface topography.

Below we review the deep controls on present architecture of sedimentary basins focussing on the know-how base for the European lithosphere.

Constraints on Basin Fill and Crustal Configuration by Seismic Imaging

Industrial Reflection Seismic and Supporting Potential Field Data

During the last decades the international oil and gas industry has widely contributed in advancing the science and technology of reflection seismology. Today, very advanced seismic recording instrumentation, processing technologies and very advanced seismic imaging algorithms generate astonishingly well-resolved and accurate images at a large depth range (0.1–10 km). Using state-of-the-art 3-D seismic imaging of today, it is possible to obtain a 3-D data cube representing the Earth below the plane of seismic measurements. 2-D depth slices through such data cubes are of such quality that geologists can readily interpret them in terms of depositional history and structural evolution. Time series of 3-D seismic cubes ("4-D time lapse seismics") can even provide information on the temporal evolution of a geological structure, e.g., as a result of hydrocarbon extraction or of CO_2 injection.

[1] These networks include ESF-EUROPROBE (Stephenson, 1993; Stephenson et al., 1996; Gee and Stephenson, 2006), ESF-Integrated Basin Studies (IBS; Cloetingh et al., 1995; Mascle et al., 1998; Durand et al., 1999), Peri-Tethys (Roure, 1994; Ziegler and Horvàth, 1996; Brunet and Cloetingh, 2003), IGCP 369 Peri-Tethyan Rift/Wrench Basins and Passive Margins (Ziegler et al., 2001), TransMED (Cavazza et al., 2004; Spakman and Wortel, 2004), the ESF-Environmental Tectonics Research Network (ENTEC; Cloetingh and Cornu, 2005), ILP TOPO-EUROPE (Cloetingh et al., 2007), ESF-EUROMARGINS, the French GDR Marges focusing on Northern Atlantic volcanic margins (Geoffroy, 2001; Callot et al., 2002) and young passive margins of the Gulf of Aden (Leroy et al., 2004; d'Acremont et al., 2005; Bellahsen et al., 2006; Huchon et al., 2007; Tibéri et al., 2007), IFP-VNIVGNI and IGIRGI scientific workshops in Moscow (Roure et al. 1996), the joint IFP-AAPG Hedberg conference in Sicily on fluid flow, deformation and diagenesis (Swennen et al., 2004), many other federative European projects dealing with sedimentary basins, as well as summer schools dedicated to the study of the lithosphere and for promoting integrated modelling approaches coupling deep and surface processes (Solheim et al., 1996; Neubauer et al., 1997; Cloetingh et al., 1998a, b; van der Wateren and Cloetingh, 1999; Chalmers and Cloetingh, 2000; Cloetingh et al., 2001, 2002a, b).

On the other hand, a priori geological information is a requisite for constraining geophysical operations. A major task of sedimentary geologists will be to formulate depositional models on the base of available borehole and seismic data that can be iteratively tested by reflection-seismic data sets. Notwithstanding the success of the seismic method, passive potential-field methods, such as gravity surveying and geomagnetic techniques, are valuable tools for prospecting deep sedimentary basins in which it is difficult to obtain high-quality reflection-seismic data. The reflection-seismic and potential-field methods complement each other. The seismic method delineates horizontal and vertical density and velocity contrasts, while gravity and geomagnetic data image only the integrated effect of vertical density and magnetization discontinuities, respectively. Moreover, the seismic and potential-field methods differ in the scale of their resolution. Depending on the wavelength of observed anomalies, potential field methods can provide first approximation information on shallow density contrast (short wavelength signal) caused for example by salt structures, on the distribution of igneous rocks, the depth to crystalline basement, the thickness of sedimentary strata (medium wavelength signal), and potentially the location of deep-seated structures (long wavelength signal).

New techniques coupling the use of passive seismology with gravity data have been successful in discriminating the distribution of salt diapirs and carbonate reservoirs in such tectonically complex environments as the Vienna Basin (Austria) and Epirus (Northern Greece) where reflection seismic method alone could hardly document the deep architecture of the prospects (Kapotas et al., 2003; Martakis et al., 2006). The first step of this new integrated work flow consists in constructing a 3D Vp/Vs velocity model, using the travel times observed between natural seismic sources (earthquakes) and a network of local seismometers. Inversion of gravity data is then used to differentiate the targeted reservoir lithologies (i.e., dense carbonate) among other high velocity bodies (e.i. light salt; Fig. 1).

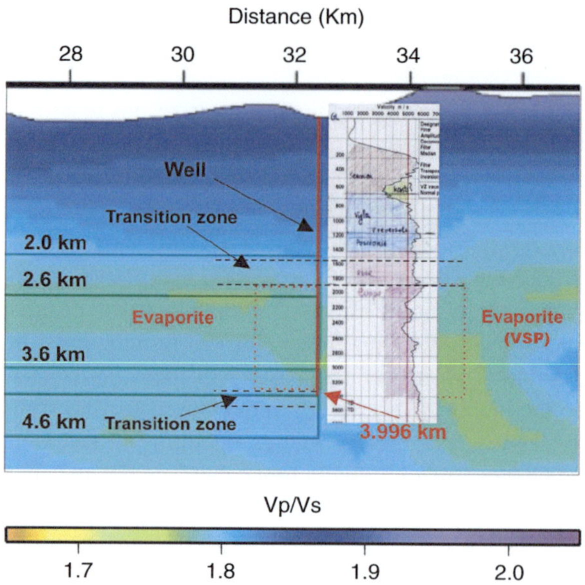

Fig. 1 The use of passive seismic and inversion of gravity data for discriminating among carbonate reservoirs and salt diapirs in Epirus (Greece; after Durham, 2003, Kapotas et al., 2003; Martakis et al., 2006; modified). Because salt and carbonates share the same high velocities, but have very distinct densities, the coupling of seismic tomography and gravity studies has been able to identify potential carbonate reservoirs in this tectonically complex salt-bearing foothills area

Crustal-Scale Reflection and Refraction Seismology

During the last decades a large number of crustal-scale reflection-seismic surveys were carried out in Europe, using many geophones in short aperture arrays, that successfully imaged the deep structure of orogenic belts, sedimentary basins, the fabric of cratonic crust, the crust/mantle boundary, and in places upper mantle reflectors (e.g., Hobbs and Klemperer, 1991; BABEL Working Group, 1993; MONA LISA Working Group, 1997; DEKORP-BASIN Research Group, 1998, 1999). In combination with extensive refraction-seismic data (EGT: Blundell et al., 1992; POLONAISE: Guterch et al., 1999; CELEBRATION: Guterch et al., 2003), providing information on the crustal and upper mantle velocity structure, and potential-field data, this has greatly advanced the understanding of the configuration and evolution of Europe's continental crust, and particularly of the transformation of orogenically destabilized crust into stabilized cratonic crust (e.g., Aichroth et al., 1992; ILIHA DSS Group, 1993; Guterch et al., 1999; Carbonell et al., 2000; Maystrenko et al., 2003; Thybo et al., 2003; Ziegler et al., 2004). A number of seismic examples are published in the Millenium Atlas (Evans et al., 2003), and in the "Atlas of the Southern Permian Basin area" (Doornenbal, 2008).

Continued technological advances, particularly in the Earth Science community's capacity to process and model seismic (and other geophysical) data, has led to many new conceptual and quantitative breakthroughs in the field of lithosphere dynamics. New data sets, collaboratively collected and interpreted in targeted multidisciplinary projects, permitted participants to integrate geophysical data with the results of geological surface and subsurface mapping.

Of particular relevance was the realization that the reactivation potential of pre-existing crustal discontinuities plays a very important role in the deformation of intraplate domains, both under compressional and extensional stress regimes. Correspondingly, identification of such crustal discontinuities is of great importance in terms of localizing neotectonic deformation and the assessment of underlying controlling processes. Surprisingly enough, and despite a flat surface topography and a horizontal Moho discontinuity, the first deep seismic profile ever recorded in France across the Paris Basin evidenced a set of Hercynian north-verging upper crustal low-angle thrusts, which root in the layered lower crust (Biju-Duval et al., 1987; Cazes and Torreilles, 1988; Fig. 2b). The same Hercynian heritage is well evidenced also in both the southern and northern forelands of the Pyrenees (Choukroune et al., 1990), and in other deep seismic profiles recorded across the Variscan fold belt of Western and Central Europe (Behr and Heinrichs, 1987). Negative inversion of such former thrust faults was first evidenced by COCORP profiles in North America, both in the Basin and Range province, where Cenozoic extension is superimposed on former Cordilleran structures, and in the internal part of the Appalachians, which has been tensionally remobilized during the rifting phase that preceded opening of the Central Atlantic Ocean. Similar extensional reactivation of former Caledonian or Hercynian thrusts has been documented in western Europe by the BIRPS, SWAT and WAM profiles (Bois et al., 1991; Hobbs and Klemperer, 1991; Lefort et al., 1991; Fig. 3), and in the northern ECORS-Pyrenees profile (Choukroune and ECORS Pyrenees team, 1989; Roure et al., 1989; Choukroune et al., 1990). Although deep seismic profiles are still missing in the Subandean basins of Venezuela and Colombia, east-verging Paleozoic thrusts have been imaged in these areas by industrial profiles beneath the regional Albian unconformity. In the Eastern Cordillera of Colombia and Merida Andes of Venezuela, reactivation of such thrusts presumably controlled the localization of Jurassic grabens and their subsequent inversion (Colletta et al., 1997; Roure, 2008).

Figure 4 outlines the crustal architecture of the main French sedimentary basins, as imaged by the ECORS programme. These sections constitute also text book illustrations for intracratonic rift systems (Rhine Graben ECORS-DEKORP profiles; Brun et al., 1991, 1992; Wenzel et al., 1991; Fig. 4a, b), asymmetric pull-apart basins (Parentis Basin-Bay of Biscaye ECORS profile; Bois and Gariel, 1994; Fig. 4), mature passive margins (Gulf of Lions ECORS-CROP profile; Burrus, 1989; De Voogd et al., 1991; Séranne et al., 1995; Séranne, 1999; Fig. 4d), but also for foreland basins resulting from deflection of the continental lithosphere under to the load of the Alps and the Pyrenees (French-Italian ECORS-CROP Alps profile; Roure et al., 1990, 1996; Fig. 4f; French-Spanish ECORS Pyrenees profile; Choukroune and ECORS team, 1989; Roure et al., 1989; Figs. 4e and e').

Continental-Scale Maps of the European Moho and Lithospheric Thickness

Dèzes and Ziegler (2004) integrated the results of crustal studies carried out since the publication of Moho depth maps by Meissner et al. (1987), Ziegler (1990b) and Ansorge et al. (1992) to obtain a better understanding of the present day crustal configuration of Western and Central Europe, and to analyze processes and their timing, which controlled the evolution of the crust in the different parts of Europe (Ziegler and Dèzes, 2006). From the Moho depth map (Fig. 5) it is evident that the stable parts of the Precambrian Fennoscandian - East-European craton are characterized by a thick crust and Moho depths ranging up to 48 km, whereas in more mobile Phanerozoic Europe Moho depths range between 24 and 48 km and no longer bear any relation to the Caledonian and Variscan orogens. On the other hand, areas underlain by the Precambrian Hebridean craton are characterized by crustal thicknesses in the range of 20–26 km, reflecting a strong overprinting by a Mesozoic rifting. By contrast, the Alpine chains, such as the Western and Central Alps, the Carpathians, Apennines, Dinarides, as well as the Betic Cordilleras and the Pyrenees are character-

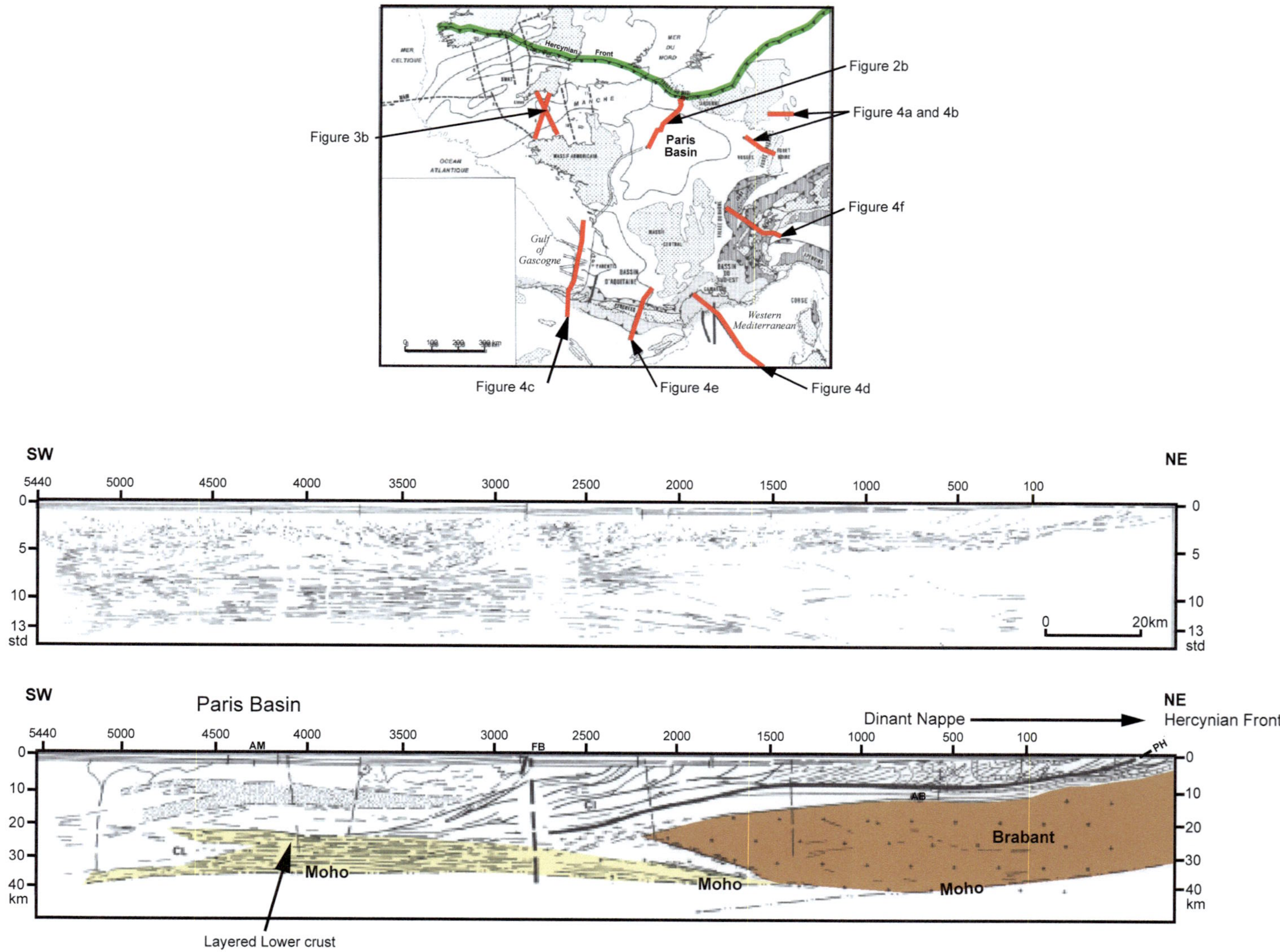

Fig. 2 **a**) Location map of the ECORS deep seismic profiles. **b**) ECORS deep seismic profile across the Paris Basin, outlining mid-crustal south-dipping reflectors inherited from the Hercynian structural grain. These Paleozoic thrusts root into a layered lower crust and flat Moho, and are unconformably overlain by poorly deformed Mesozoic and Cenozoic series (after Biju-Duval et al., 1987; Cazes and Torreilles, 1988, modified)

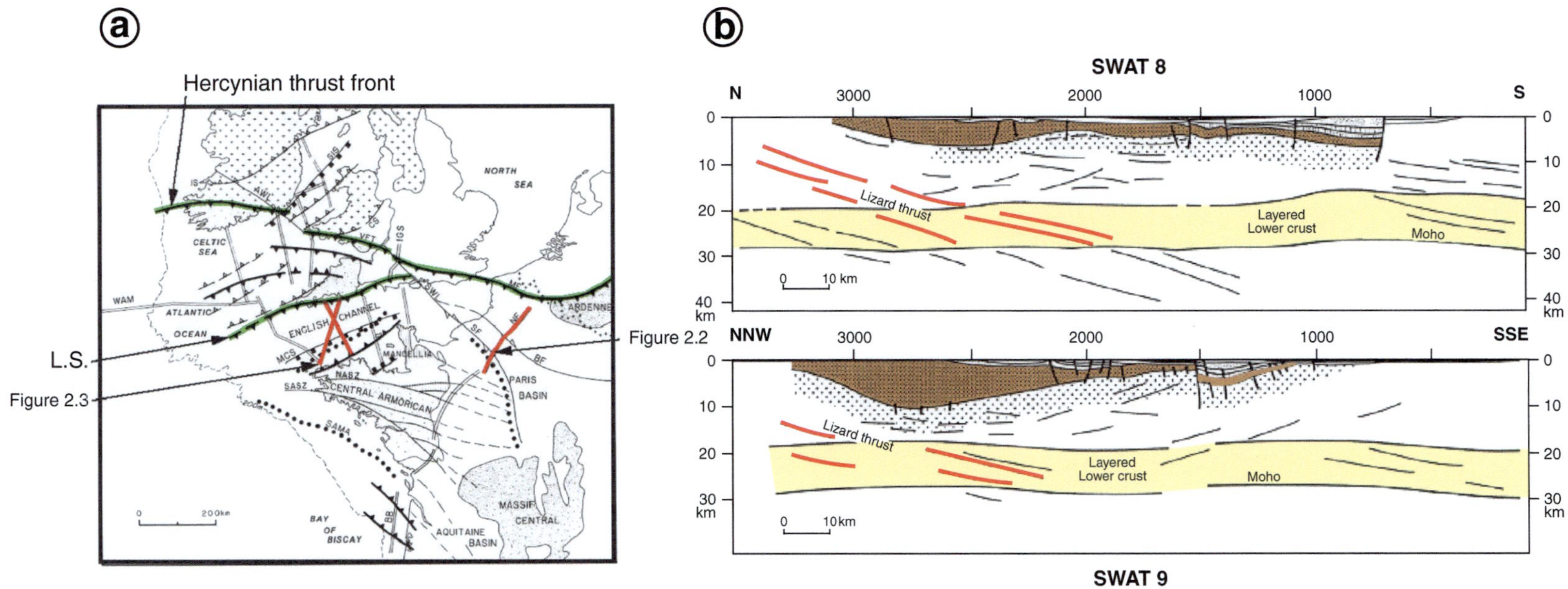

Fig. 3 Reactivation of Paleozoic thrusts in the Celtic Sea, as evidenced by SWAT and WAM deep seismic profiles (after Bois et al., 1991; Hobbs and Klemperer, 1991; Lefort et al., 1991, modified)

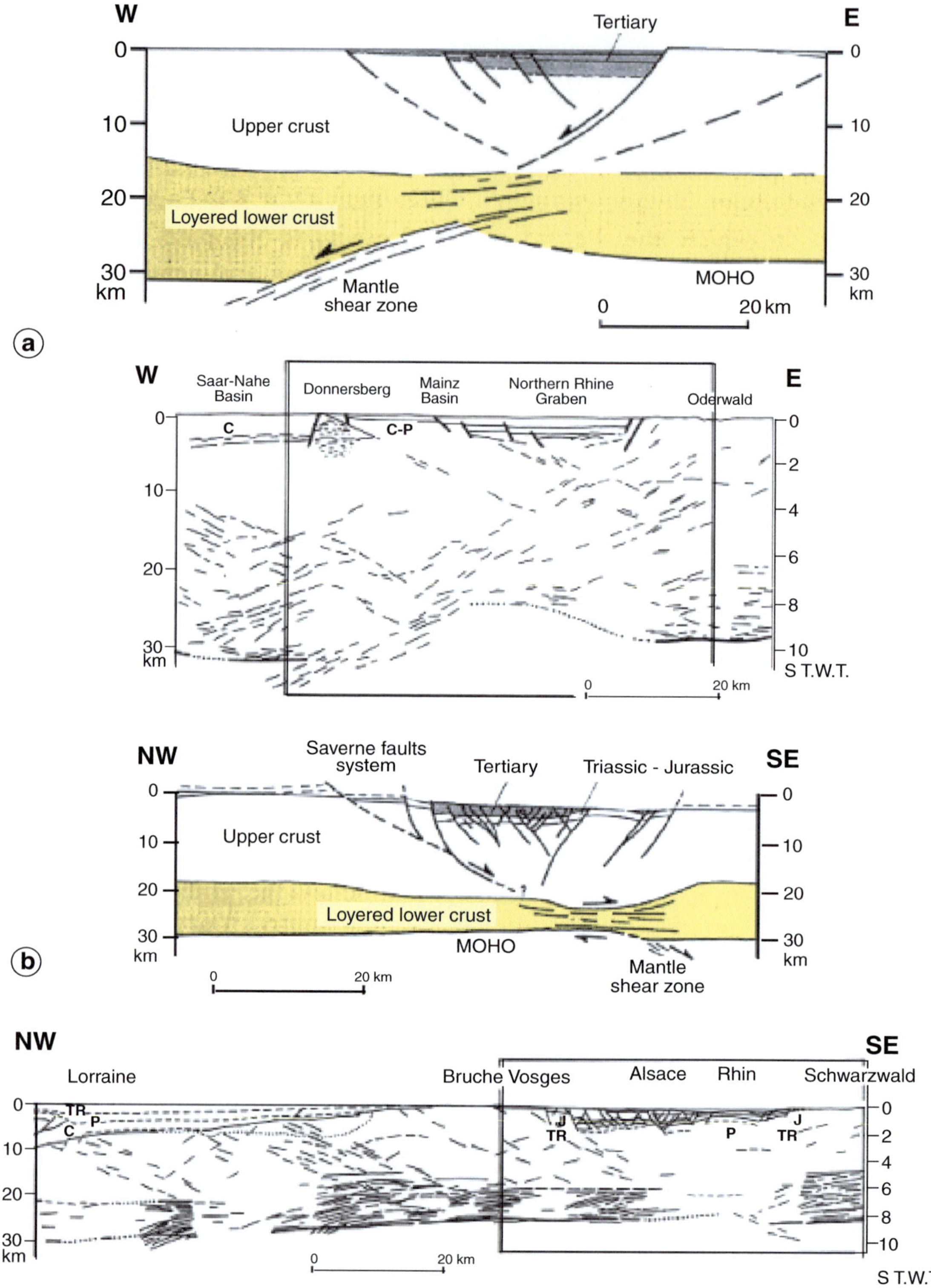

Fig. 4 An overview of the overall crustal architecture of selected French basins from ECORS, ECORS-DEKORP and ECORS-CROP deep seismic profiles. See location map of the profiles in Fig. 2: **a** and **b**) Rhine Graben (French-German ECORS-DEKORP programme; after Brun et al., 1991, 1992; Wenzel et al., 1991, modified). Notice the vertical and lateral shift of the Moho beneath the graben, and the lateral change in the asymetry of the graben, the dominant border fault being

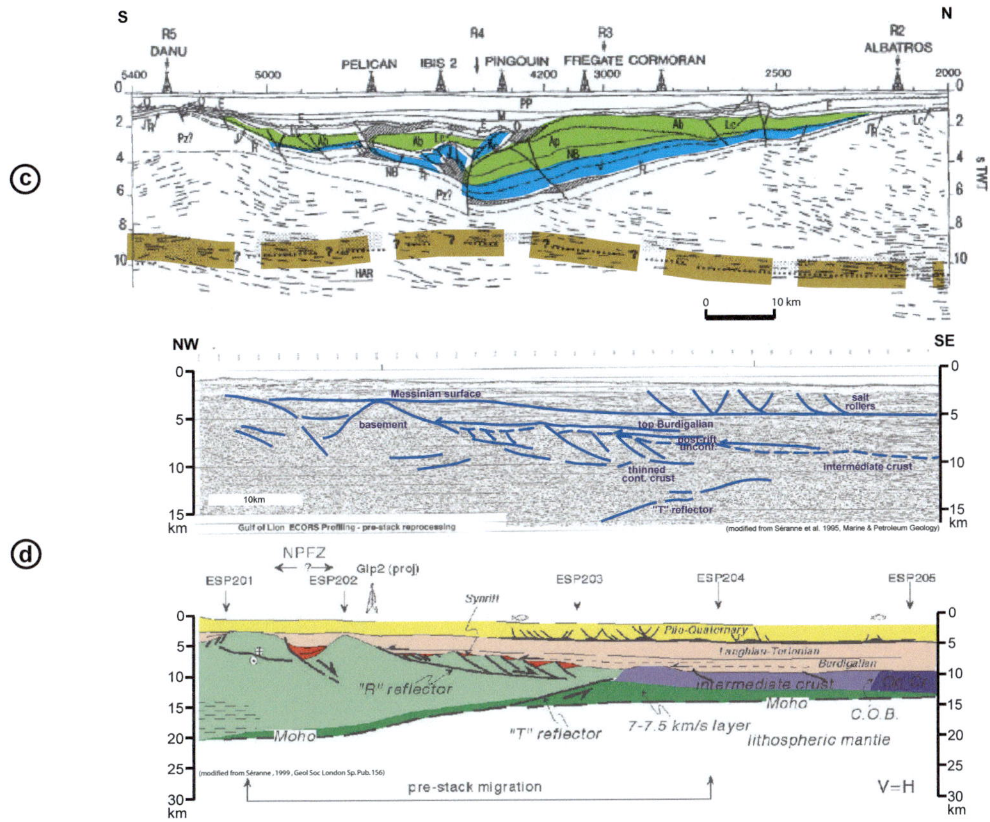

Fig. 4 (continued) located along the western side of the graben in the south, and eastern border in the north; **c**) Parentis Basin (after Bois and Gariel, 1994, modified). Notice the relative flat Moho, and the overall asymetry of the basin. Due to crustal stretching, Albian sediments rest almost directly on top of the lower crust and infra-continental mantle lithosphere. The same overall pattern can be proposed for the pre-orogenic architecture of the currently inverted Albian basins in the Pyrenees, where high temperature syn-extentional Albian metamorphism occurred (Roure et al., 1989); **d**) Gulf of Lions passive margin (after De Voogd et al., 1991; Séranne et al., 1995; Séranne, 1999, modified). Notice the progressive crustal thinning toward the southeast. Further refraction and wide-angle studies have been integrated to better precise the location of the TOC (Transition between Oceanic and Continental crusts); **e, e', f**) Foreland basins from the Western Alps (**f**), Southern Pyrenees (**e** and **e'**), and Northern Pyrenees (**e** and **e'**), as imaged by the French-Italian ECORS-CROP Alpine profile (Roure et al., 1990, 1996) and the French-Spanish ECORS Pyrenees profiles (Choukroune and ECORS team, 1989; Roure et al., 1989). Notice the bending of the European Moho and lithosphere north of the Alps (**f**), and of the Iberian Moho and lithosphere south of the Pyrenees (**e** and **e'**), thus outlining the preservation of thick crustal roots. Notice also the relatively flat European Moho and south-dipping attitude of the post-Albian unconformity beneath the northern flank of the Pyrenees (**e** and **e'**), which result from a combination of pre-orogenic Albian crustal thinning and syn-orogenic Pyrenean bending of the lithosphere. Worth to notice, the structural grain of the Iberian and European conjugate forelands of the Pyrenees is dominantly controlled by north-dipping reflections, which as best interpreted as remnants of the Hercynian orogen (including both south-verging thrusts and tilted Paleozoic strata)

ized by more or less distinct crustal roots with Moho depths attaining values of up to 60 km. The present crustal configuration of Phanerozoic Europe reflects that the crustal roots of the Caledonides and Variscides were destroyed during post-orogenic times, and that their crust was repeatedly modified by Mesozoic and Cenozoic tectonic activities. An updated version of the Moho map (Fig. 5b), incorporating recent seismic studies on e.g., the Fennoscandian Shield and the Cantabrian Mountains of Spain (Tesauro et al., 2008), confirms the main features of the map presented by Dèzes and Ziegler (2004).

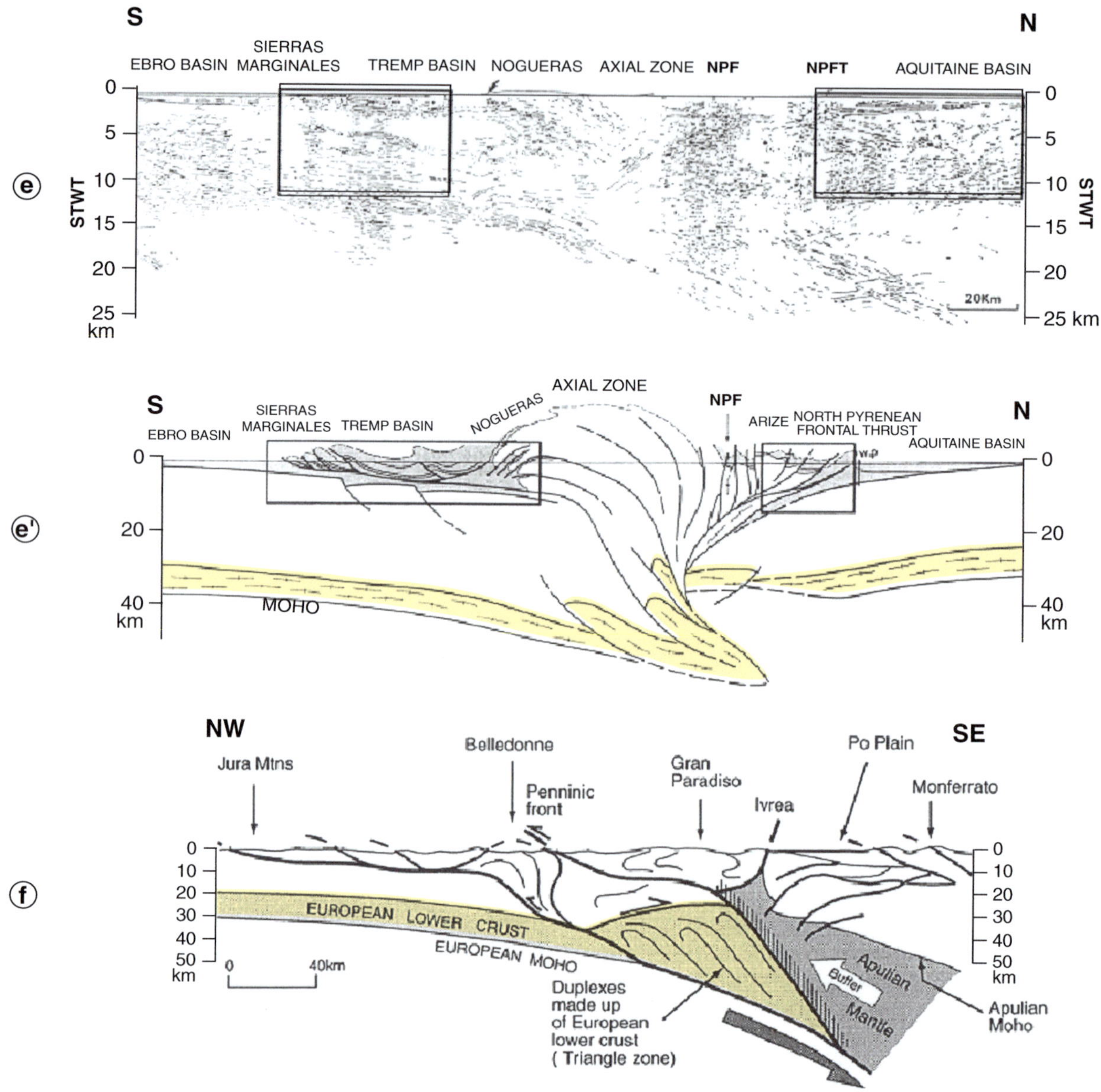

Fig. 4 (continued)

Mantle Tomography and Other Controls on Lithospheric Thickness and Mantle Heterogeneities

During the past decade, the analysis and understanding of dynamic crust-mantle processes has greatly progressed owing to major advances in the field of seismic tomography at global and regional scales (e.g., Bijwaard et al., 1998). Tomographic imaging techniques are applied to observations on body and surface waves, and provide spectacular 3-D images of mantle structures. These images can readily be linked to global plate tectonic processes, such as past and active subduction of lithospheric plates (Fukao et al., 2001). A new model parameterization technique and new 3-D ray tracing algorithms (Bijwaard and Spakman, 1999a) resulted in global mantle models that, for the first time, exhibit regional scale (60–100 km) detail (Bijwaard and Spakman, 2000, as a follow-up on Bijwaard et al., 1998). Improved focusing on lower mantle structures led to the first evidence for a whole mantle plume below Iceland (Bijwaard and Spakman, 1999b) and for upwelling of the lower mantle beneath Europe,

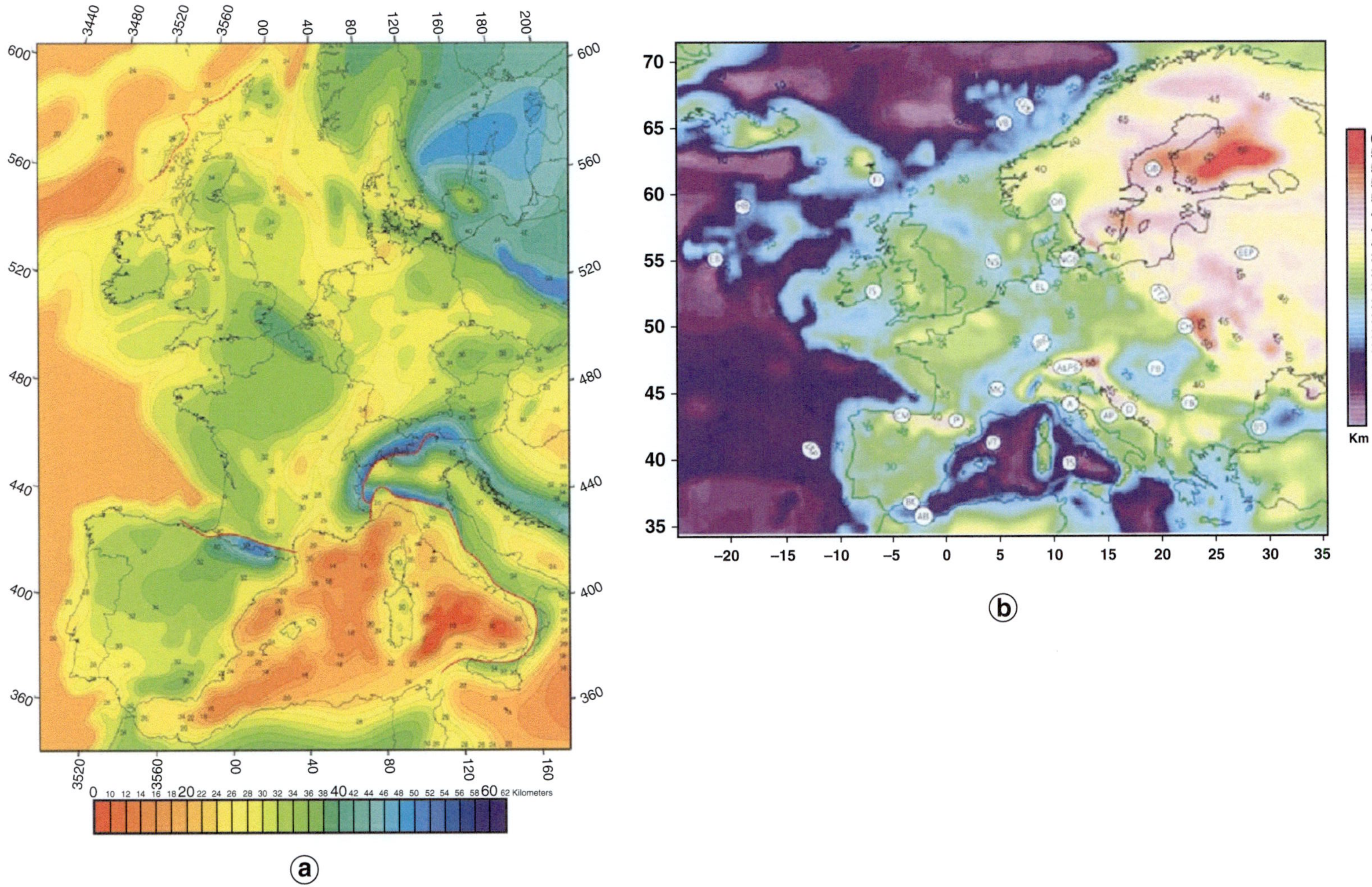

Fig. 5 **a**) Depth map of Moho discontinuity in Western Europe (2 km contour interval), constructed by integration of published regional maps (after Dèzes and Ziegler, 2004). For data sources see http://comp1.geol.unibas.ch /. *Red lines* (solid and stippled) show offsets of the Moho discontinuities. Thin oceanic crust is orange and thick crustal roots in the Alps are blue. **b**) Moho depth (km). Colour codes differ here from Fig. 5a: thin oceanic crust is blue and thick Eastern European crust pink. Abbreviations are as follows: *A*, Apennines; *AB*, Alboran Basin; *AP*, Adriatic Promontory; *BC*, Betic Cordillera; *BS*, Black Sea; *CH*, Carpathians; *CM*, Cantabrian Mountain; *D*, Dinarides; *EB*, Edoras Bank; *EL*,

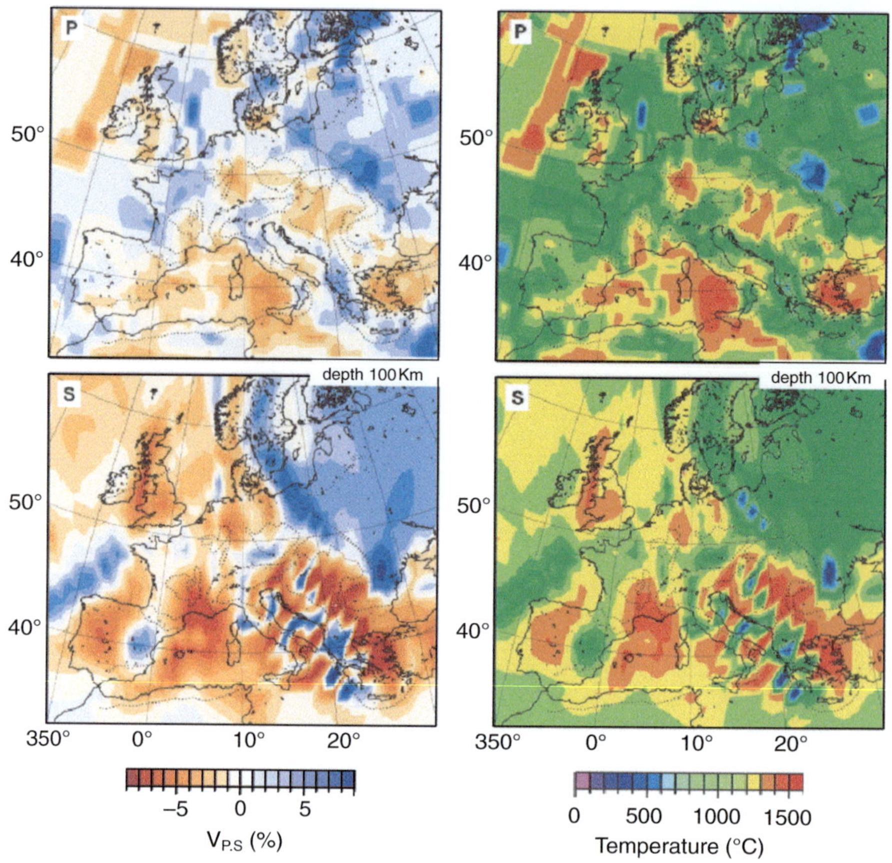

Fig. 6 Seismic velocity anomalies at 100 km under Europe for P waves (*top left*) and S waves (*bottom left*). Panels on the right show temperatures at 100 km depth estimated from the P and S velocity anomalies. The assumption that all velocity anomalies can be attributed to variations in temperature appears to be reasonable in view of the similarity in thermal structure obtained from P- and S-wave velocities (after Goes et al., 2000a)

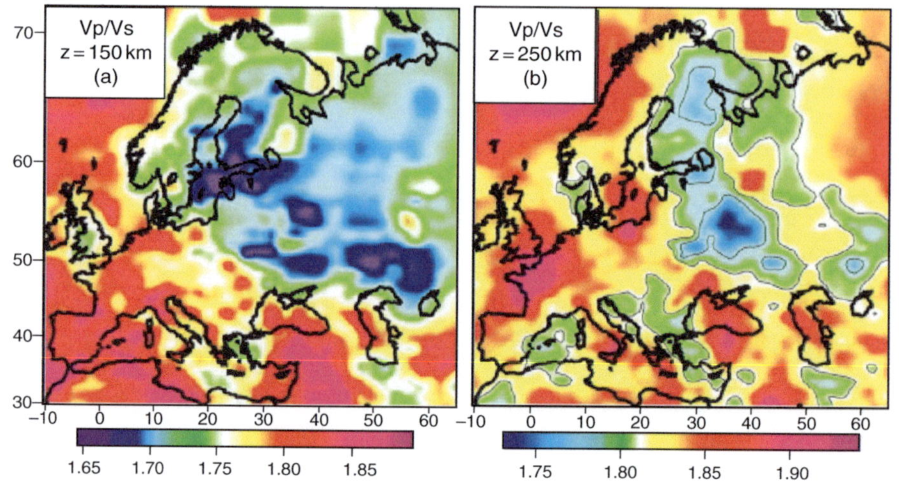

Fig. 7 Vp/Vs at (**a**) 150 km and (**b**) 250 km depth calculated from smoothed and filtered P-wave (Bijwaard and Spakman, 2000) and S-wave (Shapiro and Ritzwoller, 2002) tomography models for Europe. Variations in Vp/Vs ratio suggest differences in lithospheric composition (after Artemieva, 2006)

Fig. 5 (continued) Elbe Lineament; *EEP*, East European Platform; *FB*, Focsani Basin, *FI*, Faeroe Islands; *GB*, Gulf of Bothnia; *HB*, Hatton Bank; *IAP*, Iberian Abyssal Plain; *IS*, Iapetus Suture; *LVM*, Lofoten-Vesterålen margin; *MC*, Massif Central; *NGB*, North German Basin; *NS*, North Sea; *OR*, OsloRift; *P*, Pyrenees; *PB*, Pannonian Basin; *TS*, Tyrrhenian Sea; *TTZ*, Tesseyre-Tornquist zone; *URG*, Upper Rhine Graben; *VB*, Vøring Basin; *VT* (after Tesauro et al., 2008)

which probably underlies the longstanding Cenozoic volcanism of Western and Central Europe (Goes et al., 1999; Ziegler and Dèzes, 2007). Future developments in tomography will be directed towards a more detailed resolution (50 km) of the upper 1,000 km, although the whole mantle is addressed by such analyses (Figs. 6 and 7).

The understanding of how mantle flow relates to geomorphic or tectonic surface features is investigated applying data-assimilation methods, which permit numerical modelling of global mantle circulation (Bünge et al., 2002). In so-called variational data-assimilation models (Bünge et al. 2003) mantle heterogeneities revealed by seismic tomography can be linked to convection models via density, temperature and seismic velocity relationships (Deschamps and Trampert, 2003).

Goes et al. (2000a) developed a new inversion strategy which takes advantage of the fact that most of the variation in seismic wave velocities in the upper mantle is caused by variations in temperature. State-of-the-art seismological models and experimental data on the physical properties of mantle rocks are inverted for upper mantle temperatures, e.g., beneath Europe. Inferred mantle temperatures beneath Europe agree roughly with independent estimates from heat flow and general geological considerations (e.g., Dunai and Baur, 1995).

Beneath the French Massif Central (Granet et al., 1995), both local and global seismic tomographic studies indicate the presence of an upper mantle anomaly with a diameter of 100–300 km that rises from the 410 km mantle transition zone and involves material that is either 100–200°C hotter than the ambient mantle or contains fluids released from the lower part of the upper mantle (Lustrino and Wilson, 2007). These findings, and similar results for the Eifel region (Ritter et al., 2001), indicate diapiric upwelling of small scale, finger-like convective instabilities from the base of the upper mantle presumably that act as the main source for the Tertiary-Quaternary volcanism of Western and Central Europe. This volcanism is spatially and temporally linked to the development of a major intra-continental rift system and to domal uplift of Variscan basement massifs (Ziegler 1990a,b, 1992, 1996; Wilson and Downes, 1992, 2006; Wilson and Patterson, 2001). Tomographic images for the Upper Rhine Graben (Achauer and Masson, 2002) support the concept that it originated as a passive rift under a repeatedly changing stress field, with inherited structures playing a controlling role in its localization (Schumacher, 2002; Dèzes et al., 2004; Ziegler and Dèzes, 2007).

Tomographic evidence for mantle plumes originating at great depth (Bijwaard and Spakman, 1999b; Goes et al., 1999; Romanowicz and Gung, 2002; Montelli et al., 2004) suggests links between mantle plumes and such surface processes as intra-plate volcanism, vertical surface motions and rifting. Mantle plume activity is, however unlikely to cause on its own the development of major rifts and the splitting apart of continents (Ziegler et al., 2001; Ziegler and Cloetingh, 2004).

For the European-Mediterranean domain, recently developed tomographic models of the sub-lithospheric mantle structure (e.g., Piromallo and Morelli, 1997; Bijwaard et al., 1998; Spakman et al., 1999; Bijwaard and Spakman, 2000; Piromallo and Morelli, 2003) have greatly advanced the understanding of processes controlling the lithosphere and mantle interaction during the past and on-going evolution of the Earth's crust (Fig. 8). Conceptual models of mantle dynamics derived from tomography and analogue lab-models emphasize the role of a variety of mantle processes as driving mechanisms of major tectonic processes, the mechanical evolution of the lithosphere, and surface deformations (e.g., Wortel and Spakman, 2000; Bellahsen et al., 2003; Faccenna et al., 2003, 2004; Funiciello et al., 2004; Spakman and Wortel, 2004; Van Hinsbergen et al., 2005; Faccenna et al., 2006).

In general, continent wide scale tomographic models are based on a global observation network of seismological stations with a very heterogeneous spatial distribution. This leads to a strongly non-uniform data density and consequently to strong spatial variations in model resolution varying between 50 km to hundreds of km. In some regions of Europe, temporary seismological networks with a dense spatial distribution (30–60 km) were installed during the past decades for periods of six months to one year in order to address specific phenomena (e.g., Massif Central [Southern France], TOR [Sweden, Denmark, northern Germany], SVEKALOPCO [Finland], EIFEL [Eastern France, Western Germany], CALIXTO [Vrancea, Romania]). These and other successful experiments targeted important lithospheric transition zones, mantle plumes and subduction zones. Importantly, these experiments imaged detailed crust-mantle structures (10–30 km) associated with dynamic processes affecting surface deformation. This was only possible owing

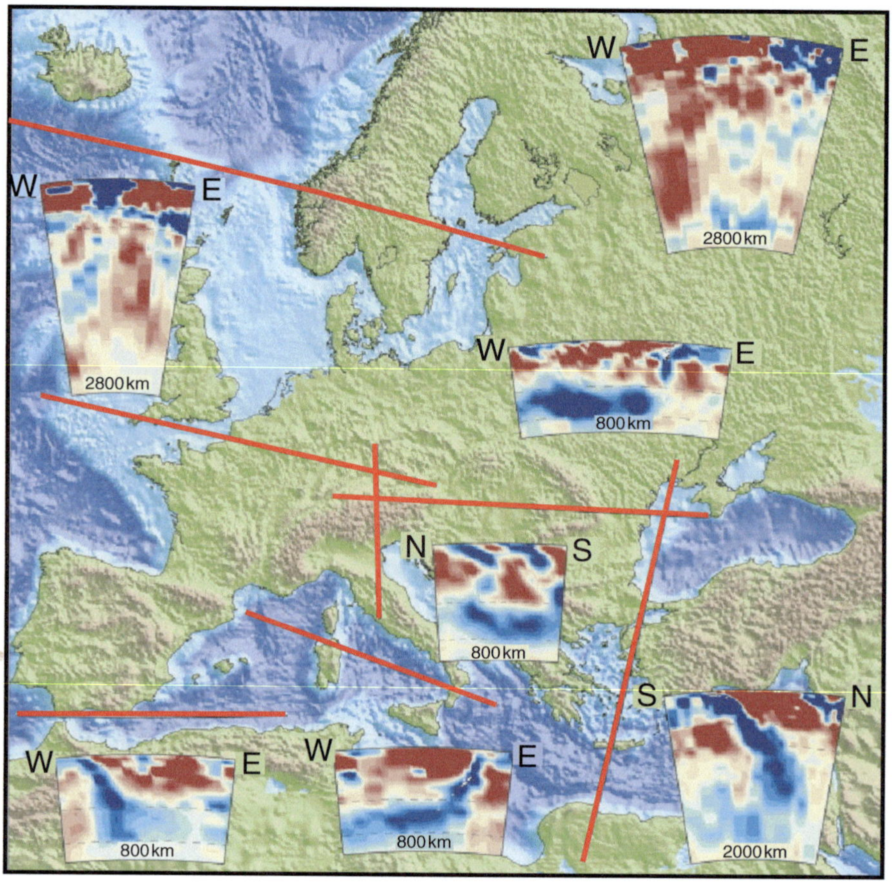

Fig. 8 Tomographic cross sections through key parts of the European continent (after Bijwaard and Spakman, 2000)

to the high resolving power that can be attained with dense observation networks. The short period of network employment, however, restricted the data volume, whilst the spatially localized nature of these experiments has prevented to obtain a contextual image of mantle structure and processes. Furthermore, by the very nature of the tomographic experiments conducted, mantle structure could only be imaged relative to an unknown background of absolute wave speed.

The shortcomings of continent-scale tomographic experiments and of local experiments with dense temporary networks can only be overcome by acquiring observations from a spatially more uniform and dense network. A considerable densification and extension of existing seismological network would permit to develop a new generation of crust-mantle models based on surface wave and body wave data and a more homogeneous spatial resolution. Moreover, this would permit to obtain models in absolute wave speeds giving strong constraints on the temperature and compositional fields of the mantle, and on associated mantle processes. Highly improved crust-mantle models (temperature, composition, presence of fluids) are a fundamental prerequisite for numerical modelling of crust-mantle processes and the resulting surface deformation both for reconstruction of topography and for establishing the current dynamic state of topography (surface motions, state of stress and strain-rate build-up). Technologically, tomographic methods are advanced enough to deal with a heterogeneous crust, ray bending effects, finite frequency effects, and even 3-D reference models of the Earth's structure. The forward leap towards much more advanced models is only hampered by the (spatial) availability of data.

A very important part of a new generation of structural models of the crust and mantle are the discontinuities in material properties that occur around the crust-mantle interface (the Moho, a velocity discontinuity interpreted as a compositional transition as well as the granulite-eclogite transition), around 410 km depth (dominated by the olivine to β-spinel transition), and around 660 km depth (dominated by the β-spinel to lower-mantle-oxides transition). Special seismological techniques can be used, and developed, to detect the topography and sharpness of these (and related) phase transitions, the precise nature of which is still a matter of active research. Receiver Function analysis of seismic data has proved to be a powerful method for the detection of the phase transition interfaces (e.g., Vinnik et al., 1996; Kind et al., 2002). The topographic configuration of these discontinuities is in fact dynamic, owing to the interaction of mantle flow (slabs, plumes) impinging on these interfaces with the physics of phase transitions. Dynamic surface topography is strongly diagnostic for the type and local nature and thermal characteristics of mantle flow. In long-wavelength mantle flow models, the dynamic surface topography is related to the dynamic topography of the internal surfaces. One of the key problems in understanding surface topography (subsidence, uplift) is the interaction between the mantle induced dynamic topography and other (shallow) topography generating processes.

Studies on mantle rocks, both xenoliths and tectonically emplaced samples, show that the mantle is heterogeneous at all observable scales, down to crystal dimensions. Recent seismological studies have shown that it is possible to image crustal and mantle structures and to determine their physical properties on a kilometric-scale or finer, much smaller than the seismic wavelength. Only recently, methods have been developed for extracting information on fine-scale heterogeneity of the crust and mantle from seismic data at scales which require a statistical representation of physical parameters (e.g., Holliger and Levander, 1994; Thybo and Perchuc, 1997; Ryberg et al., 2000; Nielsen et al., 2002; Baig and Dahlen, 2004; Shearer and Earle, 2004; Thybo, 2006; Thybo and Anderson, 2006). These methods push the attainable resolution below the usual detection limit, although they cannot provide unique solutions for the structure of the Earth.

New-generation models of the crust-mantle system can only result from concerted efforts in seismic tomography research, strong seismic-contrast and dynamic-topography research, and fine-scale imaging of crustal and mantle properties (Fig. 9). Such seismological studies can provide the principal source of information on the detailed structure of the European crust-mantle system. Data from existing global networks (FDSN, IRIS, GEOFON, EarthSCOPE) and from existing regional networks are, however, insufficient to make the necessary step forward toward the development of much more detailed crust-mantle models. This requires a much denser observation network, complementing the existing European seismological infrastructure.

Apart from its own merits, a new-generation model of crustal and mantle structure will provide the necessary input for advanced modelling of the mantle system constrained by high-resolution satellite gravity and geodetic observations of active surface deformation. This provides the "depth-to-surface" relations required for the reconstruction of mantle induced subsidence or uplift of the Earth's surface.

Lithosphere Rheology and Tectonic Controls on Basin Segmentation and Topography

The structural style of extensionally as well as compressionally deformed sedimentary sequences are controlled by their lithological composition and specifically by the distribution of competent and incompetent beds. The rheological characteristics of rock units are essentially pressure and temperature dependent and thus related to burial depth. Compressional deformation of competent, brittle beds (fracturing), such as limestones and consolidated sandstones, is characterized by the propagation of low angle faults usually cutting up-section at angles of 20–30° (ramps in thrust systems). By contrast extensional and transpressional systems are characterized by high angle faults dipping up to 60° and more. Incompetent beds, such as salt and clays, deform by ductile flow. Bed parallel compressional as well as extensional detachment (décollement) surfaces preferentially develop in evaporites, shales and over-pressured porous media. Compressional deformation of sedimentary sequences consisting of alternating competent and incompetent beds can give rise to complex patterns of thrusted flexural slip and kink folds. Changes in the rheology of sedimentary rocks, controlling their deformation are related to the evolution of pore-fluid pressures. These essentially depend on the porosity and permeability distribution within a sedimentary sequence, as well as on sedimentation-related burial, tectonic compaction and dewatering rates. Moreover, the illite-smectite transition and the transformation of kerogen to hydrocarbons may contribute to the build-up of pore-fluid pressures (Vrolijk, 1987; Vrolijk et al., 1990; Cobbold et al., 2004; De Wever, 2008). Similarly, the rheology of

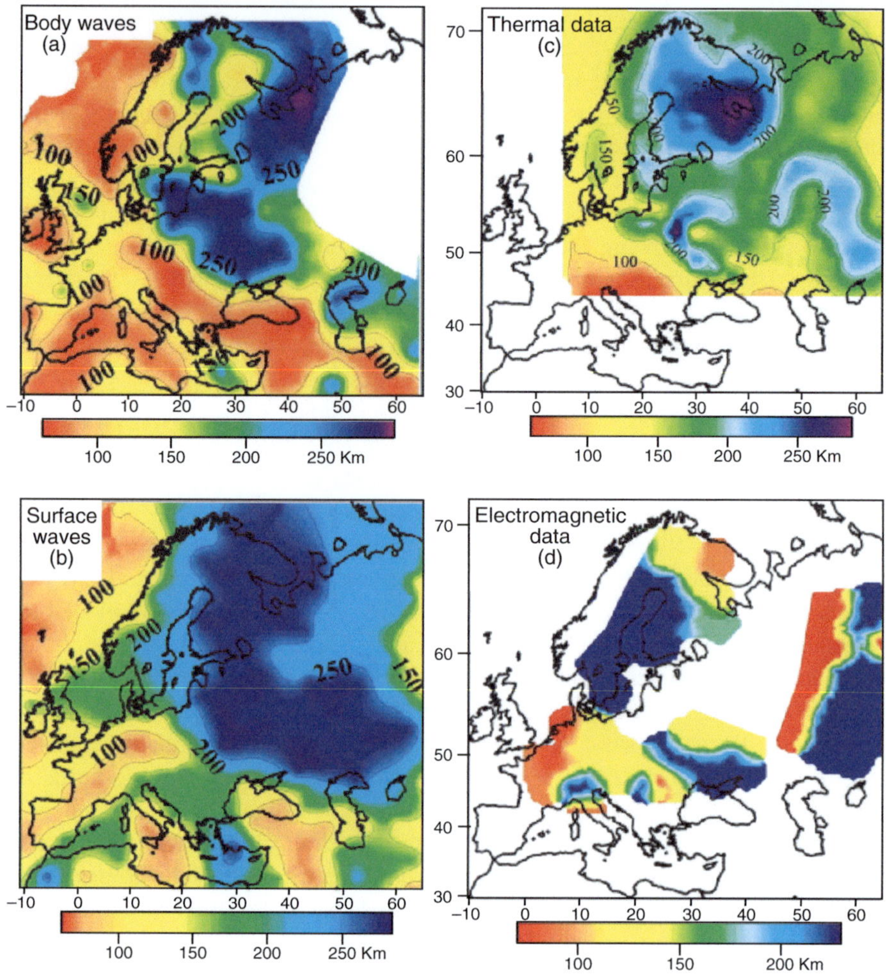

Fig. 9 Thickness of the European lithosphere as determined by (**a**) seismic tomography; (**b**) surface wave tomography; (**c**) geothermics; (**d**) magnetotellurics (after Artemieva et al., 2006)

deeper crust and mantle rocks depends on their chemical composition, temperature and pressure conditions (combination of burial and regional heat flow) and their water content. Furthermore, the crust and lithospheric mantle may be locally weakened by the occurrence of pre-existing discontinuities related to earlier deformation phases (Ziegler et al., 1998; Ziegler and Cloetingh 2004). Such inherited weakness zones, represented by e.g., crustal scale faults or eclogitized continental crust inserted into the sub-crustal mantle, may be characterized by considerably reduced strengths as compared to surrounding crustal and mantle domains.

Analogue modelling techniques were first developed, and are now routinely used by many laboratories to simulate thin-skinned deformation of sedimentary rocks (see Colletta et al., 1991, and references therein), but also of the lithosphere as a whole (Sokoutis et al., 2005, 2007, and references therein), using specific analogue materials for modelling either brittle or ductile sediments, the crust or the mantle. Further numerical developments are, however, required when investigating the effects of other parameters during basin evolution, such as pore-fluid pressure, temperature and mineralogical phase transitions.

In the following paragraphs, we shall summarize some recent advances achieved in documenting and understanding the rheology and long term behaviour of the European lithosphere, as well as a few dedicated case studies outlining (1) the incidence of deep active décollements on surface topography, (2) the respective effects of coupling and strain partitioning between the foreland and hinterland during the development of selected intramontane basins, as well as the overall dynamics of (3) intracratonic basins and (4) passive margins.

Lithosphere Strength and Deformation Mode

The strength of continental lithosphere is controlled by its depth-dependent rheological structure in which the thickness and composition of the crust, the thickness of the lithospheric mantle, the potential temperature of the asthenosphere, and the presence or absence of fluids, as well as strain rates play a dominant role. By contrast, the strength of oceanic lithosphere depends on its thermal regime, which controls its essentially age-dependent thickness (Kuznir and Park, 1987; Cloetingh and Burov, 1996; Watts, 2001; see also Burov, 2007).

Figure 10 gives synthetic strength envelops for three different types of continental lithosphere and for oceanic lithosphere at a range of geothermal gradients (Ziegler and Cloetingh, 2004). These theoretical rheological models indicate that thermally stabilized continental lithosphere consists of the mechanically strong upper crust, which is separated by a weak lower crustal layer from the strong upper part of the mantle-lithosphere that in turn overlies the weak lower mantle-lithosphere. By contrast, oceanic lithosphere has a more homogeneous composition and is characterized by a much simpler rheological structure. In terms of rheology, thermally stabilized oceanic lithosphere is considerably stronger than all types of continental lithosphere. However, the strength of oceanic lithosphere can be seriously weakened by transform faults and by the thermal blanketing effect of thick sedimentary prisms prograding onto it (e.g., Gulf of Mexico, Niger Delta, Bengal Fan; Ziegler et al., 1998).

The strength of continental crust depends largely on its composition, thermal regime and the presence of fluids, and also on the availability of pre-existing crustal discontinuities (see also Burov, 2007). Deep-reaching crustal discontinuities, such as thrust- and wrench-faults, cause significant weakening of the otherwise mechanically strong upper parts of the crust. Such discontinuities are apparently characterized by a reduced frictional angle, particularly in the presence of fluids (Van Wees, 1994). These discontinuities are prone to reactivation at stress levels that are well below those required for the development of new faults. Deep reflection-seismic profiles show that the crust of Late Proterozoic and Paleozoic orogenic belts is generally characterized by a monoclinal fabric that extends from upper crustal levels down to the layered lower crust and Moho at which it either soles out or by which it is truncated (Figs. 2, 3, 4; see Bois, 1992; Ziegler and Cloetingh, 2004). This fabric reflects the presence of deep-reaching lithological inhomogeneities and shear zones.

The strength of the continental upper lithospheric mantle depends to a large extent on the thickness of the crust but also on its age and thermal regime (see Jaupart and Mareschal, 2006). Thermally stabilized stretched continental lithosphere with a 20 km thick crust and a lithospheric mantle thickness of 50 km is mechanically stronger than unstretched lithosphere with a 30 km thick crust and a 70 km thick lithospheric mantle (compare Fig. 10b, d). Extension of stabilized continental crustal segments precludes ductile flow of the lower crust and faults will be steep to listric and propagate towards the hanging wall, i.e., towards the basin centre (Bertotti et al., 2000). Under these conditions, the lower crust will deform by distributing ductile shear in the brittle-ductile transition domain. This is compatible with the occurrence of earthquakes within the lower crust and even close to the Moho (e.g., southern Rhine Graben: Bonjer, 1997; East African rifts: Shudofsky et al., 1987).

On the other hand, in young orogenic belts, which are characterized by crustal thicknesses of up to 60 km and an elevated heat flow, the mechanically strong part of the crust is thin and the lithospheric mantle is also weak (Fig. 10c). Extension of this type of lithosphere, involving ductile flow of the lower and middle crust along pressure gradients away from areas lacking upper crustal extension into zones of major upper crustal extensional unroofing, can cause crustal thinning and thickening, respectively. This deformation mode gives rise to the development of core complexes with faults propagating towards the hanging wall (e.g., Basin and Range Province: Wernicke, 1990; Buck, 1991; Bertotti et al., 2000). However, crustal flow will cease after major crustal thinning has been achieved, mainly due to extensional decompression of the lower crust (Bertotti et al., 2000).

Generally, the upper mantle of thermally stabilized, old cratonic lithosphere is considerably stronger than the strong part of its upper crust (Fig. 10a) (Moisio et al., 2000). However, the occurrence of upper mantle reflectors, which generally dip in the same direction as the crustal fabric and probably are related to subducted oceanic and/or continental crustal material,

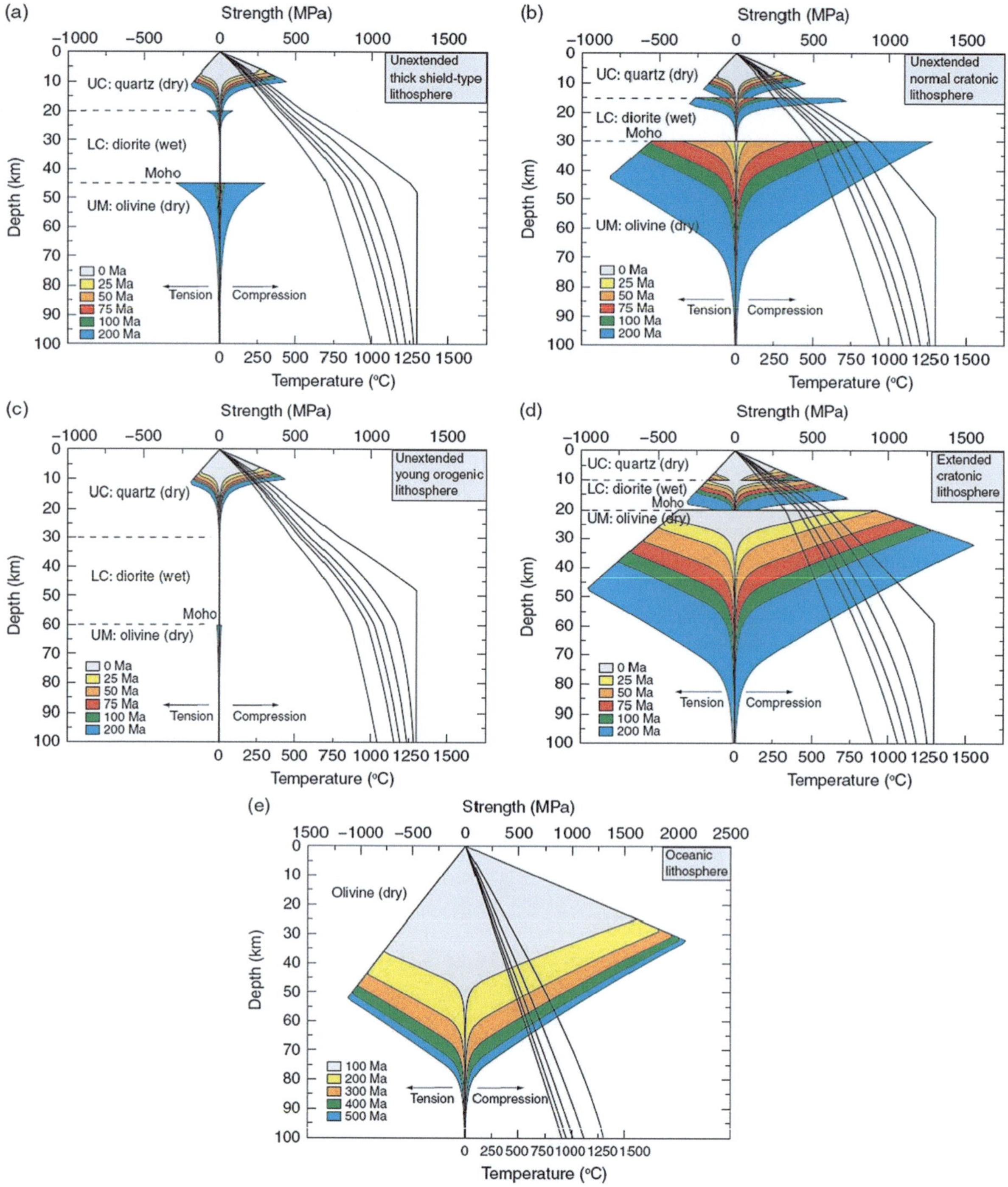

Fig. 10 Depth-dependent rheological models for various lithosphere types and a range of geothermal gradients, assuming a dry quartz/diorite/olivine mineralogy for continental lithosphere (Ziegler, et al., 1995; Ziegler et al., 2001). (**a**) Unextended, thick-shield-type lithosphere with a crustal thickness of 45 km and a lithospheric mantle thickness of 155 km. (**b**) Unextended, "normal" cratonic lithosphere with a crustal thickness of 30 km and a lithospheric mantle thickness of 70 km. (**c**) Unextended, young orogenic lithosphere with a crustal thickness of 60 km and a lithospheric mantle thickness of 140 km. (**d**) Extended, cratonic lithosphere with a crustal thickness of 20 km and a lithospheric mantle thickness of 50 km. (**e**) Oceanic lithosphere. Modified from Ziegler et al. (2001)

suggests that the continental lithospheric mantle is not necessarily homogenous but can contain lithological discontinuities that enhance its mechanical anisotropy (Vauchez et al., 1998; Ziegler et al., 1998). Such discontinuities, consisting of eclogitized crustal material, can potentially weaken the strong upper part of the lithospheric mantle. Moreover, even in the face of similar crustal thicknesses, the heat flow of deeply degraded Late Proterozoic and Phanerozoic orogenic belts is still elevated as compared to adjacent old cratons (e.g., Panafrican belts of Africa and Arabia; Janssen, 1996). This is probably due to the younger age of their lithospheric mantle and possibly also to a higher radiogenic heat generation potential of their crust. These factors contribute to weakening of former mobile zones to the end that they present rheologically weak zones within a craton, as evidenced by their preferential reactivation during the break-up of Pangea (Ziegler, 1989; Janssen et al., 1995; Ziegler et al., 2001).

Concerning rheology, the thermally destabilized lithosphere of tectonically active rifts, as well as of rifts and passive margins that have undergone only a relatively short post-rift evolution (e.g., 25 Ma), is considerably weaker than that of thermally stabilized rifts and of unstretched lithosphere (Figs. 10 and 11, Ziegler et al., 1998). In this respect, it must be realized that during rifting, progressive mechanical and thermal thinning of the lithospheric mantle and its substitution by the upwelling asthenosphere is accompanied by a rise in geotherms causing progressive weakening of the extended lithosphere. In addition, its permeation by fluids causes its further weakening (Fig. 11). Upon decay of the rift-induced thermal anomaly, rift zones are rheologically speaking considerably stronger than unstretched lithosphere (Fig. 10). However, accumulation of thick syn- and post-rift sedimentary sequences can cause by thermal blanketing a weakening of the strong parts of the upper crust and lithospheric mantle of rifted basins (Stephenson, 1989). Moreover, as faults permanently weaken the crust of rifted basins, they are prone to tensional as well as compressional reactivation and tectonic inversion (Roure et al., 1994, 1997; Ziegler et al., 1995, 1998, 2001, 2002; Brun and Nalpas, 1996; Roure and Colletta, 1996).

In view of its rheological structure, the continental lithosphere can be regarded under certain conditions as a two-layered visco-elastic beam (Fig. 12; Reston, 1990, Ter Voorde et al., 1998). The response of such a system to the build-up of extensional and compressional stresses depends on the thickness, strength and spacing of the two competent layers, on stress magnitudes and strain rates and the thermal regime (Zeyen et al., 1997; Watts and Burov, 2003). As the structure of continental lithosphere is also regionally heterogeneous, its weakest parts start to yield first once tensional as well as compressional intraplate stress levels equate their strength (Ziegler et al., 2001).

The flow properties of mantle rocks control the thickness and strength of the lithospheric plates, the degree of coupling between moving lithospheric plates and the pattern and rate of asthenospheric convection, and the rate of melt extraction at mid-ocean ridges. To be able to understand the dynamic behaviour of the outer parts of the solid Earth, notably the dynamics of lithospheric extension and associated rifting and sedimentary basin development, a detailed knowledge of the rheology and evolution of the upper mantle (30–410 km depth) and between the 410 and 670 transition zones is essential. At present, these flow properties are surprisingly poorly known. Experimental work has yielded constitutive equations describing various types of flow in mantle rocks, but it is not clearly established to what extent the experimentally observed flow mechanisms are relevant for natural crust and mantle conditions. A second problem is that trace amounts of water and melt can cause drastic weakening of mantle rocks and may cause the development of upper mantel convective instabilities (Lustrino and Wilson, 2007). Such fluid-related weakening effects are widely recognized as, for example, controlling the strength of trans-lithospheric faults in the substratum of active basins. However, only limited data are available on such effects, and a quantitative, mechanical understanding suitable for extrapolation to nature is lacking.

These problems can be addressed by means of experimental studies, scanning and transmission electron microscopy (SEM, TEM) and field studies on exposed upper mantle rocks. Integration of these approaches aims at arriving at quantitative, mechanism-based descriptions of mantle rheologies suitable for use in modelling the dynamics of the upper mantle and transition zone. Field-based studies involving structural geological and EM work on upper mantle rocks deformed in a variety of geological environments

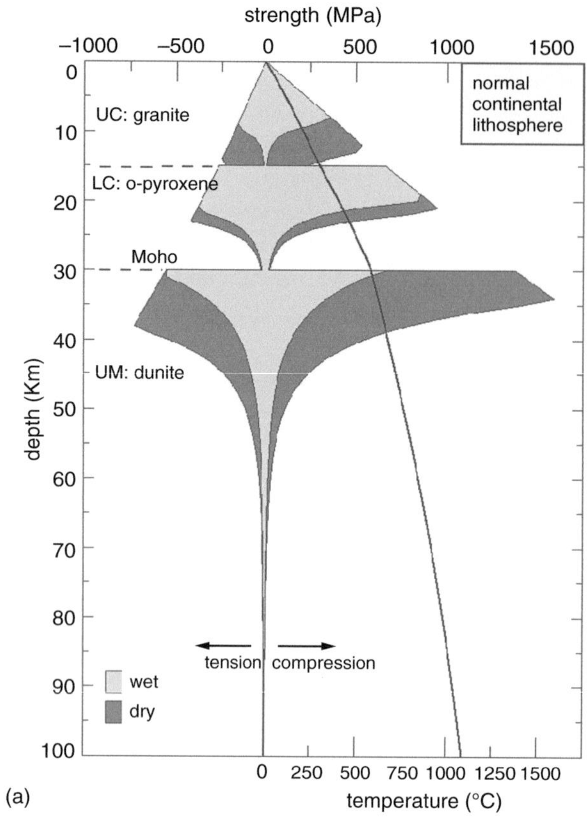

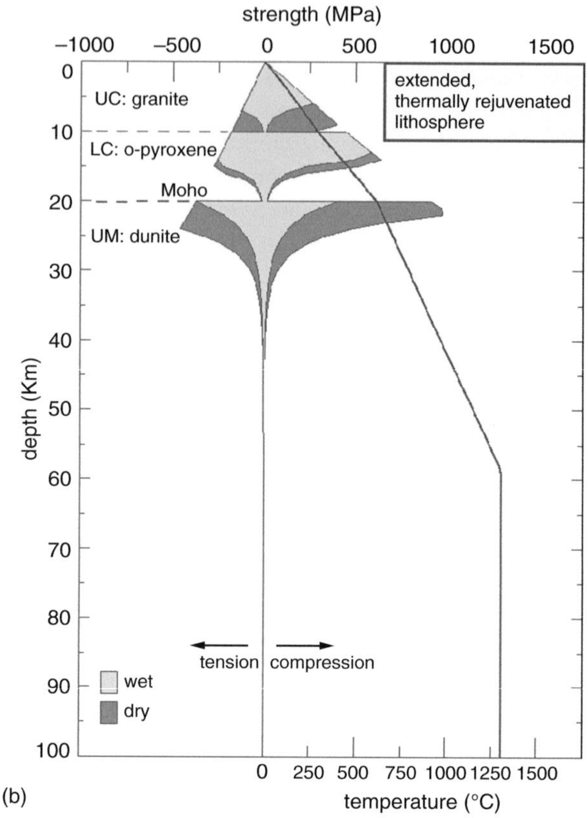

Fig. 11 Depth-dependent rheological models for dry and wet, unextended 'normal' cratonic lithosphere and stretched, thermally attenuated lithosphere, assuming a quartz/diorite/olivine mineralogy. (**a**) Unextended, cratonic lithosphere with a crustal thickness of 30 km and a lithospheric mantle thickness of 70 km. (**b**) Extended, thermally destabilized cratonic lithosphere with a crustal thickness of, 20 km and a lithospheric mantle thickness of 38 km. Modified from Ziegler et al. (2001)

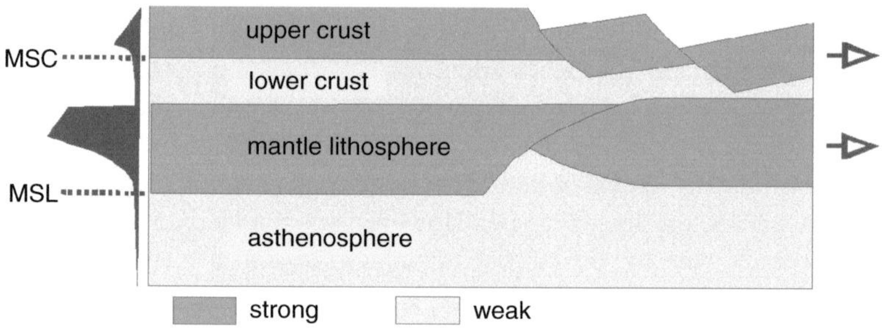

Fig. 12 Kinematic model for extension of rheologically stratified lithosphere. See strength profile on *left side* of diagram. MSC and MSL indicate the base of the mechanically strong crust and mechanically strong lithosphere, respectively. From Reston (1990)

may provide information on flow mechanisms occurring in the upper mantle. Therefore, special attention has to be paid to upper mantle rocks showing possible asthenospheric flow structures, which developed when the rocks contained some fluid or partial melts. In addition, attention has to be paid to upper mantle shear zone rocks as such shears probably control the extensional strength of the lithosphere.

Lithospheric Folding: An Important Mode of Intraplate Basin Formation

Folding of the lithosphere, involving its positive as well as negative deflection (see Figs. 13 and 14), appears to play a more important role in the large-scale neotectonic deformation of Europe's intraplate

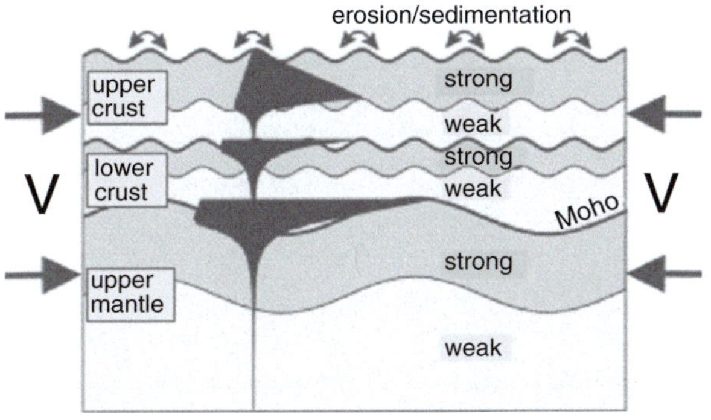

Fig. 13 Schematic diagram illustrating decoupled lithospheric mantle and crustal folding, and consequences of vertical motions and sedimentation at the Earth's surface. V is horizontal shortening velocity; upper crust, lower crust, and mantle layers are defined by corresponding rheologies and physical properties. A typical brittle-ductile strength profile (*in black*) for decoupled crust and upper mantle- lithosphere, adopting a quartz-diorite-olivine rheology, is shown for reference

domain than hitherto realized (after Cloetingh et al., 1999). The large wavelength of vertical motions associated with lithospheric folding necessitates integration of available data from relatively large areas (Elfrink, 2001), often going beyond the scope of regional structural and geophysical studies that target specific structural provinces. Recent studies on the North German Basin have revealed the importance of its neotectonic structural reactivation by lithospheric folding (Marotta et al., 2000). Similarly, the Plio-Pleistocene subsidence acceleration of the North Sea Basin is attributed to stress-induced buckling of its lithosphere (Van Wees and Cloetingh, 1996; Unternehr and van den Driessche, 2004). Moreover, folding of the Variscan lithosphere has been documented for Brittany (Bonnet et al., 2000), the adjacent Paris Basin (Lefort and Agarwal, 1996) and the Vosges-Black Forest arch (Ziegler et al., 2002; Dèzes et al., 2004; Bourgeois et al., 2007; Ziegler and Dèzes, 2007). Lithospheric folding is a very effective mechanism for the propagation of tectonic deformation from active plate boundaries far into intraplate domains (e.g., Stephenson and Cloetingh, 1991; Burov et al., 1993; Ziegler et al., 1995, 1998, 2002).

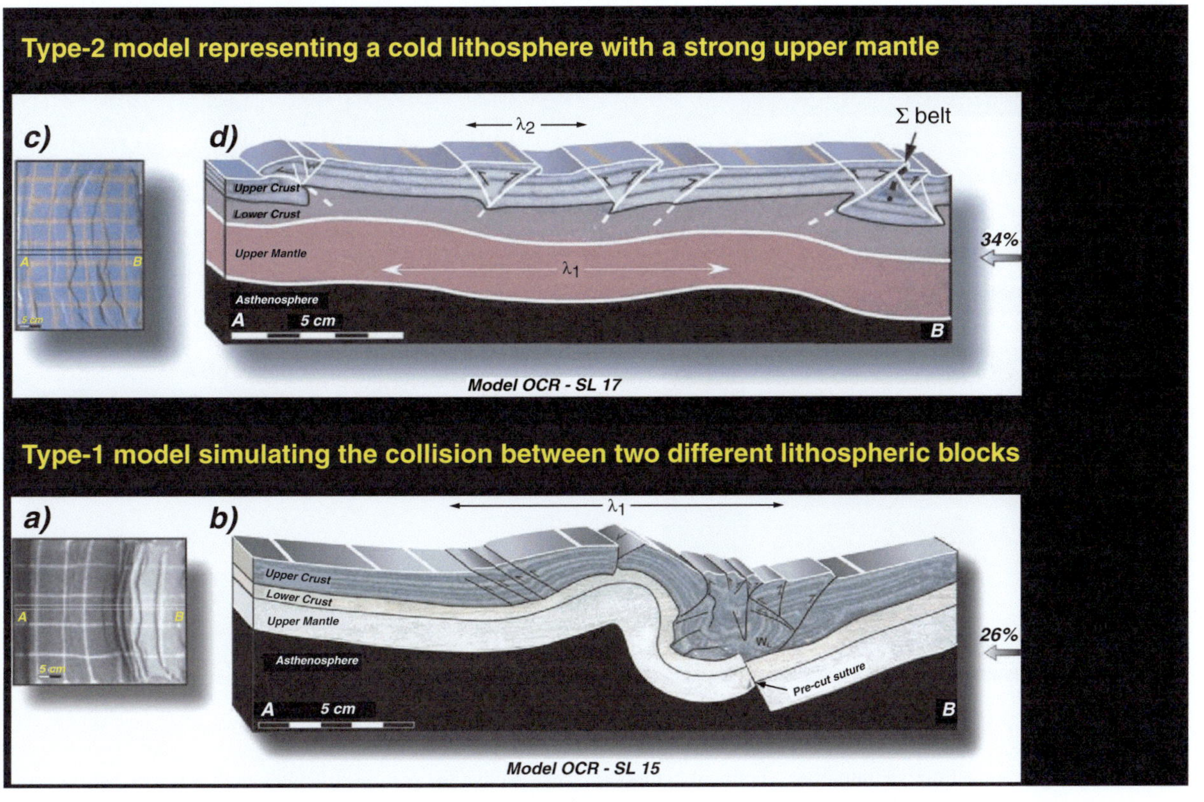

Fig. 14 Analogue tectonic modelling for continental lithosphere folding. *Top*: uniform lithosphere. *Bottom*: lithosphere blocks separated by suture zone (after Sokoutis et al., 2005)

At the scale of a micro-continent that was affected by a succession of collisional events, Iberia provides a well-documented natural laboratory for lithospheric folding and the quantification of the interplay between neotectonics and surface processes (Fig. 15; Cloetingh et al., 2002). An important factor in favor of a lithosphere-folding scenario for Iberia is the compatibility of the thermo-tectonic age of its lithosphere and the wavelength of observed deformations.

Well-documented examples of continental lithospheric folding come also from other cratonic areas. A prominent example of lithospheric folding occurs in the Western Goby area of Central Asia, involving a lithosphere with a thermo-tectonic age of 400 Ma. In this area, mantle and crustal wavelengths are 360 and 50 km, respectively, with a shortening rate of ~10 mm/year and a total amount of shortening of 200–250 km during 10–15 Myr (Burov et al., 1993; Burov and Molnar, 1998).

Quaternary folding of the Variscan lithosphere in the area of the Armorican Massif (Bonnet et al., 2000) resulted in the development of folds with a wavelength of 250 km, pointing to a mantle-lithospheric control on deformation. As the timing and spatial pattern of uplift inferred from river incision studies in Brittany is incompatible with a glacio-eustatic origin, Bonnet et al. (2000) relate the observed vertical motions to deflection of the lithosphere under the present-day NW–SE directed compressional intraplate stress field of NW Europe (Fig. 16). Stress-induced uplift of the area appears to control fluvial incision rates and the position of the main drainage divides. The area located at the western margin of the Paris Basin and along the rifted Atlantic margin of France has been subject to thermal rejuvenation during Mesozoic extension related to North Atlantic rifting (Robin et al., 2003; Ziegler and Dèzes, 2006) and subsequent compressional intraplate deformation (Ziegler et al., 1995), also affecting the Paris Basin (Lefort and Agarwal, 1996). Levelling studies in this area (Lenotre et al., 1999) also point towards its ongoing deformation.

The inferred wavelength of these neotectonic lithosphere folds is consistent with the general relationship

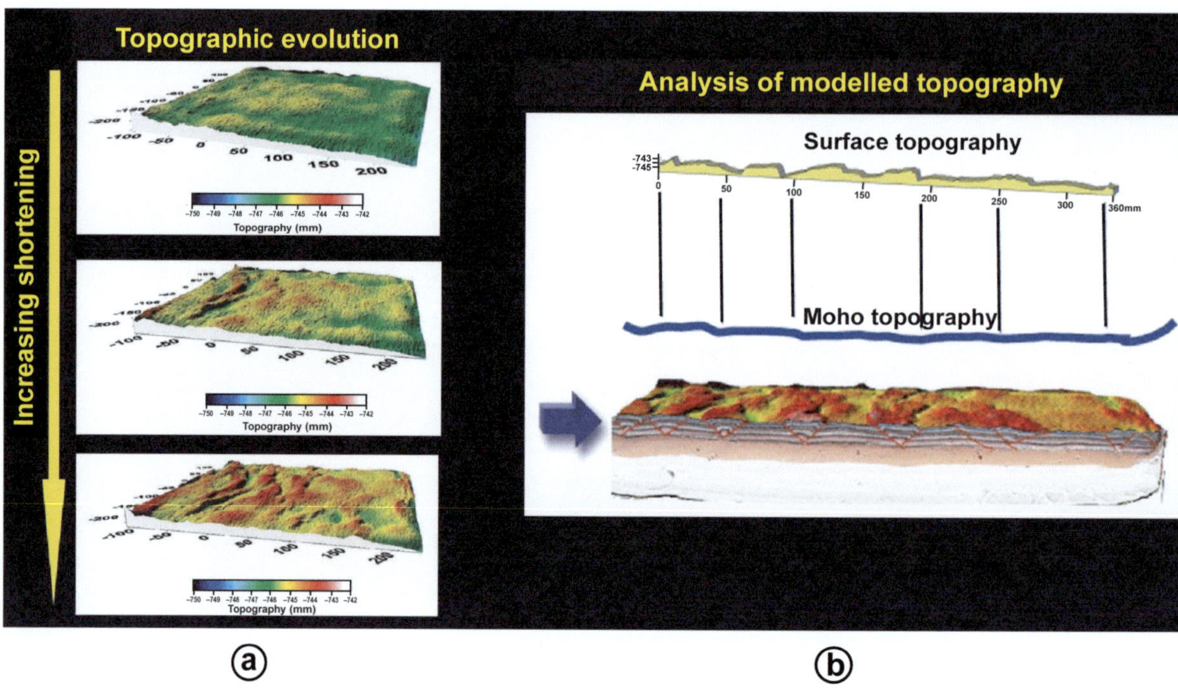

Fig. 15 Analogue modelling of intraplate continental lithosphere folding of Iberia (Fernández-Lozano et al., 2008). *Left*: incremental shortening and topographic evolution. *Top right*: 2D section of the final stage. *Bottom right*: 3D view of the final stage. Notice the pop-up structures in the upper crust (*layered*), the ductile flow of the lower crust (*orange*), and the folded mantle lithosphere (*light grey*)

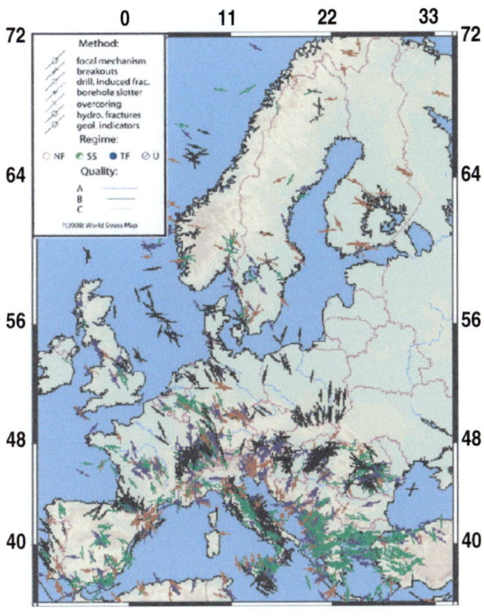

Fig. 16 Present-day stress map of Europe showing orientation of maximum horizontal stress axes (SHmax). Different symbols stand for different stress indicators; their length reflects the data quality, "A" being highest. Background shading indicates topographic elevation (*brown high*, *green low*). This map was derived from the World Stress Map database (http://www.world-stress-map.org)

that was established between the wavelength of lithospheric folds and the thermo-tectonic age of the lithosphere on the base of a global inventory of lithospheric folds (Fig. 17; see also Cloetingh and Burov, 1996; Cloetingh et al., 2005). In a number of other areas of continental lithosphere folding, also smaller wavelength crustal folds have been detected, for example in Central Asia (Burov et al., 1993; Nikishin et al., 1993).

Thermal thinning of the mantle-lithosphere, often associated with volcanism and doming, enhances lithospheric folding and appears to control the wavelength of folds. Substantial thermal weakening of the lithospheric mantle is consistent with higher folding rates in the European foreland as compared to folding in Central Asia (Nikishin et al., 1993), which is marked by pronounced mantle strength (Cloetingh et al., 1999).

Linking the Sedimentary Record to Processes in the Lithosphere

Over the last decades basin analysis has been in the forefront of integrating sedimentary and lithosphere components of previously separated fields of geology and geophysics (Fig. 18). Integrating neotectonics, surface processes and lithospheric dynamics in the reconstruction of the paleo-topography of sedimentary basins and their flanking areas is a key objective of integrated Solid-Earth science. A fully integrated approach, combining dynamic topography and sedimentary basin dynamics, is also important considering the societal importance of these basins on account of their resource potential. At the same time, most of the human population resides on sedimentary basins, often close to coastal zones and deltas that are vulnerable to geological hazards inherent to the active Earth system.

One major task of on-going research is to bridge the gap between historic and geological time scales in analyzing lithospheric deformation rates. Major progress has been made in reconstructing the evolution of sedimentary basins on geological time scales, incorporating faulting and sedimentary phenomena. From this, we have considerably increased our insights into the dynamics of the lithosphere at large time slices (millions of years). On the other hand, knowledge on present-day dynamics is rapidly growing thanks to the high spatial resolution in quantification of earthquake hypocenters and focal mechanisms, and vertical motions of the land surface. Unification, coupling and fully 3-D application of different modelling approaches to present-day observations and the geological record will permit to strengthen the reconstructive and predictive capabilities of process quantification. Particularly an intrinsically time-integrated approach will permit to assess in greater detail the importance of the geological memory of lithospheric properties on present-day dynamics. This is one of the key parameters for predicting future vertical motions.

Mechanical Controls on Basin Evolution: Europe's Continental Lithosphere

Studies on the mechanical properties of the European lithosphere revealed a direct link between its thermo-tectonic age and bulk strength (Cloetingh et al., 2005, Cloetingh and Burov, 1996; Pérez-Gussinyé and Watts, 2005). On the other hand, inferences from P and S wave tomography (Goes et al., 2000a, b; Ritter et al., 2000, 2001) and thermo-mechanical modelling (Garcia-Castellanos et al., 2000) point to a

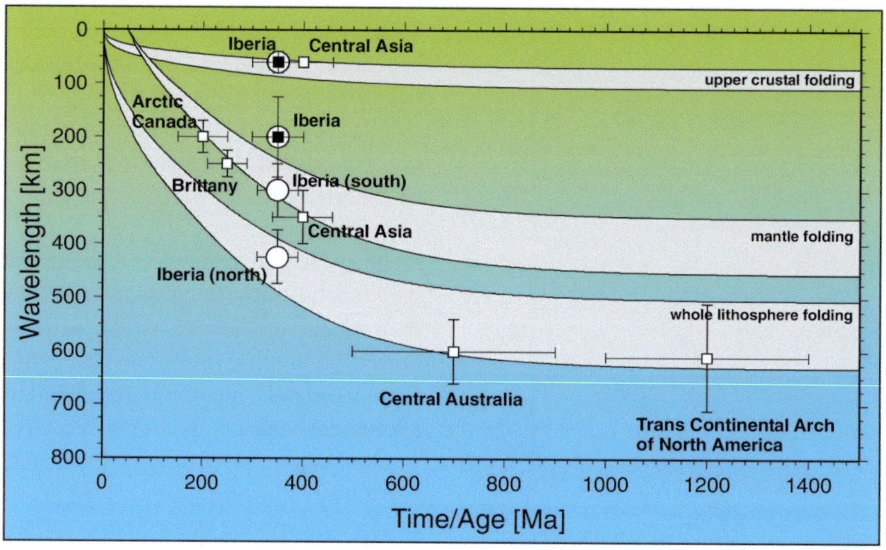

Fig. 17 Comparison of observed (*solid squares*) and modelled (*open circles*) wavelengths of crustal, lithospheric mantle and whole lithospheric folding in Iberia (Cloetingh et al., 2002c) with theoretical predictions (Cloetingh et al., 1999) and other estimates (*open squares*) for wavelengths documented from geological and geophysical studies (Stephenson and Cloetingh, 1991; Nikishin et al., 1993; Ziegler et al., 1995; Bonnet et al., 2000). Wavelength is given as a function of the thermo-tectonic age at the time of folding. Thermo-tectonic age corresponds to the time elapsed since the last major perturbation of the lithosphere prior to folding. Note that neotectonic folding of Variscan lithosphere has recently also been documented for Brittany (Bonnet et al., 2000). Both Iberia and Central Asia are characterized by separate dominant wavelengths for crust and mantle folds, reflecting decoupled modes of lithosphere folding (Cloetingh et al., 2005). Modified from Cloetingh et al. (2002)

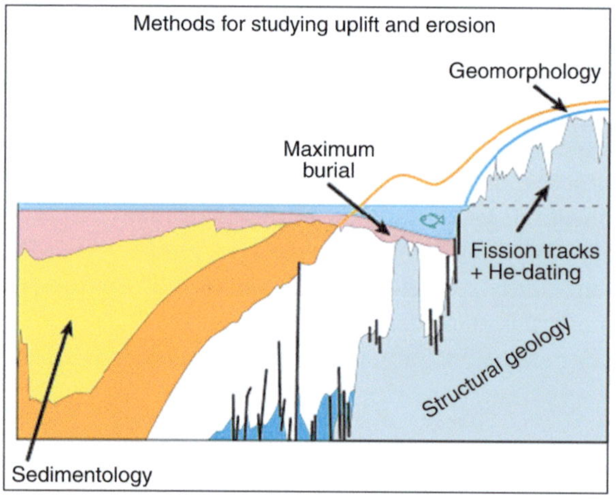

Fig. 18 Role of constraints from structural geology, geochronology, geomorphology and sedimentology in linking the sedimentary record to lithospheric processes (cartoon for coastal Norway by Japsen)

pronounced weakening of the lithosphere in the Lower Rhine area owing to high upper mantle temperatures. However, the Late Neogene and Quaternary tectonics of the Ardennes-Lower Rhine area appear to form part of a much wider neotectonic deformation system that overprints the Late Paleozoic and Mesozoic basins of NW Europe. This is supported by geomorphologic evidence and the results of seismicity studies in Brittany (Bonnet et al., 1998, 2000) and Normandy (Lagarde et al., 2000; Van Vliet-Lanoë et al., 2000), by data from the Ardennes-Eifel region (Meyer and Stets, 1998; Van Balen et al., 2000), the southern parts of the Upper Rhine Graben (Nivière and Winter, 2000), the Bohemian Massif (Ziegler and Dèzes, 2005, 2007) and the North German Basin (Bayer et al., 1999; Littke et al., 2008).

Lithosphere-scale folding and buckling, in response to the build up of compressional intraplate stresses, can cause uplift or subsidence of relatively large areas at time scales of a few My and thus can be an important driving mechanism of neotectonic processes. For instance, the Plio-Pleistocene accelerated subsidence of the North Sea Basin is attributed to down buckling of the lithosphere in response to the build-up of the present day stress field (Van Wees and Cloetingh, 1996; Unternehr and van den Driessche, 2004). Similarly, the Vosges-Black Forest arch, which at the level of the crust-mantle boundary extends from the Massif Central into the Bohemian Massif, was rapidly uplifted during the Burdigalian (±18 Ma) and since then has been maintained as a major topographic feature (Ziegler and Fraefel, 2009). Uplift of this arch is attributed to lithospheric folding controlled by compressional stresses originating at the Alpine collision zone (Ziegler et al., 2002; Dèzes et al., 2004; Ziegler and Dèzes, 2005, 2007; Bourgeois et al., 2007).

An understanding of the temporal and spatial strength distribution in the NW European lithosphere may offer quantitative insights into the patterns of its intraplate deformation (basin inversion, up thrusting of basement blocks), and particularly into the pattern of lithosphere-scale folding and buckling.

Owing to the large amount of high quality geophysical data acquired during the last 20 years in Europe, its crustal configuration is rather well known (Dèzes and Ziegler, 2004; Tesauro et al., 2008) though significant uncertainties remain in many areas about the seismic and thermal thickness of the lithosphere (Babuska and Plomerova, 1992; Artemieva and Mooney, 2001; Artemieva, 2006). Nevertheless, available data helps to constrain the rheology of the European lithosphere, thus enhancing our understanding of its strength.

So far, strength envelopes and the effective elastic thickness of the lithosphere have been calculated for a number of locations in Europe (Fig. 19, Cloetingh and Burov, 1996). However, as such calculations were made for scattered points only, or along transects, they provide limited information on lateral strength variations of the lithosphere. Although lithospheric thickness and strength maps have already been constructed for the Pannonian Basin (Lankreijer et al., 1999) and the Baltic Shield (Kaikkonen et al., 2000), such maps were until recently not yet available for all of Europe.

As evaluation and modelling of the response of the lithosphere to vertical and horizontal loads requires an understanding of its strength distribution, dedicated efforts were made to map the strength of the European foreland lithosphere by implementing 3D strength calculations (Cloetingh et al., 2005).

Strength calculations of the lithosphere depend primarily on its thermal and compositional structure and are particularly sensitive to thermal uncertainties (Ranalli and Murphy, 1987; Vilotte et al., 1993; Ranalli, 1995; Burov and Diament, 1995). For this reason, the workflow aimed at the development of a 3D strength model for Europe was two-fold: (1) construction of a 3D compositional model and (2) calculating a 3D thermal cube. The final 3D strength cube was obtained by calculating 1D strength envelopes for each lattice point (x, y) of a regularized raster covering NW-Europe (Fig. 20a). For each lattice-point the appropriate input values were obtained from a 3D compositional and thermal cube. A geological and geophysical geographic database was used as reference for the construction of the input models.

For continental realms, a 3D multi-layer compositional model was constructed, consisting of one mantle-lithosphere layer, 2–3 crustal layers and an overlying sedimentary cover layer, whereas for oceanic areas a one-layer model was adopted. For the depth to the different interfaces several regional or European-scale compilations were available that are based on deep seismic reflection and refraction or surface wave dispersion studies (e.g., Panza, 1983; Calcagnile and Panza, 1987; Suhadolc and Panza, 1989; Blundell et al., 1992; Du et al., 1998; Artemieva et al., 2006). For the base of the lower crust, we strongly relied on the European Moho map of Dèzes and Ziegler (2004) (Fig. 2.1a). Regional compilation maps of the seismogenic lithosphere thickness were used in subsequent thermal modelling as reference to the base of the thermal lithosphere (Babuska and Plomerova, 1993, 2001; Plomerova et al., 2002) (see Fig. 20b).

Figure 21a shows the integrated strength under compression of the entire lithosphere of Western and Central Europe, whereas Fig. 21b displays the integrated strength of the crustal part of the lithosphere. As evident from Fig. 21, Europe's lithosphere is characterized by major spatial mechanical strength variations, with a pronounced contrast between the strong Proterozoic lithosphere of the East-European Platform to the northeast of the Teisseyre-Tornquist Zone (TTZ) and the relatively weak Phanerozoic lithosphere of Western Europe.

A similar strength contrast occurs at the transition from strong Atlantic oceanic lithosphere to the relatively weak continental lithosphere of Western Europe. Within the Alpine foreland, pronounced northwest-southeast trending weak zones are recognized that coincide with such major geologic structures as the Rhine Graben System and the North Danish-Polish Trough, that are separated by the high-strength North German Basin and the Bohemian Massif. Moreover, a broad zone of weak lithosphere characterizes the Massif Central and surrounding areas.

In the area of the Trans-European Suture Zone, which corresponds to a zone of terranes that were

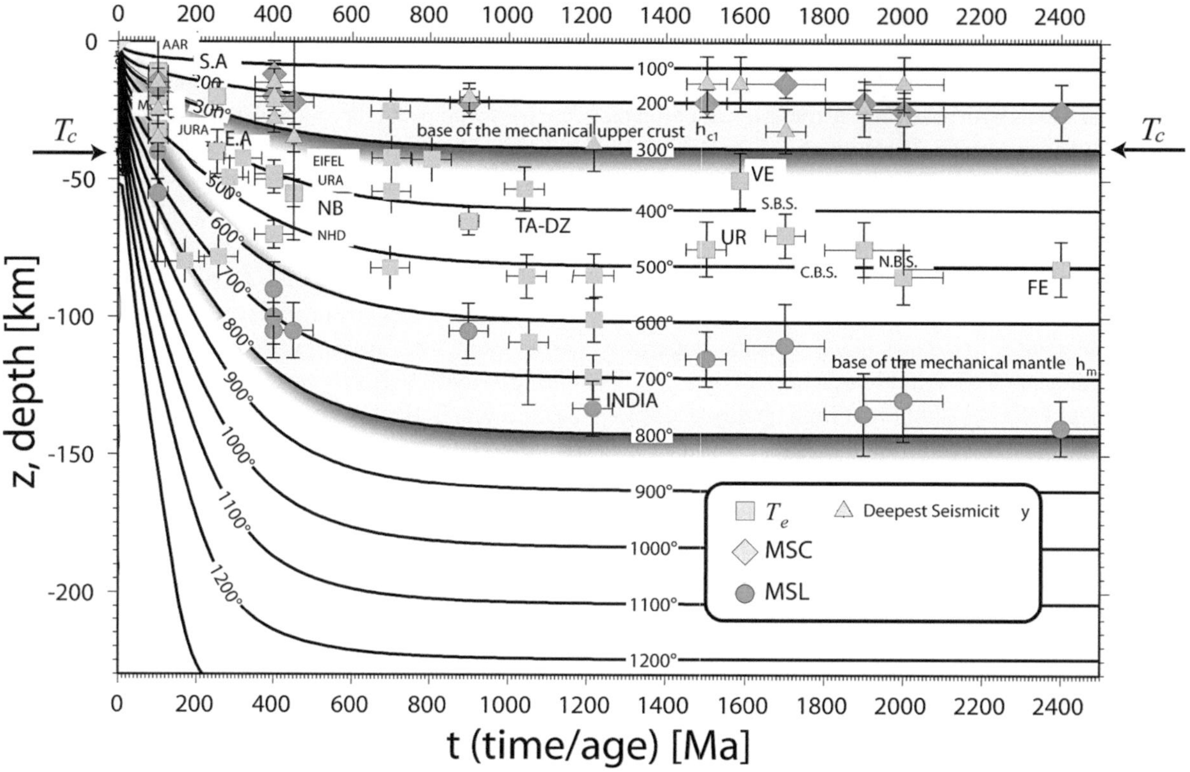

Fig. 19 Compilation of observed and predicted values of effective elastic thickness (EET), depth to bottom of mechanically strong crust (MSC), and depth to bottom of mechanically strong lithospheric mantle (MSL) plotted against the age of the continental lithosphere at the time of loading and comparison with predictions from thermal models of the lithosphere. Labeled contours are isotherms. Isotherms marked by *solid lines* are for models that account for additional radiogenic heat production in the upper crust. *Dashed lines* correspond to pure cooling models for continental lithosphere. The equilibrium thermal thickness of the continental lithosphere is 250 km. *Shaded bands* correspond to depth intervals marking the base of the mechanical crust (MSC) and the mantle portion of the lithosphere (MSL). *Squares* correspond to EET estimates, *circles* indicate MSL estimates, and *diamonds* correspond to estimates of MSC. *Bold letters* correspond to directly estimated EET values derived from flexural studies on, for example, foreland basins, *Thinner letters* indicate indirect rheological estimates derived from extrapolation of rock-mechanics studies. The data set includes (I): Old thermo-mechanical ages (1,000–2,500 Ma): northernmost (N.B.S.), central (C.B.S.), and southernmost Baltic Shield (S.B.S.); Fennoscandia (FE); Verkhoyansk plate (VE); Urals (UR); Carpathians; Caucasus, (II): Intermediate thermo-mechanical ages (500–1,000 Ma): North Baikal (NB); Tarim and Dzungaria (TA-DZ); Variscan of Europe: URA, NHD, EIFEL; and (III): Younger thermo-mechanical ages (0–500 Ma): Alpine belt: JURA, MOLL (Molasse), AAR; southern Alps (SA) and eastern Alps (EA); Ebro Basin; Betic rifted margin; Betic Cordilleras. Modified from Cloetingh and Burov (1996)

amalgamated during the Early Paleozoic, the debated occurrence of thickened crust adjacent to the Tornquist Zone (Fig. 5) was refuted by large-scale deep seismic experiments (POLONAISE, DEKORPBASIN96 and CELEBRATION; Guterch et al., 1999, 2003; BASIN Research Group, 1999). The process of terrane amalgamation was assumed to give rise to pronounced mechanical weakening of the lithosphere, particularly of its mantle part. Recent studies indicate, however, a rather steep crustal thickness gradient across the Tornquist Zone (Scheck-Wenderoth and Lamarche, 2005), which does not fit the expected configuration of an intracratonic suture zone. The Teisseyre-Tornquist Zone is now better described as the boundary zone that separates the Precambrian crust of the East European Craton from the Phanerozoic crust of Western and Central Europe, which was consolidated during the Caledonian and Variscan orogenies (Ziegler, 1990b; Berthelsen, 1998; Erlström et al., 1997; Grad et al., 2002; Guterch et al., 1999; Guterch and Grad, 2006). The northwestern prolongation of the Teisseyre-Tornquist Zone into Denmark,

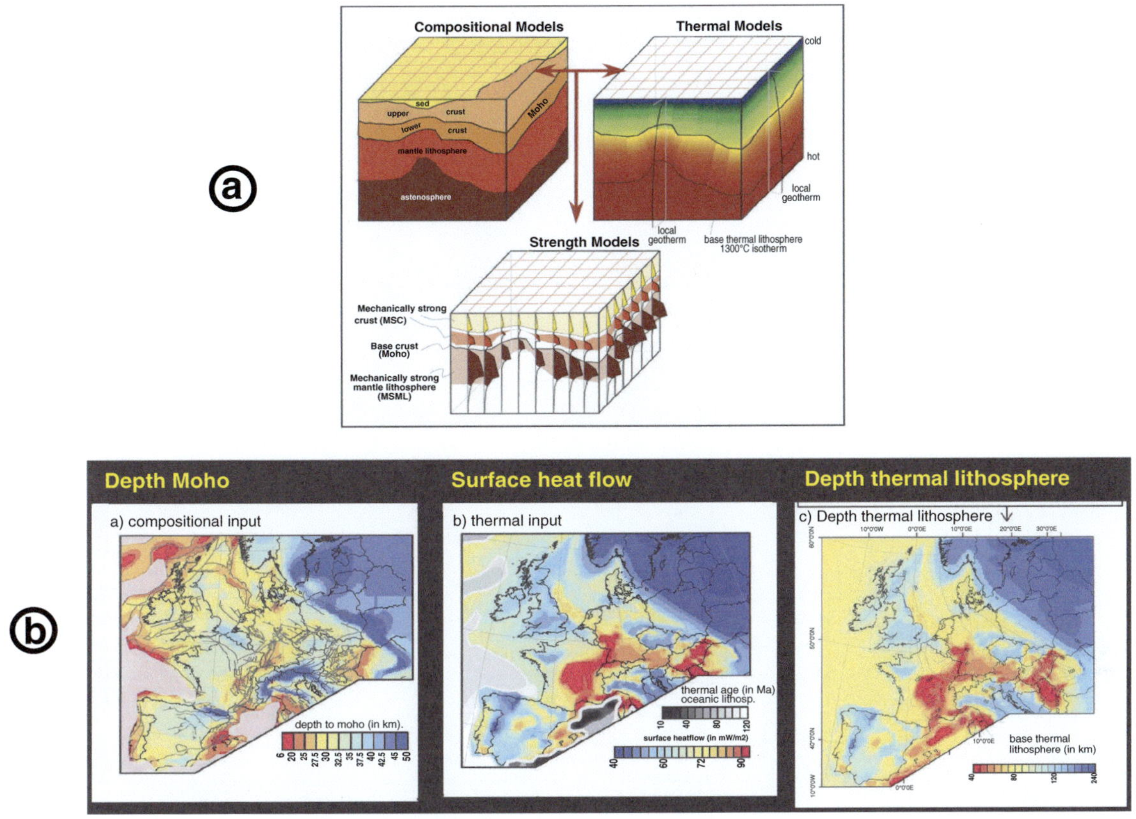

Fig. 20 **a**) From crustal thickness (*top left*) and thermal structure (*top right*) to lithospheric strength (*bottom*): conceptual configuration of the thermal structure and composition of the lithosphere, adopted for the calculation of 3-D strength models. Modified from Cloetingh et al., 2005. **b**) Heterogeneity in compressional and thermal structure in Europe's lithosphere and upper mantle. (*a*) Heterogeneity in crustal controls on lithospheric strength (from Dèzes and Ziegler, 2004). (*b*) Heterogeneity in surface heat flow. (*c*) Heterogeneity in depth (km) to the base of the lithosphere inferred from constraints from seismic tomography

where it is referred as the Sorgenfrei-Tornquist Zone (STZ), separates the stable part of the East European Craton from its weaker southwestern margin (Berthelsen, 1998; Erlström et al., 1997; Pharaoh et al., 2006).

Whereas the lithosphere of Fennoscandia is characterized by a relatively high strength, the North Sea rift system corresponds to a zone of weakened lithosphere. Other areas of high lithospheric strength are the Bohemian Massif and the London-Brabant Massif both of which exhibit low seismicity (Fig. 22).

A pronounced contrast in strength can also be noticed between the strong Adriatic indenter and the weak Pannonian Basin area (see also Fig. 21).

Comparing Fig. 12 with Fig. 13 reveals that the lateral strength variations of Europe's intraplate lithosphere are primarily caused by variations in the mechanical strength of the lithospheric mantle, whereas variations in crustal strength appear to be more modest. Variations in lithospheric mantle strength are primarily related to variations in the thermal structure of the lithosphere that can be related to thermal perturbations of the sub-lithospheric upper mantle imaged by seismic tomography (Goes et al., 2000a); lateral variations in crustal thickness play a secondary role, apart from Alpine domains which are characterized by deep crustal roots. High strength in the East-European Platform, the Bohemian Massif, the London-Brabant Massif and the Fennoscandian Shield reflects the presence of old, cold and thick lithosphere, whereas the European Cenozoic Rift System coincides with a major axis of thermally weakened lithosphere within the Northwest European Platform. Similarly, weakening of the lithosphere of southern France can be attributed to the presence of tomographically imaged upper mantle convective insta-

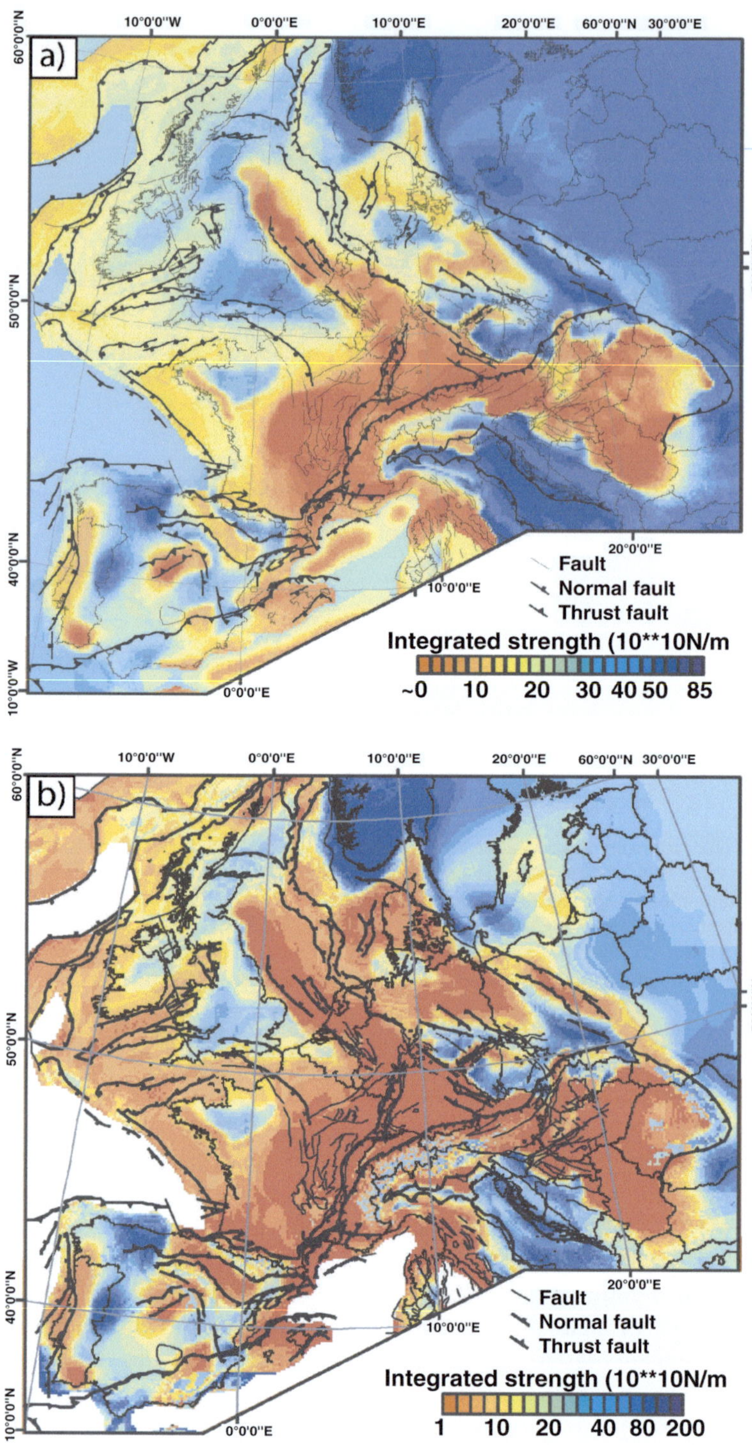

Fig. 21 Integrated strength maps for intraplate Europe. Contours represent integrated strength in compression for (**a**) total lithosphere and (**b**) crust. Adopted composition for upper crust, lower crust, and mantle is based on a wet quartzite, diorite, and dry olivine composition, respectively. Rheological rock parameters are based on Carter and Tsenn (1987). The adopted bulk strain-rate is 10^{-16} s^{-1}, compatible with constraints from GPS measurements (see text). The main structural features of Europe are superimposed on the strength maps. Modified from Cloetingh et al. (2005)

bilities rising up under the Massif Central (Granet et al., 1995; Wilson and Patterson, 2001; Lustrino and Wilson, 2007).

The major lateral strength variations that characterize the lithosphere of extra-Alpine Phanerozoic Europe are largely related to its Late Cenozoic thermal per-

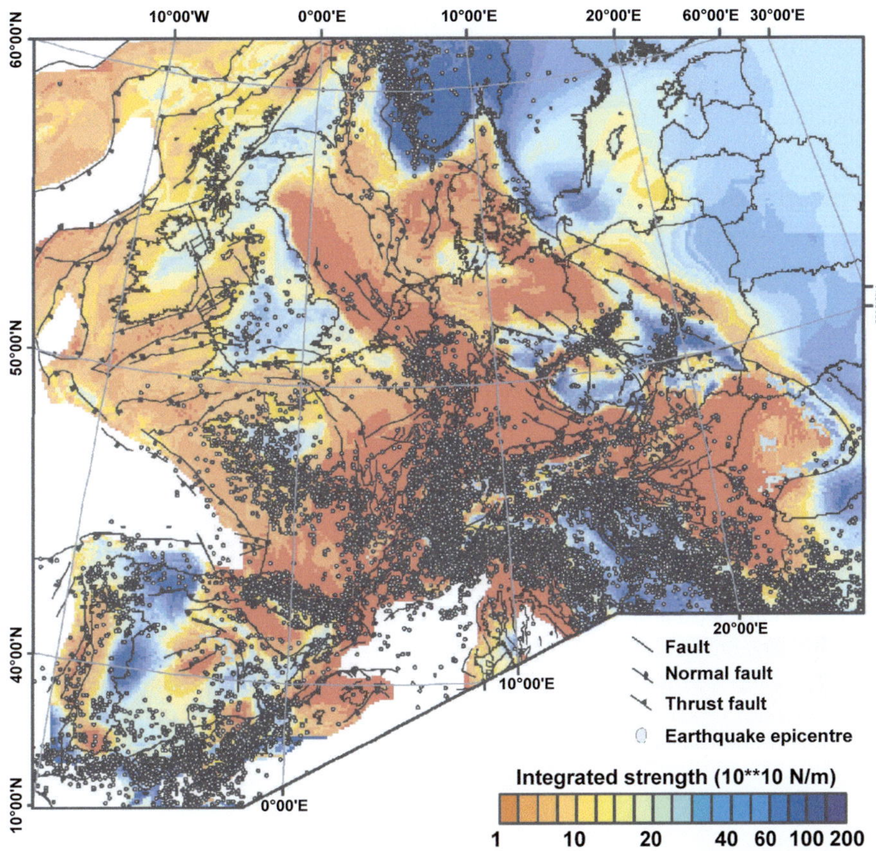

Fig. 22 Distribution of crustal seismicity superimposed on map of integrated strength for the European crust (see Fig. 21). Earthquake epicenters from the NEIC data center (NEIC, 2004), queried for magnitude >2 and focal depths <35 km

turbation as well as to Mesozoic and Cenozoic rift systems and intervening stable blocks, and not so much to the Caledonian and Variscan orogens and their accreted terranes (Dèzes et al., 2004, Ziegler and Dèzes, 2006). These lithospheric strength variations (Fig. 21) are primarily related to variations in the thermal structure of the lithosphere, and therefore, are compatible with inferred variations in the effective elastic thickness (EET) of the lithosphere (see Cloetingh and Burov, 1996; Pérez-Gussinyé and Watts, 2005).

The most important strong domains within the lithosphere of the Alpine foreland correspond to the London-Brabant, Armorican, Bohemian and West-Iberian Massifs. The strong Proterozoic Fennoscandian – East-European Craton flanks the weak Phanerozoic European lithosphere to the northeast whereas the strong Adriatic indenter contrasts with the weak lithosphere of the Alpine-Mediterranean collision zone (Cavazza et al., 2004).

Figure 22 displays on the background of the crustal strength map the distribution of seismic activity, derived from the NEIC global earthquake catalogue (USGS). As obvious from Fig. 16, crustal seismicity is largely concentrated on the presently still active Alpine plate boundaries, and particularly on the margins of the Adriatic indenter. In the Alpine foreland, seismicity is largely concentrated on zones of low lithospheric strength, such as the European Cenozoic rift system, and areas where pre-existing crustal discontinuities are reactivated under the presently prevailing NW-directed stress field, such as the South Armorican shear zone (Dèzes et al., 2004; Ziegler and Dèzes, 2007) and the rifted margin of Norway (Mosar, 2003).

It should be noted that the strength maps presented in Fig. 15 do not incorporate the effects of spatial variations in the composition of crustal and mantle layers. Future work will have to address the effects of such second order strength perturbations, adopting constraints on the composition of several crustal and mantle layers provided by seismic velocities (Guggisberg et al., 1991; Aichroth et al., 1992) and crustal and upper mantle xenolith studies (Mengel et al., 1991; Wittenberg et al., 2000).

Dynamics of Sedimentary Systems and Deformation Patterns

The largest water mass outside the ocean resides not in ice caps nor in lakes and rivers but in the pore space of the Earth's crust. By far the largest proportion of this pore space is contained in sedimentary rocks. Owing to their high porosity, sedimentary rocks are the only significant reservoirs for oil, gas and water and the most significant conduits for subsurface pollution. Therefore, predicting the architecture and properties of sedimentary rocks in the subsurface is one of the great challenges of Solid-Earth science. Progress will critically depend on successful integration of remote imaging of the subsurface and forward modelling from first principles of sedimentation, erosion and chemical reactions. Prediction includes both prediction in space ("ahead of the drill") and forecasting system behaviour in time (based on 4-D-monitoring).

Quantitative analysis of the geometries and facies patterns resulting from erosion and sediment deposition provides a key step in linking the dynamics of hinterland uplift and basin subsidence and the associated mass flux. The prospect of increasingly higher resolution in space and time will provide a better understanding of factors controlling the topographic evolution on continents and the subsidence of sedimentary basins along their margins.

During the last few years it has become increasingly evident that recent deformation has strongly affected the structure and fill of sedimentary basins. Similarly, the long-lasting memory of the lithosphere appears to play a much more important role in basin reactivation than hitherto assumed. Therefore, a better understanding of the 3-D fine structure of the linkage between basin formation and basin deformation is essential for linking lithospheric forcing and upper mantle dynamics to the dynamics of crustal uplift and erosion, and the dynamics of sedimentary systems. Structural analysis of the architecture of sedimentary basins, including paleo-stress assessment, provides important constraints on the transient nature of intra-plate stress fields.

Reconstruction of the history of sedimentary basins is a prerequisite for identifying transient processes controlling basin (de)formation. Full 3-D reconstructions, including the use of sophisticated 3-D visualization and geometric construction techniques for faulted basin architectures. 3-D back-stripping, including the effects of flexural isostasy and faulting, permits a thorough assessment of sedimentation and faulting rates and changing facies and geometries through time. The established architecture of the preserved sedimentary record serves as key input for the identification and quantification of transient processes.

Compressional Basins: Lateral Variations in Flexural Behaviour and Implications for Paleotopography

Deep seismic profiles across the Alps, the Pyrenees, the Apennines and the Carpathians have recently provided a completely new understanding on the main decoupling horizons acting during continental collision, roll-back of the subducted slab and coeval opening of back-arc basins. In many cases, the mantle lithosphere of the upper plate (Adria in the Western Alps, Europe in the Pyrenees) progressively indents the lower plate (European lithosphere in the Western Alps, Iberian lithosphere in the Pyrenees) and detaches its upper crust (Fig. 4; Roure et al., 1989, 1996). Both in the Alps and the Pyrenees, this process results in the development of foreland-propagating thrusts in the lower plate and crustal-scale antithetic back-thrusts in the upper plate, which account for the famous "crocodile" tectonics of Meissner (1989). As predicted by Laubscher with his lithophere "Verschluckung" (Laubscher, 1970, 1988, 1990), only the mantle lithosphere is proved to be recycled in the asthenosphere. Most, if not all the lower crust, which actually forms the main decoupling horizon, is progressively stacked in deeply buried duplexes and, thus, contributes to the growth of crustal roots at the base of orogens. Nevertheless lower crustal material can be subducted to depths of 55–60 km at which it enters the eclogite stability field, acquiring P-wave velocities typical for the mantle, therefore crossing the geophysically defined Moho discontinuity, and thus limits the seismically resolvable depth of crustal roots (Bousquet et al., 1997; Stampfli et al., 1998; Ziegler et al., 2001). Roure et al. (1994, 1996) and Ziegler and Roure (1996) give a detailed discussion on constraints provided by deep seismic data on the bulk geometry of Alpine belts.

Modelling of compressional basins followed essentially the same philosophy as modelling of extensional basins. Initial lithosphere-scale models focused on the role of flexural behaviour of the lithosphere during foreland basin development (e.g., Zoetemeijer et al., 1990; Van der Beek and Cloetingh, 1992). These studies drew on data sets consisting of wells, gravity data, and deep seismic profiling, such as the ECORS profile through the Pyrenees (Fig. 4), completed in the 1980s. Flexural modelling was backed up by large-scale studies on the rheological evolution of continental lithosphere (Cloetingh and Burov, 1996) that demonstrated in compressional settings a direct link between the mechanical properties of the lithosphere, its thermal structure and the level of regional intraplate stresses.

Inferences drawn from large-scale flexural modelling provided feedback for subsequent analysis on sub-basin scales. For example, modelling predictions for the presence of weak lithosphere in the Alpine belt invoke steep downward deflection of the lithosphere, favouring the development of upper crustal flexure-induced synthetic and antithetic tensional faults (Ziegler et al., 2002). Such fault systems are observed on reflection-seismic profiles in the Apennine and Sicilian foredeeps (Casero et al., 1991; Roure et al., 1991; Hippolyte et al., 1994, 1996; Casero, 2004), in the Alpine Molasse Basin of Germany and Austria (Ziegler, 1990a) and in the Carpathian foreland basin of Poland (Roca et al., 1995; Oszczypko, 2006), the Ukraine (Roure and Sassi, 1995; Izotova and Popadyuk, 1996) and Romania (Ellouz et al., 1996; Matenco et al., 1997). Such flexure-induced upper crustal normal faulting caused weakening of the lithosphere. Integrated flexural analysis of a set of profiles across the Ukrainian Carpathians and their foreland demonstrates an extreme deflection of the lithosphere, almost to the point of its failure, and very large offsets on upper crustal normal faults (Zoetemeijer et al., 1999).

Following studies on the paleo-rheology of the lithosphere, constrained by high-quality thermochronology in the Central Alps (Okaya et al., 1996) and Eastern Alps (Genser et al., 1996), the importance of large lateral variations in the mechanical strength of mountain belts became evident. This pertains particularly to a pronounced strength reduction from the external part of an orogen towards its internal parts. As a result, flexural foreland basins develop on the strong lithosphere of external zones of orogens, whereas on low strength lithosphere of their internal zones pull-apart basins can develop (Nemes et al., 1997, Cloetingh et al., 1992; Roure, 2008). The effects of lateral flexural strength variations of the lithosphere of foreland basins were explored by a modelling study that was carried out along a transect through the NE Pyrenees that is well constrained by crustal-scale seismic control and an extensive field-derived database (Verges et al., 1995). Apart from investigating the present configuration of foreland basins and quantifying the present-day mechanical structure of the lithosphere underlying the southern Pyrenees fold-and-thrust belt, the relationship between paleo-topography and flexural evolution of the orogen was analyzed (Millan et al., 1995). This novel approach led to a set of testable predictions on paleotopography and sediment supply to the foreland basin.

Topographic Expression of Compressional and Extensional Flat-Ramp Systems

In contrast to lithospheric scale folding or bending, accounting for the development of large wavelength arches and basins, and strongly asymmetric foredeep basins, reverse and normal faults control the development of narrow anticlines and steep grabens or half grabens, respectively. Such isolate structures occurring in the foreland of thrust belts or adjacent to an extensional system may be linked to them via deep seated detachment horizons, as evidenced by reflection-seismic data in the case studies presented below.

Northern Apennines Case Study

In the frontal parts of the Northern Apennines, industry-type reflection-seismic profiles document their thrusted architecture and the configuration of Pliocene synkinematic sedimentary series that were deposited in piggyback basins during the growth of anticlines and rapid subsidence of the foreland lithosphere (Fig. 23a; Pieri, 1983; Zoetemeijer et al., 1992; Doglioni and Prosser, 1997; Roure, 2008). As in

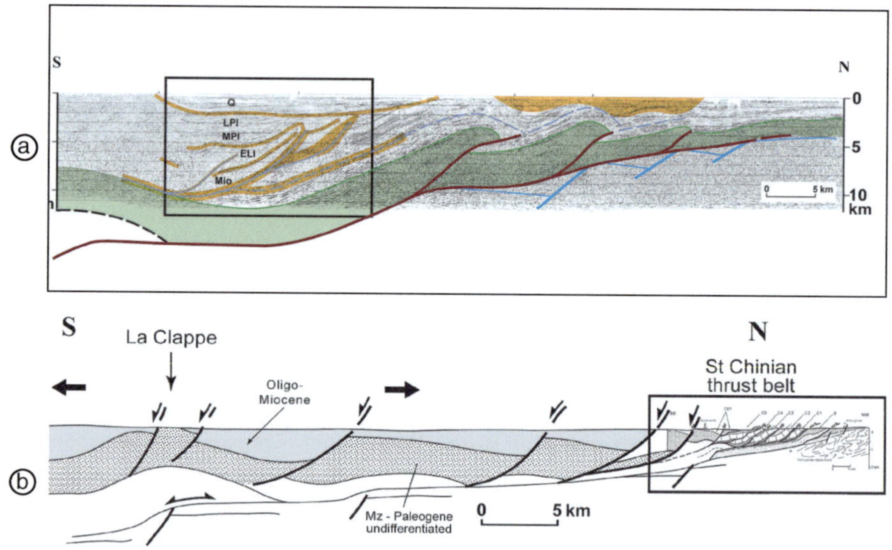

Fig. 23 Extentional and compressional flat-ramp systems and related topography: **a**) Quaternary piggy-back basin in the Northern Apennines (after Zoetemeijer et al., 1992; Roure, 2008). **b**) La Clappe roll-over anticline at the northern margin of the Gulf of Lions (Languedoc, Southeastern France, after Roure, 2008)

the southern parts of this profile Pliocene structures are unconformably overlain by thick, little deformed Quaternary sediments contained in a gentle synclinal basin, it was assumed that all tectonic contraction had stopped by the end of the Pliocene.

Earthquake focal mechanisms and GPS measurements attest, however, for a still ongoing compression and growth of structures in the vicinity of the Apennine thrust front (Picotti et al., 2007; Scrocca et al., 2007).

Since the available seismic profiles were recorded to 5 sec TwT only they do not image beneath the Quaternary basin the basal décollement of the orogenic wedge. However, by extrapolation from the northern, shallower parts of the seismic profiles and by applying cross-section balancing techniques, it is evident that the basal décollement is located within Triassic evaporites. Moreover, it is evident that the Quaternary basin takes in a piggyback position with respect to the frontal thrusted structures and is apparently located above a flat segment of the sole thrust. In the process of continued northward displacement of the orogenic wedge above ramp segments stepping up from older horizons in the south (i.e., either Paleozoic strata or crystalline basement) to progressively shallower stratigraphic horizons in thinner Mesozoic series in the north, the southern and northern flanks of the Quaternary piggy-back basin became increasingly tilted.

Further validation by coupled forward kinematic and sedimentation modelling helped to quantify the various parameters controlling the present architecture of the orogenic wedge and the accommodation space available for synkinematic trapping of Quaternary sediments. This required quantification of the overall amount of shortening, its partitioning from the basal deformation into the individual thrusts which ramp up from it, their respective velocity and timing, and the rates of bending of the lithosphere (Zoetemeijer et al. 1992).

La Clappe Case Study (Northern Margin of the Gulf of Lions)

The structural cross-section across the northern onshore segment of the Gulf of Lions margin, given in Fig. 23b, is constrained by industry-type reflection-seismic profiles. This cross-section extends from the St-Chinian fold and thrust belt in the north, representing the lateral equivalent of the North Pyrenean thrust front, across the La Clappe antiform in the south, an isolated Mesozoic carbonate massif that is flanked by tilted Oligocene and Miocene series (Roure, 2008). This cross-section, which was not investigated by ECORS deep seismic profiling, documents two phenomena, namely a negative inversion and a roll-over structure above an extensional detachment.

From the north to the south, this cross-section outlines (1) an erosional remnant of the former Late Cretaceous to Eocene Pyrenean flexural basin, (2) the thin-skinned Eocene structures of St-Chinian fold-and-thrust belt, representing a segment of the external part of the Pyrenean Orogen, and (3) a number of post-Eocene half grabens, which are controlled by arcuate listric normal faults, such as the Quarante Fault. At depth these faults sole out into the basal detachment thrust of the Pyrenean orogenic wedge. The tensional (negative) reactivation of former thrust faults can be accurately dated by the late Oligocene and early Miocene sedimentary fill of the footwall grabens, and thus coincided with the opening of the Gulf of Lions Basin (Roure et al., 1988; Séranne et al., 1995; Séranne, 1999).

The La Clappe anticline is interpreted as a textbook example of a roll-over or accomodation fold above an extensional detachment (McClay, 1989). Although seismic data evidence in the area of the La Clappe structure a vertical offset of the basement beneath the basal décollement, it is not yet clear whether this fault developed during Mesozoic rifting, controlling the development of a flat-ramp décollement during the Pyrenean orogeny, or whether it is related to the Oligo-Miocene rifting event (in a similar way as analogue models of Vendeville et al., 1987, and the infra-salt tilted blocks imaged beneath the Bresse Graben by ECORS; Bergerat et al., 1989, 1990). Both solutions can be accurately balanced, but imply very different kinematic scenarios during forward modelling.

Coupling versus Decoupling between Forelands and Orogenic Wedges and Development of Thrust-Top Pull-Apart Basins

Ziegler et al. (2002) have documented at a plate tectonic scale successive episodes of mechanical coupling and decoupling of orogenic wedges and forelands, with far-field foreland inversions reflecting periods of strong coupling between the orogen and the adjacent foreland lithosphere.

Paleostress measurements and paleomagnetic studies help to trace the coupling versus decoupling history of tectonic wedges with respect to their adjacent forelands. The inversion of microtectonic measurements in well-dated strata provides a direct control on the main horizontal stress direction at a given time, whereas paleomagnetic data are required to demonstrate whether a given tectonic unit has been rotated or not between the considered time interval and the Present. This methodology has been applied to the Southern Apennines by Hippolyte et al. (1994, 1996), documenting the successive pattern of paleostress directions in both the allochthon and the autochthon, for different stages including the Tortonian, Messinian, Lower, Middle and Upper Pliocene and Pleistocene. Surprisingly, results show periods during which paleostress directions were identical in the allochthon and the foreland, reflecting their mechanical coupling, and periods during which stress directions differed between the allochthon and the foreland, reflecting their mechanical decoupling. Similar objectives were also addressed by Malavieille (1984), Martinez et al. (2002), and McClay et al. (2004) by means of analogue models, exploring the effects of oblique convergence on strain partitioning and coupling/decoupling processes in foreland fold-and-thrust belts. Moreover, Nieuwland et al. (1999, 2000; Fig. 24) carried out a series of experiments with a sand box containing pressure sensors which showed cycles of pressure build-up and relaxation in the foreland that are directly related to cycles of thrust activation. At the onset of each cycle, a good coupling is observed between the allochthon and the autochthon, with no fault activity but with a progressive increase of the maximum horizontal compressional stress. Once this horizontal stress reaches a sufficient value, a new thrust fault nucleates, resulting in renewed decoupling of the autochthon from the allochthon and in a pressure decrease in the autochthon.

At the reservoir level, this rapid increase of the maximum compressional stress before the nucleation of a new frontal thrust is likely to account for the development of Layer Parallel Shortening (LPS) and coeval pressure-solution and quartz-cementation in sandstone reservoirs, as observed in the Sub-Andean basins of Venezuela and Colombia (Roure et al., 2003, 2005), and for the re-crystallisation and re-magnetisation of mesodolomites observed in the Alberta foreland (Robion et al., 2004; Roure et al., 2005). Deformation is thus alternatively plastic (accounting for pressure-solution and LPS in the autochthon during periods of coupling), and brittle (accounting for nucleation of a new frontal thrust and thrust propagation during the periods of decoupling).

Decoupling and strain partitioning near plate boundaries account also for the development of thrust-

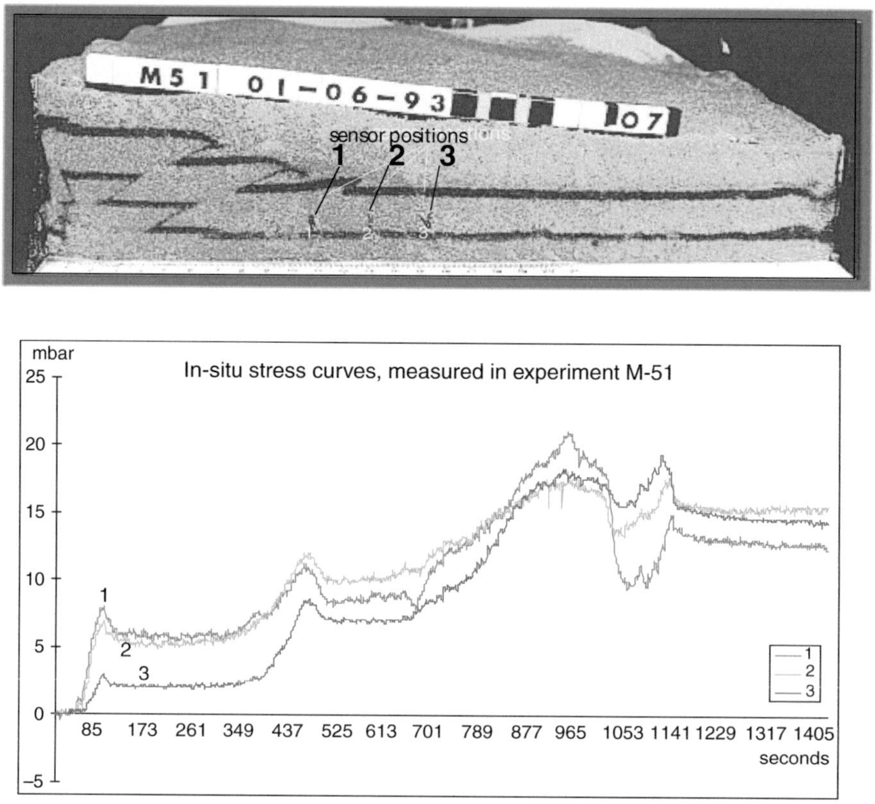

Fig. 24 Analogue experiment of thrust propagation, with pressure sensors recording the cyclic evolution of the main principal stress in the foreland autochthon, as a result of its successive coupling and decoupling with the tectonic wedge (after Nieuwland et al., 1999, 2000)

top pull-apart basins in a dominantly compressional regime. The Vienna Basin is probably the most famous and archetype of this type of basins. It developed on top of the Alpine allochthon after emplacement of its nappe systems on the foreland, in response to lateral eastward escape of the Alpine-Carpathian Block into the Carpathian embayment (Royden, 1985; Sauer et al., 1992; Seifert, 1996; Decker and Peresson, 1996; Schmid et al. 2008).

Complex piggyback basins and thrust-top pull-apart basins developed also in the internal parts of Circum-Mediterranean thrust belts in response to local and temporal changes in paleostress regime, and related strain partitioning and lateral block escape. Examples are the Sant'Arcangelo Basin in the Southern Apennines (Hippolyte et al., 1991, 1994; Di Stefano et al., 2002; Sabato et al., 2005; Monaco et al., 2007) and the Chelif Basin in North Algeria (Neurdin-Trescartes, 1995; Roure, 2008). In a very similar geodynamic setting the Gulf of Paria developed between Venezuela and Trinidad (Lingrey, 2007).

The physiography and lozenge shape of the Chelif Basin in North Algeria is clearly evident on geological maps and landsat imagery (Fig. 25; Roure, 2008). This basin is located north of the Tellian thrust front, which reached its current position during the Langhian (Frizon de Lamotte et al., 2000; Roca et al., 2004; Benaouali et al., 2006). To the north the Chelif Basin is delimited by a major east-trending lineament, known as the "Dorsale Calcaire", which separates the Kabylides crystalline basement in the north from the Tellian nappes in the south, and most likely behaved as a major strike-slip fault during the development of the Chelif Basin.

The Neogene sedimentary fill of the Chelif Basin comprises Burdigalian to Langhian syn-kinematic series, which were deposited in a piggyback position during the main southward transport of the Tellian nappes involving oblique convergence, transpression and strain-partitioning. Post-kinematic Tortonian to Pliocene series, contained in normal fault controlled depressions, overlay these syn-kinematic deposits.

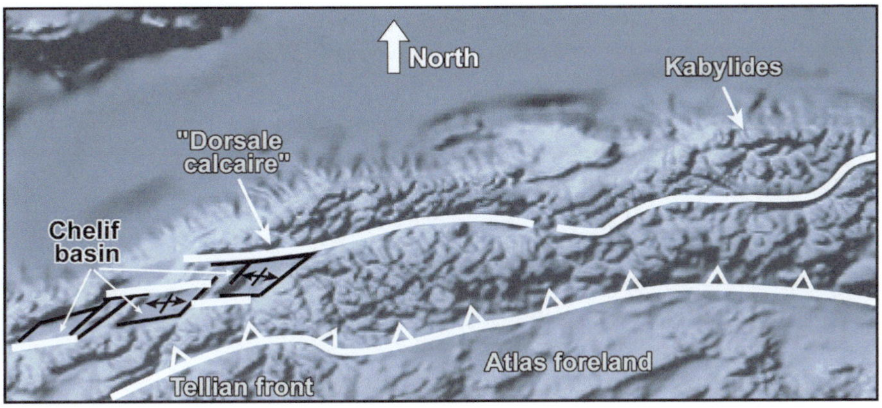

Fig. 25 Satellite image outlining the location of the Chelif thrust top pull-apart basin (North Algeria)

These normal faults, locally exposed at the surface, can be traced on seismic profiles downward into the deepest part of the basin, and trend oblique (en échelon) with respect to the Dorsale Calcaire lineament, being thus indicative of its Late Miocene-Pliocene transtensional reactivation. Comparable to the El Pilar Fault of Venezuela, the Dorsale Calcaire lineament accommodated during the Tortonian to Pliocene post-nappes transtensional episode a lateral displacement of the Kabylides relative to the Tell allochthon and the underlying African foreland crust.

Plio-Quaternary inversion of the Chelif depocentre, involving folding and erosion of Pliocene series along basin bounding faults, accounts for renewed transpression along this segment of the Dorsale Calcaire.

Intracratonic Basins

Intracratonic basins developed in the interior of continents, generally far from active plate boundaries. In cross-section they are generally saucer shaped, are characterized by a protracted subsidence history often exceeding 100 My during which there is hardly evidence for extensional faulting. As such they reflect long-term progressive crustal down warping and are generally characterized by low topographies throughout their history. The dimensions and depth of such sag basins vary considerably. Intracratonic basins are of considerable economic interest as a number of them host outstanding hydrocarbon provinces such as the West Siberian, North Sea, Williston and Michigan basins.

Driving forces controlling the subsidence of intracratonic sag basins vary considerably and are still a matter of debate. Nevertheless, considerable progress has recently been made in understanding these systems, which apart from a signal of long-lasting subsidence can be strongly influenced by structural inheritance, particularly during phases of intraplate compression. As intracratonic basins are often superimposed on a puzzle of different-aged crustal blocks that were welded together during foregoing orogenies, related crustal weakness zones tend to be reactivated throughout the basin history, causing strain localization. Prominent examples of intracratonic basins that were intensively studied during the last decade are the Central European Basin System (Ziegler, 1990a,b; Bayer et al., 1999; Van Wees et al., 2000; Scheck-Wenderoth and Lamarche, 2005; Littke et al., 2008), the Dniepr-Donetz and Donbas basins (Stephenson et al., 2001, 2006; Maystrenko et al., 2003; Stovba and Stephenson, 2003) and the West Siberian Basin (Sleep, 1976; Khain et al., 1991; Milanovsky, 1992; Hartley and Allen, 1994; Vyssotski et al., 2006).

Three end members of intracratonic sag basins are recognized, namely rift-, hot-spot- and cold-spot-driven basins (Ziegler, 1989). During the evolution of such sag basins compressional intraplate stresses can interfere with their subsidence, controlling by lithospheric folding subsidence accelerations and by reactivation of pre-existing basement discontinuities their partial inversion. Furthermore, intraplate compressional deformation can cause by disruption of sedimentary platforms the isolation of basins. Such erosional remnants of larger shelves and platforms differ from sag basins in so far as their axes do not necessarily coincide with depocentres (e.g., Paleozoic Tindouf Basin; Mesozoic Paris Basin: Ziegler, 1989, 1990).

The proto-type of a rift-driven intracratonic sag basin is the Late Cretaceous and Cenozoic North Sea Basin, which is superimposed on the Mesozoic Viking and Central grabens that transect Caledonian basement and the Northern and Southern Permian Basin. The rifting stage of the North Sea commenced in the Early Triassic and persisted intermittently into the Early Cretaceous. During the Late Cretaceous and Cenozoic post-rift thermal subsidence stage an over 4 km deep, up to 500 km wide and 1,000 km long thermal sag basin developed, widely overstepping the axial rift system (Ziegler, 1990; Kuznir et al., 1995). This type of basin essentially conforms to the lithospheric stretching model of McKenzie (1978), though its Paleocene and Plio-Quaternary subsidence accelerations can be related to deflection of the lithosphere in response to the build-up of intraplate compressional stresses (Ziegler, 1990; Van Wees and Cloetingh, 1996).

The West Siberian Basin, which extends into the Kara Sea, is up to 1,500 km wide and almost 3,000 km long and, though still debated, can be regarded as a hot-spot-driven sag basin. It evolved on a complex pattern of arc terranes and continental blocks that were assembled during the Late Paleozoic Uralian Orogeny. Following Early Permian consolidation of the Urals, their back-arc domain was affected by Late Permian-Early Triassic extension that were punctuated by major plume activity at the Permo-Triassic boundary, as evidenced by the extrusion of thick and widespread basalts. Evidence for limited crustal extension is essentially limited to the Kara Sea and the northern parts of the West Siberian Basin. Post-magmatic regional thermal subsidence of the West Siberian Basin by as much as 6,500 m is interpreted as reflecting strong thermal attenuation of the lithospheric mantle and magmatic destabilization of the crust-mantle boundary during plume impingement. Correspondingly the West Siberian Basin is interpreted as a "hot-spot" basin that, owing to limited crustal extension, does not conform to the classical stretching model of McKenzie (1978). The post-rift subsidence of the West Siberian Basin was repeatedly overprinted by the build-up of compressional stresses related to the Middle and Late Triassic folding of the northernmost Urals and Novaya Zemblya, Early Cretaceous south-verging thrusting of the South Taimyr fold belt and during the Oligocene India-Asia collision (Rudkewich, 1988, 1994; Ziegler, 1989; Peterson and Clarke, 1991; Zonenshain et al., 1993; Nikishin et al., 2002; Vyssotski et al., 2006).

Potential "cold spot" basins are the sub-circular Paleozoic Williston, Hudson Bay, Michigan and Illinois basins of North America, which have diameters of 500–750 km, vary in depth between 2 and 4.7 km and evolved on stabilized Precambrian crust. All of them lack a distinct precursor rifting or magmatic stage and are characterized by a thick crust. Subsidence of these basins began variably during the Late Cambrian to Late Ordovician and persisted into Early Carboniferous times, though subsidence rates varied through time and were not synchronous between the different basins (Quinlan, 1987; Bally 1989). As such they defy the principle of the classical thermal sag basins which develop in response to lithospheric cooling and contraction during the post-rift stage of extensional basins (McKenzie, 1978; Quinlan, 1987). During the Late Carboniferous and Permian these basins were regionally uplifted and subjected to erosion. Significantly, these intracratonic basins evolved during a period when Laurentia underwent only relatively minor latitudinal drift and was flanked by the Ordovician-Silurian Taconic-Caledonian and during the Devonian and Early Carboniferous by the Appalachian and the Antler-Inuitian subduction systems (Ziegler, 1989, 1990). It has therefore been postulated that development of these intracratonic basins was controlled by a decrease in ambient mantle temperatures related to the development of down-welling cells in the upper mantle (cold-spots), and that subsequent recovery of ambient mantle temperatures resulted in their slow uplift and erosion, unless they were incorporated into another subsidence regime, such as flexural foreland basins (Williston Basin). Subsidence of these North American intracratonic basins ceased once Laurentia started to drift northward, thus decoupling them from their cold-spots (Ziegler 1989).

The invoked cold-spot model conforms essentially to the model advanced by Heine et al. (2008), who advocates that vertical displacement of the lithosphere, controlling the development of intracratonic basins, is induced by mantle convection and related global plate kinematics. According to this model, the long-lasting subsidence of intracratonic basins and their topographic position close to sea level throughout their evolution, results from negative dynamic topography of the lithosphere that is controlled by down-welling cells of the large-scale Earth's mantle convection system. Heine et al. (2008) find that the movement of lithospheric plates relative to the underlying mantle, as well as variations in the large-scale mantle convection patterns can interfere with mantle-driven dynamic subsidence and can contribute to the creation and destruction of accommodation space in intracratonic basins. Other models for mantle-flow induced dynamic topography affecting continental domains, particularly adjacent to foreland basins, were advanced by e.g., Harper (1984), Coackley and Gurnis (1995), Burgess and Moresi (1999) and Pysklywec and Mitrovioca (2000), mainly invoking corner flow above deep reaching subducted slabs dipping beneath continental lithosphere.

The Central European Basin System (CEBS) is a complex system of intracratonic basins that extends from the British Isles to Poland over a distance of 1,500 km, is up to 1,000 km wide and contains up to 10 km thick Permian to Cenozoic sediments (Fig. 26). Owing to its hydrocarbon potential but also in terms of environmental hazard assessment, the CEBS has been the subject of extensive studies (Ziegler, 1990; Bayer et al., 1999; Van Wees et al., 2000; Scheck-Wenderoth and Lamarche, 2005; Reicherter et al. 2005; Littke et al., 2008; Stackebrandt, 2008). The CEBS is superimposed on a puzzle of crustal blocks that was amalgamated during the Caledonian orogeny and encroaches in the east onto the external parts of the Variscan orogen.

Following a latest Carboniferous-Early Permian phase of wrench faulting and magmatic activity (Neumann et al., 2004) subsidence of the CEBS commenced during the Late Permian and persisted during Mesozoic and Cenozoic times. Its long-term saucer-shaped subsidence was repeatedly overprinted by tensional and compressional stresses with strain localization along distinct linear zones corresponding to inherited crustal discontinuities. These date back to the Caledonian orogeny during which Gondwana-derived terranes were accreted to the SW margin of the East European Craton (Pharaoh et al., 2006), as well as to Permo-Carboniferous wrench tectonics (Ziegler and

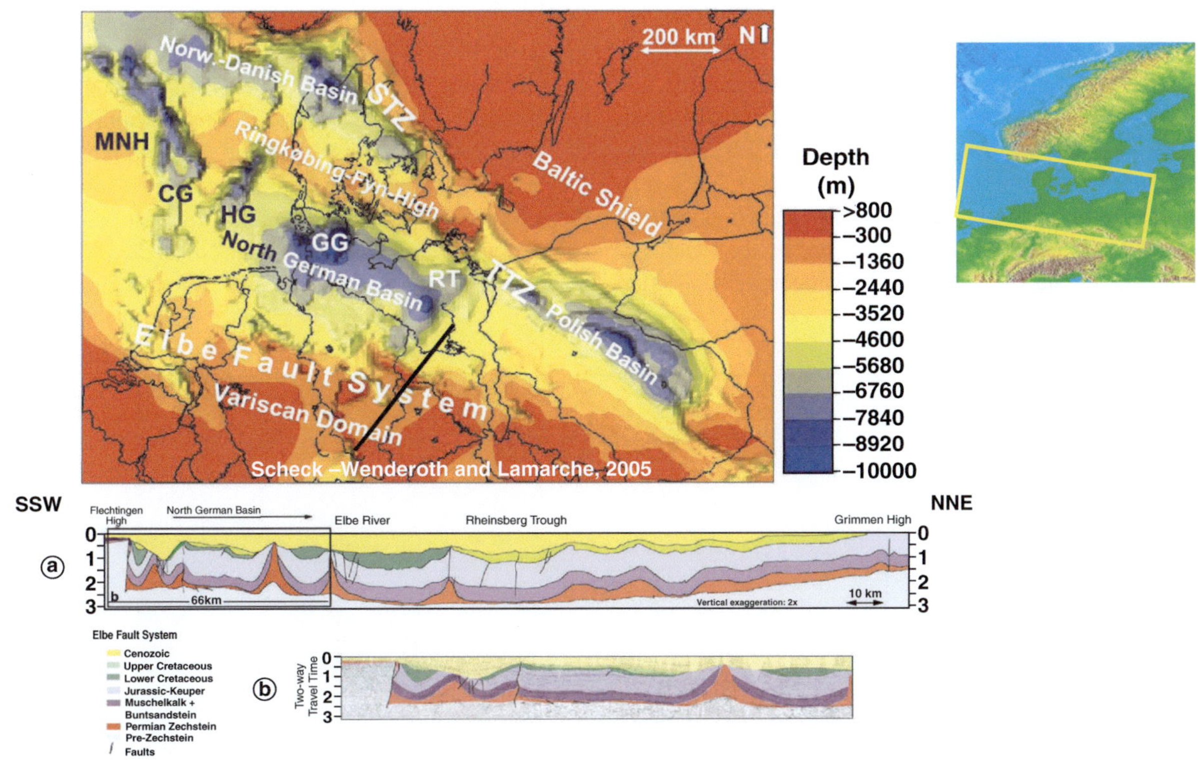

Fig. 26 Main characteristics of the Central European basin system illustrated by the depth to top Pre-Permian surface. *CG*: Central Graben; *HG*: Horn Graben; *GG*: Glückstadt Graben; *RT*: Rheinsberg Trough; *STZ*: Sorgenfrei-Tornquist Zone; *TTZ*: Teisseyre-Tornquist Zone. Black line on the map outlines the location of section a, shown with a slightly compressed horizontal scale, whereas section b outlines the southern part of section a, at a 1/1 scale

Dèzes, 2006). As such the CEBS is a good example for long-lived subsidence of a saucer-shaped basin (Fig. 26) in which the long wavelength subsidence signal is overprinted by repeated strain localization along distinct linear zones of inherited crustal weakness.

The long-wavelength subsidence pattern of the CEBS reflects the development of the large Permo-Mesozoic Norwegian-Danish and North German and Polish basins, which are separated by the Mid-North Sea Rinkøbing-Fyn chain of highs. These basins and highs are delimited by crustal-scale lineaments across which changes in crustal and lithospheric thickness and thermal properties are observed (Figs. 5 and 21; Ziegler and Dèzes, 2006; Cloetingh et al., 2007). The Teissyere-Tornquist Zone (TTZ) marks the boundary between the Precambrian East European craton (EEC) and Phanerozoic Europe. Across this zone crustal thicknesses decrease from 40–50 km in the EEC to 30–35 km in Phanerozoic Europe (Fig. 27; Guterch and Grad, 2006). In the NW prolongation of the TTZ the Sorgenfrei-Tornquist Zone (STZ) represents a Permo-Carboniferous shear zone that delimits the stable EEC from its less stable SW marginal parts (Berthelsen, 1998). The STZ coincides with a distinct step in crustal and lithosphere thickness (Gregersen et al., 2006, 2007). Another 'step' in lithosphere thickness coincides with the Elbe Fault System (EFS) delimiting the present-day southern margin of the CEBS. It is therefore not surprising that these lines exerted control on the large-scale basin evolution and on the localized deformation during the Late Cretaceous-Early Paleocene phase of intraplate compression. Along both NW-SE striking fault zones, vertical offsets related to Late Cretaceous basin inversion are in the range of several km (Scheck et al., 2002; Vejbaek and Andersen, 2002; Mazur et al., 2005).

Evolution of the CEBS commenced with the latest Carboniferous-Early Permian wrench-induced thermal destabilization of the lithosphere that was accompanied by extensive magmatic activity (Ziegler, 1990; Van Wees et al., 2000; Heeremans et al., 2004; Ziegler et al., 2004). This was followed by thermal subsidence and the establishment of two large sag basins north and south of the Mid-North Sea-Ringkøbing-Fyn chain of highs, the Northern and Southern Permian Basins (Ziegler, 1990). During this phase up to 3,500 m of Rotliegend clastics and Zechstein evaporites were deposited in the Southern Permian Basin and about 1,500 in the Northern Permian Basin. During the Triassic and Early Jurassic regional thermal subsidence of both basins continued (van Wees et al., 2000), but was overprinted by regional extensional stresses controlling the subsidence of the N–S striking Viking, Central, Horn and Glückstadt grabens and by reactivation of the NW–SE striking and Tornquist-Teisseyre zones, the subsidence of the Mid-Polish Trough. During the early Middle Jurassic, thermal uplift of the rift-related Central North Sea dome interfered with the declining subsidence of the Northern and Southern Permian Basins. Accelerated Late Jurassic and Early Cretaceous extension across the North Sea rift system was compensated along the southern margin of

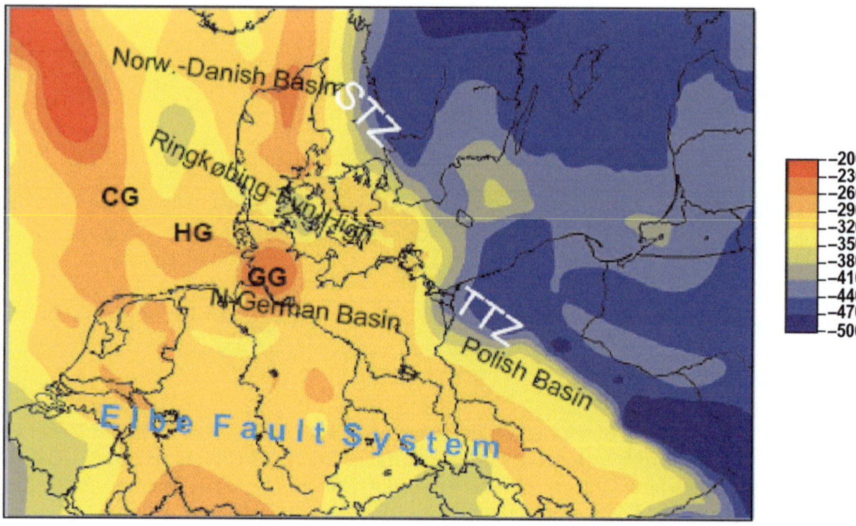

Fig. 27 Depth to the crust-mantle boundary beneath the CEBS (after Scheck-Wenderoth and Lamarche, 2005, compiled from: NEGB from Rabbel et al., 1995; Scheck and Bayer, 1999; DEKORP res. Group, 1999; NW-Germany from Giese, 1995; Polish Basin from Jensen et al., 2002; Lamarche et al., 2003: the Netherlands from Duin et al., 1995; Denmark from Thybo et al., 1999; North Sea and Great Britain from Ziegler, 1990; Chadwick and Pharaoh, 1998)

the CEBS by subsidence in a chain of NW–SE-striking transtensional basins including the Sole Pit, West and Central Netherlands, Lower Saxony and Altmark-Brandenburg basins. Likewise, Late Jurassic-Early Cretaceous tensional reactivation of the NW–SE striking STZ and TTZ is observed at the northern margin of the CEBS. During the Late Cretaceous collision-related compressional stresses built up in the Alpine foreland (Ziegler, 1990; Ziegler et al., 1998). These caused the Late Cretaceous and Paleocene inversion of NW–SE trending extensional structures flanking the CEBS, such as the West and Central Netherlands, Lower Saxony and Altmark-Brandenburg basins, the Polish Trough and the northern margin of the Danish Basin. Similarly the southern part of the North Sea Central Graben was mildly inverted (Scheck et al., 2002; Scheck-Wenderoth and Lamarche, 2005; Vejbaek and Andersen, 2002; Mazur et al., 2005; de Jager, 2007). Subsequently the western parts of the CEBS were dominated by the wide-radius thermal subsidence of the North Sea rift system, whilst its eastern parts subsided very slowly (Ziegler, 1990). The distinct Plio-Pleistocene subsidence acceleration of the North Sea Basin and the North German part of the CEBS is still enigmatic but may reflect a renewed build-up of compressional intraplate stresses (Van Wees and Cloetingh, 1996; Marotta et al., 2000).

Diapiric mobilization of the thick Permian evaporites contained in the CEBC commenced during the Triassic, giving rise to a complex network of salt-cored structures some of which remained active to the present (Fig. 28). Moreover, these evaporites provide a basin-wide decoupling horizon between pre- and post salt series that was activated during phases of crustal extension as well as during crustal shortening (Brun and Nalpas, 1996). Whilst the pre-salt series preserve the long-wavelength signal of basin subsidence (Fig. 26) the post-salt series is dominated by a short-wavelength deformation pattern dominated by salt-cored structures (Scheck-Wenderoth et al., 2008). Successive changes in tectonic regime were accompanied by distinct pulses of salt mobilisation. Triassic, E–W directed extension caused the development of N–S trending salt walls on the flanks of the large N–S striking grabens, whereas, along the margins of the CEBS, NW-SE-trending salt structures developed coeval with Late Jurassic-Early Cretaceous transtensional basin subsidence. During the latest Cretaceous-Paleocene phase of intraplate compression, crustal shortening along the southern margin of the CEBC, involving upthrusting of the Harz Mountains and the Flechtinger block, was compensated by compressional thin-skinned deformation of the post-salt series, as evidenced by the development of anticlinal structures along the northern margin of the North German Basin (e.g., Grimmen anticline; Scheck et al., 2003, 2008).

Based on the above, the Northern and Southern Permian basins of the CEBS are interpreted as hot-spot intracratonic basins, akin to the West Siberian Basin. Though the intensity of their precursor magmatic activity was not as spectacular as in Siberia, it caused considerable thinning of the mantle lithosphere and crust, thus providing the driving mechanism for the subsequent thermally driven basin subsidence (van Wees, 1994; van Wees et al., 2000 Ziegler et al., 2004). During the long-term subsidence phase of the Northern and Southern Permian basins, repeated changes in the regional stress field caused renewed destabilization of their lithosphere-asthenosphere system, as evidenced by the evolution of the North Sea rift system that transects the western parts of both basins. In the course of the Cenozoic, geotherms rose apparently resulting in a transition from stress-induced end-Mesozoic basin inversion to Plio-Pleistocene lithospheric folding-induced basin subsidence (Cloetingh et al., 2008).

The sub-circular, about 450 km wide Paris Basin, which contains up to 3 km of mainly Mesozoic sediments, is generally regarded as a typical intracratonic basin. This basin is superimposed on the deeply eroded internal parts of the Variscan orogen, the crust and lithosphere of which were thinned and thermally destabilized during the latest Carboniferous-Early Permian tectono-magmatic cycle. In the course of the Late Triassic and Early Jurassic the area of the Paris Basin subsided below the regional base level and was incorporated into a large sedimentary platform that extended from the Armorican Massif to the Bohemian Massif and from the Helvetic Tethys shelf to the CEBC. Local subsidence centres in the Paris Basin were associated to wrench faults compensating for crustal extension in the westward adjacent Western Approaches and Channel basins that remained active until end Early Cretaceous times. After an Early Cretaceous uplift and SE- ward tilting phase, the Paris Basin area was transgressed again and connections between the Helvetic Shelf and the Western Approaches and North Sea basins re-established. During the Paleocene the

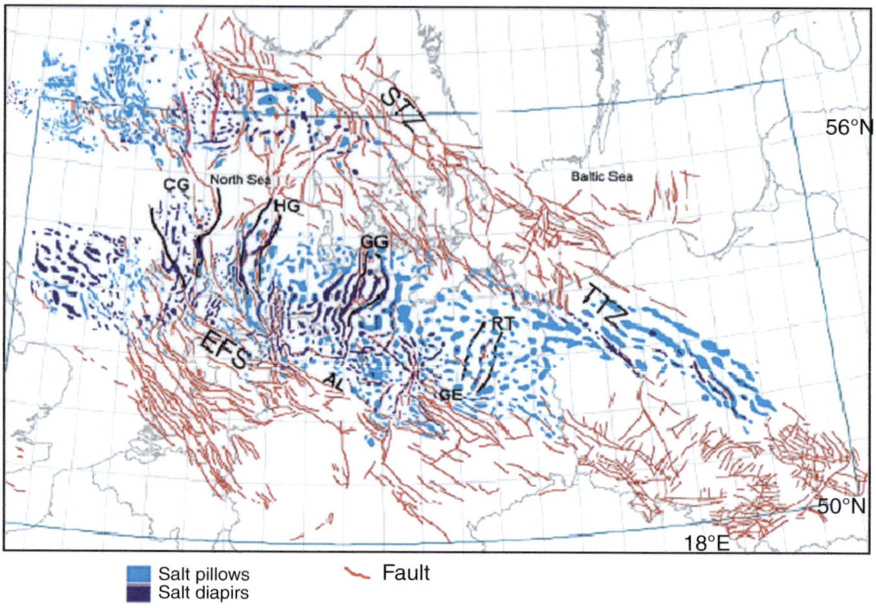

Fig. 28 Present distribution of salt structures in the CEBS illustrates the relationship between intensity of salt tectonic deformation and zones of crustal weakness

connection between the Helvetic Shelf and the Paris Basin was interrupted again owing to collision-related compressional deformation and uplift of areas flanking it to the south and southeast. From Oligocene times onward the eastern part of the Paris Basin was subjected to erosion in response to flank uplift of the Upper Rhine Graben. During the Miocene uplift of the Vosges-Black Forest arch that extends into the northern parts of the Massif Central, caused further erosion of the SE flank of the Paris Basin. Transpressional uplift of the Ardennes-Western Rhenish Massif and reactivation of the shear zone linking the Massif Central and the Armorican Massif, compensating for crustal extension across the Rhine Rift System, caused uplift and erosion of the south-western and northern margins of the Paris Basin (Ziegler, 1990; Prijac et al., 2000; Dèzes et al., 2004; Ziegler et al., 2004; Ziegler and Dèzes, 2007).

Although saucer-shaped in cross-section and semicircular in outline, the Paris Basin does not qualify as an intracratonic thermal sag basin but represents an erosional remnant of a much larger basin complex, development of which was controlled by the relaxation of thermal anomalies introduced during the Permo-Carboniferous tectono-magmatic cycle (Ziegler et al., 2004). Local subsidence centres in the Paris Basin were controlled by activity along transtensional fault systems compensating for Mesozoic crustal extension in the Western Approaches-Channel area and for Cenozoic inversion movements (Ziegler, 1990).

Passive Margins

Passive continental margin sedimentary prisms can host prolific petroleum systems such as the Atlantic shelves of Gabon, Angola, Brazil and Norway. Successful exploration of such basins depends on a reliable assessment of their heat flow regime controlling the transformation of organic matter to petroleum. Though of increasing economic relevance (White et al., 2003), the physical state of passive continental margins as well as their evolution are debated in terms of heat flow regime, crustal structure, the mode and level of isostatic compensation and the configuration of the lithosphere-asthenosphere boundary through time but also concerning factors controlling continental break-up and the post-break-up evolution of margins. This is related to different hypotheses on the geometric configuration, density, composition and thermal structure of the lithosphere below continents and oceans (Mooney and Vidale, 2003).

Integrated studies addressing at a lithospheric scale the 3D configuration of conjugate passive margins and the entire extensional system from continent to ocean to continent are still sparse. Although most research efforts on passive margins are concentrated on the very mature margins of the Northern and Southern Atlantic, and to a lesser extent on the conjugate Australia-Antarctica margins, also young passive margins are investigated, such as the Gulf of Aden, where post-rift sediments are not too thick, thus permitting to evaluate controls exerted by pre-break-up inherited structures on the segmentation of the rift (Leroy et al., 2004; d'Acremont et al., 2005; Bellahsen et al., 2006; Huchon et al., 2007; Tibéri et al., 2007).

The subsidence history of passive margins depends not only on the time that elapsed since crustal separation became effective, but also on their lithospheric configuration and its thermal disturbance at the moment of crustal separation. For instance, the crustal and remnant lithospheric mantle thickness of simple-shear upper and lower plate margins differ significantly at the crustal separation stage. This has repercussions on the post break-up subsidence and strength evolution of such conjugate margins during which their lithosphere re-equilibrates with the asthenosphere (Ziegler et al., 1998). Furthermore, in addition to mechanical stretching of the lithosphere syn-rift magmatic activity, depending on its intensity, can be accompanied by thermal attenuation of the continental lithospheric mantle and by injection of basic melts into and, by underplating of the crust, its densification. The timing and level of magmatic activity in rifts progressing to crustal separation and the opening of an oceanic basin is, however, highly variable (Ziegler and Cloetingh, 2004). For instance, in the Norwegian-Greenland rift magmatic activity flared up immediately prior to crustal separation but was preceded by a very long period of non-volcanic rifting (Ziegler, 1988) whilst the South Atlantic rift was accompanied by plume-related magmatism from its Late Jurassic inception onward. By contrast magmatic activity commenced in the in the Central Atlantic domain 15 My after the onset of rifting and terminated 15–10 My before crustal separation was achieved (Ziegler et al. 2001).

Following crustal separation, the topographic configuration of landmasses flanking passive margins and their drainage systems exert strong controls on source-sink dynamics in terms of denudation, sediment transport away from or toward the coast, and accumulation rates in offshore basins. This controls whether and to what degree a passive margin is sediment starved (e.g., West Iberian Atlantic margin: Boillot, 1984; Boillot et al., 1989) or overfilled resulting in the progradation of its post-break-up sedimentary prism onto oceanic lithosphere (e.g., Niger Delta: Doust and Omatsola, 1989; Anka et al., 2009). Obviously, the thickness of post-break-up sediments, combined with the prevailing heat flow, exerts controls on whether syn-rift or early post-rift potential hydrocarbon source rocks attain a sufficient organic maturity for hydrocabon expulsion.

Although the above-mentioned factors are pertinent to all passive margins quantification of their effects needs to be further pursued. Nevertheless, considerable progress has been made in the understanding of general passive margin characteristics.

Non-volcanic passive margins show little evidence for magmatic activity prior to the onset of sea floor spreading (Louden and Chian, 1999). The best documented example of such a non-volcanic margin is the lower plate West Galician margin of Iberia that faces the upper plate Flemish Cap margin (Boillot et al., 1987, 1995; Manatscal and Bernoulli, 1999; Manatschal et al., 2001; Witmarsh et al., 2001; Manatschal, 2004; Reston, 2007). For the Galician margin reflection seismic data, calibrated by deep-sea drilling, show that continental lithospheric mantle rocks are exposed at the sea floor owing to lithospheric extension. This "transition zone" of exhumed mantle is characterized by a p-wave velocity structure that differs from both oceanic and stretched continental crust (White et al., 2003). Acoustic velocities rise steeply to ~7 km.s–1 only 2 km beneath deformed crustal rocks, increasing gradually to normal mantle values of ~8 km.s–1 over a depth range of 6 km. This velocity variation has been interpreted as resulting from serpentinization of mantle rocks, though this issue is still being debated. Another typical characteristic of non-volcanic passive margins are high-angle detachment faults with steeper parts in the upper crust, flattening with depth and merging into a common detachment surface at the crust-mantle boundary (Reston, 2007). The coincidence of negligible magmatism, exhumed mantle and detachment surfaces is interpreted as resulting from extremely low mechanical stretching rates and thus limited thermal perturbance of the asthenosphere-lithosphere system during the rifting phases preceding crustal separation (White et al., 2003; Lavier and Manatschal, 2006).

Passive margins which have been volcanically active during the crustal separation stage have two specific attributes: (1) Sea-ward dipping reflectors (SDR) have been found along the continent-ocean transition close to the seabed that have been identified by deep-sea drilling as up to kilometre-thick piles of seaward-dipping lava flows. (2) High-velocity bodies have been imaged in deep seismic experiments at the base of the thinned crust adjacent to the continent-ocean transition for which gravity studies also require high densities. Velocities (7.2–7.6 km.s–1) and inferred densities (3,000–3,200 kg.m–3) are consistent with magnesium-rich igneous rocks generated by large-scale decompressional melting of the asthenosphere during continental break-up (Trumbull et al., 2007; White et al., 2003). Magmatism accompanying the initial phases of continental break-up is held responsible for magmatic underplating and the high-velocity-high-density bodies at the base of the stretched continental crust near the continent-ocean transition at most volcanic passive margins. Though the time of emplacement of such underplated bodies is uncertain, geophysical evidence for such underplated bodies is abundant. Well-documented examples include the African margin offshore Namibia (Stewart et al. 2000, Wyer and Watts, 2006) and South Africa (Bauer et al.; Hirsch et al. 2007), the Atlantic margin offshore Congo and Angola (Contrucci et al., 2004) and the Norwegian Margin (Ebbing et al., 2006; Fjeldskaar et al., 2002; Mjelde et al., 2005; Scheck-Wenderoth and Maystrenko, 2008). Deep crustal studies of intracratonic rifts such as the Baikal and East African rifts (Thybo et al., 2003) and the Dniepr Graben (Stephenson et al., 2001) also image high-velocity bodies in the lower crust beneath the rift axis, suggesting that underplating and permeation of the crust with mantle derived melts can also occur prior to crustal separation (Ziegler and Cloetingh, 2004).

Another characteristic feature of passive margins is a significant segmentation along strike, which indicates that margin evolution is also controlled by pre-existing structures. Such inherited lateral discontinuities in lithospheric strength are likely to have influenced the rifting process at the US East Coast passive continental margin (Wyer and Watts, 2006). In addition, modification of the lithosphere rheology induced by the rifting process may be of importance as a control on deformation and localization of the final break up (Cloetingh et al., 2005; van Wijk and Blackman, 2004). Along strike, upper plate margins may be replaced across transform faults by lower plate margin as seen in the conjugate Mid-Norwegian and Greenland margins (Mosar et al., 2002).

One of the few volcanic passive continental margins where the detailed structure of the system from the continental onshore, over the continental margin offshore and across the continent ocean transition to the deep oceanic part is well explored, is the Norwegian North Atlantic margin (Blystad et al., 1995; Breivik et al., 1999; Brekke, 2000; Nottvedt, 2000; Ebbing et al., 2006; Fernandez et al., 2005, 2004; Mjelde et al., 2005; Scheck-Wenderoth et al., 2007; Skogseid et al., 2000; Cavanagh et al., 2006; Tsikalas et al., 2005; Wandås et al., 2005). This margin experienced a long pre-break-up rifting history during which major crustal extensional detachment systems developed before crustal separation occurred 55 My ago (Gabrielsen et al., 1990, 1999; Osmundsen et al. 2002; Mosar et al., 2002). Seaward dipping reflectors, as well as lower crustal high-velocity-high-density bodies have been imaged at the continent-ocean transition. In addition to several crustal transects available for the conjugate Norwegian-Greenland margins (Faleide et al., 1988; Geoffroy, 2001; Callot et al., 2002; Mjelde et al., 2005; Tsikalas et al., 2005), a lithosphere-scale 3D study of the Vøring segment at the Norwegian margin (Fig. 29) documents the presence of a highly heterogeneous crust. Furthermore, a less dense and considerably hotter lithospheric mantle appears to be present beneath the oceanic part of the margin than beneath the stretched continental crust (Scheck-Wenderoth and Maystrenko, 2008).

A characteristic feature of passive margins is the decrease in crustal and lithospheric thickness from the continent to the ocean. In terms of the thermal regime of a passive margin, the configuration of the lithosphere-asthenosphere boundary is of special interest as it is assumed to coincide with the 1,330°C isotherm. Estimates of lithosphere thickness for the continents come from passive seismology, p-wave travel time tomography, surface wave analysis, gravity modelling and studies of the thermal field (Artemieva, 2006, Artemieva et al., 2006). The thickness of oceanic lithosphere is known to vary as a function of its age and does not exceed 125 km (Turcotte and Schubert, 2002; Mueller et al., 2008), though conflicting estimates are predicted by the half-space cooling model (Turcotte and Oxburgh, 1967), the plate-cooling model (Parsons and Sclater, 1977; Sclater, 2003; Stein and Stein, 1992) and by the analysis of surface waves (Midzi et al., 1999; Zhang and Lay, 1996).

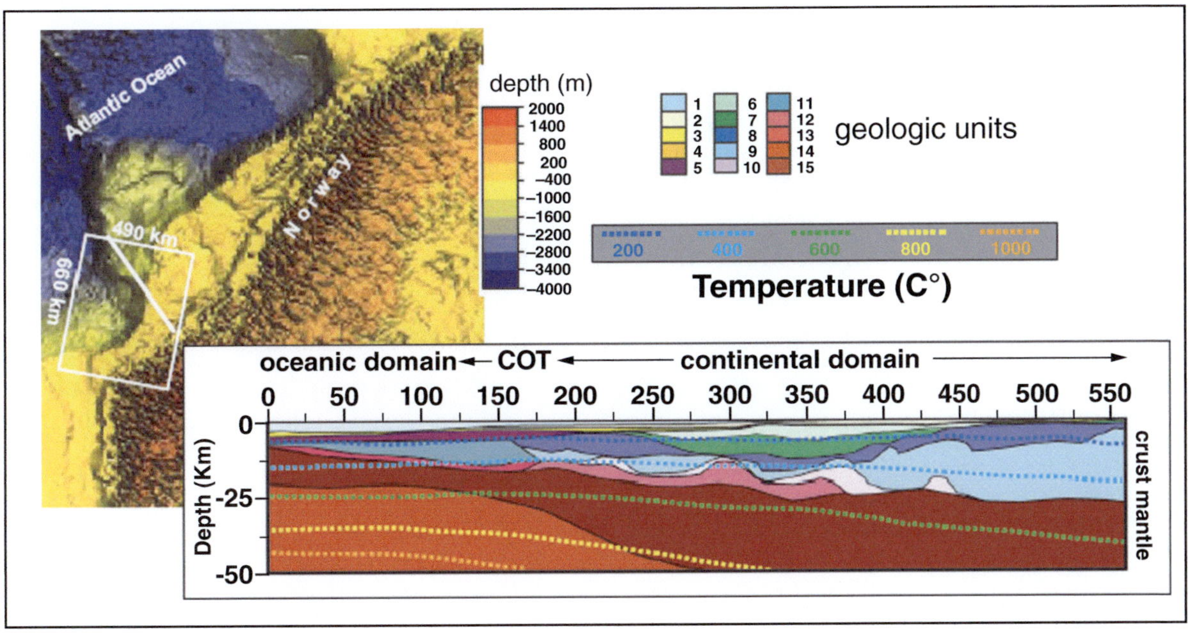

Fig. 29 Present bathymetry and structural setting at the Norwegian passive continental margin. The profile is a cross-section through a 3D model constrained by seismic and well data consistent with gravity and temperature. Numbers near geological units: *1*, water; *2–4*, post-break-up sediments (*2*, post-Middle Miocene; *3*, pre-Middle Miocene; *4*, Paleocene); *5–7*, pre-break-up sediments (*5–6*, Cretaceous; *7*, pre-Cretaceous); *8*, crystalline continental crust; *9*, high density continental lower crust; *10*, high velocity, high density lower crustal body; *11*, ocean layer 2AB; *12*, ocean layer 3A; *13*, ocean layer 3B; *14*, low density mantle; 15, normal density mantle

Studies from the North Atlantic (Breivik et al., 1999; Fernandez et al., 2004, Scheck-Wenderoth and Mystrenko, 2008) indicate that the configuration of the lithosphere-asthenosphere boundary essentially controls the present-day shallow temperature field of passive margins. This may have been of even more important in the past, particularly during the crustal separation stage before the lithosphere of the newly formed passive margin began to re-equilibrate with the asthenosphere. As a general feature, the crust is colder in the oceanic than in the continental domain whereas this trend is reversed downward. Lateral temperature differences are in the order of ~100°K at 5 km depth across the margin and increase to ~400°K at 50 km depth (Fig. 30) (Scheck-Wenderoth and Maystrenko, 2008). Higher temperatures and heat flows for the oceanic lithospheric mantle compared to the continent indicate that reduced upper mantle p-wave velocities and densities observed in the oceanic mantle lithosphere are thermally induced.

Similar results pertaining to the crustal structure have been obtained for the South Atlantic volcanic margins between Argentina and South Africa. These resulted from the Gondwana break-up that was accompanied by mantle plume activity and commenced in Late Jurassic/Early Cretaceous in the southernmost parts of the future South Atlantic and propagated northward throughout the Early Cretaceous (Clemson et al., 1997; de Wit, 2007; Keeley and Light, 1993; Light et al., 1993; Mohriak et al., 2002). For the Orange Basin on the South African and Namibian margin and for the Colorado Basin on the conjugate Argentinean margin seismic data show that the thickness of the continental crust decreases from about 40 km in unstretched onshore areas (Franke et al., 2007, 2006; Kwadiba et al., 2003; Nguuri et al., 2001; Stankiewicz et al., 2002) to less than 25 km over a distance of ~200 km in the stretched continental margin (Bauer et al., 2000; Franke et al., 2007, 2006; Mohriak et al., 2002; Stewart et al., 2000) whilst the oceanic crust is less than 10 km thick (Bauer et al., 2000; Hinz et al., 1999). On both margins the lower crust beneath the stretched margin is overlain by an up to 12 km thick sedimentary succession (Franke et al., 2006, 2007; Bauer et al., 2000; Paton et al. 2007) consisting mainly of post-break up deposits. On both

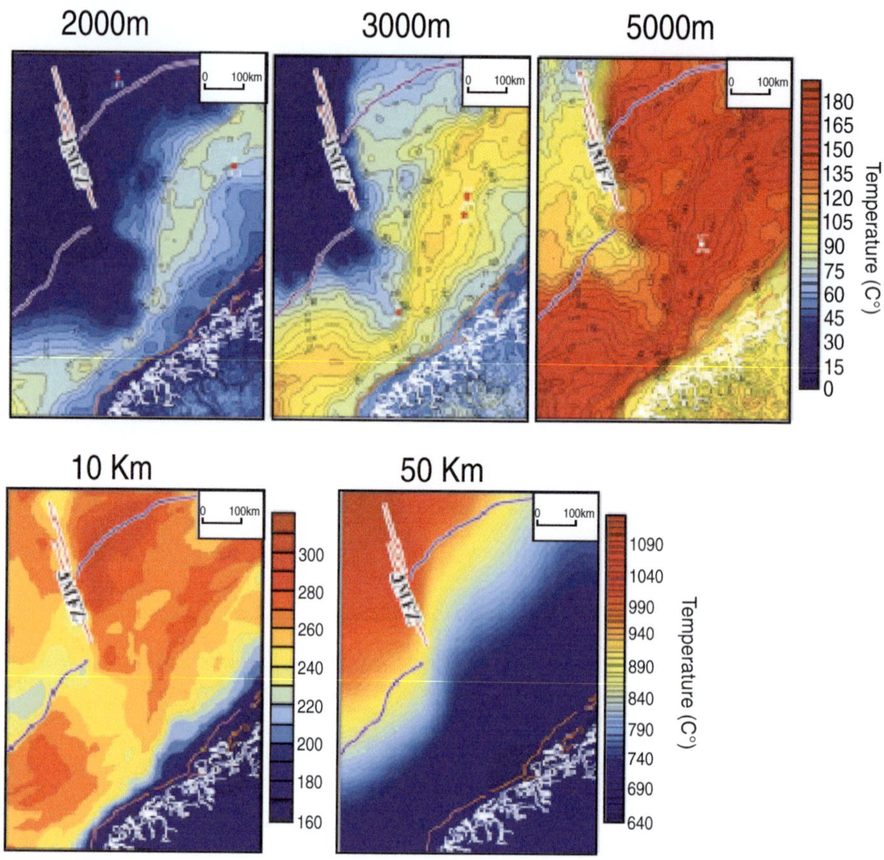

Fig. 30 Temperature variations across the Norwegian passive margin for different depths from 3D modelling of the area indicated by a white rectangle in Fig. 29 (modified after Scheck-Wenderoth and Maystrenko, 2008). White line is Norwegian coast corresponding roughly to limit between unstretched continental crust and thinned crust of margin. Red dots are well locations where measured temperatures have been compared to modelling results. JMFZ: Jan Mayen fracture zone; COT: continent-ocean transition. Coordinates are Universal Transverse Mercator Zone 33 (Northern Hemisphere)

margins the continental lower crust is characterized near the continent-ocean transition by a high-velocity-high-density-body whilst wedges of seaward dipping reflectors occurring at the base of post-break-up sediments are interpreted as syn-break-up lava flows (Gerrard and Smith, 1982; Bauer et al., 2000; Franke et al., 2007, 2006; Gladczenko et al., 1997; Hinz et al., 1999; Hirsch et al., 2007).

A number of questions remain to be addressed in the future. The duration of the rifting required to achieve crustal separation is a function of the strength (bulk rheology) of the lithosphere, the build-up rate, magnitude and persistence of the extensional stress field, constraints on lateral movements of the diverging blocks (on-trend coherence, counter acting far-field compressional stresses), and apparently not so much of the availability of pre-existing crustal discontinuities which can be tensionally reactivated (Ziegler and Cloetingh, 2004). Kinematic and dynamic modelling of passive margins will yield non-unique results as long as their structural configuration and thermal history preserved in the sediments can be reproduced by a combination of depth-dependent lithospheric stretching, phase transitions taking place in the lower crust and varying stretching velocities. Another issue of bias relates to processes held responsible for post-break magmatic activity and vertical uplift of passive margins (Ziegler and Cloetingh 2004). These questions call for a holistic 'source-to-sink' approach considering the relationship between offshore subsidence and phases of onshore exhumation.

Source – Sink Processes: Coupling of Climate, Uplift and Erosion, Subsidence and Sedimentation

The subsidence rates of sedimentary basins and the pattern of sediments accumulating in them, is controlled by multiple interacting parameters. Advanced analytical and numerical tools are required to evaluate the record of sedimentary basins and the denudation history of source areas to be able to make realistic predictions of lithofacies developments and the distribution of potential reservoirs, seals and hydrocarbon source rocks. This is a requirement not only for hydrocarbon exploration and the search for other mineral resources, but also for water management and CO_2 sequestration.

Although it has since long been realized that denudation is an important aspect of present-day morphology, there is now a growing awareness that topography development is driven by tectonics and further controlled by forces operating at depth in the Solid Earth. Moreover, it is realized that tectonic topography is modified by erosion, a process driven by external forces operating at the surface of the Solid Earth. To understand the interrelationship, interdependence and feedback mechanisms between internal and external forcing, lithospheric and surface processes need to be constrained.

Age dating of detrital minerals in sedimentary basins yields information that can be used to infer detailed spatial and temporal denudation patterns in sediment source areas (Van der Beek et al., 2006; Champagnac et al., 2008). Such data, together with the volume of sediments contained in basins, provide a powerful tool for estimating in source areas long-term regionally averaged mechanical denudation rates. Obviously, there is a clear genetic link between tectonics controlling the geomorphologic evolution of source areas and the subsidence of sedimentary basins. The chronology of denudation inferred from fission track analyses reflects the complex relationship between tectonic reactivation of major structures, development of topography and the associated geomorphologic response, related effects on local denudation rates, and other factors, such as the location of the drainage divide on the uplifted flanks of sedimentary basins (Bernet and Spiegel, 2004; Reiners and Ehlers, 2005).

Requirements for Data Acquisition and Development of New Analytical Techniques

Whilst the sedimentary record of a basin readily permits to retrace its evolution, the ability to properly date the different sedimentary sequences is of prime importance. In this respect continued efforts are required to date series that are devoid of marine fossils. A further problem in reconstructing the burial/uplift history of sedimentary basins and the unroofing of sediment source areas is the quantification of the amount of erosion. Ultimately, even if modelling approaches can provide some estimates of the past topography, it remains quite difficult to quantify paleo-elevation values as required for closer control on hydraulic heads and topography-driven fluid flow in adjacent basins. To better handle these input data and calibration points for basin modelling, and to decrease the error bars of results, new techniques have been developed or are currently under development. Some of them, among many other techniques, are presented below:

Chronostratigraphy, Sedimentation Rates and Timing of Maximum Burial

Outcrop studies, reflection-seismic and well data permit to describe the configuration and litho-stratigraphic composition of the fill of sedimentary basins. This sedimentary record must, however, be closely tied to a detailed time scale to permit construction of quantitative subsidence curves, to assess sedimentation rates, changes in subsidence rates and the development of erosional events as input for basin modelling.

Compared to the commonly used biostratigraphic methods, seismic sequence stratigraphy and magnetostratigraphy, recent advances in astronomical tuning of the sedimentary record provide for the youngest part of Earth's history higher resolution and accuracy time scales, both of the order of 10 kyr over the last 10 Myr. These are independent of radio isotopic dating and are tied to the Recent through a direct match with astronomical curves. Recent research has been directed at intercalibrating radio isotopic and astronomical time with the aim of eliminating uncertainties in mineral age dating standards (Kuiper et al., 2004). The high resolution of astronomical time scales is required to accurately date sequence boundaries, to determine whether sequence boundaries are synchronous on a regional or global scale, and to determine whether third-order sequences are tectonically controlled or related to long peri-orbital cycles. The new time scale is equally important for the dating of major tectonic events and associated changes in the sedimentary basin configuration. Recent research (Krijgsman et al., 1999; Krijgsman, 2002) aims at refining the anchored astronomical polarity time scale (APTS) for Neogene marine and continental successions, extending it through the Paleogene into the Mesozoic, developing floating astronomical time scales for parts of the Mesozoic and Paleozoic, and at the intercalibration of radio isotopic and astronomical time. Evidently, astronomical time scales – with their unprecedented accuracy and resolution – are and will be increasingly applied in research directed at sea-floor spreading histories, magnetic reversals and short polarity excursions, biochronology and evolution, sequence stratigraphy and sedimentary basin development, and at paleoclimate and paleo-oceanography, including the important link to (paleo) climate modelling (Haug et al., 2003).

Paleo-Thermometers, Rates and Timing of Unroofing

The challenge of dating geological events and reconstructing the burial history of hydrocarbon source rocks and reservoirs becomes even larger in areas, which have been uplifted and affected by erosion. For instance, maturity rank of the organic matter (Tmax and Ro) and Bottom Hole Temperatures (BHT) are usually sufficient to determine the regional heat flow and geothermal gradients and to assess the thermal evolution of a basin by means of 1D, 2D or 3D basin modelling techniques. These yield accurate results for passive margins and other sedimentary basins that have not been uplifted and eroded, and for which quantitative subsidence curves are closely constrained.

In the external parts of fold and thrust belts, where both paleo-temperatures and paleo-burial data are questionable (Fig. 31), other analytical techniques are required to constrain the PT path of hydrocarbon source rocks and reservoirs, date the episodes of their unroofing, and to quantify the amounts of erosion. For instance, measuring the paleo-temperature alone is not sufficient to derive the paleo-burial, because the same value could be reached at different paleo-depths, depending on assumptions made on the paleo-heat flow and past geothermal gradient.

On the other hand, fluid inclusion studies, as described below, can provide constraints for calibrating both the paleo-pressure and paleo-temperature of cemented reservoirs and fractures at the time of maximum burial, whereas fission track studies are used to date the crossing of a given isotherm, which can be converted into a paleoburial value, provided the geothermal gradient is constrained by other techniques.

Hydrocarbon-Bearing Fluid Inclusions and PVT Modelling

Fluid inclusions constitute isolated volume of fluids, which have been entrapped inside a mineral during its crystallization. Because their volume is constant through time, they plot, depending on their composition, on pressure/temperature diagrammes along specific curves (Fig. 32).

Aqueous fluid inclusions are common in diagenetic cements filling the pores of reservoirs and fractures.

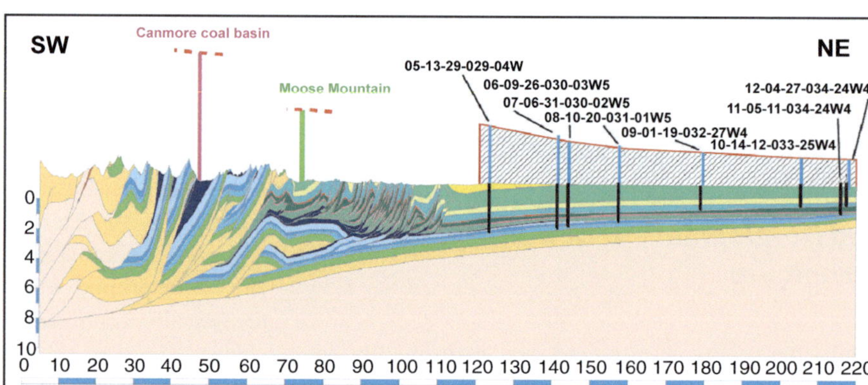

Fig. 31 Erosion estimates along the SUBTRAP Canadian transect, resulting from Tmax (peak temperatures derived from Rock-Eval pyrolysis), Ro (vitrinite reflectance) measurements and coupled 1D Genex and 2D Thrustpack kinematic and petroleum modelling (after Faure et al., 2004)

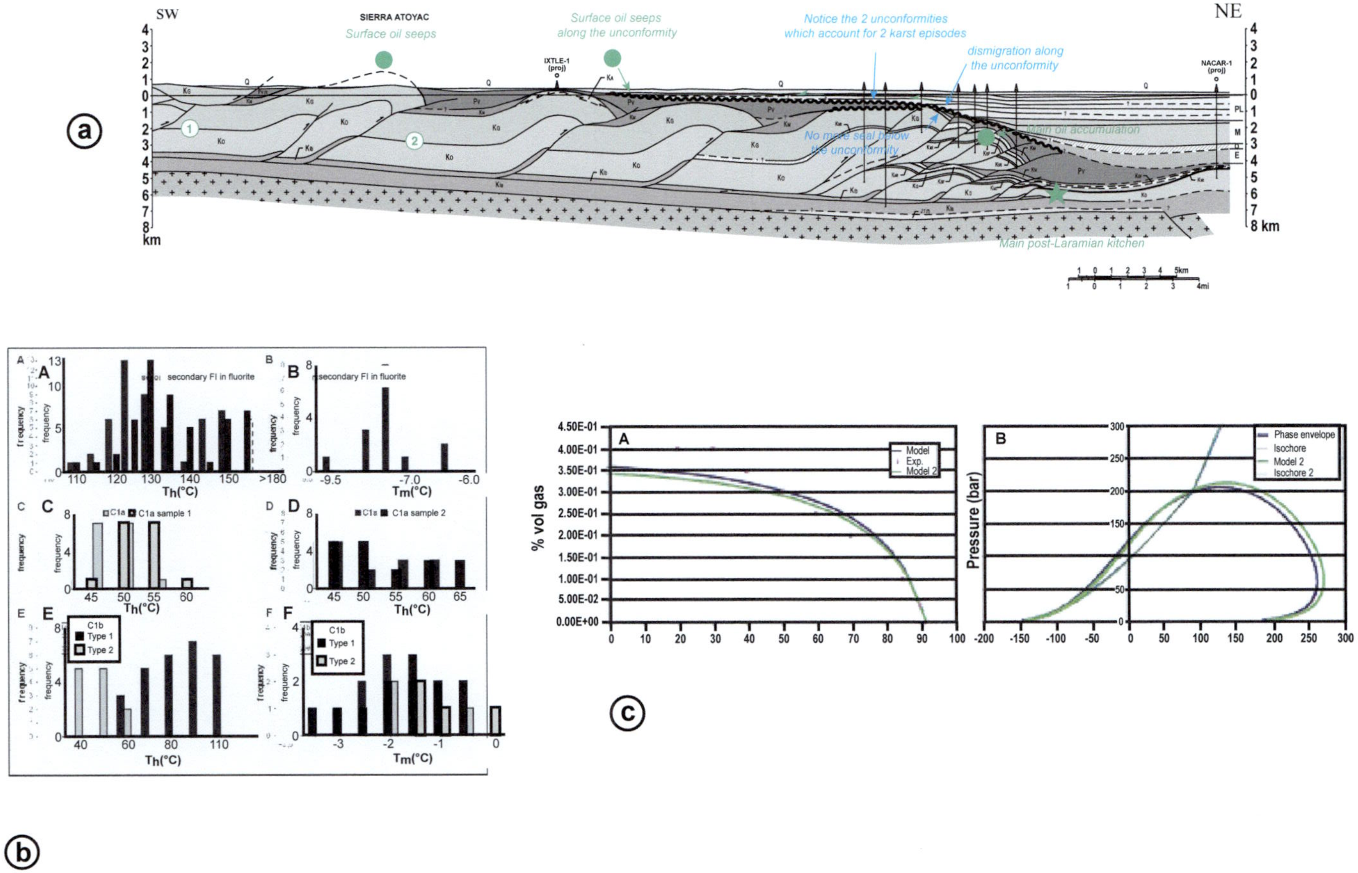

Fig. 32 Fluid inclusions microthermometry and PVT modelling. Both aqueous and hydrocarbon-bearing inclusions coexist in the same cements in Cretaceous platform carbonates of the Cordoba Platform in Eastern Mexico. Crossing of the two isochores provides both the paleo-Temperature and paleo-Pressure at the time of trapping, thus providing a direct control on the paleo-burial and paleo-geothermal gradient. **a**) Regional cross section along the Cordoba Platform-Veracruz transect; **b**) Histogram of Th (homogenization temperature) of fluid inclusions from fracture cements; **c**) PVT modelling of aqueous and HC-bearing inclusions (after Ferket et al., 2003, 2004: Roure et al., 2009)

Their homogenization temperatures (Th), measured by thermometry on thin-sections, are representative of the paleo-temperature to which the host rock was exposed at the time of cementation, assuming a thermal equilibrium between the fluids and the overburden and no advection of hot or cold fluids into the host rock.

Because hydrocarbons do not mix with water, the same cement can contain both aqueous and hydrocarbon-bearing inclusions, which have been entrapped at the same time. Therefore, the crossing point of the isochors derived from the aqueous inclusions and the hydrocarbon inclusions can be used to determine the pressure and temperature at the time of cementation, provided the composition of the hydrocarbon phase is relatively well known. Assuming a hydrostatic pressure regime, the observed paleopressure can be directly interpreted in terms of paleoburial (load of the water column).

This technique has been successfully applied on fluorites from Baluchistan and Tunisia, where hydrocarbon inclusions were big enough to allow direct sampling of the oil and its characterization by gas-chromatography mass spectroscopy (G-CMS) (Benchilla et al., 2003). When inclusions are smaller (e.g., Cordoba Platform, Mexico: Ferket et al., 2003, 2004), in situ measurements of the changes observed in oil/gas ratios and Synchrotron FTIR (Fourier transform infra-red) microanalysis may allow to determine the oil API°, to obtain a reliable shape of its isochore, thus permitting to determine the paleo-temperature and paleo-burial of the sample at the time of cementation (Fig. 31; Guilhaumou and Dumas, 2005; Roure et al., 2009).

Apatite Fission Track Technology (AFTT) and Other Low Temperature Chronometers

Age dating of detrital minerals in sedimentary basins has yielded information on sediment sources from which detailed spatial and temporal denudation patterns could be inferred. This, together with sediment volumes, provides a powerful tool for estimating regionally averaged mechanical denudation rates over long time scales. The dominant source region for sediments deposited in a basin cannot be neglected, as there is a clear genetic link between the geomorphic, tectonic and sedimentological evolution of the sediment source and sink areas (Carrapa et al., 2003). The denudation chronology inferred from fission track data reflects the complex relationship between topography development and the geomorphic response to tectonic reactivation of major structures, their effects on local denudation and the location of the drainage divide relative to the uplifted flanks of sedimentary basins (De Bruijne and Andriessen, 2002; Willingshofer et al., 2001; Braun and Van der Beek, 2004; Braun et al., 2006). Source-Sink correlations are still sparse but first studies clearly demonstrate their potential (Gallagher and Brown, 1999; Van der Beek et al., 2006; Tinker et al., 2007 b; Kounov et al., 2007).

Thermochronologic methods provide estimates of regional variations in the timing, duration and rate of uplift and erosion over long-term periods. During the last decade, AFTT has emerged as a powerful technique to unravel and quantify the denudation history of elevated basement regions, owing to its ability to constrain the low-temperature (< 120°C) cooling history of rock samples (Fig. 33) (Ehlers and Farley, 2003; Reiners and Ehlers, 2005).

Although existing fission track databases are for many areas able to quantify broad, regional-scale denudation and/or uplift patterns, they do not provide information on denudation rates at the finer scale required to constrain numerical models for surface, deeper lithospheric or asthenospheric processes. Therefore, new approaches to constrain the quantification and timing of denudation and landscape development rely on the combination of a low-temperature geothermometer (<70ºC) -apatite (U-Th)/He analyses- with a higher temperature geothermometer (120–60ºC) – AFTT, in order to obtain a finer resolution on denudation patterns and their timing. Another advantage of this approach is the distinction between regional-scale patterns of denudation based upon AFTT and more local-scale patterns based on (U-Th)/He. Future work has to focus on problems such as (a) quantification of timing and amounts of uplift and denudation, (b) finer resolutions of denudation patterns, (c) characterization of local-scale versus regional scale effects and (d) comparison of post-breakup landform evolution on conjugate margins. To further constrain vertical motions and the denudation history, results of low-temperature geochronology analyses can be complemented with techniques of exposure age dating using ^{21}Ne and ^{3}He isotopes (Meesters and Dunai, 2002a, b, c; Dunai and Wijbrans, 2000). In this context, the study of

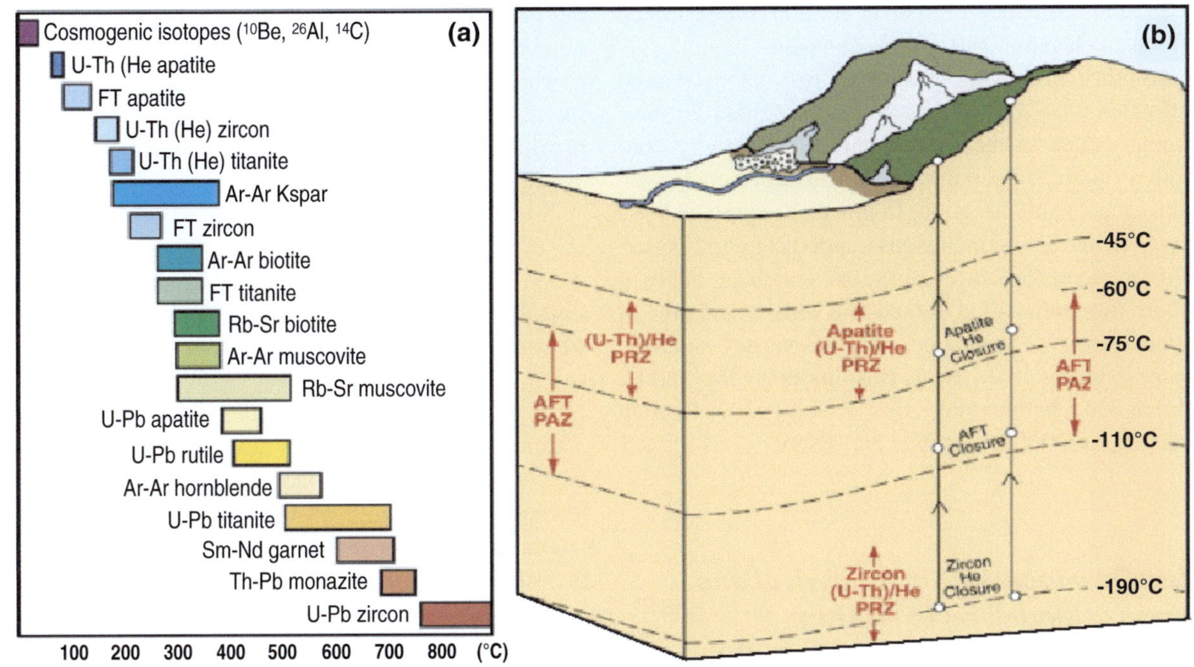

Fig. 33 **a**) Temperature range of generally used thermochronometers; **b**) Crustal depth profile showing temperature zones of several low-temperature thermochronometers as a result of surface topography and archived as time-temperature records in surface samples (after Ehlers and Farley, 2003)

paleoplanation surfaces has been often used to detect the presence of regional vertical movements, potentially related to crustal- or lithosphere-scale processes (e.g., Bonnet et al., 2000; Van Balen et al., 2000). This type of studies may significantly contribute to improve the understanding of the present-day topographic relief.

Isotope geology uses analytical methods and techniques for geochronology, such as age determinations and definition of process rates, and isotopic tracer studies to analyze transport between various chemical reservoirs. In order to be able to answer questions concerning the chemical evolution of the Solid-Earth through time and space, as well as feedback mechanisms between reservoirs contributing to the Solid-Earth, it is necessary to identify, record and quantify the underlying processes. An aspect important to the understanding of forces driving these processes is the timing and quantification of mass transfer and chemical fluxes taking place at different scales – from global to mineral lattice.

Analytical techniques in isotope geology have improved considerably, owing to advances in mass-spectrometry and chemistry, and to the use of new techniques such as laser probing. It is now possible to analyze small quantities of material, single grains and spots within a crystal with high precision. Particularly promising techniques are fission track analysis, exposure age dating using noble gas isotopes of ^{21}Ne and ^{3}He, Ar-laser probing and U-series (Sanders et al., 1999; Necea et al., 2005; Ruszkiczay-Rudiger et al., 2005; Juez-Larré and Andriessen, 2006).

Pioneer Studies for Calibrating Paleo-Elevations

Although Europe was located at low latitudes during the Carboniferous and Permian, periglacial fossil plant assemblages and sediments attest for the past high elevation of the Stephanian-Autunian depocenters of the Massif Central in France (Becq-Giraudon and van den Driessche, 1994; Becq-Giraudon et al., 1996). Similarly, pollen associations have been used to trace in Pliocene and Pleistocene deposits of Colombia (Sabana de Bogota: Hooghiemstra, 1989; Andriessen et al., 1993; Hooghiemstra and Ran, 1994; Torres et al., 2005) and the Western Alps (Fauquette et al., 2007; Jimenez-Moreno et al., 2007) the coupled effects of elevation and climate changes.

Another promising technique for paleoelevation estimates relies on stable isotopes of O and H, their average value in meteoric water being directly controlled by air temperature, which in turn depends on both climate and elevation. In this technique, the $\delta^{18}O$ and D ratios in fluid inclusions trapped in quartz minerals of cemented fractures and veins, which are likely to record the composition of meteoric water at the time of cementation, are compared with present-day meteoric water (Rowley et al., 2001; Blisniuk and Stern, 2005; Mulch and Chamberlain, 2007; Rowley, 207; Rowley and Gazione, 2007; Mulch et al., 2008).

New Developments in Numerical and Physical Modelling of Erosion, Sedimentation and Mountain Building Processes

Analogue models have been extensively used by Earth scientists to simulate the evolution of sedimentary basins at various scales, in both extensional and compressional domains, thus providing direct controls on the mechanical behaviour of rifted basins, thrust belts and salt diapirs, but also on fault nucleation and propagation, or on the gravitational collapse of passive margins (Vendeville et al., 1987; McClay, 1990; Colletta et al., 1991; Buchanan and Nieuwland, 1996; Schreurs and Colletta, 1998; Nieuwland, 2003; Buiter and Schreurs, 2006; Schreurs et al., 2006; and references therein). Attempts with analogue models have been made to simulate the fluid pressure by injecting gas into reservoirs from below (Cobbold et al., 2004), demonstrating that the main limit of analogue experiments is the constant rheology of materials during the experimental conditions, preventing simulation of transient effects of temperature, phase transition or changes in pore fluid pressure during the evolution of the model.

To overcome these limitations, numerical models were developed simulating the evolution of sedimentary basins and their deformation into thrust belts. Early kinematic models were actually quite simple, being purely geometric and deterministic with the geometry and incremental deformation along thrust faults being defined by the user. New trends in numerical modelling rely on coupling with rock mechanics to predict the localization or reactivation of faults, as well as with surface processes, in order to provide more realistic estimates of erosion, transport and deposition of sediments.

Some of these models are presented below.

Coupled Kinematic and Mechanical Models of Thrust Belt Evolution

First attempts at integrating rock mechanics in simulations of the evolution of foreland fold-and-thrust belts were undertaken in the nineties. These permitted to account for and predict overall stress patterns (i.e., lateral and vertical distribution of extensional, compressional or intermediate strike-slip domains) and the localization of zones of potential brittle failure (Zoback et al., 1993; Sassi and Faure, 1996).

Beaumont et al. (1999, 2000) and Pfiffner et al. (2000) took the rheology of the entire lithosphere into consideration in their simulations of the growth of the Pyrenees and Alpine orogens, and developed sequential evolutionary models for flexural foreland subsidence and intra-crustal decoupling, accounting for progressive lower crustal stacking and the development of antithetic backthusts. Results of these modelling efforts compared rather well with the present-day architecture of the Pyrenees and Alps, as imaged by the crustal-scale ECORS and NFP20 reflection-seismic profiles.

Stockmal et al. (2007) presented a numerical dynamic model for a regional profile crossing the Canadian Rockies and the Alberta foreland basin that accounted for strain softening, multiple layers of competent and incompetent strata, erosion and sedimentation (Fig. 34). The resolution of these simulations allowed for a direct comparison with seismic profiles.

Numerical Stratigraphic Models Coupling Erosion-Transport-Sedimentation

In less well explored frontier areas predictions of reservoir quality and hydrocarbon charge are factors that may not easily be assessed. In support of the petroleum industry, numerical tools were developed that couple various surface processes during basin modelling,

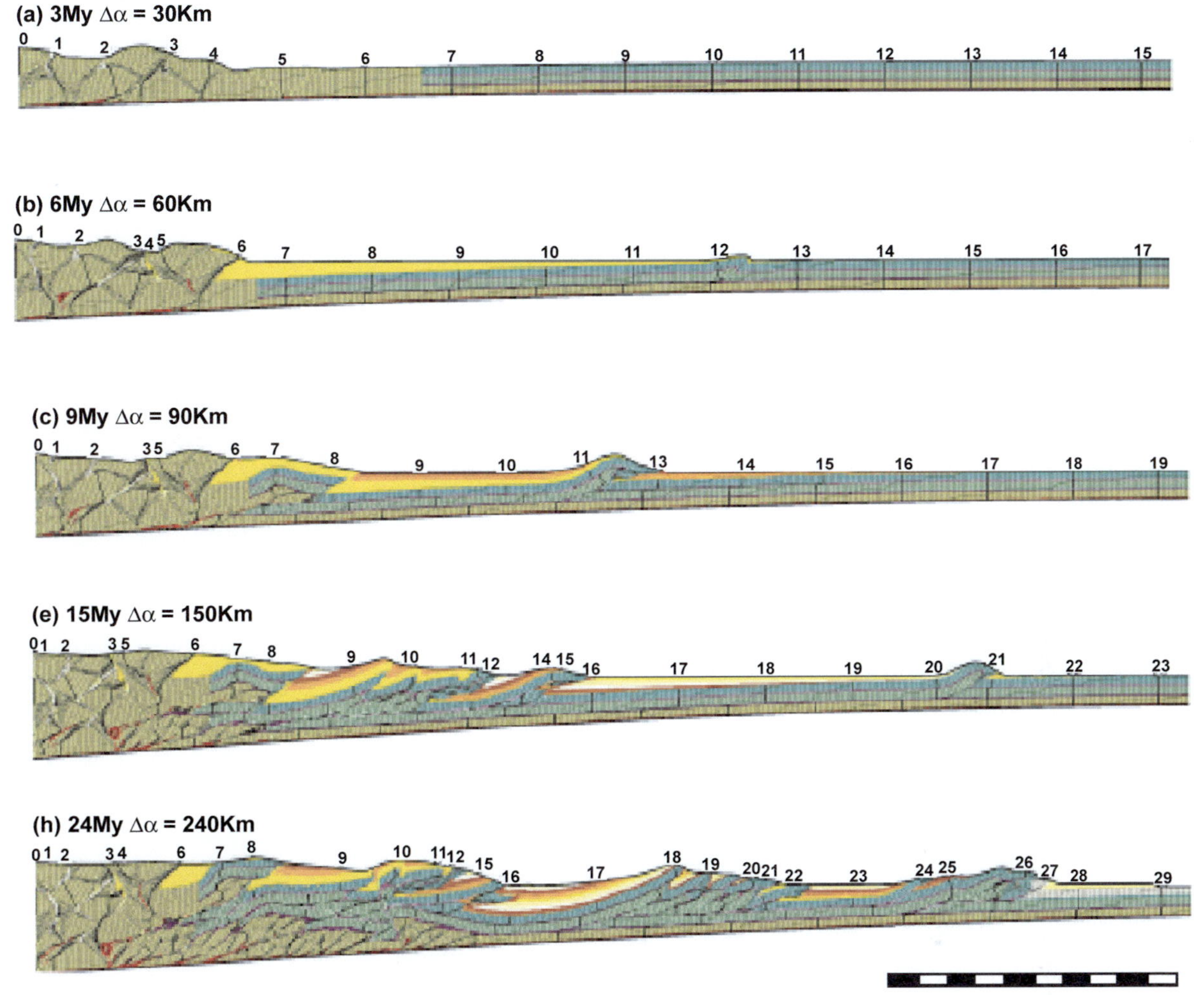

Fig. 34 Coupled kinematic and mechanic model of the Canadian Rockies thrust-belt evolution (after Stockmal et al., 2007; see also the following link to access an animation of the modelling results: ftp://rock.geosociety.org/pub/reposit/2007/2007285.pdf)

simulating the transfer and deposition of siliciclastic sediments over a wide range of environments, i.e., from fluvio-deltaic systems and coastal plains, down to deep abyssal plains. In these sedimentological models, the main input parameters are eustatic sea level fluctuations and subsidence rates controlling accommodation space development, and the sedimentary flux from source to sink areas (Garcia-Castellanos et al., 1997, 2002; Granjeon and Joseph, 1999; Granjeon, 2005; Leever et al., 2006, 2007). Output data comprise the overall configuration and architecture of the sedimentary basin fill, as well as the sand/shale ratio for individual layers. As such, these models outline main lateral and vertical lithofacies changes and the distribution of potential reservoirs and seals.

This sedimentological modelling capacity, which was particularly well suited for simulating the post-rift evolution of passive margins, was gradually adapted to cope also with more complex tectonic domains, such as fold-and-thrust belts and segments of passive margins in which activity along clay and salt detachment horizons accounts for gravitational overburden collapse.

In a case study in the Central Apennines, Albouy et al. (2002; Fig. 35) tested the coupling of the 2D Thrustpack forward kinematic tool with Dionisos, a 3D sedimentological tool, both of which had been developed by the Institut Français du Pétrole (Sassi and Rudkiewicz, 2000; Granjeon and Joseph, 1999; Barrier et al., 2005). The area was selected for its excellent surface and subsurface controls (outcrops,

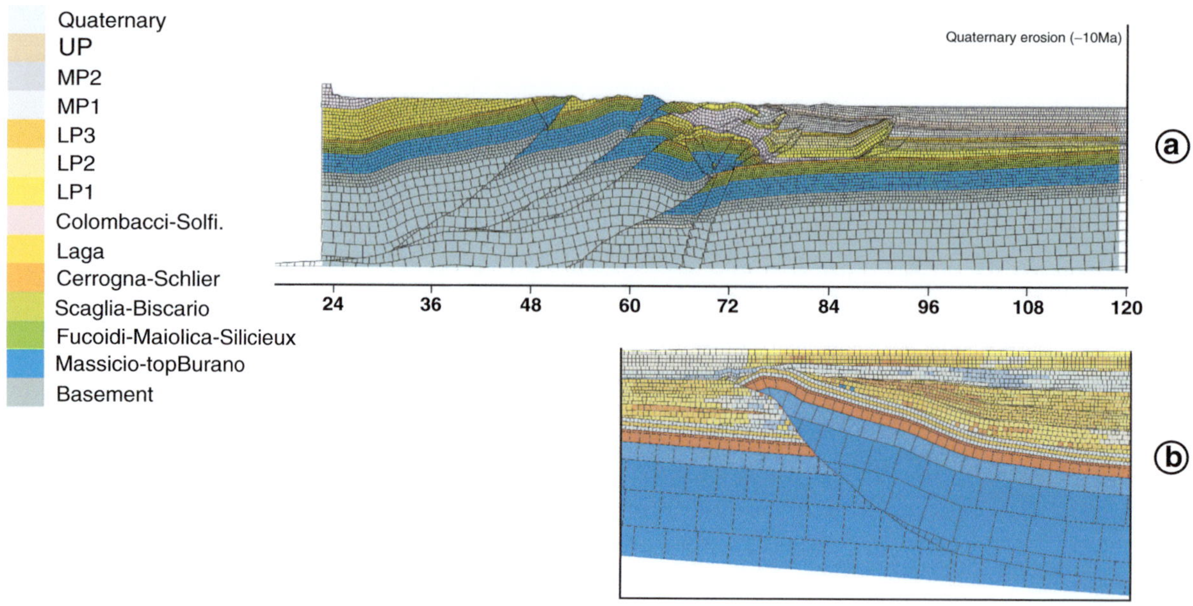

Fig. 35 Coupled Thrustpack-Dionisos modelling applied to the Central Apennines (after Albouy et al., 2002). See Further description of the workflow applied for its simulation in the text

seismic profiles and wells), and its well known lithostratigraphy, consisting of thick Miocene to Pliocene flexural and piggyback series containing the Messinian evaporite marker horizon. The workflow validated during this study comprised the following steps: (1) seismic interpretation, construction of 3 parallel structural sections crossing the thrust belt, and their palinspastic restoration to the pre-orogenic configuration; (2) 2D forward kinematic modelling of each cross-section with Thrustpack; (3) construction of the regional surface topography at incremental stages of Thrustpack deformation by lateral extrapolation between the different 2D cross-sections, and export of these surfaces into Dionisos; (4) iterative and incremental sedimentological modelling using the successive topographic surfaces constructed with Thrustpack as input data for accommodation space; and (5) export of resulting lithofacies distributions to the synkinematic layers computed by Thrustpack for the 2D regional sections.

In a case study on the western margin of the Gulf of Mexico, Alzaga et al. (2009; Fig. 36) used Thrustpack to calibrate activity on a basal detachment horizon and progressive growth of syn-extensional strata deposited between a listric normal fault located at the shelf break and a roll-over antiform that contemporaneously developed down-slope from this depocentre. Lateral displacement velocities along the listric fault and subsidence rates of the substratum were then used as input data for Dionisos modelling. Several hypotheses on sediment flux were successively tested in order to fit with the well data available on the shelf, resulting in different distribution patterns of potential reservoirs in the deeper-water parts of the margin. Unlike in the Apennines, however, no wells were available in the deeper part of this margin that could be used to evaluate the validity of the model and to further refine it.

Physical Experiments and Parameters Controlling Continental Topography and Erosion

Many academic groups have developed and still utilize the potential of the models to promote new inderstanding of dynamics and processes. However, because the petroleum industry is the primary end user of sedimentary models, such models were initially developed for documenting the progressive infill of sedimentary basins, the sink areas. It is envisaged, however, that addressing sediment source areas in a same way may materially contribute towards a better understanding of the architecture and composition of depositional systems in sedimentary basins. In this respect it ought to

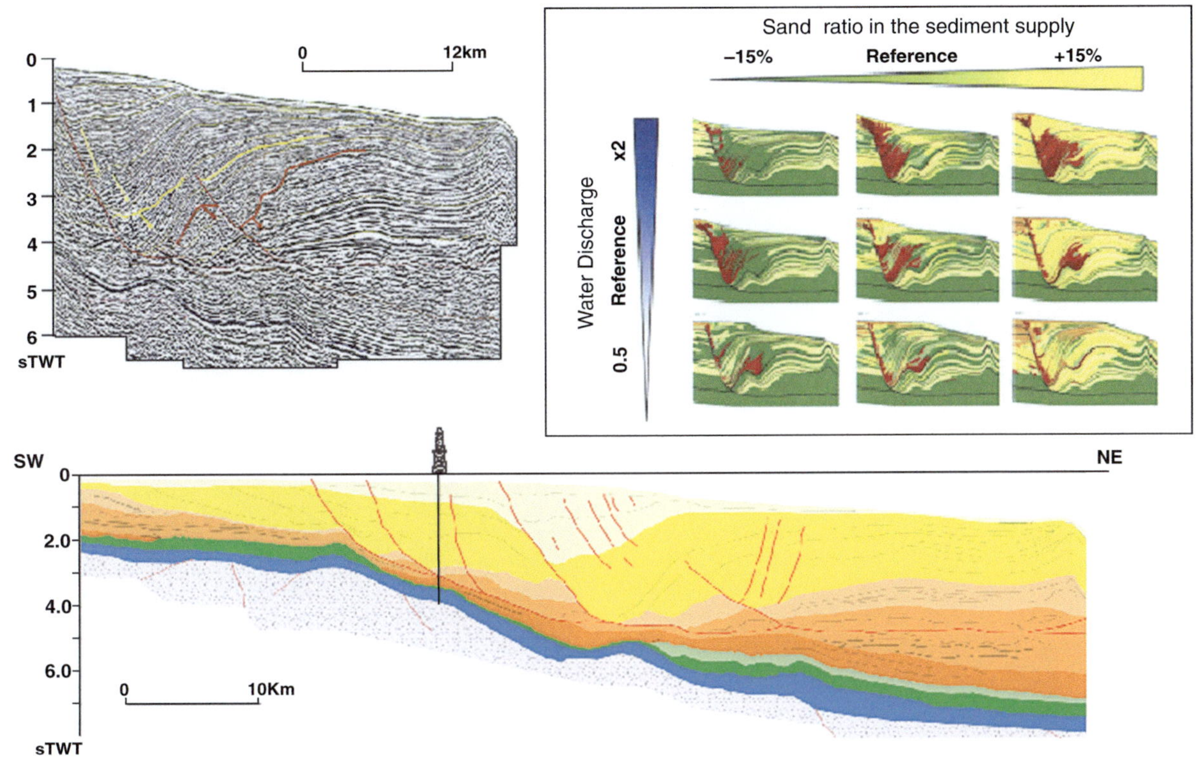

Fig. 36 Gravitational collapse of the Western margin of the Gulf of Mexico and Dionisos modelling applied for predicting the distribution of clastic reservoirs (after Alzaga-Ruiz et al., 2009; Roure et al., 2008)

be realized that surface processes, including erosion and climate, continuously degrade the topography of orogens that results from processes inherent to the deep Earth, such as mantle dynamic and flexural response of the lithosphere to tectonic accretion. Principally, surface processes control the transfer of erosion products from source to sink areas.

3D numerical models were developed to simulate the evolution of topography and drainage systems in source areas and the transport of erosion products to sink areas, thus taking a more realistic approach to mass balance than models focusing on the sink areas only. Figure 37 documents the application of these new integrated tools to the study of the Ebro Basin (Garcia-Castellanos et al., 2003). Note that this basin was fully endoreic during Oligocene to Middle Miocene times and trapped the erosion products derived from the Pyrenees in the north, the Catalan Coastal Range in the southeast and the Iberian Ranges in the southwest. Only after the Ebro River had breached the Catalan Coast Ranges by headward erosion were substantial sediment volumes transferred from the Ebro Basin to the Mediterranean Basin commencing in Late Miocene times (Gaspar-Escribano et al., 2001). The evolution of similar erosional gateways linking the Pannonian Basin to the South Carpathian foredeep, and the latter to the Black Sea, was modelled by Leever et al. (2007), outlining the steady interaction of surface and deep processes and their effect on the evolution of the Danube drainage system and the sediment transfer from the Carpathians to the Black Sea.

During the last decade new analogue models were developed and applied that simulate climate controls (rain fall) on erosion rates, the evolution of drainage systems and the resulting topography. Such models address specifically the feedback of erosion in the hinterland and sedimentation in the foreland on thrust kinematics and isostatic equilibrium of orogens (Fig. 38; Crave et al., 2000; Bonnet and Crave, 2003; Babault et al., 2005a, b; Braun, 2006).

In this new type of scaled analogue models, water is sprinkled at varying rates on top of the sand box to

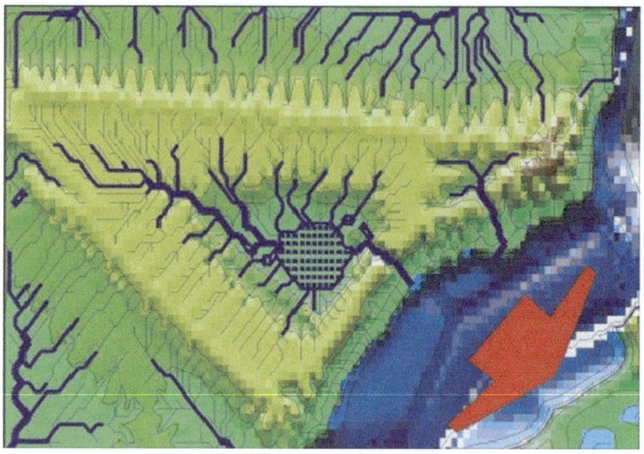

Fig. 37 3D modelling of the physiographic and topographic evolution of the Ebro Basin and its surroundings (after Garcia-Castellanos et al., 2003) See detailed description in the text

Fig. 38 Physical modelling of erosion (after Babault et al., 2005a, b, 2007) See detailed description in the text

simulate the effects of precipitation and climate changes. Various analogue materials are selected to account for both chemical and mechanical erosion.

One parameter that is difficult to simulate in such analogue model is the effect of vegetation on erosion and its long term coupling with climate. In other words, what is the impact of the biosphere on erosional rates? Do erosion rates increase under dry climates owing to the lack of vegetation, or do they increase instead during wet periods in response to increased rainfall and despite a protective vegetation cover? What is the effect of elevation-related vegetation zonation on erosion rates and on the overall shape of topographic profiles? What effects have soils and vegetation on erosional rates, particularly during periods of major climate changes, such as from glacial to interglacial, from humid to dry, and vice-versa?

Synergy between Analogue and Numerical Modelling Addressing Coupling Between Deep Earth and Surface Processes and Paleo-Topography

Feedback between Erosion and overall Structural Style of Thrust Dynamics

The interaction of erosion and tectonics is particularly evident in the evolution of fold-and-thrust belts, the near-surface structural style of which can be strongly influenced by synorogenic erosion and sediment accumulation (Beaumont et al., 1992; Avouac and Burov, 1996 Mugnier et al. 1997; Persson and Sokoutis, 2002; Babault et al., 2005a, b; Konstantinovskaia and Malavieille, 2005; Garcia-Castellanos, 2007; Mora et al., 2008). For instance, Garcia-Castellanos (2007; Fig. 38) and Mora et al. (2008) suggest that in the Andes there is a direct coupling between erosion rates, asymmetric thrust activation and the development/preservation of high plateau topography.

Newly developed methods permit to integrate analogue sandbox models of compressional deformation with numerical model for surface transport (Persson et al., 2004), further improving our understanding of the interplay between tectonics and erosion. The analogue model provides the numerical model with tectonically controlled topography whereas the numerical model calculates the evolving river network and resulting amount of erosion and sediment transport (Fig. 39). Compared to previous analogue models with symmetric (2D) erosion, this method provides a tool that incorporates process-based rates of erosion and deposition, the spatial distribution of which depends on the river network that develops on the evolving topography (Persson et al., 2004). Rivers running parallel to the evolving topography cause asymmetric erosion and deposition, thus controlling asymmetric tectonic deformation. The additional sedimentary load controls the propagation of thrust faults, as well as their vergence. High erosion rates lead not always to fewer faults within the thrust belt, but generally prolong the duration of activity along a given fault and affect its inclination.

For more refined analyses, a 4D approach is required that integrates geological, geophysical and geodetic constraints provided by high-quality modern observation systems. Important in this context is the transfer of basin modelling and 3D reflection-seismic know-how from petroleum exploration studies to the domain of environmental and Quaternary studies, which so far had hardly access to this modern geotechnology. A better understanding of active tectonic processes requires characterization of active deformation, mainly focusing on the vertical component, but also including mapping of faults, analysis of fault kinematics, and slip rates over geological and historical time scales. Data from various geoscience fields must be integrated, including data obtained by state-of-the-art monitoring systems and from geological records. Analysis of the role played by climate, erosion and tectonics on landscape evolution provides the data for quantifying the feedback between active tectonics and surface processes. Geodetic monitoring of horizontal and vertical surface motions, including GPS studies, and mapping of the subsurface, using crustal scale seismic tomography and 3D reflection-seismic data, will provide new constraints on present-day deformation patterns and related topography changes. Analogue and numerical modelling, using these new constraints as input parameters, permits validation of interpretations and provides information on dynamic processes controlling topography development in intraplate settings and adjacent orogens.

Feedback of Deep Lithosphere Processes on Vertical Movements and Topography

Novel tectonic modelling concepts and their implementation in numerical modelling software has opened new approaches to the quantification of the interplay of stresses and rheology in lithospheric deformation. Computer simulation focuses on dynamic modelling that addresses the link between basin-formation processes, basin geometries and vertical motions in space and time. Analogue modelling provides independent validation of numerical models and is particularly useful in complex settings, such as those with a pronounced 3D geometry (strike-slip systems, subduction zones). These problems are typically too complex to be treated with available numerical modelling resources. In lithospheric-scale analogue experiments, geomechanical boundary conditions and material properties can be dynamically scaled, applying the Ramberg

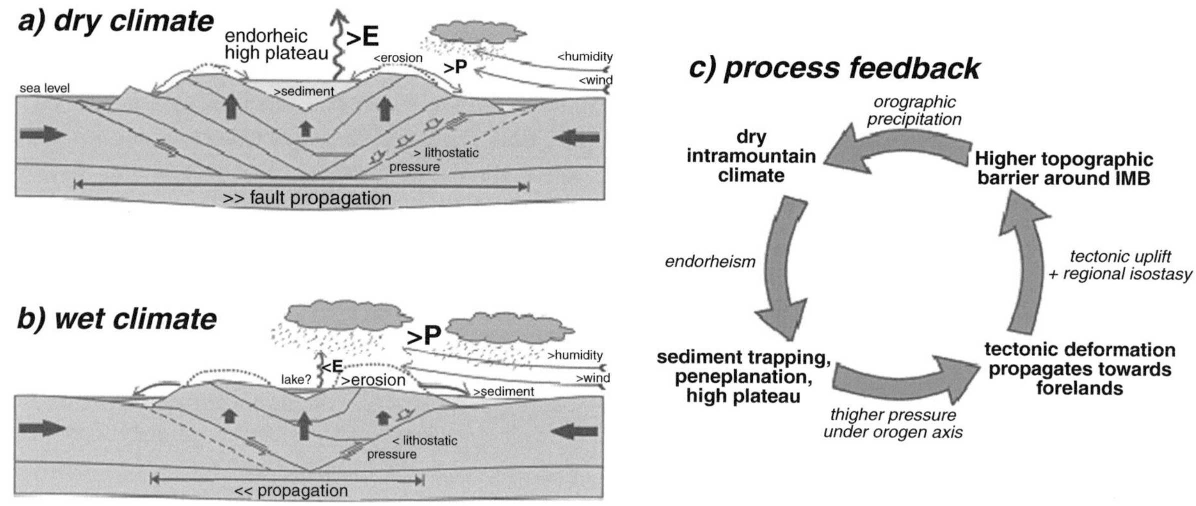

Fig. 39 Schematic diagram outlining the coupling between climate, erosion, thrust evolution and development of high Andean plateaus. **a**) Forcing forces during periods of dry climate; **b**) forcing forces during wet climate; **c**) cyclic interactions between climate and its tectonic feedback (after Garcia-Castellanos, 2007)

Number, which describes the relationship between gravitational and viscous forces (Davy and Cobbold, 1991; Burg et al., 2002).

Of particular importance has been the development of thermo-mechanical numerical modelling, addressing the physics of strain localization in the lithosphere and its consequence for poly-phase deformation and associated vertical motions (Huismans et al., 2001; Van Wijk and Cloetingh, 2002). An understanding of the role played by inherited lithospheric weakness zones during large-scale intraplate deformation, caused by collisional interaction and coupling of orogenic wedges with their foreland, has proven to be of key importance (Ziegler et al., 2002). In analogue modelling it is not yet possible to explicitly test the effects of the lithospheric thermal structure on thin- and thick-skinned deformation of the crust and its sedimentary cover. However, since the temperature structure of the lithosphere is directly related to its mechanical strength, it can be simulated in analogue models as a mechanically heterogeneous multi-layer. In analogue modelling of thermo-mechanical deformation processes, there remain, however, intrinsic limitations. As numerical modelling tools do not have these limitations, they can complement analogue models on the thermo-mechanical deformation of the crust and lithosphere. Numerical modelling has also brought insight into the degree of mechanical decoupling at depth with important implications for extensional as well as compressional tectonics.

By these means, a 4D understanding of processes controlling the interaction of the lithosphere and asthenosphere with surface dynamics is obtained. This objective is pursued through an integrated multi- and interdisciplinary approach, closely linking the development of new concepts, methodology and analytical techniques, with observation, reconstruction and modelling of Solid Earth processes. Further development of numerical 4D modelling, integrated with high-resolution analogue studies and seismic modelling, will permit to obtain a better understanding of the link between temporal and spatial scales of lithospheric processes and their expression at the Earth's surface (see Figs. 40 and 41; Garcia-Castellanos et al., 1997). The intrinsic non-linearity of these processes (Burov and Cloetingh, 1997; Kooi and Beaumont, 1996) offers a fundamental challenge requiring, by definition, the need to concentrate on localization mechanisms. Interactive 4D numerical and analogue modelling, including integration of 3D and 4D seismic data, is an avenue that will be further pursued to resolve fundamental questions on the interplay of neotectonics and surface processes in geological reconstructions.

First steps have been taken to explore in extensional and compressional settings the internal consistency of analogue and numerical modelling (Sassi and Faure,

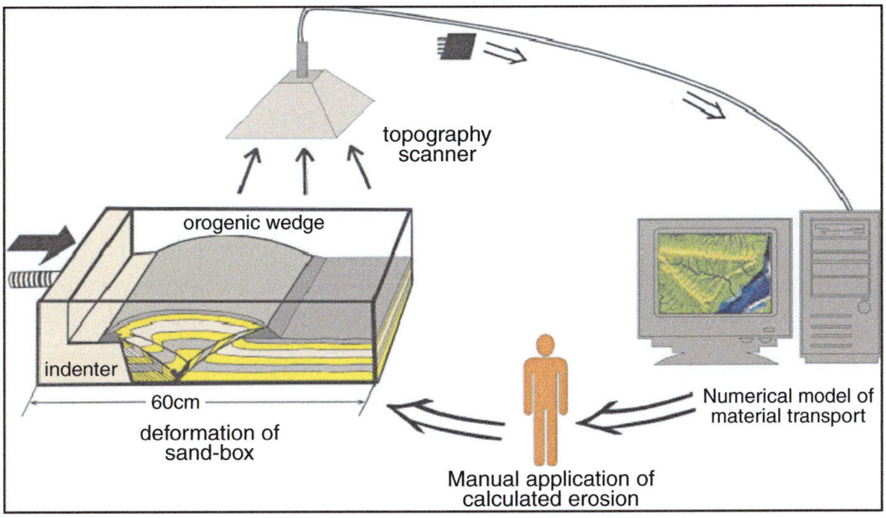

Fig. 40 Schematic diagram showing the added value of a coupled analogue-numerical modelling system such as recently applied by Persson et al. (2004). The analogue modelling facilities are used to simulate the upper crustal deformation and its dynamic response to the surface erosion and sedimentation predicted by the developed numerical models. The scanner is used to transmit the surface topography of the analogue model to the numerical model. The displayed images refer to the drainage system of the Ebro river (NE Spain) (Garcia-Castellanos et al., 2003). Subsequently, the calculated erosion/deposition is manually applied to the analogue model

1996; Ellis et al., 2004). Analogue models are very useful to evaluate in 3D the control of crustal heterogeneities on rift development and to assess the distribution and magnitude of extension in time and space (Fig. 42).

The presence or absence of a lower crustal weakness zone has a distinct effect on the width of rifts and rifted passive margins. Analogue models demonstrate the key role of lower crustal weakness zones in localizing deformation. Moreover, they indicate that the absence of lower crustal weakness zones promotes the development of wide rifted margins characterized by series of parallel trending depocentres (Figs. 42 and 43).

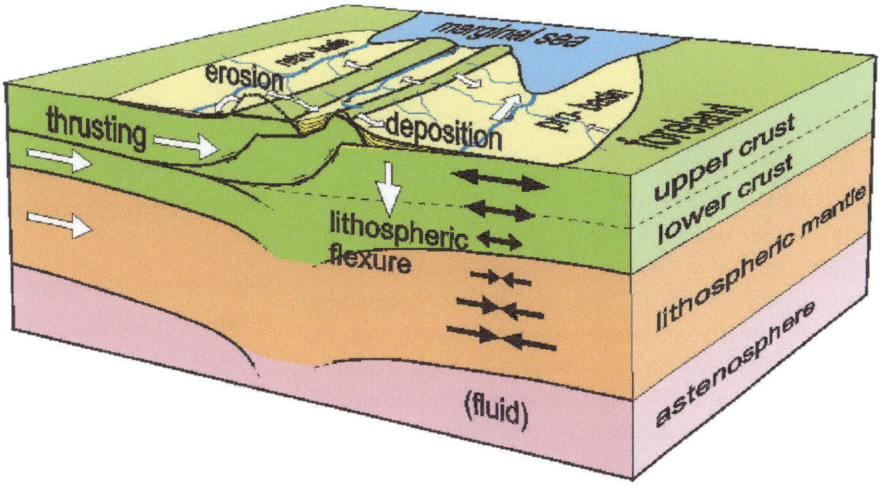

Fig. 41 Schematic diagram outlining the coupling between lithosphere rheology, foreland basin development and thrust belt evolution (Ebro Basin and Pyrenees; after Garcia-Castellanos, 2002)

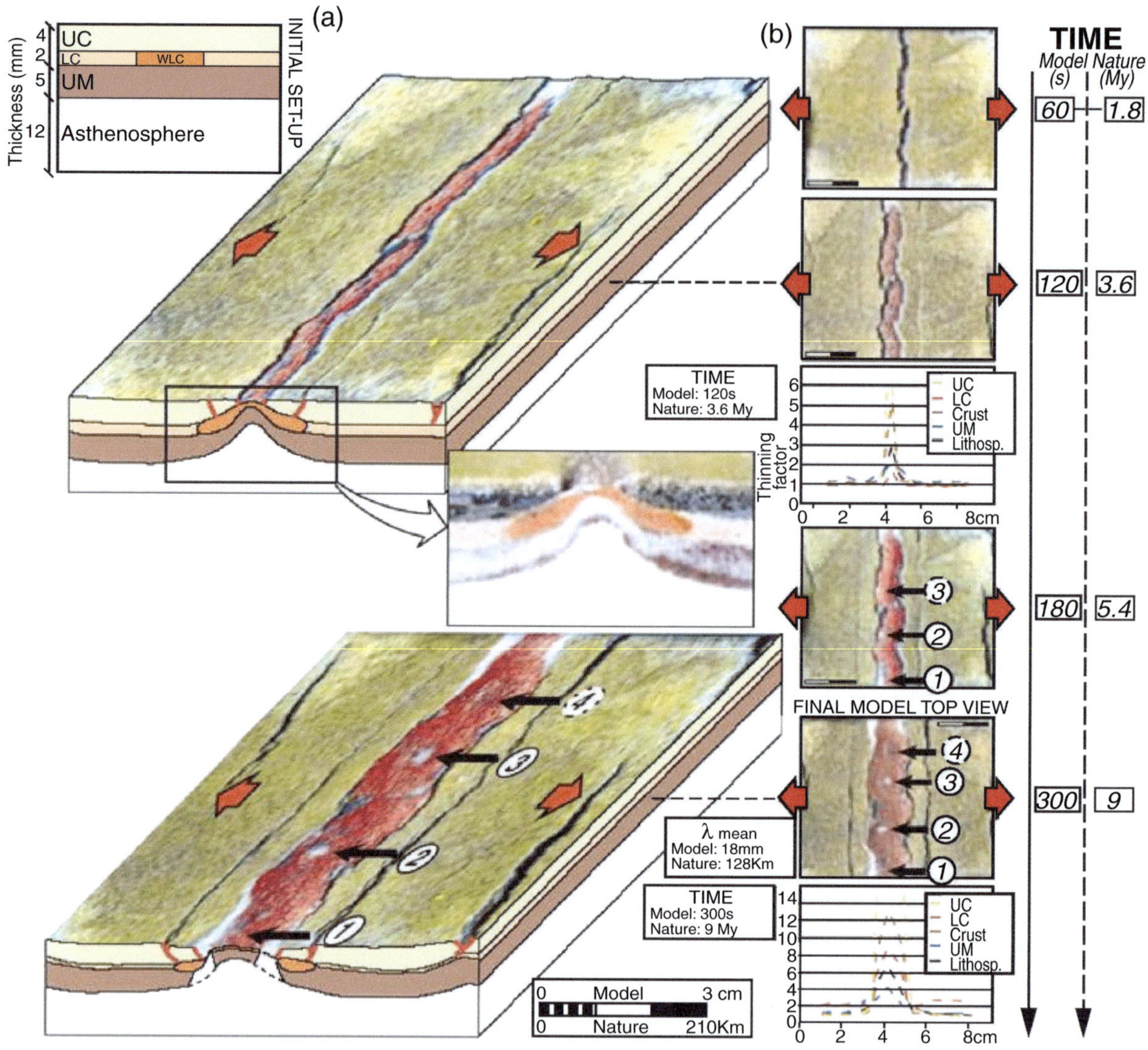

Fig. 42 Analogue tectonic model for exploring consequences of the presence of lower crustal weakness zone in extending lithosphere (from Corti et al., 2004)

These findings are of particular importance in the context of assessing "basin migration mechanisms" as documented for the Norwegian continental margin, where intermittent rifting occurred during a time span of about 280 My before continental break up finally occurred around 55 Ma and the Norwegian-Greenland Sea bean to open (Ziegler, 1988; Ziegler and Cloetingh, 2004). Results of an analogue modelling experiment on this rifted margin system are given in Fig. 44 (Sokoutis et al., 2007).

New Trends in Integrated Basin Modelling Studies and Validation

Since the mid-eighties, numerical basin models have been continuously improved, with a progressive shift from 1D to 2D and 3D models thanks to faster computers, a better understanding of physical processes and critical parameters controlling the dynamic and thermal evolution of sedimentary basins such as rifts and

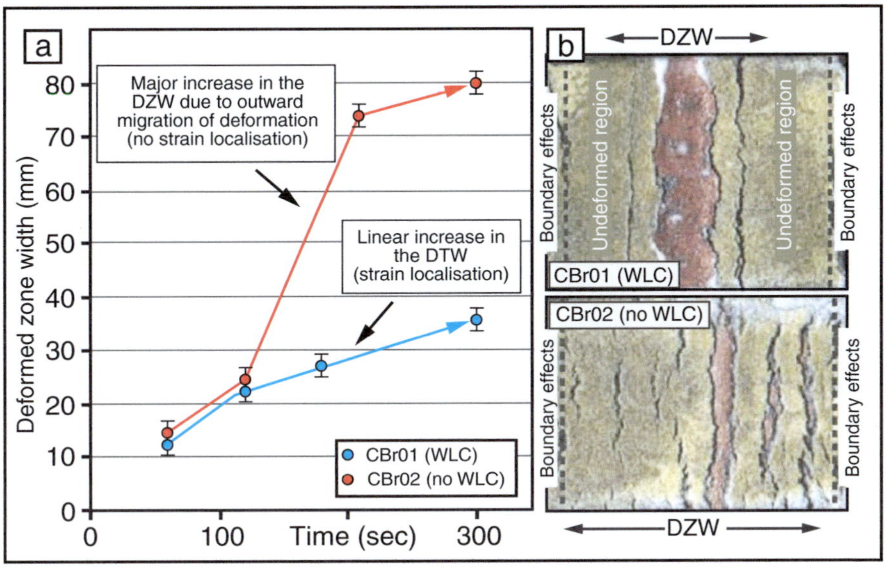

Fig. 43 Consequences of presence or absence of lower crustal weakness zones on localization of deformation in extended lithosphere. **a**) Increase of width of deformed zone during ongoing extension; **b**) Planview of extending lithosphere for two end-member models: model incorporating lower crustal weakness zone (WLC), top; model without crustal weakness zone, bottom (from Corti et al., 2003)

passive margins, foreland basins and foothills domains. Geological processes operating in sedimentary basins are too complex to be addressed by a single, multi-process numerical tool. Therefore, it is quite important to generate easy-access databases and to allow for the import and export of files from one code to the other in order to develop interactive workflows and more integrated approaches. Moreover allowance must be made for switching back and forth between basin-scale and reservoir-scale studies.

In the following, we describe such an integrated workflow, which couples analytical work and

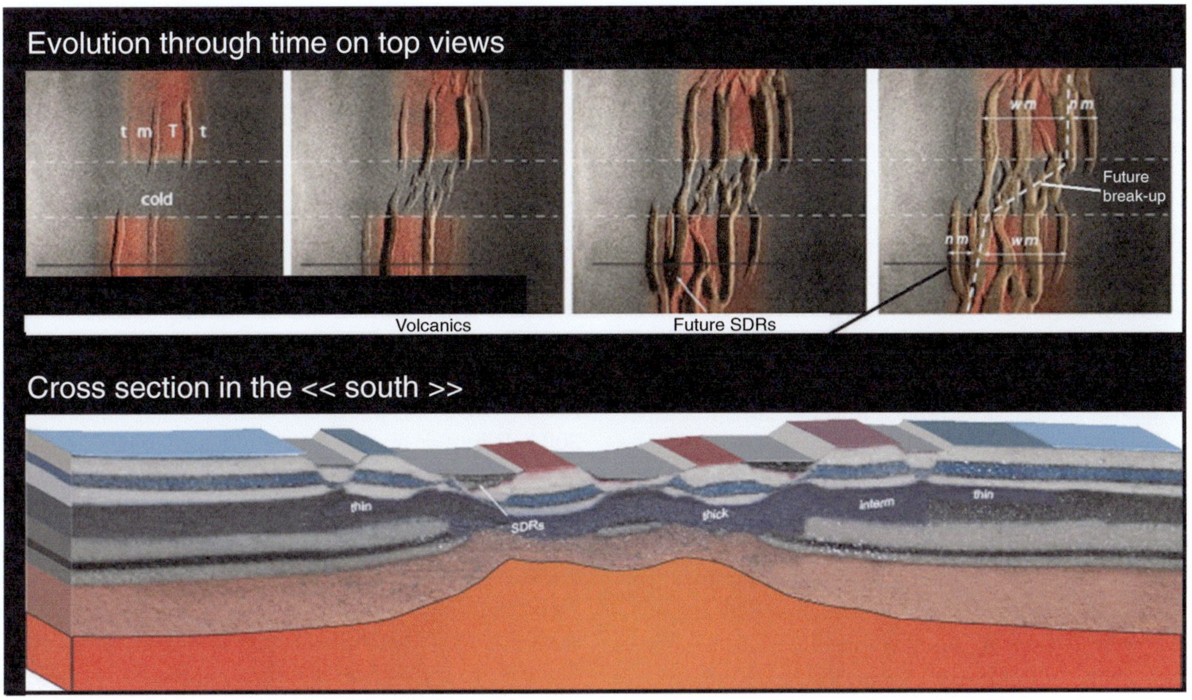

Fig. 44 Application of rifted continental margin of Mid-Norway (from Sokoutis et al., 2007)

modelling, and addresses the interactions between selected but complex geological processes operating at various temporal and spatial scales in sedimentary basins.

Dynamic Controls on Reservoir Quality in Foreland Fold-and-Thrust Belts

Integration of various datasets ranging from seismic profiles to thin-sections, analytical work and modelling is a prerequisite for the appraisal of sub-thrust sandstone reservoirs, the porosity-permeability evolution of which results from mechanical and chemical compaction, both processes interacting in response to sedimentary burial, horizontal tectonic stress and temperature.

First results of the SUBTRAP (SUB-Thrust Reservoir Appraisal) consortium studies have shown that in the Sub-Andean basins of Venezuela and Colombia the main episode of sandstone reservoirs deterioration occurred in the footwall of frontal thrusts at the time of their nucleation when the evolving thrust belt and its foreland were mechanically strongly coupled. The related build-up of horizontal tectonic stresses in the foreland induced Layer Parallel Shortening (LPS) at reservoir-scales, involving pressure-solution at detrital grain contacts, causing the in-situ mobilization of silica, rapid reservoir cementation by quartz-overgrowth and commensurate porosity and permeability reductions (Fig. 45; Roure et al., 2003, 2005). The age and duration of such quartz-cementation episodes can be roughly determined by combining microthermometric fluid inclusion studies with 1D and 2D petroleum generation modelling.

In the case of the Oligocene El Furrial sandstone of eastern Venezuela, homogenization temperatures (Th) in quartz overgrowth reflect a very narrow temperature range, averaging 110–130°C, whereas the current reservoir temperature exceeds 160°C. When plotted on burial/temperature versus time curves derived from 1D or 2D basin models calibrated against bottom hole temperatures (BHT) and the maturity rank of organic matter, it becomes obvious that cementation occurred during a short time interval, no longer than a few millions years, when the reservoir was not yet incorporated into the orogenic wedge (Roure et al., 2003, 2005).

The technique of combined microthermometry and basin modelling can also be used for dating any other diagenetic episodes, provided the reservoir was in thermal equilibrium with the overburden at the time of cementation (without advection of hot fluids). Moreover, forward diagenetic modelling at reservoir scales can benefit from such output data from basin modelling as e.g., reservoir temperature, length of the diagenetic episode and, in the case of diagenesis in an open system, fluid velocities. For the quantification of fluid-rocks interaction in the pore space of a reservoir or along open fractures transecting it, information on these parameters is indeed required. Furthermore, the composition of the fluids involved and the kinetic parameters, which control the growth or dissolution of various minerals present in the system, must be known.

Pore Fluid Pressure, Fluid Flow and Reactive Transport

When dewatering processes are slowed down by permeability barriers, which impede the vertical and lateral escape of compaction fluids, pore fluid pressures do not remain hydrostatic but can build up to geostatic levels. The build-up of excess of pore fluid pressure can impede mechanical compaction, stopping pressure-solution at quartz grain contacts, but can also cause hydraulic fracturing and failure of seals encasing a reservoir.

New basin modelling tools have been implemented for 2D simulation in tectonically complex areas of the pore fluid pressure evolution and the migration velocity water and hydrocarbons circulating in such subsurface conduits as reservoir intervals and open fractures. First tested in the Venezuelan and Canadian foothills (Schneider et al., 2002; Schneider, 2003; Faure et al., 2004, Roure et al., 2005), the CERES modelling tool (a numerical prototype for HC potential evaluation in complex areas) has now been applied in many fold-and-thrust belts around the World. It is noteworthy that the main results of fluid flow modelling in fold-and-thrust belts accounts for long episodes during which deep reservoirs behave as a closed system, whilst relatively short episodes of fast fluid expulsion are directly controlled by fold and thrust propagation (squeegee episodes). Figure 46 documents the main results of

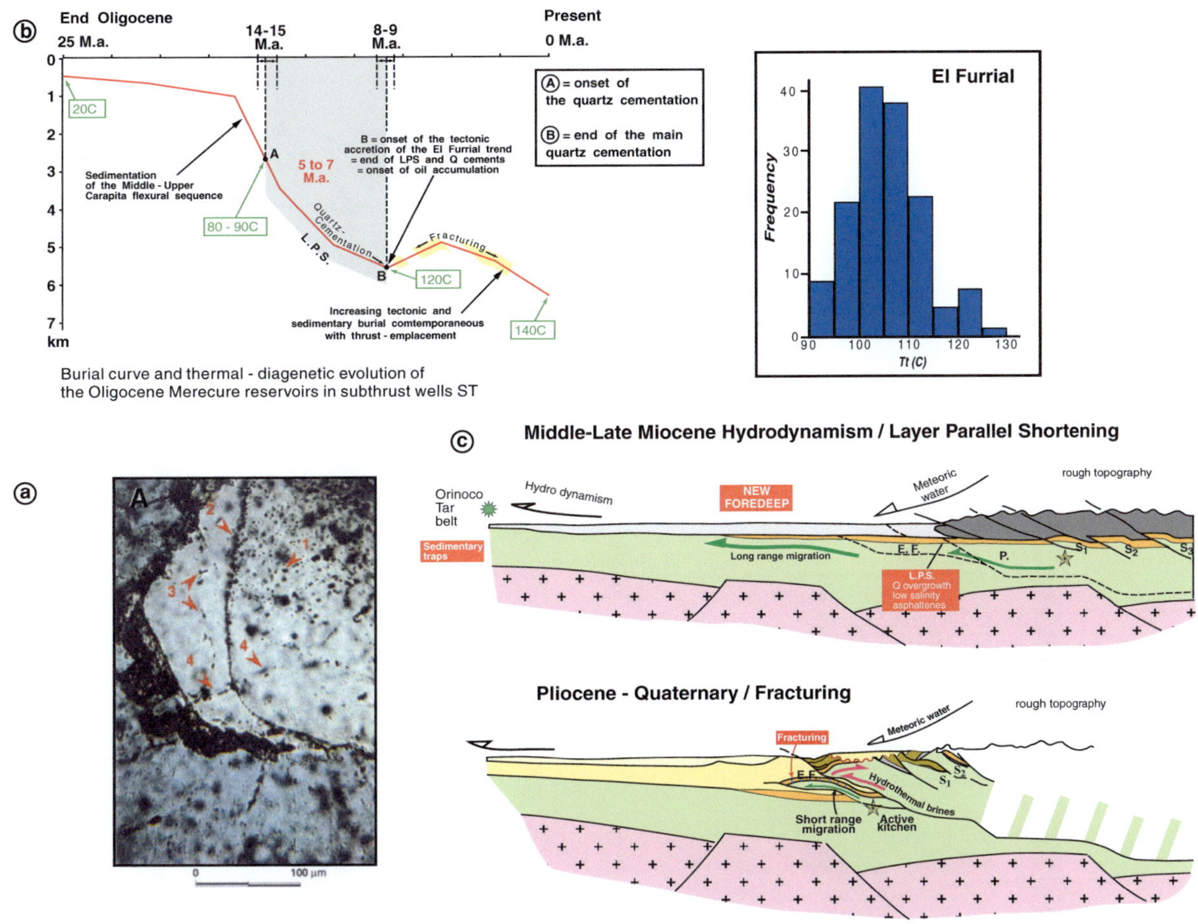

Fig. 45 Geodynamic control on quartz cementation in Sub-Andean basins (Subtrap-Venezuelan transect, after Roure et al., 2003, 2005): **a**) Thin-section evidencing various families of fluid inclusions in a detrital quartz and its diagenetic overgrowth; **b**) Diagram outlining the use of micro-thermometry (Th) and 1D thermal modelling to date the diagenetic event; **c**) Cartoon depicting the development of LPS (Layer Parallel Shortening) and quartz-cementation in the footwall of the frontal thrust

such combined kinematic and fluid flow modelling applied to a case study in the Albanian foothills (Vilasi et al., 2008).

The CERES modelling tool requires, however, modification to be able to handle the long term porosity/permeability parameters for individual faults (faults can change from non-sealing to sealing, depending on regional stresses and compaction/cementation processes), and to address these topics in 3D.

Numerical models require further improvement to properly handle reactive transport at reservoir- and basin-scales, since it probably controls the long-term porosity/permeability evolution of the main subsurface fluid circulation systems, such as porous and fractured rock units and fracture and fault systems (including hydrocarbon reservoirs).

Apart from serving the petroleum industry, new societal challenges such as CO_2 sequestration and water management also require the implementation of basin-scale reactive transport models. In such applications, basin geometries can be kept constant, whilst the time resolution required is much smaller (months or years instead of millions of years). Promising results have already been obtained in the simulation of thermo-haline circulations in the Northeast German Basin, thus accounting for the advection of saline water derived from Permian salt layers up to the surface (Fig. 47; Magri et al., 2005a, b, 2007; Magri et al., 2008).

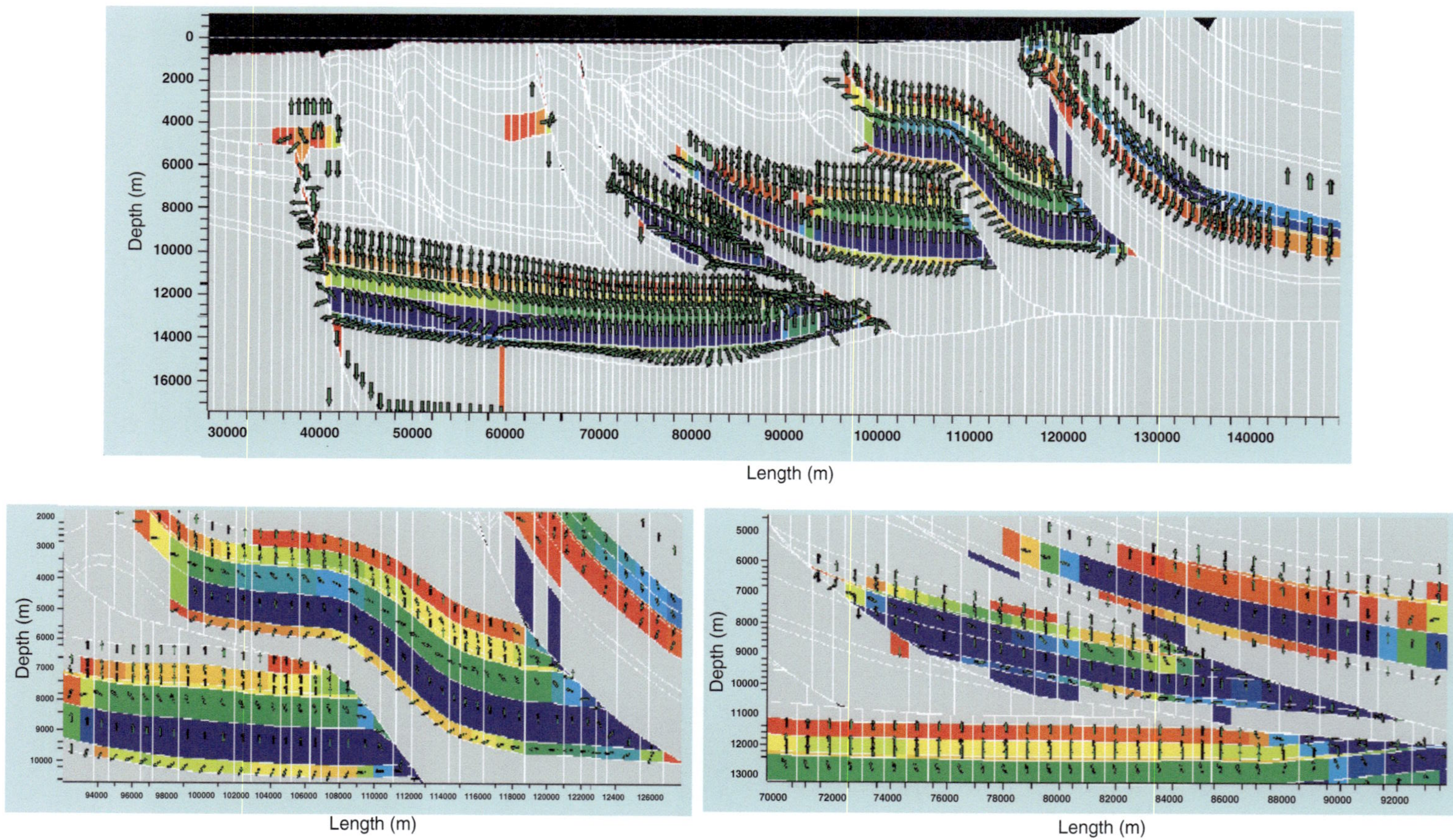

Fig. 46 Ceres fluid flow and pore fluid pressure modelling in the Albanian foreland fold-and-thrust belt (after Vilasi et al., 2009)

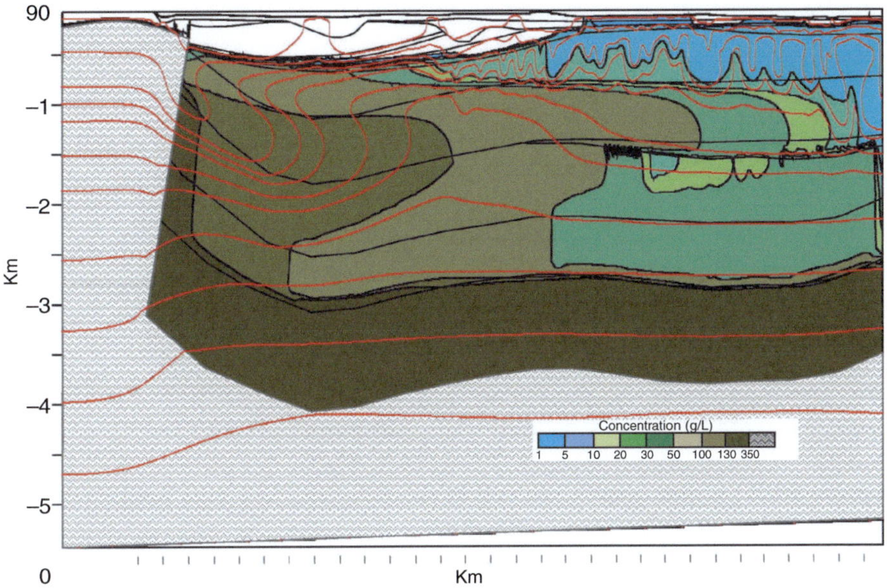

Fig. 47 Brine concentration (filled pattern, g/l) and temperature profiles (dashed lines, °C) calculated from a transient thermo-haline simulation based on a profile of the Schleswig-Holstein region (North German Basin; after Magri et al., 2005a, b, 2007, 2008)

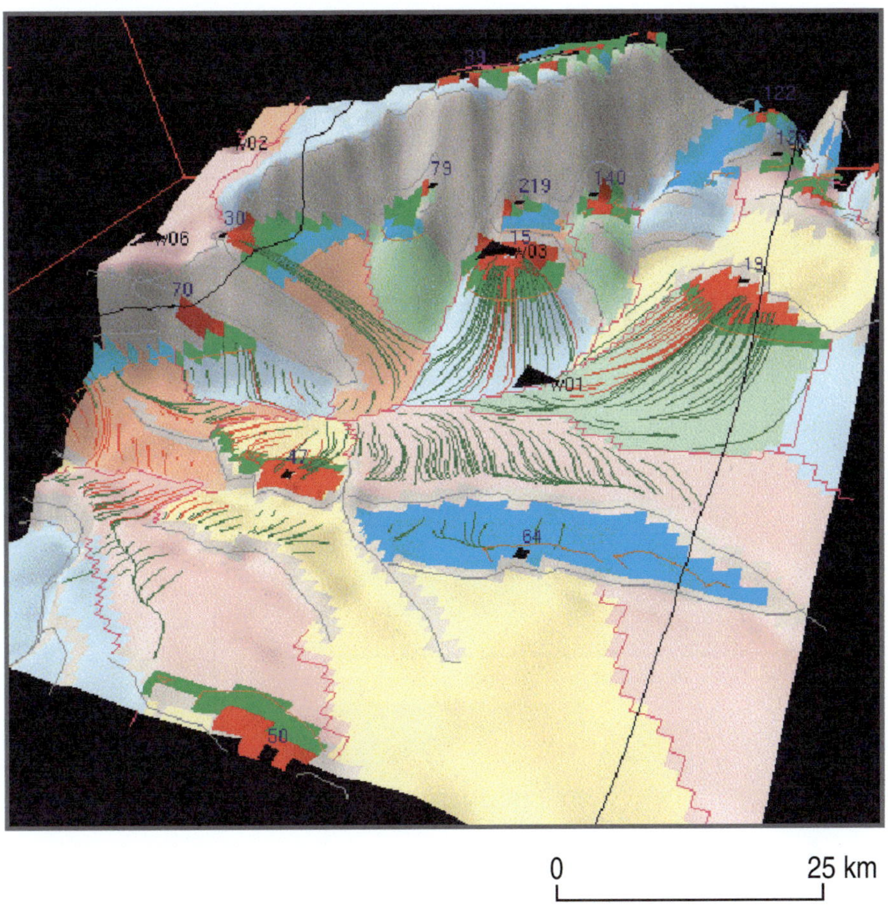

Fig. 48 Temis 3D modelling of drainage areas and fully quantitative prediction of HC trapping (after Rudkiewicz and Carpentier, 2005). Blue pattern outlines dry prospects, whereas gas (vapor phase) and oil (liquid phase) accumulations are shown in *reg and green*, respectively. Coeval migration path for gas and oil between the active kitchens (structural lows) and trapps (structural highs) are indicated with red and green lines, respectively

3D Kinematic Evolution of Complex Structures

A fully quantitative prediction of the hydrocarbon charge to a given structural or stratigraphic prospect requires 3D modelling in order to properly take into account lateral and vertical heterogeneities of the source rocks and their maturity, the drainage areas and migration conduits, and the interconnection between the various fault systems and reservoirs. One of the main limitation of current tools, however, is the over simplistic assumptions made by most models for the architecture of faults, which can hardly be handled differently than as vertical boundaries (Fig. 48). Thus, only vertical motion (subsidence and compaction) is taken into account during modelling, with the border lengths and surface areas of the models being kept constant through times, no matter whether lateral extension or contraction occurred or not.

Therefore, a major effort is currently being made to develop new tools, which are able to reconstruct the kinematics of real faults in 3D (low angle thrust faults and high-angle normal or strike-slip faults; Fig. 49; Moretti et al., 2006). This is a prerequisite for combined 3D thermal and fluid flow modelling of tectonically complex areas (Fig. 50; Baur and Fuchs, 2008).

Geomechanics, Fracturing and Reservoir Prediction

Pressure-solution related cementation and fracturing are important processes that can have repercussions of the porosity/permeability evolution of carbonate and

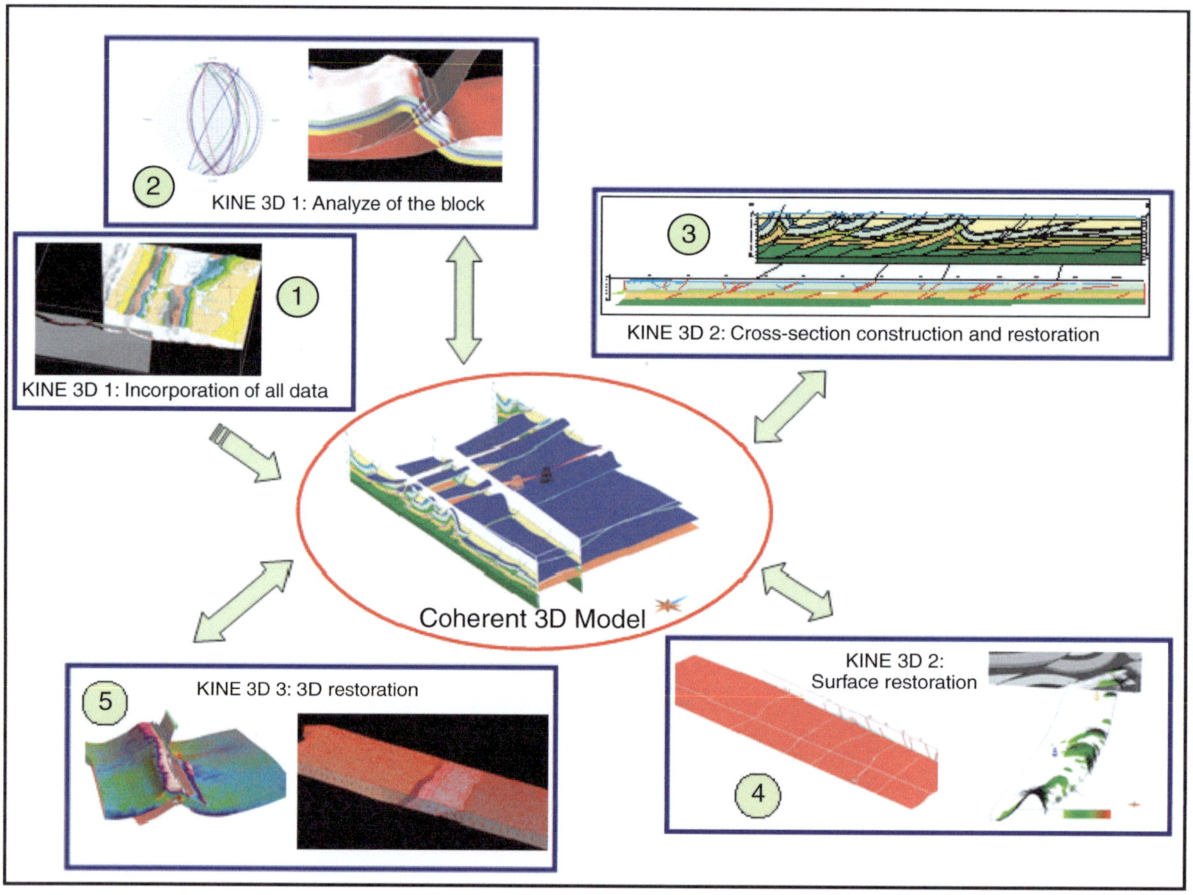

Fig. 49 Kine 3D. The workflow for 3D kinematic modelling of complex structures requires the integration of 2 and 3D seismic data, geological maps and sections when constructing the present-day architecture of the model (1), to extrapolate the fault planes from one section to the other (2), and then to proceed to the restoration of the sections (3), maps (4) or full 3D restoration (5) (after Moretti et al., 2006)

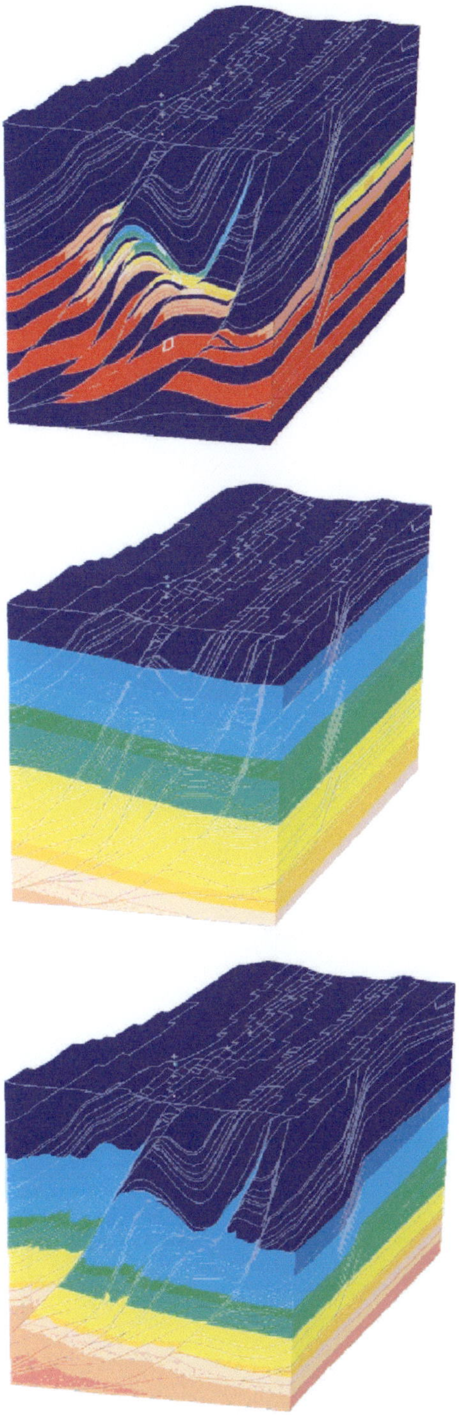

Ⓐ Ⓑ Ⓒ

Fig. 50 3D distribution of source rock maturity resulting from coupling complex kinematics with thermal modelling. Notice that current models cannot yet handle fluid flow and HC migration in complex tectonic environments. **a**) Transformation ratios (red: gas window; blue: immature); **b**) Temperatures and Ro -vitrinite reflectance- computed for present-day architecture (**c**) (petromod; after Baur and Fuchs, 2008; Bauer et al., 2009)

sandstone reservoirs, but also on the overall the long-term evolution of fluid flow and the pore-fluid pressure regime of sedimentary basins.

Once purely geometric, basin models must be progressively modified to account for more realistic physics and rock mechanics in order to better control changes induced by such processes as Layer Parallel Shortening (LPS) and stress-related opening and closure of fractures. In this respect, it is important to assess the structural fabric of a given horizon as pre-existing fractures are likely to play an important role in the pattern of fractures opening during successive tectonic episode. Nearby outcrop analogues can be used to calibrate basin-scale flow models in both frontier and mature basins, in order to properly describe the 3D architecture of sub-seismic fracture systems and complement the fragmentary information provided by cores and FMI (formation micro imager) logs (Fig. 51).

Average reservoir porosity values and directional permeability anisotropies derived from production data are currently applied in field-sized reservoir models. This information could be extrapolated to fine-tune basin-scale models.

Aspects of Future Basin Study

The feedback between methodology development and multi-scale observations is the key to validate models for tectonic controls on intraplate continental topography. In order to separate the contribution of surface and tectonic processes to the development of modern landscapes, high resolution dating of Quaternary strata must be combined with process-oriented modelling, linking the Quaternary record to long-term deep Earth processes. Some pertinent developments are in the forefront of this research domain, which is the focus of the TOPO-EUROPE Project (Cloetingh et al., 2007), one of the new challenges that has been endorsed by ILP and the European Union. Other projects addressing the evolution of continental topography, adopting similar approaches and workflows, though focusing less on its Quaternary and recent development and related societal implications, are the TOPO-ASIA Project (Himalayas and Asian intra-cratonic basins), the TOPOAFRICA Project (Guillocheau et al., 2006, 2007a, b; Braun et al., 2007) and the German SAMPLE

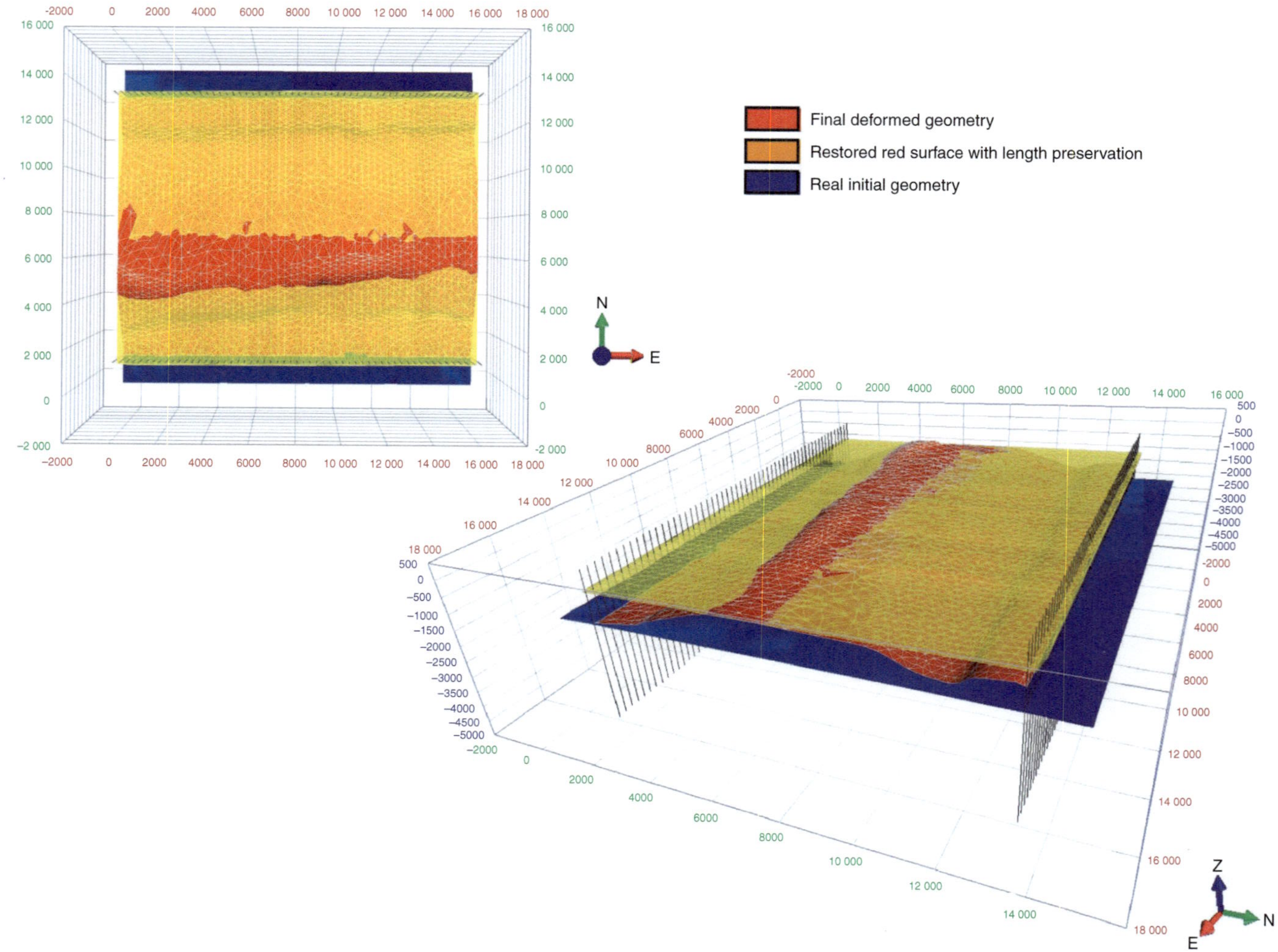

Fig. 51 Coupling of sand box experiment (basin inversion) and numerical modelling (unfolding of surfaces) (after Mattioli et al., 2007; Saeed et al., 2008). Notice that the restored surface is quite smaller (about 20% less) than the initial surface of the model. Part of this discrepancy relates to 3D strain, i.e., thickening of the sand layers (there is only little compaction operating in the sand box experiment). In natural cases, pressure-solution would also account for a change in the rock volume, the study of which requiring further investigations by means of mechanical modelling

Project (conjugate South Atlantic Margin development; Bünge et al., 2008) and ANDES Project (Oncken et al., 2006). To a large extent these integrated projects apply the analytical and modelling tools summarized in the previous paragraphs.

Although the Solid Earth has continuously changed, the record of its evolution is stored in sedimentary basins and the lithosphere. The aim of the ILP Task Force on Sedimentary Basins is to facilitate networking between the various communities (i.e., geologists and geophysicists, academy and industry) involved in the study of sedimentary basins, and to secure a wide diffusion of integrated workflows and new modelling concepts worldwide. A major challenge is to elucidate the role played by internal lithospheric processes and external forcing as controlling factors of erosion and sedimentation rates. The sedimentary cover of the lithosphere provides a high-resolution record of changing environments, and of deformation and mass transfer at the Earth surface, as well as at different depth levels in the lithosphere and sub-lithospheric mantle. Important contributions were made to explain the relationships between lithosphere-scale tectonic processes and the sedimentary record, demonstrating, for example, the intrinsic control exerted by lithospheric intraplate stress fields on stratigraphic sequences and on the record of relative sea-level change in sedimentary basins (Cloetingh et al., 1990; Guillocheau et al., 2000; De Bruijne and Andriessen, 2002; Hendriks and Andriessen, 2002; Robin et al., 2003). By now, there is a growing awareness that neotectonic processes can seriously affect the fluid flow in sedimentary basins and that fluid flow can have a major effect on the geothermal regime, and hence on calculated denudation and erosion quantities (Rowan et al., 2002; Goncalves et al., 2003; Schneider et al., 2002; Schneider, 2003; Ter Voorde et al., 2004; Vilasi et al., 2008). Monitoring of the sedimentary and deformation record provides constraints for present-day deformation rates.

Whereas in the analysis of sedimentary basins, tectonics, eustasy and sediment supply are usually treated as separate factors, an integrated approach is required that is constrained by fully 3-D quantitative subsidence and uplift history analyses. Recent work has also elucidated the control exerted by inherited mechanical weakness zones in the lithosphere on its subsequent evolution, as expressed by the geological and geophysical record of orogenic belts and sedimentary basins in intraplate domains and the related development of topography. The mechanical properties of the lithosphere depend on its temperature regime and composition (Ranally and Murphy, 1987; Ranalli, 1995; Cloetingh et al., 2003a, b; Andriessen and Garcia-Castellanos, 2004; Cloetingh et al., 2004; Cloetingh and Van Wees, 2005). Therefore, it is necessary to fully integrate geothermochronology and material property analyses in reconstructions of the evolution of the lithosphere as derived from the record of sedimentary basins. In doing so, traditional boundaries between endogen and exogen geology will be trespassed.

The sedimentary basin community, and Earth Sciences as a whole, face new societal challenges owing to on-going climate changes and the needs for CO_2 sequestration. Therefore, basin models must be adapted to new time scales, changing from the long-term resolution required for hydrocarbon resource evaluation (millions of years) towards much shorter time intervals (from less than ten to hundreds of years). In basin and reservoir models geomechanics, reactive transport and fluid-rock interactions must be taken into account to cope with accelerated subsidence and hydro-fracturing induced by hydrocarbon and water production, water injection, as well as with rapid changes in reservoir porosities and permeabilities induced either by dissolution or by pore and fracture cementation related to CO_2 injection. In this context, the 4-D geophysical survey technology can be applied for reservoir monitoring.

Sedimentary geologists and basin modellers are currently building new bridges to the Deep Earth community. The various lithospheric and sub-lithospheric mantle processes, which control the evolution of sedimentary basins, will be implemented in the numerical codes currently used by the petroleum industry. This will be of importance for investigating the heat flow and thermal evolution of rifted basins and passive margins, as well as the history of vertical movements of the Earth's surface in foreland basins and adjacent fold belts. Currently, modelling of global processes and deformation prediction of sedimentary strata, including reservoir rocks, is going through the important transition from kinematic to thermo-mechanic and dynamic modelling.

These developments cannot take place without interaction with sub-disciplines that address the Earth's structure and kinematics and the reconstructions of geological processes. In fact, the advances in structure-related research, in particular the advent of 3-D seismic velocity models, have set the stage for studies on dynamic processes within the Earth. In short, structural information is a prerequisite for modelling both sedimentary basins and Solid-Earth processes. Similarly, information on present-day horizontal and vertical motions, as well as reconstructions of past motions, temperatures or other process characteristics, is used to formulate and test hypotheses concerning dynamic processes. Inversely, the results of process modelling motivate and guide observational research. Through the emphasis on process dynamics, the full benefits of coupling at spatial and temporal scales are expected to become apparent. The scale of processes studied ranges from the planetary scale to the small scale relevant to sedimentary processes, the depth scale being reduced accordingly.

Despite the great success of plate tectonic concepts, there are still fundamental questions on the evolution of the continental lithosphere and its interaction with the sub-lithospheric mantle. At the scale of a differentiating planet, processes controlling the growth of continental lithosphere, its thickness and dynamic coupling with the underlying mantle require focused attention from a number of Earth science subdisciplines (see Artemieva, 2006). Equally important questions remain on mechanisms controlling deformation of the continents and their effects on vertical motions, dynamic topography, and the evolution and destruction of sedimentary basins. Of particular importance are the dynamics of rifting culminating in splitting of continents and the opening of oceanic basins, as well as of subduction of oceanic basins, the development of orogens (mountain building) and continent-continent collision, including their effects on continental platforms. For the quantification of Solid-Earth processes the coupling of internal and external forcing has to be addressed. Starting from large-scale mantle and lithospheric structure and processes, and going to increasingly finer scales of crustal structure, processes must be analyzed to understand the dynamics of sedimentary basins and their fill and the development of topography.

Primary and most innovative objectives of integrated sedimentary basin studies are to link lithosphere-to-surface processes and to promote 4-D approaches that will lead to integrated interpretations of existing and newly acquired geomorphologic, geologic, geophysical, geodetic, remote sensing and geotechnologic datasets. A major challenge is the incorporation of different temporal and spatial scales in the analyses of sedimentary basins, Solid-Earth and surface processes. Assessment of the roles played by climate, erosion and tectonics on landscape and basin evolution will provide key constraints for quantifying feedback mechanisms linking deep Earth and surface processes. Monitoring horizontal and vertical surface motions and mapping the subsurface, using modern geophysical, geodetic, remote sensing and geotechnical techniques, can constrain present-day deformation patterns and related topographic changes, and can provide new guidelines for investigating the past. Analogue and numerical modelling, based on such constraints, can be used to test integrated interpretations and to provide information on dynamic processes controlling subsidence and topography development in intraplate domains, such as forelands of orogens and passive margins.

The bathymetric evolution of passive margins, as well as the surface topography and morphology of continents strongly depend on the interplay of subsurface and surface processes. Erosion of growing topography has an unloading effect on the lithosphere whereas sediment accumulation has a loading effect. This is clearly demonstrated by the strong correlation between denudation and tectonic uplift rates in zones of active deformation. During collision, surface processes contribute towards the localization and growth of mountain belts and fault zones, and ensure stable growth of topography (see also Burov, this volume). During crustal extension, erosion contributes towards widening of rifted basins, so that apparent extension coefficients can increase by a factor of 1.5–2 (Fig. 52; Burov and Poliakov, 2001). Poly-phase subsidence and other deviations from time-depending asymptotic thermal subsidence can be also controlled by the feedback between surface and subsurface processes.

The topographic reaction to surface loading and unloading depends on the mechanical strength of the lithosphere as well as on the strength partitioning between the crust and lithospheric mantle. Consequently, testing different rheological profiles in areas where data on denudation and/or sedimentation rates are well constrained may provide opportunities for constraining the long-term rheology of the lithosphere (e.g., Burov and Watts, 2006).

Reliable information on (de)coupling processes at the crust-mantle and lithosphere-asthenosphere

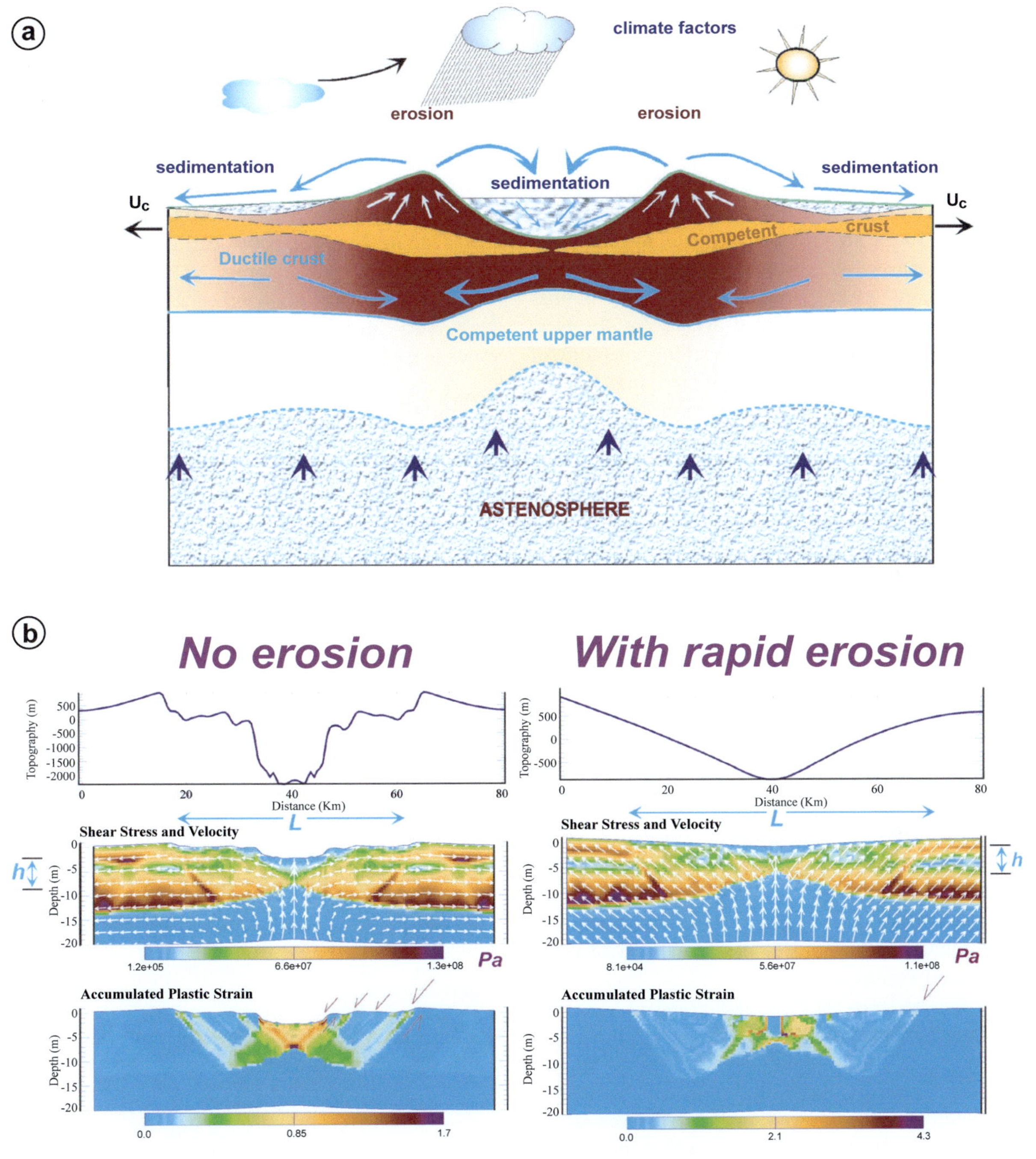

Fig. 52 (**a**) Syn- and post-rift feedback conceptual model (Burov and Cloetingh, 1997 E. Burov and S. Cloetingh, Erosion and rift dynamics: new thermo-mechanical aspects of post-rift evolution of extensional basins, Earth Planet. Sci. Lett. 150 (1997), pp. 7–26. Abstract | PDF (2335 K) | View Record in Scopus | Cited By in Scopus (58) Burov and Cloetingh, 1997); (**b**) Numerical model (Burov and Poliakov, 2001) of rift evolution with and without active surface erosion, for the same boundary and initial conditions. Erosion results in much stronger crustal thinning and a wider basin than in the case without erosion

boundaries and at the two principal phase transitions within the upper mantle (at about 410 and 660 km depth) will be of fundamental importance for modelling surface topography. Quantification of dynamic depth-to-surface relationships is a major challenge, requiring innovative approaches to 4-D modelling. The principles of available conventional fluid-dynamic modelling are robust, but require greatly increased computer power to provide adequate resolution of a mantle convection system characterized by thermal boundary layers, subducted slabs and plumes of complex structure that may evolve rapidly. New approaches must incorporate the yielding rheology of crustal and mantle materials, integrated modelling of material flow and elastic deformation (also crucial for predicting realistic topography evolution), crustal and lithospheric weakness zones and/or faults. Available large-scale mantle dynamic models may actually require modification to take instead of flow approximations elastic and plastic deformation into account when attempting to solve full stress equations with free upper surface boundary conditions, at least for the lithospheric mantle (see also Burov, this volume; Burov and Cloetingh, 2009). Mantle models need to be constrained by mantle tomography, geodetic and electromagnetic data.

A new generation of 3-D and 4-D tectonically realistic models is required for an understanding of the dynamic feedback between tectonic and surface processes, providing new insights into the evolution of tectonically active systems and related surface topography:

- Morphologically and tectonically consistent collision and exhumation models;
- Basin modelling, synthetic stratigraphy;
- Climate-coupled modelling

These future geo-modelling tools will be able to consistently treat homogeneous and heterogeneous deformation with realistic faults, so that the magnitude of uplift, subsidence, fluid flow and other types of deformation (derived from 4-D-seismic monitoring of geological markers or GPS, stress in boreholes and earthquakes) can be linked and interpreted quantitatively. The goals of 4-D modelling will be to predict and quantify (1) the overall mass transfer in sedimentary basins, including apart from sediments such fluids as water, hydrocarbons and CO_2, which circulate in porous and permeable media (reservoir horizons) at various spatial and temporal scales, and (2) the global dynamic evolution of Solid-Earth boundaries and phase transition zones, which control surface deformation. 4-D modelling will permit to define the present state of surface deformation, including its space-time gradient (a prerequisite for geoprediction) and to assess a wide range of potential geological hazards, ranging from landsides and coastal subsidence and flooding up to CO_2 storage, as well as to inventorize water and hydrocarbon resources. To achieve these goals, very high-resolution at temporal and spatial scales (e.g., 50–100 years, 5–10 km) will be required.

Acknowledgments The Task Force on Sedimentary Basins thanks ILP for its initiative and support. Jörg Negendank and Roy Gabrielsen provided helpfull reviews of the manuscript. Special thanks also to Patrick Le Foll for his graphics, and, the many colleagues who provided dedicated high resolution versions of their figures.

References

Achauer, U. and Masson, F., 2002. Seismic tomography of continental rifts revisited: from relative to absolute heterogeneities. Tectonophysics, 358, 17–37.

Aichroth, B., Prodehl, C. and Thybo, H., 1992. Crustal structure along the central segment of the EGT from seismic refraction studies. Tectonophysics, 207, 43–64.

Albouy, E., Casero, P., Eschard, R. and Rudckiewicz, J.J., 2002. Tectonics and sedimentation in the Central Apennines. Soc. Geol. It., 81° Summer Meeting, 10–12 September, 2002, Abs., 7–8.

Alzaga-Ruiz, H., Granjeon, D., Lopez, M., Séranne, M. and Roure, F., 2009. Gravitational collapse and Neogene sediment transfer across the western margin of the Gulf of Mexico: Insights from numerical models. Tectonophysics, 470, 21–41 doi: 10.1016/j.tecto.2008.06.017.

Andriessen, P. and Garcia Castellanos, D., 2004. 3D Motions of the Earth surface: from measurements to physical modelling. Phys. Chem. Earth, 29, 663–717.

Andriessen, P.A.M., Helmens, K.F., Hooghiemstra, H., Riezelos, P.A. and van der Hammen, T., 1993. Absolute chronology of the Pliocene-Quaternary sediment sequence of the Bogota area, Colombia. Quat. Sci. Rev., 12, 483–501.

Anka, Z., Séranne, M., Lopez, M., Scheck-Wenderoth, M. and Savoye, B., 2009. The long-term evolution of the Congo deep-sea fan: a basin-wide view of the interaction between a giant submarine fan and a mature passive margin (ZaiAngo project). Tectonophysics, 470, 42–56.

Ansorge, J., Blundell, D. and Mueller, St., 1992. Europe's lithosphere – seismic structure. In Blundell, D., Freeman, R. and Mueller, St., eds., A continent revealed, The European Geotraverse. Cambridge University Press, 33–69.

Artemieva, I.M., 2006. Global 1°x1° thermal model TC1 for the continental lithosphere: implications for lithosphere growth since Archean. Tectonophysics, 416, 245–277.

Artemieva, I.M. and Mooney, W.D., 2001. Thermal thickness and evolution of Precambrian lithosphere. A global study. J. Geophys. Res., 106, 16387–16414.

Artemieva, I.M., Thybo, H. and Kaban, M.K., 2006. Deep Europe today: Geophysical synthesis of upper mantle structure and lithospheric processes over 3.5 Ga. In Gee, D. and Stephenson, R.A., eds., European lithosphere dynamics. Geol. Soc., London, Mem., 32, 11–42.

Avouac, J.P. and Burov, E.B., 1996. Erosion as a driving mechanism of intracratonic mountain growth. J. Geophys. Res., 101, 17747–17769.

Babault, J., Bonnet, S., Crave, A. and van den Driessche, J., 2005a. Influence of piedmont sedimentation on erosion dynamics of an uplifting landscape: an experimental approach. Geology, 301–304. doi:10.1130/G21095.1.

Babault, J., Bonnet, S., van den Driessche, J. and Crava, A., 2007. High elevation of low relief surfacs in mountain belts: Does it equate to post-orogenic surface uplift? . Terra Nova, 19, 4, 272–277. doi:10.111/j.1365-3121.2007.00746.x.

Babault, J., van den Driessche, J., Bonnet, S., Castelltort, S. and Crave, A., 2005b. Origin of the highly elevated Pyrenean peneplain. Tectonics, 24, TC3004. doi:10.1029/2004TC001697.

BABEL Working Group, 1993. Deep seismic reflection / refraction interpretation of crustal structure along BABEL profiles A and B in the southern Baltic Sea. Geophys. J. Int., 112, 325–343.

Babuska, V. and Plomerova, J., 1992. The lithosphere in central Europe – seismological and petrological aspects. Tectonophysics, 207, 141–163.

Babuska, V. and Plomerova, J., 1993. Lithospheric thickness and velocity anisotropy – seismological and geothermal aspects. Tectonophysics, 225, 79–89.

Babuska, V. and Plomerova, J., 2001. Subcrustal lithosphere around the Saxothuringian-Moldanubian Suture Zone – a model derived from anisotropy of seismic wave velocities. Tectonophysics, 332, 185–199.

Baig, A.M. and Dahlen, F.A., 2004. Travel time biases in random media and the S-wave discrepancy. Geophys. J. Int., 158, 922–938.

Bally, A.W., 1989. Phanerozoic basins of North America. In: Bally, A.W. and Palmer, A.R. (eds.), The Geology of North America: An Overview. Geol. Soc. of Am., The Geology of North America. Vol. A, 397–446.

Barrier, L., Albouy, E., Guri, S., Rudkiewicz, J.L., Bonjakes, S., Muska, K. and Eschard, R., 2005. Coupled structural and sedimentary mass balances in the Central Albanides. Soc. Géol. France, Soc. Geol. España, ILP Conférence, Thrustbelts and foreland basins, Paris, December 2005, Abs., 47–49.

Bauer, K., Neben, S., Schreckenberger, B., Emmermann, R., Hinz, K., Fechner, N., Gohl, K., Schulze, A., Trumbull, R.B. and Weber, K., 2000. Deep structure of the Namibia continental margin as derived from integrated geophysical studies. J. Geophys. Res. B Solid Earth Planets, 105, 25829–25853.

Baur, F., Di Benedetto, M., Fuchs, T., Lampe, C. and Sciamanna, S., 2009. Integrating structural geology and petroleum systems modeling – A pilot project from Bolivian's fold and thrust belt. Mar. and Petrol. Geol., 26, 4, 573–579.

Baur, F. and Fuchs, T., 2008. 3D TecLink: a new approach in petroleum systems analysis to link basin- and structural modelling. Am. Assoc. Petrol. Geol., European Region News Letter, March 2008.

Bayer, U., Scheck, M., Rabbel, W., Krawczyk, C.M., Götze, H.-J., Stiller, M., Beilecke, Th., Marotta, A.-M., Barrio-Alvares, L. and Kuder, J., 1999. An integrated study of the NE German Basin. Tectonophysics, 314, 285–307.

Beaumont, C., Ellis, S. and Pfiffner, A., 1999. Dynamics of sediment subduction-accretion at convergent margins: Short-term modes, long-term deformation, and tectonic implications. J. Geophys. Res., 104, 17573–17601.

Beaumont, C., Fullsack, P. and Hamilton, J., 1992. Erosional control of active compressional orogens. In McClay, K.R., ed., Thrust tectonics, Chapman and Hall, London, 1–18.

Beaumont, C., Munoz, J.A., Hamilton, J. and Fullsack, P., 2000. Factors controlling the Alpine evolution of the central Pyrenees inferred from a comparison of observations and geodynamical models. J. Geophys. Res., 105, 8 121–8 145.

Becq-Giraudon, J.F., Montenat, C. and van den Driessche, J., 1996. Hercynian high-altitude periglacial phenomena in the French Massif Central: Tectonic implications. Palaeogeogr. Palaeoclimatol. Palaeoecol., 122, 1–4, 227–241.

Becq-Giraudon, J.F. and van den Driessche, J., 1994. Dépôts péri-glaciaires dans le Stéphano-Autunien du Massif Central: témoin de l'effondrement gravitaire d'un haut plateau hercynien. C. R. Acad. Sci., Paris 2, 318, 675–682.

Behr, H.J. and Heinrichs, T., 1987. Geological interpretation of DEKORP 2-S: a deep seismic reflection profile across the Saxothuringian and possible implications for late Variscan structural evolution of Central Europe. Tectonophysics, 142, 173–202.

Bellahsen, N., Faccenna, C., Funiciello, F., Daniel, J.M. and Jolivet, L., 2003. Why did Arabia separate from Africa? Insights from 3-D laboratory experiments. Earth Planet. Sci. Lett., 216, 365–381.

Bellahsen, N., Fournier, M., d'Acremont, E., Leroy, S. and Daniel, J.M., 2006. Fault reactivation and rift localization: the northeastern Gulf of Aden margin. Tectonics, 25 doi:10.1029/2004TC001626.

Benaouali-Mebarek, N., Frizon de Lamotte, D., Roca, E., Bracene, R., Faure, J.L., Sassi, W. and Roure, F., 2006. Post-Cretaceous kinematics of the Atlas and Tell systems in central Algeria: Early foreland folding and subduction-related deformation. C. R. Geoscience, Paris, 338, 115–125.

Benchilla, L., Guilhaumou, N., Mougin, P., Jaswal, T. and Roure, F., 2003. Paleoburial and pore pressure reconstruction of reservoir rocks in foothills areas: A sensivity test in the Hammam Zriba (Tunisia) and Koh-I-Maran (Pakistan) ore deposits. Geofluids, 3, 103–123.

Bergerat, F., Cazes, M., Damotte, B., Guellec, S., Mugnier, J.L., Roure, F. and et Truffert, C., 1989. Les structures distensives en Bresse d'après les données du profil sismique Jura-Bresse (programme ECORS). C. R. Acad. Sci. Paris, 309, II, 325–332.

Bergerat, F., Mugnier, J.L., Guellec, S., Truffert, C., Cazes, M., Damotte, B. and Roure, F., 1990. Extensional tectonics and ubsidence of the Bresse basin: an interpretation from Ecors data. In Roure, F., Heitzman, P. and Polino, R., eds., Deep structure of the Alps. Mém. Soc. Géol, France, 156, 145–156.

Bernet, M. and Spiegel, C., 2004. Detrital thermochronology: provenance, analysis, exhumation and landscape evolution of mountain belts. Geol. Soc. Am., Boulger, Spec. Publ., 378, 126 pp.

Berthelsen, A., 1998. The Tornquist zone northwest of the Carpathians: an intraplate pseudosuture. GFF, 20, 223–230.

Bertotti, G., Frizon de Lamotte, D., Teixell, A. and Charroud, M., eds., 2009. The geology of vertical movements. Proceedings ILP workshop, Marrakech, Tectonophysics, in press.

Bertotti, G., Podlachikov, Y. and Daehler, A., 2000. Dynamic link between the level of ductile crustal flow and style of normal faulting of brittle crust. Tectonophysics, 320, 195–218.

Bertotti, G., Schulman, K. and Cloetingh, S., eds., 2002. Continental collision and the tectono-sedimentary evolution of foreland. European Geosciences Union, Stephan Mueller Spec. Publ. Series, 1, 223 pp.

Biju-Duval, B., Bois, C., Cazes, M., Hirn, A. and Raoult, J.F., 1987. Une chaîne de montagnes sous le bassin de Paris. La Recherche, 18, 185, 242–244.

Bijwaard, H. and Spakman, W., 1999a. Fast kinematic ray tracing of first- and later-arriving global seismic phases. Geophys. J. Int., 139, 359–369.

Bijwaard, H. and Spakman, W., 1999b. Tomographic evidence for a narrow whole mantle plume below Iceland. Earth Planet. Sci. Lett., 166, 121–126.

Bijwaard, H. and Spakman, W., 2000. Non-linear global P-wave tomography by iterated linearized inversion. Geophys. J. Int., 141, 71–82.

Bijwaard, H., Spakman, W. and Engdahl, E.R., 1998. Closing the gap between regional and global travel time tomography. J. Geophys. Res., 103, (B12), , 30055–30078.

Blisniuk, P.M. and Stern, L.A., 2005. Stable isotope paleoaltimetry: A critical review. Am. J. Sci., 395, 1033–1074.

Blundell, D., Freeman, R. and Mueller, S., eds., 1992. A continent revealed, the European geotraverse. Cambridge University Press, Cambridge, 275 pp.

Blystad, P., Brekke, H., Faerseth, R.B., Larsen, B.T., Skogseid, J. and Tørurdbakken, B., 1995. Structural elements of the Norwegian continental shelf. Part 11, the Norwegian Sea region. Norwegian Petrol. Directorate Bull., 8.

Boillot, G., Beslier, M.O. and Girardeau, J., 1995. Nature, structure and evolution of the ocean-continent boundary: the lesson of the West Galicia margin (Spain). In Banda, E., Torné, M. and Talwani, M., eds., Rifted ocean-continent boundaries, Kluwer Academic, the Netherlands, 219–229.

Boillot, G., Montadert, L., Lemoine, M. and Biju-Duval, B., 1984. Les marges continentales actuelles et fossiles autour de la France. Masson, Paris, 342 pp.

Boillot, G., Recq, M., Winterer, E.L., Meyer, A.W., Applegate, J., Baltuck, M., Bergen, J.A., Comas, M.C., Davies, T.A., Dunham, K., Evans, C.A., Girardeau, J., Goldberg, G., Haggerty, J., Jansa, L.F., Johnson, J.A., Kasahara, J., Loreau, J.-P., Luna-Sierra, E., Moullade, M., Ogg, J., Sarti, M., Thurow, J. and Williamson, M., 1987. Tectonic denudation of the upper mantle along passive margins: a model based on drilling results (ODP leg 103, western Galicia margin, Spain). Tectonophysics, 132, 4, 335–342.

Boillot, G., Mougenot, D., Girardeau, J. and Winterer, E.L., 1989. Rifting processes on the West Galicia Margin, Spain. In: Tankard, A.J. and Balkwill, H.R., eds., Extensional Tectonics and Stratigraphy of the North Atlantic Margins. Am. Assoc. Petrol. Geol. Mem., 46, 363–377.

Bois, C., 1992. The evolution of the layered lower crust and the Moho through geological time in Western Europe: contribution of deep seismic reflection profiles. Terra Nova, 4, 99–108.

Bois, C., Cazes, M., Olivier, G., Lefort, J.P., Pinet, B. et al., 1991. Principaux apports scientifiques des campagnes SWAT et WAM à la géologie de la Mer Celtique, de la Manche et de la marge atlantique. In Bois, C., Gariel, O. and Pinet, B., eds., Soc. Géol. France, Mém., 159, 185–217.

Bois, C. and Gariel, O., 1994. Deep seismic investigations in the Parentis Basin (southwestern France). In Mascle, A., ed., Hydrocarbon and petroleum geology of France, European Assoc. Petrol. Geol., Spec. Publ., 4. Springer-Verlag, Berlin, 173–186.

Bonjer, K.P., 1997. Seismicity pattern and style of seismic faulting at the eastern borderfault of the southern Rhine Graben. Tectonophysics, 275, 41–69.

Bonnet, S. and Crave, A., 2003. Landscape response to climate change: Insights from experimental modelling and implications for tectonic versus climatic uplift of topography. Geology, 31, 2, 123–126.

Bonnet, S., Guillocheau, F. and Brun, J.P., 1998. Relative uplift measured using river incision: the case of the Armorican basement (France). C. R. Acad. Sci. Earth Planet. Sci., 327, 245–251.

Bonnet, S., Guillocheau, F., Brun, J.P. and Van den Driessche, J., 2000. Large-scale relief development related to Quaternary tectonic uplift of a Proterozoic-Paleozoic basement: The Armorican Massif, NW France. J. Geophys. Res., 105, 19273–19288.

Bourgeois, O., Ford, M., Diraison, M., LeCarlierdeVeslud, C., Gerbault, M., Pick, R., Ruby, N. and Bonnet, S., 2007. Separation of reifting and lithospheric folding signatures in the NW Alpine foreland. Int. J. Earth Sci. Springer, 96, 1003–1031.

Bousquet, R., Goffé, B., Henry, P., Le Pichon, X. and Chopin, C., 1997. Kinematic, thermal and petrological model of the Central Alps: Lepontine metamorphism in the upper crust and eclogitisation of the lower crust. Tectonophysics, 273, 105–127.

Braun, J., 2006. Recent advances and current problems in modelling surface processes and their interaction with crustal deformation. In Buiter, S.J.H. and Schreurs, G., eds., Analogue and numerical modelling of crustal-scale processes. Geol. Soc., London, Spec. Publ., 253, 307–325.

Braun, J., Dauteuil, O., Gallagher, K., Guillocheau, F., Robin, C., Rouby, D., Simoes, M. and Tiercelin, J.J., 2007. Topoafrica. Evolution of African topography over the last 250My. From sedimentary record to mantle dynamics. Mémoires de Géosciences, Univ. Rennes, vol. HS, 7, 94 pp.

Braun, J. and van der Beek, P.A., 2004. Evolution of passive margin escarments: what can we learn from low-temperature thermochronology. J. Geophys. Res., 109, F04009. doi: 10.1029/2004JF000147.

Braun, J., van der Beek, P.A. and Batt, G., 2006. Quantitative thermochronology: numerical methods for the interpretation of thermochronological data. Cambridge Univ. Press, 258 pp.

Breivik, A., Verhoef, J. and Faleide, J.I., 1999. Effects of thermal contrasts on gravity modelling at passive margins: results from the western Barents Sea. J. Geophys. Res., 104, 15293–15311.
Brekke, H., 2000. The tectonic evolution of the Norwegian Sea continental margin with emphasis on the Vøring and Møre basins. In Nøttvedt, A., ed., Dynamic of the Norwegian margin. Geol. Soc., London, Spec. Publ., 167, 327–378.
Brun, J.P. and Gutscher, M.A. and DEKORP-ECORS team, 1992. Deep crustal structure of the Rhine Graben from DEKORP-ECORS seismic reflection data: a summary. Tectonophysics, 208, 203–210.
Brun, J.P. and Nalpas, T., 1996. Graben inversion in nature and experiments. Tectonics, 15, 677–687.
Brun, J.P. and Wenzel, F. and ECORS-DEKORP teams, 1991. Crustal-scale structure of the southern Rhine Graben from ECORS-DEKORP seismic reflection data. Geology, 19, 758–762.
Brunet, M.F. and Cloetingh, S., eds., 2003. Integrated Peri-Tethyan basin studies. Peri-Tethys Programme, Sedim. Geol., 156, 1–10.
Buchanan, P.G. and Nieuwland, D.A., eds., 1996. Modern developments in structural interpretation, Validation and Modelling. Geological Society, Spec. Publ., 99, 369 pp.
Buck, W.R., 1991. Modes of continental lithospheric extension. J. Geophys. Res., 96, 20161–20178.
Buiter, S.J.H. and Schreurs, G., eds., 2006. Analogue and numerical modelling of crustal-scale processes. Geol. Soc, London, Spec. Publ, 253, 440 pp.
Bünge, H.P., 2006. 4-D evolution of deep mantle flow and continental topography. Geophys. Res. Abs. EGU, 8, 04153.
Bünge, H.P., Behrmann, J. and Weber, M., coordinators, 2008. SAMPLE, South Atlantic margin processes and links with onshore evolution. Proposal for establishment of a priority programme, www.sample-spp.de/.
Bünge, H.P., Hagelberg, C.R. and Travis, B.J., 2003. Mantle circulation models with variational data assimilation: inferring past mantle flow and structure from plate motion histories and seismic tomography. Geophs. J. Int., 152, 280–301.
Bünge, H.P., Richards, M.A. and Baumgardner, J.R., 2002. Mantle-circulation models with sequential data assimilation: inferring present-day mantle structure from plate-motion histories. Phil. Trans. Roy. Soc. A, 360, 2545–2567.
Burg, J.P., Sokoutis, D. and Bonini, M., 2002. Model inspired interpretation of seismic structures in collisional Central Alps: Crustal wedging and buckling at mature stage of collision. Geology, 30, 634–649.
Burg, J.P., van den Driessche, J. and Brun, J.P., 1994. Syn- to post-thickening extension: Modes and consequences. C. R. Acad. Sci. Paris II, 319, 1019–1032.
Burgess, P.M. and Moresi, L.N., 1999. Modelling rates and distribution of subsidence due to dynamic topography over subducting slabs; is it possible to identify dynamic topography from ancient strata? Basin Res., 11, 305–314.
Burgess, P., Granjeon, D., Lammers, H. and Van Oosterhout, C., 2006. Multivariable sequence stratigraphy: Tackling complexity and uncertainty with stratigraphic forward modelling, multiple scenarios and conditional frequency maps. Am. Assoc. Pet. Geol. Bull., 90, 1883–1901.
Burov, E.B., 2007. Coupled lithosphere-surface processes in collision context. In Lacombe, O., Lavé, J., Roure, F. and Vergés, J., eds., Thrust belts and foreland basins, Frontiers in Earth Sciences, Springer, 3–40.
Burov, E.B., 2009. Thermo-mechanical models for coupled lithosphre-surface processes. This volume.
Burov, E.B. and Cloetingh, S., 1997. Erosion and rift dynamics: new thermo-mechanical aspects of post-rift evolution of extensional basins. Earth Planet. Sci. Lett., 150, 7–26.
Burov, E.B. and Cloetingh, S., 2009. Interactions of mantle plumes and lithospheric folding: impact on intra plate continental tectonics. Geophysical Journal International, in press.
Burov, E.B. and Diament, M., 1995. The effective elastic thickness of continental lithosphere: What does it really mean? (constraints from rheology, topography and gravity. J. Geophys. Res., 100, 3905–3927.
Burov, E.B. and Molnar, P., 1998. Gravity Anomalies over the Ferghana Valley (central Asia) and intracratonic Deformation. J. Geophys. Res., 103, 18137–18152.
Burov, E.B., Nikishin, A.M., Cloetingh, S. and Lobkovsky, L.I., 1993. Continental lithosphere folding in central Asia (Part II): constraints from gravity and tectonic modelling. Tectonophysics, 226, 73–87.
Burov, E. and Poliakov, A., 2001. Erosion and rheology controls on syn- and post-rift evolution: verifying old and new ideas using a fully coupled numerical model. J. Geophys. Res., 106, 16461–16481.
Burov, E.B. and Watts, A.B., 2006. The long-term strength of continental lithosphere: "jelly-sandwich" or "crème-brûlée"?. Geol. Soc. Am. Today, 16, 4–10.
Burrus, J., 1989. Review of geodynamic models for extensional basins: the paradox of stretching in the Gulf of Lions (Northwest Mediterranean). Bull. Soc. Géol. France, 8, 377–393.
Calcagnile, G. and Panza, G.F., 1987. Properties of the lithosphere-asthenosphere system in Europe with a view toward Earth conductivity. Pure Appl. Geophys., 125, 241–254.
Callot, J.P., Geoffroy, L. and Brun, J.P., 2002. Development of volcanic passive margins: Three-dimensional laboratory models. Tectonics, 21, 6, 1052. doi:10.1029/2001TC901019.
Carbonell, R., Gallart, J., Perez-Estaun, A., Diaz, J., Kashubin, S., Mechie, J., Wenzel, F. and Knapp, J., 2000. Seismic wide-angle constraints on the crust of the southern Urals. J. Geophys. Res., 105, 13755–13777.
Carrapa, B., Bertotti, G. and Krijgsman, W., 2003. Subsidence, stress regime and rotation(s) of a tectonically active sedimentary basin within the western Alpine Orogen; the Tertiary Piedmont Basin (Alpine domain, NW Italy). Geol. Soc., Lond., Spec. Publ., 208, 205–227.
Carter, N.L. and Tsenn, M.C., 1987. Flow properties of continental lithosphere. Tectonophysics, 36, 27–63.
Casero, P., 2004. Structural setting of petroleum exploration plays in Italy. In Crescenti, V., D'Offizi, S., Merlino, S. and Sacchi, L., eds., Geology of Italy, Spec. vol. It. Geol. Society for the IGC32 Florence-2004. 189–200.
Casero, P., Roure, F. and Vially, R., 1991. Tectonic framework and petroleum potential of the southern Apennines. In Spencer, A., ed., Generation, accumulation and production of Europe's hydrocarbons, European Association. of Petroleum Geologists, Berlin meeting, Oxford Univ. Press, Oxford, 381–387.

Cavanagh, A.J., di Primio, R., Scheck-Wenderoth, M. and Horsfield, B., 2006. Severity and timing of Cenozoic exhumation in the southwestern Barents Sea. J. Geol. Soc., Lond., 163, 5, 761–774.

Cavazza, W., Ziegler, P.A., Spakman, W., Stampfi, G. and Roure, F., eds., 2004. The Transmed Atlas: The Mediterranean region from crust to mantle. Heidelberg, Springer-Verlag, 141 p. +CD-ROM.

Cazes, M. and Torreilles, G., eds., 1988. Etute de la croûte terrestre par sismique profonde (ECORS). Profil Nord de la France. Editions Technip.

Chadwick, R. and Pharaoh, T., 1998. The seismic reflection Moho beneath the United Kingdom and adjacent areas. Tectonophysics, 299, 255–279.

Chalmers, J.A. and Cloetingh, S., eds., 2000. Neogene uplift and tectonics around the North Atlantic. Global Planet. Change, 24, 3–4, 164–318.

Champagnac, J.P., van der Beek, P.A., Diraison, G. and Dauphin, S., 2008. Flexural isostatic response of the Alps to increased Quaternary erosion recorded by foreland basin remnants, SE France. Terra Nova, 20, 213–320.

Choukroune, P. and ECORS Pyrénées team, 1989. The ECORS Pyrenean deep seismic profile. Reflection data and the overall structure of an orogenic belt. Tectonics, 8, 1, 23–39.

Choukroune, P., Pinet, B., Roure, F. and Cazes, M., 1990. Major Hercynian structures along the ECORS Pyrénées and Biscay lines. Bull. Soc. Géol. France,

Clemson, J., Cartwright, J. and Booth, J., 1997. Structural segmentation and the influence of basement structure on the Namibian passive margin. J. Geol. Soc. Lond., 154, 477–482.

Cloetingh, S., ed., 1990. Geosphere fluctuations: short-term instabilities in the Earth's system. Contribution of Solid Earth Sciences to the International Geosphere and Biosphere Project, Global Planet. Change, 89, 177–313.

Cloetingh, S., ed., 1991. Long term sea level changes. J. Geophys. Res., 96, 6583–6949.

Cloetingh, S., 1992. Lithospheric dynamics and tectonics of sedimentary basins. Proceedings, Koninklijke Nederlandse Akademic van Wetenschappen, 95, 349–369.

Cloetingh, S., Beekman, F., Ziegler, P.A., van Wees, J.-D. and Sokutis, D., 2008. Post-rift compressional reactivation potential of passive margins and extensional basins. In: Johnson, H., Doré, A.G., Gatliff, R.W., Holdsworth, R., Lundin E., and Ritchie J.D., eds., The nature and origin of compressive margins. Geol. Soc., London, Spec. Publs., 306, 27–69.

Cloetingh, S. and Ben Avraham, Z., eds., 2002. From continental extension to collision: Africa-Europe interaction, the Dead Sea rift and analogue natural laboratories. European Geosciences Union, Stephan Mueller Spec. Publ. Series, 2, 246 pp.

Cloetingh, S., Ben Avraham, Z., Sassi, W. and Horvath, F., eds., 1996. Dynamics of extensional basins and inversion tectonics. Tectonophysics, 266, 523 pp.

Cloetingh, S., Boldreel L.O., Larsen, B.T., Heinesen, M., and Mortensen, L., eds. , 1998. Tectonics of sedimentary basin formation: Models and constraints. Tectonophysics, 300, 387 pp.

Cloetingh, S. and Burov, E., 1996. Thermo mechanical structure of European continental lithosphere: constraints from rheological profiles and EET estimates. Geophys. J. Int., 124, 695–723.

Cloetingh, S., Burov, E., Beekman, F.B.A., Andriessen, P.A.M., Garcia-Castellanos, D., De Vicente, G. and Vegas, R., 2002a. Lithospheric folding in Iberia. Tectonics, 21, 1041.

Cloetingh, S., Burov, E., Matenco, L., Toussaint, G., Bertotti, G., Andriessen, P.A.M., Wortel, M.J.R. and Spakman, W., 2004. Thermo-mechanical controls on the mode of continental collision in the SE Carpathians (Romania). Earth Planet. Sci. Lett., 218, 57–78.

Cloetingh, S., Burov, E. and Poliakov, A., 1999. Lithosphere folding: Primary response to compression? (from central Asia to Paris basin). Tectonics, 18, 1064–1083.

Cloetingh, S. and Cornu, T.G.M., 2005. Neotectonics and Quaternary fault reactivation in Europe's intraplate lithosphere. Quat. Sci. Rev, 24, 235–508.

Cloetingh, S., d'Argenio, B., Catalano, R., Horvath, F. and Sassi, W., eds., 1995a. Interplay of extension and compression in basin formation. Tectonophysics, 252, 484 pp.

Cloetingh, S., Durand, B. and Puigdefabregas, C., eds., 1995b. Integrated basin studies. Mar. Petrol. Geol. IBS Spec., Vol. 12, 787–963.

Cloetingh, S., Eldholm, O., Sassi, W., Larsen, B.T. and Gabrielsen, R., eds., 1994. Dynamics of extensional and inverted basins. Tectonophysics, 240, 1–341.

Cloetingh, S., Fernandez, M., Munoz, J.A., Sassi, W. and Horvath, F., eds., 1997a. Structural controls on sedimentary basin formation. Tectonophysics, 282, 1–442.

Cloetingh, S., Gradstein, F., Kooi, H., Grant, A. and Kaminski, M., 1990. Plate reorganization: a cause of rapid late Neogene subsidence and sedimentation around the North Atlantic? J. Geol. Soc. Lond., 147, 495–506.

Cloetingh, S., Den Bezemer, T. and Ter Voorde, M., 1998a. Sedimentary basin modelling: from the lithospheric scale to the sub-basin scale in extensional basin formation. Proceedings of the Intational School of Earth Planetary Science, Siena, 5–19.

Cloetingh, S., Boldreel, L., Larsen, B.T., Heinesen, M. and Mortensen, L., 1998b. Tectonics of sedimentary basins. Tectonophysics, 300, 1–11.

Cloetingh, S., Horvath, F., Bada, G., Lankreijer, A., eds., 2002b. Neotectonics and surface processes: the Pannonian Basin and Alpine/Carpathian systems. European Geosciences Union, Stephan Mueller Spec. Publ. Series, 3.

Cloetingh, S., Horváth, F., Dinu, C., Stephenson, R.A., Bertotti, G., Bada, G., Matenco, L., Garcia-Castellanos, D. and Group, T.W., 2003a. Probing tectonic topography in the aftermath of continental convergence in Central Europe. EOS, Transactions, Am. Geophys. Union, 84, 89.

Cloetingh, S., Lankreijer, A., Nemcok, M., Neubauer, F. and Horvath, F., eds., 2001. Sedimentary basins and hydrocarbon habitat at the margin of the Pannonian basin system. Mar. Petrol. Geol., 18, 177 pp.

Cloetingh, S., Marzo, M., Muñoz, J.A. and Vergés, J., eds., 2002c. Tectonics of sedimentary basins: from crustal structure to basin fill. Tectonophysics, 346, 1–2, 1–135

Cloetingh, S. and Podlachikov, Yu.Y., 2000. Perspectives on tectonic modelling. Tectonophysics, 320, 169–173.

Cloetingh, S., Ranalli, G. and Ricci, C.A., eds., 1996. Sedimentary basins: models and constraints. Siena, 12–30 September 1996, Proceedings, International School, Earth and Planetary Sciences, 241 pp.

Cloetingh, S., Sassi, W. and Horvath, F., eds. 1993a. The origin of sedimentary basins: Inferences from quantitative modelling and basin analysis. Tectonophysics, 226, 1–518.
Cloetingh, S., Sassi, W., Horvath, F. and Puigdefabregas, C., eds., 1993b. Basin analysis and dynamics of sedimentary basin evolution. Sediment. Geol., 86, 1–201.
Cloetingh, S. and Sassi, W. and Task Force Team, 1994. The origin of sedimentary basins – a status report from the International Lithosphere Programme task force. Mar. Petrol. Geol., 11, 659–683.
Cloetingh, S., van Balen, R.T., Ter Voorde, M., Zoetemeijer, B.P. and Den Bezemer, T.1997b. Mechanical aspects of sedimentary basin formation: development of integrated models for lithospheric and surface processes. Geol. Rundsch., 86, 226–240
Cloetingh, S., van der Beek, P.A., Van Rees, D., Roep, T.B., Biermann, C. and Stephenson, R.A.1992. Flexural interaction and the dynamics of Neogene extensional basin formation in the Alboran-Betic region. Geo-Marine Lett., 12, 66–75.
Cloetingh, S. and van Wees, J.D. 2005. Strength reversal in Europe's intraplate lithosphere; transition from basin inversion to lithospheric folding. Geology. 33 (4), 285–288.
Cloetingh, S., Ziegler, P.A., Cornu, T. and working Group, 2003b. Investigating environmental tectonics in northern Alpine foreland of Europe. EOS, Transactions, Am. Geophys. Union, 349, 356–357.
Cloetingh, S., Ziegler, P.A., Beekman, F., Andriessen, P.A.M., Matenco, L., Bada, G., Garcia-Castellanos, D., Hardebol, N., Dèzes, P. and Sokoutis, D., 2005. Lithospheric memory, state of stress and rheology: neotectonic controls on Europe's intraplate continental topography. Quat. Sci. Rev., 24, 241–304.
Cloetingh, S., Ziegler, P.A. and the Topo-Europe Working Group, 2007. Topo-Europe: The geosciences of coupled deep Earth-surface processes. Global Planet. Change, 58, 1–4, 1–118.
Coackley, B. and Gurnis, M., 1995. Far-field tilting of Laurentia during the Ordovician and constraints on the evolution of a slab under an ancient continent. J. Geophys. Res., 100, 6313–6327.
Cobbold, P.R., Mourgues, R. and Boyd, K., 2004. Mechanism of thin-skinned detachment in the Amazon fan: assessing the importance of fluid overpressures and hydrocarbon generation. Marine Petrol. Geol., 21, 8, 1013–1025.
Colletta, B., Letouzey, J., Pinedo, R., Ballard, J.F. and Balé, P., 1991. Computerized X-ray tomography analysis of sandbox models: Examples of thin-skinned thrust systems. Geology, 19, 1063–1067.
Colletta, B., Roure, F., De Toni, B., Loureiro, D., Passalacqua, H. and Gou, Y., 1997. Tectonic inheritance, crustal architecture and contrasting structural styles along the northern and southern Andean flanks. Tectonics, 16, 777–794.
Contrucci, I., Matias, L., Klingelhöfer, F., Moulin, M., Géli, L., Nouzé, H., Aslanian, D., Olivet, J.L., Réhault, J.P. and Sibuet, J.C., 2004. Study of the deep structure of the West African continental margin between 5°S and 8°S from reflexion/refraction seismics and gravity data. Geophys. J. Int., 158, 529–553.
Corti, G., Bonini, M., Sokoutis, D., Innocenti, F., Manetti, P., Cloetingh, S. and Mulugeta, G., 2003. Continental rift architecture and patterns of magma migration; a dynamic analysis based on centrifuge models. Tectonics, 23, 2, 1–20, TX 2012, doi:10.1029/2003TC001561.
Crave, A., Lague, D., Davy, P., Kermariec, J.J., Sokoutis, D., Bodet, L. and Compagnon, R., 2000. Analogue modelling of relief dynamics. Phys. Chem. Earth A, 25, 6–7, 549–553.
Davy, P. and Cobbold, P.R., 1991. Experiments on shortening of 4-layer model of the continental lithosphere. Tectonophysics, 188, 1–25.
d'Acremont, E., Leroy, S., Beslier, M.O., Bellahsen, N., Fournier, M., Robin, C., Maia, M. and Gente, P., 2005. Structure and evolution of the eastern Gulf of Aden conjugate margins from seismic reflection data. Geophys. J. Int., 160, 3, 869–890.
De Bruijne, C.H. and Andriessen, P.A.M., 2002. Far field effects of Alpine plate tectonism in the Iberian microplate recorded by fault-related denudation in the Spanish Central System. In Kohn, B.P. and Green, P.F., eds., Low temperature thermochronology; from tectonics to landscape evolution. Tectonophysics. 349, 161–184.
Decker, K. and Peresson, H., 1996. Tertiary kinematics in the Alpine-Carpathian-Pannonian system: links between thrusting, transform faulting and crustal extension. In Wessely, G. and Liebl, W., eds., Oil and gas in Alpidic thrustbelts and basins of Central and Eastern Europe, European Assoc. Geosc. Explor. Spec. Publ., 5, 69–77.
DEKORP Basin Research Group, 1998. Survey provides seismic insights into an old suture zone. EOS Trans. Am. Geophys. Union, 79, 151–159.
DEKORP Basin Research Group, 1999. The deep crustal structure of the Northeast German Basin: new DEKORP-Basin'96 deep profiling results. Geology, 27, 55–58.
DEKORP and Orogenic Processes Research Groups Krawczyk, C.M., 1999. Structure of the Saxothuringian Granulites: Geological and geophysical constraints on the exhumation of high-pressure/high-temperature rocks in the mid-European Variscan belt. Tectonics, 18, 5, 756–773.
Den Bezemer, T., Kooi, H., Podlachikov, Y. and Cloetingh, S., 1998. Numerical modelling of growth strata and grain-size distributions associated with fault-bend folding. Geol. Soc., Lond., Spec. Publ., 134, 381–401.
Deschamps, F. and Trampert, J., 2003. Mantle tomography and its relation to temperature and composition. PEPI, 140, 277–291.
de Vicente, G., Cloetingh, S., Muñoz-Martin, A., Olaiz, A., Stick, D., Vegas, R., Galindo-Zaldivar, J. and Fernandez-Lozano, J., 2008. Inversion of moment tensor focal mechanism for active stresses around the microcontinent Iberia: Tectonic implications. Tectonics, 27, TC1009. doi: 10.1029/2000/TC002093/2008.
De Voogd, B. and Nicolich, R. and the ECORS-CROP working Group, 1991. First deep seismic reflection transect from the Gulf of Lions to Sardinia (ECORS-CROP profiles in Western Mediterranean). In Continental lithosphere: Deep seismic reflections, Am. Geophys. Union, Geodynamic Series, 22, 265–274.
De Jager, J., 2007. Geological develoment. In: Wong, Th.E. Batjes D.A.J., and de Jager, J., eds., Geology of the Netherlands. Royal Academy of Art and Sciences, 5–26.

De Wever, B., 2008. Diagenesis and fluid flow in the Sicilian fold-and-thrust belt. PhD Thesis, KU-Leuven, Belgium, 184p.

De Wit, M., 2007. The Kalahari epeirogeny and climate change: Differenciating cause and effect from core to space. South Afr. J. Geol., 110, 367–392.

Dèzes, P. and Ziegler, P.A., 2004. Moho depth map of Western and Central Europe (2004) EUCOR-URGENT home page (http://www.unibas.ch/eucor-urgent).

Dèzes, P., Schmid, S.M. and Ziegler, P.A., 2004. Evolution of the European Cenozoic Rift System: interaction of the Alpine and Pyrenean orogens with their foreland lithosphere. Tectonophysics, 389, 1–33.

Di Stefano, A., Vinci, G. and Romeo, M., 2002. The Pliocene-Pleistocene deposits at the southeastern border of the Sant'Arcangelo Basin (Southern Apennines): Structural setting and biostratigraphy. Boll. Soc. Geol. It., 5, 527–537.

Doglioni, C. and Prosser, G., 1997. Fold uplift versus regional subsidence and sedimentation rate. Marine Petrol. Geol., 14, 2, 179–190.

Doornenbal, H., 2008. Overview of the Southern Permian Basin Atlas project: Management and results. EUR-09 Session, Geology of the Southern Permian Basin Area, Part I, 33rd International Geological Congress, Oslo, Abs.

Doust, H. and Omatsola, E., 1989. Niger Delta. In Edwards, J.D. and Santogrossi, P.A., eds., Divergent/Passive margin basins. Am. Assoc. Petrol. Geol. Mem. 48, 201–238.

Du, Z.J., Michelini, A. and Panza, G.F., 1998. EurID, a regionalized 3-D seismological model of Europe. Phys. Earth Planet. Inter., 105, 31–62.

Duin, E., Rijkers, R. and Remmelts, G., 1995. Deep seismic reflections in the Netherlands: an overview. Geol. Mijnbouw, 74, 191–197.

Dunai, T.J. and Baur, H., 1995. Helium, neon and argon systematics of the European subcontinental mantle: implications for its geochemical evolution. Geochim. Cosmo. Acta, 59, 2767–2783.

Dunai, T.J. and Wijbrans, J.R., 2000. Long-term cosmogenic ^{3}He production rates (152 ka-1.35 Ma) from Ar^{40}/Ar^{39} dated basalt flows at 29 degrees N latitude. Earth Planet. Sci. Lett., 176, 147–156.

Durand, B., Jolivet, L., Horvath, F. and Séranne, M., eds.,, 1999. The Mediterranean Basin: Tertiary extension within the Alpine orogen. Geol. Soc., Lond., Spec. Publ., 156.

Durham, L.S., 2003. Passive seismic. Listen: Is it the next big thing? The Explorer, Am. Assoc. Petrol. Geol., April 2003.

Ebbing, J., Lundin, E., Olesen, O. and Hansen, E.K., 2006. The Mid-Norwegian margin: a discussion of crustal lineament, mafic intrusions, and remnants of the Caledonian root by 3D density modelling and structural interpretation. J. Geol. Soc, London.

Ehlers, T.A. and Farley, K.A., 2003. Apatite (U-Th)/He thermochronometry: methods and applications to problems in tectonics and surface processes. Earth Planet. Sci. Lett., 206, 1–14.

Elfrink, N.M., 2001. Quaternary groundwater avulsions: Evidence for large-scale midcontinent folding? Association Eng. Geologists News, 44, 60.

Ellis, M.A. and Densmore, A.L., 2006. First order topography over blind thrusts. In Willett, S.D., Hovius, N., Brandon, M.T. and Fisher, D.M., eds., Tectonics, climate and landscape evolution, Geol. Soc. America, Sp. Paper, 398.

Ellis, S., Schreurs, G. and Panien, M., 2004. Comparisons between analogue and numerical models of thrust wedge development. J. Struct. Geol., 26, 9, 1659–1675.

Ellouz, N., Roure, F., Sandulescu, M. and Badescu, D., 1996. Balanced cross-sections in the Eastern Carpathians (Romania): a tool to quantify Neogene dynamics. In Roure, F., Ellouz, N., Shein, V.S. and Skvortsov, I.I., eds., Geodynamic evolution of sedimentary basins, Editions Technip, 305–325.

Evans, D.J., Graham, C., Armour, A. and Bathurst, P., 2003. The Millennium Atlas: Petroleum geology of the Central and Northern North Sea. Geol. Soc., London.

Faccenna, C., Jolivet, L., Piromallo, C. and Morelli, A., 2003. Subduction and the depth of convection in the Mediterranean mantle. J. Geophys. Res., 108, 2099.

Faccenna, C., Piromallo, C., Crespo-Blanc, A., Jolivet, L. and Rossetti, F., 2004. Lateral slab deformation and the origin of the western Mediterranean arcs. Tectonics, 23, 1–21.

Faccenna, C., Bellier, O., Martinod, J., Piromallo, C. and Regard, V., 2006. Slab detachment beneath eastern Anatolia: A possible cause for the formation of the North Anatolian fault. Earth Planet. Sci. Lett., 242, 85–97.

Faleide, J.I., Myhre, A.M. and Eldholm, O., 1988. Dipping reflectors and NE Atlantic evolution; Early Tertiary volcanism at the Western Barents Sea margin. Geol. Soc., London, Sp. Publ., 39, 135–146.

Fauquette, S., Suc, J.P., Jimenez-Moreno, G., Micheels, A., Jost, A., Favre, E., Bachiri-Taoufiq, N., Bertini, A., Clet-Pellerin, M., Diniz, F., Farjanel, G., Feddi, N. and Zheng, Z., 2007. Latitudinal climatic gradients in Western European and Mediterranean regions from the Mid-Miocene ($\sim$15 Ma) to the Mid-Pliocene ($\sim$3.5 Ma) as quantified from pollen data. Deep-time perspectives on climate change: Marrying the signal from computer models and biological proxies, The Micropaleontological Society, Geol. Soc. Lond., Spec. Paper.

Faure, J.L., Osadetz, K., Benaouali, N., Schneider, F. and Roure, F., 2004. Kinematic and petroleum modelling of the Alberta Foothills and adjacent foreland, west of Calgary. Oil Gas Sci. Technol., Revue de l'IFP, 1, 81–108.

Ferket, H., Ortuño, S., Swennen, R. and Roure, F., 2003. Diagenesis and fluid flow history in reservoir carbonates of the Cordilleran fold- and thrust- belt: The Cordoba Platform. In Bartolini, C., Burke, K., Buffler, R., Blickwede, J. and Burkart, B., eds., Mexico and the Caribbean region: plate tectonics, basin formation and hydrocarbon habitats, Am. Assoc. Petrol. Geol., Memo. 79, Ch. 10. 283–304.

Ferket, H., Swennen, R., Ortuño-Arzate, S., Cacas, M.C. and Roure, F., 2004. Hydrofracturing in the Laramide foreland fold-and-thrust belt of Eastern Mexico. In Swennen, R., Roure, F. and Granath, J., eds.,, Deformation, fluid flow and reservoir appraisal in foreland fold-and-thrust belts, Am. Assoc. Petrol. Geol. Hedberg Series, Me. 1, 133–156.

Fernandez, M., Ayala, C., Torné, M., Vergés, J., Gomez, M. and Karpuz, R., 2005. Lithospheric structure of the Mid-Norwegian margin: comparison between the Møre and Vøring margins. J. Geol. Soc. Lond., 162, 1005–1012.

Fernandez, M., Torné, M., Garcia-Castellanos, D., Vergés, J., Wheeler, W. and Karpuz, R., 2004. Deep structure of the Vøring margin: the transition from a continental shield to a young oceanic lithosphere. Earth Planet. Sci. Lett., 221, 131–144.

Fernandez-Lozano, J., Sokoutis, D., Willingshofer, E., Vicente G. de and Cloetingh, S., 2008. Analogue modelling: insights on lithospheric processes in Iberia. Bolletino di Geofysica teoretica ed applicata, 42, 54–57.
Fjeldskaar, W., Johansen, H., Dodd, T. and Thompson, M., 2003. Temperature and maturity effects of magmatic underplating in the Gjallar Ridge, Norwegian Sea. In Düppenbecker, S. and Marzi, R., eds., Multidimensional basin modelling, Am. Assoc. Petrol. Geol., Datapages Discovery Series. 7, 71–85.
Franke, D., Neben, S., Ladage, S., Schreckenberger, B. and Hinz, K., 2007. Margin segmentation and volcano-tectonic architecture along the volcanic margin off Argentina/Uruguay, South Atlantic. Marine Geol., 244, 1–4, 46–67.
Franke, D., Neben, S., Schreckenberger, B., Schultze, A., Stiller, M. and Kravczyk, C., 2006. Crustal structure across the Colorado Basin, offshore Argentina. Geophys. J. Int., 165, 850–864.
Frizon de Lamotte ,D., Saint-Bezar, B., Bracène, R. and Mercier, E., 2000. The two steps of the Atlas building and geodynamics of the West Mediterranean. Tectonics, 19, 740–761.
Fukao, Y., Widiyantoro, S. and Obayashi, M., 2001. Stagnant slabs in the upper and lower mantle transition region. Reviews of Geophysics, 2001.
Funiciello, F., Faccenna, C. and Giardini, D., 2004. Role of lateral mantle flow in the evolution of subduction system: Insights from 3D laboratory experiments. Geophys. J. Int., 157, 1393–1406.
Gabrielsen, R.H., Faerseth, R.B., Steel, R.J., Idil, S. and Klovjan, O.S., 1990. Architectural styles of basin-fill in the Northern Vicking Graben. In Blundell D.J, and Gibbs A.D., eds., Tectonic evolution of the North Sea rifts, Clarendon Press, Oxford, 158–179.
Gabrielsen, R.H., Odinsen, T. and Grunnaleite, I., 1999. Structuring of the Northern Viking Graben and the Møre Basin: the influence of basement structural grain and the particular role of the Møre-Trøndelag Fault Complex. J. Mar. Petrol. Geol., 16, 443–465.
Gallagher, K. and Brown, R., 1999. The Mesozoic denudation history of the Atlantic margins of southern Africa and southeast Brazil and the relationship to offshore sedimentation. In. Cameron, N.R., Bate, R.H. and Clure, V.S., eds., The oil and gas habitats of the South Atlantic. Geol. Soc., Lond., Spec. Publ., 153, 41–53.
Garcia-Castellanos, D., 2002. Interlay between lithospheric flexure and river transport in foreland basins. Basin Res., 14, 2, 89–104.
Garcia-Castellanos, D., 2006. Long-term evolution of tectonic lakes: Climatic controls on the development of internally drained basins. In Willett, S.D., Hovius, N., Brandon, M.T. and Fisher, D.M., eds., Tectonics, climate and landscape evolution, Geol. Soc. Am., Spec. Paper, 398.
Garcia-Castellanos, D., 2007. The role of climate in high plateau formation: Insight from numerical experiment. Earth Planet. Sci. Lett., 257, 372–390. doi:10.1016:j.epsl.2007.02.039.
Garcia-Castellanos, D., Cloetingh, S. and van Balen, R.T., 2000. Modelling the Middle Pleistocene uplift in the Ardennes–Rhenish Massif: thermo-mechanical weakening under the Eifel?. Global Planet. Change, 27, 39–52.
Garcia-Castellanos, D., Fernandez, M. and Torné, M., 1997. Numerical modelling of foreland basin formation: a program relating thrusting, flexure, sediment geometry and lithosphere rheology. Comput. Geosci., 23, 9, 993–1003.
Garcia-Castellanos, D., Vergés, J., Gaspar-Escribano, J.M. and Cloetingh, S., 2003. Interplay between tectonics, climate and fluvial transport during the Cenozoic evolution of the Ebro Basin (NE Iberia). J. Geophys. Res., 108, B7, 2347. doi:10.1029/2002JB002073.
Garcia-Castellanos, D., Fernandez, M. and Torné, M., 2002. Modelling the evolution of the Guadalquivir foreland basin (South Spain). Tectonics, 21, 3, doi: 10.1029/2001TC001339.
Gaspar-Escribano, J.M., van Wees, J.-D., ter Voorden, M., Cloetingh, S., Roca, E., Cabrera, L., Muñoz, J.A., Ziegler, P.A. and Garcia-Castellanos, D., 2001. Three dimensional modelling of the Ebro Basin. Geophys. J. Int., 145, 349–367.
Gee, D.G. and Stephenson, R.A., eds., 2006. European lithosphere dynamics. Geol. Soc. Lond. Sp. Publ., 32, 662 pp.
Genser, J., Van Wees, J.D., Cloetingh, S. and Neubauer, F., 1996. Eastern Alpine tectonometamorphic evolution: constraints from two-dimensional P-T-t modelling. Tectonics, 15, 584–604.
Geoffroy, L., 2001. The structure of volcanic margins: Some problematics from the North Atlantic/Labrador-Bafin system. Marine Petrol. Geol., 18, 4, 463–469.
Gernigon, L., Ringenbach, J.C., Planke, S. and Le Gall, B., 2004. Deep structures and breakup along volcanic rifted margins: insights from integrated studies along the outer Vøring Basin (Norway). Mar. Petrol. Geol., 21, 363–372.
Gerrard, I. and Smith, G.C., 1982. Post-Paleozoic succession and structure of the Southwestern African continental margin. In Watkins, J.S. and Drake, C.L., eds., Studies in continental margin geology, Am. Assoc. Petrol. Geol., 49–76.
Giese, P., 1995. Main features of geophysical structures in Central Europe. In dallmeyer, R.D., Franke, W. and Weber, K., eds., Pre-Permian geology of Central and Eastern Europe, Springer-Verlag, Berlin, 7–25.
Gladczenko, T.P., Hinz, K., Eldholm, O., Meyer, H., Neben, S. and Skogseid, J., 1997. South Atlantic volcanic margins. J. Geol. Soc. Lond., 154, 465–470.
Goes, S., Spakman, W. and Bijwaard, H., 1999. A lower mantle source for central European volcanism. Science, 286, 1928–1931.
Goes, S., Govers, R. and Vacher, P., 2000a. Shallow mantle temperatures under Europe from P and S wave tomography. J. Geophys. Res., 105, 11, 153–169.
Goes, S., Loohuis, J.J.P., Wortel, M.J.R. and Govers, R., 2000b. The effect of plate stresses and shallow mantle temperatures on tectonics of northwestern Europe. Global Planet. Change, 27, 23–39.
Goncalves, J., Violette, S., Robin, C., Pagel, M., Guillocheau, F., de Marsily, G., Bruel, D. and Ledoux, E., 2003. 3-D modelling of salt and heat transport during the 248 M.y. evolution of the Paris Basin: diagenetic implications. Bull. Soc. Géol. France, 174, 429–439.
Gorini, C., Le Marrec, A. and Mauffret, A., 1993. Contribution to the structural and sedimentary history of the Gulf of Lions (Western Mediterranean) from the ECORS profiles, industrial seismic profiles and well data. Bull. Soc. Géol. France, 164, 3, 353–363.
Gorini, C., Mauffret, A., Guennoc, P. and Le Marrec, A., 1994. Structure of the Gulf of Lions (Northwestern Mediterranean Sea): a review. In Mascle, A., ed., Hydrocarbon and petroleum geology of France, Eur. Assoc. Petrol. Geol., Spec. Publ., 4, Springer-Verlag, Berlin, 223–243.

Grad, M., Güterch, A. and Mazur, S., 2002. Seismic refraction evidence for crustal structure in the central part of the Trans-European Suture Zone in Poland. In: Winchester, J.A., Pharaoh, T.C. and Verniers, J., eds., Paleozoic amalgamation of Central Europe, Geol. Soc., London, Spec. Publ., 201, 295–309.

Granet, M., Wilson, M. and Achauer, U., 1995. Imaging mantle plumes beneath the French Massif Central. Earth Planet. Sci. Lett., 136, 199–203.

Granjeon, D., 2005. Stratigraphic modelling as a key to find new potentialities in exploration. OAPEC/IP Joint Seminar, Hydrocarbon reserves: Abondance or scarcity, Rueil-Malmaison, 28–30 June, 2008, Proceedings, 17 pp.

Granjeon, D. and Joseph, P., 1999. Concepts and applications of a 3D multiple lithology, diffusive model in stratigraphic modeling. In Harbaugh, J.W., Watney, W.L., Rankey, E.C., Slingerland, R., Goldstein, R.H. and Franseen, E.K., eds., Numerical experiments in stratigraphy: Recent advances in stratigraphic and sedimentologic computer simulations. SEPM Spec. Publ., 62, 197–210.

Gregersen, S., Voss, P., Shomali, Z.H., Grad, M. and Roberts, R.G. and THOR Working Group, 2006. Physical differences in the deep lithosphere of Northern and Central Europe. In: Gee, D.G. and Stephenson, R.A., eds., European lithosphere dynamics. Geol. Soc., Lond., Spec. Publ. 32, 313–322.

Gregersen, S., Wiejacz, P., Debski, W. et al., 2007. The exceptional earthquakes in Kaliningrad district, Russia, on September 21; 2004. Physics Earth Planet. Inter., 164, 1–2, 63–74.

Guggisberg, B., Kaminski, W. and Prodehl, C., 1991. Crustal structure of the Fennoscandian Shield: a traveltime interpretation of the long-range Fennolora seismic refraction profile. Tectonophysics, 195, 105–137.

Guilhaumou, N. and Dumas, P., 2005. Synchrotron FTIR hydrocarbon fluid inclusions microanalysis applied to diagenetic history and fluid flow reconstruction in reservoir appraisal. Oil Gas Sci. Technol. Revue de l'IFP, 60, 5, 1–12.

Guilhaumou, N., Larroque, C., Nicot, E., Roure, F. and Stephan, J.F., 1994. Mineralized veins resulting from fluid flow in decollement zones of the Sicilian prism: evidence from fluid inclusions. Bull. Soc. Géol. France, 165, 425–436.

Guillocheau, F., Braun, J., Simoes, M., Rouby, D., Robin, C., Dauteuil, O. and Gallagher, K., 2007a. Quantifying the evolution of the African topography over the last 205 My. In Topo-Africa: Evolution of African topography over the last 250 My. From sedimentary record to mantle dynamics, Rennes, No v. 13–16, 2007, Mémoire de Géosciences, Rennes, H.S. 7, 3 p.

Guillocheau, F., Robin, C., Allemand, P., Bourqui, S. and Brault, N., 2000. Meso-Cenozoic geodynamic evolution of the Paris Basin: 3D stratigraphic constraints. Geodinamica Acta, 13, 4, 189–246.

Guillocheau, F., Roland, N., Colin, J.P., Helin, C., Simoes, M., Rouby, D. and Robin, C., 2006. Paleogeographie de l'Afrique au Jurassique terminal-Crétacé: reconstitution des reliefs continentaux. In Réunion des Sciences de la Terre, Dijon, 12 Décembre 2006, Abs., 232.

Guillocheau, F., Roland, N., Colin, J.P., Robin, C., Rouby, D., Tiercelin, J.J. and Dauteuil, O., 2007b. Paleogeography of Africa through Meso-Cenozoic times: A focus on the continental domain evolution. In Topo-Africa: Evolution of African topography over the last 250 My. From sedimentary record to mantle dynamics, Rennes, Nov. 13–16, 2007. Mémoire de Géosciences, Rennes, H.S. 7, 3 p.

Guterch, A. and Grad, M., 2006. Lithospheric structure of the TESZ in Poland based on modern seismic experiment. Geophys. Quarterly, 50, 1, 23–32.

Guterch, A., Grad, M., Keller, G.R., Posgay, K., Vozar, J., Spicak, A., Bruckl, E., Hajnal, Z. and Thybo, H. Selvio and Celebration 2000 experiment team, 2003. Celebration 2000 seismic experiment. Stud. Geophs. Geol., 47, 3, 659–669.

Guterch, A., Grad, M., Thybo, H. and Keller, R. and POLONAISE Working Group, 1999. POLONAISE 97 – An international seismic experiment between Precambrian and Variscan Europe in Poland. Tectonophysics, 314, 101–122.

Hansen, M.B., Scheck-Wenderoth, M., Hübscher, C., Lykke-Andersen, H., Dehghani, A. and Gajewski, D., 2007. Basin evolution of the northern part of the Northeast German Basin – Insights from a 3D structural model. Tectonophysics, 437, 1–4,1–16.

Harper, J.F., 1984. Mantle flow due to internal vertical forces. Phys. Earth Plant. Inter., 36, 285–290.

Hartley, R.W. and Allen, P.A., 1994. Interior cratonic basins of Africa: relation to continental break-up and role of mantle convection. Basin Res., 6, 95–113.

Haug, G.H., Gunther, D., Peterson, L.C., Sigman, D.M., Hughen, K.A. and Aeschlimann, B., 2003. Climate and collapse of Maya civilization. Science, 14, 1731–1735.

Heine, C., Müller, R.D., Steinberger, B. and Torsvik, T.H., 2008. Subsidence of intracratonic basins due to dynamic topography. Phys. Earth Planet. Int., doi:10.1016/j.pepi.2008.05.008.

Hendriks, B.W.H. and Andriessen, P.A.M., 2002. Pattern and timing of the post-Caledonian denudation of northern Scandinavia constrained by apatite fission track termochronology. In Dore, A.G., Tuner, M.S., Stoker, J.A. and White, N.J., eds., Exhumation of the North Atlantic margin: Timing, mechanisms and implications for petroleum exploration. Geol. Soc., Lond., Spec. Publ., 196, 117–137.

Heeremans, M., Faleide, J.I. and Larsen, B., 2004. Late Carboniferous-Permian of NW Europe: an introduction to a regional map. In: Wilson M., Neumann, E.-R., Davies, G.R., Timmerman, M.J., Heerman, M. and Larsen, B.T. (eds.), Permo-Carboniferous Magmatism and Rifting in Europe. Geol. Soc., Lond., Spec. Publ., 223, 75–88.

Hinz, K., Neben, S., Schreckenberger, B., Roeser, H.A., Block, M., Gonzalvez de Souza, K. and Meyer, H., 1999. The Argentine continental margin north of 48°S: Sedimentary successions, volcanic activity during breakup. Mar. Petrol. Geol., 16, 1–25.

Hippolyte, J.C., Angelier, J., Müller, C. and Roure, F., 1991. Structure et mécanisme d'un bassin de type piggy-back": le bassin de Sant'Arcangelo (Italie méridionale). C. R. Acad. Sci., Paris, 312, 1373–1378.

Hippolyte, J.C., Angelier, J. and Roure, F., 1996. Paleostress analyses and fold-and-thrust belt kinematics in the Southern Apennines. In Roure, F., Ellouz, N., Shein, V.S. and Skvortsov, I.I., eds., Geodynamic evolution of sedimentary basins. Edition Technip, 157–169.

Hippolyte, J.C., Angelier, J., Roure, F. and Casero, P., 1994. Piggyback basin development and thrust belt evolution: structural and paleostress analyses of Plio-Quaternary basins in the Southern Apennines. J. Struct. Geol., 16, 159–171.

Hirsch, K., Bauer, K. and Scheck-Wenderoth, M., 2009. Deep structure of the western South African passive margin-results of a combined approach of seismic, gravity and isostatic investigations. Tectonophysics, in press.

Hirsch, K.K., Scheck-Wenderoth, M., Paton, D.A. and Bauer, K., 2007. Crustal structure beneath the Orange Basin, South Africa. South Afr. J. Geol., 110, 2–3, 249–260.

Hobbs, R. and Klemperer, S., 1991. The BIRPS atlas: deep seismic reflection profiles around the British Isles. Cambridge University Press, Cambridge.

Holliger, K. and Levander, A., 1994. Seismic structure of gneissic/granitic upper crust; geological and petrophysical evidence from the Strona-Ceneri Zone (northern Italy) and implications for crustal seismic exploration. Geophys. J. Int., 119, 497–510.

Hooghiemstra, H., 1989. Quaternary and Upper Pliocene glaciation and forest development in the Tropical Andes: Evidences from a long high resolution pollen record from the sedimentary basin of Bogota, Colombia. Palaeogeogr. Palaeoclimatol. Palaeoecol., 72, 11–26.

Hooghiemstra, H. and Ran, E.T.H., 1994. Late Pliocene-Pleistocene high resolution pollen sequence of Colombia: An overview of climatic change. Quat. Int., 21, 63–80.

Hoth, S., Adam, J., Kukowski, N. and Oncken, O., 2006. Influence of erosion on the kinematics of bivergent orogens: Results from sandbox simulations. In Willett, S.D., Hovius, N., Brandon, M.T. and Fisher, D.M., eds., Tectonics, climate and landscape evolution, Geol. Soc. America, Sp. Paper, 398.

Huchon, P., Manatschal, G. and Péron-Pinvidic, G., 2007. Comprendre la formation et l'évolution des marges passives: Importance de la caractérisation des transitions océan-continent. Dossier transition océan-continent, La Lettre de l'Académie des Sciences, Hiver 2007–2008, 22, 2–9.

Huismans, R.S. and Beaumont, C., 2002. Asymmetric lithospheric extension: The role of frictional plastic strain softening inferred from numerical experiments. Geology, 30, 3, 211–214.

Huisman, R.S. and Beaumont, C., 2003. Symmetric and asymmetric lithospheric extension: Relative effects of frictional-plastic and viscous strain softening. J. Geophys. Res.,

Huismans, R.S. and Beaumont, C., 2008. Roles of llithospheric strain, softening and heterogeneity in determining the geometry of rifts and continental margins. In Karner, G., Manatschal, G. and Pinheiro, L., eds., Margins, Theoretical and experimental Earth Science serie. s, Univ. Press Publ.

Huismans, R.S., Buiter, S.J.H. and Beaumont, C., 2005. The effect of plastic-viscous layering and strain-softening on mode selection during lithospheric extension. J. Geophys. Res., 110, doi: 10.1029/2004JB003114.

Huismans, R.S., Podlachikov, Y.Y. and Cloetingh, S., 2001. Transition from passive to active rifting: Relative importance of asthenospheric doming and passive extension of the lithosphere. J. Geophys. Res., 106, 11271–11291.

ILIHA DSS Group, 1993. A deep seismic sounding investigation of lithospheric heterogeneity and anisotropy beneath the Iberian Peninsula. Tectonophysics, 221, 35–51.

Izotova, T.S. and Popadyuk, I.V., 1996. Oil and gas accumulations in the Late Jurassic reefal complex of the West Ukrainian Carpathian foredeep. In Ziegler, P.A. and Horvàth, F., eds., Peri-Tethy Memoir 2, Structure and prospects of Alpine basins and forelands. Mém. Mus. natn. Hist. nat., Paris, 170, 375–390.

Janssen, M.E., Stephenson, R.A. and Cloetingh, S., 1995. Temporal and spatial correlations between changes in plate motions and the evolution of rifted basins in Africa. Geol. Soc. Am. Bull., 107, 1317–1332.

Janssen, M.E., 1996. Intraplate deformation in Africa as a consequence of plate boundary changes. PhD thesis, Vrije Universiteit, Amsterdam, 161 pp.

Jaupart, C. and Mareschal, J., 2006. Heat flow, Archean thermal regime and stabilization of cratons. EOS, Transactions, Am. Geophys. Union, December 2006, 87, Abs.

Jensen, S.L. and Thybo, P. and Polonaise 97 Working Group, 2002. Moho topography and lower crustal wide-angle reflectivity around the TESZ in southern Scandinavia and northeastern Europe. Tectonophysics, 360, 187–213.

Jimenez-Moreno, G. and Suc, J.P., 2007. Middle Miocene latitudinal climatic gradient in Western Europe: Evidence from pollen records. Palaeogeogr. Palaeoclimatol. Palaeoecol., 253, 208–225.

Juez-Larré, J. and Andriessen, P.A.M., 2006. Tectonothermal evolution of the northeastern margin of Iberia since the break-up of Pangea to present, revealed by low-temperature fission-track and (U-Th)/He thermochronology: A case history of the Catalan Coastal Ranges. Earth Planet. Sci. Lett., 243, 159–180.

Kaikkonen, P., Moiso, K. and Heeremans, M., 2000. Thermomechanical lithospheric structure of the central Fennoscandian Shield. Phys. Earth Planet. Inter., 119, 209–235.

Kapotas, S., Tselentis, G. and Martakis, N., 2003. Case study in NW Greece of passive seismic tomography: a new tool for hydrocarbon exploration. First Break, 1, 37–42.

Keeley, M.L. and Light, M.P.R., 1993. Basin evolution on the Argentine continental shelf. J. Petrol. Geol., 16, 451–464.

Khain, V.Y., 1992. The role of rifting in the evolution of the Earth's crust. Tectonophysics, 215, 1–7.

Khain, V.E., Sokolov, B.A., Kleshchev, K.A. and Shein, V.S., 1991. Tectonic and geodynamic setting of oil and gas basins of the Soviet Union. Am. Assoc. Petrol. Geol., 75, 313–325.

Kind, R., Yuan, X., Nelson, D., Sobolev, S.V., Mechie, J., Zhao, W., Kosarev, G., Ni, J., Achauer, U. and Jiang, M., 2002. Seismic images of crust and upper mantle beneath Tibet: Evidence for Eurasian plate subduction. Science, 298, 1219–1221.

Kirkwood, D., Lavoie, D., Malo, M. and Osadetz, K., eds., 2009. The history of convergent and passive margins in the Polar Realm: Sedimentary and tectonic processes, transitions and resources. Bull. Can. Petrol. Geol., in press.

Konstantinovskaia, E. and Malavieille, J., 2005. Erosion and exhumation in accretionary orogens: Experimental and geological approaches. Geochem. Geophys. Geosyst. (G3), 6, Q02006. doi 10.1029/2004GC000794.

Kooi, H., Hettema, M. and Cloetingh, S., 1991. Lithospheric dynamics and the rapid Pliocene-Quaternary subsidence in the North Sea basin. Tectonophysics, 192, 245–259.

Kooi, H. and Beaumont, C., 1996. Large-scale geomorphology: classical concepts reconciled and integrated with contemporary ideas via a surface process model. J. Geophys. Res., 101, 3361–3386.

Kounov, A., Niedermann, S., de Wit, M.J., Viola, G., Andreoli, M. and Erzinger, J., 2007. Present day denudation rates bearing on erosion processes along selected west- and south facing sections of the South African Great Escarpment and its interior derived from in situ produced cosmogenic ^{3}He and ^{21}Ne. South Afr. J. Geol., 110, 235–248.

Krijgsman, W., Hilgen, F.J., Raffi, I., Sierro, F.J. and Wilson, D.S., 1999. Chronology, causes and progression of the Messinian salinity crises. Nature, 400, 652–655.

Krijgsman, W., 2002. The Mediterranean: Mare Nostrum of Earth sciences. Earth Planet. Sci. Lett., 205, 1–12.

Kuiper, K.F., Hilgen, F.J., Steenbrink, J. and Wijbrans, J.R., 2004. 40Ar/39Ar ages of tephras intercalated in astronomical tuned Neogene sedimentary sequences in the Eastern Mediterranean. Earth Planet. Sci. Lett., 222, 583–597.

Kuznir N.J., Echtler H. and Wenzel, F., 1995. Structural and stratigraphic modelling of the Upper Rhinegraben. European Union of Geosciences, EUG VIII, Strasburg, 54.

Kuznir, N.J. and Park, R.G., 1987. The extensional strength of the continental lithosphere; its dependence on geothermal gradient, and crustal composition and thickness. Geol. Soc. Lond. Spec. Publ., 28, 35–52.

Kwadiba, M.T.O., Wright, C., Kgaswane, E.M., Simon, R.E. and Nguuri, T.K., 2003. Pn arrivals and lateral variations of Moho geometry beneath the Kaapvaal craton. Lithos, 71, 393–411.

Lacombe, O., Lavé, J., Roure, F. and Vergés, J., eds., 2007. Thrust belts and foreland basins: From fold kinematics to hydrocarbon systems. Frontiers in Earth Sciences, Springer, 487 pp.

Lacombe, O., Malandain, J., Vilassi, N., Amrouch, K. and Roure, F., 2009. From paleostress to paleoburial in fold-thrust belts: Preliminary results from calcite twin analysis in the Outer Albanides. In Bertotti, G., Frizon de Lamotte, D., Teixell, A. and Charroud, H., eds., The geology of vertical movements, Tectonophysics. in press.

Lagarde, J.-L., Baize, S., Amorese, D., Delcaillau, B. and Font, M., 2000. Active tectonics, seismicity and geomorphology with special reference to Normandy (France). J. Quat. Sci., 15, 745–758.

Lague, D., Crave, A. and Davy, P., 2003. Laboratory experiment stimulating the geomorphic response to tectonic uplift. J. Geophys. Res. Solid Earth, 108, B1, 2008.

Lamarche, J., Scheck, M. and Lewerenz, B., 2003. Heterogeneous tectonic inversion of the Mid Polish Basin related to crustal architecture, sedimentary patterns and structural inheritance. Tectonophysics, 373, 75–92.

Lankreijer, A.C., Bielik, M., Cloetingh, S. and Majcin, D., 1999. Rheology predictions across the Western Carpathians, Bohemian Massif and the Pannonian Basin: Implications for tectonic scenarios. Tectonics, 18, 1139.

Larroque, C., Guilhaumou, N., Stéphan, J.F. and Roure, F., 1996. Advection of fluids at the front of the Sicilian Neogene subduction complex. Tectonophysics, 254, 41–55.

Laubscher, H., 1970. Bewegung und Wärme in a lpinen Orogenese. Schweiz. Min. Petr. Mitt., 50, 3, 565–596.

Laubscher, H., 1988. Material balancing in Alpine orogens. Geol. Soc. Am. Bull., 100, 1313–1328.

Laubscher, H., 1990. The problem of the Moho in the Alps. Tectonophysics, 182, 9–20.

Lavier, L.L. and Manatschal, G., 2006. A mechanism to thin the continental lithosphere at magma-poor margins. Nature, Lond., 440, 7082, 324–328.

Le Douaran, S., Burrus, J. and Avedik, F., 1984. Deep structure of the Northwestern Mediterranean Basin: résults of a two ship seismic survey. Marine Geol., 55, 325–345.

Leever, K.A., Bertotti, G., Zoetemeijer, R., Matenco, L. and Cloetingh, S.A.P.L., 2006. The effects of a lateral variation in lithospheric strenght on foredeep evolution: Implications for the East Carpathians fordeedp. Tectonophysics, 421, 3–4, 251–267.

Leever, K., Garcia-Castellanos, D., Matenco, L. and Cloetingh, S., 2007. Re-establishing the connection between Central and Eastern Paratethys: incision of the Danube in the Iron Gates. EGU, 2007, Abs.

Lefort, J.P. and Agarwal, B.N.P., 1996. Gravity evidence for an Alpine buckling of the crust beneath the Paris Basin. Tectonophysics, 58, 1–14.

Lefort, J.P., Montadert, L., Pinet, B., Schroeder, I.J. and Sibuet, J.C., 1991. Structure profonde de la marge des entrées de la Manche et du plateau continental celtique: le profil WAM. In Bois, C., Gariel, O. and Pinet, B., eds., Géol. Soc. France, Mém., 159.

Lenotre, N., Thierry, P., Blanchin, R. and Brochard, G., 1999. Current vertical movement demonstrated by comparative leveling in Brittany, France. Tectonophysics, 301, 33–344.

Leroy, S., Bonneville, A., Huchon, P. and Lucazeau, F., 2004. From rifting to spreading in the eastern Gulf of Aden: a geophysical survey of a young oceanic basin from margin to margin. Terra Nova, 16, 185–192.

Light, M.P.R., Maslanyj, M.P., Greenwood, B.J. and Banks, N.L., 1993. Seismic sequence stratigraphy and tectonics offshore Namibia. Geol. Soc. Lond. Spec. Publ., 71, 163–191.

Lingrey, S., 2007. Cenozoic deformation of Trinidad: Foldbelt restoration in a region of significant strike-slip. In Lacombe, O., Roure, F., Lavé, J. and Vergés, J., eds., Thrust belts and foreland basins, Frontiers in Geosciences, Springer, 163–178.

Littke, R., Bayer, U., Gajewski, D. and Nelskamp, S., eds., 2008. Dynamic of complex intracratonic basins. The Central European basin system. Springer, 520 pp.

Louden, K.E. and Chian, D., 1999. The deep structure of non-volcanic rifted continental margins. Phil. Trans. Royal Soc. Lond. Ser. A, 357, 767–804.

Lustrino, M. and Wilson, M., 2007. The Circum-Mediterranean anorogenic Cenozoic igneous provice. Earth-Sci. Rev., 81, 1–65.

Magri, F., Bayer, U., Clausnitzer, V., Jahnke, C., Fuhrman, J., Möller, P., Pekdeger, A., Tesmer, M. and Voigt, H., 2005a. Deep reaching fluid flow close to convective instability in the North East German Basin. Results from water chemistry and numerical modelling. Tectonophysics, 397, 5–20.

Magri, F., Bayer, U., Jahnke, C., Clausmitzer, V., Diersch, H.J., Fuhrman, J., Möller, P., Pekdeger, A., Tesmer, M. and Voigt, H.T., 2005b. Fluid dynamics driving saline water in the North East German Basin. Int. J. Earth Sci., 94, 1056–1069.

Magri, F., Bayer, U., Pekdeger, A., Otto, R., Thomsen, C. and Maiwald, U., 2008. Salty groundwater flow in the shallow and deep aquifer systems of the Schleswig-Holstein area (North German Basin). Tectonophysics, doi: 10.1016/j.tecto.2008.04.019.

Magri, F., Bayer, U., Tesmer, M., Möller, P. and Pekdeger, A., 2007. Salinization problems in the North East German Basin: Results from thermohaline simulation. Int. J. Earth Sci. Geolol. Rundschau, DOI-10.1007/00531-007.0209-8.

Malavieille, J., 1984. Modélisation expérimentale des chevauchements imbriqués: application aux chaînes de montagnes. Bull. Soc. Géol. France, 7, 129–138.
Manatschal, G., 2004. New models for evolution of magma-poor rifted margins based on a review of data and concepts from West Iberia and the Alps. Int. J. Earth Sci., 93, 432–466.
Manatschal, G. and Bernoulli, D., 1999. Architecture and tectonic evolution of non-volcanic margins: Present-day Galicia and ancient Adria. Tectonics, 18, 6, 1099–1199.
Manatschal, G., Froitzheim, N., Rubenach, M.J. and Turrin, B., 2001. The role of detachment faulting in the formation of an ocean-continent transition: insights from the Iberia abyssal plain. In Wilson, R.C.L., Whitmarsh, R.B., Taylor, B. and Froitzheim, N., eds., Non-volcanic rifting of continental margins: evidence from land and sea. Geol. Soc. London, Spec. Publ., 187, 405–428.
Marotta, A.M., Bayer, U. and Thybo, H., 2000. The legacy of the NE German Basin – Reactivation by compressional buckling. Terra Nova, 12, 132–140.
Martakis, N., Kapotas, S. and Tselentis, G.A., 2006. Integrated passive seismic acquisition and methodology: Case studies. Geophys. Prospect., 54, 6, 829–847.
Martinez, A., Malavieille, J., Lallemand, S. and Collot, J.Y., 2002. Strain partitioning in an accretionary wedge, in oblique convergence: analogue modelling. Bull. Soc. Géol. France, 173, 1, 17–24.
Mascle, A., Puigdefabregas, C., Luterbacher, H.P. and Fernandez, M., eds., 1998. Cenozoic foreland basins of Western Europe. Geol. Soc. Lond. Spec. Publ., 134, 422.
Matenco, L. and Bertotti, G., 2000. Tertiary tectonic evolution of the external East Carpathians (Romania). Tectonophysics, 316, 255–286.
Matenco, L., Bertotti, G., Leever, K., Cloetingh, S., Schmid, S., Tarapoanca, M. and Dinu, C., 2007. Large-scale deformation in a locked collisional boundary: interplay between subsidence and uplift, intraplate stress, and inherited lithospheric structures in the late stage of the SE Carpathians evolution. Tectonics, doi: 10.1029/2006TC001951.
Matenco, L., Zoetemeijer, R., Cloetingh, S. and Dinu, C., 1997. Tectonic flexural modelling of the Romanian Carpathians foreland system. Tectonophysics, 282, 147–166.
Mattioni, L., Sassi, W. and Callot, J.P., 2007. Analogue models of basin inversion by transpression: role of structural heterogeneity. Geol. Soc., London, Spec. Publ., 272, 397–417.
Mauffret, A., Maillard, A., Pascal, G., Torné, M., Buhl, P. and Pinet, B., 1992. Long listening MCS profiles in the Valencia Trough (Valsis 2) and the Gulf of Lions (ECORS): a comparison. In Banda, E. and Surinach, P., eds., Geology and geophysics of the Valencia Trough, Western Mediterranean, Tectonophysics. 203, 285–304.
Mauffret, A., Pascal, G., Maillard, A. and Gorini, C., 1995. Tectonics and deep structure of the Northwestern Mediterranean Basin. Mar. Petrol. Geol., 12, 6, 645–666.
Maystrenko, Y., Bayer, U. and Scheck-Wenderoth, M., 2006. 3D reconstruction of salt movements within the deepest post-Permian structure of the Central European Basin System – the Glucckstadt Graben. Netherlands J. Geosci. Geologie en Mijnbouw, 85, 3, 181–196.
Maystrenko, Y., Stovba, S., Stephenson, R., Bayer, U., Menyoli, E., Gajewski, D., Huebscher, C., Rabbel, W., Saintot, A., Starostenko, V., Thybo, H. and Tolkunov, A., 2003. Crustal-scale pop-up structure in cratonic lithosphere: DOBRE deep seismic reflection study of the Donbas Foldbelt, Ukraine. Geology, 31, 733–736.
Mazur, S., Scheck-Wenderoth, M. and Krzywiec, P., 2005. Different modes of the Late Cretaceous – Early Tertiary inversion in the North German and Polish basins. Int. J. Earth Sci., 94, 5–6, 782–798.
McClay, K.R., 1989. Physical Models of structural styles during extension. In: Tankard, A.J. and Balkwill, H.R., eds., Extensional Tectonics and Stratigraphy of the North Atlantic Margins. Am. Assoc. Petrol. Geol., Mem., 46, 95–110.
McClay, K.R., 1990. Experimental modelling using analogue materials: deformation mechanics in analogue models of extensional fault systems. In Knipe, R.J. and Rutter, E.H., eds., Deformation mechanisms, rheology and tectonics. Geol. Soc. London, Spec. Publ., 54, 445–453.
McClay, K.R., Whitehouse, P.S., Dooley, T. and Richards, M., 2004. 3-D evolution of thrust belts formed by oblique convergence. Mar. Petrol. Geol., 21, 7, 857–877.
McKenzie, D., 1978. Some remarks on the development of sedimentary basins. Earth Planet. Sci. Lett., 40, 25–32.
Meesters, A. and Dunai, T.J., 2002a. Solving the production-diffusion equation for finite diffusion domains of various shapes; Part I, Implications for low-temperature (U-Th)/He thermochronology. Chem. Geol., 186, 333–344.
Meesters, A. and Dunai, T.J., 2002b. Solving the production-diffusion equation for finite diffusion domains of various shapes; Part II, application to cases with alpha-ejection and nonhomogeneous distribution of the source. Chem. Geol., 186, 57–73.
Meesters, A. and Dunai, T.J., 2002c. Erratum solving the production-diffusion equation for finite diffusion domains of various shapes; Part II, application to cases with alpha-ejection and nonhomogeneous distribution of the source. Chem. Geol., 186, 347–363.
Meissner, R., 1989. Rupture, creep, lamellae and crocodiles: happenings in the continental crust. Terra Nova, 1, 17–28.
Meissner, R., Wever, T.H. and Fluh, E.R., 1987. The Moho of Europe -implications for crustal development. Ann. Geophys., 5, 357–364.
Mengel, K., Sachs, P.M., Stosch, H.G., Wörner, G. and Look, G., 1991. Crustal xenoliths from Cenozoic volcanic fields of West Germany: implication for structure and composition of the continental crust. Tectonophysics, 195, 271–289.
Meyer, W. and Stets, J., 1998. Young Pleistocene to Recent uplift in the Rhenish Massif. In Neugebauer, H.J., ed., Young tectonics, magmatism, fluids; a case study of the Rhenish Massif, Schriftenreihe Sonderforschungsbereich, Bonn, Abs. 350, 74, 7–10.
Midzi, V., Singh, D.D., Atakan, K. and Havskov, J., 1999. Transitional continental-oceanic structure beneath the Norwegian Sea from inversion of surface wave group velocity data. Geophys. J. Inter., 139, 2, 433–446.
Milanovsky, E.E., 1992. Aulacogens and aulacogeosynclinal zones: regularities of their setting and tectonic development. Tectonophysics, 215, 55–68.
Millán, H., den Bezemer, T., Vergés, J., Marzo, M., Muñoz, J.A., Roca, E., Cirés, J., Zoetemeijer, R., Cloetingh, S. and Puigdefàbregas, C., 1995a. Paleo-reliefs at mountain ranges: inferences from flexural modelling in the Eastern Pyrenees and Ebro Basin. In S. Cloetingh, B. Durand, C. Puigdefàbregas, eds., Mar. Petrol. Geol., 12, 917–928.

Mjelde, R., Raum, T., Breivik, A., Shimamura, H., Murai, Y., Takanami, T. and Faleide, J.I., 2005. Crustal structure of the Vøring margin, NE Atlantic: a review of geological implications based on recent OBS data. In Doré, A.G. and Vining, B.A., eds., North west European petroleum geology and global perspectives, Proceedings of the 6th conference. Geol. Soc., London.

Mohriak, W.V., Rosendahl, B.R., Turner, J.P. and Valente, S.C., 2002. Crustal architecture of South Atlantic volcanic margins. In Menzies, M.A., Klemperer, S.L., Ebinger, C.J. and Baker, J., eds., Volcanic rifted margins, Geol. Soc. Am., Spec. Paper, 362, 159–202.

Moisio, K., Kaikkonen, P. and Beekman, F., 2000. Rheological structure and dynamical response of the DSS profile BALTIC in the SE Fennoscandian shield. Tectonophysics, 320, 175–194.

Monaco, C., Tortorici, L., Catalano, R., Paltrinieri, W. and Steel, N., 2007. The role of Pleistocene strike-slip tectonics in the Neogene-Quaternary evolution of the Southern Apennines orogenic belt: implications for oil trap development. J. Petrol. Geol., 24, 3, 339–359.

MONA LISA Working Group, 1997. Deep seismic investigations of the lithosphere in the southeastern North Sea. Tectonophysics, 269, 1–19.

Montelli, R., Nolet, G., Dahlen, F.A., Masters, G., Engdahl, E.R. and Hung, S.H., 2004. Finite-frequency tomography reveals a variety of plumes in the mantle. Science, 303, 338–343.

Mooney, W.D. and Vidale, J.E., 2003. Thermal and chemical variations in subcrustal cratonic lithosphere; Evidence from crustal isostacy. Lithos, 71, 185–193.

Mora, A., Parra, M., Strecker, M.R., Sobel, E.R., Hooghiemstra, H., Torres, V. and Jaramillo, J.V., 2008. Climatic forcing of asymmetric orogenic evolution in the Eastern Cordillera of Colombia. Geol. Soc. Am. Bull., 120, 7–8, 930–949.

Moretti, I., 2008. Working in complex areas: New restoration workflow based on quality control, 2D and 3 D restorations. Marine Petrol. Geol., 25, 205–218.

Moretti, I., Lepage, F. and Guiton, M., 2006. 3D restoration: Geometry and geomechanics. Oil Gas Sci. Technol. Revue de l'IFP, 61, 2, 277–289.

Mosar, J., 2003. Scandinavias North Atlantic passive margin. J. Geophys. Res., 108 (B8), 2360.

Mosar, J., Eide, E.A., Osmundsen, P.T., Sommaruga, A. and Torvik, T., 2002. Greenland – Norway separation: A geodynamic model for the North Atlantic. Norwegian J. Geol., 82, 281–298.

Mueller, R.D., Sdrolias, M., Gaina, C., Steinberger, B. and Heine, C., 2008. Long-term sea-level fluctuations driven by ocean basin dynamics. Science, 319, 5868, 1357–1362.

Mugnier, J.L., Baby, P., Colletta, B., Vinoue, P., Bale, P. and Leturmy, P., 1997. Thrust geometry controlled by erosion and sedimentation: A view from analogue models. Geology, 25, 427–430.

Mulch, A. and Chamberlain, C.P., 2007. Stable isotope paleoaltimetry in orogenic belts. The silicate record in surface and crustal archives. Rev. Mineral. Geochem., 66, 1, 89–118.

Mulch, A., Chamberlain, C.P., Teyssier, C. and Graham, S.A., 2008. Stable isotope paleoaltimetry: Coupled climate and surface uplift records in orogenic belts. Geophysical Research Abs, 10, EGU-2008-A-06417.

Mulch, A., Sarna-Wojcicki, A.S., Perkins, M.E. and Chamberlain, C.P., 2008. A Miocene to Pliocene climate and elevation record of the Sierra Nevada and Great Basin region. Proc. Nat. Ac. Sc, in press.

Necea, D., Fielitz, W. and Matenco, L., 2005. Late Pliocene-Quaternary tectonics in the frontal part of the Se Carpathians: Insights from tectonic geomorphology. In The evolution of the Carpathians-Pannonian system: Interaction between neotectonics, deep structure, polyphase orogeny and sedimentary basins in a source to sink natural laboratory. Tectonophysics, 410, 111–137.

NEIC, 2004. National Earthquake Information Center, neic.usgs.gov.

Nemes, F., Neubauer, F., Cloetingh, S. and Genser, J., 1997. The Klagenfurt Basin in the Eastern Alps: an intra-orogenic decoupled flexural basin? Tectonophysics, 282, 189–203.

Neubauer, F., Cloetingh, S., Dinu, C. and Mocanu, V., eds., 1997. Tectonics of the Alpine-Carpathian-Pannonian region. Tectonophysics, 272.

Neumann, E.-R., Wilson, M., Heeremans, M., Spencer, E.A., Obst, K., Timmerman, M.J. and Kirsten, L., 2004. Carboniferous- Permianm rifting and magmatism in southern Scandinavia, the North Sea and northern Germany: a review. In: Wilson M., Neumann, E.-R., Davies, G.R., Timmerman, M.J., Heerman, M. and Larsen, B.T., eds., Permo-Carboniferous magmatism and rifting in Europe. Geol. Soc., Lond., Spec. Publ., 223, 11–40.

Neurdin-Trescartes, J., 1995. Evolution du bassin néogène du Chélif (Algérie nord-occidentale). Un exemple d'interaction sédimentation-tectonique. In Arbey, F. and Lorenz, J., eds., Bassins sédimentaires africains, Géodynamique et géologie séquentielle, biominéralisation, sédimentation et organismes, 4° Colloque de géologie africaine, Pau, 25–29 Octobre 1993, 118° Congrès national des Sociétés Historiques et Scientifiques, 183–196.

Nguuri, T.K., Gore, J., James, D.E., Webb, S.J., Wright, C., Zengeni, T.G., Gwavana, O. and Snoke, J.A. and Kaapsvaal Seismic Group, 2001. Crustal structure beneath southern Africa and its implication for the formation and evolution of the Kaapvaal and Zimbabwe cratons. Geopys. Res. Lett., 28, 2501–2504.

Nielsen, L., Thybo, H. and Egorkin, A.V., 2002. Implications of seismic scattering below the 8 degrees discontinuity along PNE profile Kraton. Tectonophysics, 358, 135–150.

Nieuwland, D.A., 2003. Introduction: New insights into structural interpretation and modelling. In Nieuwland, D.A., ed., New insights into structural interpretaion and modelling. Geol. Soc., London, Sp. Publ., 212, 1–5.

Nieuwland, D.A., Leutscher, J.H. and Gast, J., 2000. Wedge equilibrium in fold-and-thrust belts. Prediction of out-of-sequence thrusting, based on sandbox experiments and natural examples. Geologie en Mijnbouw, 79, 1, 81–91.

Nieuwland, D.A., Urai, J. and Knoop, M., 1999. In-situ stress measurements in model experiments of tectonic faulting. In Lehner, F. and Urai, J., eds., Aspects of tectonic faultin. g, Springer Verlag.

Nikishin, A.M., Cloetingh, S., Lobkovsky, L. and Burov, E.B., 1993. Continental lithosphere folding in Central Asia (Part I): constraints from geological observations. Tectonophysics, 226, 59–72.

Nikishin, A.M., Ziegler, P.A., Abbott, D., Brunet, M.-F.- and Cloetingh, S., 2002. Permo-Triassic intraplate magmatism and magma dynamics. Tectonophysics, 351, 3–39.

Nivière, B. and Winter, T., 2000. Pleistocene northwards fold propagation of the Jura within the southern Rhine Graben: seismotectonic implications. Global Planet. Change, 27, 263–288.

Nottvedt A., ed., 2000. Dynamics of the Norwegian margin. Geol. Soc. London, Spec. Publ. 167.

Okaya, N., Cloetingh, S. and Mueller, S., 1996. A lithospheric cross section through the Swiss Alps (part II): constraints on the mechanical structure of a continent–continent collision zone. Geophys. J. Int., 127, 399–414.

Oncken, O., Chong, G., Franz, G., Giese, P., Götze, H.J., Ramos, V.A., Stecker, M.R. and Wiggen, P., eds., 2006. The Andes: Active subduction orogeny. Frontiers in Earth Sciences, Springer, 592 pp.

Osmundsen, P.T., Sommaruga, A., Skilbrei, J.R. and Olesen, O., 2002. Deep structure of the Mid Norway rifted margin. Norwegian J. Geol., 82, 205–224.

Oszczypko, N., 2006. Late Jurassic-Miocene evolution of the Outer Carpathian fold-and-thrust belt and its foredeep basin (Western Carpathians, Poland). Geol. Quarter., 50, 168–194.

Panza, G.F., 1983. Lateral variations in the European lithosphere and seismic activity. Phys. Earth Planet. Inter., 33, 3, 194–197.

Parson, B. and Sclater, J.G., 1977. An analysis of the variation of the ocean floor bathymetry; heat flow and with age. J. Geophys. Res., 82, 5, 803–827.

Paton, D.A., van der Spuy, D., di Primio, R. and Horsfield, B., 2007. Insights into the hydrocarbon system evolution of the Southern Orange Basin. J. South Afr. Geol., 110, 261–274.

Pérez-Gussinyé, M. and Watts, A.B., 2005. The long-term strength of Europe and its implications for plate forming processes. Nature, 436, doi:10.1038/nature03854.

Persson, K.S., Garcia-Castellanos, D. and Sokoutis, D., 2004. River transport effects on compressional belts: First results from an integrated analogue-numerical model. J. Geophys. Res., 109, 10.1029/2002JB002274.

Persson, K.S. and Sokoutis, D., 2002. Analogue models of orogenic wedges controlled by erosion. Tectonophysics, 356, 4, 323–336.

Peterson, J.A. and Clarke, J.W., 1991. Geology and hydrocarbon habitat of the West Siberian Basin. Am. Assoc. Petrol. Geol. Stud. Geol., 12, 96p.

Pfiffner, O.A., Ellis, S. and Beaumont, C., 2000. Collision tectonics in the Swiss Alps: Insight from geodynamic modelling. Tectonics, 19, 6, 1065–1094.

Pharaoh, T.C., Winchester, J.A., Verniers, J., Lassen, A. and Seghedi, A., 2006. The Western accretionary margin of the East European Craton: an overview. In: Gee, D.G. and Stephnsin, R.A., eds., European Lithosphere Dynamics. Geol. Soc., Lond., Spec. Publ., 32, 291–311.

Picotti, V., Capozzi, R., Bertozzi, G., Mosca, F., Sitta, A. and Tornaghi, M., 2007. The Miocene petroleum system of the Northern Apennines in the Central Po Plain (Italy). In: Lacombe, O., Lavé, J., Roure, F. and Vergés, J., eds., Thrust belts and foreland basins. Springer, Frontiers in Earth Sciences, 117–131.

Pieri, M., 1983. Three seismic profiles through the Po Plain. In Bally, A.W., ed., Seismic expression of structural styles. A picture and work atlas, Am. Assoc. Petrol. Geol., Studies in Geology, 15, 3.4.1/8-3.4.1/26.

Piromallo, C. and Morelli, A., 1997. Imaging the Mediterranean upper mantle by P-wave travel time tomography. Ann. Geofisica, XL, 963–979.

Piromallo, C. and Morelli, A., 2003. P wave tomography of the mantle under the Alpine-Mediterranean area. J. Geophys. Res., 108, B2, 2065.

Plomerova, J., Kouba, D. and Babuska, V., 2002. Mapping the lithosphere-asthenosphere boundary through changes in surface-wave anisotropy. Tectonophysics, 358, 175–185.

Prijac, C., Doin, M.P., Gaulier, J.M. and Guillaucheau, F., 2000. Subsidence of the Paris Basin and its bearing on the late Variscan lithosphere evolution: comparison between Plate and Chablis models. Tectonophysics, 323, 1–38.

Pysklywec, R.N. and Mitrovica, J.X., 2000. Mantle flow mechanisms of epirogeny and their possible role in the evolution of the Western Canada sedimentary basin. Can. J. Earth Sci., 37, 1535–1548.

Quinlan, G., 1987. Models of subsidence mechanism in cratonic basins and applicability to North American examples. In: Beaumont, C. and Tankard, A., eds., Sedimentary basins and basin forming mechanisms, Canadian Soc. Petrol. Geol. Memoir 12, 463–481.

Rabbel, W., Förste, K., Schulze, A., Bittner, R., Röhl, J. and Reichert, J.C., 1995. A high-velocity layer in the lower crust of the North German Basin. Terra Nova, 7, 327–337.

Ranalli, G., 1995. Rheology of the Earth. 2nd edn. Chapman and Hall, London, 413 pp.

Ranalli, G. and Murphy, D.C., 1987. Rheological stratification of the lithosphere. Tectonophysics, 132, 281–295.

Reicherter, K., Kaiser, A. and Stackebrandt, W., 2005. The post-glacial landscape evolution of the North German Basin: morphology, neotectonics and crustal deformation. Int. J. Earth Sciemces, 94, 1083–1093.

Reiners, P.W. and Ehlers, T.A., eds., 2005. Low-temperature thermochronolog: Techniques, interpretations, andaApplications. Reviews in Mineralogy & Geochemistry, 58, 662 pp.

Reston, T.J., 1990. The lower crust and the extension of the continental lithosphere; kinematic analysis of BIRPS deep seismic data. Tectonics, 9, 1235–1248.

Reston, T.J., 2007. The formation of non-volcnic rifted margins by the progressive extension of the lithosphere: example of the West Iberian margin. In: Karner, D.G., Manatschal, G. and Pinheiro, I.M., eds., Imaging, mapping and modelling continental lithospheric extension and breakup. Geol. Soc., Lond., Spec. Publ., 282, 77–110.

Ritter, J.R.R., Achauer, U. and Christensen, U.R., 2000. The teleseismic tomography experiment in the Eifel region, central Europe: design and first results. Seismol. Res. Lett., 71, 437–443.

Ritter, J.R.R., Jordan, M., Christensen, U.R. and Achauer, U., 2001. A mantle plume below the Eifel volcanic fields, Germany. Earth Planet. Sci. Lett., 186, 7–14.

Robin, C., Allemand, P., Burov, E., Doin, M.P., Guillocheau, F., Dromart, G. and Garcia, J.P., 2003. Vertical movement of the Paris Basin (Triassic-Pleistocene): from 3D stratigraphic database to numerical models. In Nieuwland, D.A., ed., New insights into structural interpretation and modelling. Geol. Soc., Lond., Spec. Publ., 225–250.

Robion, P., Faure, J.L. and Swennen, R., 2004. Late Cretaceous chemical remagnetization of the Paleozoic carbonates from the undeformed foreland of the Western Canadian Cordillera. In Swennen, R., Roure, F. and Granath, J., eds., Deformation, fluid flow and reservoir appraisal in foreland fold-and-thrust belts. Am. Assoc. Petrol. Geol., Hedberg Series, 1.

Roca, E., Bessereau, G., Jawor, E., Kotarba, M. and Roure, F., 1995. Pre-Neogene evolution of the Western Carpathians: constraints from the Bochnia-Tatra mountains section (Polish Western Carpathians). Tectonics, 14, 855–873.

Roca, E., Frizon de Lamotte, D., Mauffret, A., Bracène, R., Vergés, J., Benaouali, N., Fernandez, M., Muñoz, J.A. and Zeyen, H., 2004. Transmed Transect II (Aquitaine Basin, Pyrenees, Ebro Basin, Catalan Coastal Ranges, Valencia Trough, Balearic Promontory, Algerian Basin, Tell, Sahara Atlas, Sahara Platform). In: Cavazza, W., Roure, F., Spakman, W., Stampfli, G.M. and Ziegler, P.A., eds., Transmed Atlas. Springer-Verlag, Berlin, Heidelberg.

Romanowicz, B. and Gung, Y., 2002. Superplumes from the core-mantle boundary to the lithosphere: Implications for heat flux. Science, 296, 513–516.

Roure, F., ed., 1994. Peri-Tethyan platforms. Proceedings of the IFP/Peri-Tethys Research Conference, March 23–25 1993, Arles. Editions Technip, Paris, 294 p.

Roure, F., 2008. Foreland and hinterland basins: What is controlling their evolution? Alpine Workshop, Davos Proceedings. Swiss J. Geosci., 101.

Roure, F., Alzaga, H., Callot, J.P., Ferket, H., Granjeon, D., Gonzales, E., Guilhaumou, N., Lopez, M., Mougin, P., Ortuño, S. and Séranne, M., 2009. Long lasting interactions between tectonic loading, unroofing, postrift thermal subsidence and sedimentary transfer along the Western margin of the Gulf of Mexico: Some insights from integrated quantitative studies. In Bertotti, G., Frizon de Lamotte, D. and Teixell, A., eds., Tectonics of vertical movements, Special issue, Tectonophysic. s, in press.

Roure, F., Bordas-Lefloch, N., Toro, J., Aubourg, C., Guilhaumou, N., Hernandez, E., Lecornec-Lance, S., Rivero, C., Robion, P. and Sassi, W., 2003. Petroleum systems and reservoir appraisal in the Subandean basins (eastern Venezuela and eastern Colombian foothills). In Bartolini, C., Burke, K., Buffler, R., Blickwede, J. and Burkart, B., eds., Mexico and the Caribbean region: plate tectonics, basin formation and hydrocarbon habitats. Am. Assoc. Petrol., Spec. Publ., 79, 34, 7507#x2013;775 Geol., Mem. 79, Ch. 34.

Roure, F., Brun, J.P., Colletta, B. and Van Den Driessche, J., 1992. Geometry and kinematics of extensional structures in the French Alpine foreland (basin of SE France). J. Struct. Geol., 14, 503–519.

Roure, F., Brun, J.P., Colletta, B. and Vially, R., 1994. Multiphase extensional structures, fault reactivation, and petroleum plays in the Alpine Foreland Basin of Southeastern France. In "Hydrocarbon and petroleum geology of France". In Mascle, A., ed., Eur. Assoc. Petrol. Geol., Paris, Springer-Verlag, Spec. Publ, 4, 245–268.

Roure, F., Casero, P. and Vially, R., 1991. Growth processes and mélange formation in the southern Apennine accretionary wedge. Earth Planet. Sci. Lett., 102, 395–412.

Roure, F., Choukroune, P., Berastegui, X., Munoz, J.A., Villien, A., Matheron, P., Baryer, M., Séguret, M., Camara, P. and Déramond, P., 1989. ECORS deep seismic data and balanced cross-sections: geometric constraints on the evolution of the Pyrénées. Tectonics, Washington, 8, 1, 41–50.

Roure, F., Choukroune, P. and Polino, R., 1996. Deep seismic reflection data and new insights on the bulk geometry of mountain ranges. C. R. Acad. Sci., Paris, série IIa, 322, 345–359.

Roure, F. and Colletta, B., 1996. Cenozoic inversion structures in the foreland of the Pyrenees and Alps. In Ziegler, P. and Horvath, F., eds., Structure and prospects of Alpine basins and forelands, PeriTethys Mem. 2, Mus. Hist. Nat., Paris, 173–210.

Roure, F., Colletta, B., De Toni, B., Loureiro, D., Passalacqua, H. and Gou, Y., 1997. Within-plate deformations in the Maracaibo and East Zulia basins, Western Venezuela. Mar. Petrol. Geol., 14, 139–163.

Roure, F., Ellouz, N., Shein, S.S. and Skvortsov, I.I., eds., 1996. Geodynamic evolution of sedimentary basins. Proceedings of the international symposium, Moscow, 18–23 May 1992. Editions Technip, Paris, 480 p.

Roure, F., Heitzmann, P. and Polino, R., eds., 1990. Deep structure of the Alps. Soc. Géol. France, Paris, Mém. 156, Soc. Géol. Suisse, Zürich, Mém. 1, Soc. Geol. Italiana, Roma, Mem. 1, 350 pp.

Roure, F., Nazaj, S., Mushka, K., Fili, I., Cadet, J.P. and Bonneau, M., 2004. Kinematic evolution and petroleum systems: an appraisal of the Outer Albanides. In McKlay, K., ed., Thrust tectonics and hydrocarbon systems. Am. Assoc. Petrol. Geol., Mem. 82, Ch. 24, 474–493.

Roure, F. and Sassi, W., 1995. Kinematics of deformation and petroleum system appraisal in Neogene foreland fold-and-thrust belts. Petrol. Geosci., 1, 253–269.

Roure, F., Séguret, M. and Villien, A., 1988. Structural styles of the Pyrénées: a view from seismic reflexion to surface studies. Guide-Book Field-Trip 3, Am. Assoc. of Petrol. Geol., Mediterranean basins conference, Nice, 140 p.

Roure, F., Swennen, R., Schneider, F., Faure, J.L., Ferket, H., Guilhaumou, N., Osadetz, K., Robion, Ph. and Vendeginste, V., 2005. Incidence and importance of tectonics and natural fluid migration on reservoir evolution in foreland fold-and-thrust belts. In Brosse, E. et al., eds., Oil and Gas Science and Technology, Oil and Gas Science and Technology-Revue de l'IFP. 60, 67–106.

Royden, L.H., 1985. The Vienna Basin: a thin-skinned pull-apart basin. In Biddle, K.T. and Christie-Blick, N., eds., Strike-slip deformation, basin formation and sedimentation. SEPM, Spec. Publ., 37, 319–338.

Rowan, E.L., Goldhaber, M.B. and Hatch, J.R., 2002. Regional fluid flow as a factor in the thermal history of the Illinois Basin: Constraints from fluid inclusions and the maturity of Pennsylvanian coals. Am. Assoc. Petrol. Geol., 862, 257–277.

Rowley, D.B., 2007. Stable isotope-based paleoaltimetry: Theory and validation. Rev. Mineral. Geochem., 66, 1, 23–52.

Rowley, D.B. and Garzione, C.N., 2007. Stable isotope-based paleoaltimetry. Ann. Rev. Earth Planet. Sci., 35, 463–508.

Rowley, D.B., Pierrehumbert, R.T. and Currie, B.S., 2001. A new approach to stable isotope-based paleoaltimetry: implications for paleoaltimetry and paleohypsometry of the High Himalaya since the Late Miocene. Earth Planet. Sci. Lett., 188, 253–268.

Rudkewich, M.Y., 1988. Oil-gas complexes of the West Siberian basin. Nedra, Moscow 304p.

Rudkewich, M.Y., 1994. Evolution of the West SiberianBasin. In: Roure, F., Ellouz, N., Shein, V.S. and Skvortsov, I., eds., Geodynamic Evolution of Sedimentary Basins. Edition Technip, Paris, 123–134.

Rudkiewicz, J.L. and Carpentier, J.L., 2005. Basin modelling. IFP Short Course, Am. Assoc. Petrol. Geol., Paris Convention. Abs

Ruszkiczay-Rudiger, Z., Dunai, T., Bada, G., Fodor, L. and Horvath, E., 2005. Middle to late Pleistocene uplift rate of the Hungarian Mountain Range at the Danube Bend, (Pannonian Basin) using in situ produced 3He. Tectonophysics, 410, 157–173.

Ryberg, T., Tittgemeyer, M. and Wenzel, F., 2000. Finite difference modelling of P-wave scattering in the upper mantle. Geophys. J. Int., 141, 787–800.

Sabato, L., Pieri, P. and Tropeano, M., 2005. A lacustrine system controled by a fault-propagation fold in the Sant'Arcangelo Basin (South Apennines, Italy). Geophys. Res. Abs, 7, EGU, 02183.

Saeed, A.M., Sassi, W., Schleder, Z., Graham, S. and Schackleton, R., 2008. Quantitative analysis of the structural evolution of a 4D-analogue sandbox experiment to constrain models of palinspastic restoration. Am. Assoc. Petrol. Geol., Annual Convention, San Antonio, Abs.

Sanders, C., Andriessen, P. and Cloetingh, S., 1999. Life cycle of the Esat Carpathian Orogen: erosion history of a doubly convergent critical wedge assessed by fission track thermochronology. J. Geophys. Res., 104, 29095–29112.

Sassi, W. and Faure, J.L., 1996. Role of faults and layer interfaces on the spatial variation of stress regimes in basins: Inferences from numerical modelling. Tectonophysics, 266, 1–4, 101–119.

Sassi, W. and Rudkiewicz, J.L., 2000. Computer modelling of petroleum systems along regional cross-sections in foreland fold and thrust belts. In Geology and Petroleum Geology of the Mediterranean and Circum Mediterranean Basins, Malta, 1–4 October 2000, Proceedings, Extended Abs., C27–1–4.

Sauer, R., Seifert, P. and Wessely, G., 1992. Guidebook to excursion in the Vienna Basin and the adjacent Alpine-Carpathian thrustbelt in Austria. Mitteilungen der Osterreichischen Geologischen Gesellschaft, 85.

Scheck, M. and Bayer, U., 1999. Evolution of the Northeast German Basin – inferences from 3D structural modelling and subsidence analysis. Tectonophysics, 313, 145–169.

Scheck, M., Bayer, U., Otto, V., Lamarche, J., Banka, D. and Pharaoh, T., 2002. The Elbe Fault system in North Central Europe – a basement controlled zone of crustal weakness. Tectonophysics, 360, 281–299.

Scheck, M., Bayer, U. and Lewerenz, B., 2003. Salt movements in the Northeast German Basin and its relation to major post-Permnian tectonic phases – results from 3D structural modelling, backstripping and reflection seismic data. Tectonophysics, 361, 277–299.

Scheck-Wenderoth, M., Bayer, U. and Roure, F., eds., 2009a. Sedimentary basin dynamics. Spec. Issue, Marine Petrol. Geol., in press.

Scheck-Wenderoth, M., Faleide, J.I., Raum, T., Mjelde, R. and Horsfield, B., 2007. The transition from the continent to the ocean – a deeper view the Norwegian margin. J. Geol. Soc., Lond., 164, 4, 855–868.

Scheck-Wenderoth, M., Krzywiec, P., Maystrenko, Y., Zühlke, R. and Froitzheim, N., 2008. Permian to Cretaceous tectonics of Central Europe. In McCann, T., ed., The geology of Central Europe, Vol. 2: Mesozoic and Cenozoic. Geol. Soc., Lond., pp. 999–1030.

Scheck-Wenderoth, M. and Lamarche, J., 2005. Crustal memory and basin evolution in the Central European Basin System – new insights from a 3D structural model. Tectonophysics, 397, 1–2, 143–165.

Scheck-Wenderoth, M., Lamarche, J., Bayer, U., Lewerenz, B. and Mazur, S., 2003. Mesozoic fault systems in the Central European Basin System -new insights from a 3D structural model. In Dynamics of sedimentary systems under varying stress conditions by example of the central European basin system, Alfred-Wegener-Stiftung,73–76.

Scheck-Wenderoth, M. and Maystrenko, Y., 2008. How warm are passive continental margins? A 3D lithosphere-scale study from the Norwegian margin. Geology, 36, 5, 419–422.

Scheck-Wenderoth, M., Maystrenko, Y., Hübscher, C., Hansen, M. and Mazur, S., 2008. Dynamics of salt basins. In Littke, R., Bayer, U., Gajewski, D. and Nelskamp, S., eds., Dynamics of complex intracratonic basins: The Central European Basin System. XXIV, 520 pp.

Scheck-Wenderoth, M., Roure, F. and Bayer, U., eds., 2009b. Progress in understanding sedimentary basins. Spec. Issue, Tectonophysics, 470, 194 pp.

Schmid, S.M., Bernoulli, D., Fügenschuh, B., Matenco, L., Schefer, S., Schuster, R., Tischler, M. and Ustaszewski, K., 2008. The Alpine-Carpathian-Dinarid orogenic system: correlation and evolution of tectonic units. Swiss J. Geosci., 101, 139–183.

Schneider, F., 2003. Basin modelling in complex area: Examples from eastern Venezuelan and Canadian foothills. Oil Gas Sci. Technol. Revue de l'IFP, 58, 2, 313–324.

Schneider, F., Devoitine, H., Faille, I., Flauraud, E. and Willien, F., 2002. Ceres2D: a numerical prototype for HC potential evolution in complex area. Oil Gas Sci. Technol. Revue de l'IFP, 57, 6, 607–619.

Schreurs, G., Buiter, S.J.H. et al., , 2006. Analogue and numerical sandbox models: analogue benchmarks of shortening and extension experiments. In Buiter, S.J.H. and Schreurs, G., eds., Analogue and numerical modelling of crustal-scale processes. Geol. Soc., Lond., Spec. Publ., 253, 1–27.

Schreurs, G. and Colletta, B., 1998. Modelling transpression and transtension: analogue modelling of faulting in zones of continental transpression and transtension. Geol. Soc. London, Spec. Publ., 135, 59–79.

Schumacher, M.E., 2002. Upper Rhine Graben: the role of pre-existing structures during rift evolution. Tectonics, 21, 10.1029/2001TC900022 (6-1 to 6-17).

Sclater, J.G., 2003. Ins and outs on the ocean floor. Nature, 421, 590–591.

Scrocca, D., Carminati, E., Doglioni, C. and Marcantoni, D., 2007. Slab retreat and active shortening along the Central-Northern Apennines. In Lacombe, O., Lavé, J., Roure, F. and Vergés, J., eds., Thrust belts and foreland basins, Springer, Frontiers in Earth Sciences, 471–487.

Seifert, P., 1996. Sedimentary-tectonic development and Austrian hydrocarbon potential of the Vienna Basin. In Wessely, G. and Liebl, W., eds., Oil and gas in Alpidic thrustbelts and basins of Central and Eastern Europe, Eur. Assoc. Geosci. Eng., Spec. Publ., 5, 331–342.

Séranne, M., 1999. The Gulf of Lions continental margin (NW Mediterranean basins: Tertiary extension within the Alpine orogen. In Durand, B., Jolivet, L., Horvath, F. and Séranne, M., eds., The Mediterranean Basin: Tertiary extension within the Alpine orogen. Geol. Soc., Lond., Spec. Publ., 156, 15–36.

Séranne, M., Benedicto, A., Truffert, C., Pascal, G. and Labaume, P., 1995. Structural style and evolution of the Gulf of Lion Oligo-Miocene rifting: Role of the Pyrenean orogeny. Mar. Petrol. Geol., 12, 809–820.

Shapiro, N.M. and Ritzwoller, M.H., 2002. Monte-Carlo inversion for a global shear velocity model of the crust and upper mantle. Geophys. J. Int., 151, 1–18.

Shearer, P.M. and Earle, P.S., 2004. The Global short-period wavefield modeled with a Monte Carlo seismic phonon method. Geophys. J. Int., 158, 1103–1117.

Shudofsky, G.N., Cloetingh, S., Stein, S. and Wortel, M.J.R., 1987. Unusually deep earthquakes in east Africa: constraints on the thermo-mechanical structure of a continental rift system. Geophys. Res. Lett., 14, 741–744.

Skogseid, J. et al., , 2000. NE Atlantic continental rifting and volcanic margin formation. In Nøttvedt, A., ed., Dynamics of the Norwegian margin, Geol. Soc., Lond., Spec. Publ., 295–326.

Sleep, N.H., 1976. Platform subsidence: mechanisms and "eustatic" sea-level changes. Tectonophysics, 36, 45–56.

Sokoutis, D., Burg, J.P., Bonini, M., Corti, G. and Cloetingh, S., 2005. Lithospheric-scale structures from the perspective of analogue continental collision. Tectonophysics, 406, 1–15.

Sokoutis, D., Corti, G., Bonini, M., Brun, J.P., Cloetingh, S., Mauduit, T. and Manetti, P., 2007. Modelling the extension of heterogeneous hot lithosphere. Tectonophysics, 444, 1–4, 63–79.

Solheim, A., Riis, F., Elverhøi, A., Faleide, J.I., Jensen, L.N. and Cloetingh, S., eds., 1996. Impact of glaciations on basin evolution: data and models from the Norwegian margin and adjacent areas. Global Planet. Change, 12.

Spakman, W., Van der Lee, S. and van der Hilst, R., 1999. Travel-time tomography of the European mantle down to 1400 km. Phys. Earth Planet. Inter., 79, 3–74.

Spakman, W. and Wortel, M.J.R., 2004. A tomographoic view of the Western Mediterranean geodynamics. In Cavazza, W., Roure, F., Spakman, W., Stampfli, G.M. and Ziegler, P.A., eds., The TRANSMED Atlas – The Mediterranean region from crust to mantle. Springer-Verslag, Berlin, Heidelberg, 31–52.

Stackebrandt, W., 2008. Zur Neotektonik der Niederlasuitz, Ostdeutschland, Zeitschrift der detschen Gesellschaft für Geowissenschaften, 159, 117–122.

Stampfli, G.M., Mosar, J., Marquer, D., Marchant, R., Baudin, T. and Borell, G., 1998. Subduction and obduction processes in the Swiss Alps. Tectonophysics, 296, 159–204.

Stankiewicz, J., Chevrot, S., van der Hilst, R.D. and de Witt, M.J., 2002. Crustal thickness, discontinuity depth and upper mantle structure beneath Southern Africa: constraints from body wave conversions. Phys. Earth Planet. Inter., 130, 235–252.

Stein, C.A. and Stein, S., 1992. A model for global variation in oceanic depth and heat flow with lithosphere age. Nature, 359, 123–128.

Stephenson, R.A., 1989. Beyond first-order thermal subsidence models for sedimentary basins? In Cross, T. A, ed., Quantitative dynamic stratigraphy, Prentice-Hall, Englewood Cliffs, NJ, 113–125.

Stephenson, R.A., ed., 1993. Crustal controls on the internal architecture of sedimentary basins. Tectonophysics, 228, 140 pp.

Stephenson, R.A. and Cloetingh, S., 1991. Some examples and mechanical aspects of continental lithospheric folding. Tectonophysics, 188, 27–37.

Stephenson, R.A., Stovba, S.M. and Starostenko, V.I., 2001. Pripyat-Dniepr-Donets basin: implication for rift geodynamics and northern Peri-Tethyan tectonic history. In Ziegler, P.A., Cavazza, W. and Robertson, A.H.F., eds., PeriTethyan rift/wrench basins and passive margins, PeriTethys Mem. 6, Mém. Museum National d'Histoire Naturelle, Paris, 186, 369–406.

Stephenson, R.A., Wilson, M., De Boorder, H. and Starostenko, V.I., eds., 1996. EUROPROBE: Intraplate tectonics and basin dynamics of the Eastern European platform. Tectonophysics, 268, 1–309.

Stephenson, R.A., Yegorova, T., Brunet, M.F., Stovba, S., Wilson, M., Starostenko, V., Saintot, A. and Kusznir, N., 2006. Late Paleozoic intra- and pericratonic basins on the East European craton and its margins. In Gee, D.G. and Stephenson, R.A., eds., European Lithosphere dynamics. Geol. Soc., Lond., Spec. Publ., 32, 463–479.

Stewart, J., Watts, A.B. and Bagguley, J.G., 2000. Three-dimentional subsidence analysis and gravity modelling of the continental margin offshore Namibia. Geophys. J. Int., 141, 724–746.

Stockmal, G.S., Beaumont, C., Nguyen, M. and Lee, B., 2007. Mechanics of thin-skinned fold-and-thrust belts: insights from numerical models; dans, Whence the mountains?. In Sears, J.W., Harms, T.A., and Evenchick, C.A., eds., Inquiries into the evolution of orogenic systems: a volume in honor of Raymond A. Price, Geol. Soc. of Am., Spec. Paper, 433, 63–98.

Stovba, S.M., Maystrenko, Y.P., Stephenson, R.A. and Kusznir, N.I., 2003. The formation of the southeastern part of the Dniepr-Donets Basin: 2-D forward and reverse syn-rift and post-rift modelling. Sediment. Geol., 156, 11–33.

Stovba, S.M. and Stephenson, R.A., 2003. Style and timing of salt tectonics in the Dniepr-Donets Basin (Ukraine): implication for triggering and driving mechanisms of salt movement in sedimentary basins. Mar. Petrol. Geol., 19, 1160–1189.

Suhadolc, P. and Panza, G.F., 1989. Physical properties of the lithosphere-asthenosphere system in Europe from geophysical data. In Boriani, A., Bonafede, M., Piccardo, G.B., Vai, G.B., eds., The lithosphere in Italy -advances in earth science research. Acad. Naz. Lincei, Roma, 80, 15–40.

Swennen, R., Roure, F. and Granath, J., eds., 2004. Deformation, fluid flow and reservoir appraisal in foreland fold and thrust belts, Am. Assoc. Petrol. Geol., Hedberg Series, 1, 430 pp.

Ter Voorde, M., Van Balen, R.T., Bertotti, G. and Cloetingh, S., 1998. The influence of a stratified rheology on the flexural response of the lithosphere to un-loading by extensional faulting. Geophys. J. Int., 134, 721–735.

Ter Voorde, M., de Bruijne, C.H., Cloetingh, S. and Andriessen, P.A.M., 2004. Thermal consequences of thrust faulting: simultaneous versus successive fault activation and exhumation. Earth Planet. Sci. Lett., 223, 395–413.

Tesauro, M., Kaban, M.K. and Cloetingh, S., 2008. EuCRUST-07: A new reference model for the European crust. Geophys. Res. Lett., 35, L05313. doi:10.1029/2007GL032244.

Thybo, H., Pharaoh, T. and Guterch, A., eds., 1999. Geophysical investigation of the Trans-European suture zone. Elsevier Science, 350 pp.

Thybo, H., 2000. Crustal structure and tectonic evolution of the Tornquist Fan region as reavealed by geophysical methods. Bull. Geol. Soc. Denmark, 46, 145–160.

Thybo, H., 2006. The heterogeneous upper mantle Low Velocity Zone. Tectonophysics, 416, 53–79.

Thybo, H. and Anderson, D.L., eds., 2006. The heterogeneous mantle. Tectonophysics, 416, 309 pp.

Thybo, H., Janik, T., Omelchenko, V.D., Grad, M., Garetsky, R.G., Belinsky, A.A., Karatayev, G.I., Zlotski, G., Knudsen, M.E., Sand, R., Yliniemi, J., Tiira, T., Luosto, U., Komminaho, K., Giese, R., Guterch, A., Lund, C.E., Kharitonov, O.M., Ilchenko, T., Lysynchuk, D.V., Skobelev, V.M. and Doody, J.J., 2003. Upper lithospheric seismic velocity structure across the Pripyat Trough and the Ukrainian Shield along the EUROBRIDGE'97 profile. Tectonophysics, 371, 41–79.

Thybo, H. and Perchuc, E., 1997. The seismic 8 degrees discontinuity and partial melting in continental mantle. Science, 275, 1626–1629.

Tibéri, C., Leroy, S., d'Acremont, E., Bellahsen, N., Ebinger, C., Al-Lazki, A. and Pointu, A., 2007. Crustal geometry of the northeastern Gulf of Aden passive margin: localization of the deformation inferred from receiver function analysis. Geophys. J. Int., 168, 1247–1260.

Tinker, J., de Wit, M. and Brown, R., 2007a. Linking source and sink: evaluating the balance between onshore erosion and offshore sediment accumulation since Gondwana break-up, South Africa. Tectonophysics, doi: 10.1016/j.tecto.2007.11.040.

Tinker, J., de Wit, M. and Brown, R., 2007b. Mesozoic exhumation of the southern Cape, South Africa, quantified using apatite fission track thermochronology. Tectonophysics, doi: 10.1016/j.tecto.2007.10.009.

Torres, V., Vandenberghe, J. and Hooghiemstra, H., 2005. An environmental reconstruction of the sediment infill of the Bogota Basin (Colombia) during the last 3 million years from abiotic and biotic proxies. Palaeogeogr. Palaeoclimatol. Palaeoecol., 226, 127–148.

Travé, A., Labaume, P. and Vergés, J., 2007. Fluid systems in foreland fold-and-thrust balts: An overview from the Southern Pyrénées. In Lacombe, O., Lavé, J., Roure, F. and Vergés, J., eds., Thrust belts and foreland basins, Frontiers in Earth Sciences. Springer, 93–116.

Trumbull, R.B., Reid, D.L., de Beer, C., van Acken, D. and Romer, R.L., 2007. Magmatism and continental breakup at the west margin of southern Africa: a geochemical comparison of dolerite dikes from northwestern Namibia and the Western Cape. South African J. Geol., 110, 2–3, 477–502.

Tsikalas, F., Faleide, J.I., Eldholm, O. and Wilson, J., 2005. Late Mesozoic-Cenozoic structural stratigraphuic correlation between the conjugate mid-Norway and NE Greenland continental margins. In Doré, A.G.and Vining, B., eds., Petroleum geology: Nothwest Europe and global perspectives, Proceedings of the 6th Petrol. Geol. Conference, Geol. Soc., Lond., Spec. Publ., 785–801.

Turcotte, D.L. and Schubert, G., 2002. Geodynamics. Cambridge Univ. Press, 456 pp.

Turcotte, D.L. and Oxburgh, E.R., 1967. Finite amplitude convective cells and continental drift. J. Fluid Mech., 28, 29.

Unternehr, P. and van den Driessche, J., 2004. Lithospheric folding in the Southern North Sea: Combined effects of Alpine compression and North Atlantic opening. Geodinamica Acta, 17, 6, 401–405.

Van Balen, R.T. and Skar, T., 2000. The influence of faults and intraplate stresses on the overpressure evolution of the Halten Terrace, mid-Norwegian Margin. Tectonophysics, 320, 331–345.

Van Balen, R.T., Houtgast, R.F., Van der Wateren, F.M., Vandenberghe, J. and Bogaart, P.W., 2000. Sediment budget and tectonic evolution of the Meuse catchment in the Ardennes and the Roer Valley Rift System. Global Planet. Change, 27, 113–129.

Van Balen, R.T., Van der Beek, P.A. and Cloetingh, S., 1995. The effect of rift shoulder erosion on stratal patterns at passive margins: implications for sequence stratigraphy. Earth Planet. Sci. Lett., 134, 527–544.

Van der Beek, P.A. and Cloetingh, S., 1992. Lithospheric flexure and the tectonic evolution of the Betic Cordillers. Tectonophysics, 20, 325–344.

Van der Beek, P.A., Cloetingh, S. and Andriessen, P.A.M., 1994. Mechanisms of extensional basin formation and vertical motions at rift flanks: constraints from tectonic modelling and fission track thermochronology. Earth Planet. Sci. Lett., 121, 417–433.

Van der Beek, P.A., Robert, X., Mugnier, J.L., Bernet, M., Huyghe, P. and Labrin, E., 2006. Late Miocene? Recent exhumation of the Central Himalaya and recycling in the foreland basin assessed by apatite fission-track thermochronology of Siwalik sediments, Nepal. Basin Res., 18, 413–434.

Van Hinsbergen, D.J.J., Hafkenscheid, H., Spakman, W., Meulenkamp, J.E. and Wortel, M.J.R., 2005. Nappe stacking resulting from subduction of oceanic and continental lithosphere below Greece. Geology, 33, 325–328.

Van Vliet-Lanoë, B., Laurent, M., Everaerts, M., Mansy, J.-L. and Manby, G., 2000. Evolution néogène et quarternaire de la Somme, une flexuration tectonique active. C. R. Acad. Sci. Earth Planet. Sci., 331, 151–158.

Van Wees, J.-D., 1994. Tectonic modelling of basin deformation and inversion dynamics: the role of pre-existing faults and continental lithosphere rheology in basin evolution. Ph.D. thesis, Vrije Universiteit, Amsterdam, 164 pp.

Van Wees, J.D. and Cloetingh, S., 1996. 3D flexure and intraplate compression in the North Sea area. Tectonophysics, 266, 343–359.

Van Wees, J.D., Stephenson, R.S., Ziegler, P.A., Bayer, U., McCann, T., Dadlez, R., Gaupp, R., Narkiewicz, M., Bitzer, F. and Scheck, M., 2000. Origin of the Southern Permian Basin, Central Europe. Mar. Petrol. Geol., 17, 43–459.

Van Wijk, J.W. and Cloetingh, S., 2002. Basin migration caused by slow lithospheric extension. Earth Planet. Sci. Lett., 198, 275–288.

Van Wijk, J.W. and Blackman, D.K., 2004. Dynamics of continental rift propagation: the end-member modes. Earth Planet. Sci. Lett., 229, 3–4, 247–258.

Vauchez, A., Tommasi, A. and Barruol, G., 1998. Rheological heterogeneity, mechanical anisotropy and deformation of the continental lithosphere. Tectonophysics, 296, 61–86.

Vendeville, B., Cobbold, P.R., Davy, P., Choukroune, P. and Brun, J.P., 1987. Physical models of extensional tectonics at various scales. In Fault geometries and associated processes, Geol. Soc. Lond., Spec. Publ., 28, 95–107.

Vejbaek, O.V. and Andersen, C., 2002. Post-mid Cretaceous inversion tectonics in the Danish Central Graben -regionally synchronous tectonic events. Bull. Soc. Denmark, 49, 139–144.

Vergés, J., Millan, H., Roca, E., Munoz, J.A., Marzo, M., Cires, J., Den Bezemer, T., Zoetemeijer, R. and Cloetingh, S., 1995. Evolution of a collisional orogen: eastern Pyrenees transect and petroleum potential. Mar. Petrol. Geol., 12, 903–916.

Vilasi, N., Malandain, J., Barrier, L., Callot, J.P., Guilhaumou, N., Lacombe, O., Muska, K., Roure, F. and Swennen, R., 2009. From outcrop and petrographic studies to basin-scale fluid flow modelling: the use of the Albanian natural laboratory for carbonate reservoir characterization. Tectonophysics, 474, 367–392.

Vilotte, J.P., Melosh, J., Sassi, W. and Ranalli, G., 1993. Lithosphere rheology and sedimentary basins. Tectonophysics, 226, 89–95.

Vinnik, L.P., Green, R.W.E., Nicolaysen, L.O., Kosarev, G.L. and Petersen, N.V., 1996. Deep seismic structure of the Kaapvaal Craton. Tectonophysics, 262, 67–75.

Vrolijk, P., 1987. Tectonically driven fluid-flow in the Kodiak accretionary complex, Alaska. Geology, 15, 5, 466–469.

Vrolijk, P., Chambers, S.R., Gieskes, J.M. and O'Neil, J.R., 1990. Oxygen and hydrogen isotope ratios of interstitial waters and carbon isotope ratios of methane gases from the Northern Barbados accretionary prism, ODP Leg 110. In Moore, J.C. and Mascle, A., eds., Proceedings ODP, Scientific results 110, 189–205.

Vyssotski, A.V., Vyssotski, V.N. and Nezhdanov, A.A., 2006. Evolution of the West Siberian Basin. Mar. Petrol. Geol., 23, 93–126.

Wandås, B.T.G., Nystuen, J.P., Eide, E., Gradstein, F.M. eds., , 2005. Onshore-Offshore relationships on the North Atlantic margin. Norwegian Petrol. Soc, Spec. Publ., Elsevier, 12.

Wateren, van de, F.M. and Cloetingh, S., 1999. Feedbacks of lithospheric dynamics and environmental change of the Cenozoic West Antarctic Rift System. Global Planet. Change, 23, 1–24.

Watts, A.B., 2001. Isostasy and flexure of the lithosphere. Cambridge University Press, Cambridge, 458 pp.

Watts, A.B. and Burov, E.B., 2003. Lithospheric strength and its relationship to the elastic and seismogenic layer thickness. Earth Planet. Sci. Lett., 213, 113–131.

Wenzel, F., Brun, J.P. and the ECORS-DEKORP working group, 1991. A deep reflection seismic line across the Northern Rhine Graben. Earth Planet. Sci. Lett., 104, 140–150.

Wernicke, B., 1990. The fluid crustal layer and its implication for continental dynamics. In Salisbury, M.H. and Fountein, D.M., eds., Exposed cross-sections of the continental crust. Kluwer Academic Publishing, Dordrecht, 509–544.

White, N., Thompson, M. and Barwise, T., 2003. Understanding the thermal evolution of deep-water continental margins. Nature, 426, 334–343.

Willett, S.D., 1999. Orogeny and orography: the effects of erosion in the structure of mountain belts. J. Geophys. Res., 104, 28957–28981.

Willett, S.D., Beaumont, C. and Fullsack, P., 1993. Mechanical model for the tectonics of doubly vergent compressional orogens. Geology, 21, 371–374.

Willett, S.D., Hovius, N., Brandon, M.T. and Fischer, D.M., eds., 2006. Tectonics, climate and landscape evolution. Geol. Soc. Am., Spec. Paper, 398.

Willingshofer, E., Andriessen, P.A.M., Cloetingh, S. and Neubauer, F., 2001. Detrital fission track thermochronology of upper Cretaceous syn-orogenic sediments in the South Carpathians (Romania); inferences on the tectonic evolution of a collisional hinterland. Basin Res., 13, 379–395.

Wilson, M. and Downes, H., 1992. Mafic alkaline magmatism in the European Cenozoic rift system. Tectonophysics, 208, 173–182.

Wilson, M. and Downes, H., 2006. Tertiary-Quaternary intraplate magmatism in Europe and its relation to mantle dynamics. In: Gee, D.G., Stephenson, R.A., eds., European lithosphere dynamics. Geol. Soc., Lond. Spec. Publ. 32, 147–166.

Wilson, M. and Patterson, R., 2001. Intraplate magmatism related to short-wavelength convective instabilities in the upper mantle: Evidence from the Tertiary-Quaternary volcanic province of Western and Central Europe. Geol. Soc. Am. Bull., Spec. Publ., 352, 37–58.

Withmarsh, R.B., Minskull, T.A., Russell, S.M., Dean, S.M., Louden, K.E. and Chian, D., 2001. The role of syn-rift magmatism in the rift-to-drift evolution of the west-Iberia continental margin: geophysical observation. Geol. Soc. Lond. Spec. Publ., 187, 107–124.

Wittenberg, A., Vellmer, C., Kern, H. and Mengel, K., 2000. The Variscan lower continental crust: evidence for crustal delamination from geochemical and petrological investigations. In Franke, W., Haak, V., Oncken, O. and Tanner, D., eds., Orogenic processes: Quantification and modelling in the Variscan Belt. Geol. Soc., Lond., Spec. Publ., 179, 401–414.

Wortel, M.J.R. and Spakman, W., 2000. Subduction and slab detachment in the Mediterranean-Carpathian region. Science, 290, 1910–1917.

Wyer, P. and Watts, A.B., 2006. Gravity anomalies and segmentation at the East Coast, USA continental margin. Geophys. J. Int., 166, 1015–1038.

Yegorova, T., Bayer, U., Thybo, H., Maystrenko, Y., Scheck-Wenderoth, M. and Lyngsie, S.B., 2007. Gravity signals from the lithosphere in the Central European Basin System. Tectonophysics, 429, 1–2, 133–163.

Zeyen, H., Volker, F., Wehrle, V., Fuchs, K., Sobolev, S.V. and Altherr, R., 1997. Styles of continental rifting; crust-mantle detachment and mantle plumes. Tectonophysics, 278, 329–352.

Zhang, Y.S. and Lay, T., 1996. Global surface wave phase velocity variations. J. Geophys. Res., 101, 8415–8436.

Ziegler, P.A., 1988. Evolution of the Arctic-North Atlantic and the Western Tethys. Am. Assoc. of Petrol. Geol., Mem. 43, 198 p.

Ziegler, P.A., 1989. Evolution of Laurussia. A study in Late Paleozoic plate tectonics. Kluver Academic Publishers, Dordrecht, 102 pp.
Ziegler, P.A., 1990a. Collision related intraplate compression deformations in western and central Europe. J. Geodynam., 11, 357–388.
Ziegler, P.A., 1990b. Geological Atlas of Western and Central Europe, 2nd ed. Shell International Petroleum Mij. B.V, distributed by Geol. Soc. Publishing House Bath, 239 pp.
Ziegler, P.A., 1992. European Cenozoic rift system. Tectonophysics, 208, 91–111.
Ziegler, P.A., 1994. Cenozoic rift systems of Western and Central Europe: an overview. Geologie en Mijnbow, 73, 99–127.
Ziegler, P.A., Bertotti, G. and Cloetingh, S., 2002. Dynamic processes controlling foreland development – the role of mechanical (de)coupling of orogenic wedges and foreland. Eur. Geophys. Union, Spec. Publ., 1, 17–56.
Ziegler, P.A. and Cloetingh, S., 2004. Dynamic processes controlling evolution of rifted basins. Earth-Sci. Rev., 64, 1–50.
Ziegler, P.A., Cloetingh, S., Guiraud, R. and Stampfli, G.M., 2001. Peri-Tethyan platforms: constraints on dynamics of rifting and basin inversion. In: Ziegler, P.A., Cavazza, W., Robertson, A.H.F. and Crasquin-Soleau, S., eds., Peri-Tethys Memoir 6: Peri-Tethyan rift/ wrench basins and passive margins. Mém. Mus. Nat. Hist. Nat., Paris, 186, 9–49.
Ziegler, P.A., Cloetingh, S. and van Wees, J.-D., 1995. Dynamics of intraplate compressional deformation: the Alpine foreland and other examples. Tectonophysics, 252, 7–59.
Ziegler, P.A. and Dèzes, P., 2005. Evolution of the lithosphere in the area of the Rhine Rift System. In Behrmann, J.H., Granet, M., Schmid, S. and Ziegler P.A., eds., EUCOR-URGENT, Special Issue, Int. J. Earth Sci., 94, 594–614.
Ziegler, P.A. and Dèzes, P., 2006. Crustal evolution of Western and Central Europe. In Gee, D.G., Stephenson, R.A., eds., European Lithosphere Dynamics. Geol. Soc., Lond., Mem., 32, 43–56.
Ziegler, P.A. and Dèzes, P., 2007. Cenozoic uplift of Variscan Massifs in the Alpine foreland: Timing and controlling mechanisms. Global Planet. Change, 58, 1–4, 237–269.
Ziegler, P.A. and Fraefel, M., 2009. Response of drainage systems to Neogene evolution of the Jura fold-thrust belt and Upper Rhine Graben. Swiss J. Geosci., 102, 57–75.
Ziegler, P.A. and Horvàth, F., eds., 1996. Structure and Prospects of Alpine Basins and Forelands. Mém. Mus. Nat. Hist Nat., Paris, 547 p.
Ziegler, P.A. and Roure, F., 1996. Architecture and petroleum systems of the Alpine orogen and associated basins. In: Ziegler, P. and Horvath, F., eds., Structure and prospects of Alpine basins and forelands. Mém. Mus. Nat. Hist. Nat., Paris, 170, Peri-Tethys Mem., 2, 15–46.
Ziegler, P.A., Schumacher, M.E., Dèzes, P., van Wees, J.-D. and Cloetingh, S., 2004. Post-Variscan evolution of the lithosphere in the Rhine Graben area: constraints from subsidence modeling. In: Wilson, M., Neumann, E.-R., Davies, G.R., Timmerman, M.J., Heeremans, M., Larsen, B.T., eds., Permo-Carboniferous magmatism and rifting in Europe. Geol. Soc., Lond., Spec. Publ., 223, 289–317.
Ziegler, P.A., Van Wees, J.-D. and Cloetingh, S., 1998. Mechanical controls on collision-related compressional intraplate deformation. Tectonophysics, 300, 103–129.
Zoback, M.D., Stephenson, R.A., Cloetingh, S., Larsen, B.T., Van Hoorn, B., Robinson, A., Horvath, F., Puigdefabregas, C. and Ben-Avraham, Z., 1993. Stresses in the lithosphere and sedimentary basin formation. Tectonophysics, 226, 1–13.
Zoetemeijer, R., Deségaulx, P., Cloetingh, S., Roure, F. and Moretti, I., 1990. Lithospheric dynamics and tectonic-stratigraphic evolution of the Ebro basin. J. Geophys. Res., 95, 2701–2711.
Zoetemeijer, R., Sassi, W., Roure, F. and Cloetingh, S., 1992. Stratigraphic and kinematic modelling of thrust evolution;. northern Apennines, Italy., Geology.
Zoetemeijer, R., Tomek, C. and Cloetingh, S., 1999. Flexural expression of European continental lithosphere under the western outer Carpathians. Tectonics, 18, 843–861.
Zonenshain, L.P., Kuzmin, M.I. and Natapov, L.M., 1993. Geology of the USSR: a plate tectonic synthesis. Am. Geophys. union, Geodynamic Ser., 21, 242 p.

Recent Developments in Earthquake Hazards Studies

Walter D. Mooney and Susan M. White

Abstract In recent years, there has been great progress understanding the underlying causes of earthquakes, as well as forecasting their occurrence and preparing communities for their damaging effects. Plate tectonic theory explains the occurrence of earthquakes at discrete plate boundaries, such as subduction zones and transform faults, but diffuse plate boundaries are also common. Seismic hazards are distributed over a broad region within diffuse plate boundaries. Intraplate earthquakes occur in otherwise stable crust located far away from any plate boundary, and can cause great loss of life and property. These earthquakes cannot be explained by classical plate tectonics, and as such, are a topic of great scientific debate. Earthquake hazards are determined by a number of factors, among which the earthquake magnitude is only one factor. Other critical factors include population density, the potential for secondary hazards, such as fire, landslides and tsunamis, and the vulnerability of man-made structures to severe strong ground motion. In order to reduce earthquake hazards, engineers and scientists are taking advantage of new technologies to advance the fields of earthquake forecasting and mitigation. Seismicity is effectively monitored in many regions with regional networks, and world seismicity is monitored by the Global Seismic Network that consists of more than 150 high-quality, broadband seismic stations using satellite telemetry systems. Global Positioning Satellite (GPS) systems monitor crustal strain in tectonically active and intraplate regions. A relatively recent technology, Interferometric Synthetic Aperture Radar (InSAR) uses radar waves emitted from satellites to map the Earth's surface at high (sub-cm) resolution. InSAR technology opens the door to continuous monitoring of crustal deformation within active plate boundaries. The U.S. Geological Survey (USGS), along with other partners, has created ShakeMap, an online notification system that provides near-real-time post-earthquake maps of ground shaking intensity. These maps are especially useful for the coordination of emergency response teams and for the improvement of building codes. Using a combination of these new technologies, with paleoseismology studies, we have steadily improved the science of earthquake *forecasting* whereby one estimates the probability that an earthquake will occur during a specified time interval. A very recent development is Earthquake Early Warning, a system that will provide earthquake information within seconds of the initial rupture of a fault. These systems will give the public some tens of seconds to prepare for imminent earthquake strong ground motion. Advances in earthquake science hold the promise of diminishing earthquake hazards on a global scale despite ever-increasing population growth.

Keywords Diffuse plate boundarie · Earthquake · Global positioning syste · Paleoseismic record · Seismic tremor

Introduction

In recent years, it has been suggested that the threat of earthquakes as a global natural hazard has decreased. Statistics published in 2002 by the United Nations

W.D. Mooney (✉)
USGS, Menlo Park, CA 94025, USA
e-mail: mooney@usgs.gov

S. Cloetingh, J. Negendank (eds.), *New Frontiers in Integrated Solid Earth Sciences*, International Year of Planet Earth, DOI 10.1007/978-90-481-2737-5_6, © Springer Science+Business Media B.V. 2010

indicated that there have been a decreasing number of fatalities (as a percentage of global population) resulting from earthquakes in recent decades (United Nations, 2002). A somewhat different perspective is provided when considering the effects of the 2004 Indian Ocean earthquake and tsunami, the 2005 Kashmir, Pakistan, and the 2008 Sichuan, China earthquakes. Together, these three events caused more than 400,000 deaths and left more than 10 million people homeless.

It is increasingly recognized that mega-cities throughout the world, especially those in developing countries, are highly vulnerable to major earthquakes. While the repeat time for some large earthquakes is several centuries or longer, many large cities in earthquake hazard zones across the globe are less than a century old and thus their built environment is untested in terms of their vulnerability to earthquakes (Bolt, 2006). Moreover, the occurrence of major earthquakes away from plate boundary zones in such locations as Bhuj, India (2001, M = 7.6) and the New Madrid Zone in Missouri (1811–1812, M = 7–8) serve as a reminder that earthquakes are a global hazard. Hence, urban centers with a high population density may expect increased fatalities from future earthquakes unless necessary precautions are taken.

There are reasons to believe that earthquake losses can be reduced. Research and the development of new technologies will allow for improved earthquake forecasting through an understanding of both past earthquakes and phenomenon such as stress transfer between active faults. However, earthquake forecasting will not reduce losses unless both developing and wealthier countries focus on mitigation strategies for earthquake hazards and strengthening their existing infrastructure through retrofitting and implementing building codes.

The Science of Earthquakes – Understanding the Hazard

Background

Earthquakes are a natural phenomenon that have been documented and studied since written records began. During the fourth century B.C. a prominent Indian sage and family priest, Garga, theorized that earthquakes were caused by the sigh of the elephants that carry the Earth (Tagare, 2002). In the same century Aristotle suggested that earthquakes were caused by winds trapped in underground caves. The earthquake of 1755 that devastated Lisbon, Portugal, occurred on a major religious holiday and stirred much debate throughout Europe, thereby playing a role in The Enlightenment. Several recent earthquakes have attracted an international response and public policy discussion, as demonstrated by the response to the Mw = 9.3 December 26, 2004, Sumatra-Andaman Islands earthquake (Fig. 1) and the Mw = 7.9 Sichuan, China earthquake of May 12, 2008.

Our understanding of how and where earthquakes occur has improved significantly throughout history, most notably starting with the invention of the pen-and-paper seismograph in the 1880s. Studies of the great 1906 San Francisco earthquake played a major role in understanding the relation between earthquake ground motions and fault slip (Lawson, 1908; Reid, 1910). The advent of plate tectonics theory in the 1960s marked the greatest advancement in earthquake studies. The development of seismology, and the study of elastic waves propagating through solid and molten rock, revealed the presence of distinct layers now commonly recognized as the Earth's core, mantle, and crust. Our current theory of earthquake generation lies with the understanding that the upper, rigid outer shell of the crust and mantle, known as the lithosphere, is divided into numerous tectonic plates that continuously drift across the underlying, hot, partially molten asthenosphere. These tectonic plates meet one another at plate boundaries where they may collide, rift apart, or drag against each other, resulting in earthquakes.

The largest earthquakes occur at subduction zones where the oceanic lithosphere descends beneath the overriding continental lithosphere along what is called a megathrust fault. This follows from the fact that earthquake magnitude is proportional to the total area of the fault and the amount of slip. Whereas the Richter magnitude is familiar to the public, seismologists generally use the moment magnitude (Mw; Bolt, 2006) which is a more accurate measure of earthquake size. It is not uncommon for subduction zones to experience 10–20 m of slip, values that are rare elsewhere. Likewise, the dimensions of the fault surface area for subduction zone earthquakes can easily

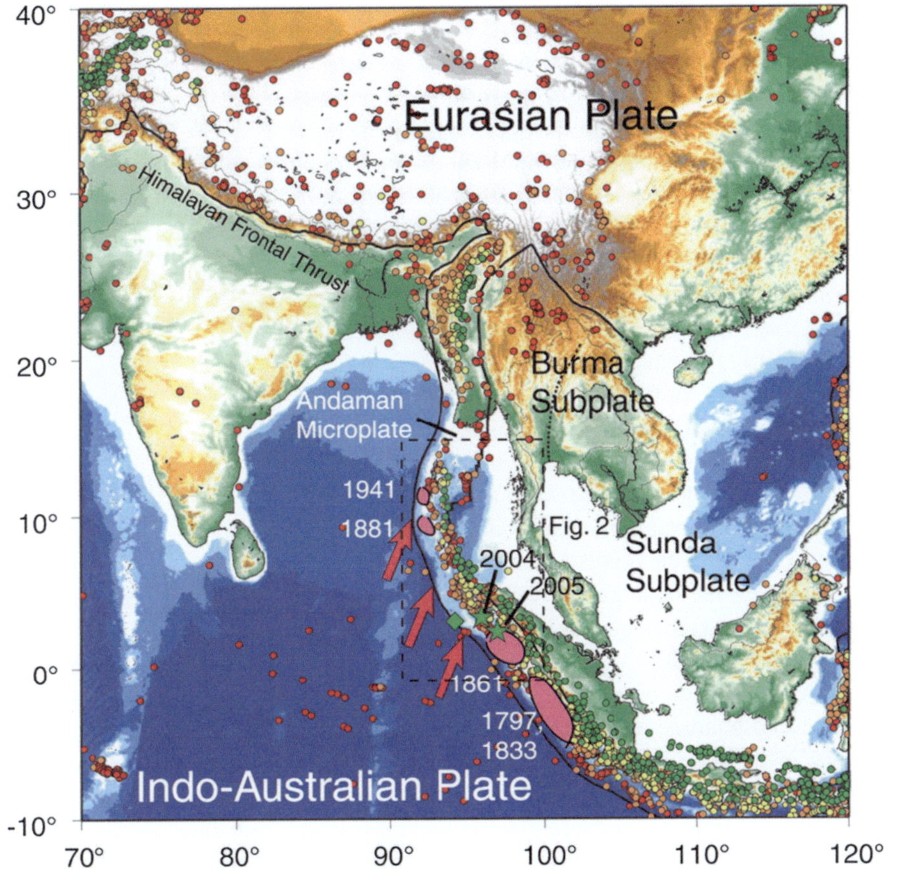

Fig. 1 Tectonic setting of Southeast Asia and the collision zone formed between the Eurasian and Indo-Australian plates. The dates of recent and historic earthquakes at the Indonesian/Andaman Islands subduction zone are shown. The Mw = 9.2 earthquake of 2004 generated a tsunami that devastated the coastline within this region

reach 1,000 km in length and 200–300 km in downdip extent. In contrast, continental strike-slip faults typically extend to a depth of only 15–30 km, and the longest mapped extent of a continental fault rupture is about 430 km (2001 Kunlun Shan, China; Xu et al., 2006). The larger area of subduction zone mega-thrust faults means that their earthquakes may have moment magnitudes in excess of Mw = 9, whereas continental earthquakes rarely exceed Mw = 8.0. In addition to megathrust earthquakes, subduction zones generate two additional types of earthquakes. Intraslab earthquakes occur to depths as great as 660 km and reach moment magnitude 8. Crustal earthquakes occur in the overriding plate due to stresses produced by the subduction process. Such crustal earthquakes are particularly hazardous due to their shallow depth and proximity to population centers. An example is the 1995 Kobe, Japan, earthquake that resulted in nearly 6,000 deaths.

One might expect that the larger subduction zone earthquakes would also be the deadliest. However, for several reasons, this is not the case. In the past 100 years, 10 of the 14 most deadly earthquakes have occurred on continental faults. The level of hazard posed by continental earthquakes can be appreciated by comparing two recent large events. The Mw = 7.9 Peru subduction zone event of Aug. 16, 2007 caused 517 deaths. In contrast, the Mw = 7.9 Sichuan, China, event of May 12, 2008 cause more than 90,000 deaths. The critical factor is that continental faults may pass directly beneath the affected communities, whereas subduction megathrusts are largely offshore, and thus strong ground shaking is attenuated.

While damage from strong ground shaking may be greater for continental earthquakes, subduction zone earthquakes pose an additional hazard from tsunamis. Ground shaking from the Mw = 9.3 Sumatra-Anadaman Islands earthquake of December 26, 2004, caused little damage, but the resulting tsunami claimed more than 230,000 lives throughout the Indian Ocean basin (Fig. 1).

Diffuse Plate Boundaries

Plate boundaries are typically conceived of as narrow contact regions between tectonic plates that can be represented as a single fault on a map. Examples of well-defined plate boundaries include the great subduction zones that circle the Pacific Ocean or the Anatolian fault in northern Turkey. Classical plate tectonics recognizes the presence of ten plates covering the Earth's surface, but this number has grown with the identification of several microplates such as the Juan de Fuca or Borneo plates. In fact, there is no consensus on the precise location of some plate boundaries as the term is conceived in classical plate tectonics. For example, the southern boundary between the Caribbean and South American plates is poorly defined, as is the boundary between the South American, Antarctic, and Scotian plates, the boundary between the Somalia and Nubia plates, and the boundary between Australian and Pacific plates (Fig. 2).

Thus, the width of a plate boundary can range from several hundred meters to several thousand kilometers. Wide plate boundaries occur where the crust and mantle lithosphere are actively deforming, and we refer to these regions as diffuse plate boundaries. It is estimated that approximately 15% of the Earth's surface is composed of these actively deforming regions (Gordon, 1995; Fig. 2). Classical tectonic theory is not sufficient to describe the kinematics that govern large scale movement at diffuse plate boundaries (Gordon and Stein, 1992). There may be hundreds of individual fault-bounded blocks separating one rigid plate from another across such wide boundary zones. Furthermore, the contact point between diffuse plate boundaries and plate interiors may also be gradual and indistinct (Gordon, 1995). Earthquakes, which are concentrated at plate boundaries, are geographically distributed throughout the diffuse plate boundary thereby spreading seismic hazards over a much broader region.

The Earthquake Cycle

The elastic rebound theory is the classical approach to explain how strain is distributed during the earthquake cycle and the mechanical process that takes place between one earthquake and the next on a particular fault (Reid, 1910). As tectonic plates gradually move past one another, there are some portions along the plate boundary where the plates are locked. At these locked areas of contact, strain accumulates and slow internal deformation takes place until the friction on the fault is exceeded and the plates suddenly slip. Global Positioning Satellite (GPS) data provide a means of monitoring crustal strain. Over a sufficiently long period of time the rate of strain energy

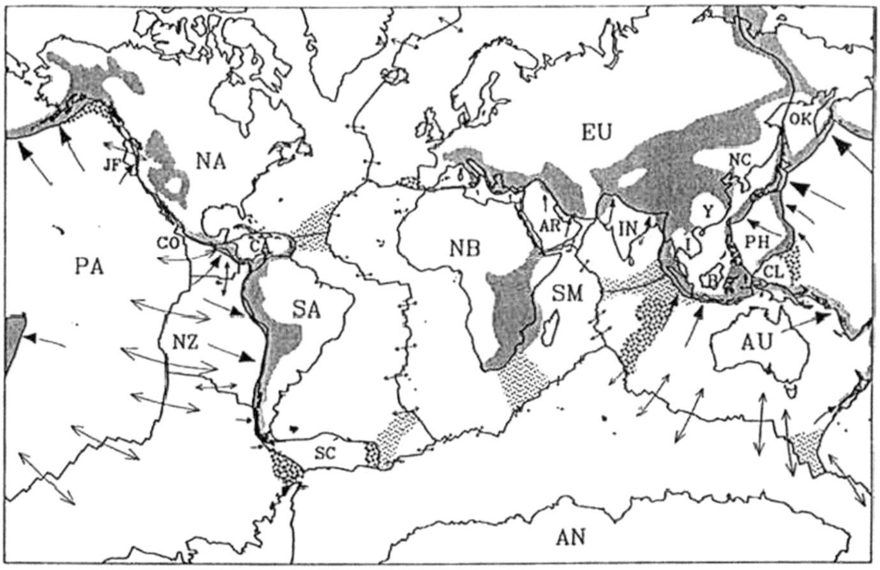

Fig. 2 Map showing idealized narrow plate boundaries, measured plate velocities, and regions of deforming lithosphere (light and dark shading), which are regarded as diffuse plate boundaries. Plate velocities are shown by arrows; their length indicates the magnitude of displacement expected in a period of 25 millions years (after Gordon, 1995)

accumulation must balance the seismic strain release in the form of earthquakes. After an earthquake nucleates at the hypocenter, fault rupture propagates along the fault at a velocity near 3 km/s second and radiates elastic energy off the fault surface. Assuming a constant stress drop, the larger the rupture area and amount of fault slip, the greater the amount of seismic energy that is released, and the more damaging is the strong ground motion. We note that some continental earthquakes, such as the Mw = 7.6 Bhuj, India, event were high-stress drop events, and therefore had a smaller fault area in comparison with other earthquakes with the same moment magnitude.

A complete description of the earthquake cycle incorporates interseismic (i.e., the time period between large seismic events) and postseismic deformation in addition to the coseismic deformation that takes place during an earthquake (Tse and Rice, 1986; Fig. 3). In these models of the earthquake cycle, slow, creeping displacement occurs in the viscous lower crust and lithosphere prior to the earthquake. This interseismic viscous creep can take place over a period of centuries for faults with long recurrence intervals. Eventually, the colder upper crust must "catch up" and release accumulated strain in a sudden coseismic displacement (an earthquake). Postseismic deformation is typically in the form of viscous creep and occurs over a period of days to weeks following an earthquake. The sum of all types of displacement over the earthquake cycle is known as the earthquake's total slip budget. Global Positioning Satellite (GPS) measurements, discussed in a later section, are able to track the progress of the interseismic, coseismic, and postseismic phases of the earthquake cycle along the Parkfield segment of the San Andreas Fault (Bakun et al., 2005; Murray and Langbein, 2006; Johanson et al., 2006).

In general, earthquakes in continental regions occur in the upper and middle crust (< 25 km depth). Earthquakes in the deeper continental crust (> 25 km depth) are rare, but some deep events are reported in old, cold crust such as the Indian craton (Raphael and Bodin, 2002; Kayal et al., 2002; Mishra and Zhao, 2003). The strength of the lithosphere depends on several factors, including pressure and temperature. The temperature in the crust typically increases by 10 to 30°C/km. The depth of continental earthquakes tells us that the upper and middle crust is in the brittle regime of failure, whereas the lower crust and mantle lithosphere undergoes creeping (ductile) deformation. Frictional strength increases linearly with pressure (depth) and ductile strength decreases exponentially with temperature. Combining both the ductile and brittle relations

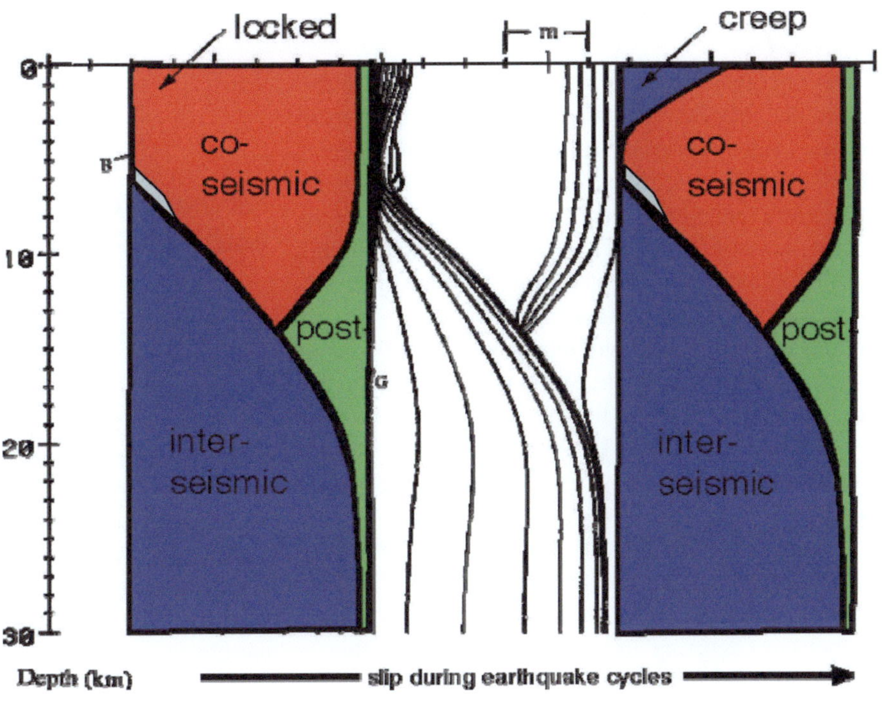

Fig. 3 The earthquake cycle, from the interseismic period, with lower crustal creep (*blue on left side*), co-seismic slip that occurs during fault rupture (*red*), and post-seismic relaxation (*green*). In this model the locked portion of the upper crust is loaded from below by the creeping middle and lower crust

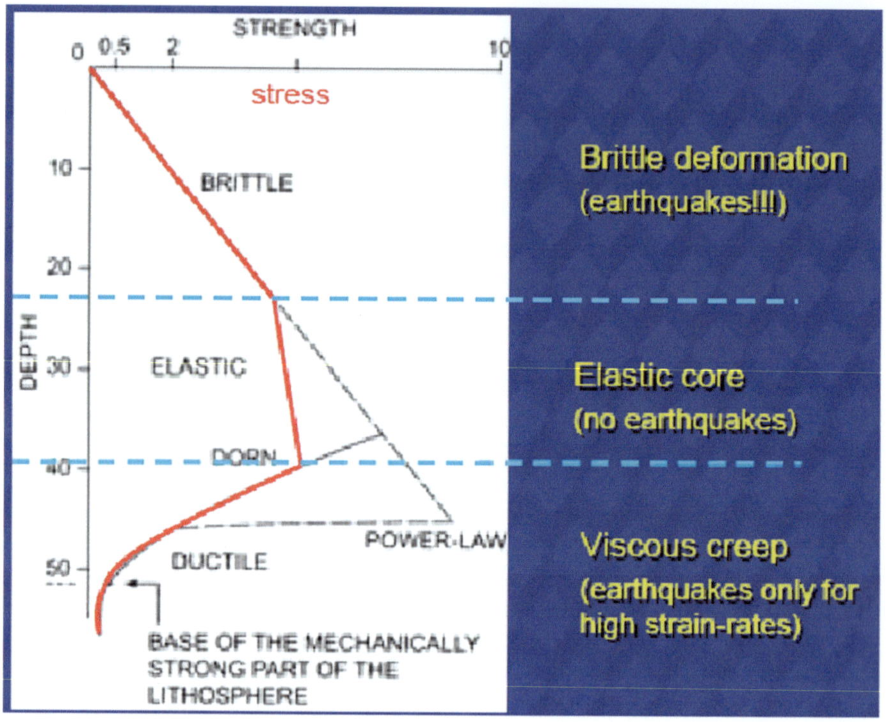

Fig. 4 Strength (stress)-depth curve for the crust and elastic lithosphere. Stress increases in the brittle crust until a mid-crustal transition zone with reduced strength is reached. At greater depth, in the ductile mantle, viscous creep occurs. Earthquakes occur primarily in the brittle upper crust (Source: Cloetingh, 1982)

yields an overall rheological strength profile for the lithosphere as a function of depth (Fig. 4). Shallow crustal earthquakes are therefore due to brittle deformation in the colder upper crust, whereas the weak, ductile lower crust and mantle lithosphere undergoes steady creep.

An important aspect of the earthquake cycle is the recurrence interval, the statistical average time period between large earthquakes along a particular segment of fault line. The recurrence interval for an earthquake of a given magnitude, say Mw = 7, is dependent on the strain rate. Since a Mw = 7 earthquake has about 4 m of slip, a low strain rate of 4 mm/year (as might occur within a diffuse plate boundary, or in an intraplate environment) will require about 1,000 years to accumulate 4 m of strain. Regions with a high strain rate of 40 mm/year, as found on the Main Boundary Fault of the Indian Himalayas, will accumulate 4 m of strain in just 100 years if these faults are fully locked.

Knowledge of the present strain rate and the recurrence interval (with standard deviation) on a fault makes it possible to estimate the probability of an earthquake occurring on a particular fault within a specific time window, such as 10 years. Such probabilistic estimates are very useful in planning seismic mitigation. Although earthquake prediction may not be feasible in the foreseeable future, earthquake *forecasting*, based on recurrence intervals, has become more sophisticated and reliable in recent years.

Earthquake Triggering: Natural and Man-Made

The stresses in the Earth's crust gradually increase due to global tectonic processes except when a fault is creeping. It is not surprising, then that many faults are already near the point of failure; such faults are referred to as critically stress *ed* faults (Zoback and Townend, 2001). Only a relatively small increase in stress will cause the fault to fail, causing an earthquake. All that is required is an extra "push" (stress increase) on a favorably oriented, critically stressed fault. Such earthquakes are called triggered earthquakes.

There are two primary types of triggered earthquakes, those induced by man-made stress increases and those induced by other earthquakes. Man-made stress increases include the filling of water reservoirs, mining activities and nuclear weapon testing (Simpson, 1986). The 1967 M = 6.5 earthquake at Koa reservoir, India, resulted in 200 deaths and 1,500 injuries (Gupta et al., 1972). This earthquake is generally considered to be a triggered earthquake.

Earthquakes cause static stress changes in the neighboring crust. The regional static Coulomb stress change due to an earthquake can be calculated if the fault geometry and slip distribution is known (King et al., 1994; Freed, 2005). These regional stress changes following a large earthquake can explain many seismic observations including both triggered earthquakes and seismic quiescence in regions that previously were seismically active (Stein et al., 1997; Stein, 1999).

Stress changes in the crust following an earthquake may be permanent (static) or transient and dynamic. As an example of dynamic triggering, it has been suggested that high-amplitude surface waves from the great (M = 9.3) Indonesian earthquake of December 2004 may have triggered micro-earthquakes as far away as Alaska (West et al., 2005). Theoretical modeling has suggested that dynamic stress is only able to cause "nearly instantaneous failures" on secondary faults, whereas there are often long delays before an earthquake occurs due to static stress changes (Belardinelli et al., 2000). However, predicting exactly how and when earthquake triggering will occur has not been possible since all earthquakes affect a very large, complex system of interacting faults across the Earth's crust.

Intraplate Earthquakes

The previous discussion has focused on earthquakes that occur either at well-defined plate boundaries, such as subduction zones, or within diffuse plate boundaries, such as the Tibetan Plateau. Earthquakes may also occur well away from a plate boundary, either in the ocean, typically on a transform fault, or in so-called stable continental crust. Such events are known as intraplate earthquakes (Fig. 5) and can cause significant loss of life when they occur near population centers. Some intraplate earthquakes at northern latitudes are caused by crustal unloading due to deglaciation. However, the ultimate cause of most intraplate earthquakes is the compressional state of stress of the Earth's crust due to the mid-ocean ridge push and asthenospheric viscous drag resistance to plate motions (Zoback and Zoback, 1980; Zoback, 1992). This stress is an effect of the gravitational potential energy of the oceanic ridges' elevated topography. The magnitude of this potential energy can be easily appreciated when one considers the towering heights of the Himalayas and Tibetan Plateau, which are the products of ridge push from the Indian Ocean ridge. Compression forces create critically stress *ed* faults within plate interiors (Zoback and Townend, 2001), and a small increase in stress may result in an earthquake. The mechanism that localizes stress perturbations and causes intraplate earthquakes in some regions, but not others, is not well understood.

The strain rate in the ductile lower crust of intraplate regions may be expressed as (Goetze, 1978; Brace and Kohlstedt, 1980):

$$\text{Strain rate}\,(\hat{e}) = A\exp(-Q/RT)(S_1 - S_3)^n \qquad (1)$$

where:

A is the flow parameter; Q is the activation energy; R is the gas constant; T is absolute temperature, and n is the stress exponent

S_1 is the maximum horizontal stress, and S_3 is the minimum horizontal stress.

Equation (1) implies that the strain rate will be at a maximum when the two horizontal stresses are at a maximum difference. The equation also implies that the strain rate will be higher for warmer thermal regimes (higher geotherm). These two implications may be used to infer that: (1) faster moving continental plates, such as India and Australia, will have higher strain rates due to the horizontal driving forces that move these plates; and (2) continental plates with thinner lithosphere will have high strain rates due to higher crustal temperatures (higher geotherm). Both of these inferences appear to be true for the Indian and Australian continental plates, which both have relatively high rates of intraplate seismicity (Fig. 5).

Continental intraplate earthquakes are widely distributed and are often associated with continental

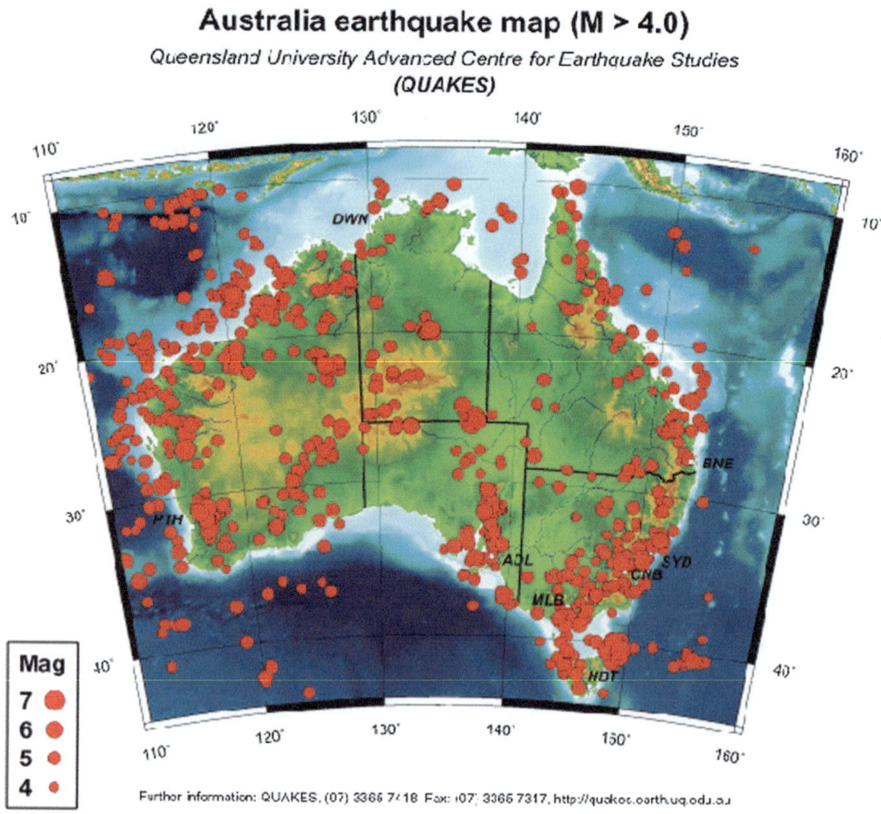

Fig. 5 Seismicity of Australia and its continental margin. Continental earthquakes in Australia are abundant, and all are intraplate earthquakes. The origin of such earthquakes that occur away from plate boundaries is a subject of considerable scientific debate

margins and interior rifts (Fig. 5). In 1811 and 1812, a sequence of three large (Mw $\geq$ 7) intraplate earthquakes occurred in New Madrid, Missouri, central USA. As summarized in Schulte and Mooney (2005), the New Madrid region is an example of intraplate seismicity associated with a paleo-rift, a correlation that has been found on a global scale in several studies (Sbar and Sykes, 1973; Sykes, 1978; Hinze et al., 1988; Johnston and Kanter, 1990; Johnston et al., 1994; Gangopadhyay and Talwani, 2003). The New Madrid earthquakes of the nineteenth century caused hundreds of casualties and were felt over an area of more than 120,000 square km. By comparison, the great 1906 (M = 7.8) earthquake of San Francisco was only felt over an area of 15,000 square km. In the eastern U.S., which is widely regarded as having a relatively low seismic hazard, a significant Mw = 7.0 to 7.3 earthquake struck Charleston, S.C., in 1886, and a large earthquake occurred off the coast of Cape Code in 1755 (the same year as the great Lisbon, Portugal, earthquake). The estimated recurrence interval for such events is estimated to be 400 years (Frankel et al., 1996).

In January 2001, a Mw = 7.7 earthquake occurred in Bhuj, western India, on a block of otherwise stable continental crust. The event caused more than 18,000 casualties, destroyed thousands of houses and critical infrastructure, and caused large ground motion more than 300 km away from the epicenter, or over a 260,000 square km region. It is evident that seismic waves emanating from earthquakes in older and colder crystalline continental crust will propagate farther, causing severe ground shaking and damage over a larger radius than those earthquakes emanating from the young, warm crust at plate boundaries. Although intraplate earthquakes contribute no more than 10% of the global released seismic energy, their widespread reach and unknown mechanism make them a cause for concern among seismologists (Bolt, 2006).

As discussed in Schulte and Mooney (2005), models that have been suggested to explain the occurrence of seismicity within continental interiors include localized stress concentration around igneous intrusions (Campbell, 1978), intersecting faults (Talwani, 1988, 1999) and ductile shear zones in the lower crust (Zoback, 1983). Fluids still present in the lower crust of ancient rift zones (Vinnik, 1989), a weak zone in the lower crust (Kenner and Segall, 2000) and stress perturbation resulting from buried rift pillows (Zoback and Richardson, 1996) have also been suggested. Liu and Zoback (1997) proposed the hypothesis that seismicity is related to elevated temperatures at depth. In their model, plate-driving forces are largely supported by the strength of the (seismogenic) upper crust; the lower crust is weakened as a result of higher temperatures. Regions with a higher geothermal gradient will have a shallower brittle-ductile transition and thus the cumulative strength of the lithosphere will be lower. Saying "generally similar" model is presented by Long (1988) who suggested that intraplate seismicity is a transient phenomenon arising from a perturbation in crustal strength as a result of a disturbance in the hydraulic or thermal properties of the lower crust. Other models call for a perturbation of the regional stress field by forces associated with lithospheric flexure after deglaciation (Stein et al., 1979; Quinlan, 1984; Grollimund and Zoback, 2001), gravitational forces at structural boundaries (Goodacre and Hasegawa, 1980; Chandrasekhar and Mishra, 2002) or sediment loading (Talwani and Rajendran, 1991). The recurrence interval for intraplate seismic zones are currently debated, with estimates ranging from relatively short (200–300 year) intervals to much longer (500–2,000 year) intervals. However, we note that the shorter recurrence intervals for major earthquakes do not readily conform to the statistics inferred from the rate of microearthquake activity (Mandal and Rastogi, 2005) nor with low (1 mm/year or less) intraplate deformation rates determined from GPS data (Calais et al., 2006).

According to Schulte and Mooney (2005), many of the models for Stable Continental Region (SCR) seismicity listed above require features that are often found in ancient rifts: numerous large faults and intrusions, an anomalous crustal structure compared with the surrounding crust and possible remnant crustal fluids. However, SCR seismicity is not unique to rift zones. In their study of intraplate seismicity, Schulte and Mooney (2005) find that: (1) 27% of all events occur within continental rifts; (2) 25% occur at rifted continental margins; (3) 36% are within non-rifted crust; and (4) 12% remain uncertain in terms of tectonic setting. The largest events ($M_w \geq 7.0$) have occurred predominantly within rifts (50%) and continental margins (48%).

Transient Aseismic Slip and Subduction Zone Seismic Tremor

Large mega-thrust events and shallow crustal earthquakes are not the only kinds of seismic events that occur at subduction zones. The recent discovery of transient aseismic slip (detected with GPS data) and seismic tremor at subduction zones suggests that the physics of faults involves an even more complicated and subtle process than previously imagined. In the last decade, observations of this slow, aseismic slip and periodic seismic tremor at many subduction zones have shown that there are alternative ways for faults to release stress than the sudden slip of an earthquake (Hirose, 1999; Rogers and Dragert, 2003). Transient aseismic slip tends to occur near the base of the locked zone on a subduction megathrust. It is unknown why certain subduction zones display transient slip and others do not. It has been hypothesized that the trigger of transient aseismic slip and seismic tremor may be pore fluid migration producing a reduction of fault-normal stress accompanied by seismic tremor within the fluid conduits (Melbourne and Webb, 2003). Aseismic slip can be quite significant and may release the equivalent of a $M_w = 7$ earthquake (Kostoglodov et al., 2003). In the Cascadia subduction zone in western North America, the aseismic slip lasts 1–2 weeks with a recurrence period of 14 months, while in the Bungo Channel of Japan, the slip lasts some 300 days with a recurrence period of 6 years (Hirose, 1999), and in Guerrero, southern Mexico, one large "silent earthquake" lasted 6–7 months (Lowry et al., 2001). Dragert et al. (2001) suggest that aseismic deep-slip events could play a key role in the cumulative stress loading and the eventual failure of shallow seismogenic zones. However, a robust connection between great earthquakes and aseismic slip events has not been made. Rubinstein et al. (this volume) provides additional discussion of transient aseismic slip and seismic tremor.

The Paleoseismic Record

Paleoseismology permits the reconstruction of the earthquake history of an area using geologic evidence of faulting and/or liquefaction in combination with radiocarbon dating (Fig. 6). This field has expanded greatly in recent years with important new

Fig. 6 *Top*: Paleoseismic expedition. Geologists are looking for evidence of faulted layers of sediments and rock, injections of liquefied sand (sand blows), and evidence for abruptly raised or lowered shorelines. *Bottom*: Buried peaty soil in British Columbia, the remains of a marsh that subsided 1.5 m during the 1700 Cascadia mega-thrust earthquake. The lighter color sand layer was then deposited by a tsunami that was caused by the earthquake

results obtained in Japan, New Zealand, China, Turkey, North America and the Mediterranean. Paleoseismic results show that earthquake recurrence intervals can range from about 20 years to more than 2,000 years.

The 1897 Great Assam (India) earthquake is the strongest instrumentally recorded intraplate earthquake, with an estimated Mw = 8.7. Mohindra and Bagati (1996) report paleoearthquakes in this region (the Shillong Plateau) dated at 1450–1650 AD, 700–1050 AD, and predating 600 AD. Together with the 1897 event, these data suggest a recurrence interval for this intraplate setting of 400–600 years. In northwestern India, a Mw = 7.6 earthquake occurred in 2001 in Gujarat (Bhuj), India, and was preceded by a similar strong event some 150 km to the NW in 1801. Paleoseismic investigations and evidence from ancient temples indicate that no large earthquakes have occurred in this region since the ninth century AD. Thus, the faults in this region appear to have recurrence intervals greater than 1,000 years.

As previously discussed, the New Madrid seismic zone of the central United States experienced three earthquakes of Mw greater than Mw = 7.5 in 1811–1812. Paleoliquifaction studies have determined similar large historic events at about 1450 AD, 900 AD 300 AD and 2350 BC (each date has an uncertainty of about 150 years; Tuttle et al., 2005). These results indicate a recurrence interval of about 500 years, which is about the same value as estimated for the 1897 Great Assam earthquake. However, these short recurrence intervals are enigmatic in view of the low regional strain rates measured with GPS data (Calais et al., 2006).

Paleoseismic techniques have also been applied to subduction zone settings. Seismic hazards in the Pacific Northwest are due to the subduction of the Juan de Fuca plate beneath the North American plate (the Cascadian subduction zone). This subduction is also responsible for the volcanic activity throughout the Pacific Northwest including the eruption of Mount St. Helens in 1980. A significant portion of the plate boundary between the Juan de Fuca and North America plates is locked (Savage et al., 1981; Dragert and Hyndman, 1995) raising the possibility that the Cascadian subduction can produce a megathrust earthquake (Mw > 8.0) that could rupture along its entire 1,100 km length and generate a tsunami comparable to the 2004 Indian Ocean tsunami. Indeed, by using paleoseismology, one such earthquake (and the resulting tsunami) was recently determined to have occurred in 1700, causing significant devastation to harbors in Japan and no doubt throughout many other coastal communities in the Pacific region (Nelson et al., 1995; Satake and Atwater, 2007).

Lessons from the Earthquake Record

A Survey of Earthquake Hazards

The primary hazardous effect of an earthquake is strong ground motion, which is measured in terms of acceleration and velocity. The strong ground motions of concern to civil engineers are usually those occurring near to the earthquake source, typically from 1 km to about 25 km, depending on the earthquake magnitude and regional seismic attenuation relations. Ground motions near the earthquake source are important because these motions are strong enough to endanger the built environment. Near-field ground motion data are also the most valuable data concerning the detailed properties of the source mechanism of the earthquake, i.e., geometry of the fault plane, the amount of slip as a function of space and time, and the rupture velocity. Each of these factors will affect the energy that is radiated during the rupture process. In order to design more earthquake resistant structures, a major effort has been made to model strong ground motion data including the duration, frequency content and fault radiation patterns. Earthquake strong ground motion data are thus of mutual interest to seismologists and engineers. In recent years it has been shown that most large (fault dimension greater than 100 km) earthquakes are usually complex multiple events (Kanamori, 2005), often making it difficult to model short period (less than 1 sec) ground motions. Progress in modeling near-field strong motion data has recently advanced due to the installation of significantly denser strong motion recording networks.

In addition to damaging strong ground motion, earthquakes may generate even more destructive secondary effects, such as liquefaction, landslides, tsunamis and fires. There are many examples of such effects, including April 18, 1906, Mw = 7.8 Great San Francisco earthquake that produced powerful ground

Fig. 7 Photographs taken after the 1906 Great San Francisco earthquake and fire. *Left*: damage to houses on the east side of Howard Street near Seventeenth Street. *Right*: The remains of the Hibernia bank on Market Street. The fire caused more destruction than the ground shaking of the earthquake

shaking that affected all of coastal northern California (Freed et al., 2007). Property damage amounted to $524 million and the number of fatalities reached 3,000 (Fig. 7). Broken water mains prevented firefighters from battling the numerous fires that appeared all over the city. In this case, the fires that raged in the city for three days caused more damage and loss of life than the earthquake (Cutcliffe, 2000). This earthquake and the 1934 Long Beach, California, event led to the adoption of better building codes in the State of California.

The Great Kanto (Tokyo region, Japan) earthquake of 1923 is another example of an earthquake with deadly secondary effects. This event resulted in more than 100,000 deaths and two million homeless, largely due to fires. The greater Tokyo region is home to about a quarter of the Japanese population of 130 million people. Paleoseismic evidence indicates that there have been seventeen $Mw \geq 8$ earthquakes in the Tokyo region in the past 7,000 years (Shishikura, 2003). This implies a long-term average recurrence interval of about 400 years for large events, with the last two $Mw = 8$ events having occurred in 1703 and 1923. The probability of a repeat of the 1923 Great Kanto earthquake is estimated to be just 0.5% in the next 30 years (Stein et al., 2006). However, $Mw = 7.3$ earthquakes occur more frequently than every 400 years, the last such event having occurred in 1855 (the Ansei-Edo event). According to the Japanese government, a $Mw = 7.3$ earthquake beneath Tokyo, similar to the 1855 event, could cause more than 11,000 deaths and destroy 240,000 or more homes. The estimates losses would amount to US$1 trillion. Stein et al. (2006) estimate that an event with a similar magnitude and location as the 1855 event has a twenty percent likelihood within a 30 year period.

The Japanese population was exposed to a dangerous secondary hazard on July 16, 2007, when a $Mw = 6.6$ earthquake occurred in close proximity to Kashiwazaki-Kariwa (KKNPP), the third largest nuclear power plant in the world (Fig. 8). Strong ground shaking caused a minor fire to break out, knocked over 100 barrels of low-level nuclear waste and required the power plant to be shut down. The plant, owned by Tokyo Electric Power Company (TEPCO), has seven boiling water reactors sitting on approximately 4.2 km^2 of land, and is home to seven of Japan's fifty five nuclear reactors, and produces

Fig. 8 The Kashiwazaki-Kariwa nuclear power plant (KKNPP), located about 10–20 km from the epicenter in the Niigata prefecture. This power plant was shut down after the July 16, 2007, earthquake caused damage to the plant

over thirty percent of the nation's power. All Japanese nuclear facilities have been engineered to withstand earthquakes of up to Mw = 6.5. In this instance, implementation of earthquake building codes in Japan's nuclear facilities almost certainly saved lives.

- Tsunamis are another secondary effect of earthquakes. In one well known case, the Mw = 9.2 earthquake that struck the coast of Sumatra, Indonesia, in December of 2004 triggered an Indian Ocean tsunami that devastated several countries separated by more than 4,000 miles, from Southeast Asia to Africa. The tsunami death toll exceeded 230,000 and led to the displacement of millions of people.
- A Mw = 7.9 earthquake struck eastern Sichuan, China, on May 12, 2008, and resulted in the death of some 89,000 people and left over a million homeless. This earthquake occurred within the Longmen Shan region which is located at the boundary between the high topography of the Tibetan Plateau to the west and the relatively stable Sichuan Basin to the east (Fig. 9; Burchfiel et al., 1995). The ground shaking was felt over much of central, eastern, and southern China (Fig. 9). The earthquake led to numerous landslides that buried villages and complicated rescue efforts by blocking transportation routes. Medical supplies, water, and food may not reach isolated communities affected by the disaster and the inability to distribute critical supplies may dramatically increase the casualties.

Earthquake Engineering and Building Codes

The design of buildings to sustain earthquake strong ground motions is a critical step in reducing the loss

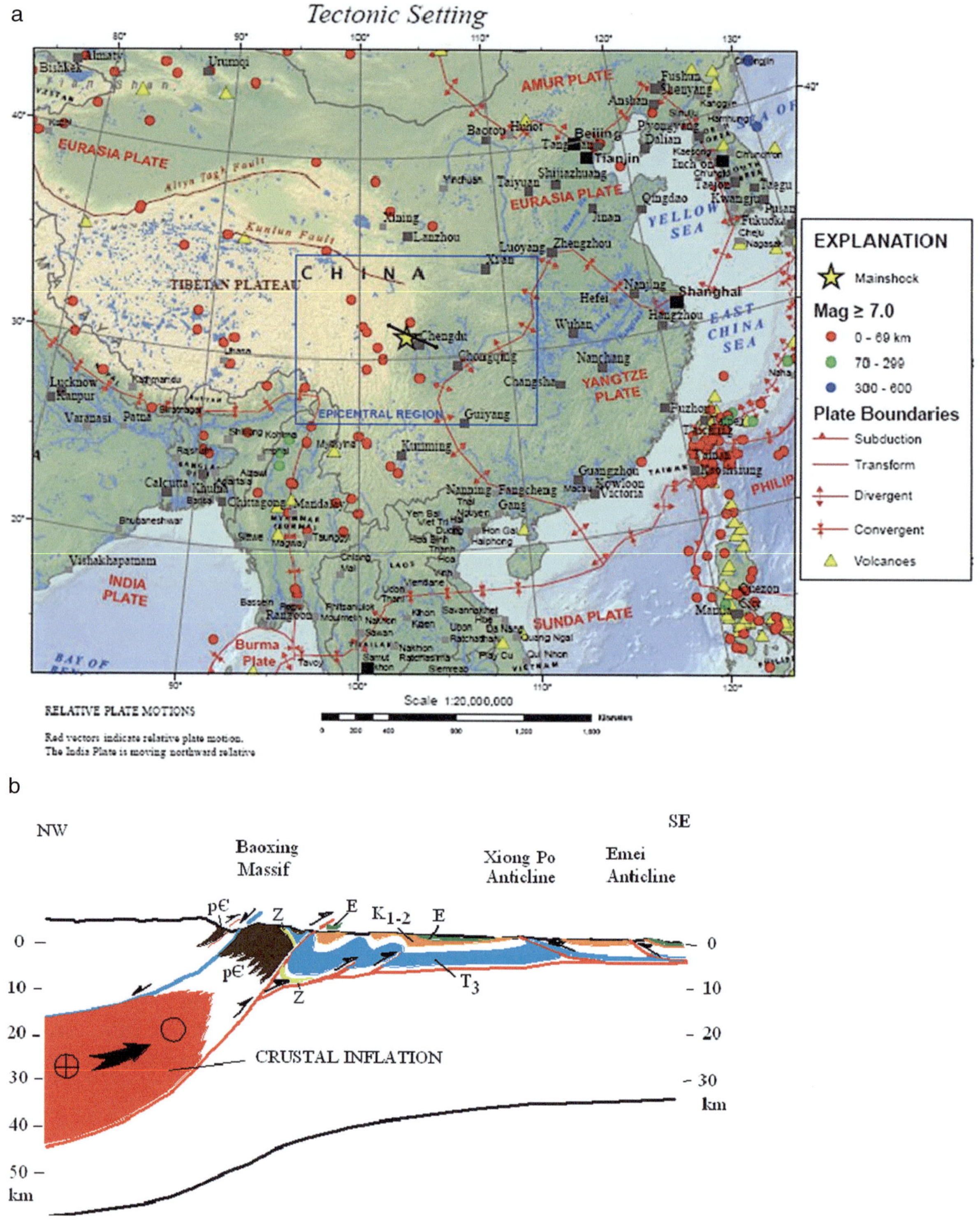

Fig. 9 **A**: Location map of China and neighboring countries. Star in center of map marks the location of the Mw = 7.9 Wenchuan (Sichuan Province) earthquake. The epicenter is on the eastern flank of the Tibetan Plateau. Black line near star marks the location of cross-section in part **B**: Crustal cross section at the hypocentral location of the Wenchuan, China, earthquake. The thicker crust of the Tibetan Plateau is being thrust eastward over the neighboring Sichuan basin

Fig. 10 Rescue workers and local residents search for survivors in the rubble following the August 15, 2007, Mw = 8.0 Pisco, Peru earthquake. Many of the deaths and injuries occurred in homes constructed with highly vulnerable adobe bricks

of life. The importance of building codes was highlighted by the August 15, 2007, earthquake in Pisco, Peru (USGS, 2007). Peru is a country where traditional and modern building designs are found in close proximity. Adobe buildings account for 65% of all buildings in rural areas and nearly 35% of all buildings in urban areas. Adobe bricks are indigenous, sun-dried building materials consisting of sand (50–70%), clay (15–30%), and silt (0–30%), that are often mixed with a binding material, such as straw. Adobe brick walls are highly vulnerable to collapse when subjected to severe ground shaking. When the Mw = 7.9 Pisco earthquake struck, many of the adobe houses in Pisco and Ica collapsed, whereas the modern reinforced concrete buildings were only superficially damaged (Fig. 10). There were more than 500 fatalities due to the Pisco earthquake, and an estimated 58,000 homes (80% within the city of Pisco) were destroyed, leaving more than 250,000 people without shelter (Fig. 10).

Disaster struck Iran in 2003, when a Mw = 6.6 earthquake ruptured along the Bam Fault in central Iran. The earthquake caused 43,000 fatalities, most of these due to building collapse (Eshghi and Zaré, 2004). Like Peru, the Bam area of Iran also utilizes traditional housing constructed from adobe. The tectonic setting of the Bam, Iran, earthquake is crustal compression and reverse faulting, as confirmed by earthquake focal mechanisms and analogue stress models of this continental collision zone (Fig. 11; Eshghi and Zaré, 2004; Sokoutis et al., 2003).

It is not always the case that traditional structures are weaker than modern designs. In the 2005 Mw = 7.6 Kashmir earthquake in Pakistan, western-style construction such as concrete block and brick masonry structures suffered more intense damage than the traditional timber-brick masonry typically used in this region (Naseem et al., 2005). In this case, buildings constructed using traditional styles and timber materials responded much better to ground shaking than all other building types. Traditional wood-framed buildings in Indonesia also perform much better than modern brick or unreinforced concrete building. A comparison of the 2005 Kashmir earthquake to the Pisco and Bam earthquakes indicates the importance of creating a building code appropriate for each specific region.

Future Directions in Earthquake Science

Enhanced Seismic Monitoring

Seismic monitoring systems have undergone tremendous growth during the past twenty-five years. The Global Seismic Network (GSN) was initiated by the

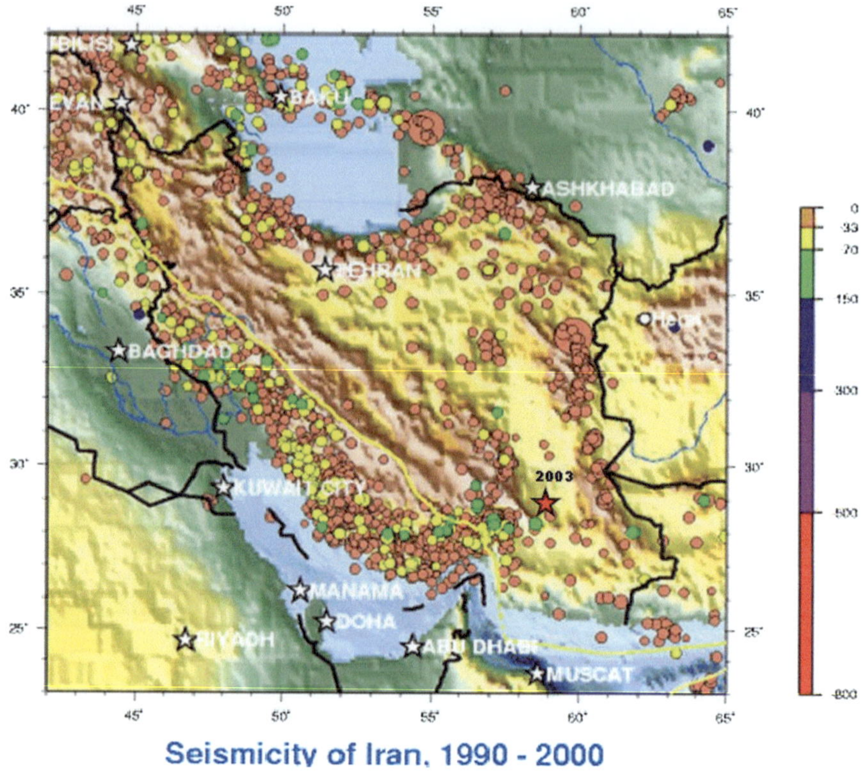

Fig. 11 Seismicity map of Iran, with location of the Mw = 6.6 Bam earthquake (*red star*) of 2003 that caused some 43,000 fatalities. The recurrence interval for large earthquakes in this region is estimated to be more than 1,000 years. However, even regions with long recurrence intervals may be highly vulnerable to earthquake disasters

Incorporated Research Institutions for Seismiology (IRIS) and now has more than 150 high-quality, broadband seismic stations (Fig. 12). This system is operated in collaboration with the US Geological Survey and the University of California-San Diego. Some 75% of these stations are available in realtime using satellite telemetry systems.

Many national seismographic systems have also been upgraded. The disastrous 1995 Kobe earthquake in Japan led to major upgrades in the seismic monitoring systems in that country. These include a high-sensitivity seismic array with 698 stations, a broadband array with 74 stations (F-net) called Hi-net and a strong-motion network with 1,043 accelerometers. The high-sensitivity array can rapidly and accurately locate earthquakes; the broadband array provides data on the earthquake source; and the strong motion array provides earthquake engineering data (as well as information about the source). A similar program of network upgrades has been completed in Taiwan. In mainland China, there are more than two thousand short-period seismographs, two hundred broadband stations and more than four hundred accelerometers. In Europe, a federation of national seismic systems, and international data collection program (e.g., ORFEUS and GEOSCOPE) provide abundant realtime data. In the United States, the Advanced National Seismic System (ANSS) is a comprehensive system that provides realtime seismic data from seismic sensors located in the free field and in buildings. Similar to other national networks, instrumentation includes a network of broadband sensors, accelerometers and high-gain seismic stations. The total number of sensors exceeds 7,000 in number, and the system automatically broadcasts information when a significant event occurs. Significant network upgrades have taken place in Mexico, Thailand, and Malaysia.

Global Positioning Systems (GPS)

Global Positioning Satellite (GPS) technology can detect minute motions of the Earth's crust that increase the stress on active faults and eventually leads to earthquakes (Segall and Davis, 1997). This technology

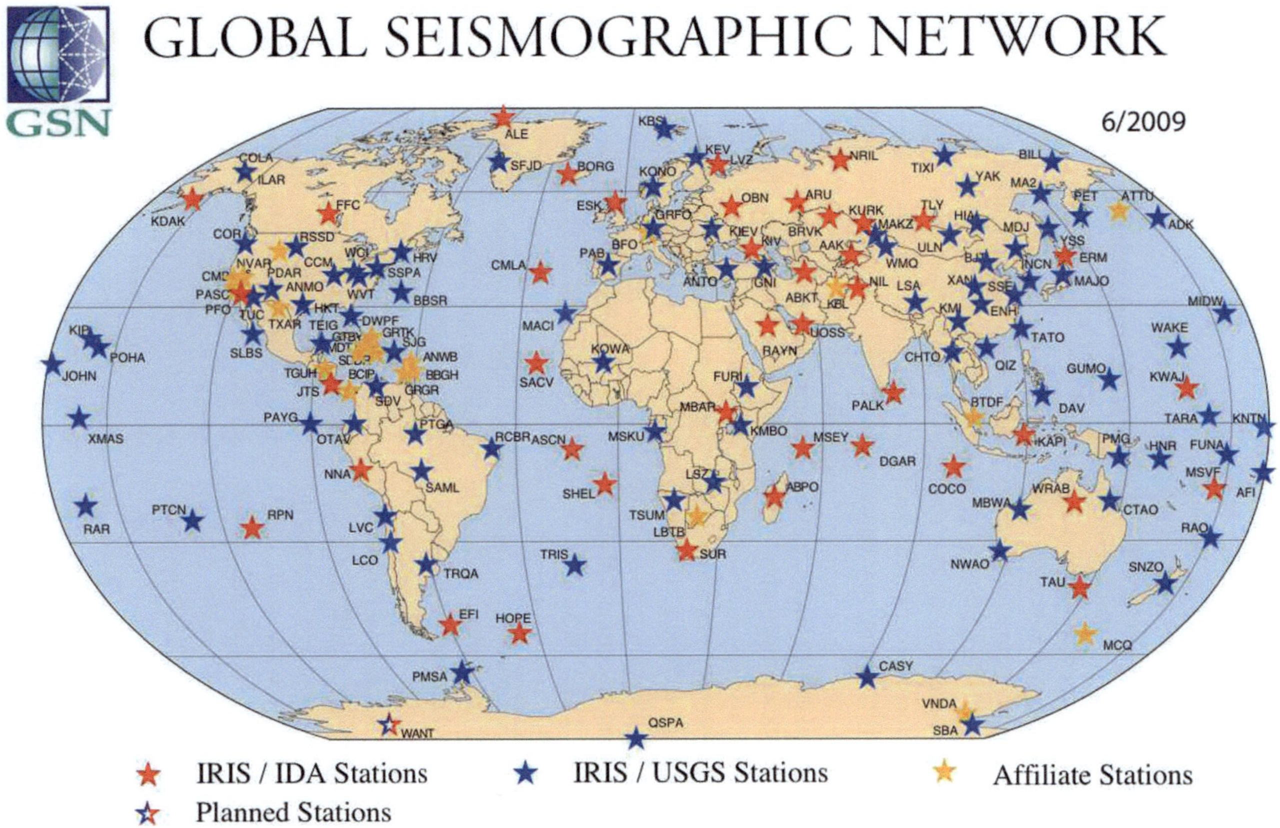

Fig. 12 Map of station locations of the Global Seismographic Network that monitors worldwide seismicity. All stations are located on continents and oceanic islands. Further sea-bottom stations are planned, but face technical challenges

provides an excellent picture of how slip (or ground displacement) can accumulate on faults throughout the earthquake cycle (e.g., Bakun et al., 2005). Satellites deployed across the globe emit precisely timed radio signals to tracking stations on the ground that record both gradual, aseismic motion as well as sudden displacements during earthquakes. GPS networks may be deployed in campaign (temporary) and permanent modes, but the decreasing cost and widespread use of this technology has been shifting more deployments to permanent status (Jordan, 2003). These data help in estimating earthquake potential, identifying active blind thrust faults and determining the potential response of major faults to the regional change in strain. As well, the ability of GPS technology to provide a measurement of the total slip caused by an earthquake complements traditional seismological methods of determining earthquake magnitude.

GPS measurements of crustal deformation are available for nearly all active tectonic environments. These data provide new and more accurate maps of the present crustal deformation field, a fundamental measurement of active continental tectonics. GPS data are important for studies of the earthquake source process since the measurement of surface displacement is mathematically related to a dislocation on a fault in an elastic medium. This relation permits the inversion of the geometry of the earthquake rupture. Such an inversion is more reliable when performed using near-field strong motion data (e.g., Bakun et al., 2005)

GPS data are also useful for the study of postseismic processes. The 1989 Loma Prieta, California, earthquake showed postseismic strain with a characteristic decay transient of 1.4 years (Savage et al., 1994). These authors report, contrary to expectations, that the transient parallel to the fault is smaller than the transient perpendicular to the fault. The interpretation of this observation is still debated.

GPS and older geodetic data have been used in a search for precursory crustal deformation prior to large earthquakes. Slow precursors were found for eight convergent margin earthquakes, including the 1960 9.2 M Chile, 1964 9.2 M Prince William Sound, Alaska, and the 1,700 Cascadian earthquakes (Roeloffs, 2006). On the other hand, no pre-seismic deformation was detected for the following terrestrial earthquakes: 2004 6.0 M Parkfield, 1992 7.3 M Landers, 2003 8.1 M Tokachi-oki (Irwan et al., 2004), and 1999 7.1 M Hector Mine earthquakes (Mellors et al., 2002). Since slow creep can go entirely undetected unless high quality GPS array data are available, it is presently inconclusive how often earthquakes are preceded by slow aseismic slip. This is an important research topic.

Interferometric Synthetic Aperture Radar (InSAR)

InSAR is a recent, innovative technology that permits the imaging of earthquake (crustal) deformation down to the millimeter scale (Wright et al., 2001a). Similar to GPS measurements, radar waves are emitted from satellites across the globe to the Earth's surface. In the case of InSAR, these radio waves are reflected from the ground surface and returned to the satellite. The satellite is sensitive to both: (1) the intensity of the returning electromagnetic wave, which has a different signature depending on the nature of the ground material, and (2) the phase of the returning wave, which will have been altered if ground displacement has taken place between successive passes of the satellite over the same location. This technology opens the door to continuously mapping deformation along active plate boundaries over larger areas and in greater detail than can practically be monitored by GPS measurements. InSAR derived interferograms have successfully been used to acquire a rapid map of surface deformation after an earthquake, such as the 1999 Izmit earthquake and in tracking interseismic strain accumulation along a large section of the Northern Anatolian Fault through minute measurements of surface displacement over a nearly decadal timescale (Wright et al., 2001b; Fig. 13).

InSAR techniques are also effective in measuring deformation on active volcanoes and landslides, both of which are significant geological hazards. For example, magma movement can be detected at otherwise apparently dormant volcanoes. As more InSAR satellites come into orbit, the capability has emerged to make measurements more frequently, and thereby make greater use of the technique as a monitoring tool. InSAR measurements of fault slip complement determinations made using seismic and GPS measurements, and generally cover a wider geographic area.

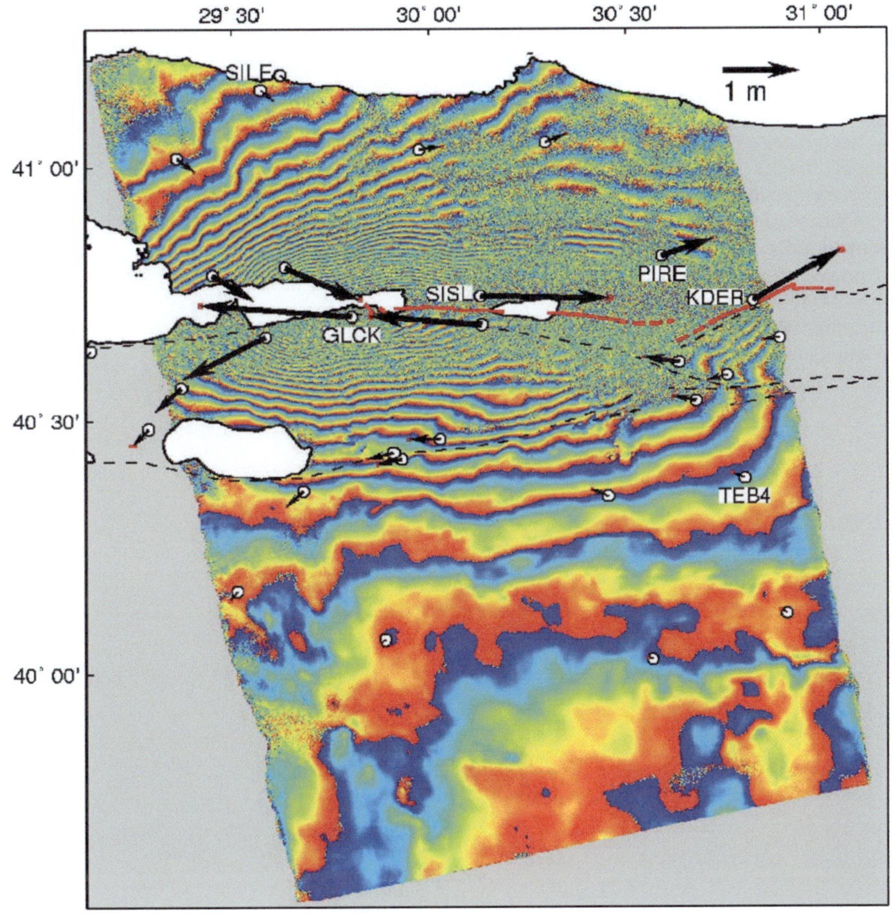

Fig. 13 Radar interferogram for the Izmit earthquake (data copyright ESA) revealing the surface displacements, measured in the satellite's line-of-sight, in the 35-day period between the two image acquisitions. Each interference fringe is equivalent to 28 mm of displacement in the satellite line-of-sight, or approximately 70 mm if caused by pure horizontal motion. *Red lines* are the mapped surface rupture [Barka, 1999] and the *dashed lines* are previously mapped segments of the North Anatolian Fault [Şaroglu et al., 1992]. (*after* Wright et al., 2001a)

Shakemaps of Seismic Intensities

Seismic intensity is a measurement of the severity of an earthquake's effects at different sites. The Modified Mercalli Intensity (MMI) scale ranges from Roman numeral I to XII, the highest level being total destruction. The MMI scale predates instrumental recordings, and is derived from field observations of damage. The intensity for historical earthquakes can also be determined from newspaper accounts, diaries, and other documents. The local intensity of an earthquake is of greater importance than the earthquake magnitude to those who manage emergency response because the intensity directly relates to damage effects.

A recent key development by the U.S. Geological Survey and its partners is an online system that provides near-real-time post-earthquake information regarding ground shaking. Shakemap (Wald et al., 2003) provides a map view of the ground shaking intensity in the region of an earthquake based on measurements from seismometers. Whereas an earthquake has a unique location and magnitude, the intensity of ground shaking it produces depends on such factors as the distance from earthquake, local site conditions and seismic wave propagation effects due to complexities in the structure of the Earth's crust. Shakemap software produces near real-time intensity maps for earthquakes, such as the May 12, 2008 Mw 7.9 Eastern Sichuan earthquake in China (Fig. 14). The widespread availability of such maps through the internet is valuable for the coordination of emergency response teams. The ground-shaking of hypothetical future earthquakes can also be evaluated, as well as the damage that would be associated with them today. ShakeMap thus serves as a useful, predictive tool by simulating the seismic intensity related to hypothetical future earthquakes.

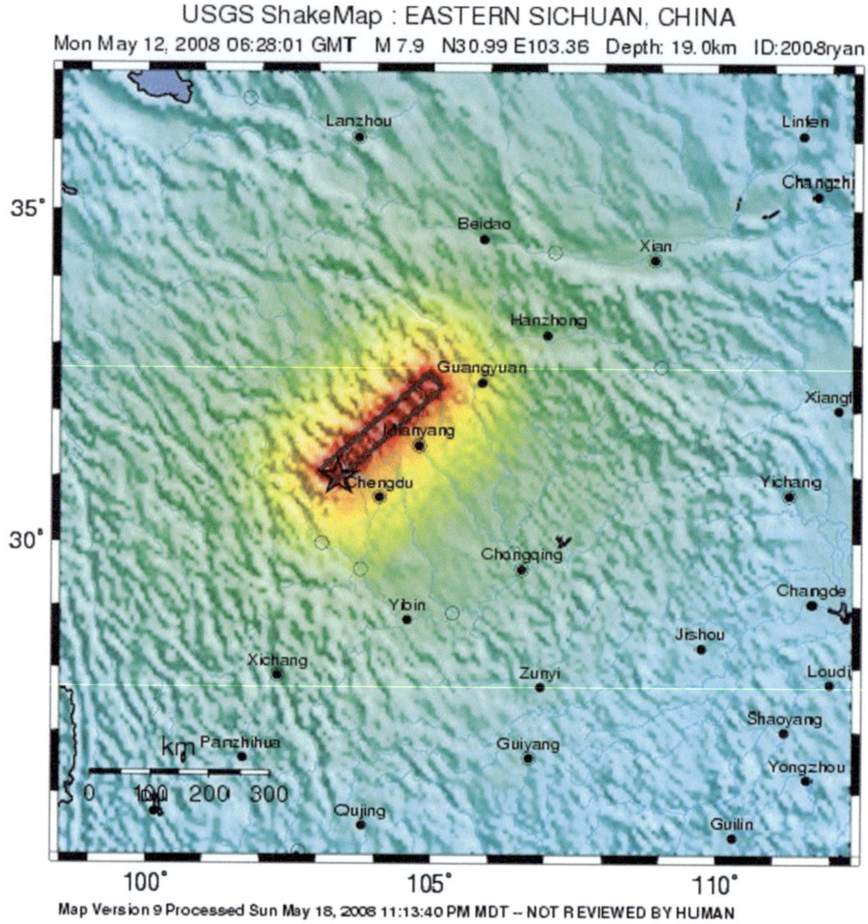

PERCEIVED SHAKING	Not felt	Weak	Light	Moderate	Strong	Very strong	Severe	Violent	Extreme
POTENTIAL DAMAGE	none	none	none	Very light	Light	Moderate	Moderate/Heavy	Heavy	Very Heavy
PEAK ACC.(%g)	<.17	.17-1.4	1.4-3.9	3.9-9.2	9.2-18	18-34	34-65	65-124	>124
PEAK VEL.(cm/s)	<0.1	0.1-1.1	1.1-3.4	3.4-8.1	8.1-16	16-31	31-60	60-116	>116
INSTRUMENTAL INTENSITY	I	II-III	IV	V	VI	VII	VIII	IX	X+

Fig. 14 Seismic shaking intensity map produced by the USGS shortly after the Mw = 7.9 May 12, 2008, Wenchuan, China. The map correctly indicated that a high population density NW of the epicenter were subjected to violent-to-extreme ground shaking intensities. Such maps, which are produced by processing data from local seismographs, are useful in planning earthquake emergency response

Earthquake Forecasting vs. Earthquake Prediction

Earthquake *prediction* refers to the ability to calculate the specific magnitude, place and time for a particular future earthquake, similar to how meteorologists can now forecast an oncoming hurricane or tornado on a short timescale. The current state of earthquake science precludes any ability to truly predict specific future earthquakes. Earthquake *forecasting*, refers to modeling the probabilities that earthquakes of specified magnitudes, and faulting types will occur during a specified time interval (usually several years) on a specific fault segment. Such probability estimates, when calculated over a specific time interval, are known as time-*dependent* earthquake forecasting. Time-*independent* forecasting, also known as long-term forecasting, is a general assessment of the likelihood of faults to rupture, not over a specific timeframe, and does not take into account whether earthquakes have occurred recently on particular faults. Time-independent forecasting is frequently used to evaluate building codes and developments or projects that must be sustainable in the long-term. A comparison between short-term and time-independent forecasting models can be found in Helmstetter et al. (2006). In order to calculate either type of earthquake probability forecast, a variety of data are assembled and analyzed, including earthquake recurrence intervals from paleoseismic, historical and instrumental records, deformation and slip rates from GPS and InSAR and long-term plate-tectonic models.

In 2007, the Working Group on California Earthquake Probabilities (WGCEP) developed a state-wide rupture (time-dependent) forecast called the Uniform California Earthquake Rupture Forecast (UCERF). This probability map specifies the likelihood of a Mw > 6.7 earthquake striking California over the next 30 years (Field et al., 2008; Fig. 15). Such probability maps are critical to ensure public safety in regions of high seismic hazard such as California or Alaska. The UCERF forecast will be used by the California Earthquake Authority (CEA) to analyze potential earthquake losses, set earthquake insurance premiums and develop new building codes.

Earthquake Early Warning

It is evident from the preceding review that much progress has been made in understanding earthquakes. Nevertheless, routine short-term earthquake prediction has not been achieved. Indeed, it will likely require many decades of additional research to address this problem. Therefore, it is useful to ask if it is feasible to provide an early warning of impending strong ground

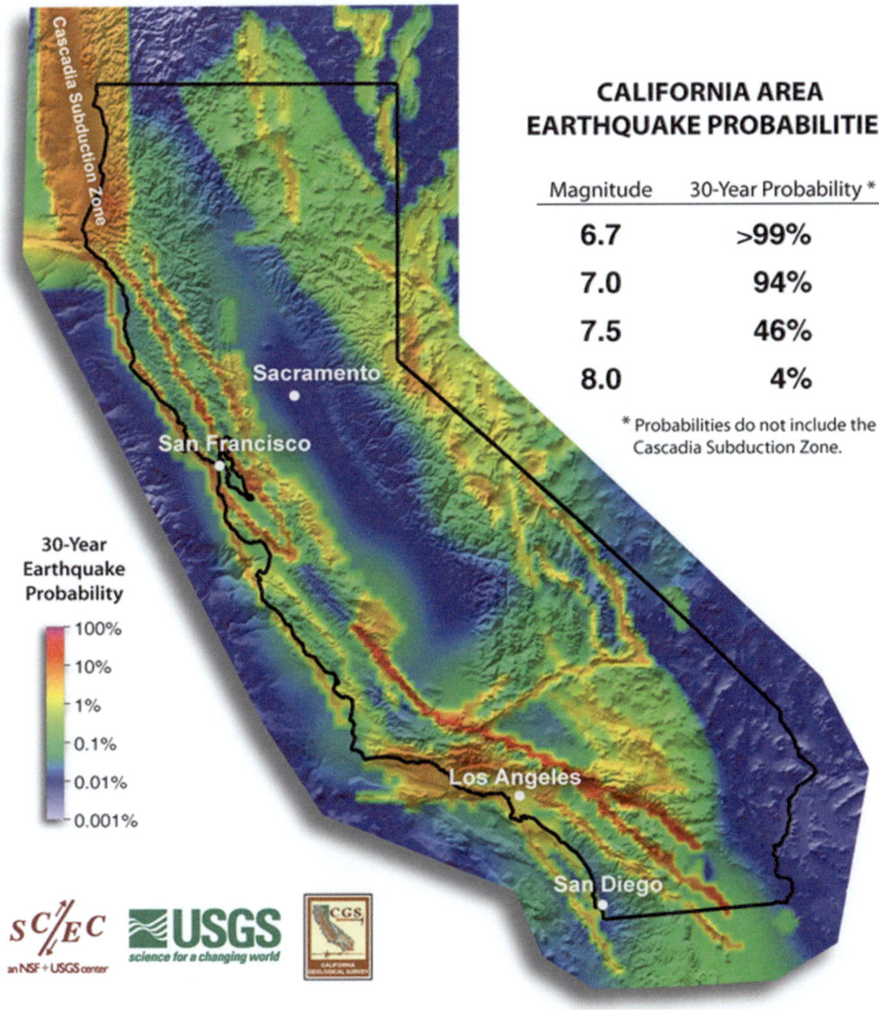

Fig. 15 Probabilistic earthquake hazard map for the State of California, USA, showing in yellow and orange those regions with higher probabilities for an earthquake with Mw $\geq$ 6.7 in the next 30 years. Boxes outlined in white located the Greater San Francisco and Greater Los Angeles areas with high seismic risk

motion based on an automated earthquake monitoring system, much as the lowered gates and flashing red lights at a railroad crossing announce the imminent arrival of a train.

Rapid earthquake notification is distinct from early warning systems. The former is a broadcast system that exists in many seismic networks that provides earthquake information within minutes after an earthquake occurs. In contrast, an early warning system provides an alert within seconds of the initial rupture of a significant (Mw $\geq$ 5) earthquake, indicating that strong ground shaking can be expected. A warning that provides only some tens of seconds of advanced notice of incoming strong ground motion may appear inconsequential, but in fact it would allow enough warning time for critical systems (e.g., high-speed trains) to be shut down as well as for mobile individuals to take protective cover from falling objects. The physical basis for such a system, first realized by Cooper (1868), is the fact that electromagnetic signals (radio and internet communications) travel faster than elastic waves. Additionally, the first arriving P-waves have much lower ground motions than the later arriving surface waves.

Earthquake early warning systems need to estimate the potential magnitude of an earthquake within the first few seconds of the rupture process (Ellsworth and Beroza, 1995; Beroza and Ellsworth, 1996). The feasibility of such a system requires that there be sufficient information in the first-arriving compressional-wave (P-wave) at local seismic stations to estimate the potential size of the earthquake using empirical relations (Allen and Kanamori, 2003; Kanamori, 2005). Test cases show that there is a strong correlation between earthquake magnitude and the frequency content of the initial few seconds of the seismogram. Early warning systems use the information contained in the initial portions of the seismic waveforms (the P-wave arrival) to estimate the eventual magnitude of the earthquake. This method of waveform analysis, as well as other methods (Cua and Heaton, 2003), can provide robust earthquake early warnings, especially in densely instrumented regions, such as Japan, Taiwan, Europe, and California.

Earthquake early warning systems have already been successfully operated. Mexico successfully issued an early warning to the public with their Seismic Alert System (SAS) during the Mw = 7.3 September 14, 1995 Copala earthquake that occurred on the subduction zone at the west coast, some 300 km from Mexico City. Over 4 million people in the city were warned. The success was in part due to the fact that the earthquake occurred during the day, when the majority of people were awake and had access to radios (Lee and Espinosa-Aranda, 1998). The Mexican Seismic Alert System consists of four units: seismic detection, telecommunications, central control, and early warning. The field stations are located 25 km apart, each monitor a region 100 km in diameter, and can estimate the magnitude of an earthquake within 10 s of its initiation. Other early warning systems have been installed in several other countries, including Japan, Taiwan, and Turkey (Lee et al., 1998). In view of the difficulty of achieving short-term earthquake predictions, earthquake early warning, like improved building codes, can be expected to play an increasingly important role in mitigating earthquake affects.

Closing Comments

Progress in the science of earthquakes, and the mitigation of earthquake effects has often been the results of knowledge gained from a devastating event. Examples include the earthquakes of 1906 in San Francisco, California; 1923 in the Kanto District, Japan; 1976 in Tangshan, China; and 2004 in Sumatra-Andaman Islands in Indonesia and India. We have highlighted some key concepts such as the earthquake cycle and recurrence intervals that are used in describing the underlying cause of earthquake. We have summarized some lessons learned from the earthquake record, such as the larger geographical area that experiences high seismic intensities for earthquakes that occur in continental interiors. Finally, we have described five important advances that have been made that have greatly enhanced our ability to monitor, report, and respond to large, damaging earthquakes. These five advances are: (1) enhanced seismic monitoring and notification; (2) GPS and (3) InSar monitoring; (4) the introduction of Shakemaps, and (5) progress in earthquake forecasting and early warning.

Have these steps succeeded in reducing earthquake hazards? The answer is certainly "Yes". Will these steps ensure a reduction in worldwide losses for the foreseeable future? The answer to this question is "Maybe". The reason for this equivocal answer is multifold. As shown by the 2005 Kashmir, Pakistan, and 2008, Sichuan, China, earthquakes, rapid population growth is resulting in the urbanization of formerly remote, seismically hazardous regions. These two earthquakes were in relatively isolated mountainous regions, but both caused more than 70,000 deaths, with millions left homeless. Due to the mountainous topography, massive landslides and rock falls were an additional, secondary effect, in addition to the severe ground shaking. Secondary hazards pose a severe threat to all affected by an earthquake. The December 26, 2004 tsunami is a prime example; the tsunami wave was a secondary hazard produced by the $Mw = 9.2$ earthquake that occurred at the Sumatra subduction zone. Another example occurred in Chimbote, Peru in 1971, when a $Mw = 7.9$ earthquake struck off the coast of Peru, and a landslide buried the city of Yungay and thousands of its residents. These examples should serve as reminders that one should prepare not only for earthquake strong ground motion, but also its secondary repercussions for example.

Remote and mountainous regions are not the only population centers at heightened risk. The growth of the world's mega-cities also presents a great challenge. There are twenty-two cities that have a population greater than 10 million; the list of cities at risk from earthquakes includes: Tokyo, Lima, Los Angeles, Mexico City, Beijing, Dhaka, Istanbul, Mumbai, Karachi, and Tehran. These ten cities comprise more than 120 million residents. Landslides and rock falls are not the main hazard in these megacities, building collapse and fires are. For this reason, the mitigation of earthquakes hazards in these cities depends on the adoption and enforcement of adequate building codes and fire safety. These new codes will require a very significant financial investment, one that in many cases may be beyond the reach of the local or national government.

Although some earthquakes have long recurrence intervals ($\geq$ 1,000 years), they are still extremely dangerous – possibly even more than those with short recurrence intervals because these communities may fail to plan for earthquakes which they believe will not happen in their lifetime. Unfortunately, this was the case in the Sichuan province in China on May 12, 2008. The Longmen Shan fault system has a recurrence interval of about 2,000 years. As a result, the public was not prepared to deal with an earthquake of such great magnitude (Fig. 16).

Contrary to expectations, the death toll caused by an earthquake does not always correspond with the magnitude. Such was the case for the December 26, 2003 ($Mw = 6.6$) Bam, Iran earthquake that killed approximately 31,000 people, injured 30,000, and left 75,600

Fig. 16 Ruins of an acient temple in Hanwang-Mianzhu, Sichuan, China. The destruction of the Wenchuan, China, earthquake reinforces the point that a long recurrence interval is not equivalent to safety. This temple withstood hundreds of years of environmental forces but was destroyed by the $Mw = 7.9$ earthquake on May 12, 2008. Source: Sarah Bahan, USGS

homeless. Some 85% of buildings in Bam were damaged or destroyed as a result of severe ground shaking, the main cause of damage to buildings and infrastructure. Furthermore, the proximity of a region to the seismic source is very important. The closer to the epicenter, the stronger the ground shaking will be.

Technological improvements hold the promise for reducing losses due to earthquakes. These improvements can be divided into two types: (1) better monitoring and risk assessment before an earthquake happens, and (2) improved reporting and response after the event occurs. Better seismic monitoring and GPS systems provide critical data for improved risk assessments. After an earthquake occurs, Shakemaps can be used to quickly and efficiently alert the government, media, and general public of the hazard. InSAR maps, if available quickly, provide a comprehensive picture of the region affected. Such technological advances will make it possible to continue to improve seismic hazard assessments, monitoring, mitigation and response.

References

Allen R.M. and H. Kanamori, 2003, The potential for earthquake early warning in southern California. *Science* 300, 786–789.

Bakun W, Aagaard B, Dost B, Ellsworth W, Hardbeck J, Harris R, Ji C, Johnston M, Langbein J, Lienkaemper J, Michael A, Nadeau R, Reasenburg P, Reichle M, Roeloffs E, Shakai A, Simpson R and Waldhauser F, 2005, Implications for prediction and hazard assessment from the 2004 Parkfield earthquake. *Nature* V. 437, p. 969–974.

Barka A., 1999, The 17 August 1999 Izmit earthquake. *Science* 285, 1858–1859.

Belardinelli M.E., Bizzarri A., Cocco M., 2000, Earthquake triggering by static and dynamic stress changes. *Eos Trans. AGU*. Fall Meet. Suppl, 81, 48.

Beroza G.C. and W.L. Ellsworth, 1996, Properties of the seismic nucleation phase. *Tectonophysics* 261, 209–227.

Bolt B., 2006, *Earthquakes*. Fifth edition. W. H. Freeman and Company, New York.

Brace W., and Kohlstedt D., 1980, Limits on lithospheric stress imposed by laboratory measurements. *J. Geophys. Res.* 85, 6248–6252.

Burchfiel B.C., Z. Chen, Y. Liu, L.H. Royden, 1995, *Tectonics of the Longmen Shan and adjacent regoins, central China. Int. Geol. Rev.* 37, 8, edited by W.G. Ernst, B.J. Skinner, L.A. Taylor.

Calais E., Han J.Y., DeMets C., d Nocquet J.M., 2006. Deformation of the North American plate interior from a decade of continuous GPS measurements. *J. Geophys. Res.* 111, doi: 10.1029/2005JB004253.

Campbell D.L., 1978. Investigation of the stress-concentration mechanism for intraplate earthquakes, *Geophys. Res. Lett.*, 5(6), 477–479.

Chandrasekhar D.V. and Mishra D.C., 2002. Some geodynamic aspects of Kutch basin and seismicity: An insight from gravity studies, *Curr. Sci. India*, 83(4), 492–498.

Cloetingh S., 1982. Evolution of passive margins and initiation of subduction zones, Ph.D. thesis, Utrecht Univ., Netherlands.

Cooper J.D., 1998. Letter to the Editor, San Francisco Daily Evening Bulletin, Nov. 3, 1868 (as quoted in Lee and Espinosa-Aranda, 1998).

Cua G. and T.H. Heaton, 2003. An envelope-based paradigm for seismic early warning, (abstract) *Trans. Am. Geophys. Union* 84, F1094–1095

Cutcliffe C. H. 2000. Earthquake resistant building design codes and safety standards: The California experience. *Geophys. J.*, 51, 259–262.

Dragert H. and Hyndman R., 1995. Continuous GPS monitoring of elastic strain in the northern Cascadia subduction zone. *Geophys. Res. Lett.* 22:755–758.

Dragert H; Wang K; James TS, 2001. A Silent Slip Event on the Deeper Cascadia Subduction Interface. *Science* 292(5521): 1525–1528.

Ellsworth W.L. and G.C. Beroza, 1995. Seismic evidence for an earthquake nucleation phase. *Science* 268, 851–855.

Eshghi S. and Zaré M. 2004. Preliminary observations on the Bam, Iran, earthquake of December 26, 2003. *EERI Special Earthquake Report.*

Field E.H., Dawson T.E., Felzer K.R., Frankel A.D., Gupta V., Jordan T.H., Parsons T., Petersen M.D., Stein R.S., Weldon II R.J., Wills C.J., 2008. The Uniform California Earthquake Rupture Forecast, Version 2 (UCERF 2), U.S. Geological Survey Open-File Report 2007-1337, CGS Special Report 203, SCEC Contribution #1138.

Frankel A., Mueller C., Barnhard T., Perkins D., Leyendecker E., Dickman N., Hanson S., Hopper M., 1996. National Seismic Hazard Maps: Documentation June 1996, U.S. Geological Survey Open-File Report 96-532, Denver, CO, 111 pp.

Freed A.A., 2005, Earthquake triggering by static, dynamic, and postseismic stress transfer, *Ann. Rev. Earth Plant. Sci.*, 33, 335–367.

Freed A. M., Ali S. T. and Burgmann R. 2007. Evolution of stress in Southern California for the past 200 years from coseismic, postseismic and interseismic stress changes. *Geophys. J. Int.*, 169, 1164–1179.

Gangopadhyay A. and Talwani P., 2003. Symptomatic features of intraplate earthquakes, *Seism. Res. Lett.*, 74, 863–883.

Goetze C., 1978, The mechanisms of creep in olivine, *Phil. Trans. Roy. Soc. Lond. A.*, 288, 99–119.

Goodacre A.K. and Hasegawa H.S., 1980. Gravitationally induced stresses at structural boundaries, *Can. J. Earth Sci.*, 17, 1286–1291.

Gordon R.G., 1995. Plate motions, crustal and lithospheric mobility, and paleomagnetism: prospective viewpoint. *J. Geophys. Res.*, v. 100, p. 24367–24392.

Gordon R.G. and Stein S., 1992. Global tectonics and space geodesy. *Science*, v. 256, p. 333–342.

Gupta H.K., Rastogi B.K., and Narain H., 1972, Common features of the reservoir-associated seismic activities, *Bull. Seis. Soc. Am.*, 62, 481–492.

Grollimund B. and Zoback M.D., 2001. Did deglaciation trigger intraplate seismicity in the New Madrid seismic zone?, *Geology*, 29(2), 175–178.

Helmstetter A., Kagan Y.Y., and Jackson D.D., 2006. Comparison of Short-Term and Time-Independent Earthquake Forecast Models in Southern California. *Bull. Seismol. Soc. Am.*, v. 96, no. 1, p. 90–106.

Hinze W.J. et al., 1988. Models for Midcontinent tectonism: an update, *Rev. Geophs.*, 26(4), 699–717.

Hirose H., et al. 1999. A slow thrust slip event following the two 1996 Hyuganada earthquakes beneath the Bungo Channel, southwest Japan. *Geophs. Res. Lett.* 26(21): 3237–3240.

Irwan M., Kimata F., Hirahara K., Sagiya T., Yamagiwa A., 2004. Measuring ground deformations with 1-Hz GPS data: the 2003 Tokachi-oki earthquake (preliminary report). *Earth Planets Space*, 56: 389–393. http://news. thomasnet.com/IMT/archives/2007/07/earthquake_japan_nuclear_kashiwazaki-kariwa_plant_industry_automotive_production_halt.html

Johanson I.A., Fielding E.J., Rolandone F., and Buergmann R., 2006. Coseismic and postseismic slip of the 2004 Parkfield earthquake from space-geodetic data. *Bull. Seismol. Soc. Am.*, 96(4B):S269–S282.

Johnston A.C. and Kanter L.R., 1990. Earthquakes in stable continental crust, *Sci. Am.*, 262(3), 68–75.

Johnston A. C., Coppersmith K. J., Kanter L.R. and Cornell C.A., 1994. The earthquakes of stable continental regions: assessment of large earthquake potential, TR-102261, Vol. 1–5, ed. Schneider J.F., *Electric Power Research Institute (EPRI)*, Palo Alto, CA.

Jordan T. 2003. *Living on an Active Earth: Perspectives on Earthquake Science*. The National Academies Press: Washington, D.C.

Kanamori H., (2005). Real-time seismology and earthquake damage mitigation. *Ann. Rev. Earth Planet. Sci.* 33, 195–214.

Kayal J.R., Zhao D., Mishra O.P., De R., Singh O.P., 2002, The 2001 Bhuj earthquake: Tomographic evidence for fluids at the hypocenter and its implications for rupture nucleation, *Geophys. Res. Lett.*, 29, 2152–2155.

Kenner S.J. and Segall P., 2000. A mechanical model for intraplate earthquakes; application to the New Madrid seismic zone, *Science*, 289(5488), 2329–2332.

King G.C.P., Stein R.S. and Lin J., 1994, Static stress changes and the triggering of earthquakes, *Bull. Seismol. Soc. Am.*, 84, 935–953.

Kostoglodov V., et al., 2003. A large silent earthquake in the Guerrero seismic gap, *Mexico. Geophs. Res. Lett. Col.* 32(15): 1807, doi:10.1029/2003GL017219.

Lawson A.C. (Ed.), 1908, The California earthquake of April 18, 1906. Report of the State Earthquake Investigation Commission. (reprinted in 1969 by the Carnegie Institution of Washington, D.C.).

Lee W. H. K. and Espinosa-Aranda J. M., 1998. Earthquake Early Warning Systems: Current Status and Perspectives in Early Warning Systems for Natural Disaster Reduction. J. Zschau and A. N. Kuppers, Eds. Springer, New York: 834 pp.

Lee W.H.K. and Espinosa-Aranda J.M., 2003. Earthquake early warning systems: Current status and perspectives: in 'Early Warning Systems for Natural Disaster Reduction', J. Zschau and A. N. Kuppers, Eds. Springer, Berlin: pp. 409–423.

Liu L. and Zoback M.D., 1997. Lithospheric strength and intraplate seismicity in the New Madrid seismic zone, *Tectonics*, 16(4), 585–595.

Long L.T., 1988. A model for major intraplate continental earthquakes, *Seism. Res. Lett.*, 59(4), 273–278.

Lowry AR; Larson KM; Kostoglodov V; Bilham R. Transient fault slip in Guerrero, southern Mexico. *Geophys. Res. Lett.*, vol. 28, no.19, pp. 3753–3756, 2001.

Mandal P., and Rastogi B.K., 2005, Self-organized fractal seismicity and b value of aftershocks of the 2001 Bhuj earthquake in Kutch (India): *Pure Appl. Geophys.*, v. 162, no. 1, p. 53–72.

Melbourne TI, and FH Webb., 2003, Slow but not quite silent. *Science* 300: 1886–1887.

Mellors R.J., Sichoix L., Sandwell D.T., 2002. Lack of precursory slip to the 1999 Hecot Mine, California, earthquake as constrained by InSAR. *Bull. Seis. Soc. Am.* 92: 1443–1449.

Mishra O.P. and ZhaoD., 2003, Crack density, saturation rate and porosity at the 2001 Bhuj, India, earthquake hypocenter: a fluid-driven earthquake?, *Earth Planet. Sci. Lett.*, 212 (3–4), 393–405.

Mohindra R. and Bagati T.N., 1996, Seismicially induced soft-sediment deformation structures (seismites) around Sumdo in the lower Spiti valley (Tethys Himalaya), *Sediment. Geol.*, 101, 69–83.

Murray J. and Langbein J., 2006. Slip on the San Andreas Fault at Parkfield, California, over two earthquake cycles, and the implications for seismic hazard. *Bull. Seismol. Soc. Am.*, 96(4B):S283–S303.

Naseem A., Ali Q., Javed M., Hussain Z., Naseer A., Ali S. M., Ahmed I., Ashraf M., Scawthorn C. 2005. First report on the Kashmir earthquake of October 8, 2005. *EERI Special Earthquake Report*.

Nelson A.R., Atwater B.F., Bobrowsky P.T., Bradley L.A., Clague J.J., et al., 1995. Radiocarbon evidence for extensive plate-boundary rupture about 300 years ago at the Cascadian subduction zone. *Nature* 378: 371–374.

Quinlan G., 1984. Postglacial rebound and the focal mechanisms of eastern Canadian earthquakes, *Can. J. Earth Sci.*, 21, 1018–1023.

Raphael A., and Bodin P., 2002, Relocating aftershocks of the 26 January 2001 Bhuj earthquake in western India. *Seis. Res. Lett.* 73, 417–418

Reid H.F., 1910, The Mechanics of the Earthquake, The California Earthquake of April 18, 1906, Report of the State Investigation Commission, Vol.2, Carnegie Institution of Washington, Washington, D.C.

Roeloffs E, May 2006, Evidence for Aseismic Deformation Rate Changes Prior to Earthquakes. *Annual Review of Earth and Planetary Sciences*. Vol. 34: 591–627. (doi:10.1146/annurev.earth.34.031405.124947)

Rogers G., and Dragert H., 2003. Episodic Tremor and Slip on the Cascadia Subduction Zone: The Chatter of Silent Slip. *Science* 300(5627): 1942–1943.

Şaroglu F., Ő. Emre, and I. Kuşçu, Active Fault Map of Turkey, General Directorate of Mineral Research and Exploration (MTA), Eskişehir Yolu, 06520, Ankara, Turkey, 1992.

Satake K. and Atwater B.F., 2007. Long-Term Perspectives on Giant Earthquakes and Tsunamis at Subduction Zones. *Annu. Rev. Earth Planet. Sci.* 35:349–374.

Savage J.C., Lisowski M., Prescott W.H., 1981. Geodetic strain meansurements in Washington. *J. Geophys. Res.* 86: 4929–4940.

Savage J.C., Lisowski M., Svarc J., 1994, Postseismic deformation following the 1989 (M = 7.1) Loma Prieta, California, earthquake, *J. Geophys. Res.*, 99, 13757–13765.

Sbar M.L. and Sykes L.R., 1973. Contemporary compressive stress and seismicity in eastern North America: An example of intra-plate tectonics, *Geol. Soc. Am. Bull.*, 84, 1861–1882.

Schulte S.M. and Mooney W.D., 2005. An updated global earthquake catalogue for stable continental regions: reassessing the correlation with ancient rifts. *Geophys. J. Int.* 161, 707–721.

Segall P. and Davis J.L., 1997, GPS applications for geodynamics and earthquake studies, *Ann Rev. Earth Planet. Sci.*, 25, 301–336.

Shishikura M., 2003, Cycle of interpolate earthquakes along the Sagami Trough deduced from tectonic geomorphology, *Bull. Earthquake Res. Inst. Univ. Tokyo* 78, 245–254.

Simpson D.W., 1986, Triggered earthquakes, *Ann. Rev. Earth Plant. Sci.*, 14, 21–42.

Sokoutis D. et al. Insights from scaled analogue modeling into the seismotectonics of the Iranian region. *Tectonophysics* Volume 376, Issues 3–4, 4 December 2003. pp. 137–149

Stein R.S., Barka A.A., and Dieterich J.H., 1997, Progressive failure on the North Anatolian fault since 1939 by earthquake stress triggering, *Geophys. J. Int.*, 128, pp. 594–604.

Stein R.S., 1999, The role of stress transfer in earthquake occurrences, *Nature*402, 605–609.

Stein R.S., Toda S., Parsons T. And Grunewald E., 2006, A new probabilitstic seismic hazard assessment for greater Tokyo, Phil. Trans. R. Soc. A, 1965–1988, doi:10.1098/rsta.2006.1808.

Stein S., Sleep N., Geller R.J., Wang S.C. and Kroeger G.C., 1979. Earthquakes along the passive margin of eastern Canada, *Geophys. Res. Lett.*, 6(7), 537–540.

Sykes L.R., 1978. Intraplate seismicity, reactivation of preexisting zones of weakness, alkaline magmatism, and other tectonism postdating continental fragmentation, *Rev. Geophys.*, 16(4), 621–688.

Tagare G. V. 2002. Earthquakes: Some ancient speculations. Speech given at the Institute for Oriental Study, Thane, India. http://www.orientalthane.com/speeches/gvtagare/1.html

Talwani P., 1988. The intersection model for intraplate earthquakes, *Seism. Res. Lett.*, 59(4), 305–310.

Talwani P., 1999. Fault geometry and earthquakes in continental interiors, *Tectonophysics*, 305, 371–379.

Talwani P. and Rajendran K., 1991. Some seismological and geometric features of intraplate earthquakes, *Tectonophysics*, 186, 19–41.

Tuttle M.P., Schweig III, E.S., Campbell J., Thomas P.M., Sims J.D., Jafferty III, R.H., 2005, Evidence for New Madrid earthquakes in

Tse S., and Rice J., 1986. Crustal Earthquake Instability in Relation to the Depth Variation of Frictional Slip Properties, *J. Geophys. Res.*, 91(B9), 9452–9472.

United Nations Population Division. World population prospects: the 2002 Revision.http://www.un.org/popin/data.html

USGS Earthquake Hazard Summary, Magnitude 8.0 – NEAR THE COAST OF CENTRAL PERU, 2007. http://earthquake.usgs.gov/eqcenter/eqinthenews/2007/us2007gbcv/#summary

USGS earthquake summary for the July 16, 2007 earthquake in Japan. http://earthquake.usgs.gov/eqcenter/eqinthenews/2007/us2007ewac/

Vinnik L.P., 1989. The origin of strong intraplate earthquakes, *translated from* O prirode sil'nykh vnutrilplitovykh zemletyaseniy, Doklady Akademii Nauk SSSR, 309(4), 824–827.

Wald D., Wald L., Worden B., and Goltz J., 2003. ShakeMap – A tool for earthquake response, US Geological Survey Fact Sheet FS-087-03. (available at: http://pubs.usgs.gov/fs/fs-087-03/)

West M., Sanches J.J., McNutt S.R., 2005. Periodically Triggered Seismicity at Mount Wrangell, Alaska, After the Sumatra Earthquake. *Science* Vol. 308, No. 5725, 1144–1146.

Wright T.J., E.J. Fielding and B.E. Parsons, 2001a. Triggered slip: observations of the 17 August 1999 Izmit (Turkey) earthquake using radar interferometry, *Geophys. Res. Lett.*, 28(6), 1079–1082.

Wright T.J., Parsons B.E. Fielding E.J., 2001b. Measurement of interseismic strain accumulation across the North Anatolian Fault by satellite radar interferometry, *Geophys. Res. Lett.* 28(10), 2117–2120.

Xu X., Yu G., Klinger Y., Tapponnier P. & Van Der Woerd J., 2006. Reevaluation of surface ruion of the 2001 Kunlunshan earthquake (M w 7.8), northern Tibetan Plateau, China, *J. Geophys. Res.*, **111**, B05316, doi:10.1029/2004JB003488.

Zoback M.L. and Zoback M.D., 1980, State of stress in the conterminous United States, *J. Geophys. Res.*, 85, 6113–6156.

Zoback M.D., 1983. Intraplate earthquakes, crustal deformation and in-situ stress, in *A workshop on 'The 1886 Charleston, South Carolina, earthquake and its implications for today'*, Open-File Report, No. 83–843, pp. 169–178, eds Hays, W.W., Gori, P.L. and Kitzmiller, C., US Geological Survey, Reston, VA.

Zoback M.D. and Townend J., 2001, Implications of hydrostatic pore pressure and high crustal strength for the deformation of intraplate lithosphere, *Tectonophysics* 336, 19–30.

Zoback M.L., 1992, First- and second-order patterns of stress in the lithosphere: the world stress map project. *J. Geophys. Res.*, 97, 11703–11728

Zoback M.L. and Richardson R.M., 1996. Stress perturbation associated with the Amazonas and other ancient continental rifts, *J. Geophys. Res. B Solid Earth Planets*, 101(3), 5459–5475.

Passive Seismic Monitoring of Natural and Induced Earthquakes: Case Studies, Future Directions and Socio-Economic Relevance

Marco Bohnhoff, Georg Dresen, William L. Ellsworth, and Hisao Ito

Abstract An important discovery in crustal mechanics has been that the Earth's crust is commonly stressed close to failure, even in tectonically quiet areas. As a result, small natural or man-made perturbations to the local stress field may trigger earthquakes. To understand these processes, Passive Seismic Monitoring (PSM) with seismometer arrays is a widely used technique that has been successfully applied to study seismicity at different magnitude levels ranging from acoustic emissions generated in the laboratory under controlled conditions, to seismicity induced by hydraulic stimulations in geological reservoirs, and up to great earthquakes occurring along plate boundaries. In all these environments the appropriate deployment of seismic sensors, i.e., directly on the rock sample, at the earth's surface or in boreholes close to the seismic sources allows for the detection and location of brittle failure processes at sufficiently low magnitude-detection threshold and with adequate spatial resolution for further analysis. One principal aim is to develop an improved understanding of the physical processes occurring at the seismic source and their relationship to the host geologic environment. In this paper we review selected case studies and future directions of PSM efforts across a wide range of scales and environments. These include induced failure within small rock samples, hydrocarbon reservoirs, and natural seismicity at convergent and transform plate boundaries. Each example represents a milestone with regard to bridging the gap between laboratory-scale experiments under controlled boundary conditions and large-scale field studies. The common motivation for all studies is to refine the understanding of how earthquakes nucleate, how they proceed and how they interact in space and time. This is of special relevance at the larger end of the magnitude scale, i.e., for large devastating earthquakes due to their severe socio-economic impact.

Keywords Earthquakes · Passive Seismic monitoring · Borehole Seismology · Crustal mechanics · Physics of Faulting

Introduction

Global monitoring of seismicity detects the occurrence of earthquakes down to about M = 4. The resulting pattern of their distribution traces the plate boundaries and highlights the most active intraplate seismic zones. In many parts of the globe the detection threshold is lower because of the presence of regional and local seismic networks. Within regions such as the west coast of North America, Japan and Western Europe regional thresholds on the order of M = 1–2 have been achieved. Such well-designed local seismic networks not only record earthquake activity at low magnitude detection thresholds but also resolve the focal depths and focal mechanisms of earthquakes. The tools of modern Passive Seismic Monitoring, referred to as PSM in the following, allows the refinement of earlier

M. Bohnhoff (✉)
Department of Geophysics, Stanford University, Stanford, CA 94305-2215, USA
e-mail: marcob@stanford.edu

Marco Bohnhoff now at Helmholt - Centre Potsdam GFZ (bohnhoff@gfz-potsdam.de)

S. Cloetingh, J. Negendank (eds.), *New Frontiers in Integrated Solid Earth Sciences*, International Year of Planet Earth, DOI 10.1007/978-90-481-2737-5_7, © Springer Science+Business Media B.V. 2010

seismotectonic models pioneered in the 1970s through the application of modern methods for determining crustal structure, locating earthquakes and determining focal mechanisms. These include 3-D mapping of active faults and fault systems, routine moment tensor determination of source processes, analysis of earthquake interaction, high-resolution characterization of active faults within hydrocarbon and geothermal reservoirs, and investigation of the systematics of the earthquake cycle for large magnitude earthquakes along plate-bounding faults.

The concepts behind PSM date back to the 1930s and 1940s and were accompanied by the quantification of the earthquake phenomenon. The introduction of the earthquake magnitude scale for regional events in California by C.F. Richter (1935) and the discovery of the earthquake frequency-magnitude relation by B. Gutenberg and C.F. Richter (1941) are certainly the most prominent and relevant examples. At about the same time, the idea of developing earthquake studies from regional and local seismic networks was introduced to study aftershock sequences in Japan (Imamura et al., 1932).

Based on technical developments in the 1960s that permitted the low-power operation and recording of many seismic stations on a common time base, new seismic networks and processing methods were developed that permitted the routine analysis of earthquakes of $M < 2$, commonly referred to as "microseismicity". A comprehensive review of principles and applications of microearthquake networks of this period was given by Lee and Stewart (1981). Using a similar approach extensive research using local arrays of seismic stations was undertaken to implement a nuclear test ban treaty. Over the last two decades, progress in the field of PSM has occurred primarily as a result of advances in seismic instrumentation and computation facilities to store and serve large data sets. The transition to digital high-frequency full waveform data acquisition systems with increased dynamic range also stimulated the development of more sophisticated analysis schemes allowing refinement of existing models at local, regional and plate-boundary scale.

A key objective of modern PSM is to collect data that can be used to resolve earthquake source processes in space and time during the rupture process. Seismic waves observed at a receiver carry information about the source process, but are also modified by propagation through the earth. When earth structure is sufficiently well known, it is possible to correct the observed waveforms for many propagation effects. As the waves propagate in the heterogeneous and inelastic earth, information about the source process is lost due to scattering and attenuation. Once lost, this information cannot be recovered, and so the solution is to place the receiver "close enough" to the source. Because the losses are greatest at the highest frequencies, to record a signal with a ~1 m wavelength the sensor must be within less than 1 km of the source, assuming anelastic and scattering losses corresponding to a damping factor of Q ~500. Clearly, if one intents to understand earthquake processes on a specific scale, one needs to record close to the source.

In this paper we summarize the principal objectives involved in PSM and review selected examples and future directions from key-locations representing various environments such as hydrocarbon and geothermal reservoirs, seismically quiet intra-plate regions, and large-scale transform faults and subduction zones. We also consider rock-deformation experiments in the laboratory thus giving examples that cover rupture length scales from millimeter to hundreds of km. Table 1 gives an overview on the different magnitude ranges and the relevant scales of rupture dimension, displacement, dominant frequency and seismic moment for the different environments discussed in this paper. We highlight that a comprehensive understanding of the physical processes responsible for brittle failure requires investigations that span this large spatial and frequency range. The effort to monitor these processes using adequately designed receiver geometries is important, especially with regard to socio-economic implications as shown in the cases of subduction megathrusts and large earthquakes along plate-bounding transform faults.

Quantifying the Earthquake Process

Earthquakes are the vibratory motion of the earth created by the sudden release of energy within the solid rock mass of the planet. Most earthquakes are caused by slip on faults, and as a consequence the term "earthquake" is commonly used to refer to the earthquake source process rather than the seismic waves it causes. Because the waves travel to great distances through the earth even for small underground disturbances they provide a powerful observational basis for studying the location, strength and fundamental nature of the earth-

Table 1 Overview on different earthquake magnitude ranges and relevant scales for rupture length, displacement, dominant frequency and seismic moment. Length and displacement scales are approximate and appropriate for crustal earthquakes with stress drops of 3 MPa. Note that ranges given may overlap between earthquake class depending on source-receiver distances and type of wave recorded

Magnitude range	Class	Length scale	Displacement scale	Frequency scale	Seismic moment*
8–10	Great	100–1,000 km	4–40 m	0.001–0.1 Hz	1 KAk–1 MAk
6–8	Large	10–100 km	0.4–4 m	0.01–1 Hz	1 Ak–1 KAk
4–6	Moderate	1–10 km	4–40 cm	0.1–10 Hz	1 mAk–1 Ak
2–4	Small	0.1–1 km	4–40 mm	1–100 Hz	1 μAk–1 mAk
0–2	Micro**	10–100 m	0.4–4 mm	10–1,000 Hz	1 nAk–1 μAk
−2–0	Nano	1–10 m	40–400 μm	0.1–10 kHz	1 pAk–1 nAk
−4 to −2	Pico	0.1–1 m	4–40 μm	1–100 kHz	1 fAk–1 pAk
−6 to −4	Femto	1–10 cm	0.4–4 μm	10–1,000 kHz	1 aAk–1 fAk
−8 to −6	Atto	1–10 mm	0.04–0.4 μm	1–100 MHz	1 tAk–1 aAk

* 1 Aki (Ak) is defined as 10^{18} Nm. The unit is named after Keiiti Aki, who pioneered the use of seismic moment in theory and practice. The International Association of Seismology and Physics of the Earth's Interior recommended in 2007 the adoption of the Aki as the standard unit of earthquake size.

** The term "microearthquake" traditionally refers to earthquakes $M < 3$. The earthquake class names used here are a compromise between the SI naming conventions, which would require that a microearthquake had a magnitude between $M = 2$ and $M = 4$, and traditional practice.

quake source. Seismology is the science of the analysis of these waves, and over the past century it has become a deep and sophisticated branch of mathematical physics (e.g., Aki and Richards, 2002). From the waves it is possible, in theory, to extract a detailed description of the earthquake source process in space and time. The study of the earthquake source, however, is of necessity an empirical science, as we have little control over when and where earthquakes occur. Aside from analog experiments performed in the laboratory or earthquakes induced by industrial modification of underground conditions, the seismologist must be prepared at all times to capture the earthquake when it happens.

Modern seismological instruments are designed to record the wide range of frequencies and amplitudes contained in the seismic waves of particular interest. Successful PSM also requires a geographic distribution of instruments that encircle the source. Only by recording waves from a range of azimuths and distances is it possible to accurately determine the location of the earthquake source (hypocenter) and determine its basic source properties (moment tensor, focal mechanism, etc.). For most natural earthquakes recorded by surface stations, this requires a network of instruments with at least one station within a focal depth of each earthquake. When these conditions are met, the initial point of rupture in an earthquake can be determined to a precision of a few hundred meters. Substantially higher precision locations can be obtained using the seismic wave field (Rubin et al., 1999; Waldhauser and Ellsworth, 2000).

Obtaining an accurate geographic description of where earthquakes occur is among the most basic steps toward developing a tectonic understanding of a region. Other physical measures of the complex mechanical event producing the earthquake take many forms, including the dimensions of the faulted region, the direction and amount of slip in both space and time, as well as traditional measures based on the amplitudes of the radiated elastic waves. To relate the characteristics of one event to another, the observed quantities must generally be summarized through the use of either an empirical relation, such as magnitude, or a quantity derived from a physical model, such as seismic moment. Based on recordings from the Southern California Seismic Network that initially consisted of ~10 stations at 100 km spacing, Richter (1935) developed the local magnitude scale as a first approach to quantify the earthquake size in a physical sense on an instrumental basis. He defined

$$M_l = \log_{10} A - \log_{10} A_0(\Delta)$$

where M_l is the event magnitude, A is the maximum amplitude recorded by the Wood-Anderson seismograph, and $\log_{10} A_0$ is the reference term used to account for amplitude attenuation with epicentral distance (Δ). This concept was further developed and generalized by Gutenberg (1945) then including also teleseismic events.

Based on Richter's results, Ishimoto and Ida (1939) as well as Gutenberg and Richter (1941) discovered a systematic relation between the magnitude of earthquakes and their frequency resulting in

$$\log N = a - b^* M$$

where N is the number of events and a and b are constants representing the overall level of seismic activity and the ratio between small and large earthquakes, respectively. To date, numerous studies of earthquake statistics extending over as much as 34 magnitude units from phonon emission in crystals to devastating earthquakes have confirmed this relation and determined the value b to be close to 1 on average (Davidson et al., 2007). Consequently, the number of earthquakes increases by about a factor of 10 for each decrease of one magnitude unit. This simple relation exemplifies that the number of seismic events detected by a seismic network entirely depends on its magnitude-detection threshold or magnitude of completeness (M_c) for which the network detects all events in the target area. M_c is directly linked to the average source-receiver distance of a seismic network and station density. The value of the resulting earthquake catalog for scaling-related studies clearly increases with decreasing M_c.

Of similar importance as the quantitative description of earthquake magnitude are physical descriptions of the source process. The lowest-order approximation of an earthquake is a double-couple point source. Describing the source by force equivalents leads to elegant mathematical approaches for propagation of elastic waves (Aki and Richards, 2002). Further refinement of the source processes replaces the point model by a spatially extensive description, such as a circular crack, distribution of point sources, or more complex geometries. Digital waveforms with sufficient dynamic range allow routine measurement of the properties of a seismic source such as the seismic moment tensor, rupture duration and stress drop.

It is beyond the scope of this paper to provide an introduction into the theory of faulting and the propagation of elastic waves. Rather, we briefly describe the role of stress as a key-parameter for the generation of earthquakes. Following the great 1906 San Francisco earthquake, H.F. Reid (1910) developed the theory of elastic rebound that has been largely confirmed as the basis of earthquakes by modern GPS measurements. Reid concluded that the slip during the earthquake resulted from the release of previously accumulated elastic strain energy stored in the crust astride the San Andreas Fault. Once the applied stress exceeded the rock strength the stored energy was released by rapid slip on the fault. We now know that the elastic strain energy accumulated over centuries due to plate motion, and the forces are building up today for another great earthquake, as this seismic cycle of accumulation and release continues. The fault is a plane of weakness in the earth that will be re-activated when the applied shear stress (τ) exceeds a critical value. This value is primarily a function of the effective normal stress representing the difference between normal stress (σ_n) and fluid pressure (P_F) across the fracture plane and the roughness of the fracture surfaces. Assuming a constant normal stress, the behavior remains elastic until the peak-strength is reached and slip occurs. Amongst various formulations for the peak shear strength (T), the simplest one is the linear Mohr-Coulomb criterion

$$T = C_o + (\sigma_n - P_F) * \tan\phi$$

where ϕ is the friction angle and C_o is cohesion. In the case of a natural earthquake, shear failure results from shear stress accumulation through tectonically driven loading. In the case of induced seismicity the effective normal stress and the frictional resistance to sliding can be lowered by increasing the pore fluid pressure through fluid injection at depth as well as contraction in geothermal fields and depleted oil reservoirs. If the fracture is subject to shear stress greater than the product of the effective normal stress and the coefficient of friction, the rocks will slip and generate an earthquake (e.g., Raleigh et al., 1972). Earthquakes generally may be classified into natural and induced events where induced microseismicity is meant to be associated with a wide range of engineering activities. These include deep well activities related to hydrocarbon production, hydro-fracturing treatments, or dam impounding. A review on different types of triggered earthquakes was given by McGarr et al. (2002).

In the following we review selected examples and case studies representing the present state-of-the-art for the respective environments.

Case Studies

Monitoring the Failure Process: Acoustic Emission Activity and Fracturing in the Laboratory

The lowermost end of magnitude scales systematically observed using PSM approaches is represented by rock-deformation experiments in the laboratory under controlled conditions. Non-destructive testing methods using advanced analysis of acoustic emissions (AE) are being used more routinely to investigate processes of brittle fracturing and frictional sliding in situ in the laboratory. The technique exploits the radiation of elastic waves from fractures that may propagate as shear or tensile cracks in a stressed rock sample. AE analysis is an indispensible tool to analyze fracture nucleation and propagation on the sample scale. However, the technique is also used to analyze very high frequency microseismic activity associated with excavation damage and rockburst in underground mining (e.g., Young and Collins, 2001; Plenkers et al., 2008). Typically, the AEs are recorded by piezoceramic transducers that are attached to the rock surface converting the recorded elastic waves into a voltage signal (Fig. 1a). Suitably designed P- and S-wave sensors have resonant frequencies that range from a few hundred kHz to several MHz. Modern multichannel transient recording systems allow storage of pre-amplified full waveform signals with broad bandwidth (up to 16 bit) and high sampling rates (>10 MHz). Figure 1b shows typical waveforms of AE signals recorded during a rock-deformation experiment. The numbers of events that are being recorded largely depend on the material and the characteristics of the recording system. They may range between a few hundred to several thousand AEs for a single test. Consequently, signal processing, such as location of AE hypocenters and source type analysis have to build on automated procedures (Zang et al., 1996, 2000; Stanchits et al., 2006). The recorded AEs typically represent <5% of the experimentally-induced cracks (Lockner, 1993; Zang et al., 2000) but have been found to provide very accurate spatial images of brittle deformation structures such as shear and compaction bands (Fortin et al., 2006).

Shear fracture nucleation was observed in triaxial compression tests performed on low-porosity igneous rocks and porous sandstones at confining pressures <100 MPa (Lockner et al., 1991; Lockner and Byerlee, 1991, 1992; Lei et al., 2000, Stanchits et al., 2006) and in experiments performed using an indenter to induce a shear fracture (Zang et al., 2000). In general, location of AE hypocenters in these experiments permits precise monitoring of the formation of microcrack clusters that precede unstable growth of a macrofracture (Lockner et al., 1991). During loading of the specimens randomly distributed AEs develop into planar nucleation spots with increasing stress (Lockner and Byerlee, 1991; Stanchits and Dresen, 2003). The nucleation patches consist of spatially correlated events contained in a volume of about a cubic centimeter from which future shear bands develop. In most experiments, fracture nucleation has been found to be associated with a burst in AE activity, a decrease in ultrasonic velocities along ray paths crossing the future fault trace (Stanchits et al., 2006) and a significant drop in b-values that develop at 60–80% peak load prior to sample failure (Scholz, 1968; Stanchits and Dresen, 2003).

Propagation of macroscopic fractures in rock from nucleation spots involves evolution of a crack damage zone of finite width surrounding the fracture tip. In the brittle deformation regime the dominant dissipative mechanisms operating in the fracture process zone include cracking and concomitant formation and closure of pore space. In experiments, the distribution of AE hypocenters typically shows a close spatial correspondence with the macroscopic fracture trace (Fig. 1c) with crack damage increasing strongly towards the fault (Zang et al., 2000). However, the width of the process zone in cross section as defined by the distribution of AEs or microcracks depends on the resolution of the method and may vary considerably (Janssen et al., 2001).

Dense spatial coverage of transducers surrounding the sample allows determination of focal mechanisms for larger amplitude AEs with a dominantly double couple radiation pattern. Although AE hypocenters commonly align with the fault trace, their focal mechanisms often reveal a complex orientation pattern of nodal planes. This likely reflects the corrugated path of the macroscopic fracture on the scale of individual grains (Zang et al., 1998). AE first motion polarities and full waveform analysis of larger AEs allow separating the events in dominantly tensile, shear and implosive (pore collapse) sources (Lei et al., 1992; Zang et al., 1998; Graham et al., 2009). With an

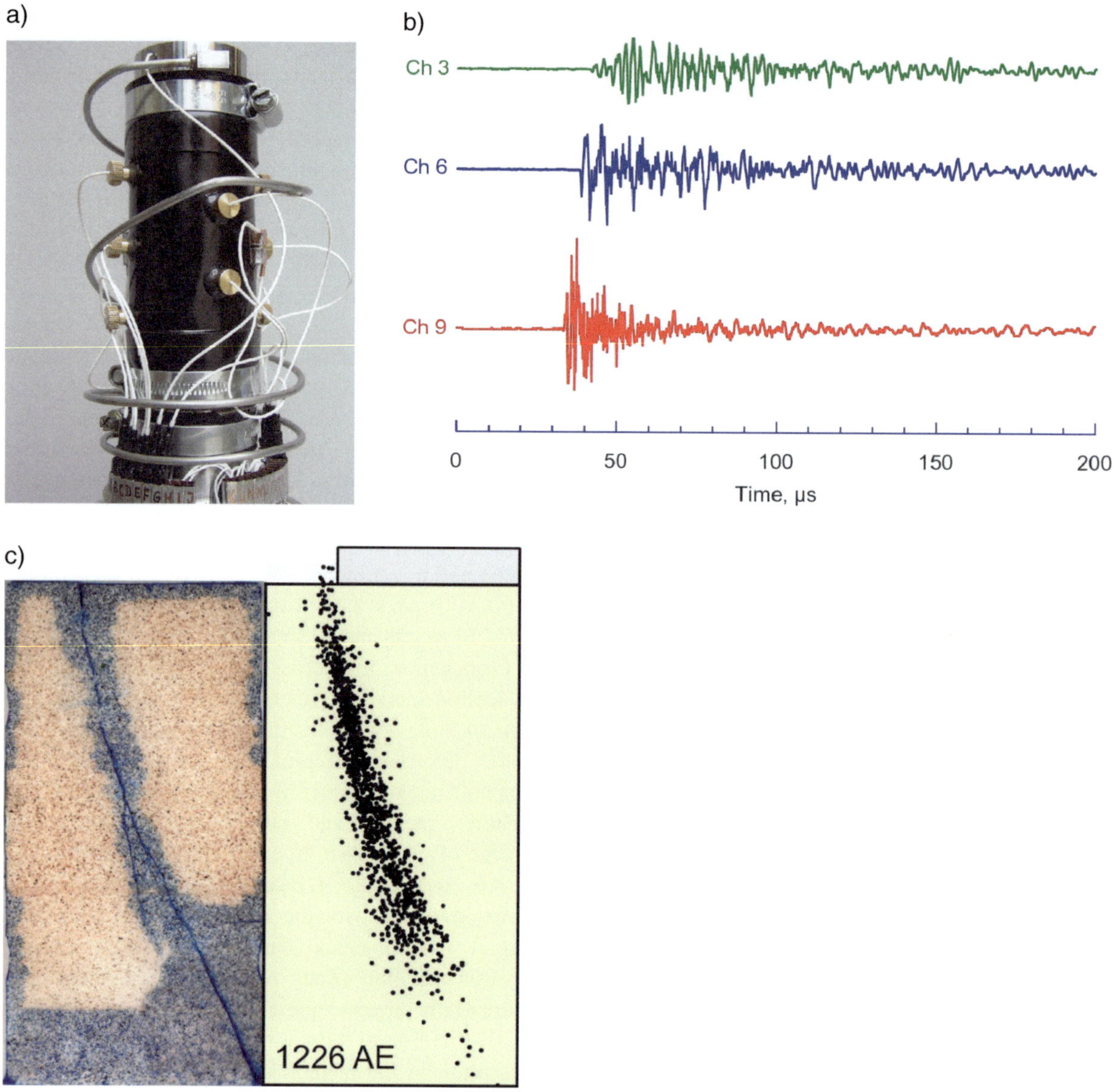

Fig. 1 **a**) Experimental setup to monitor acoustic emissions (AE) during rock-deformation experiments in the laboratory. The cylindrical sample is jacketed in rubber sleeve clamped to steel endcaps to protect specimen from oil confining medium. Sensors are piezoceramic transducers in brass housings and glued directly to the sample surface. **b**) Typical waveforms of acoustic emission (AE) signals recorded by a piezoceramic transducer during a rock-deformation experiment. **c**) *Left*: Shear fracture in low porosity sandstone. Length of sample is 100 mm. Specimen was recovered after the test, impregnated with blue epoxy and cut in half along its axis. The shear fracture initiated at an indenter placed at the top of the specimen and propagated towards the bottom. *Right*: Black dots represent located AE hypocenters. Note the close spatial correspondence of hypocenter cloud and fracture trace

increase in loading and specimen damage a shift in dominant AE source types is commonly observed. In dense brittle rock like granite, the relative proportion of shear and collapse events increases at the expense of tensile sources as the sample approaches peak stress. In porous rocks like sandstone deformed at elevated confining pressures cracking is dominated by shear and implosive source types.

Detailed analysis of AE activity in rocks contributes significantly to unravel the microphysics of nucleation, shear zone formation and brittle failure in laboratory experiments. In addition it is currently used successfully to study laboratory-scale compaction bands in porous rocks and the mechanisms involved in the formation of borehole breakouts.

Tracking the Hydro-Frac: Passive Seismic Monitoring in Hydrocarbon Reservoirs

At the scale of hydrocarbon reservoirs, the PSM of microseismicity induced by stress changes during hydraulic fracturing through massive fluid-injection is a key method to directly image fracture growth and permeability enhancements in hydrocarbon or geothermal reservoirs. Microseismic data are also crucial in other rock engineering applications, such as mining activity, excavation stability in nuclear waste repositories or geotechnical stability studies. Moreover, similar techniques are currently developed and tested to enhance underground storage of carbon dioxide into different target formations such as depleted oil and gas reservoirs, saline aquifers and coal-bearing mines.

Systematic fluid-injection was pioneered at the Rangely Oil Field, Colorado, confirming the hypothesis that earthquakes may be triggered by the increase of fluid-pressure (Raleigh et al., 1972). The injection was seismically monitored by a temporary local seismic array at the surface and at regional distance during a 7-year period detecting almost 1,000 events. Changes in the number of earthquakes were correlated with changes in the fluid-injection rates over the years (Gibbs et al., 1973) confirming earlier findings by Evans (1966) and Healy et al. (1968) who studied the Denver earthquakes in the context of nearby-injection of chemical waste. In the Geothermal Industry, field efforts began with the pioneering work at Los Alamos National Laboratory in the early 1970s at the Fenton Hill site then referred to as the "Hot Dry Rock" (HDR) project. Later, this term was replaced by "Enhanced Geothermal Systems" (EGS) to more correctly reflect the quality of resource.

PSM of hydraulic fracturing experiments aims at following the spatiotemporal growth of the created hydro-fractures by imaging the associated small-scale seismicity (e.g., Rutledge et al., 1994). Since the Rangely experiment, an increasing number of PSM field campaigns have been conducted in the petroleum industry confirming the direct correlation between injection flow rate and pressure and the rate of induced seismicity (e.g., Kovach, 1974; Oppenheimer and Iyer, 1980; Albright and Pearson, 1982; McGarr, 1991; Philips et al., 1998; 2002). An inherent need for a low magnitude detection threshold monitoring close to the seismic source are additional monitoring wells in direct vicinity to the injection well. The monitoring wells are usually equipped with geophones to detect elastic waves and to determine the hypocenter location and source parameters. The quality of the hydraulic fracture characterization is directly linked to the source-receiver geometry and the quality of the velocity model for the source area. In that respect, the study by Rutledge et al. (2004) exemplifies that an adequate data set of downhole seismic recordings forms the base for further evaluation such as interpretation of faulting mechanisms (Fig. 2).

However, using adequately spaced sensors at reservoir depth is not a simple task due to the enormous costs involved in deep drilling and the fact that the downhole sensors are operated somewhat remote-controlled with a single cable of several kilometer lengths being the only connection between sensor and surface equipment. In that respect, an illustrative exemplification for downhole PSM may be using high-tech equipment in outer space where also no direct access exists to the actual sensor and its periphery. In both cases a single dysfunction within the whole monitoring system is crucial to shut-down the entire operation.

A general observation in monitoring of fluid-injection induced microseismicity was the alignment of the seismic cloud along the direction of the maximum principal stress. Furthermore, most events cluster into well-defined small-scale geometrical patterns (e.g., Phillips et al., 2002). In recent years the enhanced use of such data sets has significantly moved forward applying collapsing methods (Jones and Stewart, 1997) and relative relocation techniques (Waldhauser and Ellsworth, 2000). During hydraulic fracturing experiments the seismicity follows the generation and growth of newly created fractures as discussed above. Alternatively, induced slip on pre-existing fault planes may reflect linear and planar features indicating networks of intersecting fractures or fracture containment between stratigraphic layers of differing mechanical properties or states of stress. This is further discussed and elaborated in the following chapter. Today, PSM is a standard tool applied in the hydrocarbon (and geothermal) industry to image permeability enhancements at depth

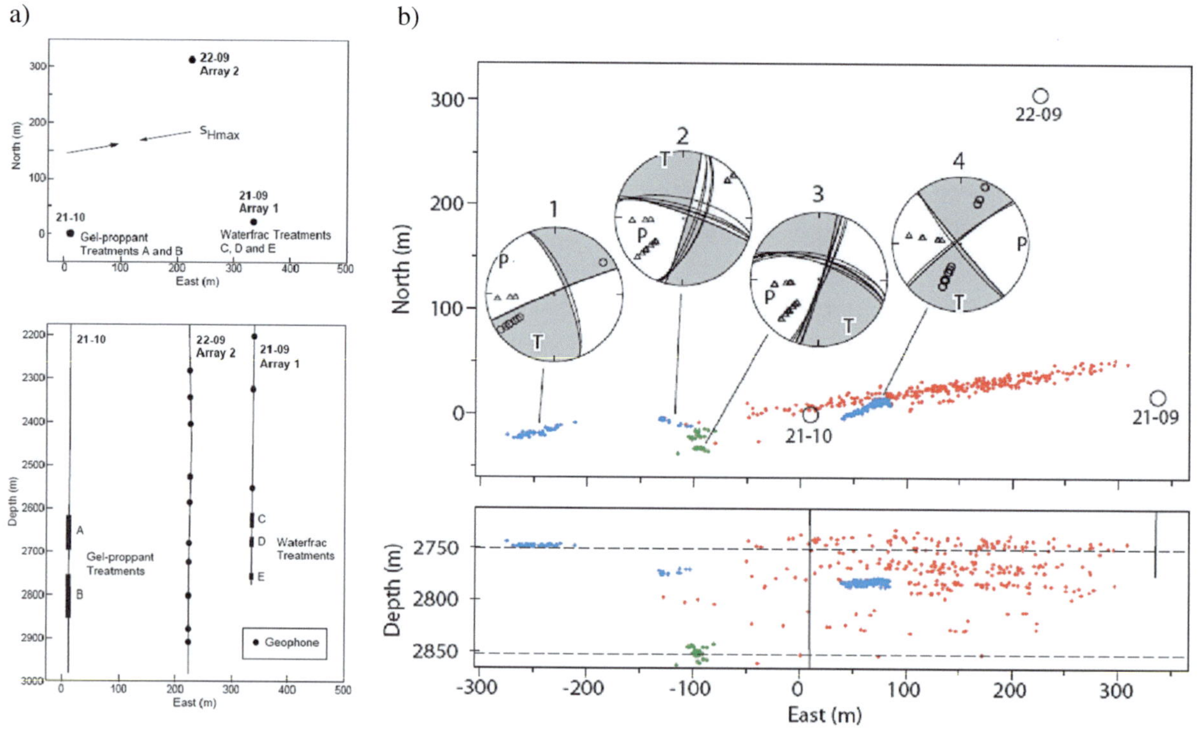

Fig. 2 **a**) Downhole seismic array during a hydraulic fracture experiment in the Carthage Cotton Valley Gas Field, Texas, in plan and depth view. Downhole geophones are placed in boreholes in close vicinity to the injection point. The injection intervals are marked by the heavier lines along the boreholes. **b**) Source locations of induced microseismic events and composite focal mechanisms for cluster events. Color refers to the individual event clusters. The three wells are indicated as shown in **a**). Figure reproduced with authors and journal permission from Rutledge et al. (2004)

in the frame of developing potential reservoirs or to improve their productivity by enhanced oil and gas recovery.

Induced Seismicity at Crustal Depth: The KTB Deep Borehole Observatory

In the previous chapter we discussed induced seismicity as one principal phenomenon to characterize hydrocarbon and geothermal reservoirs, i.e., PSM as a commercial tool. However, specially designed fluid-injection and seismic monitoring campaigns for purely scientific purposes are likewise important towards a better understanding of the physical processes related to induced microseismicity at reservoir or crustal depth. In that respect, the German Continental Deep Drilling Program (KTB) in SE Germany offers the unique opportunity to address such attempts. The KTB consists of two deep boreholes and was designed to study the rheology and properties of the continental crust at depth. The project was conducted in distinct phases including drilling of a 4 km deep pilot borehole in 1990. During the main phase a 200 m apart superdeep borehole was drilled which reached its final depth of 9.1 km in 1994. Being located at the margin of the Bohemian Massif and the contact zone of the Saxothuringian with the Moldanubian, both drillholes mainly penetrated metabasites and gneisses reaching a temperature of 265°C at the final depth of the main hole. A number of challenging studies were performed at the KTB that were summarized by Emmermann and Lauterjung in a "Journal of Geophysical Research" special volume (1997). Among the outstanding results are a continuous profile of the complete stress tensor that was determined from hydraulic fracturing experiments, compressional and tensile failures of the borehole wall, and faulting mechanisms of induced seismic events (Barton and Zoback, 1994; Brudy et al., 1997; Bohnhoff et al., 2004).

Two fluid-injection experiments have been carried out at the KTB drilling site with the aim of triggering microearthquakes at depth in a presumed stable and aseismic interplate continental crust. The experiments also investigated the consistence between measured stresses and frictional faulting theory. During a short-term hydro-fracturing experiment in 1994, microseismicity was induced deeper than 8 km depth. A total of 200 m^3 of heavy brine was injected into the main hole within a period of 48 h at rates of up to 600 l/min (Zoback and Harjes, 1997; Jost et al., 1998). The seismic activity during the experiment was monitored by a seismic network consisting of a borehole sonde (geophone) in the pilot hole at 4 km depth and a surface network of 73 stations that were clustered in subarrays. About 400 microearthquakes were detected at the borehole sonde whereas the strongest 100 were also recorded at the surface. The largest induced event with $M_l = 1.2$ was seen even by stations of the regional network at 150 km epicentral distance and magnitudes of all other events were below zero. The hypocentral depth was limited to 9.1 km. The principal conclusion of the experiment was that very small pressure perturbations are able to trigger seismicity, i.e., that Byerlee's law is valid to great crustal depth and that the crust is in a state of incipient frictional failure even at depths and temperatures approaching the brittle-ductile transition (Zoback and Harjes, 1997).

Based on these results a long-term fluid-injection experiment was carried out in 2000 aiming at injecting 4,000 m^3 of fresh water over 60 days at low flow rates (30–70 l/min). The objective was to investigate whether the limited hypocentral depth observed in 1994 reflected a change in rheological behaviour or the limited range of pore-pressure increase during short-term injection. Principal results were presented by Baisch et al. (2002), Baisch and Harjes (2003), and Bohnhoff et al. (2004). PSM was achieved by deploying a seismic network to detect the induced microseismicity at low-magnitude detection threshold. The network included a downhole sensor in the pilot hole and a 39-station surface network (Fig. 3a) where stations were equally spaced within an aperture of 25 km to allow determining reliable faulting mechanisms for the stronger events. Almost 2,800 microearthquakes were detected at the borehole sensor and the 237 strongest events could be located with a precision of a few tens of meters using a sophisticated location procedure (Baisch et al., 2002). 125 focal mechanisms were determined that showed predominantly strike-slip faulting. Relating the hypocentral distribution to the faulting mechanisms and the fault structure at the KTB, it was concluded that the larger faults (unfavourably oriented to the present stress field) did not produce any seismicity but, instead, act as pathways for the injected fluid. The brittle failure occurs on their asperities and nearby smaller faults both oriented in agreement with the present stress field (Bohnhoff et al., 2004). Due to several leaks in the borehole casing that were unknown before, seismicity occurred at distinct depth levels between 3 and 9 km depth. Two events occurred at 10 km and 15 km depth, respectively. The spatio-temporal distribution of hypocenters showed strong clustering and a large number of events had highly similar waveforms. Regions which were seismically active at a certain time often showed reduced or no activity at later times indicating local shear stress relaxation. Furthermore, the area around the injection point that was seismically active in 1994 became active in 2,000 only after the downhole pressure of the earlier experiment was exceeded (Fig. 3b). This behaviour was interpreted to represent a crustal "memory effect" also known as the Kaiser effect in material sciences (Baisch et al., 2002). The confirmed overall limitation of hypocentral depths shallower than 9.1 km suggests changes of rheological properties of the upper crust and thus supports an onset of the transition from the brittle towards ductile deformation at this depth. Single events at depths >10 km were observed and might indicate the existence of critically stressed fractures even at temperature >300°C in agreement with findings of Dresen et al. (1997) based on microstructural analysis of cuttings.

The Parkfield Earthquake Experiment

The Parkfield segment of the San Andreas Fault is located near the midpoint of the 1,200 km-long transform boundary between the North American and Pacific plates in California. It defines the transition zone between the locked southern and the creeping central sections of the fault (Fig. 4). To the south, the San Andreas Fault moves in large-magnitude earthquakes, the most recent being the great M 7.9 Fort Tejon earthquake of 1857. This southern segment

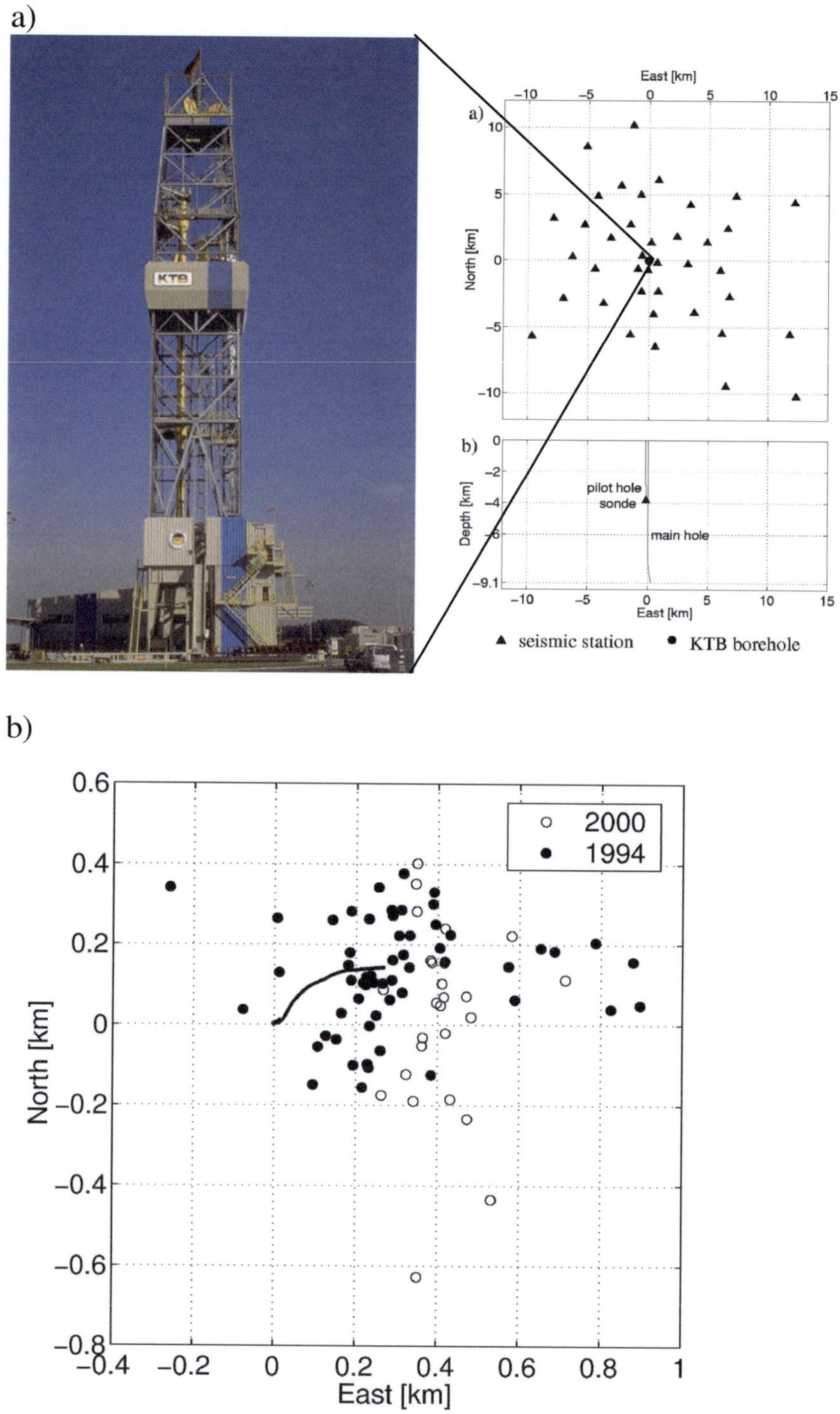

Fig. 3 **a**) *Left*: View of the KTB main hole drill rig. *Right upper part*: Map view of the temporary seismic network operated during the 2000 injection experiment. Locations of the 39 surface stations (*triangles*) are given with respect to the KTB main hole that is indicated by the black dot. Right lower part: Position of the borehole seismometer that was operated at 3,827 m depth in the pilot hole at about 200 m distance to the main hole (view from South). Fluid injection was performed through the main hole at 9,030 m depth. **b**) Map view of hypocenter locations for induced events near the open hole section (hypocentral depth>8.8 km) during the 2000 (*open circles*) and 1994 (*filled circles*) injection experiments, respectively. Coordinates are given with respect to the top of the main hole (the trajectory of which is indicated by the *black line*). Nearly all events during 2000 occurred at the edge of the volume which was seismically active during 1994. Figure reproduced with authors and journal permission from Baisch et al. (2002)

produces almost no microearthquakes ($M \geq 0$). In contrast, the central creeping segment to the north generates thousands of microearthquakes with magnitudes up to $M = 5$ annually (Hill et al., 1990). Displacement on this central segment of the fault keeps pace with the plate motion and little or no shear strain is accumulating along it (Lisowski and Prescott, 1981). The abundant microearthquakes that occur on it only contribute a few percent to the slip budget, with almost all of the displacement occurring aseismically.

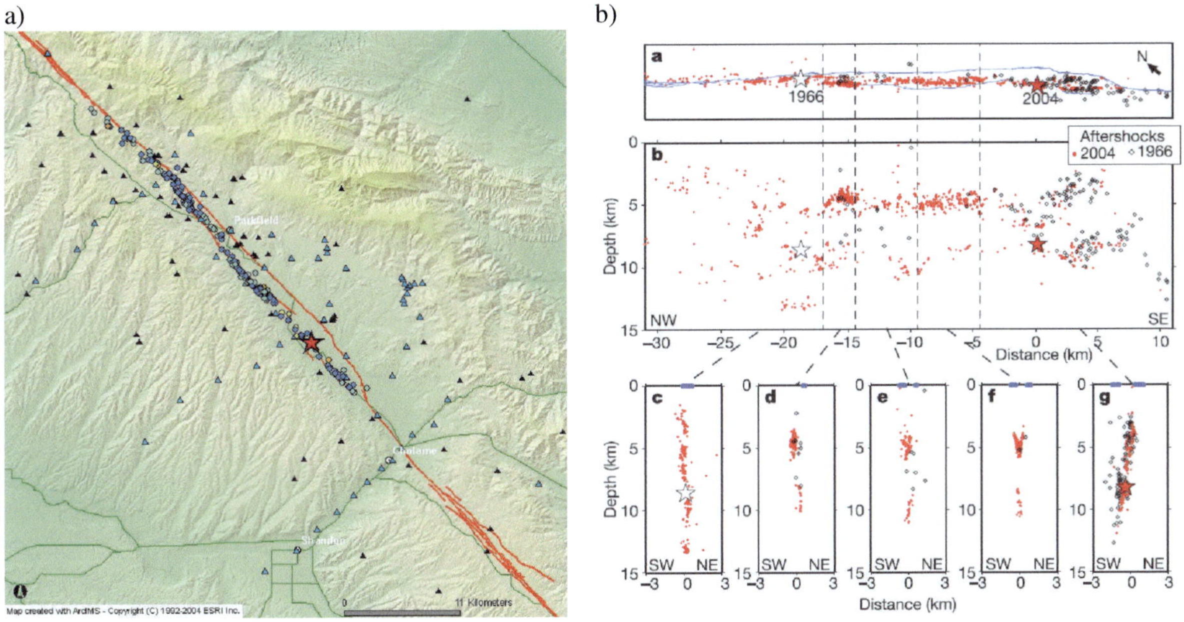

Fig. 4 **a**) Surface trace of the San Andreas Fault in the Parkfield area (*red lines*) and aftershock locations of the 2004 Parkfield earthquake (epicenter is the star). *Black triangles* mark continuously recording seismic stations (surface and borehole) and the blue triangles are the strong motion recorders operated in event-triggered mode. **b**) Spatial distribution of Parkfield aftershocks (1966: black diamonds, 2004: *red dots*). The events were relocated using a three-dimensional velocity model and the double-difference relocation technique. San Andreas fault traces (*purple lines*) and the 1966 (*open star*) and the 2004 (*red star*) main shock hypocenters are also shown. Upper part (indicated by the bold *a*): Map view, middle (indicated by the bold *b*): Along fault section, lower (indicated by the bold *c–g*): Cross sections. Figure reproduced with authors and journal permission from Bakun et al. (2005)

The transition between fully locked and fully creeping behavior occurs along the 40-km-long Parkfield Segment (Murray and Langbein, 2006). Within the segment the fault is locked over a portion of the seismogenic zone between 5 and 10 km depth which retards the creep rate of the fault at shallower depth. Abundant microearthquake activity occurs within the transitional regions, both above and below the principal zones of locking (Waldhauser et al., 2004). Six times since the 1857 earthquake this locked region has ruptured in $M = 6$ earthquakes, at an average rate of one event every 25 years with the most recent recurrence in 2004. The 2004 earthquake, together with the preceding decades of fault slip, strain accumulation and seismicity, and its postseismic response and aftershocks were captured in exquisite detail by the Parkfield Earthquake Experiment (see the special volume of Bulletin of the Seismological Society of America edited by Harris and Arrowsmith in 2006). Begun in 1985 as a collaborative project of the U.S. Geological Survey (USGS) and the California Division of Mines and Geology (now California Geological Survey (CGS)) as a prototype earthquake prediction experiment, the observation networks were designed to identify geological and geophysical techniques that would be generally useful for understanding fault behavior, and that could be used in an attempt at short-term warning based on specific behavior observed in previous Parkfield earthquakes (Bakun et al., 1987). The rationale for the prediction experiment was based upon the characteristic earthquake hypothesis supported by the series of $M = 6$ earthquakes, identical $M = 5$ foreshocks to the 1934 and 1966 earthquakes, and anomalous fault zone deformation observed shortly before the 1966 shock (Bakun and McEvilly, 1979; Bakun and McEvilly, 1984, Bakun and Lindh, 1985). The instrumental network included strong motion accelerometers, surface and borehole seismometers, borehole strainmeters, magnetometers, resistivity arrays, 2-color electronic distance measurements, and GPS.

The experiment built upon nearly 2 decades of intensive investigations that began with the 1966 earthquake. The seismological investigations, in particular, played an important role in establishing methods that are in wide use today and in advancing our understanding of the association between earthquakes and faults. Prior to the 1966 Parkfield earthquake, local seismographic networks were virtually non-existent and the large distances between stations in regional networks did not allow the routine determination of focal depth or precise determination of epicenter. A detailed investigation of the aftershocks of the 1966 Parkfield earthquake was conducted by Eaton et al. (1970) beginning just 3 days after the mainshock using a dense network of portable seismograph stations deployed throughout the epicentral region (Fig. 4). This pioneering study employed independent crustal velocity models, calibrated by refraction profiling, for the North American and Pacific Plate sides of the fault and iterative least-squares Geiger's equations with station adjustments to determine hypocenters with horizontal and vertical precision of less than 1 km. The spatial distribution of aftershocks aligned beneath the surface rupture revealed small-scale features of the fault geometry. Virtually all of the earthquake focal mechanisms indicated right-lateral strike slip displacement. It left no doubt that the San Andreas Fault has ruptured in the earthquake and that the aftershocks were almost exclusively associated with the coseismic fault.

Within six months, a local telemetered network capable of resolving focal depths was in operation at Parkfield. Beginning there and in the San Francisco Bay region, telemetered local networks were systematically expanded to provide end-to-end coverage of the entire San Andreas Fault system by 1986. Today, the California Integrated Seismic Network (www.cisn.org), operated by 13 institutions, integrates data from more than 200 broad band, 500 short period, 50 borehole and 1,000 strong motion stations. The extensive public archives of digital waveforms (www.ncedc.org and www.data.scec.org) form the backbone for a very wide range of research projects, including the detailed studies of seismicity. Advances in waveform analysis, including the systematic cross correlation of the waveform catalog to measure differential travel times with sub-sample precision (Schaff et al., 2004) and high-resolution earthquake location algorithms such as the double-difference method (Waldhauser and Ellsworth, 2000) and double-difference tomography (Zhang and Thurber, 2003) have revolutionized our ability to image subsurface fault structure. The differences between traditional single event locations and tomographic double-difference locations at Parkfield, in particular (Thurber et al., 2006) are redefining our understanding of fault geometry and its influence on structural evolution and topography (Simpson et al., 2006). Along the creeping fault, it was found that most of the microearthquakes are repeating characteristic events that exhibit regular recurrence behavior (Waldhauser and Ellsworth, 2002; Schaff et al., 2002). At Parkfield, the transition between predominantly repeating and non-repeating behavior of the microearthquakes coincides with the transition from creeping to locked behavior of the fault (Waldhauser et al., 2004).

As part of the Parkfield Earthquake Experiment, 11 digitally-recording, 3-component seismometers were cemented into boreholes at depths between 75 m and 275 m below the surface to record high-frequency waves from microearthquakes (Malin et al., 1989). The network was fully operational in 1987 and continues today as the U.C. Berkeley High Resolution Seismographic Network (HRSN). By emplacing the seismometers beneath the shallow weathering layer, the high frequency content of the seismogram was enhanced and the stability of the observation point was improved over nearby surface sites. Comprehensive studies of repeating earthquakes and their temporal variations by Nadeau et al. (1995) led to the discovery of correlations between variations in recurrence interval and variations in tectonic loading. Repeated surveys with a mechanical vibrator established new limits on the precision of travel time measurements, producing evidence of temporal variations in the velocity of the low-velocity waveguide associated with the shallow fault (Korneev et al., 2000). The HRSN was also instrumental in the discovery of nonvolcanic tremor beneath the San Andreas Fault immediately to the southeast of Parkfield (Nadeau and Dolenc, 2005).

The San Andreas Fault Observatory at Depth

A new chapter in earthquake studies at the Parkfield segment was opened in 2004 when work on the San Andreas Fault Observatory at Depth (SAFOD) was initiated with the drilling of the 2.2 km deep pilot hole into the granitic basement 1.8 km southwest of the main fault trace (Hickman et al., 2004). SAFOD is a comprehensive project to drill through the San Andreas Fault at Parkfield, carry out a comprehensive program of downhole measurements, obtain core samples of the fault, and install seismic instruments in the near-field of repeating microearthquakes. This project is being carried out under the National Science Foundation's EarthScope Project in cooperation with USGS. Preparatory work began several years earlier, including potential field, active seismic, magneotelluric, and fault-zone guided wave surveys (see special collection of papers in Geophys. Res. Lett. with introduction by Hickman et al., 2004). As part of pre-drilling surveys for SAFOD, the USGS telemetered network and UCB HRSN were augmented by up to 59 3-component instruments as part of the Parkfield Area Seismic Observatory (PASO) experiment with instrument spacing of ~1 km designed to refine our understanding of 3-D crustal structure and the seismically defined active faults (Thurber et al., 2004; Roecker et al., 2004).

The vertical SAFOD pilot hole was also instrumented with a 32-element array of 15 Hz 3-component geophones (Chavarria et al., 2004). With a group interval of 40 m, the 1,240 m-long array provided a first look at the spatially unaliased wavefield from local earthquakes and explosions. Secondary phases between the direct P- and S-waves and following the S-wave arrival displayed remarkable coherence across the array. Chavarria et al. (2003) determined that these phases were reflected and scattered off of near vertical faults near the well. Imanishi and Ellsworth (2006) used the high frequency content of the seismograms and close proximity to the microearthquakes to determine source parameters for earthquakes ranging in size from M –0.2 to 2.1. Stable estimates of the spectral ratio for pairs of events were obtained by stacking the seismograms from the 32-element array. They found that static stress drop and the apparent stress (radiated energy/seismic moment) showed no deviation from moment-independent scaling laws reported for moderate and large earthquakes, indicating that the breakdown in stress scaling reported in other studies was likely an artifact of inadequate recording bandwidth as proposed by Ide and Beroza (2001). Peak stress drops for these Parkfield repeating microearthquakes exceeded 50 MPa in some cases, approaching the inherent rock strength given by Byerlee's Law.

The main SAFOD hole was drilled in stages in 2004 and 2005 along a diagonal path that crosses the divide between Salinian basement accreted to the Pacific Plate and Cretaceous sediments of North America. It provides a portal into the inner workings of this major plate boundary fault. Following drilling and casing of the main hole in the summer of 2005 to a total vertical depth of 3.1 km it was possible to conduct spatially extensive and long-duration observations of active tectonic processes within the actively deforming core of the San Andreas Fault. Observatory instrumentation consists of retrievable seismic, deformation and environmental sensors deployed inside the casing and a fiber optic strainmeter installed behind casing in the main hole (Ellsworth et al., 2007b). By using retrievable systems deployed with instrumentation that takes maximum advantage of advances in sensor technology on either wire line or rigid tubing, the hole can be used for a wide range of scientific purposes.

To meet the scientific and technical challenges of building the observatory, borehole instrumentation systems developed for use in the petroleum industry and by the academic community in other deep research boreholes have been adapted for use at SAFOD. These systems included 15 Hz omnidirectional and 4.5 Hz gimbaled seismometers, micro-electro-mechanical accelerometers, tiltmeters, and a fiber optic interferometric strainmeter. A 1,200-m-long, 3-component 80-level clamped seismic array was also operated in the main hole for 2 weeks of recording in May of 2005, collecting continuous seismic data at 4,000 samples per second.

A new type of seismic wave was observed when a 3-component downhole seismograph was installed at a depth of 2,700 m, approximately 40 m from the principal active strands of the San Andreas Fault (Ellsworth et al., 2007a). Sonic log velocities in the core of the fault are very low, with Vp and Vs averaging as little as 3.2 km/s and 1.8 km/s within a ~20-to-30 m-wide zone. This 20–25% reduction from surrounding velocities forms a channel that is embedded in a broader ~200-m-wide low velocity zone. The velocity reductions of both zones are large enough that different types of fault zone guided waves should propagate along them. In fact, the seismograms with the new wave type also contain the well-known, SH-type fault guided waves. The new wave arrives between the direct P and S waves and appears to be a "leaky mode" fault zone guided P wave. The characteristics of this phase give special insights into the structure and multi-stranding of the San Andreas Fault zone. Analytical and numerical acoustical models of the observed fault core suggest that its velocity reductions and thickness must extend along the entire 2-to-6 km source-receiver distance, providing support for the hypothesis that multiple, narrow, low velocity fault zones extend deep into the seismogenic crust.

The same seismometer recorded a recurrence of a Mw = 1.8 repeating earthquake at 100 m range on August 11, 2006 (Ellsworth et al., 2007a). Despite the clipping, a dynamic stress drop of 4–7 MPa during the first 0.5–1.0 ms of rupture could be measured using the Kostrov (1964) solution for a dynamically growing circular crack. This is comparable to the average stress drop for the entire event of 10 MPa determined with seismograms from nearby PASO surface and HRSN shallow borehole stations. An aftershock sequence of extremely small events was also observed in the SAFOD main hole (Fig. 5), but went undetected in the SAFOD pilot hole, and at the HRSN and PASO stations. The first detected aftershock was noted as soon as the seismograms returned on scale at +2.5 s,

a)

b)

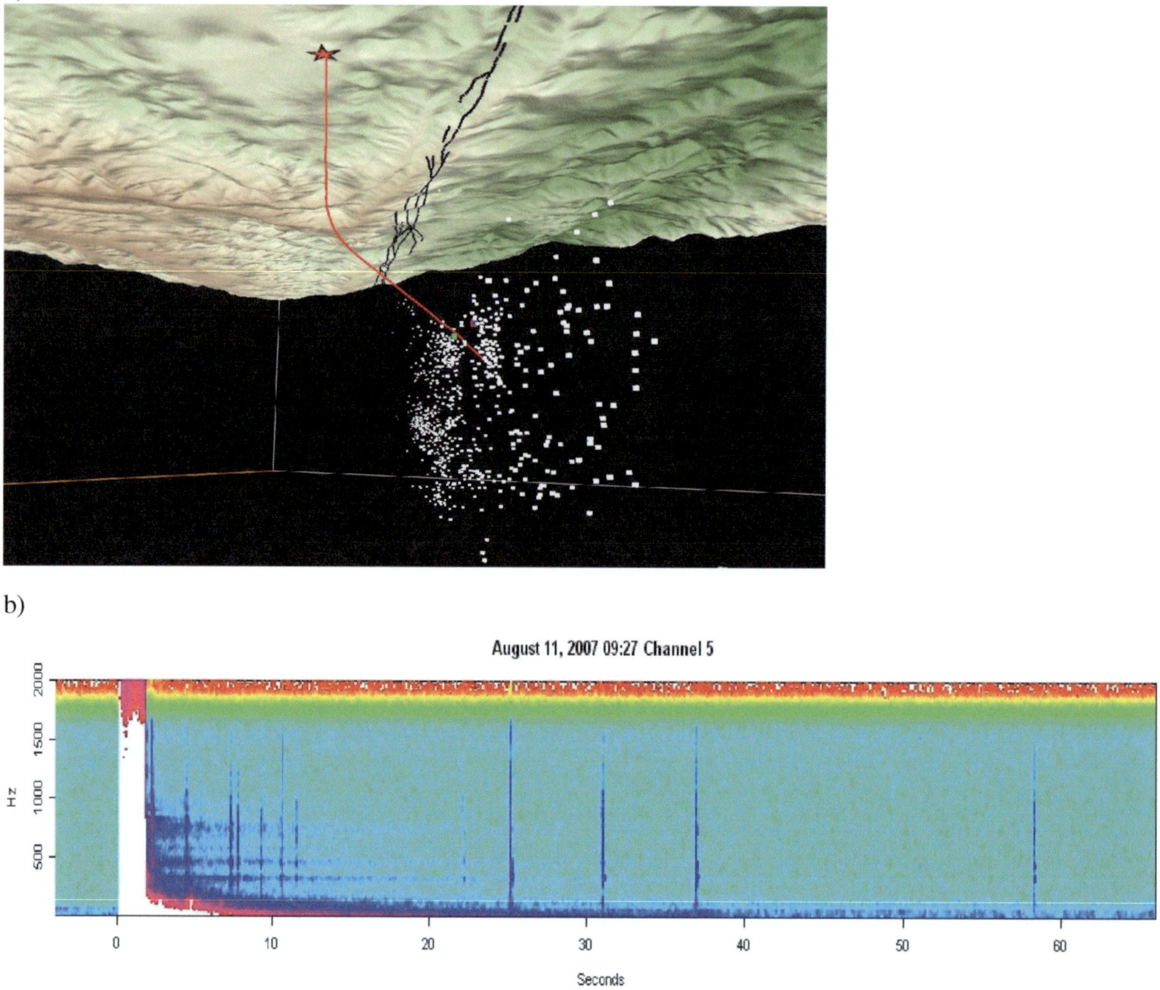

Fig. 5 **a**) Seismicity (*white dots*) and trajectory of the SAFOD main hole (*red line*) at the Parkfield segment of the San Andreas Fault seen from a vantage point in the Earth looking to the northwest. The surface track of the fault is shown in black draped over by topography. Figure reproduced with authors and journal permission from Hickman et al. (2007). **b**) Spectrogram of the M 1.8 "Hawaii" SAFOD target earthquake of August 11, 2006 recorded at 2,700 m depth in the SAFOD main hole (named due to its relative vicinity to Hawaii compared with other nearby clusters). The recording was clipped due to the high level of ground motion for the first 1.8 s. Vertical blue lines on the spectrogram that appear after the recording returned on scale are aftershocks. The strongest are about M = –2 (nanoearthquakes), while the weakest are as small as M = –3.5 (picoearthquakes)

and the activity rate declined following Omori's Law. Aftershock moment magnitudes range from M = –1.9 (nanoearthquake) to below M = –3.5 (picoearthquake). Aftershock locations determined using the P-wave polarizations and S-P intervals (16–23 ms) define a northwest trending zone located 100 m below the borehole fault crossing at a depth of 2.7 km. Their very narrow depth range suggests that they define a "streak" as has been commonly observed along creeping faults in California (Rubin et al., 1999; Waldhauser et al., 1999). The aftershocks coincide with the shallower of the two faults that have deformed the casing in the SAFOD main hole that was successfully cored in 2007. As of 2008 the SAFOD experiment has entered the stage where observation of microearthquake activity in the near field became integrated with the broader Parkfield Experiment networks. All data collected at SAFOD as part of the EarthScope project are open and freely available through the EarthScope data portal forming a profound state-of-the-art data base allowing researchers to address numerous scientific objectives in near future.

Outlook and Future Directions

In the following we will give an overview on two ongoing large-scale campaigns where PSM is embedded in mutiparamter approaches to monitor natural seismicity at plate boundary scale offshore of the Kii Penninsula in Japan and at the North Anatolian Fault Zone in northwestern Turkey.

NanTroSEIZE: Monitoring of a Locked Segment Along the Convergent Plate Boundary Offshore of Japan

The PSM efforts undertaken at the Parkfield segment as introduced in the previous chapter represent the leading edge in terms of seismic monitoring technology at an active transform plate boundary located onshore. A similar impact for marine and downhole seismic monitoring can be expected from the Nankai Trough Seismogenic Zone Experiment (NanTroSEIZE) at the convergent plate boundary offshore of the Kii Peninsula in Japan (Tobin and Kinoshita, 2006). As the December 2004 Sumatra earthquake and Indian Ocean tsunami so tragically demonstrated, great subduction earthquakes represent one of the greatest natural hazards on the planet. Accordingly, drilling into and instrumenting an active interplate seismogenic zone is a very high priority in the Initial Science Plan of the Integrated Ocean Drilling Program (IODP). NanTroSEIZE attempts to drill, sample, and instrument the seismogenic portion of a subduction zone plate boundary fault, a so-called megathrust. The fundamental goal of NanTroSEIZE is building up a distributed observatory spanning the up-dip limit of seismogenic and tsunamigenic behavior at a location where great subduction earthquakes have occurred in the past and are expected to occur in near future. PSM will be a dominant part but the project will also include observing the geodetic and hydrogeologic behavior of subduction megathrusts and the aseismic to seismic transition of the megathrust system from the shallow onset of the plate interface to depths where earthquakes occur (Fig. 6). At this location representing the rupture region of the M = 8 Tonankai earthquake of 1944, the plate interface and active mega-splay faults that triggered tsunamis in the past are accessible to drilling.

The most ambitious objective of NanTroSEIZE clearly is to access and instrument the Nankai plate interface right in the seismogenic zone. The science plan entails sampling and long-term instrumentation of the inputs to the subduction conveyor belt: faults that splay from the plate interface to the surface and that may accommodate a major portion of coseismic and tsunamigenic slip, and the main plate interface at depths of ~4 to 6 km. Data acquired within this project are expected to substantially quantify conditions pertinent for stable versus unstable sliding (which define seismic vs. aseismic behavior) and the frictional strength of likely fault zone material. NanTroSEIZE will sample fault rocks over a range of pressure and temperature conditions across the aseismic to seismogenic transition. The composition of faults and fluids and associated pore pressure and state of stress will be studied and strain partitioning at the décollement and splay faults will be investigated. One of the main efforts of the project will be to install borehole observatories below the seafloor to achieve permanent PSM and to provide in-situ recordings of critical parameters such as strain, tilt, pressure, and temperature in order to test whether interseismic variations or detectable earthquake precursory phenomena exist.

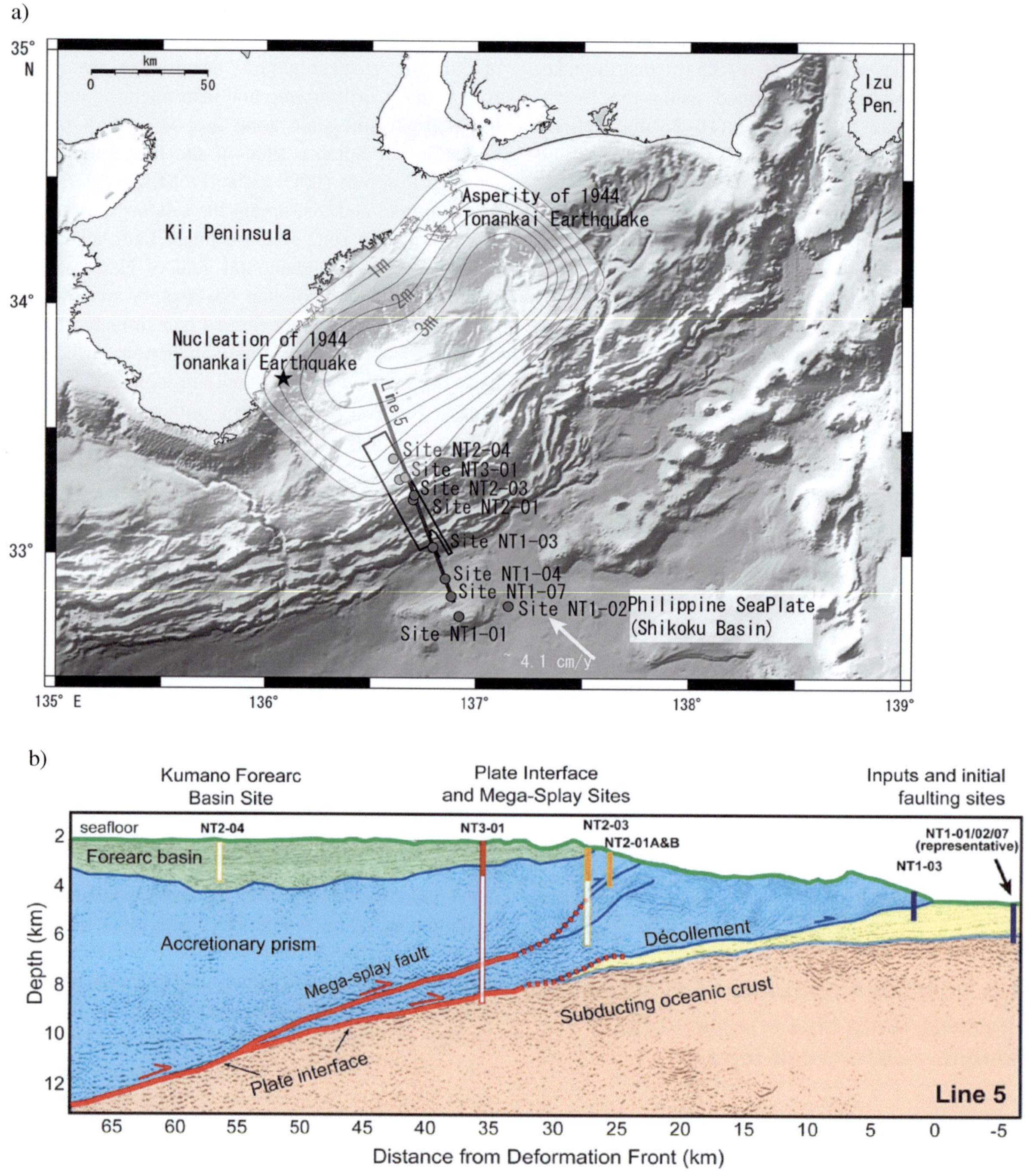

Fig. 6 **a**) Index map showing the location of planned NanTroSEIZE drilling sites. Rectangle encloses the area for 3D-seismic survey. The contour shows the coseismic slip of the 1944 Tonankai earthquake as determined by Kikuchi et al. (2003). **b**) Interpretation of seismic Line KR0108-5 (Park et al., 2002) showing locations of Stage 1 sites proposed for logging while drilling, coring, and downhole measurement, after Tobin and Kinoshita (2006). *Solid rectangles* = proposed drilling depths for Stage 1, *empty rectangles* = proposed depths for future operations. *Blue* = Stage 1, *orange* = Stage 2, *red* = Stage3

The following topical hypotheses will be addressed: (a) Systematic, progressive material and state changes control the onset of seismogenic behavior on subduction thrusts; (b) Subduction zone mega-thrusts are weak faults; (c) Within the seismogenic zone, relative plate motion is primarily accommodated by coseismic frictional slip in a concentrated zone; (d) Physical properties, chemistry, and state of the fault zone change systematically with time throughout the earthquake cycle; (e) The megasplay thrust fault system slips in discrete events which may include tsunamigenic slip during great earthquakes. We refer to the webpage www.iodp.org/NanTroSEIZE for a detailed discussion of these hypotheses. The overall drilling targets involve a total of eight drill sites (Fig. 6) in different tectonic regimes. Two sites target the pre-subducting plate section, one will sample the frontal thrust of the accretionary wedge, three sites target the megasplay fault system at different depths, one site will sample the megasplay uplift history recorded in the forearc basin sediments, and one ultra-deep site targets the plate interface in the seismogenic zone. Drilling is planned to be conducted in four stages: Stage 1 calls for drilling and sampling at six of the sites down to ~1 km below sea floor. Comprehensive coring and logging of the boreholes is planned, including extensive use of logging-while-drilling (LWD) technology to obtain high-quality logs. At least one borehole observatory installation is planned to monitor pore-fluid pressure, strain, temperature, and seismicity above the plate boundary. Stage 1 was accomplished in 2007/2008. Stage 2 will involve drilling the first deep riser hole, targeting the megasplay fault zone at ~3–3.5 km below the seafloor at site NT2-03 (see Fig. 6). Extensive coring, downhole experiments to measure pore pressure and seismic properties, and an initial, retrievable, long-term monitoring package are all planned for this site. Stage 3 will focus on 5.5–6 km deep drilling into the seismogenic zone and across the plate interface into subducting crust at site NT3-01 (see Fig. 6). Stage 4 will be concerned with installing the final long-term observatory systems into the two ultra-deep boreholes. Monitoring efforts will be planned as much as possible for robust, long-duration deployment, such that data pertinent to the behavior and evolution of the plate interface fault system during a significant portion of the seismic cycle can be recorded. In Japan, a seafloor fiber-optic network is being developed for seismic monitoring in the Kumano Basin region. One exciting possibility is that the NanTroSEIZE boreholes ultimately could be connected to this network, allowing real-time access to the entire set of recorded data. It is expected that the NanTroSEIZE mission will be completed in 2012/2013.

Clearly, one of the most challenging opportunities given by the NanTroSEIZE project is the aim to install a seismological observatory at seismogenic depth across a megathrust system. Scientific needs for long term in situ monitoring at and in the vicinity of splay faults and plate boundary faults include the monitoring of low-frequency and slow seismic events, the detection and quantification of strain partitioning during the interseismic period as well as near-source seismic, hydrological and geodetic observations. One of the most essential parts of the NanTroSEIZE monitoring system is a distributed, multi-level multi-sensor system that was proposed by Tobin and Kinoshita (2006). Based on this concept a possible plan for a sensor-downhole telemetry system interface and system installation was developed (Ito 2007; Namba et al., 2008). The major technological challenges in the installation of the NanTroSEIZE monitoring system are certainly the great depth for sensor deployments and the telemetry-system, and the high temperatures (80–100°C at 3.5 km, 170–180°C at 6 km depth) under which long-term monitoring is planned. Further important issues are long-term reliability and stability, the coupling of the sensors to the formation, geodetic and pore pressure measurements at multiple intervals, vertical drilling and core sampling at the fault interval, broadband high dynamic range and high resolution recording, the telemetry system through a "Christmas tree" wellhead system of riser drill holes and the real-time monitoring through seafloor cables. To date no such system has successfully been installed anywhere in the world. Data transfer is planned to be organized through a borehole telemetry system. The system should be reliable and has to work for many years at high temperatures. At the same time, the system should be able to deliver high-quality data as the fundamental base for a number of pioneering scientific studies. However, there are a number of fundamental requirements such as long-term reliability, long mean time before failure (MTBF), redundancy, protection against failure, high dynamic range, continuous recording, a maximum of two penetrator connections through pressure controlled well head and a low power consumption. The monitoring system is designed to basically consist of three sections: (1) A downhole module array that digitizes seismic signals continuously and transmits the data to a recorder on sea floor, (2) a sub-sea recorder that receives data from downhole and stores data, and (3) a communication unit that transmits commands from sea surface to the subsea recorder to check the status of the downhole systems and that receives QC data.

The NanTroSEIZE deep borehole observatory system will be deployed as a vertical sensor array. It will be linked to a network of sea floor cables to integrate the shallow borehole observatories in the system of ocean bottom observatories. Furthermore, it is planned to integrate the entire NanTroSEIZE offshore system into the existing Japanese seismic onshore network to establish a long-term amphibious monitoring system. Such a combined installation would then cover the entire seismogenic zone from the active continental margin down to the decoupled part of the subduction zone.

The Istanbul/Marmara Branch of the North Anatolian Fault Zone in NW Turkey: Locked or Creeping?

The examples discussed in the two previous chapters focused on the seismic monitoring efforts along two of the most prominent large-scale active plate boundaries involving transform faulting along the San Andreas Fault in California and thrust faulting at the Nankai Trough offshore Japan. Here, we focus on an example for an ongoing campaign involving PSM at the North Anatolian Fault Zone (NAFZ) in northwestern Turkey. There the densely populated greater Istanbul region is presently facing a high probability for a large earthquake in the near future. The main branch of the NAFZ extends below the northern Sea of Marmara within less than 20 km distance to the Istanbul metropolitan region with its >13 million inhabitants (Fig. 7a). The NAFZ represents a 1,600 km long transform plate boundary that slips at an average rate of 20–30 mm/y (e.g., Barka, 1992; Reilinger et al., 2006). It has developed in the framework of the northward moving Arabian plate (in the East) and the Hellenic subduction zone where the African lithosphere is subducting below the Aegean. Comparison of long-term and GPS-derived slip rates indicates a recently accelerated westward movement of the Anatolian plate with respect to stable Eurasia (e.g., Muller and Aydin, 2005). During the 20th century, the NAFZ has ruptured over 900 km of its length by a series of westward migrating mainshocks starting in 1939 near Erzincan in Eastern Anatolia (Barka, 1999; Ambraseys, 2002). The only major segment of the NAFZ that has not been activated since 1766 is located below the Sea of Marmara between the 1912 Ganos and 1999 Izmit ruptures (e.g., Stein et al., 1997; Barka et al., 2002; Armijo et al., 2005; see Fig. 7a). This part of the NAZF is interpreted to represent a "seismic gap", i.e., a region within a seismically active area with a longer lasting low level of seismicity. The estimated 30-year probability for an event $M \geq 7$ at the seismic gap below the Sea of Marmara is 35–70% (Parsons, 2004).

The Sea of Marmara is interpreted to represent a pull-apart basin in the framework of the transtensional regime formed by dextral strike-slip along the NAFZ and the southward rollback of the Hellenic subduction zone (e.g., Flerit et al., 2004). Recent multi-channel seismic and bathymetric surveys provided state-of-the-art information on the shallow Pleistocene-Holocene stratigraphy. However, geological interpretation of the data yielded two opposing views of the structural setting and current faulting activity. Le Pichon and co-workers have suggested that a uniform Marmara block is delimited by the NAFZ to the north, which displays pure strike-sip displacement (Le Pichon et al., 2001). Instead, Armijo et al. (2002; 2005) suggested that the Marmara Sea area constitutes a complex network of pull-apart basins bounded by active normal faults and strike-slip transfer segments. Numerical modeling of different active fault patterns (Muller and Aydin, 2005) shows that a series of pull-apart basins along a master strike-slip fault reproduce best the observed deformation pattern within the Marmara Sea. The opposing interpretations of fault kinematics in the Marmara segment of the NAFZ have important consequences for the seismic hazard assessment pertinent to the Istanbul area. For example Armijo et al. (2005) suggested a maximum seismic risk for a well defined 70–100 km long fault segment between the 1912 and 1999 ruptures. Whether this fault segment operates largely as a normal or strike-slip fault remains unclear as described above.

Current seismic activity in the eastern Sea of Marmara indicates a complex fault network active at the transition between the western end of the Izmit

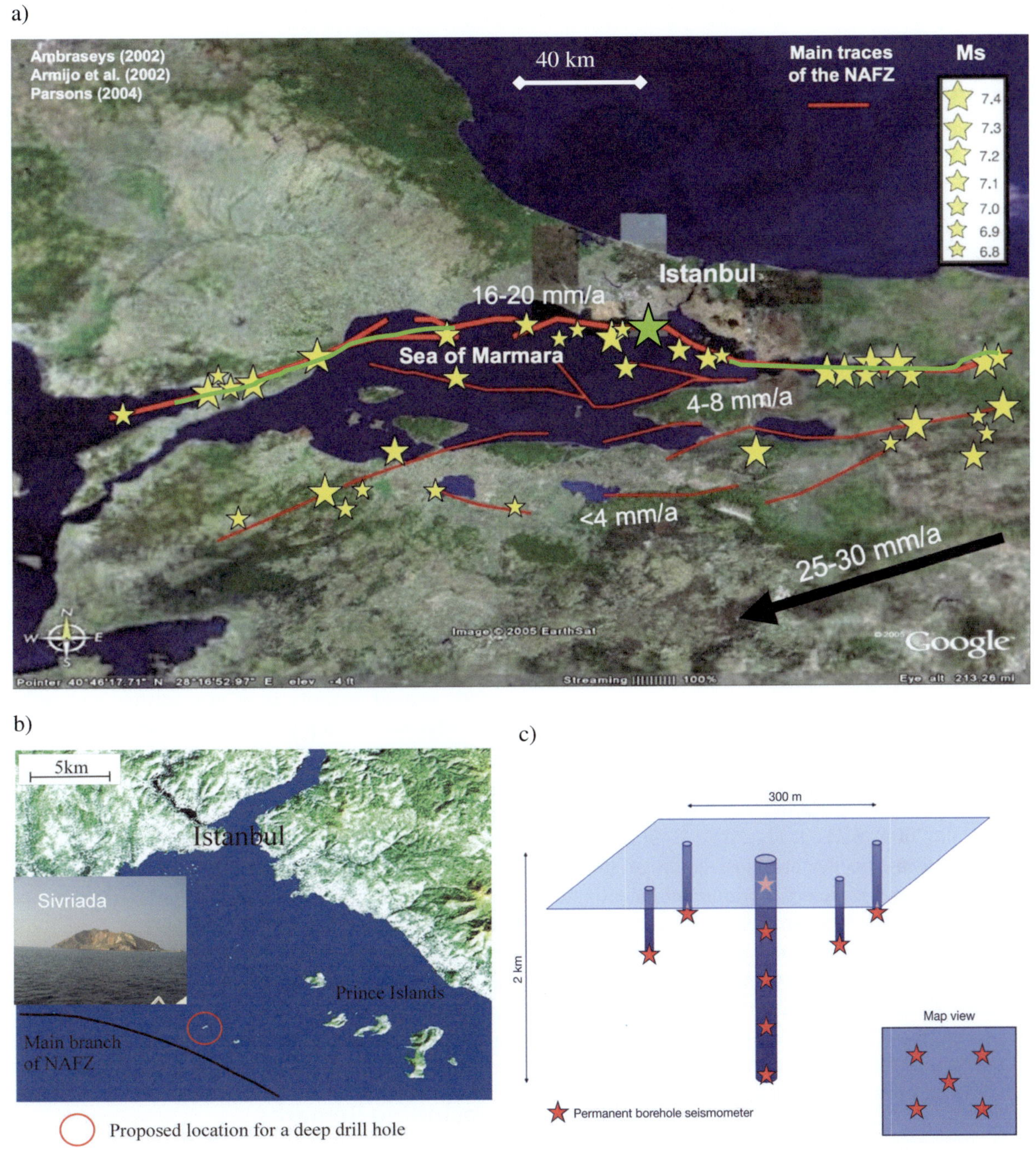

Fig. 7 **a**) Google Earth ® view of the Istanbul/Sea of Marmara region. Red lines indicate major segments of the North Anatolian Fault Zone (NAFZ). Stars indicate main shocks M > 6.8 that occurred in the last 2000 years (after Ambraseys 2002; Parsons 2004). The NAFZ off-shore Istanbul below the Sea of Marmara last ruptured in a major earthquake in 1766. Its presumable epicenter is represented by the *green star* and its rupture extending along >100 km below the Sea of Marmara is interpreted as a seismic gap (see text for details). **b**) Proposed location for a deep drill hole at the NAFZ on the island of Sivriada to install a deep geophysical observatory (after Dresen et al., 2007). **c**) Planned seismic downhole array on Sivriada consisting of sensors in a 2 km deep and several nearby shallow boreholes

earthquake rupture and the assumed seismic gap south of Istanbul. The majority of focal mechanism solutions indicate both strike-slip motion as well as normal faulting activity for Izmit aftershocks in that area (Karabulut et al., 2002; Bohnhoff et al., 2006). However, the results of PSM efforts during the past decades to image the quantity and spatio-temporal distribution of (micro)seismicity along the main branch of the fault below the Sea of Marmara lack the spatial resolution required for an accurate definition of seismic versus aseismic zones due to an $M_c = 2.8$ for the region. Therefore, the NAFZ in the eastern Marmara region and in particular the Istanbul area is a preferred location where PSM at low-magnitude detection threshold can substantially contribute to the ongoing debate whether the fault is locked or creeping. Furthermore, an adequately dense station distribution would allow determining the dominant type of faulting (normal versus strike-slip) with regard to the current but diverse seismotectonic models. Should a significant portion of normal faulting be expected this would have a direct impact on the tsunami-potential of the expected major earthquake in the region where historical documents report a non-negligible tsunami hazard (Yalciner et al., 2002).

During recent years there has been an increasing effort to improve the PSM of the Marmara region in general and the area south of Istanbul in particular (Özalaybey et al., 2002; Karabulut et al., 2002; Sato et al., 2004; Baris et al., 2007; Bohnhoff et al., 2007). Focusing on the area just south of Istanbul, the Prince Islands represent a prominent location near (2–3 km) the vicinity of the presumed main branch of the NAFZ. There, a permanent seismic array (PIRES) was installed on the two outermost islands of Sivriada and Yassiada in 2006 to achieve high-precision seismic monitoring down to $M_c = 1.5$ (Bohnhoff et al., 2007 Bulut et al., 2009). PIRES thereby locally enhances the data base for microseismicity at the fault by a factor of more than 10. The PIRES array will be extended by drilling shallow boreholes that will be equipped with chains of borehole seismometers to be cemented into the wells to achieve high-quality recordings on a long-term base. The downhole sensors will virtually extend the aperture of the PIRES array that is limited by the size of the islands (see Fig. 7b,c). In addition, plans to deploy permanent sea-floor instrumentation within the Cinarcik Basin in the eastern Sea of Marmara are currently being developed. As a long-term perspective to achieve multi-parameter monitoring at the NAFZ in direct vicinity to Istanbul it is planned to install a deep geophysical observatory in a deep vertical borehole on Sivriada (Dresen et al., 2007). Besides a further improved PSM close to the expected hypocentral region at the "seismic gap", the deep-borehole observatory will allow to measure in-situ stress and heat and fluid flow at the fault prior to a major earthquake for the first time. This will assist estimation of the in-situ strength of a major plate boundary fault. Fluid pressure in close proximity to active fault strands will be monitored. The borehole observatory will address fundamental questions such as (a) Is the NAFZ in northwestern Turkey a weak or a strong plate boundary? (b) What is the orientation of the principal stresses at the fault zone? (c) Do stresses vary with depth and time? (d) How does the structural heterogeneity affect the earthquake nucleation process? (e) How is the deformation partitioned between aseismic creep and seismic slip at the termination of a major earthquake rupture? (f) What is the continuation of the fault at depth; is there a deep-seated master fault in the Sea of Marmara south of Istanbul?

Recordings from the above mentioned PSM campaigns as well as those from new seismic and multi-parameter networks presently under construction will supply seismic near-fault recordings, partly in real-time. Those recordings are available to be included in the seismic early-warning system for the greater Istanbul area. However, the benefit from such a system in the case of Istanbul is naturally limited due to the close vicinity of the fault to the population center. Therefore, detailed microzonation studies (Bindi et al., 2007; Parolai et al., 2007) to estimate local peak-ground accelerations become even more important to ensure the optimization of emergency plans and earthquake preparedness for Istanbul.

Concluding Remarks

The case studies and future directions introduced in the previous chapters stress the power of Passive Seismic Monitoring (PSM) as a tool to record, analyze and quantify processes associated with the occurrence of earthquakes over a broad range of magnitudes. The magnitude scales considered extend from acoustic emissions generated during rock-deformation experiments in the laboratory ($M \sim -4$), over natural and induced microseismicity monitored by downhole seismic instrumentation in hydrocarbon and geothermal reservoirs and at crustal depth ($-3 < M < 1$), towards the occurrence of seismicity at major plate-bounding fault systems ($M > -3.5$) including devastating large and great earthquakes with significant socio-economic impact ($M > 7$). As shown for these distinct environments, only adequately designed PSM networks can record the brittle-failure processes and allow capturing the entire range of rupture surface dimensions extending from millimeter to plate-boundary scale. A comprehensive evaluation of the acquired data sets then addresses various scientific objectives including scale-dependence of the relevant parameters over the entire range of magnitudes observed.

On the lower end of the magnitude scale the ability to run specially designed experiments to generate brittle failure under controlled conditions with sensors right on the pressurized rock sample certainly forms the base to further develop and refine physical models of the failure processes including earthquake nucleation. On the larger end, the technology and instrumentation to adequately monitor actively deforming plate boundaries that are expected to host large and devastating earthquakes in the near future close to densely populated areas has undergone substantial progress in recent years. What is needed in the future is a systematic transfer of know-how gained at either end of the magnitude scale. This would allow to address societal relevant issues and allow to handle if not control the seismic risk in a responsible way. In that respect the study by Rutledge et al. (2004) may serve as one good example (see also Fig. 2). The authors analyzed precisely located microearthquake data detected during hydraulic fracture treatments and observed that anomalous event counts and moment release within dense clusters delineates bends or jogs in the fracture zones. They compare high-moment asperities within the reservoirs to dense patches of seismicity observed along creeping sections of the San Andreas Fault, which have been attributed to localized zones of strength or stress concentration. These asperity zones are surrounded by larger regions undergoing stable, aseismic slip (e.g., Nadeau et al., 1995). Rutledge et al. conclude that this similarity and the large moment deficits in terms of volume injected suggests a large component of aseismic slip induced by the massive fluid injections as observed along creeping sections of the San Andreas Fault in California.

Generating an improved understanding of seismic versus aseismic deformation and, more generally, both the earthquake nucleation process and the ability to perform deterministic earthquake prediction are desirable tasks. Many seismologists agree that deterministic earthquake prediction is not a realistic target to be achievable in the near future. However, an improved understanding of the earthquake-triggering process, the interaction between faults and thus a better understanding of the spatio-temporal distribution of earthquakes along plate boundaries will help prepare people for future earthquakes. Earthquakes cannot be prevented from occurring but society will have to live with the seismic hazard. Therefore, all of the examples discussed in the previous sections demonstrate that the use of high-level technology and development of pioneering experiments will generate high-quality data sets to improve fundamental knowledge of earthquake science. The upgraded know-how then needs to be transferred and applied to improve preparedness of population centers facing earthquakes.

Physical losses in the form of buildings and infrastructure caused by earthquakes are not only damaging in the short term, but they also cause losses such as a slow-down in economic development, investors and consumer confidence. Such losses become apparent only over the time as they can hardly be quantified during the first months after the earthquake. Bird and Bommer (2004) reviewed 50 recent earthquakes worldwide and found that the primary cause of building damage is ground shaking. Therefore, any effort to use the refined understanding of seismic source processes for improving the knowledge about the influence of local geology for estimations of ground motions in regions pertinent to severe seismic hazard is needed to locate areas that are facing high peak-ground accelerations in the case of a large earthquake. Likewise, building up efficient earthquake early-warning systems to make optimal use of the time window between the occurrence of a major earthquake and the time the destructive shear and surface waves reach the population center is of major importance to allow for maximum warning time. This is of major relevance especially if the potential hypocenter area is located close to metropolitan regions as is the case e.g., in the greater Los Angeles and San Francisco Bay areas (California), in Tokyo and Nagoya (Japan) or around Istanbul (Turkey), and Bucharest (Romania) in Europe. In all these cases the length of the time window is a few seconds only. Therefore, only a state-of-the-art seismic monitoring system with online data transmission and near real-time data processing is capable to ensure building up an rapid-response early-warning systems immediately after the earthquake occurred. Most existing seismological processing methods have not yet been optimized for real-time applications as required for early warning. Consequently, the further development of real-time determination of hypocenter and source parameters as well as modeling and simulation methods is one of the major challenges of today's seismology. By the year 2025 more than 5,500 million people around the world will live in cities. A significant proportion of the largest of these cities are located close to regions of known seismic hazard. As a consequence, there is clearly a need to be prepared and prevent natural disasters resulting in catastrophic damage and loss. The deployment and implementation of optimized PSM systems for each individual region can serve to limit or even minimize the impact of large future earthquakes. Besides the use of PSM for applied technology such as earthquake early-warning systems, ongoing active and creative research is needed using data sets acquired in the different environments as discussed in this paper. This will contribute to develop, constrain, and refine our understanding of the physical processes being responsible for the occurrence of small-scale brittle-failure and major earthquakes in both space and time.

References

Aki, K., Richards, P. (2002) Quantitative Seismology. 2nd Edition, University Science Books.

Albright, J.N., Pearson, C.F. (1982) Acoustic Emissions as a Tool for Hydraulic Fracture Location: Experience at the Fenton Hill Hot Dry Rock Site. *SPE J.* 22, 4, 523–530.

Ambraseys, N.N. (2002) The seismic activity of the Sea of Marmara region over the last 2000 years. *Bull Seismol Soc Am.* 92, 1–18.

Armijo, R., Meyer, B., Navarro, S., King, G., Barka, A. (2002) Asymmetric slip partitioning in the Sea of Marmara pull-apart: A clue to propagation processes of the North Anatolian fault? *Terra Nova.* 14, 2, 80–86.

Armijo, R. et al. (2005) Submarine fault scarps in the Sea of Marmara pull-apart (North Anatolian fault): Implications for seismic hazard in Istanbul. *Geochem Geophysics Geosystems.* doi:10.1029/2004GC000896.

Baisch, S., Harjes, H.P. (2003) A model for fluid-injection-induced seismicity at the KTB, Germany. *Geophys J Int.* 152, 160–170.

Baisch, S., Bohnhoff, M., Ceranna, L., Tu, Y., Harjes, H.P. (2002) Probing the crust to 9 km depth: Fluid injection experiments and induced seismicity at the KTB superdeep drilling hole. *Bull Seismol Soc Am.* 92, 2369–2380.

Bakun, W.H., McEvilly, T.V. (1979) Earthquakes near Parkfield, California: Comparing the 1934 and 199 Sequences. *Science.* 205, 4413, 1375–1377.

Bakun, W.H., McEvilly, T.V. (1984) Recurrence models and Parkfield, California, earthquakes. *J Geophys Res.* 89, B5, 3051–3058.

Bakun, W.H., Lindh, A.G. (1985) The Parkfield, California, Earthquake Prediction Experiment. *Science.* 229, 4714, 619–624.

Bakun, W.H. et al. (1987) Parkfield earthquake prediction scenarios and response plans. USGS Open-File Report 87–192.

Bakun, W.H. et al. (2005) Implications for prediction and hazard assessment from the 2004 Parkfield earthquake. *Nature.* doi:10.1038/natures04067.

Baris, S., Irmak, T.S., Grosser, H., Ozer, M.F., Woith, H., Ulutas, E., Tuncer, M.K. (2007) Monitoring seismicity in the eastern Marmara: the Armutlu Network. *Geophys Res Abstr.* 9, 10198.

Barka, A. (1992) The North Anatolian fault zone. *Annal Tectonicae.* 6, 164–195.

Barka, A. (1999) The 17 August Izmit earthquake. *Science.* 285, 1858–1859.

Barka, A. et al. (2002) The Surface Rupture and Slip Distribution of the 17 August 1999 Izmit Earthquake (M 7.4). *Bull Seismol Soc Am.* 92, 1, 43–60.

Barton, C.A., Zoback, M.D. (1994) Stress perturbations associated with active faults penetrated by boreholes: Possible evidence for near-complete stress drop and a new technique for stress magnitude measurement. *J Geophys Res.* 99, 9373–9390.

Bindi, D., Parolai, S., Grosser, H., Milkereit, C., Durukal, E. (2007) Empirical ground-motion prediction equations for northwerstern Turkey using the aftershocks of the 1999 Kocaeli earthquake. *Geophys Res Lett.* 34, L08305.

Bird, J.F., Bommer, J.J. (2004) Earthquake Losses Due to Ground Failure. *Eng Geol.* doi:10.1016/j.enggeo.05.006.

Bohnhoff, M., Baisch, S., Harjes, H.P. (2004) Fault mechanisms of induced seismicity at the superdeep German Continental Deep Drilling Program (KTB) borehole and their relation to fault structure and stress field. *J Geophys Res.* doi:10.1029/2003JB002528.

Bohnhoff, M., Grosser, H., Dresen, G. (2006) Strain Partitioning and Stress Rotation at the North Anatolian Fault Zone from aftershock focal mechanisms of the 1999 Izmit Mw = 7.4 Earthquake. *Geophys J Int.* 166, 373–385.

Bohnhoff, M., Bulut, F., Aktar, M., Childs, D.M., Dresen, G. (2007) The North Anatolian Fault Zone in the broader Istanbul/Marmara region: Monitoring a ‚seimic gap‘. AGU Fall Meeting T51C-0688.

Brudy, M., Zoback, M.D., Fuchs, K., Rummel, F., Baumgärtner, J. (1997) Estimation of the complete stress tensor to 8 km depth in the KTB scientific drill holes: Implications for crustal strength. *J Geophys Res.* 102, B8, 18453–18475.

Chavarria, J.A., Malin, P., Catchings, R.D., Shalev, E. (2003) A Look Inside the San Andreas fault at Parkfield Through Vertical Seismic Profiling. *Science.* 302, 5651, 1746–1748.

Chavarria, J.A., Malin, P.E., Shaley, E. (2004) The SAFOD Pilot Hole seismic array: Wave propagation effects as a function of sensor depth and source location. *Geophys Res Lett*. doi:10.1029/2003GL019382.

Davidson, J., Stanchits, S., Dresen, G. (2007) Scaling and Universality in Rock Fracture. *Phys Rev Lett*. doi: 10.1103/PhysRevLett.98.125502.

Dresen, G., Duyster, J., Stoeckhert, B., Wirth, R., Zulauf, G. (1997) Quartz dislocation microstructure between 7000 m and 9100 m depth from the Continental Deep Drilling Program KTB. *J. Geophys. Res.* 102, B8, 18443–18452.

Dresen, G., Aktar, M., Bohnhoff, M., Eyidogan, H. (2007) Drilling the North Anatolian Fault. *Sci Drill Spec Iss*. doi:10.2204/iodp.sd.s01.17.2007.

Eaton, J.P., O'Neill, M.E., Murdock, J.N. (1970) Aftershocks of the 1966 Parkfield-Cholame, California, earthquake: a detailed study. *Bull Seismol Soc Am*. 60, 1151–1197.

Ellsworth, W.L., Hickman, S.H., Zoback, M.D., Imanishi, K., Thurber, C.H., Roecker, S.W. (2007a) Micro- Nano- and Picoearthquakes at SAFOD: Implications for Earthquake Rupture and Fault Mechanics. AGU S12B-05.

Ellsworth, W.L., Malin, P.E., Imanishi, K., Roecker, S.W., Nadeau, R., Oye, V., Thurber, C.H., Waldhauser, F., Boness, N.L., Hickman, S.H., Zoback, M.D. (2007b) Seismology inside the Fault Zone: Applications to Fault-Zone Properties and Rupture Dynamics. *Sci Drill Spec Iss*. doi:10.2204/iodp.sd.s01.04.2007.

Emmermann R. Lauterjung J. (Eds.) (1997) The German Continental Deep Drill Program KTB: Overview and major results. *J Geophys Res* 102(B8):18179–18201.

Evans, D. (1966) Denver area earthquakes and the Rocky Mountain Arsenal disposal well. *Mountain Geologist*. 3, 1, 23–26.

Flerit, F., Armijo, R., King, G., Meyer, B. (2004) The mechanical interaction between the propagating North Anatolian Fault and the back-arc extension in the Aegean. *Earth Plant Sci Lett*. 224, 347–362.

Fortin, J., Stanchits, S., Dresen, G., Guéguen, Y. (2006) Acoustic emission and velocities associated with the formation of compaction bands in sandstone. *J Geophys Res*. doi:10.1029/2005JB003854.

Gibbs, J.F., Healy, J.H., Raleigh, C.B., Coakley, J.M. (1973) Seismicity in the Rangely, Colorado, area: 1962–1970. *Bull Seismol Soc Am*. 63, 1557–1570.

Gutenberg, B. (1945) Amplitudes of surface waves and magnitudes of shallow earthquakes. *Bull Seismol Soc Am*. 35, 3–12.

Gutenberg, B., Richter, C.F. (1941) Seismicity of the Earth. *Geol Soc Am Spec Pap*. 34, 1–133.

Graham, C.C., Stanchits, S., Main, I.G., Dresen, G. (2009) Comparison of polarity and moment tensor inversion methods for source analysis of acoustic emission data, Int. J. Rock Mech. Min. Sci. doi:10.1016/j.ijrmms.2009.05.002'.

Harris, R.A., Arrowsmith, J.R. (2006) Introduction to the Special Issue on the 2004 Parkfield Earthquake and the Parkfield Earthquake Prediction Experiment. *Bull Seismol Soc Am*. doi:10.1785/0120050831.

Healy, J.H., Rubey, W.W., Griggs, D.T., Raleigh, C.B. (1968) The Denver Earthquakes. *Science*. 161, 3848, 1301–1310.

Hickman, S.H., Zoback, M.D., Ellsworth, W.L. (2004) Introduction to special sections: Preparing for the San Andreas Fault Observatory at Depth. *Geophys Res Lett*. doi:10.1029/2004GL020688.

Hickman, S.H., Zoback, M.D., Ellswoth, W.L., Boness, N., Malin, P., Roecker, S., Thurber, C. (2007) Structure and Properties of the San Andreas Fault in Central California: Recent Results from the SAFOD Experiment. *Sci Drill*. doi:10.2204/iodp.sd.s01.39.2007.

Hill, D.P., Eaton, J.P., Jones, L.M. (1990) 5. Seismicity, 1980–86, in The San Andreas Fault System, California. Wallace RE (ed.) U.S. Geological Survey Professional Paper 1515, United States Government, Printing Office, Washington.

Ide, S., Beroza, G.C. (2001) Does apparent stress vary with earthquake size, Geophys. Res. Lett. 28, 3349–3352.

Imamura, A., Kodaira, T., Imamura, H. (1932) The earthquake swarms of Nagusa and vicinity. *Bull Earthquake Res Inst Univ Tokyo*. 10, 636–648.

Imanishi, K., Ellsworth, W.L. (2006) Scaling Relationships of Microearthquakes at Parkfield, CA, Determined Using the SAFOD Pilot Hole Seismic Array. Abercrombie RE McGarr A Di Toro G Kanamori H (eds.) Earthquakes: radiated energy and the physics of faulting, American Geophysical Union Monograph 170.

Ishimoto, M., Ida, K. (1939) Observations sur les seismes enrigstres par le microsismographe construit dernierement (I). Bull Earthquake Inst Univ Tokyo 17:443–478 (in Japanese with French abstract).

Ito, H. (2007) Long-term monitoring in deep boreholes in the Nankai subduction zone. *Sci Drill Spec Iss*. 1, 117–119.

Janssen, C., Wagner, C., Zang, A., Dresen, G. (2001) Fracture Process Zone in Granite-a Microstructural Analysis. *Int J Earth Sci*. 90, 46–59.

Jones, R.H., Stewart, R.C. (1997) A method for determining significant structures in a cloud of earthquakes. *J Geophys Res*. 102, B4, 8245–8254.

Jost, M.L., Büsselberg, T., Jost, Ö., Harjes, H.P. (1998) Source parameters of injection-induced microearthquakes at 9 km depth at the KTB Deep Drilling site, Germany. *Bull Seismol Soc Am*. 88, 815–832.

Karabulut, H., Bouin, M.P., Bouchon, M., Dietrich, M., Cornou, C., Aktar, M. (2002) The Seismicity in the Eastern Marmara Sea after the 17 August 1999 Izmit Earthquake. *Bull Seismol Soc Am*. 92, 1, 387–393.

Kikuchi, M., Nakamura, M., Yoshikawa, K. (2003) Fault asperity of large earthquakes in Japan inferred from low-gain historical seismograms. *Earth Plan Space*. 55, 159–172.

Korneev, V.A., McEvilly, T.V., Karageorgi, E.D. (2000) Seismological Studies at Parkfield VIII: Modeling the Observed Travel-Time Changes. *Bull Seismol Soc Am*. 90, 702–708.

Kostrov, B.V. Selfsimilar problems of propagation of shear cracks (Tengential rupture crack propagation in medium under shearing stress). *PPM-J Appl Meth Mech*. 28, 5, 1077–1087. (1964).

Kovach, R.L. (1974) Source mechanisms for Wilmington oil field, California, subsidence earthquakes. *Bull Seismol Soc Am*. 64, 699–711.

Lee WHK Stewart SW (1981) Principles and Applications of Microearthquake Networks. Academic Press, New York.

Lei, X., Nishizawa, O., Kusunose, K., Satoh, T. (1992) Fractal Structure of the Hypocenter Distributions and Focal Mechanism Solutions of Acoustic Emission in Two Granites of Different Grain Sizes. *J Phys Earth*. 40, 617–634.

Lei, X., Kusunose, K., Rao, M.V.M.S., Nishizawa, O., Satoh, T. (2000) Quasi-static fault growth and cracking in homogeneous brittle rock under triaxial compression using acoustic emission monitoring. *J Geophys Res 105*. 6127–6139.

LePichon, X. et al. (2001) The active Main Marmara Fault. *Earth Plant Sci Lett*. 192, 595–616.

Lisowski, M., Prescott, W.H. (1981) Short-range distance measurements along the San Andreas fault system in central California, 1975 to 1979. *Bull Seismol Soc Am*. 71, 1607–1624.

Lockner, D.A. (1993) The Role of Acoustic Emission in the Study of Rock Fracture. *Int J Rock Mech Min Sci Geomech*. 30, 7, 883–899.

Lockner, D.A., Byerlee, J.D., Kuksenko, V., Ponomarev, A., Sidorin, A. (1991) Quasi-static fault growth and shear fracture energy in granite. *Nature*. 350, 6313, 39–42.

Lockner, D.A., Byerlee, J.D. (1991) Precursory AE patterns leading to rock fracture. 5th Conf Acoust Emiss Geol Struct Mat 45–58 Trans Tech Publications Clausthal-Zellerfeld-Germany and Pennsylvania State University.

Lockner, D.A., Byerlee, J.D. (1992) Fault growth and acoustic emissions in confined granite. Proceedings of the 22nd Midwestern Mechanics Conf 165–173 Appl Mech Rev Rolla Missouri.

Malin, P.E., Blakeslee, S.N., Alvarez, M.G., Martin, A.J. (1989) Microearthquake Imaging of the Parkfield Asperity. *Science*. 244, 4904, 557–559.

McGarr, A. (1991) On a possible connection between three major earthquakes in California and oil production. *Bull Seismol Soc Am*. 81, 948–970.

McGarr, A., Simpson, D., Seeber, L. (2002) Case Histories of Induced and Triggered Seismicity. In: International handbook of earthquake and engineering seismology Part A (International geophysics, Vol.81 A), p 647–661.

Muller, J.R., Aydin, A. (2005) Using mechanical modelling to constrain fault geometries proposed for the northern Marmara sea. *J Geophys Res*. doi:10.1029/2004JB003226.

Murray, J., Langbein, J. (2006) Slip on the San Andreas Fault at Parkfield, California, over Two Earthquake Cycles, and the Implications for Seismic Hazard. *Bull Seismol Soc Am*. 96, S283–S303.

Nadeau, R.M., Dolenc, D. (2005) Nonvolcanic tremor Deep Beneath the San Andreas Fault. *Science*. 307, 5708, 389.

Nadeau, R.M., Foxall, W., McEvilly, T.V. (1995) Clustering and periodic recurrence of microearthquakes on the San Andreas fault at Parkfield, California. *Science*. 267, 503–507.

Namba, Y., Ito, H., Kato, K., Higuchi, K., Kyo, M. (2008) Engineering specifications on LTBMS telemetry system for NanTroSEIZE 3.5 km riser hole. *JAMSTEC Rep Res Dev*. 7, 43–58.

Oppenheimer, D.H., Iyer, H.M. (1980) Frequency-wavenumber analysis of geothermal microseisms at Norris Geyser basin, Yellowstone National Park, Wyoming. *Geophysics*. 45, doi:10.1190/1.1441099.

Özalaybey, S., Ergin, M., Aktar, M., Tapirdamaz, C., Bicmen, F., Yörük, A. (2002) The 1999 Izmit Earthquake Sequence in Turkey: Seismological and Tectonic Aspects. *Bull Seismol Soc Am*. 92, 1, 376–386.

Park, J.-O., Tsuru, T., Kodaira, S., Cummins, P.R., Kaneda, Y. (2002) Splay fault branching along the Nankai subduction zone, Science 297, 1157–1160.

Parolai, S., Richwalski, S.M., Zschau, J., Durukal, E., Özel, O., Birgören, G., Ansal, A., Erdik, M. (2007) Project 'Megacity Istanbul': Estimation of site effects and ground motion scenarios. *Geophys Res*. Abstr 8 07038.

Parsons, T. (2004) Recalculated probability of M≥7 earthquakes beneath the Sea of Marmara, Turkey. *J Geophys Res*. doi:10.1029/2003JB002667.

Phillips, W.S., Fairbanks, T.D., Rutledge, J.T., Anderson, D.W. (1998) Induced microearthquake patterns and oil-producing fracture systems in the Austin chalk. *Tectonophysics*. doi:10.1016/S0040-1951(97)00313-2.

Phillips, W.S., Rutledge, J.T., House, L.S., Fehler, M.M.C. (2002) Induced Microearthquake Patterns in Hydrocarbon and Geothermal Reservoirs: Six Case Studies. *PAGEOPH*. 159, 345–369.

Plenkers, K., Kwiatek, G. JAGUARS-group (2008) JAGUARS-Project: Spectral analysis of microseismicity and acoustic emission in a deep South African gold mine. *Seismol Res Lett*. 79, 330.

Raleigh, C.B., Healy, J.H., Bredehoeft, J.D. (1972) Faulting and crustal stress at Rangely, Colorado. Flow and Fracture of rocks. *Am Geophys Union Geophys Monograph*. 16, 275–284.

Reid, H.F. (1910) The mechanics of the earthquake. Carnegie Institution of Washington.

Reilinger, R. et al. (2006) GPS constraints on continental deformation in the Africa-Arabia-Eurasia continental collision zone and implications fort he dynamics of plate interactions. *J Geophys Res*. doi:10.1029/2005JB004051.

Richter, C.F. (1935) An instrumental earthquake magnitude scale. *Bull Seismol Soc Am*. 25, 1–32.

Roecker, S., Thurber, C., McPhee, D. (2004) Joint inversion of gravity and arrival time data from Parkfield: New constraints on structure and hypocenter locations near the SAFOD drill site. *Geophys Res Lett*. doi:10.1029/2003GL019396.

Rubin, A.M., Gillard, D., Got, J.L. (1999) Streaks of microearthquakes along creeping faults. *Nature*.doi:10.1038/23196.

Rutledge, J.T., Phillips, W.S., Roff, A., Albright, J.N., Hamilton-Smith, T., Jones, S., Kimmich, K. (1994) Subsurface fracture mapping using microearthquakes detected during primary oil production, Clinton County, Kentucky, paper SPE 28384, Soc. of Petro. Eng. Ann. Tech. Conf.

Rutledge, J.T., Phillips, W.S., Mayerhofer, M.J., (2004) Faulting Induced by Forced Fluid Injection and Fluid Flow Forced by Faulting: An Interpretation of Hydraulic-Fracture Microseismicity, Carthage Cotton Valley Gas Field, Texas. Bull Seismol Soc Am doi:10.1785/012003257.

Sato, T., Kasahara, J., Taymaz, T., Ito, M., Kamimura, A., Hayakawa, T., Tan, O. (2004) A study of microearthquake seismicity and focal mechanisms within the Sea of Marmara (NW Turkey) using ocean bottom seismometers (OBSs. *Tectonophysic*. 391, 303–314.

Schaff, D.P., Bokelmann, G.H.R., Beroza, G.C. (2002) High-resolution image of Calaveras Fault seismicity. *J Geophys Res*. doi:10.1029/2001JB000633.

Schaff, D.P., Bokelmann, G.H.R., Ellsworth, W.L., Zanzerkia, E., Waldhauser, F., Beroza, G.C. (2004) Optimizing Correlation Techniques for Improved Earthquake Location. *Bull Seosmol Soc Am*. doi:10.1785/0120020238.

Scholz, C.H. (1968) The Frequency – Magnitude Relation of Microfracturing in Rock and its Relation to Earthquakes. *Bull Seismol Soc Am*. 58, 1, 399–415.

Simpson, R.W., Barall, M., Langbein, J., Murray, J.R., Rymer, M.J. (2006) San Andreas Fault Geometry in the Parkfield, California. *Region Bull Seismol Soc Am*. doi:10.1785/0120050824.

Stanchits, S., Vinciguerra, S., Dresen, G. (2006) Ultrasonic velocities, acoustic emission characteristics and crack damage of basalt and granite. *PAGEOPH*. 163, 974–993.

Stanchits, S., Dresen, G. (2003) Separation of Tensile and Shear Cracks Based on Acoustic Emission Analysis of Rock Fracture. Proc Int Symposium: Non-Destructive Testing in Civil Engineering (NDT-CE) Berlin 107.

Stein, R.S., Barka, A., Dieterich, J.H. (1997) Progressive failure ob the North Anatolian fault since 1939 by earthquake stress triggering. *Geophys J Int*. 128, 594–604.

Thurber, C., Roecker, S., Zhang, H., Baher, S., Ellsworth, W.L. (2004) Fine-scale structure of the San Andreas fault zone and location of the SAFOD target earthquakes. *Geophys Res Lett*. doi:10.1029/2003GL019398.

Thurber, C., Zhang, H., Waldhauser, F., Hardebeck, J., Michael, A., Eberhart-Phillips, D. (2006) Three-Dimensional Compressional Wavespeed Model, Earthquake Relocations, and Focal Mechanisms for the Parkfield, California, Region. Bull. Seismol. Soc. Am. doi:10.1785/0120050825.

Tobin, H., Kinoshita, M. (2006) The IODP Nankai Trough seismogenic zone experiment. *Sci Drill*. doi:10.2204/iodp.sd.2.06.2006.

Waldhauser, F., Ellsworth, W.L., Cole A. (1999) Slip-parallel seismic lineations on the Northern Hayward Fault, California. Geophys. Res. Lett. 26, 3525–3528.

Waldhauser, F., Ellsworth, W.L. (2000) A Double-Difference Earthquake Location Algorithm: Method and Applications to the Northern Hayward Fault, California. *Bull Seismol Soc Am*. doi:10.1785/0120000006.

Waldhauser, F., Ellsworth, W.L. (2002) Fault structure and mechanics of the Hayward Fault, California, from double-difference earthquake locations. *J Geophys Res*. doi: 10.1029/2000JB000084.

Waldhauser, F., Ellsworth, W.L., Schaff, D.P., Cole, A. (2004) Streaks, multiplets, and holes: High-resolution spatio-temporal behavior of Parkfield seismicity. *Geophys Res Lett*. doi:10.1020/2004GL020649.

Yalciner, A.C., Alpar, B., Altinok, Y., Özbay, I., Imamura, F. (2002) Tsunamis in the Sea of Marmara: Historical documents for the past, models for the future. *Mar Geol*. 190, 445–463.

Young, R.P., Collins, D.S. (2001) Seismic studies of rock fracture at the Underground Research Laboratory, Canada. *Int J Rock Mech Mining Sc*. 38, 787–799.

Zang, A., Wagner, F.C., Stanchits, S., Janssen, C., Dresen, G. (2000) Fracture process zone in granite. *J Geophys Res*. 105, 23651–23661.

Zang, A., Wagner, F.C., Stanchits, S., Dresen, G., Andresen, R., Haidekker, M.A. (1998) Source analysis of acoustic emissions in granite cores under symmetric and asymmetric compressive load. *Geophys J Int*. 135, 1113–1130.

Zang, A., Wagner, F.C., Dresen, G. (1996) Acoustic emission, microstructure, and damage model of dry and wet sandstone stressed to failure. *J Geophys Res*. 101, 17507–17521.

Zhang, H., Thurber, C.H. (2003) Double-Difference Tomography: The Method and Its Application to the Hayward Fault, California. *Bull Seismol Soc Am*. doi:10.1785/0120020190.

Zoback , M.D., Harjes , H.P. (1997) Injection-induced earthquakes and crustal stress at 9 km depth at the KTB deep drilling site, Germany. *J Geophys Res*. 102, B8, 18477–18491.

Non-volcanic Tremor: A Window into the Roots of Fault Zones

Justin L. Rubinstein, David R. Shelly, and William L. Ellsworth

Abstract The recent discovery of non-volcanic tremor in Japan and the coincidence of tremor with slow-slip in Cascadia have made earth scientists reevaluate our models for the physical processes in subduction zones and on faults in general. Subduction zones have been studied very closely since the discovery of slow-slip and tremor. This has led to the discovery of a number of related phenomena including low frequency earthquakes and very low frequency earthquakes. All of these events fall into what some have called a new class of events that are governed under a different frictional regime than simple brittle failure. While this model is appealing to many, consensus as to exactly what process generates tremor has yet to be reached. Tremor and related events also provide a window into the deep roots of subduction zones, a poorly understood region that is largely devoid of seismicity. Given that such fundamental questions remain about non-volcanic tremor, slow-slip, and the region in which they occur, we expect that this will be a fruitful field for a long time to come.

Keywords Tremor · ETS · Slow earthquakes · Slow-slip

Introduction

The analysis of seismic waves provides a direct, high-resolution means for studying the internal structure of the Earth and the geological processes that are actively reshaping it today. The subject of this chapter is a newly discovered source process, non-volcanic tremor, that appears to be most common in the lower crust and upper mantle. Non-volcanic tremor has great potential for shedding light on the dynamics of regions of the Earth that have been difficult to study before.

Beginning in the late nineteenth century, the first seismologists had the challenge of understanding the origin of the feeble motions or "tremors" recorded on the early seismographs. By the turn of the twentieth century, they had established that distant earthquakes, or teleseisms, were the source of the most obvious eventful signals and that the earthquakes were distributed in well-defined belts around the globe. The study of those belts of earthquakes and the nature of the earthquake source itself continued as the principal focus of seismological research during the twentieth century and continues to be a central topic today.

Seismologists were also busy developing a comprehensive understanding of the other signals seen on the seismogram. These ranged from the impulsive signature of nearby earthquakes and explosions to the continuous microseisms generated by waves in the ocean, the ringing of the Earth itself in free oscillations, volcanic tremors created by subterranean movement of magma, sonic booms caused by meteors and aircraft, and cultural noise produced by power plants, highways, trains, etc.

It is in this context that the discovery by Kazushige Obara in 2002 of a new source of seismic waves that had been overlooked for a century took the Earth sciences by surprise. Obara (2002) found that long-duration trains of weak seismic motions were originating from a band following the contour of the subducting Philippine Sea plate beneath southwestern Japan, at depths near 30 km (Fig. 1). These "non-volcanic

J.L. Rubinstein (✉)
United States Geological Survey; Menlo Park, CA 94025, USA
e-mail: jrubinstein@usgs.gov

S. Cloetingh, J. Negendank (eds.), *New Frontiers in Integrated Solid Earth Sciences*, International Year of Planet Earth, DOI 10.1007/978-90-481-2737-5_8, © Springer Science+Business Media B.V. 2010

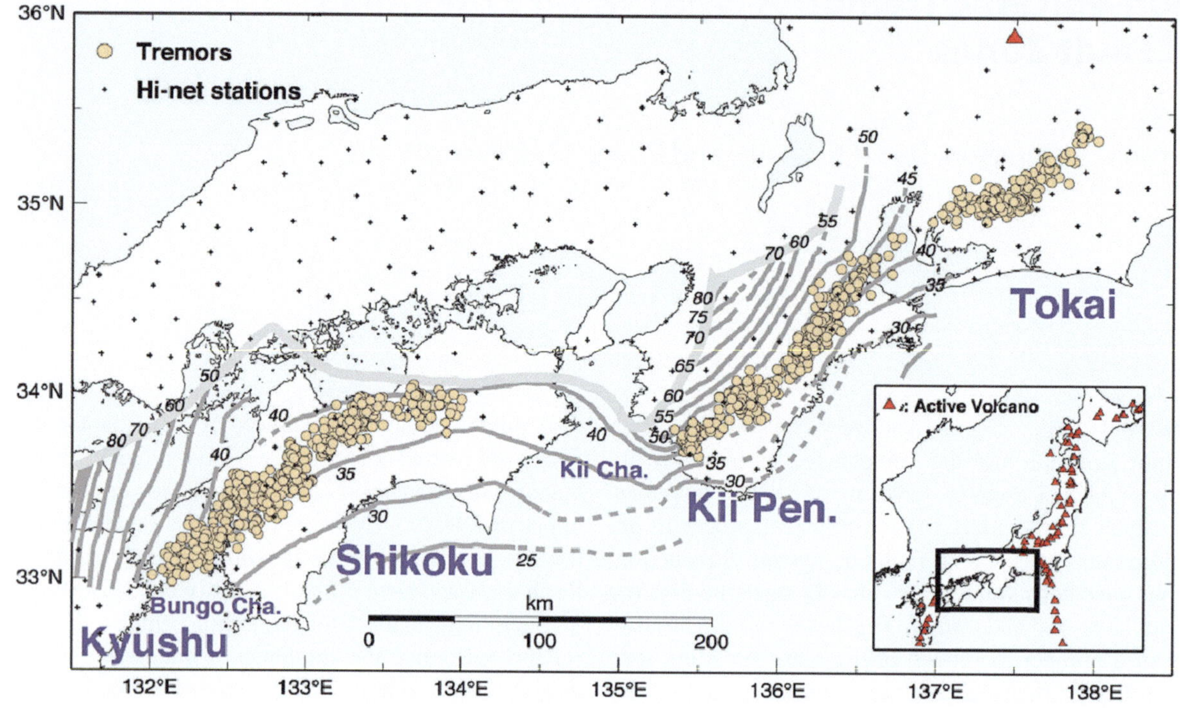

Fig. 1 Map of Southwestern Japan showing distribution of Hi-net stations (*dots*), isodepths of the subduction interface, and locations of tremor (*filled circles*) identified by Obara (2002). Active volcanoes shown by *red triangles*. Figure from Obara (2002)

tremors" were unlike anything previously known to seismologists. Something new had been discovered.

Active periods of non-volcanic tremor lasted for a week or more and migrated laterally along the subduction zone near the seismic-aseismic transition on the plate interface. Individual tremor episodes typically lasted for tens of minutes with dominant seismic frequencies of 1 to 10 Hz. The tremor propagated at the *S* wave velocity, suggesting that these waves were *S* body waves. They typically emerged slowly from the background noise and lacked any regular structure, with peaks in amplitude occurring at random during the episode and without any easily identifiable seismic body wave arrivals. Their appearance is similar to that of a distant train passing by the seismometer.

Episodic Tremor and Slip

The previous year, another important discovery had been made across the Pacific in the Cascadia subduction zone by Dragert et al. (2001). Using Global Positioning Satellite (GPS) measurements, they deduced that a slow earthquake had occurred over a period of about a week beneath Vancouver Island in British Columbia and Puget Sound in Washington State. Slow earthquakes are like regular earthquakes in that they represent slip on a fault, but unlike regular events, they rupture very slowly and take place over many hours/days such that they do not radiate strong high frequency energy like regular earthquakes. Dragert et al. (2001) inferred that event probably ruptured the plate interface downdip of the seismogenic zone. The moment magnitude of the event was 6.7. Miller et al. (2002) expanded this discovery with the remarkable observation that slow earthquakes had been occurring there for years at semi-regular intervals of 13–16 months. But were these transient slip events detected by GPS related in some way to non-volcanic tremor?

Following the report of tremor in southwest Japan by Obara (2002) and the suggestion that it might relate to the prior observation of Cascadia slow slip (Julian, 2002); Rogers and Dragert (2003) demonstrated that tremor indeed accompanied slow slip in Cascadia and that slow-slip and tremor appeared to be coming from the same location. They termed the coupled phenomenon episodic tremor and slip (ETS), showing that each slip event was accompanied by a level of tremor activity far exceeding that observed at other times (Fig. 2). In Cascadia, this phenomenon is highly regular; the coefficient of variation of ETS recurrence times for the 25 year interval (1982–2007) documented by Rogers (2007) is 0.1, far less than 0.5 which is typical for earthquakes (Ellsworth et al., 1999).

Similarly, Obara et al. (2004) reported the discovery of slow slip events accompanying the previously observed tremor in western Shikoku. Presumably because of the smaller size of the events (approximately moment magnitude 6), the slow slip was not readily observed by GPS instruments and was instead recognized using borehole tiltmeters. Slow slip and tremor were observed to migrate together, with a recurrence interval of approximately 6 months between major episodes. Hirose and Obara (2006) reported similar episodes in the Tokai region.

With improving geodetic networks, slow slip events, or periods of accelerated creep, have been recognized to be relatively common in subduction zones. These events can occur at a variety of depths, with durations ranging at least from days (e.g. Obara et al., 2004) to years (Hirose et al., 1999). Strong correlations between tremor and slow slip have now been established in Cascadia and southwest Japan, and evidence is building that similar correlations may exist in Alaska (Ohta et al., 2006; Peterson and Christensen, 2009), Mexico (Payero et al., 2008), and Costa Rica (Brown et al., 2005; Schwartz et al., 2008) (Fig. 3). On the other hand, slow slip has been seen in some places without generating detectible tremor (Delahaye, et al., 2009; Segall et al., 2006; Ozawa et al., 2002, 2003; Kawasaki et al., 1995, 2001; Montgomery-Brown et al., 2009). Thus the two phenomena are not always linked. Slow-slip and non-volcanic tremor were also recently been reviewed by Schwartz and Rokosky (2007).

Consistent, slow-slip at the rates that many of these events are slipping (mm/h) should not radiate the high frequencies (>1 Hz) that we observe associated with tremor. Slip rates that are many orders of magnitude larger are necessary to radiate such high frequencies (Aki and Richards, 2002). The slow slip events that do have tremor associated with them may have brief periods of accelerated slip that produce this tremor, or perhaps have smaller regions within the larger, slow-slipping region that slip much more rapidly. Alternatively, tremor may come from some other process that is coincident with the slow-slip.

New Opportunities

Individually or in concert, non-volcanic tremor and slow-slip represent a great research opportunity for geoscientists. The recent discovery of these phenomena offers many opportunities to learn about the

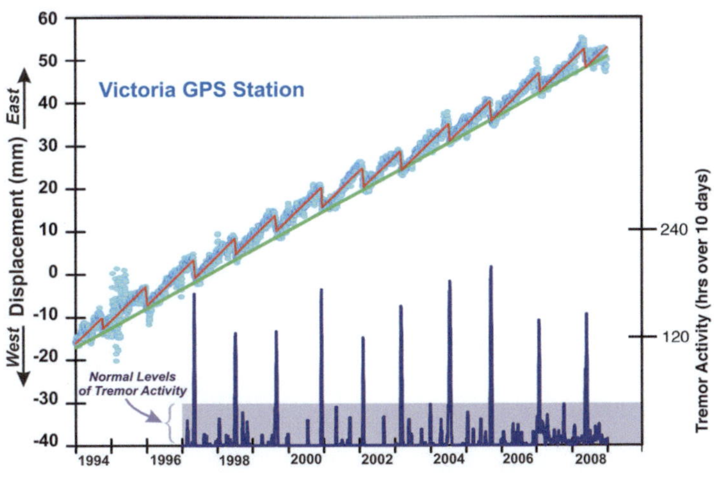

Fig. 2 Comparison of hours of tremor activity to GPS observations of displacement. *Blue dots* represent daily location GPS solutions for the east component of ALBH GPS station. Every ~14 months the trend of the GPS locations reverses direction for ~2 weeks away from the secular trend. These 2 week excursions represent the slow slip events. *Red line* segments represent a linear fit to the trend of the daily locations within periods where the displacement has not reversed. *Green line* represents a linear fit to the entire trend of the displacement. *Black lines* represent hours of non-volcanic tremor recorded in the same region. There is a remarkable correlation between the times when the GPS displacement has a reversal and periods of strong tremor activity. Figure modified from Rogers and Dragert (2003)

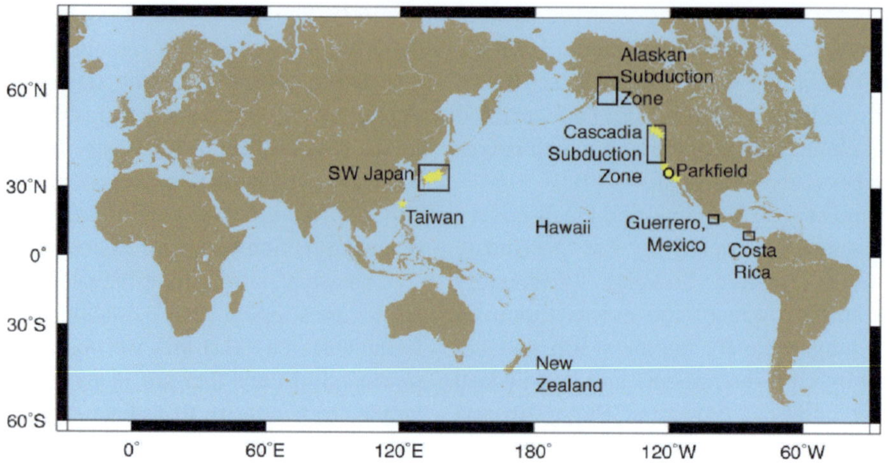

Fig. 3 Map of the world showing regions where non-volcanic tremor and associated slip have been observed. Regions where non-volcanic tremor and coincident slow-slip have observed are indicated by *black boxes*. We note that presently there is only data that indicates there is and episodic relationship between slow-slip in Cascadia and SW Japan, but in the other boxed locations slow-slip and tremor have been observed to be coincident at least once. The *circle* indicates Parkfield, California where non-volcanic tremor has been observed but no associated slow-slip has been observed. The *yellow stars* indicate locations of non-volcanic tremor that was triggered by earthquakes – note that there are many locations where triggered tremor has been observed but spontaneous tremor has yet to be observed. Other locations that are mentioned in the text are also indicated on the map without symbols

rheology and deformation mechanisms in the tremor source region, and it is accelerating research in related topics.

The global distribution of non-volcanic tremor is still being ascertained, but it has been found in a number of different tectonic environments. While most observations of tremor, to date, have been in the Cascadia (e.g., Rogers and Dragert, 2003; Szeliga et al., 2004; McCausland et al., 2005) and the Nankai subduction zones (Obara, 2002), tremor has also been identified in the subduction zones of Alaska (Peterson and Christensen 2009), and Mexico (Payero et al., 2008), but is absent in others including the Japan Trench in northern Honshu (Obara, 2002). It has also been observed along the strike-slip plate boundary in California (Nadeau and Dolenc, 2005; Gomberg et al., 2008), the region beneath the Western Tottori earthquake in Japan (Ohmi et al., 2004), and the central Ranges of Taiwan, within Taiwanese collision zone (Peng and Chao, 2008). Repeating slow-slip events, without tremor, have also been identified in New Zealand (Delahaye et al., 2009; McCaffrey et al., 2008).

Tremor and slow-slip can be used to learn about the conditions in the deep root zone of some major faults. In strike-slip-dominated regions (Gomberg et al., 2008; Nadeau and Dolenc, 2005) and collisional environments (Peng and Chao, 2008) tremor is found below the seismogenic zone, a poorly sampled and understood region. In subduction zones, tremor is typically found in depth ranges with little seismicity (Kao et al., 2005; Shelly et al., 2006), leaving it poorly sampled as well.

In addition to facilitating the study of regions that produce tremor and slow-slip, these phenomena have led to a frenzied development of new techniques and more careful analysis of many data sources. As a result new phenomena are being discovered, including: Very Low Frequency earthquakes (VLF) with durations of ~10 s (Ito et al., 2007) and slow earthquakes with one hundred second duration (Ide et al., 2008). With the continued pace and fervor of research and development in this field, we fully expect that there will be many more breakthroughs in this young and exciting field.

Fundamental Properties of Tremor

Non-volcanic tremor is typically observed as long-duration, low-amplitude, non-impulsive seismic radiation most readily seen in the 1–10 Hz frequency band (Fig. 4, 5). The lack of easily identifiable features makes it difficult to distinguish from cultural or environmental noise and likely prevented an earlier identification of tremor. The unremarkable nature of its

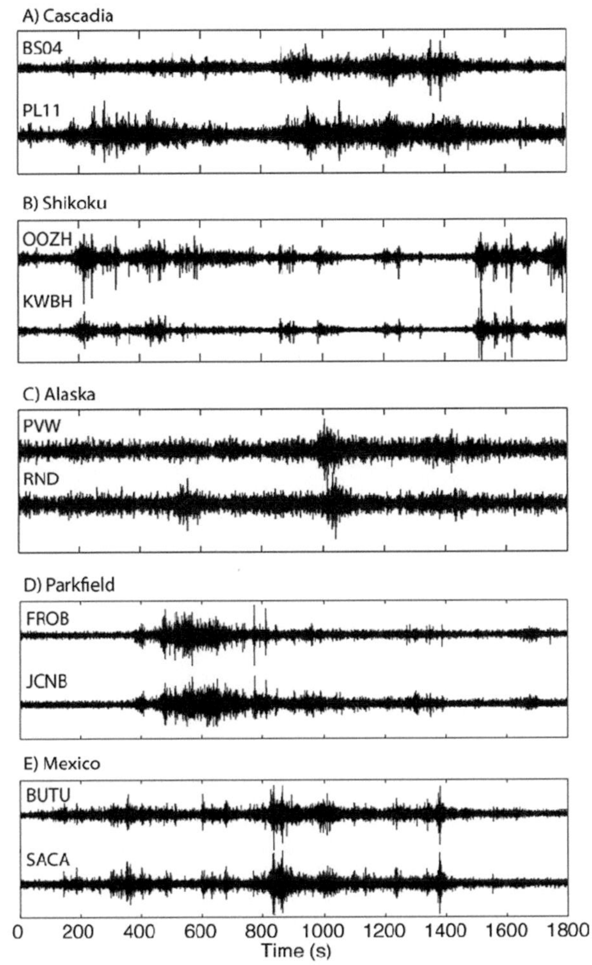

Fig. 4 Recordings of non-volcanic tremor in (**a**) the Cascadia subduction zone (**b**) the Nankai Trough (**c**) the Alaska subduction zone (**d**) Parkfield, California on the San Andreas strike-slip fault and (**e**) the Mexican subduction zone. Records are bandpass filtered at 1–8 Hz. (**b**) is modified from Shelly et al. (2007a)

waveforms poses a challenge for those trying to identify it. Most use very simple methods based on envelope amplitude like those that Obara (2002) used to initially identify tremor, although more complex, automated methods to identify tremor are starting to be developed (Kao et al., 2007a; Wech and Creager, 2008 Suda et al., in press). The absence of easily identified body wave arrivals also contributes to the difficulty in locating non-volcanic tremor. Methods used to locate earthquakes largely depend on the impulsive nature of their body wave phases, rendering them rather ineffective for locating tremor. The issue of tremor location is more fully explored in section "Locating Non-volcanic Tremor".

While non-volcanic tremor usually lacks distinguishable arrivals, impulsive arrivals in Japanese tremor have been observed (Katsumata and Kamaya, 2003). These arrivals are typically *S* waves, but *P* waves have also been found (Shelly et al., 2006). These body wave arrivals are regularly identified and cataloged by the Japanese Meteorological Agency (JMA) as Low Frequency Earthquakes (LFEs). These observations are made primarily on the Hi-Net in Japan, a nationwide network of high-sensitivity borehole seismometers (Obara et al., 2005). The unprecedented density and low noise of the instruments in the Hi-net facilitates the detection of weak signals. LFEs are only rarely identified in regions with tremor outside of Japan (e.g. Kao et al., 2006, Sweet et al., 2008). It is unclear if this difference represents a real variation in tremor activity or simply a limitation in the observation capabilities of networks outside of Japan.

At many time-scales tremor can appear to be very stable, maintaining a fairly constant amplitude for significant amounts of time (Fig. 4) with some waxing and waning of tremor amplitude. At other times, tremor is rather spasmodic, with many bursts that have significantly higher amplitude than the ongoing background tremor (Fig. 4). These bursts can range from less than one minute to tens of minutes. The maximum amplitude of tremor is always relatively small, but appears to vary somewhat from region to region.

Tremor duration is also highly variable. The duration of tremor can range from discrete bursts that last only minutes to ongoing sources that last hours or days (Rogers and Dragert, 2003). During an ETS episode, tremor activity sometimes may continue for days uninterrupted or may also turn on and off erratically throughout the episode. Minor episodes of tremor are routinely observed outside of times of major ETS events. This is also true in California near the town of Parkfield, where correlated slip has not been observed despite excellent detection capabilities provided by borehole strainmeters (Johnston et al., 2006; Smith and Gomberg, in press), in that it is very infrequent that a week goes by without tremor being observed in the Parkfield area.

Watanabe et al. (2007) examined the relationship between duration and amplitude of tremor in southwest Japan, comparing exponential and power law models. They found that the exponential model provided a much better fit, suggesting that tremors, unlike earthquakes, must be of a certain size. As a result, they propose that tremor is generated by fluid processes of a fixed size, or alternatively, that tremor is generated by shear slip on a fault patch of fixed size with variable stress drop.

The spectral content of non-volcanic tremor clearly distinguishes it from earthquakes (Fig. 5), although, at times, non-volcanic tremor can look similar to volcanic tremor. Relative to local earthquakes, tremor is deficient in high frequency energy, in that it has a much steeper drop off of amplitude with increasing frequency. Because of the presence of low-frequency noise and attenuation and smaller source spectra at high frequencies, tremor is most easily identified in a narrow frequency band ranging from approximately 1–10 Hz (Obara, 2002). While energy from tremor undoubtedly extends to a wider frequency range, it is in this frequency range where tremor typically has its highest signal to noise ratio.

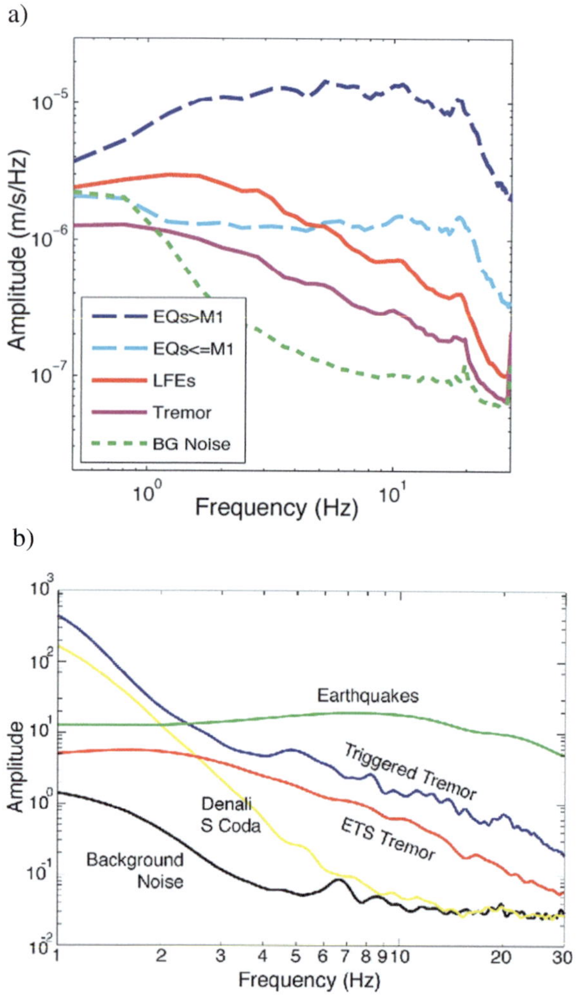

Fig. 5 Velocity spectrum of tremor in Shikoku, Japan (**a**) and Vancouver Island, Canada (**b**). Tremor and local earthquakes have significantly different spectral shape. Triggered tremor (**b**) also has a similar spectral shape as ambient tremor. Figures from Shelly et al. (2007a) (**a**) and Rubinstein et al. (2007) (**b**). We note in (**a**) that the tremor falls below the noise at the lowest frequencies, this is because the noise and tremor were measured at different times and the level of noise during the period of measured tremor was much lower

The tremor wavefield is believed to be dominated by shear waves because it propagates at the S wave velocity and shows higher amplitudes on horizontal components of motion (Obara, 2002; La Rocca et al., 2005). Furthermore, polarization analysis of tremor indicates that tremor is largely composed of shear waves (La Rocca et al., 2005; Wech and Creager, 2007; Payero et al., 2008; Miyazawa and Brodsky, 2008). It seems likely that tremor is generated by a shear source, although fluid based sources can produce shear waves as well (e.g., Chouet, 1988).

Tremor is also highly repeatable with respect to location. Within an individual ETS episode, highly-similar bursts of tremor repeat many times, suggesting that tremor radiates from an individual location many times (Shelly et al., 2007a). From ETS episode to ETS episode, tremor also typically occurs in the same locations (Shelly et al., 2007a; Kao et al., 2006), whereby much of the area where tremor occurs is the same from event to event. Ambient tremor occurring outside ETS events is typically found in these same locations as well.

Most tremor episodes occur spontaneously, but it also can be triggered when the source region is being dynamically stressed by large amplitude teleseismic surface waves (e.g., Miyazawa and Mori, 2005, 2006; Rubinstein et al., 2007; Gomberg et al., 2008). While triggered tremor has been frequently identified in regions where ambient tremor exists, e.g., Parkfield, Vancouver Island, and Japan, it also has been identified in regions where tremor has not previously been identified, e.g., Taiwan and Southern California. It should be noted however, that the existence of ambient tremor in these regions cannot be ruled out because the appropriate studies have not yet been conducted. Similarly, ambient tremor has been found in many regions where triggered tremor has yet to be seen. These incongruities may imply that there are fundamental differences between these regions or processes, or simply that the data in these regions has yet to be thoroughly analyzed.

Locating Non-volcanic Tremor

The very features of the tremor wavefield that make it such a rich phenomena – including the long duration of the source process and absence of distinct body wave arrivals in the seismogram – also make it very difficult to determine where these waves originate. Standard earthquake location methods, like those described below, rely on picking body wave arrivals and most often cannot be used because impulsive arrivals are difficult to find within tremor. Thus, a wide and sometimes novel suite of techniques to locate the tremor source has been developed to exploit some of the unique characteristics of the tremor wave field. These methods largely reproduce the same epicentral locations for tremor, but often have significant differences in the depths (Hirose et al., 2006), whereby some methods suggest that tremor is largely confined to the plate interface in Japan (e.g., Shelly et al., 2006) and other methods indicate that tremor is distributed within a volume of more than 40 km depth in Cascadia (e.g., Kao et al., 2005). The drastic difference in depth distributions of tremor produced by these methods requires significantly different mechanical models to produce tremor in Cascadia and Japan. Thus, precise location of the tremor source in both space and time is a critical step in understanding the mechanics of tremor generation. Doing this will allow us to determine the appropriate physical model for tremor and whether the differences in depth distribution of tremor are real or if they are driven by differences in methodology or data quality.

In general, we can describe the observed seismogram as a convolution of the source process in both space and time with the impulse response of the earth (Green's function) that connects the source positions with the receiver. The resulting seismogram contains a mix of direct body wave arrivals, converted phases and waves scattered by the complex 3D structure of the earth. If the source process has an impulsive beginning it is usually possible to measure the arrival time of the direct P- and S-waves on the seismogram. For earthquakes, this is typically the case and it is then straightforward to estimate the location of the waves' source as is the point that yields the smallest discrepancy between the observed arrival times and those predicted by an appropriate earth model. This is the location of the initial rupture, or hypocenter. Essentially all earthquakes are located in this manner. Commonly, this is done using an iterative least-squares algorithm based on "Geiger's method", the Taylor series expansion of the travel time about a trial hypocenter (Shearer, 1999). This method is attractive, as it only depends on travel time calculations which can be done quickly and efficiently using ray theory. Typically this method cannot be applied to tremor because it often does not have impulsive arrivals that coherently observed at many stations. At the Japan Meteorological Agency, analysts have sometimes been successful in identifying S-waves (and occasionally P-waves) from "low frequency" earthquakes (LFEs) embedded in tremor episodes and locating their hypocenters using these standard methods (Katsumata and Kamaya, 2003).

Waveform Envelope Location Methods

One of the most successful and widely used approaches to locate tremor uses the envelope of the tremor signal to determine the relative arrival times of the waves across a network of stations. First employed by Obara (2002), this method takes advantage of the station to station similarity of smoothed waveform envelopes of high-pass filtered tremor seismograms. Using cross-correlation, one can compute the delay between the envelopes at a pair of stations. The relative arrival times across the network can then be used to locate the tremor source. The errors in the envelope correlation measurements are typically larger than those involved in picking arrival times of earthquakes. Consequently, the location uncertainty is fairly large, particularly for the focal depth, which can exceed 20 km. This method and variants on it are the most commonly used methods to locate non-volcanic tremor (e.g., McCausland et al., 2005; Wech and Creager, 2008; Payero et al., 2008).

Amplitude Based Location Methods

Envelope cross correlation works because the energy output of the tremor source varies with time, waxing and waning on time scales that vary from seconds to minutes. It is reasonable to consider that short-duration periods of high amplitude represent either the constructive interference of waves being radiated from multiple locations in the tremor source or particularly strong radiation from a specific location. In the latter case, it should be possible to exploit both the arrival time and amplitude information to localize the source. Kao and Shan (2004) developed a "source scanning algorithm" to determine the hypocenter by back projection of the observed absolute amplitudes onto the source volume. When the summed wave amplitudes from a network of stations achieve a maximum at a particular location in both space and time, the event hypocenter has been found. The method is closely related to the back projection reconstruction of rupture kinematics of Ishii et al. (2005) used to image the 2004 Sumatra-Andaman Island earthquake. Kao and Shan (2004) have shown that the method compares favorably with conventional methods for locating earthquakes. Since the source scanning algorithm only requires the computation of travel times, and not their partial derivatives, it can be readily implemented in 3D velocity models using an eikonal solver (Vidale, 1988). The epicentral locations computed using this method are similar to those from other methods, with the majority of tremor in Cascadia lying between the surface projections of the 30 and 45 km depth contours of the subduction interface (Kao et al., 2005). They also find tremor at a wide range of depths (>40 km), with errors estimated to be on the order ± 3 and ± 5 km for the epicenters and depth.

Small Aperture Seismic Array Based Location Methods

Seismic arrays (Capon, 1969; Filson, 1975; Goldstein and Archuleta, 1987) offer an attractive alternative to regional seismic networks for making use of the phase and amplitude information in the wavefield to study the tremor source as they have been used to locate earthquakes and study earthquake rupture propagation (Spudich and Cranswick, 1984; Fletcher et al., 2006). Following this logic, many seismic arrays have been deployed to record non-volcanic tremor. The ETS episode of 2004 was well recorded by three small arrays deployed above the tremor source region in the northern Puget Sound region in British Columbia and Washington (La Rocca et al., 2005, 2008). Even with just 6 or 7 stations, the arrays proved capable of measuring the backazimuth and apparent velocity of the dominant signal in the 2–4 Hz band. Triangulation for the source location using the 3 arrays provided rough estimates of the source position that were comparable to those determined from envelope correlation (McCausland et al., 2005). Significantly, P-wave energy was also detected on the arrays arriving at different velocities than the S-wave energy.

Phase Based Location Methods

If discrete phase arrivals could be identified in the tremor seismogram and correlated across a network of seismic stations, it would be possible to apply standard earthquake location methods (e.g., Geiger's method) to locate the tremor source. Using LFEs that have some phase picks, Shelly et al. (2006) improved the LFE locations in southwestern Japan using waveform cross-correlation with a double-difference technique. These well-located events were then used as templates in a systematic cross-correlation-based search of tremor episodes in southwestern Japan (Shelly et al., 2007a). These authors found that a significant portion of the tremor seismogram could be explained by multiple occurrences of LFEs. This result is discussed in greater detail in section "Low Frequency Earthquakes". This procedure of cross correlating a known event with another time interval has also been used with great success in studying earthquakes (Poupinet et al., 1984) and has led to the recognition that many earthquakes are "doublets" or repeating earthquakes (e.g. Nadeau et al., 2004; Waldhauser et al., 2004; Uchida et al., 2007). It should be noted that imperfect matches are still useful, as the relative delay between the reference event and match across the network of stations can be used to locate the two events relative to one another (see Schaff et al., 2004), potentially providing a very high resolution image of the tremor source region. The search for template events outside of Japan is an area of ongoing effort by a number of research groups. As of this writing, these efforts have met with limited success. We should note that current templates do not explain all of the tremor signals in Japan either. Brown et al. (2008) has worked to address these limitations using an autocorrelation technique to identify repeating tremor waveforms to use as templates.

Another opportunity to improve tremor locations is to identify *P* waves or compute *S-P* times, as most methods purely use *S* wave arrivals. La Rocca et al. (2009) retrieve *S-P* times by cross-correlating the vertical component of recordings of tremor against the horizontal components. This method relies on the assumption that the tremor arrives at near-vertical incidence so that the *P* waves are predominantly recorded on the vertical component and the *S* waves are predominantly on the horizontal component. Using these newly computed *S-P* times, La Rocca et al. (2009) dramatically improve the vertical resolution of tremor locations in Cascadia. For the events that they locate, tremor appears to lie on or very close to the subduction interface.

The Future of Tremor Location

Despite the progress being made in localizing the tremor source, much work remains to be done. With the exception of locations based on template events and *S-P* times, the location uncertainties are currently much larger than those routinely achieved for earthquakes. In general, the tremor epicenters are much better determined than the focal depths, but even epicentral estimates provided by the different methods do not necessarily agree. Other opportunities would include trying to locate tremor as a line or areal source. While much remains to be done, there are ample opportunities for improving upon the existing analysis methods, implementing new techniques, and gathering data in better ways.

Ideally, we would like to image the tremor source process in both space and time as is now commonly done for earthquakes (Hartzell and Heaton 1983). However, the use of the full waveform for studying the tremor source process is hampered by inadequate knowledge of the path Green's function at the frequencies represented in non-volcanic tremor. Knowledge of this information would allow correcting for the Green's function and determining the true source-spectrum of tremor. Learning about the true source spectrum, would undoubtedly teach us a lot about the source processes of non-volcanic tremor.

Developing a Physical Model for Tremor

In this section, we aim to elucidate the physical processes underlying non-volcanic tremor. There are two predominant models to explain the mechanics of non-volcanic tremor: (1) tremor is a result of fluid-flow and fluid processes at the plate interface and within the overlying plate; and (2) tremor is a frictional process that represents failure on a fault with rupture speeds that are much lower than earthquakes. In the following section we will first discuss the evidence for the fluid based model for non-volcanic tremor. We then present two case studies, examining where and why tremor occurs. The evidence from these case studies suggests that the frictional model, explains some attributes of non-volcanic tremor that the fluid-flow model does not. We note that the frictional models, often still appeal to high fluid pressures and the presence of fluids to explain their observations.

In the first case study, we focus our attention on Japan, where diverse and active subduction along with high-quality data has provided an excellent natural laboratory. These conditions have helped lead to the identification and location of tremor and other slow events on a variety of times scales in southwestern Japan. Growing evidence suggests that these events represent plate convergence shear failure on the subduction interface in the transition zone.

In the second case study, we examine tremor activity triggered by tiny stress perturbations from tides and distant earthquakes. These observations can tell us about the conditions under which tremor occurs, and they indicate a sensitivity to stress far beyond what is seen for earthquakes at comparable depths. This argues that tremors probably occur on faults that are very close to failure, which might be achieved if expected high confining pressures are mitigated by near-lithostatic pore fluid pressures.

The Fluid Flow Model for Non-volcanic Tremor

At the time he discovered non-volcanic tremor, Obara (2002) argued that tremor might be related to the movement of fluid in the subduction zone. The depths at which tremor is believed to occur is consistent with depths where significant amounts of subduction related dehydration from basalt to eclogite is occurring (Peacock and Wang, 1999; Julian 2002; Yoshioka et al., 2008), so large amounts of fluid could be present at or near the plate interface. High fluid pressures could then change the fracture criterion of the rock, thus causing hydraulic fracturing, which would radiate the tremor (Obara, 2002). Obara (2002), then goes on to suggest that long-durations of tremor could be a sequence of fractures that are opening as a chain reaction. Other work, examining the stress regime in which tremor is occurring supports the notion that tremor is a product of hydraulic fracturing (Seno, 2005). Others have argued that non-volcanic tremor is caused by brine resonating the walls of fluid conduits near the plate interface (Rogers and Dragert, 2003). This is quite similar to fluid oscillation models for tremor seen at volcanoes (Chouet, 1988; Julian, 2000). Considering the similarities between non-volcanic and volcanic tremor, we expect that much can be learned by comparing the two processes.

Focal mechanism analysis of one burst of non-volcanic tremor in Japan showed that the tremor appeared to be the result of a single-force type source mechanism, which is consistent with fluid flow and not frictional slip (Ohmi and Obara, 2002). This is in contrast with studies of low frequency earthquakes that indicate that tremor appears to be a double-couple source (i.e. shear on a plane) (Ide et al., 2007a; Shelly et al., 2007a).

Additional evidence that non-volcanic tremor is related to fluid flow comes from the distribution of depths where tremor is identified. Studies from both Japan and Cascadia have determined that tremor depths range more than 40 km (e.g, Kao et al., 2005; Nugraha and Mori, 2006). The locations where the tremor is generated in Cascadia correspond well with high-reflectivity regions believed to have fluids (Kao et al., 2005). If tremor is distributed at this wide range of depths, fluid movement seems a much more viable mechanism to produce tremor than slip, as it seems much more likely for there to regions of fluid distributed widely than regions of slip. As discussed in section "Locating Non-volcanic Tremor" and later in section "Tremor Locations: A Broad Depth Distribution in Some Areas?", other studies suggest that tremor is being radiated from the plate interface and does not have a large depth distribution (La Rocca et al., 2009; Shelly et al., 2006; Brown et al., in press). Clearly, precisely determining tremor locations is critical for our understanding of the source processes of tremor.

Case Study I: Non-volcanic Tremor in Japan

Since its discovery in southwest Japan (Obara, 2002), non-volcanic tremor has been extensively studied using high-quality data from the Hi-net borehole seismic network, operated by the National Research Institute for Earth Science and Disaster Prevention (NIED) (Obara, 2005). Hi-net data is supplemented by numerous surface stations operated by the Japan Meteorological Agency (JMA), individual universities, and other agencies. Using Hi-net data, Obara (2002) located the tremor source by waveform envelope cross-correlation and found that the epicenters occurred in a band corresponding to the 35–45 km depth contours of the subducting Philippine Sea Plate in the Nankai Trough (Fig. 1). This band extends from the Bungo Channel in the southwest to the Tokai region in the northeast. Gaps in this band, such as that beneath the Kii Channel, may correspond to where a fossil ridge is being subducted resulting in an area that lacks hydrated oceanic crust (Seno and Yamasaki, 2003).

Following the discovery of ETS in Cascadia (Rogers and Dragert, 2003), Obara et al. (2004) established a similar relationship between tremor and slow slip in Nankai Trough using precise measurements of tilt (Obara et al., 2004). Based on these measurements, slow slip events were modeled to occur on the plate interface, downdip of the seismogenic zone, with durations of ~1 week and equivalent moment magnitudes near 6.0. The locations of slip matched with epicentral locations of tremor, but it was not clear whether the depth of the tremor source matched the depth of slow slip.

Low Frequency Earthquakes

The discovery of low-frequency earthquakes (LFEs) in Southwest Japan (Katsumata and Kamaya, 2003) has led to significant progress in our understanding of tremor processes, including markedly reducing the uncertainty in tremor depths. In Japan, LFEs are routinely identified by the JMA and included in the seismic event catalog. Although some of these events are volcanic, many come from regions far from active volcanoes and are, in fact, relatively strong and isolated portions of non-volcanic tremor. Using mostly S-wave arrival times (few P-wave arrivals are determined for LFEs), JMA estimates the hypocenter and origin time for each event, although the locations generally have large uncertainty, especially in depth. Based on these catalog locations, it was unclear whether the tremor was emanating from the megathrust, within the Wadati-Benioff zone immediately below, or within the upper plate. Drawing from analogies with volcanic

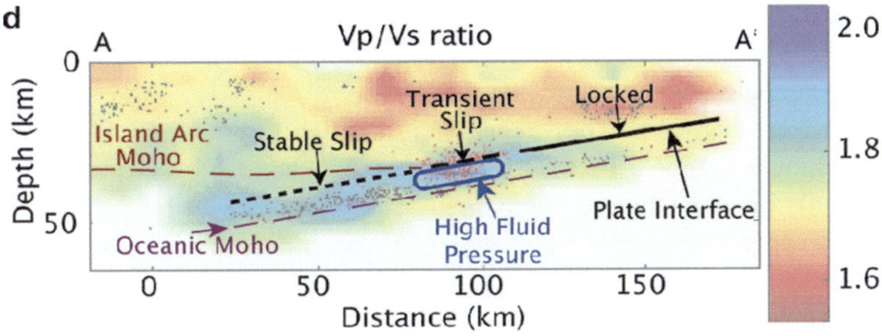

Fig. 6 Cross-section showing hypocenters, Vp/Vs ratios, and structures in western Shikoku. *Red dots* represent LFEs while *black dots* are regular earthquakes. Figure from Shelly et al. (2006)

tremor, initial models of tremor generation proposed that tremor and LFEs might be due to fluid flow near the upper plate Moho (Julian, 2002; Katsumata and Kamaya, 2003; Seno and Yamasaki, 2003).

Shelly et al. (2006) located LFEs and tectonic earthquakes in western Shikoku using waveform cross-correlation and double-difference tomography (Zhang and Thurber, 2003). They found that waveform similarity among LFEs was strong enough to provide accurate differential time measurements, and thus very good focal depth determinations in this region. These locations showed LFEs occurring in a narrow depth range, approximately on a plane dipping with the expected dip of the subducting plate (Fig. 6). These events located 5–8 km shallower than the Wadati-Benioff zone seismicity, and were interpreted as occurring on the megathrust. Based on these locations and the observed temporal and spatial correspondence between tremor and slow slip, Shelly et al. (2006) proposed that LFEs were likely generated directly by shear slip as part of much larger slow slip events, rather than being generated by fluid flow as had been previously suggested.

Support for this hypothesis was provided by Ide et al. (2007a), who determined a composite mechanism for LFEs in western Shikoku using two independent methods. Although the small size of LFEs would normally prevent such an analysis, Ide et al. (2007a) stacked LFE waveforms to improve the signal-to-noise ratio and also utilized waveforms of intraslab earthquakes of known mechanism. Results from an empirical moment tensor using S-waves as well as the mechanism from P-wave first motions both showed motion consistent with slip in the plate convergence direction (Fig. 7). Thus, the kinematics of LFEs appeared to be very similar to regular earthquakes.

Although the above analyses provided strong evidence for the mechanism of LFEs, the relationship between LFEs and continuous tremor was uncertain. Shelly et al. (2007a) argued that the extended duration of tremor could be explained by many LFEs occurring in succession. To identify this correspondence, they used waveforms of catalog LFEs as templates in a matched filter technique applied simultaneously

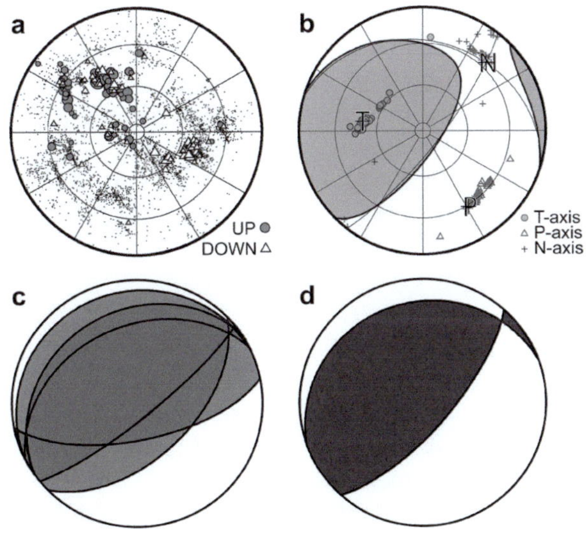

Fig. 7 Comparison of LFE, slow slip event, and megathrust earthquake mechanisms. (**a**) P-wave first motions determined by Ide et al. (2007a) for low frequency earthquakes by cross correlation-based first motion determination. Solid circles and open triangles indicate compressional and dilatational first motions for LFE *P* waves, respectively. SNR for most observations (*small dots*) is too low to determine the polarity. (**b**) Moment tensor inversion results from empirical Green's function analysis of LFE *S* waves. T-, P-, and N-axes are shown together with symbols showing uncertainty and corresponding *P*-wave first motion distribution. (**c**) Overlay of the mechanism for three slow slip events near the study area. (**d**) Mechanism of the 1946 Nankai earthquake, which is the most recent megathrust earthquake in this region and representative of relative plate motion between the Philippine Sea Plate and the overriding plate on the dipping plate interface of the Nankai Trough subduction zone. All these figures are shown in equal area projection of lower focal hemisphere. Figure from Shelly et al. (2007a)

across multiple stations and components (Gibbons and Ringdal, 2006). They found that significant portions of tremor could be matched by the waveforms of a previously recorded LFE. They concluded that, like LFEs, continuous tremor in southwest Japan is also generated directly by shear slip as a component of the larger slow slip events. Importantly, this technique also provided a means to locate this tremor more precisely in space and time.

The successful matching of LFE and tremor waveforms implies that tremor recurs in the same location (or very nearby) during a single ETS episode. Analyzing a two week long ETS episode in western Shikoku, Shelly et al. (2007b) showed that even during a given episode, tremor is generated repeatedly in roughly the same location. In particular, certain patches of the fault, where clusters of LFEs locate, appear to radiate strong tremor in intermittent bursts. The authors suggested that the region of the fault surrounding these patches may slip in a more continuous fashion during an ETS event, driving the LFE patches to repeated failure in a model somewhat analogous to that proposed for repeating earthquakes (Schaff et al., 1998; Nadeau and McEvilly, 1999).

Tremor Migration

Several studies have examined the spatial and temporal evolution of tremor in southwest Japan and found that systematic migration is common. Obara (2002) reported migration of the tremor source along the subduction strike direction at rates of 9–13 km/day, over distances approaching 100 km. Tremor and slip were later seen to migrate together along strike, always at rates of ~10 km/day (Obara et al., 2004; Hirose and Obara, 2005). Along-strike migration directions do not appear to be consistent and migration sometimes occurs bilaterally or activity appears to stall or jump. Similar along-strike migration characteristics have also been reported in Cascadia (Dragert et al., 2004; Kao et al., 2007b).

In addition to relatively slow, along-strike migration, a much faster tremor migration, occurring primarily in the subduction dip direction, was reported by Shelly et al. (2007a, b). Locating tremor by the template LFE method (described above) greatly improved the temporal resolution of tremor locations, allowing locations on a timescale of seconds. Activity was seen to repeatedly migrate up to 20 km at rates of 25–150 km/h, orders of magnitude faster than the observed along-strike migration rates, yet still orders of magnitude slower than typical earthquake rupture velocities. As with the along-strike migration, no preferential direction was observed for along-dip migration. Tremor activity could be seen to propagate updip, downdip, and bilaterally. The downdip migration examples, coupled with relatively fast migration rates, make it unlikely that fluid flow accompanies the tremor. Although it is unclear what generally prevents similar migration velocities in the along-strike direction, a subtle segmentation of the plate boundary, perhaps due to a corrugation in the slip direction, was suggested as a possibility (Shelly et al., 2007b). A similar hypothesis has been proposed to explain streaks of seismicity on faults (Rubin et al., 1999).

A Wide Range of Slow Events

Ito et al. (2007) discovered another new source process occurring along the southwest Japan subduction zone using long period, 20–50 s waveforms. These events, with estimated durations of ~10 s and seismic moment magnitudes of 3.1–3.5, were termed very low frequency (VLF) earthquakes. Timing of these events corresponded with tremor and slow slip. In fact, each VLF was accompanied by a tremor burst in the 2–8 Hz frequency band, but not all tremor bursts were accompanied by detectible VLF events. Focal mechanisms showed thrust faulting, leading to the conclusion that VLFs were also generated by shear slip in the plate convergence direction.

Given the growing number of kinds of shear slip events that occur in the transition zone in southwest Japan (Fig. 8), Ide et al. (2007b) proposed that these events, ranging in duration from ~1 s (LFEs) to years (long-term slow slip), belonged to a single family. This family was unified by a scaling law in which moment scales linearly with duration, rather than as duration cubed as for ordinary earthquakes (Fig. 9). While observations constrain the region between slow events and ordinary earthquakes to be essentially empty, events slower than the proposed scaling relation for a given magnitude might exist beyond the current limits of detection. After this relation was proposed, Ide et al. (2008) detected events predicted by

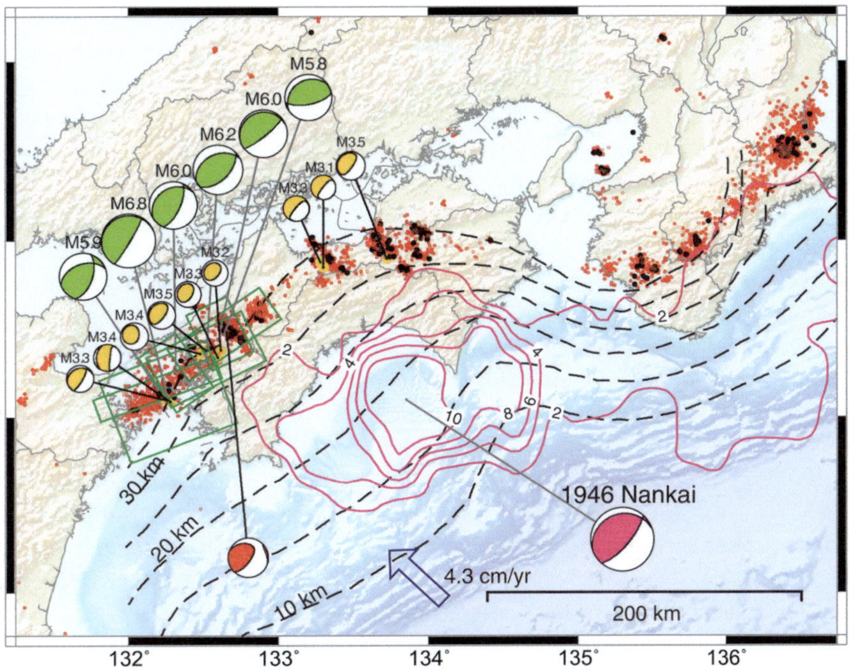

Fig. 8 Various types of earthquakes and their mechanisms along the Nankai Trough, western Japan. *Red dots* represent LFE locations determined by Japan Meteorological Agency. *Red and orange beach balls* show the mechanism of LFEs and VLFs, respectively. *Green rectangles* and *beach balls* show fault slip models of SSE. *Purple contours* and the *purple beach ball* show the slip distribution (in meters) and focal mechanism of the 1946 Nankai earthquake (M8). The top of the Philippine Sea Plate is shown by *dashed contours*. *Blue arrow* represents the direction of relative plate motion in this area. Figure from Ide et al. (2007b)

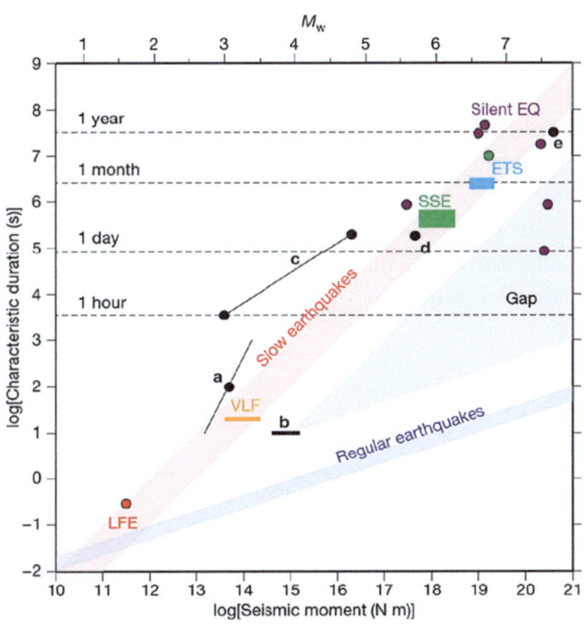

Fig. 9 LFE (*red*), VLF (*orange*), and SSE (*green*) occur in the Nankai trough while ETS (*light blue*) occur in the Cascadia subduction zone. These follow a scaling relation of M_0 proportional to t, for slow earthquakes. *Purple circles* are silent earthquakes. Black symbols are slow events. *a* Slow slip in Italy, representing a typical event (*circle*) and proposed scaling (*line*). *b*, VLF earthquakes in the accretionary prism of the Nankai trough. *c*, Slow slip and creep in the San Andreas Fault. *d*, Slow slip beneath Kilauea volcano. *e*, Afterslip of the 1992 Sanriku earthquake. Typical scaling relation for shallow interplate earthquakes is also shown by a *thick blue line*. Figure from Ide et al. (2007b)

the scaling law with a source duration of 20–200 s and moment magnitude 3–4 under the Kii Peninsula. Such events at these long durations may be common but are difficult to detect due to noise levels and the domination of near-field terms that decay with squared distance. These ~100 s events exhibit a close correspondence between moment rate and high-frequency radiated energy, providing a link between the larger, longer-duration events detected geodetically and smaller shorter-duration events detected seismically.

Case Study II: Stress Interactions of Tremor with Other Earth Processes

Since the discovery of non-volcanic tremor, authors have been interested in the stress interactions between non-volcanic tremor and other earth processes. The periodic nature of ETS makes it easy to connect earth processes to it. For example, the 14-month periodicity of ETS in Northern Cascadia has the same periodicity as the Chandler Wobble (also called the pole-tides). Based on this connection, some have argued that the small gravitation changes associated with the Chandler Wobble are responsible for the periodicity of ETS in Cascadia (Miller et al., 2002; Shen et al., 2005). Similar claims have been made for ETS in Mexico and

Japan, where climatic loading has been argued as the source of the ~12 and ~6 month periodicities of ETS in those locations respectively (Lowry, 2006). However the wide range of dominant ETS periods, from 3 to 20 months in different regions, suggests that outside forcing is, at most, a secondary factor.

A much clearer impact on tremor activity results from small stress changes from distant and local earthquakes as well as the earth and ocean tides. With the aim of elucidating the physical processes underlying non-volcanic tremor, we examine these weak stress perturbations and their effect upon non-volcanic tremor and ETS activity.

Earthquakes Influencing Tremor

Strong evidence suggests that non-volcanic tremor can be influenced by local and distant earthquakes both dynamically, where it is instantaneously triggered by the passage of seismic waves, and in an ambient sense, where periods of active tremor appear to be started or stopped by an earthquake.

Along with the discovery of non-volcanic tremor, Obara (2002) identified the interaction of self-sustaining tremor and local earthquakes. Specifically, periods of active tremor are observed to both turn on and turn off shortly following local and teleseismic earthquakes (Obara, 2002, 2003). An increase in tremor rates is also seen following two strong earthquakes in Parkfield, CA (Nadeau and Guilhem, 2009). A similar observation has been made in Cascadia, where ETS episodes that are "late" appear to be triggered by teleseismic earthquakes (Rubinstein et al., 2009). The interpretation of these observations is complex. For local and regional events, the change in the static stress field caused by the earthquake could be large enough to either start or stop a period of enhanced tremor activity. For teleseismic events, the changes in static stress will be negligible, such that the dynamic stresses associated with them must somehow start or stop a period of enhanced tremor. Rubinstein et al. (2009), propose that when a region is particularly loaded, the small nudge that the dynamic stresses from a teleseismic earthquake provide are enough to start an ETS event going. No satisfactory model has been proposed to explain how a teleseismic event might stop a period of active tremor.

The other mode in which tremor can be influenced by earthquakes is instantaneous triggering by the strong shaking of an earthquake. The first observations of instantaneous triggering of tremor come from Japan, where high-pass filtering broadband records of teleseismic earthquakes showed that there is tremor coincident with the large surface waves (Obara, 2003). Further study identified that tremor was instantaneously triggered by a number of different earthquakes in Japan (Miyazawa and Mori, 2005; 2006). Most observations of triggered tremor are triggered by surface waves, but in at least one case tremor has been observed to have been triggered by teleseismic P waves (Ghosh et al., in press(a)). While triggered tremor is typically larger than self-sustaining tremor, the spectrum of triggered tremor is very similar to that of regular tremor, suggesting that they are the same process (Rubinstein et al., 2007; Peng et al., 2008).

Careful analysis of the phase relationship between the surface waves from the Sumatra earthquake and the tremor it triggered in Japan shows that the tremor is very clearly modulated by surface waves. The tremor turns on when there are positive dilatations associated with the Rayleigh waves and turns off when the dilatation is negative (i.e. during compression) (Miyazawa and Mori, 2006) (Fig. 10). Miyazawa and Mori (2006) interpret this to mean that tremor is related to pumping of fluids from changes in pore space, which might induce brittle fracture and thus generate tremor. Observations of tremor on Vancouver Island triggered by the

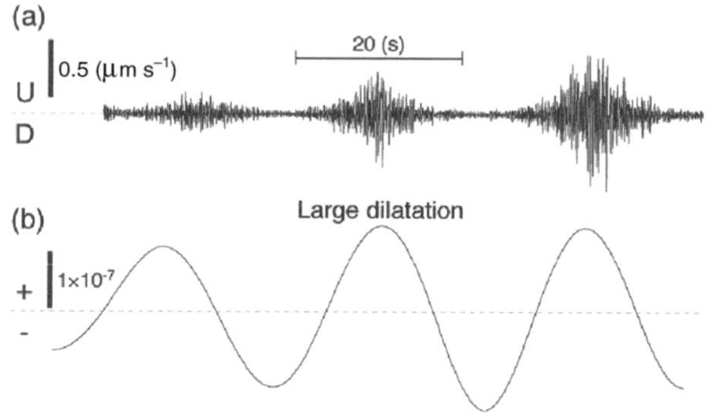

Fig. 10 Figure comparing non-volcanic tremor triggered by the Sumatra earthquake (**a**) to dilatations from the Rayleigh waves from that same earthquake (**b**). Traces have been adjusted to reflect the timing and cause and effect relationship between the surface waves and the tremor. Figure modified from Miyazawa and Mori (2006)

Denali earthquake show instead that tremor is clearly triggered by the Love waves, which have no dilatational component (Rubinstein et al., 2007) (Fig. 11). Rubinstein et al. (2007) offer an alternative explanation for this process, that increased coulomb failure stress from the teleseismic waves promotes slip on the plate interface. They show that when shear stress from the Love waves encourages slip on the plate interface, tremor turns on and when it discourages slip, tremor turns off. This also is supported by the apparent modulation of the triggered tremor amplitude by the shear stress amplitude (Rubinstein et al., 2007), which is predicted by modeling of Coulomb based triggering (Miyazawa and Brodsky, 2008). This behavior is not observed for all observations of triggered tremor (e.g., Peng et al., 2008; Rubinstein et al., 2009). Rubinstein et al. (2007) also argue that this model can explain the observations of tremor being modulated by dilatation (Miyazawa and Mori, 2006), in that increased dilatation also results in a reduction of the Coulomb Failure Criterion on the fault, and should thus encourage slip. Further study of the tremor triggered by the Sumatra earthquake in Japan shows that either model, frictional failure or pumping of fluids can explain the phasing of the tremor with the surface waves (Miyazawa and Brodsky, 2008). It has further been suggested that the difference in triggering behaviors in Cascadia and Japan may be related to the effective coefficient of friction, implying that fluid pressure may be higher in Cascadia than in southwest Japan (Miyazawa et al., 2008)

Tremor triggered at teleseismic distances by large earthquakes offers a powerful tool for identifying additional source regions. For example, tremor was triggered in 7 locations in California by the 2002 Denali earthquake (Gomberg et al., 2008) and underneath the Central Range in Taiwan by the 2001 Kunlun earthquake (Peng and Chao, 2008). With the exception of the tremor triggered in the Parkfield region of California, these observations of triggered tremor are in locations where tremor had never been observed previously. Notably, none of these source regions are in subduction zones. This suggests that tremor is much a much more common process than previously thought and is not limited to subduction zones. Furthermore, these findings indicate that the necessary conditions for producing non-volcanic tremor must exist in a wide variety of tectonic environments.

Computations of shear stress change imparted by teleseismic earthquakes are on the order of tens of kPa (Hill, 2008), or about 10^5 times smaller than the expected confining pressures at these depths. Based on this, some have argued that tremor, at least in its triggered form, occurs on faults that are extremely close to

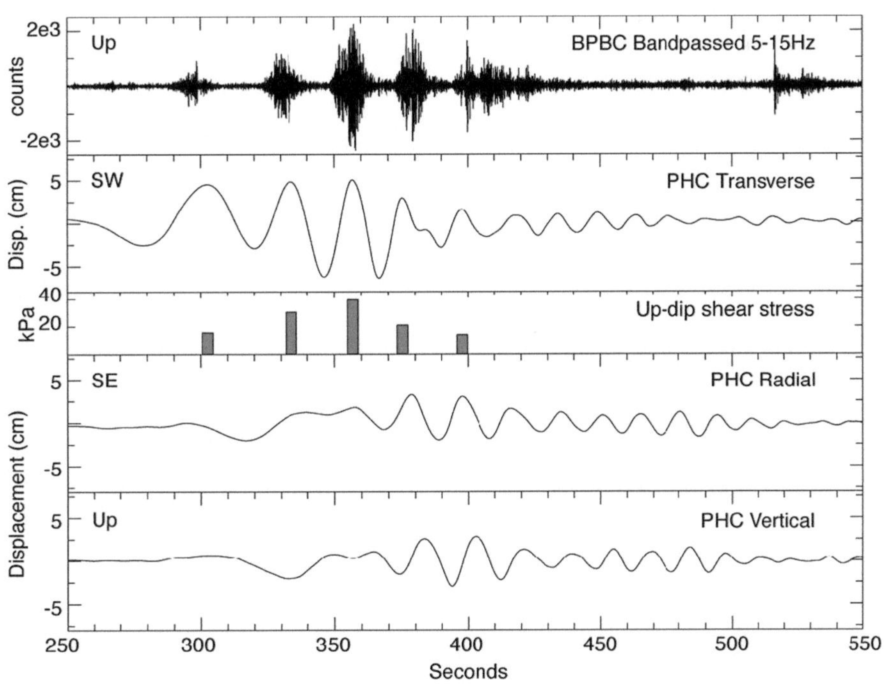

Fig. 11 Figure comparing non-volcanic tremor triggered by the Denali earthquake (a) to surface waves from that same earthquake (b, d, e). Traces have been adjusted to reflect the timing and cause and effect relationship between the surface waves and the tremor. Middle panel reflects the approximate peak shear stress on the plate interface enhancing slip in a subduction sense from the five largest Love wave pulses. Figure modified from Rubinstein et al. (2007)

failure, possibly because of near lithostatic fluid pressures (Miyazawa and Mori, 2006; Rubinstein et al., 2007; Peng and Chao, 2008).

Despite the incremental stresses associated with teleseismic earthquakes, the presence of triggered tremor appears to be strongly controlled by the amplitude of the triggering waves in Parkfield (Peng et al., 2009) and less so on Vancouver Island (Rubinstein et al., 2009). While the amplitude of triggering waves is clearly important in determining whether tremor will be triggered, many other factors are likely important. These include the presence of an ongoing ETS episode or elevated levels of tremor (Rubinstein et al., 2009), frequency content, and azimuth of the earthquake. We also note that while amplitudes and therefore dynamic stresses associated with local and regional, medium-magnitude events are of similar amplitude as those from teleseismic earthquakes, tremor triggered by these events, if it occurs, cannot be observed easily as it is obscured by the larger, high frequency energy associated with the body waves and coda from these events (Rubinstein et al., 2009).

The Tides Influencing Tremor

The periodic changes in gravitation caused by the moon and the sun (the lunar and solar tides) are frequently employed by the earth science community as a way to better understand earth processes. It seems quite logical that when the small stresses associated with the tides encourage slip on faults, seismicity should increase and conversely it should decrease when the tidal stresses discourage slip on these same faults. While a number of studies have identified a very weak correlation between the tides and seismicity rates in particularly favorable conditions (e.g., Tanaka et al., 2002; Cochran et al., 2004, Wilcock, 2001), careful studies of large data sets find no significant correlation of the tides and earthquakes (e.g. Vidale et al., 1998, Cochran and Vidale, 2007).

In contrast to the results from earthquakes, non-volcanic tremor in Japan, Cascadia and Parkfield have been seen to respond strongly to tidal forcing. A comparison of the hourly tremor durations in eastern Shikoku for two ETS events shows that tremor duration is strongly periodic at the two strongest tidal forcing periods of 12.4 and 24 h (Nakata et al., 2008). Examining LFEs in the same location, Shelly et al. (2007b) also determined that non-volcanic tremor is strongly periodic with the lunar tide of 12.4 h and more weakly periodic with the lunisolar tide of 24–25 h. Similarly, a study of non-volcanic tremor in Cascadia showed that the amplitude of tremor in three ETS episodes was strongly periodic at both the 12.4 and 24–25 h tidal periods (Rubinstein et al., 2008). Nadeau et al. (2008), also identify a periodicity to non-volcanic tremor in Parkfield that indicates that it is influenced by the tides.

The periodicity of tremor is such that in both Japan and Cascadia, it is more energetic with high water (Shelly et al., 2007b; Rubinstein et al., 2008). Lambert et al. (2009) similarly show that tremor levels in Cascadia are highest when the normal stress on the plate interface is highest, although this time also corresponds to the time where shear stresses encouraging thrust slip are largest. Neither of the papers that identify this correlation of tremor with water level compute the specific stresses on the fault plane, but they comment that the stresses induced by the tides are miniscule compared to the confining stress of the overburden. Rubinstein et al. (2008) estimates the confining pressures to be approximately 10^5 times larger than ~10 kPa stresses induced by the tides. Nakata et al. (2008) estimate the peak change in Coulomb stress from the solid-earth tides to be ~1 kPa with a maximum rate of ~10 kPa/day, assuming that it occurs as shear slip on the plate interface. Using this computation they find that the temporal behavior of tremor strongly parallels the predictions of a rate-and-state model that predicts seismicity rate changes given a changing stress field. They also note that the stressing rate from the slow-slip event is comparable to the stressing rate from the tides and argue that the tides only should affect tremor if the slow-slip stressing rate is similar in amplitude as the tidal stressing rate.

All of these observations support the argument that tremor is being produced by faults that are very close to failure because they are extremely weak or under near-lithostatic fluid pressures (Nakata et al., 2008, Shelly et al., 2007b, Rubinstein et al., 2008). This parallels the observation that the faults that produce tremor must be at least an order of magnitude more sensitive to triggering than regions where earthquake swarms are produced (Nakata et al., 2008).

Theoretical Models of Slow Slip (and Tremor)

Several studies have attempted to model subduction zone slow slip using a variety of theoretical models in order to constrain the underlying physical mechanisms. Most of these studies do not attempt to model tremor, but rather focus on simulating slow slip. In order for the event to remain slow, the frictional resistance of the sliding surface must increase as the slip velocity increases. In other words, some form of velocity-strengthening friction must put the brakes on slip to keep it from accelerating and becoming an earthquake. The models discussed below all simulate slow slip behavior, but do so through different mechanisms.

Yoshida and Kato (2003) reproduced episodic slow-slip behavior using a two-degree of freedom block-spring model and a rate-and-state friction law. They argue that temporal and spatial variation in stress and frictional properties are necessary conditions for ETS. Similarly, Kuroki et al. (2004) and Hirose and Hirahara (2004) use more complex numerical simulations of slip, and find that they require spatial heterogeneity in frictional properties to be able to reproduce ETS.

Shibazaki and Iio (2003) imposed a rate and state dependent friction law with a small cutoff velocity, such that behavior in the transition zone is velocity weakening at low slip velocity and velocity strengthening at high slip velocity. Such a slip law naturally generates slow slip behavior and may be supported by laboratory data for halite (Shimamoto, 1986) and quartz gouge (Nakatani and Scholz, 2004), under certain conditions. However, it's unclear whether more realistic lithologies, temperatures, and slip speeds behave the same way (Liu and Rice, 2005). Shibazaki and Shimamoto (2007) used a similar approach to specifically model short-term slow slip events. They successfully reproduced slow slip events with propagation velocities of 4–8 km/day, similar to what is observed in Cascadia and southwest Japan and find that this propagation velocity scales linearly with slip velocity in their models.

Liu and Rice (2005, 2007) took a somewhat different approach to achieving slow slip behavior in their rate-and-state-based models. They were able to reproduce transients with a recurrence interval of about a year using laboratory-based friction values with temperature dependence and inserting a region of width W with very high pore pressures updip from the stability transition. They found that the slip behavior primarily depended on the value of the parameter W/h^*, where h^* represents the maximum fault size that produces stable sliding under conditions of velocity-weakening friction. This is similar to the findings of Hirose and Hirahara (2004), who are able to produce slow-slip in a rate and state based model and find a dependence of the slip behavior on the ratio of the width of the slipping region to its lateral dimension. In modeling of Liu and Rice (2007), the recurrence interval of slow slip events decreases with increasing effective normal stress. An effective stress of ~2–3 MPa produces a recurrence interval of 14 months, corresponding to that observed in northern Cascadia.

Another alternative is that dilatant stabilization may play an important role in regulating slow slip and/or tremor behavior, as proposed by Segall and Rubin (2007) and Rubin (2008). They argue that the fault size constraints of the model of Liu and Rice (2005, 2007) may be too specific given the apparent abundance of slow slip events in subduction zones (Rubin and Segal, 2007). Dilatancy that accompanies shear slip will tend to create a suction and thus reduce pore fluid pressure in the fault zone. Depending on the slip speed and permeability, dilatant strengthening could allow slip to occur at slow speeds but prevent it from reaching dynamic speeds typical of earthquakes. If pore fluid pressures in the fault zone approach lithostatic, as has been suggested, the effect of dilatancy becomes relatively more important in controlling slip behavior. In this model, regions of particularly high permeability could slip faster than those with lower permeability, potentially generating tremor.

Many properties of slow slip and tremor can also be explained with a Brownian walk model, where the radius of a circular fault expands and contracts according to this random process (Ide, 2008). Although this model does not address the underlying physical mechanisms, it successfully reproduces the observed frequency content, migration, and scaling of tremor and slow slip, predicting a slight modification to the scaling law proposed by Ide et al. (2007b).

Few laboratory experiments designed to simulate tremor and slow slip have been performed thus far. One recent study by Voisin et al. (2008) examined the effect of cumulative slip on a NaCl sample designed to emulate the frictional conditions in a subduction zone. Although it's unclear how closely this analog represents real conditions of a subduction zone, the experiment succeeded in producing a transition from stick slip behavior, to slow slip, and finally to steady-state creep with increasing cumulative displacement. In addition, they recorded a seismic signal that was qualitatively very tremor-like. They note that the change in behavior with the evolution of their sample is consistent with some features of the modeling discussed above, namely near-neutral stability (Yoshida and Kato, 2003; Liu and Rice, 2005) and a large slip-weakening distance (Shibazaki and Iio, 2003; Kuroki et al., 2004). Tremor-like signals have also been observed in dehydration experiments (Burlini et al., 2009), suggesting that tremor may arise from fluid induced micro-crack propagation and fluid interaction with crack walls, or that metamorphic dehydration reactions supply the fluid necessary to reduce effective pressure and allow tremor to occur.

Clearly additional laboratory experiments specifically designed to study non-volcanic tremor would be important. Considering that laboratory studies have helped reveal many new facets of earthquakes and brittle failure, we expect that laboratory studies will also allow for great insight into the physical processes underlying non-volcanic tremor and slow-slip. Particularly useful will be laboratory simulations that explore the varying conditions expected where tremor is generated (lithology, temperature, pressure, fluid pressure). This parameter space hasn't been thoroughly explored because earthquakes are not abundant in these conditions.

Discussion and Outstanding Questions

We are only beginning to understand the mechanism and environment that produces tremor. Many questions remain unanswered. Following is a discussion of some of the outstanding issues that are topics of ongoing research.

Understanding Why Tremor Occurs in Certain Places

By now, we are beginning to constrain where tremor does and does not occur. By examining the physical conditions in each of these regions including the depth, temperature, mineralogy and metamorphic state, we may succeed in deducing those conditions that are essential for tremor and thereby learn about the source process. We first compare two different tectonic environments where tremor is observed (subduction and strike-slip faults). We then compare the two similar tectonic environments – southwest and northeast Japan – one where tremor is observed and the other where it is not.

An interesting comparison can be made between strike-slip and subduction tremor-hosting environments. The best-documented strike-slip examples are beneath the San Andreas Fault near Parkfield in central California (Nadeau and Dolenc, 2005) and rare instances of activity beneath the source region of the 2000 Western Tottori earthquake in southwest Japan (Ohmi and Obara, 2002; Ohmi et al., 2004). Tremor triggered by teleseismic waves from the Denali earthquake has been observed in several places in California in addition to Parkfield (Gomberg et al., 2008), as discussed above, but tremor has not yet been investigated at other times at these other locations.

Although the subduction and strike-slip environments that generate tremor may appear quite different, some common features are clear. In each case tremor activity occurs below the crustal seismogenic zone of the major fault. These regions appear to correspond to the transitions from stick slip (earthquake-generating) to stable sliding portions of the fault. In subduction zones, the region of tremor and slow slip corresponds to depths where fluids are expected to be liberated from the subducting slab through metamorphic reactions (e.g. Hacker et al., 2003; Yamasaki and Seno, 2003), although varying thermal structures between different regions suggests that tremor does not correspond to a single metamorphic reaction (Peacock, 2009). Seismic studies support the existence of elevated fluid pressures near the tremor in southwest Japan (Kodaira et al., 2004, Shelly et al., 2006, Nugraha and Mori, 2006, Wang et al., 2006, Matsubara et al., 2009), Cascadia (Audet et al., 2009), and Mexico (Song et al., 2009). Furthermore, some have argued that two prominent gaps in tremor in Japan are due to the lack of dehydration reactions and the associated high fluid pressures above them (Seno and Yamasaki, 2003; Wang et al., 2006). Indeed, numerical models of slow slip (see below) often invoke near-lithostatic fluid pressures and thus very low effective stress. Unlike subduction zones, strike-slip faults do not necessarily have an obvious source of fluids. At least for the San Andreas Fault, however, Kirby et al. (2002) have proposed that the fossil slab from previous subduction in this region-may still provide a fluid source. Although fluids might be a necessary condition, they do not appear to be sufficient. For example, no tremor has been reported in hydrothermal areas such as the Geysers, California, Long Valley, California, and Coso Geothermal Field, California.

Indeed, identifying where tremor does not occur is equally important for understanding the underlying mechanisms. While tremor is widespread in the Nankai Trough subduction zone of southwest Japan, it is demonstrably absent at similar levels in the Japan Trench subduction zone of northeastern Japan. Despite the lack of tremor, slow slip is sometimes observed in northeast Japan, often as a large afterslip following an interplate earthquake (e.g. Heki et al., 1997). A major difference between NE and SW Japan subduction is the thermal structure of the subducting plate. In the southwest, the relatively young Philippine Sea Plate subducts at a moderate rate, while in the NE, the much older Pacific plate subducts at a faster rate. Thus the conditions are much colder at a given depth in NE Japan than they are in the SW. This difference significantly influences the seismicity of these regions (Peacock and Wang, 1999); intraslab earthquakes extend to 200 km in NE Japan and only to 65 km depth in the SW. It seems probable that this variability would affect tremor generation as well. If fluids from metamorphic reactions are important in the tremor generation process, they would be released at much greater depth in the NE than in SW. This effect, though, could be negated by advection of fluids to the depths where tremor is believed to originate. Studies of b-values in Tohoku – a region devoid of tremor – suggest that this indeed has happened, leaving the region of 40–70 km depth low in fluids (Anderson, 1980) and less likely to produce tremor. In SW Japan, it has been suggested that the downdip limit of tremor may correspond to where the downgoing slab intersects the island arc Moho, possibly due to the ability of the mantle wedge to absorb fluids though serpentinization (Katsumata and Kamaya, 2003). In NE Japan, however, similar fluid-releasing reactions would take place at a depth of approximately 100 km, long after the slab was in contact with the island arc mantle, preventing the fluids from rising to the depths where tremor is generated. Others have suggested that the segmentation of tremor distribution in Japan is due to stress conditions there, arguing that the stress state of the forearc mantle wedge in NE is compressional and prevents tremor, while in SW Japan the mantle wedge is in tension allowing for hydro-fracture, which they believe to be responsible for tremor (Seno, 2005).

Although new reports come in frequently, thus far only limited locations and times have been searched for tremor. New observations are enabled both by new analyses and by new instrumentation. How does the currently reported distribution of tremor relate to the "true" distribution?

One factor arguing that tremor is widespread, but at levels at or near the noise level, is the variation in the strength of tremor in the currently-identified regions. Some of the strongest tremor may be generated in western Shikoku, where LFEs can be identified and located using methods similar to those for regular earthquakes. Although the Hi-net borehole network certainly assists in this, fewer LFEs are identified in other parts of southwest Japan despite similar station quality and density.

While the maximum amplitude of tremor varies from place to place, it is clearly limited to be relatively small. It is very likely that tremor occurs at or below the noise level of current instrumentation in many places and may evade detection. In other words, the currently recognized distribution of tremor sources should probably be thought of not as the regions that generate tremor, but rather the regions that generate *strong* tremor. Improved seismic instrumentation, increased seismometer density, and addition of low noise seismic sites (e.g., boreholes) would greatly help in identifying tremor in new locations, as well assist in characterizing tremor in locations where tremor has already been seen.

Tremor Locations: a Broad Depth Distribution in Some Areas?

The locations of tremors are fundamental to understanding the underlying processes. A broad distribution of tremor has been reported by several sources for Cascadia (McCausland et al., 2005; Kao et al., 2005; 2006). A similar result has been reported for Mexico (Payero et al., 2008). Although previous studies have argued that tremor is distributed in depth in Japan (e.g., Nugraha and Mori, 2006), these findings and those from other subduction zones contrast with recent results showing that tremor in Japan is concentrated in depth at the plate interface (Shelly et al., 2006; Ohta and Ide, 2008). Does this difference represent a real variation?

A broad depth distribution in tremor would be most easily explained by a fluid-flow mechanism. However, a moment tensor solution in southwest Japan (Ide et al., 2007a) and polarization analysis in Cascadia (Wech and Creager, 2007) argue strongly that tremor is generated by shear slip in both locations. In this case, the broad depth distribution might represent shear slip distributed in depth (Kao et al., 2005). While it's possible to imagine multiple slip interfaces in the subduction zone (e.g. Calvert, 2004), it's perhaps more difficult to imagine these slip zones distributed over a depth range of several 10s of kilometers.

One possibility that must be considered is that not all tremor is generated by the same process. Since "tremor" describes any low-amplitude, extended duration seismic signal, there is no requirement that all tremor be alike. In this scenario, the broad depth distribution and polarization results from Cascadia could be explained if most tremor is generated by shear slip on the plate interface and a smaller component generated at shallower depth by fluid flow (or some other mechanism) in the overlying crust. These tremor sources, while they could be distinct, would still need to be linked as they happen synchronously in episodes of ETS. Although volcanic tremor is believed to arise from multiple processes (McNutt, 2005) so far no evidence has been reported suggesting distinct types of non-volcanic tremor.

Another possibility is that location uncertainty and/or selection bias of different may explain depth discrepancies. No tremor location method locates every part of the signal – to varying degrees, methods either locate only part of the signal or obtain some sort of average over longer periods of time. Methods like source scanning (Kao and Shan, 2004) and LFE location (Shelly et al., 2006) fall into the former category, locating only relatively impulsive events within tremor, while waveform envelope methods fall into the latter category, obtaining some average location over a longer time period. This difference might in part explain the lack of consistency in depth determinations using different location methods in Cascadia (Royle et al., 2006; Hirose et al., 2006). However, the broad tremor depth distributions could also be the result of large location uncertainties. In particular, amplitude-based methods such as source-scanning could be strongly affected by multiple simultaneous sources, as interference of waves from multiple sources could alter the timing of amplitude peaks. This uncertainty would most strongly affect depth estimation. New locations from Cascadia based on *S-P* times (La Rocca et al., 2009) as well as locations from Cascadia and Costa Rica based on waveform cross-correlations (Brown et al., in press) show events localized near the plate interface. This may indicate that, as in southwest Japan, tremor in these areas tracks the plate interface, although again, selection bias must be considered.

Clearly, further studies are needed in order to reduce location uncertainty and resolve this debate, confirming either a broad or narrow tremor depth distribution. One promising avenue for improved locations is the use of seismic arrays. We could learn a great deal about tremor from the installation of multiple large seismic arrays, like the one installed in Washington to record an ETS episode in 2008 (Ghosh et al., in press(b)). Besides providing greatly improved signal-to-noise, such arrays would be capable of distinguishing and locating multiple simultaneous sources, decomposing the complex wavefield in a way that has not thus far been possible.

Relationship Between Tremor and Slow Slip

The precise relationship between slow slip and tremor is still uncertain. Mounting evidence suggests that where tremor is generated by shear failure at the plate interface in the plate convergence direction its distribution in space and time is closely tied to slow slip. Even within this framework, multiple models can be envisioned. One end member would be the idea that slow slip is simply the macroscopic sum of a great many small tremor-generating shear failures (e.g. Ide et al., 2008). In this model, slow slip cannot occur without tremor. This idea may be supported by the linear relationship observed between hours of tremor and slow slip moment (Aguiar et al., 2009), the close correspondence between moment rate and tremor energy for 100 s events (Ide et al., 2008), and the linear relationship between cumulative tremor amplitude measured in reduced displacement and moment measured from strain records of slow-slip events (Hiramatsu et al., 2008). However, this model fails to explain regions that exhibit slow slip without tremor, and the energy radiated through tremor appears to be extremely low compared to the geodetic moment (and slip) of the slow slip events (Ide et al., 2008). Additionally, having many small sources poses a problem of coherence for generating low-frequency energy. An alternative model might be that tremor is only generated at limited locations on the plate boundary, where changes in frictional properties (as a result of geometric, petrologic, or pore pressure heterogeneity) lead to locally accelerated rupture and radiation of seismic waves above 1 Hz while the slow-slip is accommodated elsewhere on the plate boundary. A third, intermediate model might have tremor accompanying slow slip everywhere, with its amplitude varying according to local frictional properties, so as to be undetectable in many locations.

At least in southwest Japan, slow slip events of a week or so are accompanied by (or composed of) slow shear slip events of a range of sizes and durations, at least from tremor/LFEs (~1 s duration) to VLFs (~10 s) (Ito et al., 2007) to 100 s events (Ide et al., 2008) and possibly 1000 s events (Shelly et al., 2007b). While it is clear that these events all contribute to the weeklong slow event, more work needs to be done to clarify their relationships and interactions.

While a clear deformation signal has been observed associated with tremor in the Cascadia and southwest Japan subduction zones, no deformation has yet been detected associated with tremor beneath the San Andreas Fault near Parkfield (Johnston et al., 2006). This could argue for a different mechanism of tremor in this region. However, recent results suggest that at least a portion of the tremor in this zone occurs on the deep extension of the fault, similar to southwest Japan (Shelly et al., 2009). Likewise, based on correlations with small seismic velocity variations following the 2004 Parkfield earthquake, Brenguier et al. (2008) suggest that the Parkfield tremor relates to slow slip at depth. Therefore it's plausible that the basic mechanism is the same as in subduction zones, but the deformation signal is too small to resolve with current instrumentation. A major difference between Cascadia/Japan and Parkfield is the distribution of tremor in time. While the majority of tremor in these subduction zones is concentrated in episodes of relatively intense activity lasting one to a few weeks, tremor near Parkfield appears more diffusely in time. In Parkfield, there are still periods of intense activity, but their intensity relative to the background rate is much smaller than that observed for periods of ETS in Cascadia and Japan. In Cascadia and Japan, a deformation signal is usually not detected until after a few days of active tremor (e.g. Szeliga et al., 2008; Wang et al., 2008). If the tremor and slip beneath the San Andreas are occurring relatively continuously, the associated deformation could be absorbed into the normal interseismic strain signal. Nevertheless, work is ongoing to detect a geodetic complement to tremor in this region; recently, a long-baseline strainmeter has been installed that should offer improved resolution over current instrumentation.

Seismic Hazard Implications

Another important avenue of research is to understand the seismic hazard implications of non-volcanic tremor and ETS. It has been argued that the seismic hazard during an ETS is higher than it is during periods that are quiescent (e.g., Rogers and Dragert, 2003). This is frequently used as a practical justification as to why ETS and tremor should be studied, although we have yet to see a great subduction zone earthquake preceded by an ETS event. Whether the conjecture that ETS elevates seismic hazard is correct is dependent upon the relationship between the area slipping in slow-slip and the seismogenic zone (Iglesias et al., 2004); if the slow-slip event extends into the seismogenic zone, one would expect it to bleed off some of the accumulated strain energy and therefore decrease the hazard (e.g., Yoshioka et al., 2004; Kostoglodov et al., 2003; Larson et al., 2007; Ohta et al., 2006), but if the slow-slip event terminates below the down-dip extent of the seismogenic zone it would effectively load the region (e.g., Brudzinski et al., 2007; Dragert et al., 2001; Lowry et al., 2001). In the loading case, the affect of ETS on seismic hazard may be negligible as the stresses will be quite small. The utility of this information has yet to be fully realized. Indeed, Mazzoti and Adams (2004) used statistical methods to estimate that the probability of a great earthquake is 30 to 100 times higher during an ETS episode than it is at other times of the year, but it is difficult to see how this could be used by emergency managers or the general public as this happens every 14 months in the Seattle region and more frequently elsewhere. If we further consider that any plate boundary where ETS is occurring has more than 1 ETS generating region on it (at least 7 on the Cascadia boundary (Brudzinski and Allen, 2007)), we find hazard estimation even more difficult as all of the ETS generating regions would contribute to the hazard on the entire subduction zone at different times. The problem of hazard estimation based on ETS is further complicated by our poor understanding of the physical and frictional properties at these depths. Knowledge of the physical and frictional properties of the subduction zone is necessary to understand how ETS will affect the earthquake producing region up-dip of it.

There is other information we have learned from tremor and slow-slip which has been useful for better characterizing hazard in subduction zones. Prior to the discovery of non-volcanic tremor and slow-slip, hazard models for subduction zones typically determined the area of the locked zone (i.e. the region expected to slip in a megathrust earthquake) using temperature profiles for the subducting slab as a guide to when it will slip in stick-slip vs creep-slip. Slow-slip events provide a new tool to map the strength of coupling on the plate interface, which in turn can be used to estimate seismic potential (Correa-Mora et al., 2008). Meade and Loveless (2009) offer an alternative interpretation of coupling, suggesting that observations of apparent, partial elastic coupling may actually indicate that an ongoing $M_W>=8$ slow earthquake is occurring with a duration of decades to centuries. Similarly, McCaffrey et al. (2008) used slow-slip events and the geodetically observed transition from fault locking to free slip at the Hikurangi subduction zone in New Zealand to show that the locked/partially-locked region in this subduction zone is much larger than predicted. Similar work in Cascadia has shown that the locked zone in the Cascadia subduction zone is both larger than expected by thermal models, but also closer to and therefore more dangerous to the major population centers of the region (e.g., Seattle and Vancouver) (McCaffrey, 2009; Chapman, 2009). This method can easily be applied to any subduction zone with slow-slip event and geodetic coverage, which allow seismologists to better characterize the region that will slip in a major earthquake and the hazards associated with it.

There is additional evidence that hazard assessment based on slow-slip is promising. Specifically, we note that slow-slip events in Hawaii (Segall et al., 2006; Brooks et al., 2006, 2008; Wolfe et al., 2007), New Zealand (Delahaye et al., 2009; Reyners and Bannister, 2007), Tokai (Yoshida et al., 2006), and Mexico (Larson et al., 2007; Liu et al., 2007) do appear to have triggered earthquakes. While none of the triggered earthquakes were large enough to pose a hazard to people, the fact that events were triggered demonstrates that the stresses associated with the slow-slip events are large enough to influence earthquakes and therefore affect seismic hazard. While this clearly indicates that there is a relationship between slow-slip events and earthquakes, this is still a difficult problem, as recurrence times of large earthquakes are quite long and therefore makes testing the significance of any prediction very difficult. Another avenue which may be promising is the suggestion of frictional models that the behavior of ETS in a region may change as the region gets closer to catastrophic failure, as hinted by some numerical models (e.g. Liu and Rice, 2007; Shibazaki and Shimamoto, 2007). Similary, Shelly (in press) suggested that changes in tremor migration patterns near Parkfield in the months before the 2004 M 6.0 earthquake might have reflected accelerated creep beneath the eventual earthquake hypocenter. If further observations solidify these hints of a connectionn between ETS and earthquakes, measurements of tremor and slow slip could become powerful tools to forecast large earthquakes.

An additional complication with the earthquakes triggered by the slow slip events in Hawaii and New Zealand is the question as to whether the slow-slip is the same in these events as they are in ETS. The slow-slip in Hawaii and New Zealand that triggers earthquakes occurs in the demonstrable absence of strong tremor, which may imply that different physical processes are occurring. This is another important avenue of future research, clarifying whether the slow-slip events in Hawaii and New Zealand are members of the same family of events that ETS based slow-slip events are. It is certainly possible that these events are producing tremor, only very weakly. Further study of these events and the physical conditions in which these occur should help understand the physics of ETS and slow-slip.

Because very little is known about tremor in continental regimes, it's hazard implications are poorly understood at present, but it does stand to reason that if there is slow-slip associated with the tremor seen in continental regions, that tremor would raise the likelihood of earthquakes. As we learn more about tremor in continental regions and subduction zones we expect that more can be said about the hazard is poses in continental regions.

Summary

We have already learned a great deal about non-volcanic tremor, but the field is still in its infancy. Investigation up to this point has mostly concentrated on understanding the tremor source. No doubt much work remains, but as our understanding of the source progresses, we will begin to find tremor to be an effective tool to study the conditions of deep deformation at various locations in the earth. Through new instrumentation and analysis, and well as new modeling and laboratory experiments, we expect progress to continue at a rapid pace. While tremor and other slow-slip processes may occur in the deep roots of fault zones, we expect that these discoveries will add to our knowledge of tectonic processes in a broad sense, eventually feeding back to aid our understanding of earthquakes.

Acknowledgements The authors would like to thank Roland Burgmann, Joan Gomberg, Jeanne Hardebeck, Stephanie Prejean, Tetsuzo Seno, John Vidale, and an anonymous reviewer for their thorough reviews. We also thank Chloe Peterson, Doug Christensen, Xyoli Perez-Campos, and Vladimir Kostoglodov for their help in procuring sample tremor data for Fig. 4. For Fig. 4: data from Mexico was part of the MesoAmerican Subduction Experiment (MASE) project; data from Alaska comes from the Broadband Experiment Across Alaskan Ranges (BEAAR) experiment; data from Parkfield comes from the High Resolution Seismic Network (HRSN); data from Cascadia comes from the Cascadia Arrays For Earthscope experiment (CAFE); and the data from Shikoku, Japan is from the High Sensitivity Seismic Network (Hi-Net).

References

Aguiar, A.C, T.I. Melbourne, and C.W. Scrivner (2009), Moment release rate of Cascadia tremor constrained by GPS, *J. Geophys. Res.*, 114, B00A05, doi:10.1029/2008JB005909.

Aki, K. and P.G. Richards (2002), Quantitative Seismology, 2nd Edition, University Science Books, Sausalito.

Anderson, R.N. (1980), Phase changes and the frequency-magnitude distribution in the upper plane of the deep seimic zone beneath Tohoku, Japan, *J. Geophys. Res.*, 85, 1389–1398.

Audet, P., M.G. Bostock, N.I. Christensen, and S.M. Peacock (2009), Seismic evidence for overpressured subducted oceanic crust and megathrust fault sealing, *Nature*, 457, 76–78, doi:10.1038/nature07650.

Brenguier, F, M. Campillo, C. Hadziioannou, N.M. Shapiro, R.M. Nadeau, and E. Larose (2008), Postseismic relaxation along the San Andreas fault investigated with continuous seismological observations, *Science*, 321, 1478–1481.

Brooks, B.A., J.H. Foster, M. Bevis, L.N. Frazer, C.J. Wolfe and M. Behn, (2006), Periodic slow earthquake on the flank of Kilauea volcano, Hawai'I, *Earth Planet. Sci. Lett.*, 246, 207–216.

Brooks, B.A., J. Foster, D. Sandwell, C.J. Wolfe, P. Okubo, M. Poland, and D. Myer (2008), Magmatically triggered slow slip at Kilauea Volcano, Hawaii, *Science*, 321, 1177, doi:10.1126/science.1159007.

Brown, J.R., G.C. Beroza, and D. R. Shelly (2008), An autocorrelation method to detect low frequency earthquakes within tremor, *Geophys. Res. Lett.*, 35, L16305, doi:10.1029/2008GL034560.

Brown, J.R., G.C. Beroza, S. Ide, K. Ohta, D.R. Shelly, S.Y. Schwartz, W. Rabbel, M. Thorwart, and H. Kao, Deep low-frequency earthquakes in tremor localize to the plate interface in multiple subduction zones, *Geophys. Res. Lett.*, in press.

Brown, K.M., M.D. Tryon, H.R. DeShon, L.M. Dorman, S.Y. Schwartz (2005). Correlated transient fluid pulsing and seismic tremor in the Costa Rica subduction zone, *Earth Planet. Sci. Lett.*, 238, 189–203.

Brudzinski, M.R. and R.M. Allen (2007), Segmentation in episodic tremor and slip all along Cascadia, *Geology*, 35. 907–910.

Brudzinski, M., E. Cabral-Cano, F. Correa-Mora, C. Demets, and B. Marquez-Azua (2007), Slow slip transients along the Oaxaca subduction segment from 1993 to 2007, *Geophys. J. Intl.*, 171, 523–538, doi:10.1111/j1365-246X.2007.03542.x.

Burlini, L., G. Di Toro, and P. Meredith (2009), Seismic tremor in subduction zones, *Rock Phys. Evidence, Geophys. Res. Lett.*, 36, L08305, doi:10.1029/2009GL037735.

Calvert A. (2004) Seismic reflection imaging of two megathrust shear zones in the northern Cascadia subduction zone, *Nature*, 428, 163–167.

Capon, J. (1969). Investigation of long-period noise at the large aperture seismic array, *J. Geophys.* Res., 74, 3182–3194.

Chapman, J. (2008), M.S. Thesis, Central Washington University, Ellensburg, Washington.

Chouet, B. (1988). Resonance of a fluid driven crack: Radiation properties and implications for the source of long-period events and harmonic tremor, *J. Geophys. Res.*, 93, 4375–4400.

Cochran, E.S., J.E. Vidale, and S. Tanaka (2004). Earth tides can trigger shallow thrust fault earthquakes, *Science*, 306, 1164–1166.

Cochran, E.S. and J.E. Vidale (2007), Comment on tidal synchronicity of the 26 December 2004 Sumatran earthquake and its aftershocks, *Geophys. Res. Lett.*, 34, l04302, doi:10.1029/2006GL028639.

Correa-Mora, F., C. DeMets, E. Cabral-Cano, O. Diaz-Molina, and B. Marquez-Azua (2008), Interplate coupling and transient slip along the subduction interface beneath Oaxaca,

Mexico, *Geophys. J. Int.* 175, 269–290, doi:10.1111/j.1365-246X.2008.03910.x

Delahaye, E.J., J. Townend, M.E. Reyners, and G. Rogers (2009), Microseismicity but no tremor accompanying slow slip in the Hikurangi subduction zone, New Zealand, *Earth Planet. Sci. Lett.*, 277, 21–28.

Dragert, H., K. Wang, and T.S. James (2001), A silent slip event on the deeper Cascadia subduction interface. *Science*, 292, 1525–1528.

Dragert, H., K. Wang and G. Rogers (2004), Geodetic and seismic signatures of episodic tremor and slip in the northern Cascadia subduction zone, *Earth Planet. Space*, 56, 1143–1150.

Ellsworth, W.L., M.V. Matthews, R.M. Nadeau, S.P. Nishenko, P.A. Reasenber, R.W. Simpson (1999), A Physically-based earthquake recurrence model for estimation of long-trerm earthquake probabilities, *U.S. Geol. Surv. Open-File Rept.* 99–522.

Filson, J. (1975), Array seismology. *Ann. Rev. Earth Plaet. Sci.*, 3, 157–181.

Fletcher, J.B., P. Spudich, and L. Baker (2006), Rupture propagation of the 2004 Parkfield, California earthquake from observations at the UPSAR, *Bull. Seismol. Soc. Am.*, 96, 129–142.

Ghosh, A., J.E. Vidale, Z. Peng, K.C. Creager, and H. Houston, Complex non-volcanic tremor near Parkfield, California, triggered by the great 2004 Sumatra earthquake, *J. Geophys. Res.*, in press (a).

Ghosh, A., J.E. Vidale, J.R. Sweet, K.C. Creager, and A.G. Wech, Tremor patches in Cascadia revealed by seismic array analysis, *Geophys. Res. Lett.* (in press (b)).

Gibbons, S.J. and F. Ringdal (2006), The detection of low magnitude seismic events using array-based waveform correlation, *Geophys. J. Int.*, 165, 149–166.

Goldstein, P. and R.J. Archuleta (1987), Array analysis of seismic signals, *Geophys. Res. Lett.*, 14, 13–16.

Gomberg, J., J.L. Rubinstein, Z. Peng, K.C. Creager, J.E. Vidale, and P. Bodin, (2008) Widespread triggering on non-volcanic tremor in California, *Science*, 319, 713.

Hacker, B.R., Peacock, S.M., Abers, G.A., and Holloway, S., (2003), Subduction Factory 2. Intermediate-depth earthquakes in subducting slabs are linked to metamorphic dehydration reactions. *J. Geophys. Res.*, 108, doi:10.1029/2001JB001129.

Hartzell, S.H. and T.H. Heaton (1983), Inversion of strong ground motion and teleseismic waveform data for the fault rupture history of the 1979 Imperial Valley, California, earthquake, *Bull. Seismol. Soc. Am.*, 73, 1553–1583.

Heki, K, S. Miyazaki, and H. Tsuji (1997), Silent fault slip following an interplate thrust earthquake at the Japan Trench, *Nature*, 386, 595–598.

Hill, D.P (2008), Dynamic stress, coulomb failure, and remote triggering, *Bull. Seismol. Soc. Am.*, 98, 66–92.

Hiramatsu, Y., T. Watanabe, and K. Obara (2008), Deep low-frequency tremors as a proxy for slip monitoring at plate inteface, *Geophys. Res. Lett.*, 35, L13304, doi:10.1029/2008GL034342.

Hirose, H. and K. Hirahara (2004). A 3-D quasi-static model for a variety of slip behaviors on a subduction fault, *PAGEOPH* , 161, 2417–2431.

Hirose, H. and K. Obara (2005), Repeating short- and long-term slow slip events with deep tremor activity, around the Bungo channel region, southwest Japan, *Earth Planet. Space*, 57, 961–972.

Hirose, H. and K. Obara (2006), Short-term slow slip and correlated tremor episodes in the Tokai region, central Japan, *Geophys. Res. Lett.*, 33, L17311, doi:10.1029/2006GL026579.

Hirose, H., K. Hirahara, F. Kimata, N. Fujii, and S. Miyazaki (1999), A slow thrust slip event following the two 1996 Hyuganada earthquakes beneath the Bungo Channel, southwest Japan. *Geophys. Res. Lett.* 26, 3237–3240.

Hirose, H., H. Kao, and K. Obara (2006). Comparative study of nonvolcanic tremor locations in the Cascadia subduction zone using two different methods, Eos Trans. AGU, 87, *Fall Meet. Suppl.* Abstract T41A-1533.

Ide, S., D. R. Shelly, and G. C. Beroza (2007a), Mechanism of deep low frequency earthquakes: Further evidence that deep non-volcanic tremor is generated by shear slip on the plate interface, *Geophys. Res. Lett.*, 34, L03308, doi:10.1029/2006GL028890.

Ide, S., G. C. Beroza, D. R. Shelly, and T. Uchide (2007b), A new scaling law for slow earthquakes, *Nature*, 447, 76–79.

Ide, S. (2008), A Brownian walk model for slow earthquakes, *Geophys. Res. Lett.*, 35, doi:10.1029/2008GL034821.

Ide S., Imanishi K, Yoshida Y., Beroza G.C., and Shelly D.R. (2008), Bridging the gap between seismically and geodetically detected slow earthquakes, *Geophys. Res. Lett.*, 35, L10305, doi:10.1029/2008GL034014.

Iglesias, A., S.K. Singh, A.R. Lowry, M. Santoyo, V. Kostoglodov, K.M. Larson, S.I. Fracno-Sanchez (2004), The silent earthquake of 2002 in the Guerrero seismic gap, Mexico (Mw=7.6): Inversion of slip on the plate interface and some implications, *Geofisica Int.*, 43, 309–317.

Ishii, M., P. M. Shearer, H. Houston, and J. E. Vidale (2005), Extent, duration and speed of the 2004 Sumatra-Andaman earthquake imaged by the Hi-Net array, *Nature*, 435, doi: 10.1038/nature03675.

Ito Y., K. Obara, K. Shiomi, S. Sekine, and H. Hirose (2007), Slow earthquakes coincident with episodic tremors and slow slip events, *Science*, 315, 503–506, doi:0.1126/science.1134454.

Johnston, M.J.S., R.D. Borcherdt, A.T. Linde, and M.T. Gladwin (2006), Continuous borehole strain and pore pressure in the near field of the 28 September 2004 M 6.0 Parkfield, California, earthquake: Implications for nucleation, fault response, earthquake prediction, and tremor, *Bull. Seismol. Soc. Am.*, 96, S56–S72.

Julian, B.R. (2000), Period doubling and other nonlinear phenomena in volcanic earthquakes and tremor, *J. Volcanol. Geothermal Res.*, 101, 19–26.

Julian, B. R. (2002), Seismological Detectic of Slab Metamorphism, *Science*, 296, 1625–1626.

Kao, H. and S-J. Shan (2004), The source-scanning algorithm: Mapping the distribution of seismic sources in time and space. *Geophys. J. Intl.* 157, 589–594.

Kao, H., S. Shan, H. Dragert, G. Rogers, J. F. Cassidy, and K. Ramachandran (2005), A wide depth distribution of seismic tremors along the northern Cascadia margin, *Nature*, 436, 841–844.

Kao, H. S-JShan, H. Dragert, G. Rogers, J.F. Cassidy, K. Wang, T.S. James, and K. Ramachandran (2006), Spatial-temporal patterns of seismic tremors in northern Cascadia. *J. Geophys. Res.*, 111, doi:10.1029/2005JB003727.

Kao, H., P. J. Thompson, G. Rogers, H. Dragert, and G. Spence (2007a), Automatic detection and characterization of seismic tremors in northern Cascadia, *Geophys. Res. Lett.*, 34, L16313, doi:10.1029/2007GL030822.

Kao, H., S.-J. Shan, G. Rogers, and H. Dragert (2007b), Migration characteristics of seismic tremors in the northern Cascadia margin, *Geophys. Res. Lett.*, 34, L03304, doi:10.1029/2006GL028430.

Katsumata, A., and N. Kamaya (2003), Low-frequency continuous tremor around the Moho discontinuity away from volcanoes in the southwest Japan, *Geophys. Res. Lett.* 30, doi:10.1029/2002GL015981.

Kawasaki I., Asai Y., Tamura Y., Sagiya T., Mikami N., Okada Y., Sakata M., and Kasahara M., (1995), The 1992 Sanriku-Oki, Japan, ultra-slow earthquake, *J. Phys. Earth*, 43, 105–116.

Kawasaki, I., Y.Asal, and Y. Tamura, (2001), Space-time distribution of interplate moment release including slow earthquakes and the seismo-geodetic coupling in the Sanriku-oki region along the Japan trench, *Tectonophysics*, 330, 267–283.

Kirby, S., K. Wang, and T. Brocher (2002), A possible deep, long-term source for water in the Northern San Andreas Fualt System: A ghost of Cascadia subduction past? Eos Trans. AGU, 83, *Fall Meet. Suppl.*, Abstract S22B-1038.

Kodaira, S., T. Iidaka, A. Kato, J.-O. Park, T. Iwassaki, and Y. Kaneda (2004), High pore fluid pressure may cause silent slip in the Nankai Trough, *Science*, 304, 1295–1298.

Kostoglodov, V., S.K. Singh, J.A. Santiago, S.I. Franco, K.M. Larson, A.R. Lowry, and R. Bilham (2003), A large silent earthquake in the Guerrero seismic gap, Mexico, *Geophys. Res. Lett.*, 30, doi:10.1029/2003GL017219.

Kuroki, H., H.M. Ito, H. Takayama, and A. Yoshida (2004). 3-D simulation of the occurrence of slow slip events in the Tokai region with a rate- and state-dependent friction law, *Bull. Seismol. Soc. Am.*, 94, 2037–2050.

La Rocca, M., W. McCausland, D. Galluzo, S. Malone, G. Saccorotti, and E. Del Pezzo (2005). Array measurements of deep tremor signals in the Cascadia subduction zone, *Geophys. Res. Lett.*, 32, L21319, doi:10.1029/2005GL023974.

La Rocca, M., D. Galluzzo, S. Malone, W. McCausland, G. Saccorotti, E. Del Pezzo (2008). Testing small-aperture array analysis on well-located earthquakes, and application to the location of deep tremor, *Bull. Seismol. Soc. Am.*, 93, 620–635.

La Rocca, M., K.C. Creager, D. Galluzzo, S. Malone, J.E. Vidale, J.R. Sweet, and A.G. Wech (2009), Cascadia tremor located near plate interface constrained by *S* minus *P* wave times, *Science*, 323, 620–623, doi:10.1126/science.1167112.

Lambert, A., H. Kao, G. Rogers, and N. Courtier (2009), Correlation of tremor activity with tidal stress in the northern Cascadia subduction zone, *J. Geophys. Res.,* 114, B00A08, doi:10.1029/2008JB006038.

Larson, K.M., V. Kostoglodov, S. Miyazaki, and J.A.S. Santiago (2007), The 2006 aseismic slow slip event in Guerrero, Mexico: New results from GPS, *Geophys. Res. Lett.*, 34, L13309, doi:10.1029/2007GL029912.

Liu, Y. and J. R. Rice (2005), Aseismic slip transients emerge spontaneously in three-dimensional rate and state modeling of subduction earthquake sequences, *J. Geophys. Res.*, 110, doi:10.1029/2004JB003424.

Liu, Y. and J. R. Rice (2007), Spontaneous and triggered aseismic deformation transients in a subduction fault model, *J. Geophys. Res.*, 12, B09404, doi:10.1029/2007JB004930.

Liu, Y., J.R. Rice, and K. M. Larson (2007), Seismicity variations associated with aseismic transients in Guerrero, Mexico, 1995–2006, *Earth Planet. Sci. Lett.*, 262, 493–504.

Lowry, A.R. (2006). Resonant slow fault slip in subduction zones forced by climatic load stress, *Nature*, 442, 802–805.

Lowry, A.R., K.M. Larson, V. Kostoglodov, and R. Bilham (2001), Transient fault slip in Guerrero, Southern Mexico, *Geophys. Res. Lett.*, 28, 3753–3756.

Matsubara, M., K. Obara, and K. Kashara (2009), High-V_P/V_S zone accompanying non-volcanic tremors and slow-slip events beneath southwestern Japan, *Tectonophysics,* 472, 6–17, doi:10.1016/j.tecto.2008.06.013.

Mazzoti, S. and J. Adams (2004), Variability of near-term probability for the next great earthquake on the Cascadia subduction zone, *Bull. Seismol. Soc. Am.*, 94, 1954–1959.

McCausland, W., S. Malone and D. Johnson (2005). Temporal and spatial occurrence of deep non-volcanic tremor: From Washington to Northern California, *Geophys. Res. Lett.* 32, doi:10.1029/2005GL024349.

McCaffrey R, Wallace L.M., and Beavan J (2008), Slow slip and frictional transition at low temperature at the Hikurangi subduction zone, *Nat. Geosci.*, 1, 316–320.

McCaffrey, R. (2009), Time-dependent inversion of three-component continuous GPS for steady and transient sources in northern Cascadia, *Geophys. Res. Lett.,* 36, L07304, doi:10.1029/2008GL036784.

McNutt S.R. (2005), Volcanic Seismology, *Ann. Rev. Earth Plaet. Sci.,* 32:461–491, doi:10.1146/annurev.earth.33.092203.122459.

Meade, B.J. and J.P. Loveless (2009), Predicting the geodetic signature of MW>=8 slow slip events, *Geophys. Res. Lett.,* 36, L01306, doi:10.1029/2008GL03634.

Miller, M. M., Melbourne, T., Johnson, D. J. & Sumner, W. Q. (2002) Periodic slow earthquakes from the Cascadia subduction zone. *Science* 295, 2423.

Miyazawa, M. and E. E. Brodsky (2008), Deep low-frequency tremor that correlates with passing surface waves, *J. Geophys. Res.*, 113, B01307, doi:10.1029/2006JB004890.

Miyazawa, M., E.E. Brodsky, and J. Mori (2008), Learning from dynamic triggering of low-frequency tremor in subduction zones, *Earth Planets Space*, 60, e17–e20.

Miyazawa, M. and J. Mori, (2005), Detection of triggered deep low-frequency events from the 20032005 Tokachi-oki earthquake, *Geophys. Res. Lett.*, 32, doi:10.1029/2005GL022539.

Miyazawa, M. and J. Mori (2006), Evidence suggesting fluid flow beneath Japan due to periodic seismic triggering from the 2004 Sumatra-Andaman earthquake, *Geophys. Res. Lett.,* 33, doi:10.1029/2005GL025087.

Montgomery-Brown, E.K., P. Segall, and A. Miklius (2009), Kilauea slow slip events: Identification, source inversions, and relation to seismicity, *J. Geophys. Res.,* 114, B00A03, doi:10.1029/2008JB006074.

Nadeau, R., A. Thomas, and R. Burgmann (2008), Tremor-tide correlations at Parkfield, CA, *Eos Trans. AGU,* 89, Fall Meet Suppl., Abstract U33A-0054.

Nadeau, R.M. and A. Guilhem (2009), Nonvolcanic tremor evolution and the San Simeon and Parkfield, California, Earthquakes, *Science,* 325, 191–193, doi:10.1126/science.1174155.

Nadeau, R.M. and T. V. McEvilly (1999), Fault slip rates at depth from recurrence intervals of repeating microearth-

quakes, *Science* 285, 718–721, DOI: 10.1126/science.285.5428.718.
Nadeau, R.M., A. Michelini, R.A. Uhrhammer, D. Dolenc, and T.V. McEvilly (2004). Detailed kinematics, structure, and recurrence of micro-seismicity in the SAFOD target region, *Geophys. Res. Lett.*, 31, L12S08, doi:10.1029/2003GL019409.
Nadeau, R. M. & Dolenc, D. (2005) Nonvolcanic tremors deep beneath the San Andreas fault. *Science* 307, 389; published online 9 December 2004 (10.1126/science.1107142).
Nakata, R., N. Suda, and H. Tsuruoka, (2008), Non-volcanic tremor resulting from the combined effect of Earth tides and slow slip events, *Nat. Geosci.*, 1, 676–678, doi:10.1038/ngeo288.
Nakatani, M., and C. H. Scholz (2004), Frictional healing of quartz gouge under hydrothermal conditions: 1. Experimental evidence for solution transfer healing mechanism, *J. Geophys. Res.*, 109 B07201, doi:10,1029/2001JB001522.
Nugraha, A.D. and J. Mori (2006). Three-dimensional velocity structure in the Bungo Channel and the Shikoku area, Japan, and its relationship to low-frequency earthquakes, *Geophys. Res. Lett.*, 33, L24307, doi:10.1029/2006GL028479.
Obara, K. (2002), Nonvolcanic deep tremor associated with subduction in southwest Japan. *Science*, 296, 1679–1681.
Obara, K. (2003), Time sequence of deep low-frequency tremors in the Southwest Japan Subduction Zone: Triggering phenomena and periodic activity, *Chigaku Zasshi (J. Geogr.)*, 112, 837–849 (in Japanese).
Obara, K., H. Hirose, F. Yamamizu, and K. Kasahara (2004), Episodic slow slip events accompanied by non-volcanic tremors in southwest Japan subduction zone. *Geophys. Res. Lett.* 31, doi:10.1029/2004GL020848.
Obara, K., K. Kasahara, S. Hori, and Y. Okada (2005), A densely distributed highsensitivity seismograph network in Japan: Hi-net by National Research Institute for Earth Science and Disaster Prevention. *Rev. Sci. Instrum.* 76, doi:10.1063/1.1854197.
Ohmi, S. and K. Obara (2002), Deep low-frequency earthquakes beneath the focal region of the Mw 6.7 2000 Western Tottori earthquake, *Geophys. Res. Lett.*, 29, doi:10.1029/2001GL014469.
Ohmi, S., I. Hirose, and J. Mori (2004), Deep low-frequency earthquakes near the downward extension of the seismogenic fault of the 2000 Western Tottori earthquake, *Earth Planets Space*, 56, 1185–1189.
Ohta, Y., J. T. Freymueller, S. Hreinsdóttir, and H. Suito, (2006), A large slow slip event and the depth of the seismogenic zone in the south central Alaska subduction zone, *Earth Planet. Sci. Lett.*, 247, 108–116.
Ohta, K., and S. Ide (2008), A precise hypocenter determination method using network correlation coefficients and its application to deep low frequency earthquakes, *Earth Planets. Space*, 60, 877–882.
Ozawa, S., M. Murakami, M. Kaidzu, T Tada, T. Sagiya, Y. Hatanaka, H. Yarai, and T. Nishimura (2002), Detection and monitoring of ongoing aseismic slip in the Tokai region, central Japan, *Science*, 298, 1009–1012.
Ozawa, S., S. Miyazaki, Y. Hatanaka, T. Imakiire, M. Kaidzu, M. Murakami (2003), Characteristic silent earthquakes in the eastern part of the Boso peninsula, Central Japan, *Geophys. Res. Lett.*, 30, doi:10.1029/2002GL016665.
Payero, J.S., V. Kostoglodov, N. Shapiro, T. Mikumo, A. Iglesia, X. Perez-Campos, R.W. Clayton (2008), Nonvolcanic tremor observed in the Mexican subduction zone, *Geophys. Res. Lett.*, 35, L07305, doi:10.1029/2007GL032877.
Peacock, S.M. (2009), Thermal and metamorphic environment of subduction zone episodic tremor and slip, *J. Geophys. Res.*, 114, B00A07, doi:10.1029/2008JB005978.
Peacock, S. M. & Wang, K. (1999) Seismic consequences of warm versus cool subduction metamorphism: Examples from southwest and northeast Japan. *Science* 286, 937–939.
Peng, Z. and K. Chao (2008), Non-volcanic tremor beneath the Central Range in Taiwan triggered by the 2001 MW7.8 Kunlun earthquake, *Geophys. J. Int.*, 175, 825–829, doi:10.1111/j.1365-246X.2008.03886.x.
Peng, Z., J.E. Vidale, K.C. Creager, J.L. Rubinstein, J. Gomberg, and P. Bodin (2008), Strong tremor near Parkfield, CA excited by the 2002 Denali Fault earthquake, *Geophys. Res. Lett.*, 35, L23305, doi:10.1029/2008GL036080.
Peng, Z., J.E. Vidale, A.G. Wech, R.M. Nadeau, and K.C Creager (2009), Remote triggering of tremor along the San Andreas Fault in central California, *J. Geophys. Res.*, 114, B00A06, doi:10.1029/2008JB006049.
Peterson, C.L. and D.H. Christensen (2009). Possible relationship between nonvolcanic tremor and the 1998–2001 slow slip event, south central Alaska, *J. Geophys. Res.*, 114, B06302, doi:10.1029/2008JB006096.
Poupinet, G., W. L. Ellsworth, and J. Fréchet (1984), Monitoring velocity variations in the crust using earthquake doublets: An application to the Calaveras Fault, California, *J. Geophys. Res.*, 89, 5719–5731.
Reyners, M. and S. Bannister (2007), Earthquakes triggered by slow slip at the plate interface in the Hikurangi subduction zone, New Zealand, *Geophys. Res. Lett.*, 34, L14305, doi:10.1029/2007GL030511.
Rogers, G. and H. Dragert (2003), Episodic tremor and slip on the Cascadia subduction zone: The chatter of silent slip. *Science*, 300, 1942–1943.
Rogers, G. (2007), Episodic Tremor and Slip in Northern Cascadia – Going Back in Time, paper presented at the 2007 Seismol. Soc. Am. Annual Meeting, Waikoloa, Hawaii., 11–13 April.
Royle G.T, Calvert A.J., Kao H (2006), Observations of non-volcanic tremor during the northern Cascadia slow slip event in February 2002, *Geophys. Res. Lett.*, 33, L18313, doi10.1029/2006GL027316.
Rubin, A.M., D. Gillard, and J.-L. Got (1999), Streaks of microearthquakes along creeping faults. *Nature*, 400, 635–641.
Rubin, A.M. and P. Segall (2007), Episodic slow-slip transients and rate-and-data friction, *Eos Trans. AGU*, 88, Fall Meet. Suppl., Abstract T21A-0374.
Rubin, A.M. (2008), Episodic slow slip events and rate-and-state friction, *J. Geophys. Res.*, 113, B11414, doi:10.1029/2008JB005642.
Rubinstein, J.L., J.E. Vidale, J. Gomberg, P. Bodin, K.C. Creager and S.D. Malone (2007), Non-volcanic tremor driven by large transient shear stresses, *Nature*, 448, doi:10.1038/nature06017, 579–582.
Rubinstein, J.L., M. La Rocca, J.E. Vidale, K.C. Creager, and A.G. Wech (2008), Tidal modulation of non-volcanic tremor, *Science*, 319, 186–189.

Rubinstein, J.L., J. Gomberg, J.E. Vidale, A.G. Wech, H. Kao, K.C. Creager, and G. Rogers (2009), Seismic wave triggering of nonvolcanic tremor, episodic tremor and slip, and earthquakes on Vancouver Island, *J. Geophys. Res.,* 114, B00A01, doi:10.1029/2008JB005875.

Schaff, D. P., G. C. Beroza, and B. E. Shaw (1998), Postseismic response of repeating aftershocks, *Geophys. Res. Lett.*, 25, 4549–4552.

Schaff, D.P., G.H.R. Bokelmann, W.L. Ellsworth, E. Zanzerkia, F. Waldhauser, and G.C. Beroza (2004), Optimizing correlation techniques for improved earthquake location, *Bull. Seism. Soc. Am.,* 94, 705–721, doi:10.1785/0120020238.

Schwartz, S.Y., J.I. Walter, T.H. Dixon, K.C. Psencik, M. Protti, V. Gonzalez, M. Thorwart, and W. Rabbel (2008), Slow slip and tremor detected at the northern Costa Rica seismogenic zone, *Eos Trans. AGU,* 89, Fall. Meet. Suppl., Abstract U31B-06.

Schwartz, S.Y. and J.M. Rokosky (2007), Slow slip events and seismic tremor at circum-pacific subduction zones, *Rev. Geophys.* 45, RG3004, doi:10.1029/2006RG000208.

Segall, P. and A.M. Rubin (2007), Dilatency stabilization of frictional sliding as a mechanism for slow slip events, *Eos Trans. AGU*, 88, Fall Meet. Suppl., Abstract T13F-08.

Segall, P., E.K. Desmarais, D. Shelly, A. Miklius, and P. Cervelli (2006), Earthquakes triggered by silent slip events on Kilauea volcano, Hawaii, *Nature*, 442, 71–74.

Seno, T. and T. Yamasaki (2003), Low-frequency tremors, intraslab and interplate earthquakes in Southwest Japan – from a viewpoint of slab dehydration. *Geophys. Res. Lett.* 30, doi:10.1029/2003GL018349.

Seno, T. (2005), Variation of downdip limit of the seismogenic zone near the Japanese islands, Implications for the serpentinization mechanism of the forearc mantle wedge, *Earth Planet. Sci. Lett.*, 231, 249–262.

Shearer, P.M. (1999), Introduction to Seismology, Cambridge University Press, Cambridge.

Shelly, D.R., Possible deep fault slip preceding the 2004 Parkfield earthquake, inferred from detailed observations of tectonic tremor, *Geophys. Res. Lett.* (in press).

Shelly, D. R., G. C. Beroza, S. Ide, and S. Nakamula (2006), Low-frequency earthquakes in Shikoku, Japan and their relationship to episodic tremor and slip. *Nature* 442, 188–191.

Shelly, D. R., Beroza, G. C. & Ide, S. (2007a), Non-volcanic tremor and low frequency earthquake swarms. *Nature*, 446, 305–307.

Shelly, D.R., G.C. Beroza, and S. Ide (2007b), Complex evolution of transient slip derived from precise tremor locations in western Shikoku, Japan. *Geochem. Geophys. Geosyst.*, 8, Q10014, doi:10.1029/2007GC001640.

Shelly, D.R., W.L. Ellsworth, T. Ryberg, C. Haberland, G.S. Fuis, J. Murphy, R.M. Nadeau, and R. Burgmann (2009), Precise location of San Andreas Fault tremors near Cholame, California using seismometer clusters: Slip on the deep extension of the fault?, *Geophys. Res. Lett.,* 36, L01303, doi:10.1029/2008GL036367.

Shen, Z-K., Q. Wang, R. Burgmann, Y. Wan, and J. Ning (2005). Pole-tide modulation of slow slip events at circum-Pacific subduction zones. *Bull. Seismol. Soc. Am.*, 95, 2009–2015.

Shibazaki, B., and Y. Iio (2003), On the physical mechanism of silent slip events along the deeper part of the seismogenic zone, *Geophys. Res. Lett.*, 30(9), 1489, doi:10.1029/2003GL017047.

Shibazaki B, Shimamoto T (2007) Modelling of short-interval silent slip events in deeper subduction interfaces considering the frictional properties at the unstable-stable transition regime. *Geophys. J. Intl.* 171, 191–205.

Shimamoto, T., (1986). Transition between frictional slip and ductile flow for Halite shear zones at room temperature, *Science,* 231, 711–714.

Smith, E.F. and J. Gomberg, A search in strainmeter data for slow slip associated with triggered and ambient tremor near Parkfield, California, *J. Geophys. Res.*, in press.

Song, T.-R.A, D.V. Helmberger, M.R. Brudzinski, R.W. Clayton, P. Davis, X. Perez-Campos, S.K. Singh (2009), Subducting slab ultra-slow velocity layer coincident with silent earthquakes in southern Mexico, *Science,* 324, 502–505, doi:10.1126/science.1167595.

Spudich, P. and E. Cranswick (1984), Direct observation of rupture propagation during the 1979 Imperial Valley earthquake using a short baseline accelerometer array, *Bull. Seismol. Soc. Am.*, 74, 2083–2114.

Suda, N.R., R. Nakata, and T. Kusumi, An automatic monitoring system for non-volcanic tremors in southwest Japan, *J. Geophys. Res.,* in press.

Sweet, J., K. Creager, J. Vidale, A. Ghosh, M. Nichols, T. Pratt, and A. Wech (2008), Low Frequency Earthquakes in Cascadia, paper presented at 2008 IRIS Workshop, Stevenson Washington, 9 June 2008.

Szeliga, W., T.I. Melbourne, M.M. Miller, and V.M. Santillan (2004), Southern Cascadia episodic slow earthquakes, *Geophys. Res. Lett.*, L16602, doi:10.1029/2004GL020824.

Szeliga, W., T. Melbourne, M. Santillan, and M. Miller (2008), GPS constraints on 34 slow slip events within the Cascadia subduction zone, 1997–2005, *J. Geophys. Res.*, 113, B04404, doi:10.1029/2007JB004948.

Tanaka, S., M. Ohtake, and H. Sato (2002), Evidence for tidal triggering of earthquakes as revealed from statistical analysis of global data, *J. Geophys. Res.*, 107(B10), 221, doi:10.1029/2001JB001577.

Uchida, N., T. Matsuzawa, W. L. Ellsworth, K. Imanishi, T. Okada, and A. Hasegawa (2007), Source parameters of a M4.8 and its accompanying repeating earthquakes off Kamaishi, NE Japan - implications for the hierarchical structure of asperities and earthquake cycle, *Geophys. Res. Lett.*, 34, doi:10.1029/2007GL031263.

Vidale, J.E. (1988). Finite-difference travel time calculation, *Bull. Seismol. Soc. Am.*, 78, 2062–2076.

Vidale, J.E., D.C. Agnew, M.J.S. Johnston, and D.H. Oppenheimer (1998). Absence of earthquake correlation with Earth tides: An indication of high preseismic fault stress rate, *J. Geophys. Res.*, 103, 7247–7263.

Voisin, C., J-R. Grasso, E. Larose, and F. Renard (2008), Evolution of seismic signals and slip patterns along subduction zones: Insights from a friction lab scale experiment, *Geophys. Res. Lett.,* 35, L08302, doi:10.1029/2008GL033356.

Waldhauser, F., W. L. Ellsworth, D. P. Schaff, and A. Cole (2004), Streaks, multiplets, and holes: High-resolution spatio-temporal behavior of Parkfield seismicity. *Geophys. Res. Lett.*, 31, doi:10.1029/2004GL02069.

Wang, Z., D. Zhao, O.P. Mishra, and A. Yamada (2006), Structural heterogeneity and its implications for the low frequency tremors in Southwest Japan, *Earth. Planet. Sci. Lett.*, 251, 66–78.

Wang, K., H. Dragert, H. Kao, and E Roeloffs (2008), Characterizing an "uncharacteristic" ETS event in northern Cascadia, *Geophys. Res. Lett.*, 35, L15303, doi:10.1029/2008GL034415.

Watanabe T, Hiramatsu Y, and Obara K (2007) Scaling relationship between the duration and the amplitude of non-volcanic deep low-frequency tremors, *Geophys. Res. Lett.*, 34, L07305, doi:10.1029/2007GL029391.

Wech A. G., K. C. Creager (2007), Cascadia tremor polarization evidence for plate interface slip, *Geophys. Res. Lett.*, 34, L22306, doi:10.1029/2007GL031167.

Wech, A.G. and K.C. Creager (2008), Automated detection and location of Cascadia tremor, *Geophys. Res. Lett.*, 35, L20302, doi:10.1029/2008GL035458.

Wilcock, W.S.D. (2001). Tidal triggering of microearthquakes on the Juan de Fuca Ridge, *Geophys. Res. Lett.*, 28, 3999–4002.

Wolfe, C.J., B.A. Brooks, J.H. Foster, and P.G. Okubo (2007), Microearthquake streaks and seismicity triggered by slow ear-thquakes on the mobile south flank of Kilauea Volcano, Hawai'I, *Geophys. Res. Lett.*, 34, L23306, doi:10.1029/2007GL031625.

Yamasaki T. and T. Seno (2003), Double seismic zone and dehydration embrittlement, *J. Geophys. Res.*, 108, doi:10.1029/2002JB001918.

Yoshida, S. and N. Kato (2003). Episodic aseismic slip in a two-degree-of-freedom block-spring model, *Geophys. Res. Lett.*, 30, doi:10.1029/2003GL017439.

Yoshida, A., K. Hososno, T. Tsukakoshi, A. Kobayashi, H. Takayama, and S. Wiemer (2006), Change in seismic activity in the Tokai region related to weakening and strengthening of the interplate coupling, *Tectonophysics*, 417, 17–31.

Yoshioka, S., T. Mikumo, V. Kostoglodov, K.M. Larson, A.R. Lowry, and S.K. Singh (2004), Interplate coupling and a recent aseismic slow slip event in the Guerrero seismic gap of the Mexican subduction zone, as deduced from GPS data inversion using a Bayesian information criterio, *Phys. Earth Planet. Interior.*, 146, 513–530.

Yoshioka, S., M. Toda, and J. Nakajima (2008), Regionality of deep low-frequency earthquakes associated with subduction of the Philippine Sea plate along the Nankai Trough, southwest Japan, *Earth Planet. Sci. Lett.*, 272, 189–198.

Zhang, H. & Thurber, C. H. (2003) Double-difference tomography: The method and its application to the Hayward fault, California. *Bull. Seismol. Soc. Am.*, 93, 1875–1889.

Volcanism in Reverse and Strike-Slip Fault Settings

Alessandro Tibaldi, Federico Pasquarè, and Daniel Tormey

Abstract Traditionally volcanism is thought to require an extensional state of stress in the crust. This review examines recent relevant data demonstrating that volcanism occurs also in compressional tectonic settings associated with reverse and strike-slip faulting. Data describing the tectonic settings, structural analysis, analogue modelling, petrology, and geochemistry, are integrated to provide a comprehensive presentation of this topic. An increasing amount of field data describes stratovolcanoes in areas of coeval reverse faulting, and shield volcanoes, stratovolcanoes, and monogenic edifices along strike-slip faults, whereas calderas are mostly associated with pull-apart structures in transcurrent regimes. Physically-scaled analogue experiments simulate the propagation of magma in these settings, and taken together with data from subvolcanic magma bodies, they provide insight into the magma paths followed from the crust to the surface. In several transcurrent tectonic plate boundary regions, volcanoes are aligned along both the strike-slip faults and along fractures normal to the local least principal stress (σ_3). At subduction zones, intra-arc tectonics is frequently characterised by contraction or transpression. In intra-plate tectonic settings, volcanism can develop in conjunction with reverse faults or strike slip faults. In most of these cases, magma appears to reach the surface along fractures striking parallel to the local σ_1. In some cases, there is a direct geometric control by the substrate strike-slip or reverse fault: magma is transported beneath the volcano to the surface along the main faults, irrespective of the orientation of σ_3. The petrology and geochemistry of lavas erupted in compressive stress regimes indicate longer crustal residence times, and higher degrees of lower crustal and upper crustal melts contributing to the evolving magmas when compared to lavas from extensional stress regimes. Small volumes of magma tend to rise to shallow crustal levels, and magma mixing is common in the compressional regimes. In detailed studies from the Andes and Anatolia, with geographic and temporal coverage with which to compare compressional, transcurrent and extensional episodes in the same location, there do not appear to be changes to the mantle or crustal source materials that constitute the magmas. Rather, as the stress regime becomes more compressional, the magma transport pathways become more diffuse, and the crustal residence time and crustal interaction increases.

Keywords Compressional tectonics · Reverse faults · Strike-slip faults · Volcanism · Magma transport

Introduction

Volcanism has been thought to require regional extensional tectonics because this stress state favours magma upwelling along vertical fractures perpendicular to the regional least principal stress (σ_3) (Anderson, 1951; Cas and Wright, 1987; Watanabe et al., 1999, and references therein). Because the greatest principal stress (σ_1) is horizontal in compressional settings, the resulting hydraulic fractures are horizontal (Hubbert

A. Tibaldi (✉)
Department of Geological Sciences and Geotechnologies, University of Milan-Bicocca, Italy
e-mail: alessandro.tibaldi@unimib.it

S. Cloetingh, J. Negendank (eds.), *New Frontiers in Integrated Solid Earth Sciences*, International Year of Planet Earth, DOI 10.1007/978-90-481-2737-5_9,

and Willis, 1957; Sibson, 2003) and therefore sills should form in compressional settings, without associated surface volcanism. Based on these arguments, many authors have concluded that volcanic activity should be rare or absent in compressional settings (e.g. Williams and McBirney, 1979; Glazner, 1991; Hamilton, 1995; Watanabe et al., 1999). However, active volcanoes are common at rapidly convergent margins, which are under horizontal compression. A good example is the Andean margin of South America which is actively shortening under a horizontal σ_1, according to state of stress data and Global Positioning System measurements (Schafer and Dannapfel, 1994; Norabuena et al., 1998; Kendrick et al., 2001). Although there are local variations in crustal stress state along the Andes and other compressional settings, some volcanoes are clearly situated over zones of active compression. In the present paper the term "compression" is related to the state of stress where σ_1 is horizontal; this can produce contractional deformation (i.e. reverse faulting) if σ_3 is vertical, or transcurrent deformation (strike-slip faulting) if σ_3 is horizontal.

To explain the occurrence of volcanism at convergent margins, Nakamura (1977) proposed that the overall tectonics of the arcs should be strike-slip instead of reverse. Although compression is the simplified stress state of convergent margins, transpression and strike-slip faulting can initiate when convergence is only 10% off of perpendicular (Fitch, 1972; McCaffrey, 1992; Busby and Bassett, 2007). A transcurrent regime would allow magma to ascend through vertical dykes parallel to the direction of σ_1 (Nakamura and Uyeda, 1980). This idea is consistent with numerical models of Hill (1977) and Shaw (1980) that suggest composite systems of tensional and shear fractures for dyke propagation, and it is also consistent with field data (e.g. Tibaldi and Romero-Leon, 2000; Pasquarè and Tibaldi, 2003; Lara et al., 2006) and geophysical data (Roman et al., 2004). In some strike-slip fault settings, volcanism has been associated with local dilation occurring at releasing bends and pull-apart basins (Pasquarè et al., 1988; Petrinovic et al., 2006; Busby and Bassett, 2007).

A true contractional tectonic environment with reverse faulting has been considered a highly unfavourable setting for volcanism (Cas and Wright, 1987; Williams and McBirney, 1979; Glazner, 1991; Hamilton, 1995; Watanabe et al., 1999); rather, intrusive emplacement of plutonic rocks is expected (e.g. Cas and Wright, 1987). Based on field mapping of plutonic rocks, it has been suggested that transpressional tectonics can be an efficient mechanism for moving magma through the lithosphere (Saint Blanquat et al., 1998). More recently, Marcotte et al. (2005) suggest that transpression can act as a mechanism for magma upwelling, but it results in the movement of only a small volume of magma to the surface.

Thus, there is growing evidence for volcanism in compressional settings, although the specific mechanisms are still debated. It is timely to use this review of the relevant data on volcanism in compressional settings, including strike-slip and reverse fault tectonics, to help focus the debate. The review is interdisciplinary in considering structural analysis, petrologic and geochemical data, and results from scaled physical models. The focus of this review is convergent margins and intra-plate volcanism; it does not address the relationship between compressional tectonics and plutonism, which has been the focus of a great deal of recent studies (e.g. Olivier et al., 1999; Rosenberg, 2004; Weinberg et al., 2004, and references therein), nor does it address the relationship between strike-slip tectonics and volcanism in ocean ridge or other divergent margin settings.

Understanding whether magma can reach the surface in a zone of regional compression is not only a leading edge matter of debate; there are also significant implications to the mitigation of natural hazards and to natural resources development. The evaluation of volcanic and seismic hazards involves the reconstruction of the structural architecture and the stress state of the volcano and the surrounding basement. Moreover, when conducting hydrogeologic and geothermal resource evaluations in volcanic areas, it is crucial to understand the state of stress in the crust.

Reverse Fault Tectonics and Volcanism

Field Examples

This section summarizes data for active volcanoes whose tectonic settings have a demonstrated correlation between reverse faults and volcanism. In conducting our survey of the literature, we distinguish a true regional contractional tectonic environment with that derived from local compressional deformation resulting from gravitational spreading of substrata under the volcano load, such as at Socompa in Chile (van Wyk de Vries et al., 2001). In some cases the temporal coincidence between contractional tectonics and volcanism is equivocal (e.g. Taapaca in Chile, Clavero et al., 2004), whereas in other cases it has been shown that local volcanic loading can induce a strain partitioning causing deflection and flattening of regional compressive structures (Branquet and van Wyk de Vries, 2001). There are many cases in the literature, however, where field geological and structural data show clear correlation between volcanism and reverse faults. This section divides these several localities and geodynamic settings into intra-plate volcanism and subduction zone volcanism.

Intra-Plate Setting

In the Mojave Desert area (USA), Glazner and Bartley (1994) found that much of the late Cenozoic alkali basalt flows and cinder cones were erupted during crustal contraction caused by oblique convergence. This compressional stress state was expressed in reverse faults and transpressional faults. These authors indicate that other alkali basaltic fields of similar age in the Southwestern USA may be associated with extensional tectonics. Glazner and Bartley (1994) conclude that buoyancy of the alkali basalt magmas provided sufficient overpressure to drive them upward through the actively contracting crust, and that an association of alkali basalts and coarse clastic sedimentary rocks need not indicate an extensional stress state.

Other volcanoes in intra-plate contractional tectonic settings are found in the region between the Caucasus mountain belt and Armenia (Fig. 1). In the Great Caucasus, the Quaternary Elbrus volcano (Russia) is located in the core of a fault and fold region subjected to strong compression and uplift (Kostyuchenko et al., 2004). In Armenia, the Aragats, Gegham and Ararat Plio-Quaternary volcanoes are located adjacent to a series of Quaternary and active faults, showing primarily reverse motions based on detailed field measurements (Westaway, 1990; Mitchell and Westaway, 1999).

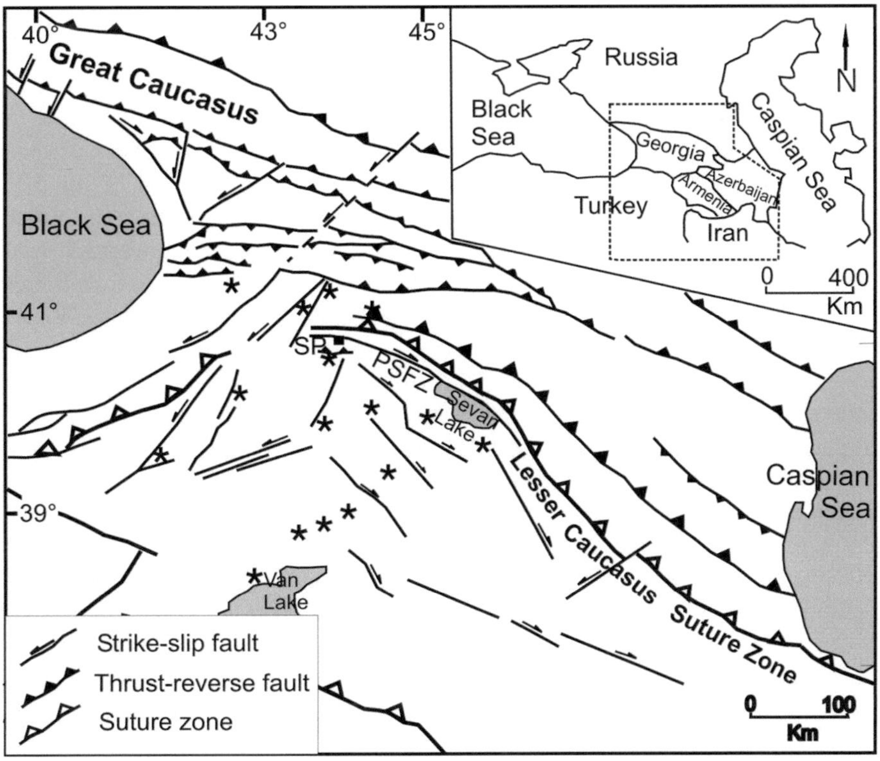

Fig. 1 Main Quaternary faults and volcanoes of the Great Caucasus – Lesser Caucasus zone. The orogenic zone s.s. is affected mostly by reverse faults, whereas south of the mountain belt strike-slip faults dominate. Volcanoes are present both in the transcurrent zones as well as in the northern area affected by pure contractional deformations. SP = town of Spitak; PSFZ = Pambak–Sevan Fault Zone. Modified after Koçyigit et al. (2001)

Subduction Zones

Volcanic arcs overlying subduction zones are the classical locus of compressional tectonics and volcanism. Although arcs contain back-arc zones of extension and transpressional and strike-slip tectonics along the volcanic front, there are several examples of volcanism in zones of contraction. In the Andes of South America, contractional structural settings are commonly found, as described in the following.

In the Northern Volcanic Zone (NVZ) of the Andes (Alemán and Ramos, 2000), the Late Oligocene change in plate convergence rates renewed mountain building by folding and thrusting; including the development of the Cauca depression (Colombia) and the Interandean Valley (Ecuador) between the Cordillera Occidental and Central/Real (Fig. 2), as well as the beginning of an important cycle of magmatic activity (Barberi et al., 1988; Lavenu et al., 1992). Some movement was accommodated by reactivation of strike-slip faults (Litherland and Aspden, 1992; Ferrari and Tibaldi, 1992). In the Northern Central Andes crustal shortening and uplift in a compressional setting occurred during the following time intervals: Aymara (28–26 Ma), Pehuenche (25–23 Ma), Quechua 1 (17–15 Ma), Quechua 2 (9–8 Ma), Quechua 3 (7–5 Ma) and Diaguita (3–5 Ma) events (Allmendinger et al., 1990, 1997; Jordan and Gardeweg, 1989; James and Sacks, 1999; Ramos et al., 2002). Magmatic activity occurred during these episodes, initially in the eastern Cordillera Oriental and subsequently broadening to encompass the currently active magmatic belt.

Looking more closely at the NVZ of the Andes, a stress tensor analysis of the 1998–1999 tectonic swarm of earthquakes that took place north of Quito, Ecuador, showed reverse fault mechanisms (Legrand et al., 2002). These reverse faults occurred in the area of the active Guagua Pichincha stratovolcano (Fig. 2). Field structural analysis of this area shows Quaternary reverse faults linked to an approximate E–W directed shortening (Ego et al., 1996). The active Cotopaxi volcano is located tens of kilometres south of Guagua Pichincha, in proximity to a transition between reverse faults and strike-slip faults (Tibaldi, 2005a) (Fig. 2). The faults surrounding both volcanoes belong to the contractional fault tectonics characterizing the western part of the northern Interandean valley.

Further south in the Interandean Valley in the Latacunga area, a series of Pliocene-Quaternary stratovolcanoes are chronologically linked to the presence of reverse faults (Winkler et al., 2005) (Fig. 2), including a location above the hanging wall of the Pisayambo reverse fault along the eastern side of the valley. Another volcano in the Latacunga area is located in the block bounded to the west by the west-dipping La Victoria reverse fault; and to the east by the east-dipping Pisayambo fault; this crustal architecture indicates that magma found its way to the surface across a full-ramp compressional basin (Cobbold et al., 1993).

In the southernmost part of the Interandean Valley (Cuenca area), several late Miocene intrusions and a few effusive lavas are present (Fig. 2). The dioritic Cojitambo intrusion forms the prominent peak of Cojitambo. Radial columnar cooling structures suggest that the intrusion penetrated the sediments at shallow depth (Hungerbuhler et al., 2002). Two ages of 5.4±0.6 and 7.8±0.8 Ma have been obtained from the intrusion (Steinmann, 1997). The younger 5.4±0.6 Ma age, obtained from a large dacite block displaying flow structures, may suggest a later extrusive phase (Hungerbuhler et al., 2002). From 8 to 9 Ma onward the southern Interandean Valley has been characterised by the development of a contractional reverse fault tectonics, thus contemporaneous to volcanism and subvolcanic intrusions (Steinmann, 1997; Hungerbuhler et al., 2002).

Five large, active stratovolcanoes are present in the compressional Subandean region of the Ecuadorian portion of the NVZ (Reventador, Sumaco, Pan de Azucar, Cerro Negro and Sangay, Fig. 2). Reventador stratovolcano is the best studied; the currently active volcano commenced eruptions at 0.32 Ma BP (INECEL, 1988). N–S- to NNE-striking reverse faults cut this area in the Quaternary (Tibaldi and Ferrari, 1992); during late Pleistocene-Holocene activity of Reventador, some reverse faults with the same orientations were still active, locally with a component of right-lateral strike-slip motions. Seismicity shows widespread events with intensity up to 6.9 Mw and compression (Tibaldi, 2005). Some of these faults cut the volcano creating differential uplift along N–S- to NNE-trending tectonic blocks. Hypocenter swarms and field data indicate east-vergent basal thrusting under the volcano and N–S-trending fault-propagation folds to the east. Thus, field and geophysical data demonstrate that the entire history of development of

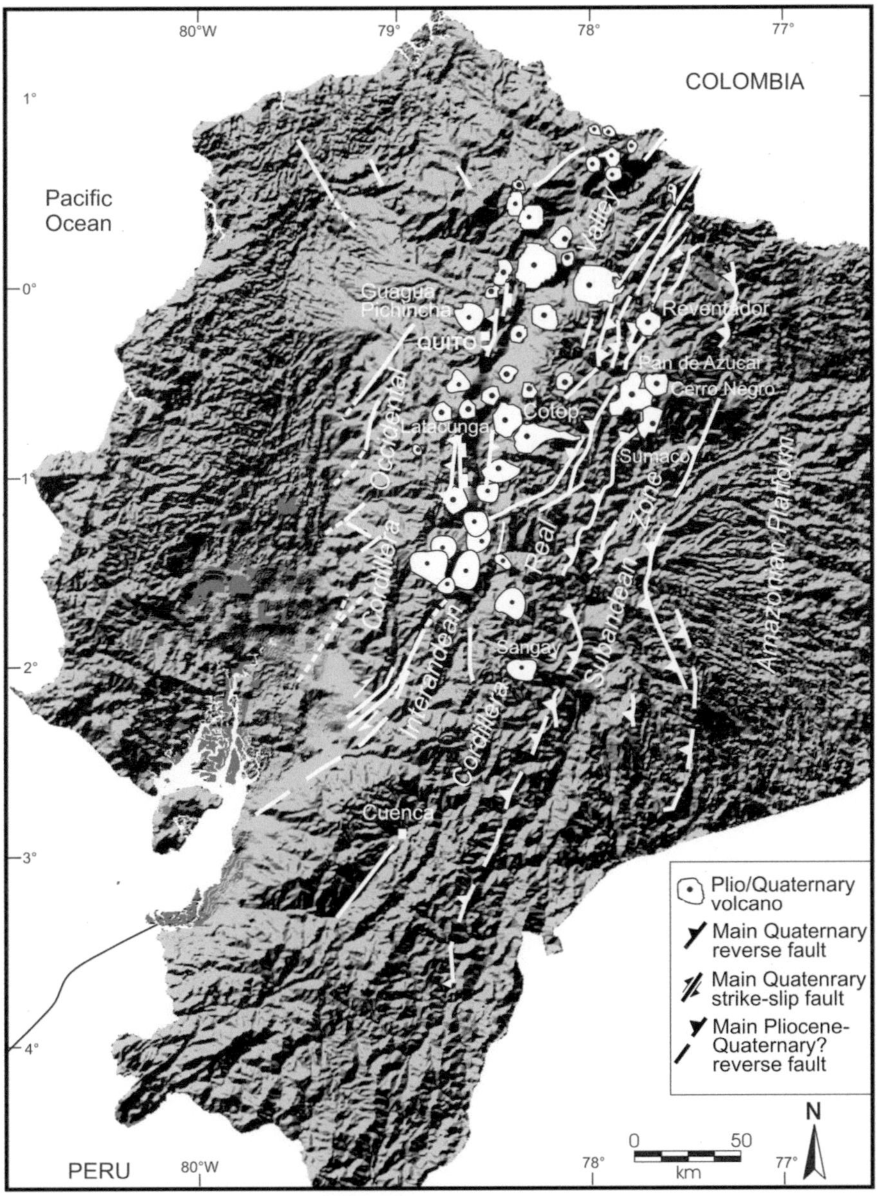

Fig. 2 Main Plio-Quaternary volcanoes and faults in the Ecuadorian volcanic belt, NVZ of the Andes. Note the presence of volcanoes along or very close to reverse faults. The volcano names refer to cases described in the text. Based on data from Barberi et al. (1988), Tibaldi (2005a, b), Tibaldi and Ferrari (1992), Winkler et al. (2005), Tibaldi et al. (2007)

the El Reventador volcano occurred within a contractional tectonic settings with reverse to reverse-oblique faulting (Tibaldi, 2005b). The Quaternary Sumaco and Sangay, and the Pliocene Pan de Azucar and Cerro Negro are all located within the Napo uplift, a block defined by two major west-dipping reverse faults that underwent major uplift and compression during Plio-Quaternary times (Pasquarè et al., 1990; Spikings et al., 2000).

Magmatic activity in the Central Volcanic Zone (CVZ) of the Andes in the time period from Late Oligocene to Recent times has been extensively studied (Coira et al., 1982, 1993; Petford and Atherton, 1985; Solar and Bonhomme, 1990; Hammerschmidt et al., 1992; Peterson, 1999; James and Sacks, 1999; Kay et al., 1999; Wörner et al., 2000b; Victor et al., 2004). The Calchaquì valley in Argentina contains several volcanoes associated with reverse faults (Guzmàn et al., 2006). Los Gemelos and El Saladillo are both monogenetic, strombolian, basaltic-shoshonitic volcanoes that constitute the easternmost recognized examples of mafic Plio-Quaternary volcanism in the southern CVZ (Fig. 3). Two regional faults delimit the borders of the Calchaquí valley, as thrusts with opposite vergence: the eastern Calchaquí fault and the western Toro Muerto fault. While Los Gemelos are set in the hanging wall of Calchaquí back-thrust fault, El Saladillo is set in the footwall of Toro Muerto fault (Fig. 3). Fault kinematic data indicate right strike-slip movement in the vicinity of Los Gemelos during the Pleistocene-Holocene (Marrett et al., 1994; Guzmàn et al., 2006). The emplacement of these volcanoes should be related with a reverse to transpressional tectonic zone parallel to the valley, where the alignment of the cones is outlining the trend of conjugated faults.

In the CVZ immediately east of the Atacama Desert in Chile, thrust faults are related to compressional neotectonics (Flint et al., 1993). A small unnamed volcanic cone is trapped in the thrust system; Quaternary ignimbrites (underlying the cone) are deformed by west-verging thrusts striking N–S (Fig. 4B; Branquet and Van Wyk de Vries, 2001). The thrust traces display a westward arcuate deflection at the western cone foot, whereas to the south one of the thrusts curves into a NW–SE-striking left lateral strike-slip oblique thrust. This cone lies above the hanging wall of a reverse main fault.

The Southern Volcanic Zone (SVZ) of the Andes, between latitude 30°S and 47°S has been extensively studied with an emphasis on the relationship between tectonics and magmatism (Lopez-Escobar et al., 1977, Hickey et al., 1986, Futa and Stern, 1988, Hildreth and

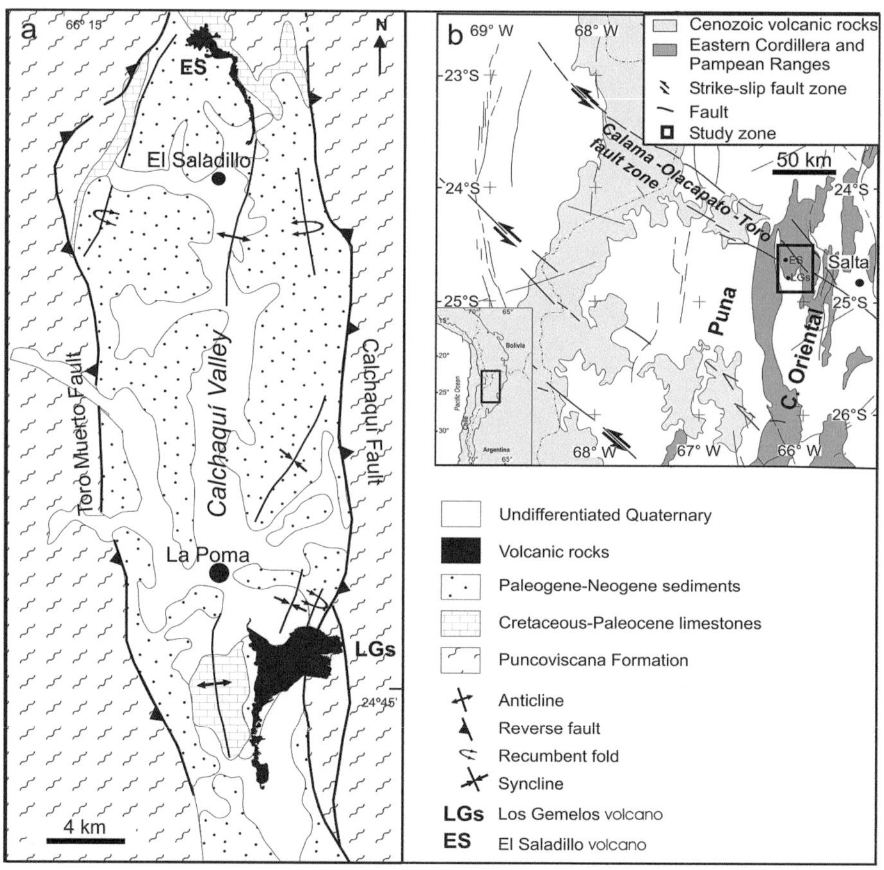

Fig. 3 Los Gemelos and El Saladillo are both monogenetic, strombolian, basaltic-shoshonitic volcanoes that constitute the easternmost recognized examples of mafic Plio-Quaternary volcanism in the southern CVZ. Two regional faults delimit the borders of the Calchaquí valley, as thrusts with opposite vergence: the eastern Calchaquí fault and the western Toro Muerto fault. While Los Gemelos are set in the hanging wall of Calchaquí back-thrust fault, El Saladillo is set in the footwall of Toro Muerto fault (after Guzmàn et al., 2006)

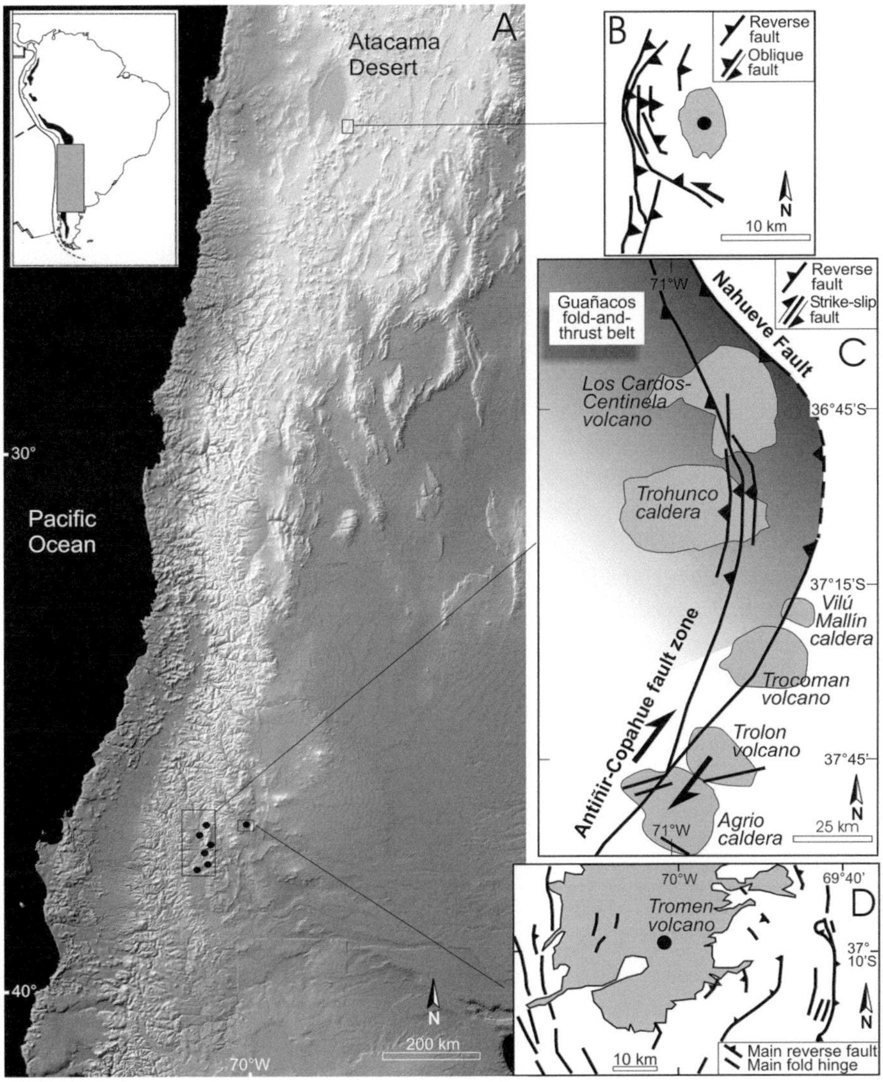

Fig. 4 Location and structures of selected centers of the Argentina-Chile volcanic arc discussed in the text, showing the association of reverse faults with coeval volcanism based on field data. Case *B* has been redrawn after Branquet and Van Wyk de Vries (2001); case *C* after Tibaldi (2008); case *D* after Galland et al. (2007a)

Moorbath, 1988, Tormey et al., 1991, Ramos et al., 1996; Dungan et al., 2001, Folguera et al. 2006a,b, Kay and Ramos 2006). The arc has a well-defined volcanic front, zone of back-arc extension, and transition zones between the two. The Liquine-Ofqui Fault Zone is the controlling fault for activity along the volcanic front between 37 and 47 S. North of this zone, the Agrio Fold and Thrust Belt and the Malargue Fold and Thrust Belt mark a transition to a transpressional and contractional zone (Ramos et al., 1996; Folguera et al., 2006).

In the Late Oligocene to Miocene extension prevailed (Stern, 2004), and magmatic activity occurred well west of the current position of the magmatic front, within the Coastal Cordillera at the northern end of the Southern Andes (33°S; Wall and Lara, 2001), and to as far west as the Pacific coast at latitude 42°S (Muñoz et al., 2000), where magmatic activity also extended as far east as the Atlantic coast. Extension ended by the Early to Middle Miocene (Stern, 2004). The sediments and volcanic rocks deposited in the Late Oligocene to Early Miocene basins were deformed and uplifted in the Middle to Late Miocene when the magmatic arc returned to its current location in the Main Cordillera (Godoy et al., 1999; Muñoz et al., 2000; Jordan et al., 2001).

Looking more closely at the SVZ, Tromen volcano lies in the foothills in Argentina, near the western edge of the Neuquen basin (Fig. 4D). The edifice is more than 3,000 m high and more than 30 km in diameter. The main period of volcanic activity was late Pliocene to Holocene (Zollner and Amos, 1973; Holmberg, 1975; Llambıas et al., 1982; Galland et al., 2007a). However, Tromen is still active and may have erupted in 1822 (Marques and Cobbold, 2002; Simkin et al., 1981). Its main volcanic products are widespread basaltic lava flows and a few domes of andesite or dacite (Groeber, 1929; Zollner and Amos, 1973; Holmberg, 1975; Galland et al., 2007a). According to recent $^{39}Ar/^{40}Ar$ data, Tromen has been almost continuously active since 2.5 Ma (Galland et al., 2007a). The Mesozoic substratum of Tromen has shortened and thickened tectonically, although it is unclear whether the shortening was due to thick-skinned or thin-skinned tectonics (Kozlowski et al., 1996; Zapata et al., 1999). The main structure is an eastward-verging thrust system (Tromen thrust) at the eastern foot of the volcano. In its hanging wall, a series of thrust faults and anticlines have brought Jurassic evaporites to outcrop (Galland et al., 2007a).

In the Eastern Neuquén Andes of Argentina the Trohunco Plio-Pleistocene volcanic complex (Fig. 4C) has an average diameter of 15 km, a height of 2,700 m a.s.l., and its substrate crops out at an altitude of 1,500 m. Andesitic lavas and breccias dip radially outwards from a 15-km-wide semicircular depression interpreted as a caldera (Folguera et al., 2006a). Miocene intracaldera volcanic breccias and tuffs outcrop on the hanging wall blocks of two west-dipping reverse faults. These faults are confined within the Miocene rocks and are interrupted by the Pliocene deposits of the Trohunco volcano, suggesting that the two faults moved prior to the emplacement of the volcano. On the eastern caldera floor, Miocene volcanic breccias are thrust onto Miocene sedimentary rocks along a series of west-dipping planes. Some of these planes are confined within the Miocene rocks, but others also cut the Pliocene or the Plio-Quaternary deposits indicating faulting coeval to the emplacement of the Trohunco volcanic complex. The orientation, kinematics and location of this fault swarm are consistent with the regional trace of the Quaternary Antiñir-Copahue fault zone (Fig. 4C) (Miranda et al., 2006).

The Los Cardos-Centinela volcanic complex is located about 15 km northeast of the Trohunco caldera (Fig. 4C). The volcano has an average diameter of 19 km, a height of 2,841 m a.s.l., and its substrate crops out at an altitude of 1,500 m. The rocks consist of basaltic lava flows, breccias and pyroclastic deposits (Miranda et al., 2006). The older rocks are composed of basal breccias and lavas dated 3.2–26 Ma (Rovere, 1993). A sequence of volcanoclastic deposits and lava flows of Quaternary age overlie these deposits, with some eruptive centers still recognizable, mantled by lavas erupted from other small centers. The age of these eruptive centers has been assigned to the late Pliocene or Pleistocene by Folguera et al. (2006a) and, at least in part, younger than 30 ka by Miranda et al. (2006). These authors also indicate the presence of a kink fold affecting the eastern slope of the Los Cardos-Centinela volcanic complex. Here the Pliocene lavas have a primary dip of 15°, but a kink increases the average strata dip to 25°. Folguera et al. (2006a) suggest that this kink fold is associated with a west-dipping thrust fault exhuming the core and substrate of the volcanic complex. The main frontal thrust of the Plio-Quaternary Guanacos fold-and-thrust belt, known as Nahueve Fault, is located farther east (Fig. 4C). This fault strikes NW–SE and dips to the SW. A NNW-striking reverse fault affects the western flank of the volcanic complex, confined within the Miocene rocks, whereas a series of Plio-Quaternary folds with NNW-trending hinge lines are present farther west. Another fault dips WSW to W and offsets Plio-Quaternary deposits, suggesting it might belong to the Quaternary Antiñir-Copahue fault zone (Fig. 4C) (Miranda et al., 2006). In spite of these data, another interpretation recently suggests that these volcanoes are located in a zone affected by normal faults in the Quaternary (Folguera et al., 2008), thus living open the debate.

The Japanese volcanic arc also has studies linking contractional tectonics and magmatism. Field and offshore data indicate that the structural evolution of the eastern Japan Sea margin and northern Honshu Island has been characterised by a Late Miocene-Quaternary compressional plate-tectonic regime (van der Werff, 2000; Yoshida, 2001) (Fig. 5A), resulting in widespread basement inversion of previous normal faults into reverse faults and low angle thrusts. Particularly during Pliocene-Recent times, mainly east-dipping reverse faults extended along the base of the continental slope and are associated with small active arc volcanoes. The relation between reverse faulting and coeval volcanism is also expressed by onshore geophysical data. Iwate stratovolcano is located in northern Honshu and has been active since the 1997 (Fig. 5B). Although the volcano did not erupt, on

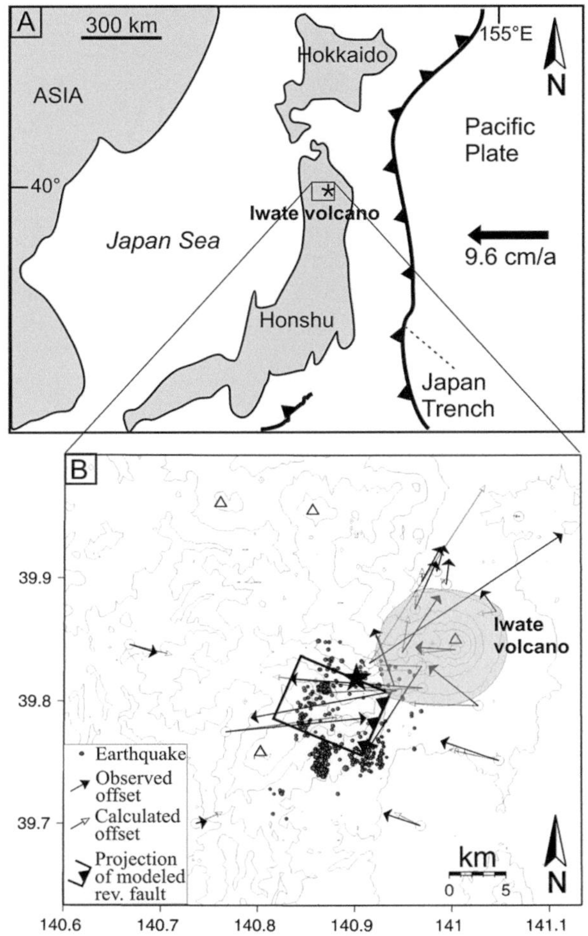

Fig. 5 (**A**) Location of the Iwate Volcano in the Japan Volcanic Arc, Northern Honshu Island. This area is under a Quaternary compressional state of stress. The *black line* with *triangles* indicates the Japan subduction trench. *Arrow* indicates subduction rate after Nakanishi (1989). (**B**) Coseismic displacements associated with the M 6.1 compressional tectonic earthquake of September 3, 1998. Black and white arrows indicate observed and calculated displacements, respectively. The rectangle fault model was estimated from inversion analyses and represents the projection of the seismogenetic reverse fault. Black triangles on the hanging wall block. The black star gives the earthquake epicenter, and the open triangles denote Quaternary/active volcanoes. Modified after Miura et al. (2000)

September 3, 1998, a large earthquake with a M = 6.1 hit Shizukuishi-town that is located southwest of the summit of the volcano (Nakahara et al., 2002). The earthquake focal mechanism is given by a reverse fault dipping to the west (Earthquake Information Center, University of Tokyo, 1998), and a part of the fault reached the ground surface at the northeastern part of the focal area (Koshiya and Ohtani, 1999). Analyses of the local GPS network data confirm that the reverse motions occurred on a tectonic fault that averagely slipped 0.37 m on a plane striking N203° and dipping 35° (Miura et al., 2000) (Fig. 5B).

Analogue Modelling Data

Analogue modelling of volcanic systems provides important insights into the mechanics of compressional stress and magma intrusion. These experiments are scaled physical models in which deformation is synchronous with fluid injection. In the sandbox experiments of Roman-Berdiel (1999), stiff silicones were used to represent magmas of high viscosity (10^{18} Pa s), and the resulting intrusions were thus wide. In the experiments of Benn et al. (1998), a single piston controlled the rates of both shortening and fluid injection, coupling them artificially. In these experiments fluid viscosity was too large for hydraulic fracturing to occur, and thus to simulate the propagation of magma paths. The first analogue experiments on hydraulic fracturing were designed to investigate borehole breakouts in oilfield development (Hubbert and Willis, 1957). The material representing host rock was gelatine, which failed readily in axial tension.

To study interactions between magmatic and regional tectonic processes in a brittle crust, Galland et al. (2003) developed an experimental method in which boundary stresses and an injecting fluid of low viscosity lead to hydraulic fracturing in suitably weak model material. From the shape and orientation of intrusions, these authors suggest that hydraulic fracturing was a viable emplacement mechanism in compressional settings. The experiment al results led Galland et al. (2003) to conclude that magmas in orogenic belts can rise along thrust faults, and in turn unconsolidated intrusions may strongly influence thrust formation.

Galland et al. (2007a) also showed movements of magma in the crust along thrust planes, demonstrating that magma can propagate to the surface in a tectonic regime characterised by contractional deformation (Fig. 6A). If the thrust in the substrate is arcuate in plan view, vertical tension fractures can form in the shallower portion of the hanging wall subparallel to the imposed shortening (Galland et al., 2007a). These structures can be attributed to superposition of

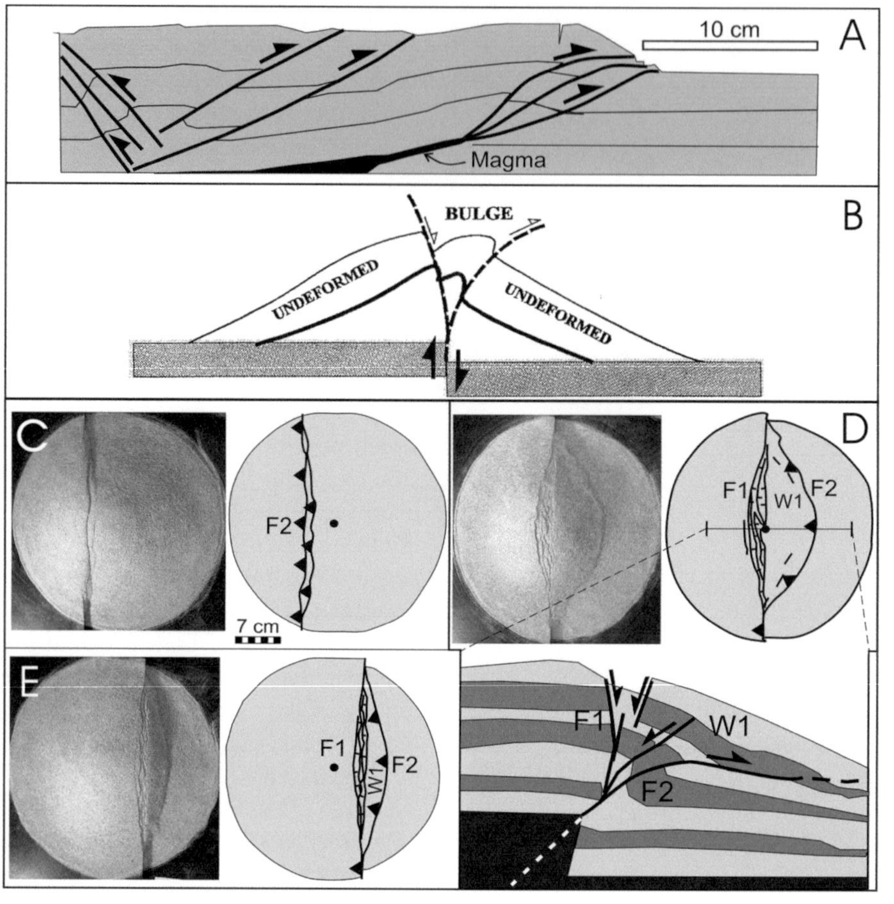

Fig. 6 Examples of experiments where volcanoes in contractional settings have been modelled, apart from case *B.* where deformation in a volcano lying above a vertical fault has been simulated. In (**A**) Galland et al. (2007a) showed movements of magma in the crust along thrust planes. In (**B**) Merle et al. (2001) studied the role played by the reactivation of a vertical basement fault in deforming the overlying volcano, resulting in two splay faults, one normal and one reverse. In (**C, D,** and **E**) photos in plan view (*to the left*) and fault traces (*to the right*) of experiments of volcanic cones lying above substrate reverse faults, dipping to the left, with different position with respect to the cone (from Tibaldi, 2008); (**C**) cone summit located on the footwall block, (**D**) above: the cone summit above the surface trace of the substrate fault, (**D**) below: section view of the deformation within the cone; (**E**) cone summit on the hanging wall block

a local load, due to the uplifted area, on the regional stress field (Johnson, 1970).

Deformation of the volcanic cone itself has been investigated in several experimental studies, but few focus on compressional stress. Merle et al. (2001) studied the role played by the reactivation of vertical basement faults in the destabilisation process of overlying volcanic cones. Results show that basement fault reactivation induces the formation of two splay faults within the volcano; these faults, one normal and one reverse, define a central block (Fig. 6B). Experimental models of volcanic cones located above brittle substratum undergoing regional compressive deformation showed that the volcanic loading induces a strain partitioning involving deflection and flattening of regional compressive structures (Branquet and Van Wyk de Vries, 2001). These authors suggest that although anticlinal thrust ridges, observed around many volcanoes, have generally been interpreted as being due to gravitational spreading, their study shows that this is not necessarily the case, as they can also be a symptom of regional compression.

A series of experiments have been exclusively devoted to reconstruct the deformation pattern in a volcano overlying reverse faults with different dip and position respect to the cone (Tibaldi, 2008). In cases where the volcano summit was located on the footwall block (in Fig. 6C–E the substrate reverse fault dips to the left), the substrate fault cuts across the cone flank creating essentially a single deformation zone characterised by reverse motions (Fig. 6C). This pattern was found to be the same for all the investigated fault dips, from 25° to 40°. In the experiments where a cone overlies the surface trace of the substrate fault (Fig. 6D), the substrate fault splays into a subvertical normal fault F1 and a subhorizontal thrust F2. These findings are similar to those of Vidal and Merle (2000) and Merle et al. (2001), although they refer to a substrate vertical fault and hence not to a contractional setting. In section view (Fig. 6D, below), it can be clearly seen that the layering within the volcano is offset along a series of extensional structures F1 to the left and along the gently inclined reverse fault F2 to the right; these two fault zones bound the block W1. In cases where the volcano summit was located on the hangingwall block, the substrate fault cuts across the cone flank creating two zones of deformation, similar to the previous case (Fig. 6E). The outcomes of Tibaldi (2008) indicate that in most cases a zone of local extension forms across the cone, whereas contractional deformations are concentrated along a shear zone located at lower elevations in the cone flank and with an arcuate trace in plan view. The zone of local extension is the prime candidate to allow magma upwelling inside the cone.

Magma Paths

Although it is clear that volcanoes overlie some zones of active compression, there are few field or experimental data that indicate which plane of principal stress constitutes the magma path in contractional settings. At Iwate volcano (NE-Japan, Fig. 5) during the 1998 crisis (http://hakone.eri.u-tokyo.ac.jp/vrc/erup/iwate.html), a large earthquake of M=6.1 occurred at 16:58 (JST) on 3 September 1998 about 10 km southwest of Iwate Volcano, which was followed by small earthquakes scattered around the first one. According to Tohoku University, the hypocenter of the largest earthquake is located just south of the pressure source of the deformation that had continued at Iwate Volcano during 1998. The focal mechanism analysis indicates a reverse faulting with the principal compressive stress axis in the E–W direction. N–S-striking reverse active faults are present in this area; the activity of one of them, the Nishine-fault, was estimated to be about 0.7 m/1000 years by the Japan Active-Fault Research Group. GPS surveys showed a steady and continuous extension between southern and northern sites of the volcano in 1998 (Miura et al., 2000). Based on these results, it has been suggested that intrusion of an E–W-striking dyke at around 10 km-depth below the summit of the volcano, thus resulting in a magma path parallel to the direction of σ_1.

At Tromen volcano (Galland et al, 2007a), lying (Fig. 4D) above the hanging wall of a coeval west-dipping thrust fault, most of the eruptions have been from the upper part of the edifice. In the central part of the volcano, the magmatic conduits are in the form of subvertical andesitic dykes, striking almost E–W. On the eastern and southern flanks, E–W alignments of volcanic domes probably follow underlying fractures (Galland et al., 2007a, b). Also in this case dykes intruding the volcano are parallel to the direction of σ_1.

In the experiments carried out by the same authors (Fig. 6A), fluid erupted from arcuate thrusts at the leading edges of the plateaus, the conduits being inclined dykes. Vertical tension fractures formed on the surface of the plateau, in directions almost parallel to the imposed shortening. This situation has been attributed by Galland et al. (2007a) to superposition of a local load, because of the uplifted area, on the regional stress field. At a shallow depth, the stress field was mainly due to the load of the uplifted area, so that the greatest stress was vertical, whereas at deeper levels the stress was mainly due to regional compression, so that the greatest stress was horizontal. In any case, the vertical tension fractures did not serve as conduits in these experiments.

The experiments of Tibaldi (2008) do not show fractures parallel to the contraction direction, probably because the reverse/thrust faults affecting the substrate are not arcuate in plan view as those simulated in the experiments (Fig. 6A) of Galland et al. (2007a). In Tibaldi's (2008) experiments (Fig. 6C–E), the propagation of the rectilinear (in plan view) reverse faults of the substrate across the volcanic cone results in a local extensional field with σ_3 parallel to the general contraction direction, suggesting that in a contractional tectonic setting, below a volcano, magma might migrate along the substrate reverse fault and then, inside the volcano, along the steeper-dipping fault zone F1. This magma migration along the steeper fault is facilitated by the very local extension and is supported by some field findings of the eruptive vents close to or along the F1 zones, such as at Trohunco caldera and the Los Cardos-Centinela volcanic complex. Magma migration along faults is consistent with several cases found in different geodynamic settings and spanning from evidence at deep crustal level (e.g. Rosenberg, 2004), to near surface level (e.g. Tibaldi et al., 2008), and to emplacement of dykes and vents along pre-existing fractures (Tibaldi, 1995; Ziv and Rubin, 2000; Corazzato and Tibaldi, 2006).

At Guagua Pichincha volcano, Ecuador (Legrand et al., 2002), and Miyakejima volcano, Japan (Fujita et al., 2004), where the tectonic settings are also compressional, seismic data suggest subhorizontal magma transport at depth, followed by subvertical transport nearer the surface. This two-part magma ascent, deep transport along inclined sheets along reverse/thrust faults and shallow transport in vertical dykes, can therefore explain the magma paths in contractional settings. The orientation of dykes within the volcano still requires clarification, and other geometries are permitted by the data.

Strike-Slip Fault Tectonics and Volcanism

Field Examples

Volcanism in transcurrent tectonic settings is more common that that in contractional settings, and examples worldwide are available. However, the presence of magmatism along major continental transform faults and their role as conduits between the upper mantle and the Earth's surface have been seldom discussed in the literature (Skulski et al., 1987; 1991; Xu et al., 1987; Aydin et al., 1990). Similar to the summary of compressional tectonics, we will document first the intraplate setting and then the subduction setting. Since the examples include different volcano types, we subdivide the discussion between composite volcanoes and minor volcanic edifices located within the above structural settings. The relationship between strike-slip faulting and calderas will be described in a dedicated subsection.

Intra-Plate Setting

Anatolia is one of the best regions worldwide to study volcanism associated to strike-slip faulting; volcanism is associated with a lithospheric block escaping westward as a consequence of the Cenozoic, N–S directed convergence between Eurasia and Africa-Arabia (Dewey et al., 1986). The tectonic extrusion occurs among two major strike-slip faults: the right-lateral North Anatolian Fault (NAF) and the left-lateral East Anatolian Fault (EAF). Adiyaman et al. (2001) illustrate Neogene and Quaternary volcanic activity along the NAF, which took place at transtensional segments of the fault that formed pull-apart basins in locations such as the Galatia Massif, the Niksar Basin and the Erzincan Basin. Quaternary volcanic activity in the Niksar pull-apart basin has been investigated also by Tatar et al. (2007). In the Erzincan Basin, the occurrence of a dozen andesitic volcanic cones and many hot springs along the traces of the master faults bordering the basin was reported by Aydın and Nur (1982).

Adiyaman et al. (1998) has analyzed satellite images and DEM in selected volcanic areas of Eastern Anatolia; this area, located between the Taurus suture zone and the Caucasus belt, is characterized by the Arabia-Eurasia convergence. This study identifies the Nurettin Daglar complex, an example of volcanism related to a horsetail-shaped fault tip: An E–W strike-slip fault, the Akdogan fault, ends to the east into a set of subparallel fractures that gradually turn NE and NNE, hence producing a horsetail. Radar images (Adiyaman et al., 1998) also show a group of edifices, the Kandildag volcanoes, emplaced within the releasing bend of a 270-km-long, Tutak-Hamur-Caldiran strike-slip fault system. Toprak (1998) analyzed the distribution of polygenetic and monogenetic volcanoes of the Neogene-Quaternary Cappadocian Volcanic Province (CVP, Central Anatolia) to identify the relationship between the location of vents and regional tectonic lineaments. This work elucidates the different role played by two different tectonic lineaments in controlling volcanism in the province. The late Miocene-Quaternary Tuzgolu-Ecemis fault system and the CVP fault system, active from Mid- Miocene to Late Pliocene times, generated the NE–SW elongated depression which hosts the volcanic province. According to Toprak (1998), the CVP fault system controls the general location of the volcanic edifices in the province since the depression was produced by this fault system. However, the exact location of the volcanoes is mostly influenced by the Tuzgolu and Ecemis, strike-slip fault zones, as illustrated by the distribution of polygenetic volcanoes, grouped in two clusters along the two lineaments. The intersection of the two strike-slip fault zones appears an important factor, at least in the case of Quaternary Erciyes and Hasandag polygenetic volcanoes. As concerns the monogenetic centers, these are aligned along to the older CVP fault system in the westernmost sector of the province, but as they approach the Tuzgolu fault, they tend to parallel it; Toprak (1998) interprets this evidence in terms of feeder-dykes using the recent fractures of the Tuzgolu fault zone to reach the surface and feed the monogenetic cones.

Tectonic manifestations of the recent collision between the Arabian plate and the Eurasian plate are present in the Armenian Upland region, structurally dominated by the North Armenian active fault arc, composed of steeply-dipping, oblique to strike-slip faults. Within this setting, the Syunik rhombus-shaped structure is regarded (Karakhanian et al., 1997) as a typical pull-apart basin formed along the active, right-lateral Khanarassar fault zone, due to the *en-echelon* overstepping of two strike-slip segments. Late-Pleistocene to Holocene andesitic-basaltic volcanoes are located within and along the borders of the Syunik pull-apart structure. North of Armenia, in the Georgian Lesser Caucasus, Rebai et al. (1993) documented in the Aboul-Samsar volcanic range, the existence of a set of volcanic cones and composite volcanoes, clearly associated with the activity of NE-striking, N-striking and NW-striking faults, most of which show evidence of strike-slip activity (Fig. 1).

In the Tibetan Plateau, Quaternary volcanism has been volumetrically limited but geographically widespread (e.g. Deng, 1993; Turner et al., 1996). Cooper et al. (2002) analysed Quaternary volcanic rocks from the Ashikule Basin (AKB) in northwestern Tibet, in order to gain insight into the origin of the magmas produced in this continental collision setting. The basin contains several silicic domes (450–500 and 250–300 ka) and trachyandesitic flows (250–300 and 66–120 ka). From a structural point of view, this area corresponds to the intersection of three major strike-slip systems: the Altyn Tagh, the Ghoza-Longmu Co and the Karakax. Cooper et al. (2002) propose that the distribution of volcanic rocks in northern and central Tibet is spatially associated with strike-slip faults. Isotopic and chemical data for young (<120 ka) trachyandesites suggest that lithospheric thinning across a releasing bend in a strike-slip fault could have produced melts similar to the AKB lavas. Farther north, in Central Mongolia, an area of Holocene volcanism situated in the Taryat Basin was triggered by left-lateral, strike-slip displacements on an E–W magma-tapping fault (Chuvashova et al., 2007).

Subduction Zones

Oblique subduction has been shown to strongly favor the development of strike-slip fault systems; strain is partitioned into a component normal to the plate boundary and a wrench component taken up by strike-slip deformation within the overriding plate (e.g., Fitch, 1972; Beck, 1983; Glazner, 1991; Scheuber and Reutter, 1992). A clear example of this geodynamic arrangement is the oblique convergence between the Philippine Sea Plate and the Philippine archipelago (Fitch, 1972), forming the left-lateral Philippine Fault System (PFS). Lagmay et al. (2005) focus on the justly famous Mayon Volcano (2,463 m), located on the Bicol Peninsula in the south-eastern part of Luzon Island. Their structural analysis reveals that the growth of the volcano is most likely linked to a major releasing bend in the PFS, characterized by substantial transtensional deformation. The influence of the PFS on the volcanoes of the north-western Bicol Volcanic Arc (Luzon, Philippines) has also been observed in two composite volcanoes, Mt. Labo and Mt. Caayunan volcanoes (Pasquarè and Tibaldi, 2003). The authors link the existence of these two volcanoes to the tectonic control exerted both by regional NW-directed main faults (sub-parallel to the PFS) and by NE-trending secondary faults roughly orthogonal to the PFS (Fig. 13B).

Oblique subduction also occurs approximately 2,000 km north at the convergence between the Philippine Sea Plate and SW Japan. Right-lateral, strike-slip movements have taken place during the Quaternary along the Median Tectonic Line (MTL) on Shikoku Island (Kanaori et al., 1994). Gutscher and Lallemand (1999) identified one more, active dextral strike-slip fault further north, in western Honshu Island, which they called the North Chugoku Fault Zone (NCFZ). Several volcanoes are aligned both along the MTL and the NCFZ; in particular, Unzen volcano (Fig. 7) is

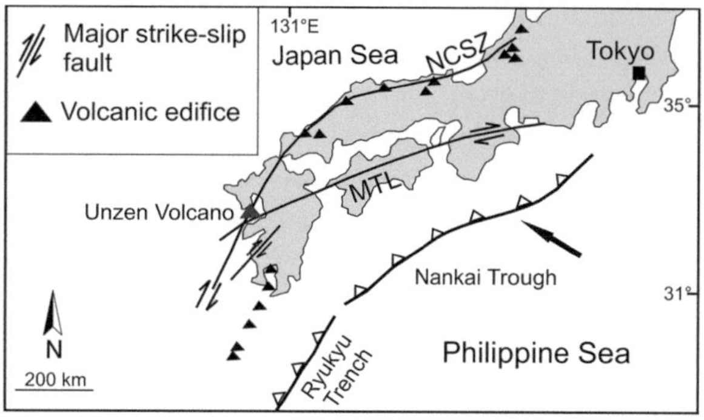

Fig. 7 Tectonic setting of SW Japan with the oblique, NW-directed subduction of the Philippine Sea Plate and active volcanoes. The dextral strike-slip movement is partitioned along the North Chugoku Shear Zone (NCSZ) and the Median Tectonic Line (MTL). Modified after Gutscher and Lallemand (1999)

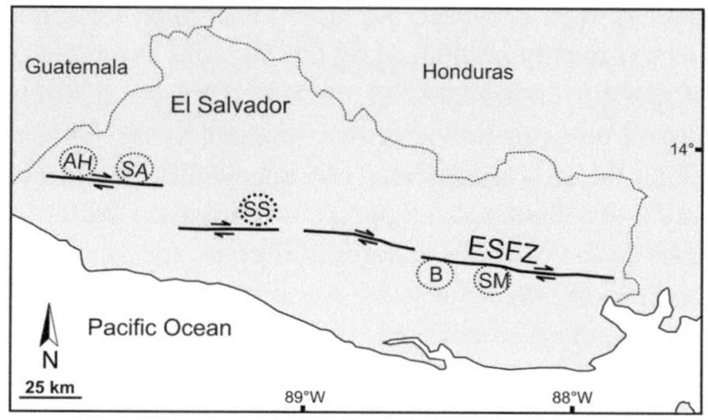

Fig. 8 Schematic representation of the main fault segments of the El Salvador Fault Zone (ESFZ). The main segments display active volcanism, but the intervening regions are devoid of volcanoes. AH, Ahuachapàn area; B, Berlın Volcano; SA, Santa Ana Volcano; SM, San Miguel Volcano; SS, San Salvador Volcano. Modified after Agostini et al. (2006)

adjacent to the intersection of these two major strike-slip faults (Gutscher and Lallemand, 1999).

In the Solomon Island Arc, oblique convergence has generated left-lateral transtension, as discussed by Auzende et al. (1994). Based on the interpretation of geophysical profiles, the authors demonstrate block tilting, rhombohedral deformation and widespread volcanic activity in the Mborokua Basin, coinciding with a pull-apart formed between two WNW-trending, left-lateral strike-slip lineaments.

In Antarctica, the Ross Sea Region has been affected, from 30 Ma on, by the reactivation of NW–SE trending, regional, strike-slip faults (Salvini et al., 1997). N–S extensional faults occur between NW–SE faults and indicate E–W extension during the Cenozoic, induced by NW–SE right-lateral strike-slip motions together with regional crustal extension. Volcanic activity concentrates along N–S fault alignments in a belt between Victoria Land and the Ross Sea. Central volcanoes lie at intersections of these alignments with the regional, strike-slip faults.

The Cocos Plate subducts obliquely beneath the Caribbean Plate; this motion is accommodated in continental El Salvador by strike-slip motions along the E–W trending El Salvador Fault Zone (ESFZ). Corti et al. (2005) and Agostini et al. (2006), analysing the regional framework of the ESFZ, document the fragmented nature of this fault system (Fig. 8), made of main segments separated by pull-apart structures. The main fault segments display active volcanism, but the intervening pull-apart basins are devoid of volcanics. The strike-slip fault segments act as preferential pathways for magma ascent (Corti et al., 2005; Agostini et al., 2006); both studies highlight that more detailed analyses should be conducted, to interpret the absence of volcanic activity within pull-apart basins, where extension is generally considered to favour volcanism.

About 600 km NW of El Salvador, El Chichòn volcano is located within the 450-km long volcanic gap, between the eastern end of the Trans-Mexican Volcanic Belt and the northwestern end of the Central American Volcanic arc. Its presence in a wide area devoid of volcanoes is related to the activity of the Transcurrent Fault Province of southern Mexico (Meneses-Rocha, 1991). A structural study by Garcìa-Palomo et al. (2004) documents the location of this volcano sitting at the junction of three main structures: (i) the extensional zone called Chapultenango Fault System, forming a half-graben structure; (ii) the Buena Vista syncline with *en echelon* pattern; and (iii) the left-lateral, strike-slip E–W San Juan Fault, transecting the volcano. The authors speculate that the growth of El Chichòn volcano was favoured by the structural junction between the Chapultenango Fault System and the San Juan Fault.

In Southern Mexico, the Oligocene volcanism in Taxco Volcanic Field (TVF) was synchronous (Alaniz-Alvarez et al., 2002) with strike-slip faulting and located within an extensional basin formed in the right-handed step-over between the Chichila and Tetipac dextral faults. The activity of another major Mexican volcano, Nevado de Toluca, has been related to strike-slip faulting (Bellotti et al., 2006); the edifice is dissected by three main fault systems that join close to the volcanic edifice: from older to younger, the Taxco-Querétaro, San Antonio and Tenango fault systems. The younger one, the E–W Tenango fault system, has been active since Late Pleistocene. This fault system is characterized by transtensive left-lateral strike-slip movement, and is partly coeval with the latest eruptive phase, the Nevado Supersynthem, which formed the present summit cone (Bellotti et al., 2006). In partial disagreement with the above are the conclusions of Garcìa-Palomo et al. (2000), who claim the present stress regime is governed by N–S extension forming E–W faults which also controlled the occurrence of two distinct flank collapses of the volcano prior to 40,000 yr ago.

The Andes display complex relationships between tectonics and magmatism, as already pointed out in Section "Field Examples". In the NVZ of Ecuador, Tibaldi and Ferrari (1992) documented the occurrence of the Cayambe stratovolcano close to the trace of the right-lateral strike-slip Cayambe-Chingual Fault, which predated the development of the volcano (Ferrari and Tibaldi, 1989). The authors propose an active role of the fault in favouring the rise of magma feeding the volcano; they also suggest that the absence of evidence of Holocene motions of the fault in the vicinity of the volcano may be related to the thermal anomaly induced by the volcano or by volcanic loading.

About 150 km north-east of Cayambe volcano in Colombia, the Galeras volcanic complex is traversed by the NE-striking Buesaco right-lateral strike-slip fault (Tibaldi and Romero-Leon, 2000). The tectonic control of this fault on the volcanism of the Galeras complex is testified by the alignment in a NE–SW direction, of the Cerro La Guaca cinder cone, the southern scarp of the youngest Galeras sector collapse, the active main crater and two parasitic vents which have outpoured lava flows. Moreover, according to Rovida and Tibaldi (2005), all the main recent structures on the flanks of Galeras strike NE.

In the southern portion of the CVZ of the Andes, Cenozoic magmatism developed both in the N–S-trending volcanic arc and in four, NW–SE directed transverse volcanic belts. These belts are associated with major fault zones, seismically active fracture zones (Schurr et al., 1999; Hanus et al., 2000). One of these zones, the Calama–Olacapato–Toro (COT) fault zone extends for more than 300 km east of the arc. Its tectonic activity within the Eastern Cordillera is testified by the NW–SE alignment of Miocene magmatic centres (Matteini et al., 2002a, b). The Eastern Cordillera contains the plutonic complexes Las Burras and Acay (13–14 Ma) and the volcanic complexes Almagro and Negra Muerta (6–7 Ma) (Riller et al., 2001; Matteini et al., 2005a, b; Hauser et al., 2005; Petrinovic et al., 2005; Acocella et al., 2007).

To gain more insight into the volcanic and tectonic activity of the COT volcanic belt in Pleistocene time, Petrinovic et al. (2006) analysed petrological and structural data collected at the Tocomar volcanic centre, and the San Jeronimo and the Negro de Chorrillos (SJ–NCH) monogenetic volcanoes. They observe that vertical strike-slip faults are spatially associated with mafic magmas of the San Jeronimo and the Negro de Chorrillos volcanoes, whereas normal faults seem to have controlled the emission of acidic pyroclastic rocks like those erupted at Tocomar volcanic centre. Hence, the eruption of highly viscous rhyolite magmas appears to have been facilitated in a tectonic setting characterized by horizontal extension, whereas the emission of less viscous basaltic andesites to shoshonites was controlled by subvertical strike-slip faults.

Portions of the SVZ of the Andes have an oblique subduction setting. Quaternary deformation in the southern SVZ is localized in the intra-arc zone along the arc-parallel, strike-slip Liquine-Ofqui fault zone (LOFZ), an approximately 1,100 km long, dextral strike-slip system that accommodates part of the margin-parallel component of oblique subduction (Nelson et al., 1994; Cembrano et al., 1996; Lavenu and Cembrano, 1999; Rosenau et al., 2006). Several authors showed that the Quaternary southern volcanic zone is strongly controlled by the LOFZ south of 38° (e.g., Hervé, 1994; Lopez-Escobar et al., 1995; Lavenu and Cembrano, 1999; Rosenau, 2004). Most stratovolcanoes are directly emplaced over the main trace of the LOFZ; several minor eruptive centers are spatially related to a major center or stratovolcano, but others occur in clusters directly emplaced over the LOFZ. The Caviahue-Copahue complex (38°S), is a striking example of the structural control of transcurrent tectonics on volcanic activity; Melnick et al. (2006a,b) propose this volcano was built within a horsetail-like structure (Fig. 9A) that developed at the northern tip of the LOFZ.

Transtensional conditions suitable for the development of volcanism occur also at intra-arc rifts such as

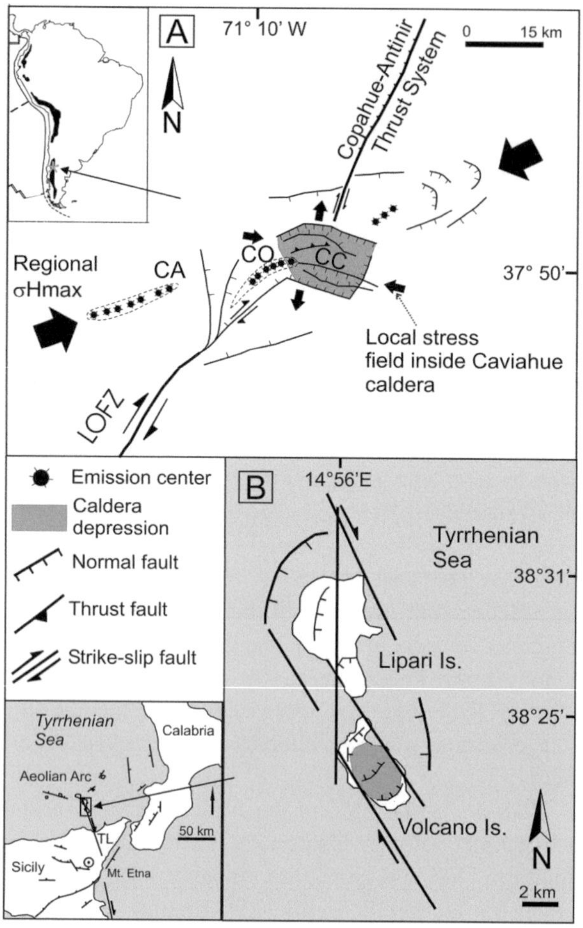

Fig. 9 Examples of field cases illustrated in the main text. In (**A**) Melnick et al. (2006a) suggest that Copahue volcano developed within part of the horsetail-like structure formed at the end of the Liquine-Ofqui fault zone. CO, Copahue volcano; CA, Callaqui volcano; CC, Caviahue caldera. Redrawn after Melnick et al. (2006a). In (**B**) the structural model for the Lipari-Vulcano Volcanic Complex pull-apart system is illustrated. Modified after Ventura et al. (1999) and Gioncada et al. (2003)

the Taupo Volcanic Zone (TVZ) in the North Island of New Zealand, as documented by Spinks et al. (2005) based on remote sensing and field-based structural data. This study shows that segments with the highest degree of extension correspond to the highly active, rhyolitic Taupo and Okataina Caldera Complexes, while segments with a higher degree of dextral transtension correspond to less voluminous andesitic stratovolcanoes.

The structural evolution in the central portion of the Aeolian Arc (Southern Italy) appears to be coherent with the evolution observed in zones of transition from arc-related to rift volcanism (Ventura et al., 1999). Together with Salina, Lipari and Vulcano islands form a NNW–SSE alignment which is orthogonal to the volcanic arc. The development of Lipari and Vulcano is linked to an active crustal discontinuity (Barberi et al., 1994; Ventura, 1994), corresponding to the NNW–SSE-trending, dextral strike-slip Tindari-Letojanni fault system (Ghisetti, 1979). The strike-slip motions along this fault resulted in the formation of pull-apart (Fig. 9B) basin structures (Ventura et al., 1999; Mazzuoli et al., 1995). The horizontal motions along the strike-slip system are accommodated by N–S to NE–SW trending normal faults where pure extension occurs (Mazzuoli et al., 1995, Gioncada et al., 2003).

Jové and Coleman (1998) describe an area within the trace of the Calaveras and Hayward strike-slip faults, northern California USA, where small volumes of alkali basalts were erupted in Pliocene time. These two fault segments are major splays of the San Andreas Fault (SAF) system in the San Francisco bay region. Jové and Coleman (1998) interpret the occurrence of mafic alkaline volcanism along the trace of the Calaveras Fault system near the Coyote Lake reservoir, northern California, in terms of stepping-fault interaction. According to the authors, the right-stepping *en echelon* fault geometry, coupled with right lateral motion of the San Andreas Fault, was responsible for the development of localized volcanic activity.

Field and age data by Busby et al. (2007) and Putirka and Busby (2007) show that the high-K_2O Little Walker Center, in the central Sierra Nevada, is among the largest Miocene volcanic areas in the mountain chain (Fig. 10). The authors propose it formed at a releasing step-over on dextral transtensional faults at the beginning (at about 10 Ma) of the Walker Lane transtensional faulting which tapped high-K partial melts. The three-dimensional arrangement of volcanic deposits in strike-slip basins is not only the product of volcanic processes, but also of tectonic processes. Busby and Bassett (2007) focus on a strike-slip basin within the Jurassic arc of southern Arizona to elaborate a facies model for strike-slip basins dominated by volcanism. Late Jurassic, intra-arc, strike-slip basins developed along the Sawmill Canyon fault zone, probably a strand of the Mojave-Sonora megashear system (Busby et al., 2005). The spatial distribution of volcanic lithofacies and vents

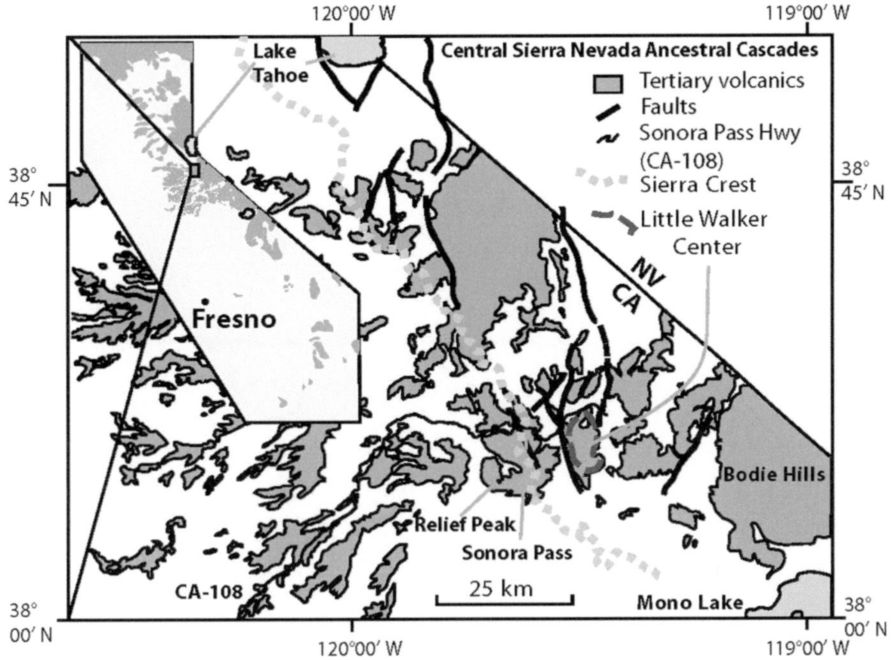

Fig. 10 Tertiary volcanic and intrusive rocks and the frontal fault system of the central Sierra Nevada in California, USA, taken from Putirka and Busby (2007). The faults form part of the dextral transtensional Walker Lane belt at the western edge of the extensional Basin and Range province. High K_2O magmatism is associated with this zone of transtension

was controlled by the structure of the basin (Santa Rita Glance Conglomerate) which the authors infer to be a releasing-bend strike-slip basin (Fig. 11). Small, multivent, polygenetic eruptive centers formed along releasing bends in the strike-slip fault; the deposits within the bends interfinger with ignimbrite sheets emitted from calderas at releasing step-overs produced at other locations along the Sawmill Canyon fault zone (Busby et al., 2005). According to the Busby and Bassett (2007), the best modern analogue to the Santa Rita Glance Conglomerate is probably the Sumatra volcanic arc in Indonesia. Small volcanic centers (in Sumatra, mainly rhyolite domes) occur along releasing bends, whereas caldera complexes develop in pull-apart basins formed along releasing step-overs (Bellier and Sebrier, 1994; Bellier et al., 1999). This observation introduces our section on calderas, which have been observed associated with strike-slip faulting in the most diverse tectonic settings.

Transcurrent Faults and Calderas

According to Holohan et al. (2007), regional-tectonic settings with active strike-slip deformation and associated calderas include:

(1) Major shear zones, along which calderas may form in: (i) zones of lateral shear, e.g. Calama - Olacapato - El Toro fault zone - Negra Muerta caldera (Riller et al., 2001); (ii) pull-apart gräben formed at releasing step-overs of transurrent fault systems such as the Great Sumatran Fault Zone (Toba and Ranau calderas, Aldiss and Ghazali, 1984; Bellier and Sebrier, 1994; Bellier et al., 1999, Fig. 12), and in Guatemala (Burkhart and Self, 1985);
(2) Transfer zones between segments of orthogonal or oblique rifts, e.g. Campi Flegrei caldera (Acocella et al., 1999);
(3) Oblique rifts with variable transtensive components, e.g. Taupo Volcanic Zone (Rotorua and Kapenga calderas, Spinks et al., 2005), Ethiopian Rift (Fantale, Gariboldi and Gedemsa calderas, Acocella et al., 2002). Some authors attribute volcanism in strike-slip fault zones to local extensional processes at pull-apart and releasing bend basins (Sylvester, 1988; Aydin et al., 1990; Mann, 2007).

We provide some details of calderas formed by different structural processes in these settings from the most representative works on the topic. An outstanding example of the association between pull-apart basins in strike-slip zones and large volcanic caldera complexes

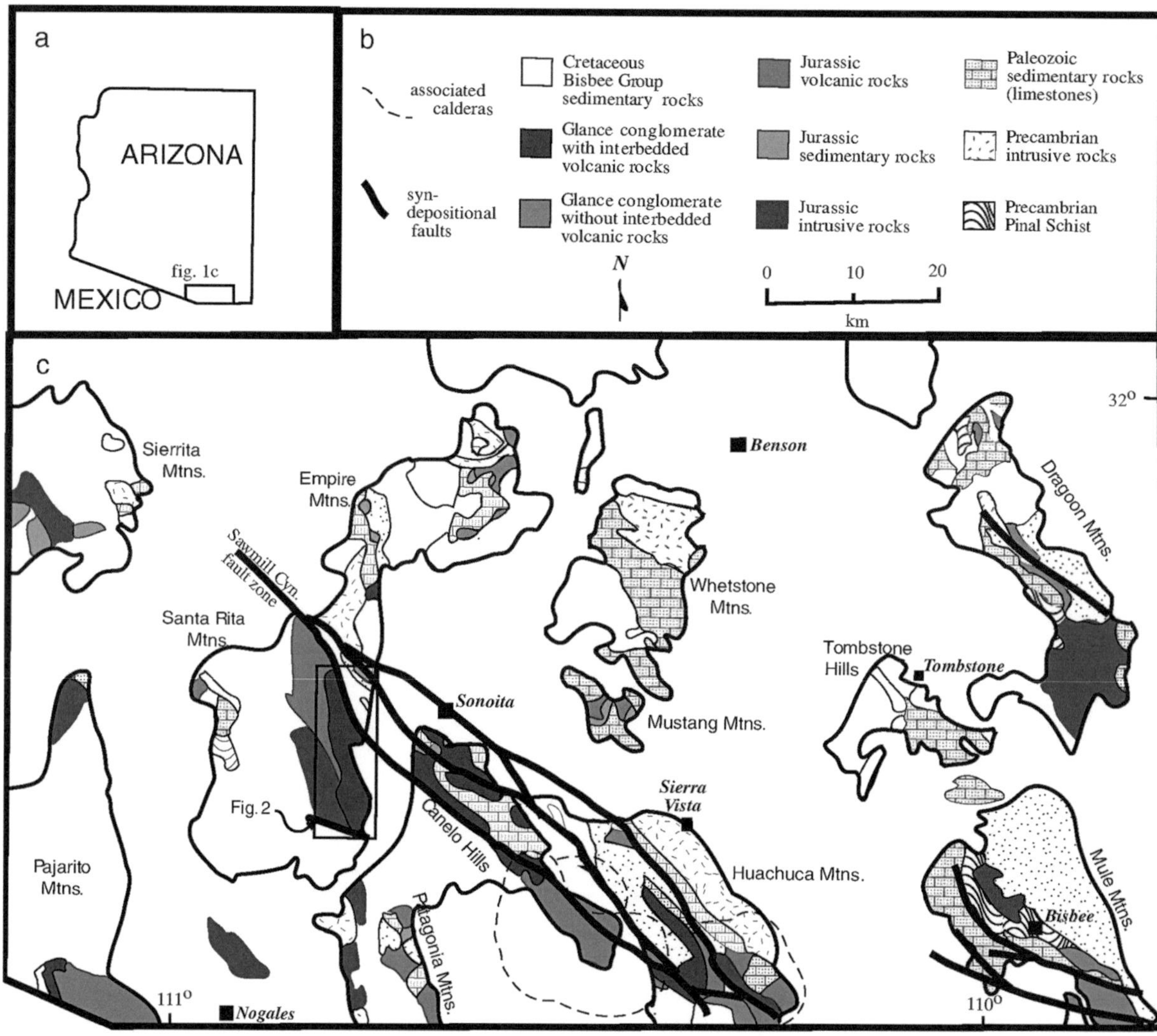

Fig. 11 Location of the Glance Conglomerate in the Santa Rita Mountains, southern Arizona, from Busby et al. (2007). The conglomerate is interbedded with volcanic rocks along a releasing-bend strike-slip basin. Small, multivent, polygenetic eruptive centers formed along releasing bends in the strike-slip fault; the deposits within the bends interfinger with ignimbrite sheets emitted from calderas at releasing step-overs produced at other locations along the Sawmill Canyon fault zone

is provided by Girard and van Wyk de Vries (2005) in their study of the Las Sierras-Masaya volcanic complex (Nicaragua), a large basaltic lava and ignimbrite shield with nested calderas. The volcanic complex is enclosed within the Managua Graben, whose location in a strike-slip zone (the Nicaraguan Depression) lead to a pull-apart basin. The authors suggest that the graben and volcano are linked tectonically, and that the graben started forming in response to a regional stress field modified around the dense, ductile intrusive complex underlying the volcanic complex.

Sumatra (Indonesia) consists of oblique convergence with partitioning of plate convergence into a component normal to the plate boundary and a wrench component accommodated by strike-slip deformation within the overriding plate, mainly localised along the Great Sumatran Fault (GSF, Fitch, 1972; Jarrard, 1986). The GSF runs parallel to the trench (Fig. 12) and follows approximately the magmatic arc, where major calderas are localized. From patterns of active and inactive (recent) faulting, Bellier and Sebrier (1994) inferred that the 0.074 Ma Toba and 0.55 Ma

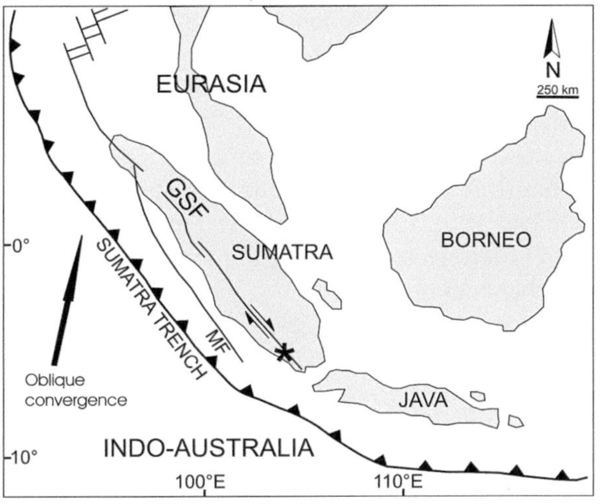

Fig. 12 Sketch of the regional geodynamic framework of Sumatra showing the role of the Great Sumatran Fault (GSF) in accommodating the convergence off the Sumatra Trench. MF = Mentawai fault. Asterisk shows location of the Ranau Caldera. Modified after Bellier et al. (1999)

Ranau calderas formed in presently inactive pull-apart gräben. In particular, the Ranau caldera (Fig. 12) evolution can be summarised as follows (Bellier et al., 1999). A pull-apart basin formed in a wide releasing step-over along the strike-slip fault system, with normal faults bounding the future location of the Ranau caldera. In this 12 × 16.5 km step-over, volcanic and geothermal activity developed, followed by the development of minor calderas in the step-over zone. The increased volcanic activity then produced incremental caldera growth. The ultimate episode of caldera evolution was most likely a paroxysmal pyroclastic eruption which caused the collapse of the block delimited by the boundary faults. Sieh and Natawidjaja (2000) cast some doubts on the relationship between tectonism and volcanism along the GSF; according to the authors, only 9 of the 50 young Sumatran volcanoes are located within 2 km of the trace of the GSF. They believe the close association of these volcanoes with the GSF does not, by itself, demonstrate a genetic relationship. They accept, however, that the close association of six volcanic centers with dilatational step-overs suggests that these structures are indeed controlling the locations of a few of the arc's volcanic centers.

Located about 2,400 km east of Ranau, the Tondano caldera (North Sulawesi, Indonesia) is one of the largest calderas in the world; the area is characterized by an active, sinistral strike-slip fault zone trending ENE–WSW (Lécuyer et al., 1997). This faulting accommodates the N–S movement of the Celebes Sea plate and represents a transfer fault zone between the Celebes Sea subduction zone and the Moluccas sea subduction zone. The authors point out the similarity existing between the development of the Tondano caldera and the Toba and Ranau ones. The Tondano caldera was clearly controlled by a currently-inactive pull-apart structure developed in the left step-over between ENE-trending, left-lateral strike-slip faults.

In Italy, the Campi Flegrei Volcanic District (CFVD) shows both active tectonics and volcanism along a predominant NE–SW trend, and represents a key site to assess the role of transverse fractures on volcanic activity. The results of analogue and numerical modelling by Acocella et al. (1999) suggest that the CFVD is characterized by subvertical transfer faults connecting adjacent, NW–SE regional faults parallel to the Tyrrhenian margin. In such a framework, volcanic activity along the preferential NE–SW direction at the CFVD would be favoured by the sub-vertical dip of the transfer fractures. The transfer faults might also be responsible for the emission of the most primitive products, as confirmed by petrochemical data. According to Acocella et al. (1999), narrow rifts in other locations worldwide also show volcanic activity at transfer fault zones, such as on the western branch of the African Rift System, where volcanic activity occurs at transfer fault zones associated with strike-slip faults (Ebinger, 1989). The Rio Grande Rift also displays volcanic activity in correspondence with strike-slip faults at the main transfer fault zone (Aldrich, 1986).

Moore and Kokelaar's (1998) study of the Caledonian Glencoe volcano, Scotland, describes the influence of regional-tectonic faulting in caldera development. Within a probably left-lateral strike-slip regime, extension and/or transtension at Glencoe occurred along several orthogonally intersecting regional-tectonic faults. The authors inferred a complex history of differential and incremental volcano-tectonic subsidence (i.e. piecemeal collapse) of caldera floor blocks along the regional-tectonic faults.

The formation of late Miocene to Recent collapse calderas on the Puna plateau in the CVZ of the Andes is generally interpreted in terms of the kinematic framework of the Nazca and South American Plates. Riller et al. (2001) present evidence that caldera dynamics and associated ignimbrite volcanism are genetically linked to the activity of NW–SE-striking zones of left-lateral transtension. In fact, most calderas in the southern central Andes are associated with NW–SE-striking transcurrent fault systems such as the Lipez, Calama-Olacapato-El Toro, Archibarca and Culampaja (Salfity, 1985) which define four major transverse volcanic zones. The recognition of a genetic relationship between caldera dynamics and regional, left-lateral transtension is strengthened by the detailed analysis of the tectono-magmatic history of the Negra Muerta Caldera, which has recently been the subject of other studies (Petrinovic et al., 2005; Ramelow et al., 2006). Riller et al. (2001) explain the formation of this partially eroded and asymmetric caldera in terms of an evolution in two successive increments, driven by left-lateral strike shear and fault-normal extension on the prominent Calama Olacapato-El Toro fault zone.

Analogue Modelling

In the last decade, analogue modelling in scaled experiments has been used to test the control exerted by strike-slip faulting on volcanic activity. van Wyk de Vries and Merle (1998) used analogue modelling to evaluate the effect of volcanic loading in strike-slip zones, as well as the effect of regional strike-slip faults on the structure of volcanic edifices. Their analogue models indicate that volcanoes in strike-slip zones develop extensional pull-apart structures. A feedback mechanism can arise, in which loading-related extension enables increased magma ascent, eruptions, and hence increased loading. The authors suggest that the Tondano caldera (North Sulawesi) may be the result of feedback between volcano loading and faulting. Other major volcano-tectonic depressions such as Toba, Ranau (Sumatra), and Atitlan (Guatemala) might have a similar origin.

Holohan et al. (2007) made scaled analogue models to study the interactions between structures associated with regional-tectonic strike-slip deformation and volcano-tectonic caldera subsidence. Their results show that while the magma chamber shape mostly influences the development and geometry of volcano-tectonic collapse structures, regional-tectonic strike-slip faults may have a strong influence on the structural evolution of calderas. Considering the case of elongate magma chamber deflation in strike-slip to transtensional regimes, they show that regional-tectonic structures can control the development of calderas. In fact, regional strike-slip faults above the magma chamber may form a pre-collapse structural grain that can be reactivated during subsidence. The experiments of Holohan et al. (2007) show that such faults preferentially reactivate when they are coincident with the chamber margins.

Based on previous experiments reproducing the formation of transfer zones and transform faults (Courtillot et al., 1974; Elmohandes, 1981; Serra and Nelson, 1988), Acocella et al. (1999) use analogue models to demonstrate that the occurrence of volcanic activity at Campi Flegrei may be related to the subvertical dip of NE–SW transfer fractures. The analogue experiments confirm that the NE–SW transverse fractures at Campi Flegrei and on the Tyrrhenian margin are transfer fault zones between adjacent NW–SE normal faults. The experiments also show that the transfer faults are steeper than the adjacent normal faults.

Girard and van Wyk de Vries (2005) have tested the effect of intrusions on strike-slip fault geometries. Their analogue models, reproducing the Las Sierras-Masaya intrusive complex in the strike-slip tectonic context of the Nicaraguan Depression, show that pull-apart basin formation around large volcanic complexes within strike-slip tectonics can be caused by the presence of an underlying ductile intrusion. To generate a pull-apart basin in this context, both transtensional strike-slip motion and a ductile intrusion are required. Their experiments reveal how strike-slip motion, even transtension, does not produce pull-aparts with no intrusion. A shield-like volcanic overload has no effect either. They conclude that the pull-apart that is forming at Las Sierras-Masaya volcanic complex is produced by the transtensive regional deformation regime and by the presence of the dense, ductile intrusive complex underlying the volcanic area.

A series of centrifuge analogue experiments were performed by Corti et al. (2001) with the purpose of modelling the mechanics of continental oblique extension (in the range of 0° to 60°) in the presence of underplated magma at the base of the continental crust. The main conclusions of their modelling are the following: (i) the structural pattern is characterised by the presence of *en echelon* faults, with mean trends not perpendicular to the stretching vector and a component of movement varying from pure normal to strike-slip;

(ii) the angle of obliquity controlling the ratio between the shearing and stretching component of movement strongly affects the deformation pattern of the models. In nature, this pattern results in magmatic and volcanic belts which are oblique to the rift axis and arranged *en echelon*, in agreement with field examples in continental rifts (i.e. Main Ethiopian Rift) and oceanic ridges.

Recently, emphasis has been placed on the effects of faulting on the lateral instability of volcanic edifices. Two key studies address transcurrent settings using analogue models. Lagmay et al. (2000) conducted analogue sand cone experiments to study instability generated on volcanic cones by basal strike-slip movement. Their results demonstrate that edifice instability may be generated when strike-slip faults beneath a volcano move as a result of tectonic adjustments. The instability is localised on the flanks of the volcano above the strike-slip shear, manifested (Fig. 13A) as a pair of sigmoids composed of one reverse and one normal fault. Two destabilised regions are created on the cone flanks between the traces of the sigmoidal faults. Lagmay et al. (2000) compare their results to two examples of volcanoes on strike-slip faults: Iriga volcano (Philippines) which was subjected to non-magmatic collapse, and Mount St. Helens (USA).

Norini and Lagmay (2005) built analogue models of volcanic cones traversed by strike-slip faulting and analysed the cones to assess the resulting deformation. Their study shows that symmetrical volcanoes that have undergone basal strike-slip offset may be deformed internally without showing any change whatsoever in their shape. Moreover, slight changes in the

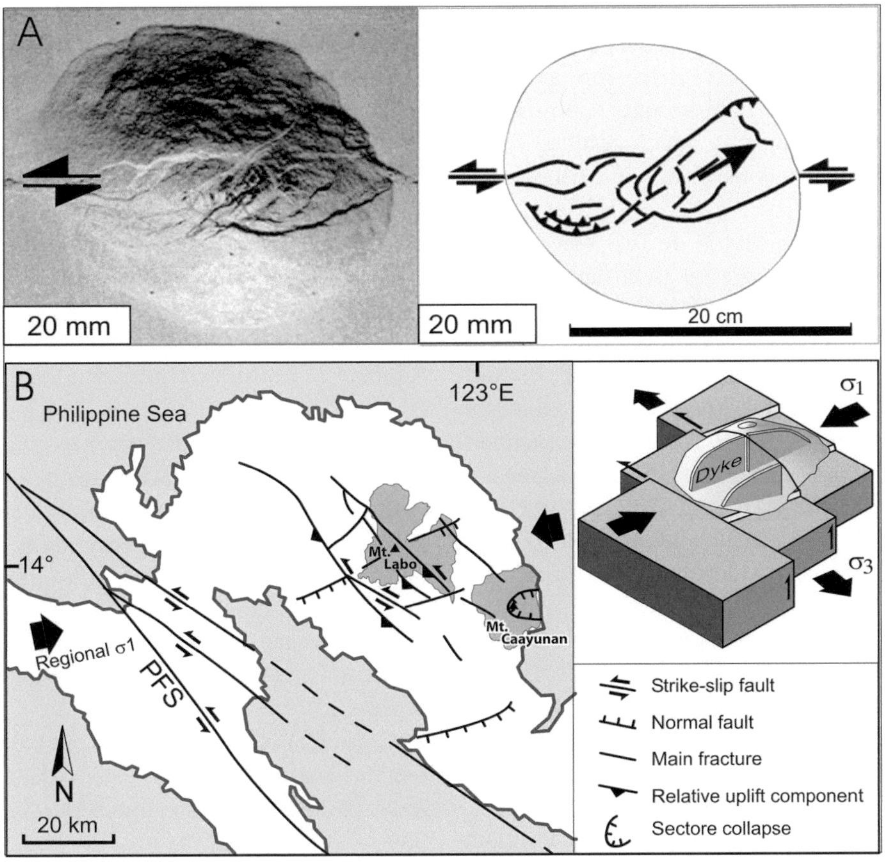

Fig. 13 (**A**) Surface deformation of analogue cones subjected to basal strike-slip faulting. To the left the photograph shows the superficial structures formed after 20 mm basal displacement. To the right, the sketch depicts the superficial features formed. Modified after Norini and Lagmay (2005). (**B**) Sketch of the main structures and Quaternary state of stress of the north-western Bicol Volcanic Arc. Main faults strike NW and secondary faults strike NE. The block diagram shows that the near-surface magma paths (dyking) followed the NE-striking fractures that are nearly parallel to σ_1 and perpendicular to σ_3. PFS = Philippine Fault System. Modified after Pasquarè and Tibaldi (2003)

basal shape of the cone induced by strike-slip movement can be restored by faster reshaping processes due to the deposition of younger eruptive products. The authors report the case of the perfectly symmetrical Mayon volcano (Philippines), suggesting that it may already be internally deformed and its faultless appearance might be misleading in terms of risk assessment.

Magma Paths

A few authors dealing with volcanism in a strike-slip tectonics setting have addressed, by field data or analogue modelling, the problem of identifying the paths through which magma reaches the surface to feed eruptions. In their study on analogue modelling dealing with strike-slip faulting and flank instability, Lagmay et al. (2000) consider also the case of Mount St. Helens, set on a right-lateral strike-slip fault; their experiments show that the fault strongly controlled the path of the intruding magma which resulted in the emplacement of a cryptodome prior to the catastrophic 1980 collapse.

Pasquarè and Tibaldi (2003) on two volcanoes of the Bicol Peninsula, observe by field data and analogue models, that the elongation of single edifices, apical depressions of domes and alignment of multiple centres, as well as all secondary faults in the studied area, trend NE–SW, i.e. perpendicularly to the main fault trend in the region, which is roughly perpendicular to the Philippine Fault System (PFS). Pasquarè and Tibaldi (2003) hypothesize that, at depth, magma probably used the main NW-striking regional faults because they are the deepest and widest crustal vertical structures, whereas the near-surface magma paths (dyking) followed the NE-striking fractures which are nearly parallel to σ_1 and perpendicular to σ_3 (Fig. 13B). The authors also point out that an upward change of orientation of magma-feeding fractures has been noticed in other transcurrent zones such as at Galeras volcano (Colombia, Tibaldi and Romero-Leon, 2000).

Holohan et al. (2007), who analysed by analogue modelling the interactions between structures associated with regional-tectonic strike-slip deformation and volcano-tectonic caldera subsidence, suggest a similarity between the roof-dissecting Riedel shears and Y-shears appearing in their models and the regional strike-slip faults that dissect the central floors of the Negra Muerta (Riller et al., 2001; Ramelow et al., 2006) and Hopong calderas. According to the authors, these fault systems might be regarded as preferential pathways in nature for the ascent of magma and other fluids before, during, or after caldera formation.

Busby and Bassett (2007) document that the intrabasinal lithofacies of the Santa Rita Glance Conglomerate record repeated intrusion and emission of small volumes of magma along intrabasinal faults. The interfingering of the eruptive products indicates that more than one vent was active at a time; hence the name "multivent complex" is applied. They propose that multi-vent complexes reflect the proximity to a continuously active fault zone, whose strands frequently tapped small batches of magma, emitted to the surface at releasing bends. Dacitic domes growing just outside the basin, were probably fed by the master, strike-slip fault, just as modern dome chains are commonly located on faults (Bailey, 1989; Bellier and Sebrier, 1994; Bellier et al., 1999).

Marra (2001) on the Mid-Pleistocene volcanic activity in the Alban Hills (Central Italy) documents two nearly contemporaneous eruptions of lava flows and ignimbrites in the Alban Hills as produced by two distinct tectonic triggers, tapping different depths of a magma reservoir. The geometries of the main structural dislocations in Quaternary strata indicate a structural pattern which is consistent with local strain partitioning in transpressive zones along strike-slip fault bends, superimposed on regional extension. Based on this analysis, Marra (2001) suggests that a local, clockwise block rotation between parallel N–S strike-slip faults might have generated local crustal decompression, enabling volatile-free magma to rise from deep reservoirs beneath the Alban Hills and feeding fissure lava flows. In contrast, the main ignimbrite eruptions appear to have tapped shallow, volatile-rich magma reservoirs and to have been controlled by extensional processes.

Chiarabba et al. (2004), on the basis of a shallow seismic tomography of Vulcano Island (Aeolian Arc, Italy) observe that at shallow depth (i.e. <0.5 km), the plumbing system of the volcano is mainly controlled by N–S striking faults, whereas at a depth >0.5 km, the rise of magma is controlled by NW–SE fractures associated with the activity of the NW–SE striking, right-lateral strike-slip to oblique-slip, Tindari-Letojanni fault system (Mazzuoli et al., 1995). This implies that magma intrudes along the NW–SE strike-slip faults but its ascent to the surface is controlled by N–S to NNW–SSE tensional structures (normal faults and tension fractures), which are orthogonal to the regional extension. Chiarabba et al. (2004) conclude that also Aydin et al. (1990) observed that in strike-slip zones, magma preferentially rises at the surface along the extensional structures rather than the main strike-slip fault segments. Also Corti et al. (2001) showed that magma emplaces at depth along faults parallel to the main shear zone but upraises to the surface along cracks that are orthogonal to the orientation of the extension.

Finally, Rossetti et al. (2000) illustrate how the effusive and intrusive rocks belonging to the McMurdo Volcanic Group (Antarctica) were emplaced along the western shoulder of the Ross Sea during the Cenozoic. The Mc Murdo dykes are widespread in the coastal sector of Victoria Land, along the western shoulder of the Ross Sea. Based on field evidence, Rossetti et al. (2000) propose that the intrusion of the Mc Murdo dykes was triggered along a crustal-scale, non-coaxial transtensional shear zone where the strike-slip component increased over time.

Petrologic and Geochemical Effects

The classic view of a convergent margin is that arc-like lavas erupt along the volcanic front, and alkalic basalts with no arc signature erupt in the back arc (Gill, 1974). However, structural analysis has shown that within an overall convergent margin setting, arc-like magmas erupt in areas of local compression, transpression, transtension, and extension. This summary paper does not compare the petrology and geochemistry of arc lavas to rift lavas, or even lavas of the volcanic front to those in the backarc. The focus is on smaller scale variations in stress state within the arc front of the convergent margin. The approach minimizes changes to petrology and geochemistry due to differences in the mantle source region, and instead allows us to compare petrology and geochemistry among magmas where the principal variable is the state of stress in the continental crust. This focus also emphasizes that interdisciplinary studies that link detailed structural information with petrology and geochemistry are relatively rare.

The SVZ of the Andes between latitude 30 S and 47 S has been used as a natural laboratory for studying the relationship between tectonics and continental magmatism for many years (Lopez Escobar et al., 1977; Hickey et al., 1986; Futa and Stern, 1988; Hildreth and Moorbath, 1988; Tormey et al., 1991; Dungan et al., 2001). This portion of the arc provides systematic variation in the age of the subducting slab, angle of subduction, volume of sediments in the trench, crustal thickness, and tectonic style. The arc also has a well-defined volcanic front, zone of back-arc extension, and transition zones between the two. These features also vary with time, as described in a recent compilation volume (Kay and Ramos, 2006). Considering present-day volcanic activity, the Liquine-Ofqui Fault Zone (LOFZ) is the controlling fault for activity along the volcanic front between 37 and 47 S (Hervé, 1994; Lopez-Escobar et al., 1995; Lavenu and Cembrano, 1999; Rosenau, 2004). The LOFZ is a greater than 1,100 km intra-arc strike slip zone that merges into the foreland fold and thrust belt at about 37 S (Ramos et al., 1996). Compared to volcanic rocks in the more compressional and transpressional area north of the LOFZ, the compositions of eruptive products in the LOFZ are primarily basalt-dacite or andesitic, with little evidence for upper crustal contamination or extensive residence time (Lopez-Escobar et al., 1977; Hickey et al., 1986; Futa and Stern, 1988; Hildreth and Moorbath, 1988; Tormey et al., 1991; Dungan et al., 2001). Lava composition is primarily controlled by mantle and lower crustal processes; the strike-slip LOFZ appears to allow more rapid passage through the crust and lesser occurrence of assimilation or magma mixing compared to the more contractional setting further north in the SVZ.

North of the LOFZ, the Agrio Fold and Thrust Belt and the Malargue Fold and Thrust Belt (Ramos et al., 1996; Folguera et al., 2006b) mark a transition to a transpressional and compressional zone. Within the zone of compression, basalts and basaltic andesites are rare, and the mineral assemblage becomes more hydrous. Hornblende andesite is the predominant rock type in northern centers of the SVZ, with subordinate biotite. In the compositional interval from andesite to rhyolite, crustal inputs cause Rb, Cs, and Th enrichment and isotopic variability indicating both lower crustal and upper crustal melts commingling with the ascending magma (Hildreth and Moorbath, 1988, Tormey et al., 1991, Dungan et al., 2001). These features are absent in the eruptive products controlled by the strike-slip LOFZ further south. The evolution from basalt to andesite occurs in the lower crust; there is enrichment of La/Yb as well as Rb, Cs, and Th. The most probable lower crustal protolith is a young, arc-derived garnet granulite (Tormey et al., 1991). In the northern part of the SVZ, with a greater prevalence of compression and transtension, petrologic and geochemical variations indicate predominantly andesitic systems with compositional variations indicating relatively low degrees of mantle melting, high degrees of mixing and assimilation of lower to mid crustal materials, and an overlay of upper crustal contamination evident in upper crustal rocks (Hildreth and Moorbath, 1988, Tormey et al., 1991, Dungan et al., 2001). The contrast between the petrology and geochemistry of volcanic rocks in the northern part of the SVZ (compression and transtension) compared to the strike-slip LOFZ-controlled portion of the SVZ have been attributed to a shallowing of the subducted slab and increasing crustal thickness in the north. In addition, the lithology and age of the continental crust in the north also exert a control on magma compositions. The thickening of the continental crust in the more compressional setting may be related to the transition from the dominantly strike-slip environment of the LOFZ.

Kay et al. (2005) evaluate the temporal trends in petrologic and geochemical effects in the Andean Arc between 33 and 36 S over a 27-million year period of record. The detailed study is used to compare the temporal trends at a single region to the present day north to south geographic trends among Holocene centers of the volcanic front just described. In the arc segment studied by Kay et al. (2005), the crustal stress regime is transtensional from 27 to 20 Ma; abundant mafic rocks with relatively flat REE patterns erupted, suggesting higher degrees of mantle melting. More evolved compositions have petrologic and geochemical variation indicating relatively low degrees of upper crustal contamination. From 19 to 7 Ma, the stress regime becomes compressional, with a significant increase in the amount of plutonic rocks. The lavas that did erupt in this compressional regime have steep REE patterns suggesting lower to mid-crustal fractionation of an amphibole-rich mineral assemblage. Geochemical data also indicate increasing degrees of upper crustal contamination. In general, as the compressional stress regime develops, there appears to be a longer crustal residence time, leading to a greater amount of plutonism, higher degrees of crustal contributions to developing magmas, and a hydrous fractionating mineral assemblage. The petrologic and geochemical features of these lavas are very similar to the characteristics of Holocene activity in the northern part of the SVZ. From 6 to 2 Ma, the dip of the subducting slab decreases, leading to a waning of magmatic activity as the volume of mantle melts decreases.

The SVZ of the Andes includes a belt of silicic volcanism, both ignimbrites and flows, between 35 and 37 S (Hildreth et al., 1999). The systems appear to have initially developed in a compressional state of stress in the crust. Voluminous eruptions of silicic magma, however, appear to coincide with a transition from compression to transpressional or even extensional conditions. During the compressional phase, there appears to have been extensive interaction with the lower and upper crust. Small batches of magma appear to have incorporated crustal melts and been subject to periodic magma mixing. As the compressional state of stress relaxed, shallow crustal melts coalesced, ultimately erupting to form the surface deposits (Hildreth et al., 1999).

In their study of the geology of a portion of the Peruvian Andes in the CVZ, Sebrier and Soler (1991) noted that during a transition from extensional to compressional states of crustal stress, there was not a corresponding change in the petrology or geochemistry of the erupted magmas. They found that calc-alkaline magmas of similar composition were the dominant eruptive product, independent of the state of stress in the crust.

Anatolia is characterised by widespread post-Oligocene volcanism associated with compression, strike-slip, and extensional crustal stress regimes. In western Anatolia, volcanic activity began during the Late Oligocene – Early Miocene in a compressional regime. Andesitic and dacitic calc-alkaline rocks are preserved, with some shallow granitic intrusions. An abrupt change from N–S compression to N–S stretching in the middle Miocene was accompanied by a gradual transition to alkali basaltic volcanism (Yilmaz, 1990). In eastern Anatolia, the collision-related compressional tectonics and associated volcanic activity began in the Late Miocene to Pliocene and continued almost without interruption into historical times (Yilmaz, 1990; Pearce et al., 1990; Yilmaz et al., 1998). Volcanism on the thickened crust north of the Bitlis Thrust Zone varies from the mildly alkaline volcano, Nemrut, and older Mus volcanics in the south, through the transitional calc-alkaline/alkaline volcanoes Bingöl and Süphan and the alkaline volcano Tendürek to the calc-alkaline volcano Ararat and older Kars plateau volcanics in the north (Pearce et al., 1990; Yilmaz et al., 1998; Coban, 2007). After initial phases of alkaline lavas, there were widespread eruptions of andesitic and dacitic calc-alkaline rocks during the Pliocene. A second, larger-volume phase of volcanism, partly overlapped with the initial phase, involving alkaline and transitional lavas; this phase began during the Quaternary and is ongoing (Pearce et al., 1990).

The calc-alkaline lavas of both Anatolian regions were erupted at a time when the compressional regime led to crustal thickening, as observed in the Andes. Petrology and geochemistry of the lavas from the compressional regime display many geochemical and isotopic signatures indicating extensive crustal contamination, and polybaric crystallization (Yilmaz, 1990; Coban, 2007). As found in the northern part of the Andean SVZ, rare earth elements are depleted in the heavier elements, indicating the importance of hornblende crystallization at depth in the calc-alkaline series lavas, in contrast to the consistently anhydrous crystallization sequences of the alkaline lavas (Yilmaz, 1990; Coban, 2007).

In the multi-vent complexes of the Santa Rita Mountains (Arizona, USA), the volcanic and subvolcanic rocks appear to record small-volume eruptions controlled by the complex faulting in the developing strike-slip basin (Busby and Bassett, 2007). Similarly, in a study of lavas from Mt. Rainier (Washington, USA) erupted during a compressional phase, Lanphere and Sisson (2003) suggest that the primary effect of compression is to lower the magma supply rate. Eruptive products at Mt. Rainier do not bear a recognizable signature of the compressive stress regime, other than smaller volume flows.

In their study of alkali basalts formed in an intraplate compressive state of stress, Glazner and Bartley (1994) note that other alkali basal fields in the southwestern USA also formed in an extensional and strike-slip state of stress. There do not appear to be petrologic or geochemical variations that correlate with the different states of stress. A relatively uniform alkali basaltic magma appears to have reached the surface in variable states of crustal stress without significant alteration in composition or other chemical characteristics.

Although focused studies on the relationship between crustal state of stress and petrology and geochemistry of eruptive products are uncommon, there are several traits of the petrologic and geochemical characteristics of magmas in compressional or weakly transpressional systems (Fig. 14). In general, plutonism tends to be favored over volcanic activity. The composition of volcanic rocks suggests longer crustal residence times, and higher degrees of lower crustal and upper crustal contributions to the magmas. Small volumes of magma tend to rise to shallow crustal levels (Marcotte et al., 2005, Busby and Bassett, 2007). In detailed studies with geographic to temporal coverage with which to compare compressive, transpressional and extensional episodes, there do not appear to be changes to the source materials that constitute the magmas. Rather, the change in crustal stress regime governs the magma transport pathway, and the crustal residence time. As the stress regime becomes more compressional, the magma transport pathways become more diffuse, and the crustal residence time increases. As a result, there are greater amounts of crustal melting and assimilation, greater degrees of magma mixing, and lower eruptive volumes as compression increases. Taken to its limits, these conditions lead to the often cited feature that compressional stress regimes tend to favour plutonism over volcanism. In the case of the silicic volcanic belt between 35° and 37 S in the Andes, the development of a plutonic belt in a compressive setting appears to have been interrupted by a transition in the state of stress of the crust from extension to transpressional or extensional, leading to large-volume eruption of dominantly rhyolitic magmas.

Conclusions

Volcanism occurs in compressional tectonic settings comprising both contractional and transcurrent deformation. The data include field examples worldwide encompassing subduction-related volcanic arcs and intra-plate volcanic zones. Moreover, several experiments conducted using scaled models demonstrate magma ascent under horizontal crustal shortening. In contractional settings, reverse faults can serve as magma pathways, leading to emplacement of volcanoes at the intersection between the fault plane

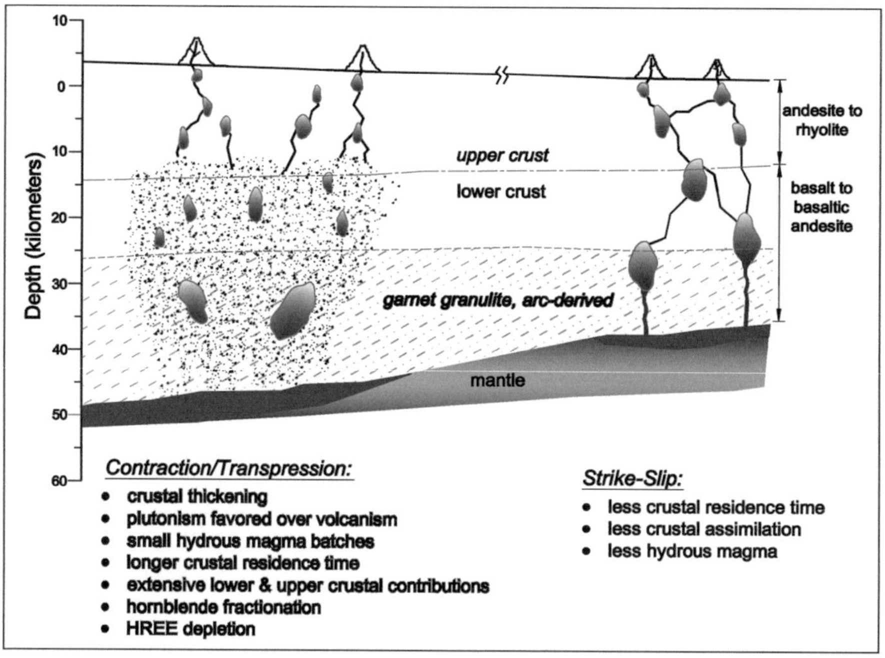

Fig. 14 Schematic petrogenetic summary diagram depicting in cross-sectional view the controls exerted by crustal stress state on contractional-derived volcanics (*left*) and strike-slip-derived volcanics (*right*), drawn based upon conditions in the Southern Volcanic Zone of the Andes and Eastern Anatolia. The cross section is not continuous between the two crustal stress states. Rough stippled pattern represents zone of lower and mid crustal partial melting and dark grey represents coalesced magma bodies. The source areas (mantle, lower crust, upper crust) and processes (fractional crystallization, assimilation, magma hybridization, mixing) occur within both crustal states, but the relative proportions vary significantly between the two states

and the topographic surface (Fig. 15A). Alternatively, magma can ascend along reverse faults and then vertically migrate, giving rise to the emplacement of volcanoes above the hanging wall fault block (Fig. 15B). The geometry of dykes feeding magma to the surface in these cases is still not clear, although it seems that within volcanic cones in contractional settings most dykes are parallel to the σ_1. The edifice type is most frequently stratovolcanoes and satellite monogenetic cones. In strike-slip fault zones, volcanic activity is primarily related to local extensional processes occurring at pull-apart basins, which form at a releasing stepover (Fig. 15C) between en echelon segments of a strike-slip fault, or at releasing bend basins, which form along a gently curved (Fig. 15D) strike-slip fault. Volcanoes can also develop directly above the trace (Fig. 15E) of strike-slip faults and hence be related to purely lateral shear processes without associated extension. Less frequently, volcanic activity can develop along extensional structures at the tips of main strike-slip faults (horsetail structures, Fig. 15F). Stratovolcanoes, shield volcanoes, pyroclastic cones and domes may occur at all these types of strike-slip fault structures, whereas calderas are preferentially located within pull-apart basins. The petrology and geochemistry of lavas erupted in compressive stress regimes suggest longer crustal residence times, and higher degrees of lower crustal and upper crustal contributions to the magmas. Small volumes of magma tend to rise to shallow crustal levels. There do not appear to be significant changes in the mantle or crustal source materials for magmas; rather, the type of crustal stress regime governs the magma transport pathway and crustal residence time. As the stress regime becomes more compressional, the magma transport pathways become more diffuse and the crustal residence time and crustal contribution to the magmas increases.

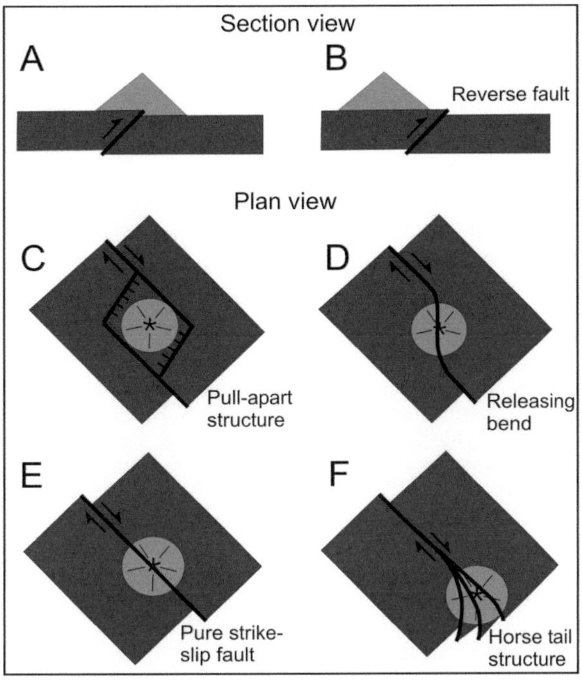

Fig. 15 Sketch of the most frequent location of surface volcanic features in compressional tectonic settings. In a contractional environment with reverse faults, most volcanoes are placed at the intersection between the fault plane and the topographic surface (**A**) or above the hanging wall fault block (**B**). They are most commonly stratovolcanoes and satellite monogenetic cones. In strike-slip fault zones, volcanism can occur at pull-apart basins (**C**); at releasing bend structures (**D**); directly along rectilinear strike-slip faults (**E**); and at the tips of main strike-slip faults (horsetail structures, **F**). Stratovolcanoes, shield volcanoes, pyroclastic cones and domes may occur at all the above types of strike-slip fault structures, whereas calderas are preferentially located within pull-apart basins

Acknowledgements C.J. Busby is greatly acknowledged for her useful suggestions on a previous version of the manuscript. This is a contribution to the International Lithosphere Programme – Task II project "New tectonic causes of volcano failure and possible premonitory signals".

References

Acocella V, Korme T, Salvini F, Funiciello R (2002). Elliptic calderas in the Ethiopian Rift: control of pre-existing structures. J Volcanol Geotherm Res 119:189–203.

Acocella V, Salvini F, Funiciello R, Faccenna C (1999). The role of transfer structures on volcanic activity at Campi Flegrei (Southern Italy). J Volcanol Geotherm Res 91:123–139.

Acocella V, Vezzoli L, Omarini R, Mattini M, Mazzuoli R (2007). Kinematic variations across Eastern Cordillera at 24°S (Central Andes): Tectonic and magmatic implications. Tectonophysics 434:81–92.

Adiyaman O, Chorowicz J, Arnaud ON, Gündogdu N, Gourgaud A (2001). Late Cenozoic tectonics and volcanism along the North Anatolian Fault: new structural and geochemical data. Tectonophysics 338:135–165.

Adiyaman Ö, Chorowicz J, Köse O (1998). Relationships between volcanic patterns and neotectonics in Eastern Anatolia from analysis of satellite images and DEM. J Volcanolol Geotherm Res 85:17–32.

Agostini S, Corti G, Doglioni C, Carminati E, Innocenti F, Tonarini S, Manetti P, Di Vincenzo G, Montanari D (2006). Tectonic and magmatic evolution of the active volcanic front in El Salvador: insight into the Berlín and Ahuachapán geothermal areas. Geothermics 35:368–408.

Alaniz-Álvarez SA, Nieto-Samaniego AF, Morán-Zenteno DJ, Alba-Aldave L (2002). Rhyolitic volcanism in extension zone associated with strike-slip tectonics in the Taxco region, southern Mexico. J Volcanol Geotherm Res 118:1–14.

Aldrich MJ, Jr. (1986). Tectonics of the Jemez lineament in the Jemez Mountains and Rio Grande rift. J Geophys Res 91: 1753–1762

Aldiss DT, Ghazali SA (1984). The regional geology and evolution of the Toba volcano-tectonic depression, Indonesia. J Geol Soc London 141:487–500.

Alemán A, Ramos VA (2000). Northern Andes. In: Cordani UG, Milani EJ, Thomaz Filho A, Campos DA (Eds.), Tectonic Evolution of South America International Geological Congress, 31, 453–480. Río de Janeiro.

Allmendinger RW, Figueroa D, Snyder D, Beer J, Mpodozis C, Isacks BL (1990). Foreland shortening and crustal balancing in the Andes at 30°S latitude. Tectonics 9(4) : 789–809.

Allmendinger R, Jordan T, Kay SM, Isacks B (1997). The evolution of the Altiplano-Puna plateau of the Central Andes. Annu Rev Earth Planet Sci 25 : 139–174.

Aldrich MJ, Jr. (1986). Tectonics of the Jemez lineament in the Jemez Mountains and Rio Grande rift. J Geophys Res 91 : 1753–1762.

Anderson EM (1951). The Dynamics of Faulting. Oliver and Boyd, Edinburgh.

Auzende J-M, Collot J-Y, Lafoy Y, Gracia E, Géli L, Ondréas H, Eissen J-P, Olisukulu C, Tolia D, Biliki N, Larue MB (1994). Evidence for sinistral strike-slip deformation in The Solomon Island arc. Geo-Marine Lett 14:232–237.

Aydin, A, Nur, A (1982). Evolution of pull-apart basins and their scale independence. Tectonics 1:91–105.

Aydin A, Schultz RA, Campagna D (1990). Fault-normal dilatation in pull-apart basins: implications for relationship between strike-slip fault and volcanic activity. In: Boccaletti M, Nur A (Eds.), Active and Recent Strike-Slip Tectonics. Ann. Tectonicae Special Issue, pp. 45–52.

Bailey RA (1989). Quaternary volcanism of Long Valley caldera, and Mono-Inyo Craters, Eastern California: 28th International Geological Congress, Field trip Guidebook T313. American Geophysical Union, Washington, DC.

Barberi F, Coltelli M, Ferrara G, Innocenti F, Navarro JM, Santacroce R (1988). Plio-Quaternary volcanism in Ecuador. Geol Mag 125:1–14.

Barberi F, Gandino A, Gioncada A, La Torre P, Sbrana A, Zenucchini C (1994). The deep structure of the Eolian arc (Filicudi-Panarea-Vulcano sector)in light of gravity, magnetic and volcanological data. J Volcanol Geotherm Res 61:189–206.

Branquet Y, van Wyk de Vries B (2001). Effects of volcanic loading on regional compressive structures: New insights from natural examples and analogue modelling. Comptes Rendu de l'Académie des Sciences 833:455–461.

Beck ME (1983). On the mechanism of tectonic transport in zones of oblique subduction. Tectonophysics 93:1–11.

Bellier O, Bellon H, Sebrier M, Sutanto MRC (1999). K-Ar age of the Ranau tuffs; implications for the Ranau Caldera emplacement and slip-partitioning in Sumatra (Indonesia). Tectonophysics 312:347–359.

Bellier O, Sebrier M (1994). Relationship between tectonism and volcanism along the Great Sumatran Fault zone deduced by SPOT image analyses. Tectonophysics 233:215–231.

Bellotti F, Capra L, Groppelli G, Norini G (2006). Tectonic evolution of the central-eastern sector of Trans Mexican volcanic belt and its influence on the eruptive history of the Nevado de Toluca Volcano (Mexico). J Volcanol Geotherm Res 158: 21–36.

Benn K, Odonne F, de Saint Blanquat M (1998). Pluton emplacement during transpression in brittle crust: new views from analogue experiments. Geology 26:1079–1082.

Burkhart B, Self S (1985). Extension and rotation of crustal blocks in northern Central America and effect on the volcanic arc. Geology 13:22–26.

Busby CJ, Bassett KN (2007). Volcanic facies architecture of an intra-arc strike-slip basin, Santa Rita Mountains, Southern Arizona. Bull Volcanol 70:85–103.

Busby CJ, Bassett K, Steiner MB, Riggs NR (2005). Climatic and tectonic controls on Jurassic intra-arc basins related to northward drift of North America. In: Anderson TH, Nourse JA, McKee JW, Steiner MB (Eds.), The Mojave-Sonora Megashear Hypothesis: Development, Assessment, and Alternatives. Geol Soc Am Spec Pap 393, pp. 359–376.

Busby CJ, Hagan J, Putirka K, Pluhar C, Gans P, Rood D, De Oreo S, Skilling, I, Wagner, D (2008). The ancestral Cascades arc: Implications for the development of the Sierran microplate and tectonic significance of high-K2O volcanism. Geol Soc Am Spec Pap, Hopson Volume 438 : 331–378.

Cas RAF, Wright JV (1987). Volcanic Successions. Allen & Unwin, London, 528 pp.

Clavero, JE, Sparks RS, Pringle MS, Polanco E, Gardeweg MC (2004). Evolution and volcanic hazards of Taapaca Volcanic Complex, Central Andes of Northern Chile. J Geol Soc (London) 161:603–618.

Cembrano J, Hervè F, Lavenu A (1996). The Liquine Ofqui fault zone: A long-lived intra-arc fault system in southern Chile. Tectonophysics 259:55–66.

Chiarabba C, Pino NA, Ventura G, Vilardo G (2004). Structural features of the shallow plumbing system of Vulcano Island Italy. Bull Volcanol 66:477–484.

Chuvashova IS, Rasskazov SV, Yasnygina TA, Saracina EV, Fefelov NN (2007). Holocene Volcanism in Central Mongolia and Northeast China: Asynchronous Decompressional and Fluid Melting of the Mantle. J Volcanol Seismol 1:372–396.

Coban H (2007). Basalt magma genesis and fractionation in collision and extension related provinces: A comparison between eastern, central, and western Anatolia. Earth Sci Rev 80:219–238.

Cobbold PR, Davy P, Gapais D, Rossello EA, Sadybakasov E, Thomas JC, Tondji Biyo JJ, De Urreiztieta M (1993). Sedimentary basins and crustal thickening. Sediment Geol 86(1–2) : 77–89.

Coira B, Davidson J, Mpodozis C, Ramos VA (1982). Tectonic and magmatic evolution of the Andes of northern Argentina and Chile. Earth Sci Rev 18 : 303–332.

Coira B, Kay SM, Viramonte JG (1993). Upper Cenozoic magmatic evolution of the Argentine Puna-a model for changing subduction geometry. Int Geol R Review 35 : 677–720.

Cooper KM, Reid MR, Dunbar NW, McIntosh WC (2002). Origin of mafic magmas beneath northwestern Tibet: Constraints from 230Th-238U disequilibria. Geochem Geophys Geosyst doi:10.1029/2002GC000332.

Corazzato C, Tibaldi A (2006). Basement fracture control on type, distribution, and morphology of parasitic volcanic cones: an example from Mt. Etna, Italy. In: Tibaldi A, Lagmay M (Eds.), Interaction between Volcanoes and their Basement, Journal of Volcanology and Geothermal Research, Special issue, 158, pp. 177–194.

Corti G, Bonini M, Innocenti F, Manetti P, Mulugeta G (2001). Centrifuge models simulating magma emplacement during oblique rifting. J Geodyn 31:557–576.

Corti G, Carminati E, Mazzarini F, Garcia MO (2005). Active strike-slip faulting in El Salvador (Central America). Geology 33:989–992.

Courtillot V, Tapponier P, Varet J (1974). Surface features associated with transform faults: a comparison between observed examples and an experimental model. Tectonophysics 24:317–329.

Deng W (1993). Study on trace element and Sr, Nd isotopic geochemistry of Cenozoic potassic volcanic rocks in north Tibet. Acta Petrol Sin 9:379–387.

Dewey JF, Hempton MR, Kidd WSF, Saroglu F, Sengör AMC (1986). Shortening of continental lithosphere; the neotectonics of eastern Anatolia, a young collision zone. In: Coward MP, Ries AC (Eds.), Collision Tectonics. Geol Soc Spec Pub 19, pp. 3–36.

Dungan M, Wulff A, Thompson R (2001). Eruptive stratigraphy of the Tatara–San Pedro Compelx, 36°S, Southern Volcanic Zone, Chilean Andes: Reconstruction method and implications for magma evolution at long-lived arc volcanic centers. J Petrol 42:555–626.

Ebinger CJ (1989). Geometric and kinematic development of border faults and accommodation zones, Kivu-Rusizi Rift, Africa. Tectonics 8:117–133.

Ego F, Sebrier M, Lavenu A, Yepes H, Eguez A (1996). Quaternary state of stress in the northern Andes and the restraining bend model for the Ecuadorian Andes. Tectonophysics 259:101–116.

Elmohandes SE (1981). The central european graben system: rifting imitated by clay modelling. Tectonophysics 73:69–78.

Ferrari L, Tibaldi A (1989). Seismotectonics of northeastern Ecuadorian Andes (abstract). Ann Geophys 39.

Ferrari L, Tibaldi A (1992). Recent and active tectonics of the North-Eastern Ecuadorian Andes. J Geodynamics 15(1/2) : 39–58.

Fitch TJ (1972). Plate convergence, transcurrent faults, and internal deformation adjacent to southeast Asia and the western Pacific. J Geophys Res 77:4432–4460.

Flint S, Turner P, Jolley EJ, Hartley AJ (1993). Extensional tectonics in convergent margin basins: an example from the Salar de Atacama, Chilean Andes. Geol Soc Am Bull 105:603–617.

Folguera A, Bottesi G, Zapata T, Ramos VA (2008). Crustal collapse in the Andean backarc since 2 Ma: Tromen volcanic plateau, Southern Central Andes (36°40′–37°30′S). Tectonophysics 459:140–160.

Fujita E, Ukawa M, Yamamoto E (2004). Subsurface cyclic magma sill expansions in the 2000 Miyakejima volcano eruption: Possibility of two-phase flow oscillation. J Geophys Res, doi:10.1029/2003JB002556.

Futa K, Stern C (1988). Sr and Nd isotopic and trace element compsitions of Quaternary volcanic centers of the southern Andes. Earth Planet Sci Lett 88:253–263.

Galland O, de Bremond d'Ars J, Cobbold PR, Hallot E (2003). Physical models of magmatic intrusion during thrusting. Terra Nova, doi: 10.1046/j.1365-3121.2003.00512.x.

Galland O, Cobbold PR, de Bremond d'Ars J, Hallot E (2007a). Rise and emplacement of magma during horizontal shortening of the brittle crust: Insights from experimental modelling. J Geophys Res, doi:10.1029/2006JB004604.

Galland O, Hallot E, Cobbold PR, Ruffet G, de Bremod d'Ars J (2007b). Volcanism in a compressional Andean setting: A structural and geochronological study of Tromen volcano (Neuquen province, Argentina). Tectonics, 26, TC4010, doi:10.1029/2006TC002011.

García-Palomo A, Macias JL, Espindola JM (2004). Strike-slip faults and K-alkaline volcanism at El Chichon volcano, southeastern Mexico. J Volc Geotherm Res 136: 247–268.

García-Palomo A, Macías JL, Garduno VH (2000). Miocene to Recent structural evolution of the Nevado de Toluca volcano region, central Mexico. Tectonophysics 318:281–302.

Ghisetti F (1979). Relazioni tra strutture e fasi trascorrenti e distensive lungo i sistemi Messina-Fiumefreddo, Tindari-Letojanni e Alia-Malvagna (Sicilia nord-orientale): uno studio microtettonico. Geol Rom 18:23–56.

Gill JB (1974). Role of underthrust oceanic crust in the genesis of a Fijian talc-alkaline suite. Contr Mineral Petrol 43:29–45.

Gioncada A, Mazzuoli R, Bisson M, Pareschi MT (2003). Petrology of volcanic products younger than 42 ka on the Lipari-Vulcano complex (Aeolian Islands, Italy): an example of volcanism controlled by tectonics. J Volcanol Geotherm Res 122:191–220.

Girard G, van Wyk de Vries B (2005). The Managua Graben and Las Sierras-Masaya volcanic complex (Nicaragua); pull-apart localisation by an intrusive complex: results from analogue modelling. J Volcanol Geotherm Res 144:37–57.

Glazner AF (1991). Plutonism, oblique subduction, and continental growth: An example from the Mesozoic of California. Geology 19:784–786.

Glazner AF, Bartley JM (1994). Eruption of alkali basalts during crustal shortening in southern California. Tectonics 13:493–498.

Godoy E, Yáñez G, Vera E (1999). Inversion of an Oligocene volcano-tectonic basin and uplifting of its superimposed Miocene magmatic arc in the Chilean Central Andes: first seismic and gravity evidences. Tectonophysics 306 : 217–236.

Groeber P (1929). Lıneas fundamentales de la geologıa del Neuquen, sur de Mendoza y regiones adyacentes, 110 pp., Ministerio de Agricultura, Direccion General de Minas, Geologıa y Hidrologıa, Buenos Aires.

Gutscher MA, Lallemand S (1999). Birth of a major strike-slip fault in SW Japan. Terra Nova 11:203–209.

Guzmán SR, Petrinovic IA, Brod JA (2006). Pleistocene mafic volcanoes in the Puna–Cordillera Oriental boundary, NW-Argentina. In: Tibaldi A, Lagmay AFM (Eds.), Interaction between volcanoes and their basement. J Volcanol Geotherm Res 158, pp. 51–69.

Hamilton WB (1995). Subduction systems and magmatism. In: Smellie JR (Ed.), Volcanism Associated with Extension to Consuming Plate Margins. Geol Soc London Spec Publ 81, pp. 3–28.

Hammerschmidt K, Döbel R, Friedrichsen H (1992). Implication of 40Ar/39Ar dating of Early Tertiary volcanic rocks from the north-Chilean Precordillera. Tectonophysics 202(1): 55–81.

Hanus V, Vanek J, Spicak A (2000). Seismically active fracture zones and distribution of large accumulations of metals in the central part of Andean South America. Miner Depos 35:2–20.

Hauser N, Matteini, M, Omarini R, Mazzuoli R, Vezzoli L, Acocella V, Uttini A, Dini A, Gioncada A (2005). Aligned extrusive andesitic domes in the southern sector of the Late Miocene Diego de Almagro Volcanic Complex, Salta, Argentina: evidence for transtensive tectonics in the Central Andes. Proceedings of the XVI Congreso Geologico Argentino, vol. II, pp. 153–158.

Hervè F (1994). The southern Andes between 39° and 44°S latitude: the geological signature of a transpressive tectonic regime related to a magmatic arc. In: Reutter KJ, Scheuber E, Wigger PJ (Eds.), Tectonics of the Southern Central Andes. Springer, Berlin, pp. 243–248.

Hickey R, Frey F, Gerlach D, Lopez-Escobar L (1986). Multiple sources for basaltic arc rocks from the southern volcanic zone of the Andes (34 to 41S): Trace element and isotopic evidence for contributions from subducted oceanic crust, mantle, and continental crust. J Geophys Res 91:5963–5983.

Hildreth W, Fierstein J, Godoy E, Drake R, Singer B (1999). The Puelche volcanic field: Extensive Pleistocene rhyolite lava flows in the Andes of central Chile. Rev Geol de Chile 26:275–309.

Hildreth W, Moorbath S (1988). Crustal contributions to arc magmatism in the Andes of central Chile. Contributions Mineral Petrol 98:455–489.

Hill DP (1977). A model for earthquake swarms. J Geophys Res 82:1347–1352.

Holmberg E (1975). Descripcion geologica de la Hoja 32c, Buta Ranquil (Prov. Mendoza-Neuquen), Bull. 152, 71 pp., Serv. Nac. Min. Geol., Buenos Aires.

Holohan EP, van Wyk de Vries B, Troll VR (2007). Analogue models of caldera collapse in strike-slip tectonic regimes. Bull Volcanol, doi 10.1007/s00445-007-0166–x.

Hubbert MK, Willis DG (1957). Mechanics of hydraulic fracturing in Structural Geology, MK Hubbert (Ed.),, Macmillan, New York, pp. 175–190.

Hungerbuhler D, Steinmann M, Winkler W, Seward D, Eguez A, Peterson DE, Helg U, Hammer C (2002). Neogene stratigraphy and Andean geodynamics of southern Ecuador. Earth-Sci Rev 57:75–124.

INECEL (by Aguilera E, Almeida E, Balseca W, Barberi F, Ferrari L, Innocenti F. Pasquarè G, Tibaldi A) (1988). Mapa Geologico del Volcan El Reventador y Estudio Vulcanologico del El Reventador, Ministerio de Energia y Minas, Quito, Ecuador, 117 pp.

James DE, Sacks IS (1999). Cenozoic formation of the Central Andes: A geophysical perspective. In : Skinner BJ (Eds.) Geology and Ore Deposits of the Central Andes. Society of Economic Geology, Special Publication, 7, 1–26.

Jarrard RD (1986). Terrane motion by strike-slip faulting of forearc slivers. Geology 14:780–783.

Johnson AM (1970). Physical Processes in Geology. W. H. Freeman, New York, 592 pp.

Jordan T, Gardeweg M (1989). Tectonic evolution of the late Cenoizoic Central Andes (20–33°S). In : Abrahams B (Eds.), The Evolution of the Pacific Ocean Margins. Oxford University Press, New York, 193–207.

Jordan T, Burns W, Veiga R, Pángano F, Copeland F, Kelley S, Mpodozis C (2001). Extensional basin formation in the southern Andes caused by incresed convergence rate: a mid-cenozoic trigger for the Andes. Tectonics 20(3) : 308–324.

Jové CF, ColemanRG (1998). Extension and mantle upwelling within the San Andreas fault zone, San Francisco Bay area, California. Tectonics 17:883–890.

Kanaori Y, Kawakami S, Yairi K (1994). Seismotectonics of the Median Tectonic Line in southwest Japan: Implications for coupling among major fault systems. Pure Appl Geophys 142:589–607.

Karakhanian AS, Trifonov VG, Azizbekian OG, Hondkarian DG (1997). Relationship of Late Quaternary tectonics and volcanism in the Khanarassar active fault zone, the Armenian Upland. Terra Nova 9:131–134.

Kay S, Godoy E, Kurtz A (2005). Episodic arc migration, crustal thickening, subduction erosion, and magmatism in the south-central Andes. Geol Soc Amer Bull 117:67–88.

Kay SM, Mpodozis C, Coira B (1999). Neogene magmatism, tectonism and mineral deposits of the Central Andes (22°–33°S Latitude). In : Skinner B (Ed.), Geology and ore deposits of the Central Andes. Society of Economic Geology, Special Publication, 7, 27–59.

Kay SM, Ramos VA (Eds.) (2006). Evolution of an Andean margin: a tectonic and magmatic view from the Andes to the Neuquen Basin. Geol Soc Am Special Paper 407 : 343 pp.

Kendrick E, Bevis M, Smalley Jr R, Brooks B (2001). An integrated crustal velocity field for the central Andes. Geochem Geophys Geosyst 2 : XX, 2001GC000191.

Koçyigit A, Yılmaz AY, Adamia S, Kuloshvili S (2001). Neotectonics of East Anatolian Plateau (Turkey) and Lesser Caucasus: implication for transition from thrusting to strike-slip faulting. Geodinamica Acta 14:177–195.

Koshiya S, Ohtani M (1999). Earthquake fault of the M6.1 earthquakes occurred at the northern part of Iwate Prefecture on September 3, 1998. Chikyu, 21, 307–311, (in Japanese).

Kostyuchenko SL, Morozov AF, Stephenson RA, Solodilov LN, Vedrentsev AG, Popolitov KE, Aleshina AF, Vishnevskaya VS, Yegorova TP (2004). The evolution of the southern margin of the East European Craton based on seismic and potential field data. Tectonophysics 381:101–118.

Kozlowski EE, Cruz CE, Sylwan CA (1996). Geologıa estructural de la zona de Chos Malal, Cuenca Neuquina, Argentina, paper presented at XIII Congreso Geologico Argentino y III Congreso de Exploracion de Hidrocarburos, 15–26.

Lagmay AMF, Tengonciang A, Uy H (2005). Structural setting of the Bicol Basin and kinematic analysis of fractures in Mayon Volcano, Philippines. J Volcanol Geotherm Res 144:23–36.

Lagmay AMF, van Wyk de Vries B, Kerle N, Pyle DM (2000). Volcano instability induced by strike-slip faulting. Bull Volcanol 62:331–346.

Lanphere M, Sisson T (2003). Episodic Volcano Growth at Mt. Rainier, Wasthington: A product of tectonic throttling? Geol. Soc. Am. Abstracts with Programs, vol. 35, no. 6, 644 pp.

Lara LE, Lavenu A, Cembrano J, Rodríguez C (2006). Structural controls of volcanism in transversal chains: Resheared faults and neotectonics in the Cordón Caulle–Puyehue area (40.5°S), Southern Andes. In: Tibaldi A, Lagmay AFM (Eds.), Interaction Between Volcanoes and Their Basement, Spec. Issue, J Volcanol Geotherm Res 158, pp. 70–86.

Lavenu A, Noblet C, Bonhomme MG., Egüez A, Dugas F, Vivier G (1992). New K-Ar age dates of Neogene and Quaternary volcanic rocks from the Ecuadorian Andes: implications for the relationship between sedimentation, volcanism, and tectonics. J South Am Earth Sci 5 : 309–320.

Lavenu A, Cembrano J (1999). Compressional and transpressional stress pattern for Pliocene and Quaternary brittle deformation in forearc and intra-arc zones (Andes of Central and Southern Chile). J Struct Geol 21:1669–1691.

Lécuyer F, Bellier O, Gourgaud A, Vincent PM (1997). Tectonique active du nord-est de Sulawesi (Indonesie) et controle structural de la caldeira de Tondano: Paris, Acadáme des Sciences, Comptes Rendues 325: 607–613.

Legrand D, Calahorrano A, Guillier B, Rivera L, Ruiz M, Villagomez D, Yepes H (2002). Stress tensor analysis of the 1998-1999 tectonic swarm of northern Quito related to the volcanic swarm of Guagua Pichincha volcano, Ecuador. Tectonophysics 344:15–36.

Litherland M, Aspden JA (1992). Terrane-boundary reactivation: a control on the evolution of the Northern Andes. J South Am Earth Sci 5(1): 71–76.

Llambıas EJ, Palacios M, Danderfer JC (1982). Las erupciones holocenas del volcan Tromen (Provincia del Neuquen) y su significado en un perfil transversal E-O a la latitud de 37°S, paper presented at Quinto Congreso Latinoamericano de Geologia, Buenos Aires, pp. 537–545.

Lopez-Escobar L, Cembrano J, Moreno H (1995). Geochemistry and tectonics of the Chilean Southern Andes Quaternary volcanism (37°–46°S). Rev Geol de Chile 22:219–234.

Lopez-Escobar L, Frey F, Vergara M (1977). Andesites and High-alumina basalts from the central-south Chilean high Andes: Geochemical evidence bearing on their petrogeneis. Contrib Mineral Petrol 63:199–228.

Mann P (2007). Global catalogue, classification and tectonic origins of restraining- and releasing bends on active and ancient strike-slip fault systems. Geol Soc London Spec Publ 290:13–142.

Marcotte SB, Klepeis KA, Clarke GL, Gehrels G, Hollis JA (2005). Intra-arc transpression in the lower crust and its relationship to magmatism in a Mesozoic magmatic arc. Tectonophysics 407:135–163.

Marques MO, Cobbold P (2002). Topography as a major factor in the development of arcuate thrust belts: Insights from sandbox experiments. Tectonophysics 348:247–268.

Marra F (2001). Strike-slip faulting and block rotation: A possible triggering mechanism for lava flows in the Alban Hills? J Struct Geol 23:127–141.

Marrett RA, Allmendiger RW, Alonso RN, Drake RE (1994). Late Cenozoic tectonic evolution of the Puna Plateau and adjacent foreland, northwestern Argentine Andes. J South Am Earth Sci 7:179–207.

Matteini M, Mazzuoli R, Omarini R, Cas R, Maas R (2002a). The geochemical variations of the upper Cenozoic volcanism along the Calama-Olocapato-El Toro transversal fault system in central Andes (24°S): petrogenetic and geodynamic implications. Tectonophysics 345:211–227.

Matteini M, Mazzuoli R, Omarini R, Cas R, Maas R (2002b). Geodynamical evolution of the central Andes at 24°S as inferred by magma composition along the Calama-Olocapato-El Toro transversal volcanic belt. J Volcanol Geotherm Res 118:225–228.

Matteini M, Gioncada A, Mazzuoli R, Acocella V, Dini A, Guillou H, Omarini R, Uttini A, Vezzoli L, Hauser N (2005a). The magmatism in the easternmost sector of the Calama-Olocapato-El Toro transversaul fault system in the Central Andes at 24°S: Geotectonic significance. Proceedings of the 6th International Symposium on Andean Geodynamics, Barcelona, Spain, pp. 499–501.

Matteini M, Acocella V, Vezzoli L, Dini A, Gioncada A, Guillou, H, Mazzuoli R, Omarini R, Uttini, A, Hauser, N (2005b). Geology and petrology of the Las Burras-Almagro magmatic complex, Salta Argentina. Proceedings of the XVI Congreso Geologico Argentino I, pp. 479–484.

Mazzuoli R, Tortorici L, Ventura G (1995). Oblique rifting in Salina, Lipari and Vulcano islands (Aeolian islands, southern Italy). Terra Nova 7:444–452.

McCaffrey KJW (1992). Igneous emplacement in the transpressive shear zone; Ox Mountains igneous complex. J Geol Soc London 149 : 221–235.

Meneses-Rocha JJ (1991). Tectonic Development of the Ixtapa Graben, Chiapas, Mexico. PhD, University of Texas, Austin, 308 pp.

Merle O, Vidal N, van Wyk de Vries B (2001). Experiments on vertical basement fault reactivation below volcanoes. J Geophys Res 106:2153–2162.

Miranda F, Folguera A Leal PL, Naranjo JA, Pesce A (2006). Upper Pliocene to Lower Pleistocene volcanic complexes and Upper Neogene deformation in the south-central Andes (36°30'–38°S). Geol Soc Am Spec Paper 407:287–298.

Mitchell J, Westaway R (1999). Chronology of Neogene and Quaternary uplift and magmatism in the Caucasus: constraints from K-Ar dating of volcanism in Armenia. Tectonophysics 304:157–186.

Miura S, Ueki S, Sato T, Tachibana K, Hamaguchi H (2000). Crustal deformation associated with the 1998 seismo-volcanic crisis of Iwate Volcano, Northeastern Japan, as observed by a dense GPS network. Earth Planet Space 52:1003–1008.

Moore I, Kokelaar P (1998). Tectonically controlled piecemeal caldera collapse; a case study of Glencoe Volcano, Scotland. Geol Soc Amer Bull 110:1448–1466.

Muñoz J, Troncoso R, Duhart P, Crignola P, Farmer L, Stern CR (2000). The relationship of the mid-Tertiary coastal magmatic belt in south-central Chile to the late Oligocene increase in plate convergence rate. Revista Geológica de Chile 27(2) : 177–203.

Nakahara H, Nishimura T, Sato H, Ohtake M, Kinoshita S, Hamaguchi H (2002). Broad-band source process of the 1998 Iwate Prefecture, Japan, earthquakes as revealed from inversion analyses of seismic waveforms and envelopes. Bull Soc Seismol Am 92:1708–1720.

Nakamura K (1977). Volcanoes as possible indicators of tectonic stress orientation: principle and proposal. J Volcanol Geotherm Res 2:1–16.

Nakamura K, Uyeda S (1980). Stress gradient in arc–back arc regions and plate subduction. J Geophys Res 85:6419–6428.

Nakanishi M (1989). Mesozoic magnetic anomaly lineations and sea-floor spreading of the NW Pacific. J Geophys Res 94:15, 437–15,446.

Nelson MR, Forsythe R, Arit I (1994). Ridge collision tectonics in terrane development. J South Am Earth Sci 7:271–278.

Norabuena E, Leffler-Griffin L, Mao A, Dixon T, Stein S, Sacks S, Ocola L, Ellis M (1998). Space geodetic observations of Nazca-South America convergence across the Central Andes. Science 279:358–362.

Norini G, Lagmay AMF (2005). Deformed symmetrical volcanoes. Geology 33:605–608.

Olivier P, Ameglio L, Richen H, Vadeboin F (1999). Emplacement of the Aya Variscan granitic pluton (Basque Pyrenees) in a dextral transcurrent regime inferred from a combined magneto-structural and gravimetric study. J Geol Soc London156:991–1002.

Pasquarè G, Poli S, Vezzoli L, Zanchi A (1988). Continental arc volcanism and tectonic setting in Central Anatolia, Turkey. Tectonophysics 146:217–230.

Pasquarè FA, Tibaldi A (2003). Do transcurrent faults guide volcano growth? The case of NW Bicol Volcanic Arc, Luzon, Philippines. Terra Nova 15:204–212.

Pasquarè G, Tibaldi A, Ferrari L (1990). Relationships between plate convergence and tectonic evolution of the Ecuadorian active Thrust Belt. In: Agusthithis SS (Ed.), Critical Aspects of Plate Tectonic Theory, Theophrastus Publications, pp. 365–387.

Pearce JA, Bender JF, De Long SE, Kidd WSF, Low PJ, Güner Y, Saroglu F, Yilmaz Y, Moorbath S Mitchell JG (1990). Genesis of collision volcanism in Eastern Anatolia, Turkey. J Volc Geoth Res 44:189–229.

Peterson U (1999). Magmatic and metallogenic evolution of the Central Andes. In: Skinner B. (Ed.), Geology and ore deposits of the Central Andes. Society of Economic Geology, Special Publication, 7, 109–153.

Petford N, Atherton MP (1995). Crustal segmentation and the isotopic significance of the Abancay Deflection Northern Central Andes, 9°–20°S. Revista Geológica de Chile 22 : 235–243.

Petrinovic IA, Riller U, Brod JA (2005). The Negra Muerta Volcanic Complex, southern central Andes: geochemical characteristics and magmatic evolution of an episodically active volcanic centre. J Volcanol Geotherm Res 140:295–320.

Petrinovic IA, Riller U, Brod JA, Alvarado G, Arnosio M (2006). Bimodal volcanism in a tectonic transfer zone: Evidence for tectonically controlled magmatism in the southern Central Andes, NW Argentina. J Volcanol Geotherm Res 152:240–252.

Putirka K, Busby CJ (2007). The tectonic significance of high-K_2O volcanism in the Sierra Nevada, California. Geology 35:923–926.

Ramelow J, Riller U, Romer RL, Oncken O (2006). Kinematic link between episodic trapdoor collapse of the Negra Muerta Caldera and motion on the Olacapato-El Toro Fault Zone, southern central Andes. Int J Earth Sci (Geol Rundsch) 95:529–541.

Ramos V, Cegarra M, Cristallini E (1996). Cenozoic tectonics of the high Andes of west-central Argentina (30º–36.5ºS). Tectonophysics 259:185–200.

Ramos V, Cristallini E, Pérez D (2002). The Pampean flat-slab of the Central Andes. J South Am Earth Sci 15 : 59–78.

Rebai S, Philip H, Dorbath L, Borissoff B, Haessler H, Cisternas A (1993). Active tectonics in the Lesser Caucasus: coexistence of compressive and extensional structures. Tectonics 12:1089–1114.

Riller U, Petrinovic I, Ramelow J, Strecker M, Oncken O (2001). Late Cenozoic tectonism, collapse caldera and plateau formation in the central Andes. Earth Planet Sci Lett 188:299–311.

Roman DC, Moran SC, Power JA, Cashman KV (2004). Temporal and spatial variation of local stress fields before and after the 1992 eruptions of Crater Peak vent, Mount Spurr volcano, Alaska. Bull Seismol Soc Am 94:2366–2379.

Roman-Berdiel T (1999). Geometry of granite emplacement in the upper crust: Contribution of analogue modeling. In: Castro A, Fernandez C, Vigneresse JL (Eds.), Understanding Granites: Integrating New and Classical Techniques. Geol Soc Lond Spec Publ 174:77–94.

Rosenau M (2004). Tectonics of the southern Andean intra-arc zone (38°–42°S). Ph.D. Thesis, Freie Universitat Berlin, 159 pp.

Rosenau M, Melnick D, Echtler H (2006). Kinematic constraints on intra-arc shear and strain partitioning in the southern Andes between 38°S and 42°S latitude. Tectonics, doi:10.1029/2005TC001943.

Rosenberg CL (2004). Shear zones and magma ascent: A model based on a review of the Tertiary magmatism in the Alps. Tectonics, doi:10.1029/2003TC001526.

Rossetti F, Storti F, Salvini F (2000). Cenozoic noncoaxial transtension along the western shoulder of the Ross Sea, Antarctica, and the emplacement of McMurdo dyke arrays. Terra Nova 12:60–66.

Rovere E (1993). K/Ar ages of magmatic rocks and geochemical variations of volcanics from South Andes (37° to 37°15'S-71°W). Proceedings 2nd Japan Volcanological Society Congress, 107.

Rovida A, Tibaldi A (2005). Propagation of strike-slip faults across Holocene volcano-sedimentary deposits, Pasto, Colombia. J Struct Geol 27:1838–1855.

Saint Blanquat M, Tikoff B, Teyssier C, Vigneresse JL (1998). Transpressional kinematics and magmatic arcs. In: Holdsworth RE, Strachan RA, Dewey JF (Eds.), Continental Transpressional and Transtensional Tectonics. Geol Soc London Spec Publ 135, pp. 327–340.

Salfity JA (1985). Lineamentos transversales al rumbo andino en el noroeste argentino, IV Congreso Geologico Chileno, 2:119–137.

Salvini F, Brancolini G Busetti M, Storti F, Mazzarini F, Coren F (1997). Cenozoic geodynamics of the Ross Sea region, Antarctica: Crustal extension, intraplate strike-slip faulting, and tectonic inheritance. J Geophys Res 102:24, 669–24,696.

Schafer KH, Dannapfel M (1994). State of in situ Stress in Northern Chile and in Northwestern Argentina. In: Reuter KJ, Scheuber E, Wigger PJ (Eds.), Tectonics of the Southern Central Andes. Structure and Evolution of an Active Continental Margin. Springer, New York, pp. 103–110.

Scheuber E, Reutter K (1992). Magmatic arc tectonics in the Central Andes between 21° and 25°S. Tectonophysics 205:127–140.

Schurr B, Asch G, Rietbrock A, Kind R, Pardo M, Heit B (1999). Seismicity and average velocities beneath the Argentine Puna plateau. Geophys Res Lett 26:3025–3028.

Sebrier M, Soler P (1991). Tectonics and Magmatism in the Peruvian Andes from Late Oligocene to Present. Geol Soc Am Spec Paper 265:259–278.

Serra S, Nelson RA (1988). Clay modeling of rift asymmetry and associated structures. Tectonophysics 153:307–312.

Shaw HR (1980). The fracture mechanisms of magma transport from the mantle to the surface. In: Hargraves RB (Ed.), Physics of Magmatic Processes. Princeton University Press, Princeton, NJ, pp. 201–264.

Sibson RH (2003). Brittle-failure controls on maximum sustainable overpressure in different tectonic regimes. Am Assoc Pet Geol Bull 87:901–908.

Sieh K, Natawidjaja D (2000). Neotectonics of the Sumatran fault, Indonesia. J Geophys Res 105:28,295–28,326.

Simkin T, Siebert L, McClelland L, Bridge D, Newhall C, Latter JH (1981). Volcanoes of the world: A regional directory, gazetteer, and chronology of volcanism during the last 10,000 years. U.S. Hutchinson Ross Publishing, 232 pp.

Skulski T, Francis D, Ludden JN (1987). The Tertiary lavas of SW Yukon and NW British Columbia; transform fault related magmatism? Geol Assoc Canada Programs Abstr 12:89.

Skulski T, Francis D, Ludden JN (1991). Arc transform magmatism in the Wrangell volcanic Belt. Geology 19:11–14.

Soler P, Bonhomme M (1990). Relations of magmatic activity to plate dynamics in central Perú from Late cretaceous to Present. In: Kay SM, Rapela CW (Eds.), Plutonism from Antarctica to Alaska . Geological Society of America, Special Paper, 241, 173–191.

Spikings R, Seward D, Winkler W, Ruiz G (2000). Low temperature thermochronology of the northern Cordillera Real, Ecuador: Tectonic insights from zircon and apatite fission track analysis. Tectonics 19:649–668.

Spinks KD, Acocella V, Cole JW, Bassett KN (2005). Structural control of volcanism and caldera development in the transtensional Taupo Volcanic Zone, New Zealand. J Volcanol Geotherm Res 144:7–22.

Steinmann M (1997). The Cuenca Basin of Southern Ecuador: tectono-sedimentary history and the Tertiary Andean evolution. PhD Thesis, Swiss Federal Institute of Technology, Zurich, n. 12297, 185 pp.

Stern CR (2004). Active Andean volcanism: its geologic and tectonic setting. Rev Geol de Chile 31:161–206.

Sylvester AG (1988). Strike-slip faults. Geol Soc Am Bull 100:1666–1703.

Tatar O, Yurtmen S, Temiz H, Gursoy H, Kocbulut F, Mesci BL, Guezou JC (2007). Intracontinental Quaternary Volcanism in the Niksar Pull-Apart Basin, North Anatolian Fault Zone, Turkey. Turkish J Earth Sci 16:417–440.

Tibaldi A (1995). Morphology of pyroclastic cones and tectonics. J Geophys Res 100:24,521–24, 535.

Tibaldi A (2005a). Quaternary compressional deformation around the Cotopaxi Volcano, Ecuador. AGU Chapman conference on "The Effects of Basement, Structure, and Stratigraphic Heritages on Volcano Behaviour", 16–20 November 2005, Taal volcano, Tagaytay City, Philippines.

Tibaldi A (2005b). Volcanism in compressional settings: is it possible? Geophys Res Lett, doi:10.1029/2004GL021798.

Tibaldi A (2008). Contractional tectonics and magma paths in volcanoes. J Volcanol Geotherm Res, in press.

Tibaldi A, Corazzato C, Rovida A (2007). Late Quaternary kinematics, slip-rate and segmentation of a major Cordillera-parallel transcurrent fault: The Cayambe-Afiladores-Sibundoy system, NW South America. J Struct Geol 29:664–680.

Tibaldi A, Ferrari L (1992). Latest Pleistocene-Holocene tectonics of the Ecuadorian Andes. Tectonophysics 205:109–125.

Tibaldi A, Romero-Leon JL (2000). Morphometry of Late Pleistocene- Holocene faulting and volcano-tectonic relationships in the southern Andes of Colombia. Tectonics 19:358–377.

Tibaldi A, Vezzoli L, Pasquarè FA, Rust D (2008). Strike-slip fault tectonics and the emplacement of sheet-laccolith systems: The Thverfell case study (SW Iceland). J Struct Geol 30:274–290.

Tibaldi A (2008). Contractional tectonics and magma paths in volcanoes. J Volcanol Geotherm Res 176 : 291–301.

Toprak, V (1998). Vent distribution and its relation to regional tectonics, Cappadocian Volcanics, Turkey. J Volcanol Geotherm Res 85:55–67.

Tormey DR, Hickey-Vargas R, Frey F, Lopez-Escobar L (1991). Recent lavas from the Andean volcanic Front (33 to 42S): Interpretations of along-arc compositional variations. Geol Soc Am Spec Paper 265:57–78.

Turner S, Arnaud N, Liu J, Rogers N, Hawkesworth C, Harris N, Kelley S, van Calsteren P, Deng W (1996). Post-collision, shoshonitic volcanism on the Tibetan plateau: Implications for convective thinning of the lithosphere and the source of ocean island basalts. J Petrol 37:45–71.

Van der Werff W (2000). Backarc deformation along the eastern Japan Sea margin, offshore northern Honshu. J Asian Earth Sci 18:71–95.

van Wyk de Vries B, Merle O (1998). Extension induced by volcanic loading in regional strike-slip zones. Geology 26:983–986.

van Wyk de Vries B, Self S, Francis PW, Keszthelyi L (2001). A gravitational spreading origin for the Socompa debris avalanche. J Volcanol Geotherm Res 105:225– 247.

Ventura G (1994). Tectonics, structural evolution and caldera formation on Vulcano island (Aeolian archipelago, southern Tyrrhenian Sea). J Volcanol Geotherm Res 60:207–224.

Ventura G, Vilardo G, Milano G, Pino NA (1999). Relationships among crustal structure, volcanism and strike-slip tectonics in the Lipari-Vulcano Volcanic Complex (Aeolian Islands, Southern Tyrrhenian Sea, Italy). Phys Earth Planet Inter 116:31–52.

Victor P, Oncken O, Glodny J (2004). Uplift of the western Altiplano plateau: Evidence from the Precordillera between 20° and 21°S (northern Chile). Tectonics, 23, 4, TC4004 10.1029/2003TC001519.

Vidal N, Merle, O (2000). Reactivation of basement fault beneath volcanoes: a new model of flank collapse. J Volcanol Geotherm Res 99:9–26.

Wall RW, Lara LE (2001). Lavas Las Pataguas: volcanismo alcalino en el antearco andino del Mioceno Inferior, Chile central. Revista Geológica de Chile 28(2) : 243–258.

Watanabe T, Koyaguchi T, Seno T (1999). Tectonic stress controls on ascent and emplacement of magmas. J Volcanol Geotherm Res 91:65–78.

Weinberg RF, Sial AN, Mariano G (2004.) Close spatial relationship between plutons and shear zones. Geology 32: 377–380.

Westaway R (1990). Seismicity and tectonic deformation rate in Soviet Armenia: Implications for local earthquake hazard and evolution of adjacent regions. Tectonics 9:477–503.

Williams H, McBirney A (1979). Volcanology. Freeman, Cooper & Co., 397pp.

Winkler W, Villagomez D, Spikingsc R, Abegglend P, Toblere St, Eguezb A (2005). The Chota basin and its significance for the inception and tectonic setting of the inter-Andean depression in Ecuador. J South Am Earth Sci 19:5–19.

Wörner G, Hammerschmidt K, Henjes-Kunst F, Lezaun J, Wilke H (2000b). Geochronology (40Ar/39Ar, K-Ar and He-exposure ages) of Cenozoic magmatic rocks from northern Chile (18–22°S): implications for magmatism and tectonic evolution of the central Andes. Revista Geológica de Chile 27: 205–240.

Xu J, Zhu G, Tong W, Cui K, Liu Q (1987). Formation and evolution of the Tancheng-Lujiang wrench fault system: a major shear system to the northwest of the Pacific ocean. Tectonophysics134:273–310.

Yilmaz Y (1990). Comparison of young volcanic associations of western and eastern Anatolia formed under a compressional regime: a review. J Volcanol Geotherm Res 44:69–87.

Yılmaz Y, Guner Y, Saroglu F (1998). Geology of Quaternary volcanic centers of east Anatolia. J Volcanol Geotherm Res 85:173–210.
Yoshida T (2001). The evolution of arc magmatism in the NE Honshu arc, Japan. Tohoku Geophys J 32:131–149.
Ziv A, Rubin AM (2000). Stability of dyke intrusion along pre-existing fractures. J Geophys Res 105:5947–5961.
Zapata TR, Brisson I, Dzelalija F (1999). La estructura de la faja plegada y corrida andina en relacion con el control del basamento de la Cuenca Neuquina, Boletın de Informaciones Petroleras, December 1999, pp. 112–121.
Zollner W, Amos AJ (1973). Descripcion geologica de la Hoja 32b, Chos Malal (Prov. Neuquen), Bull. 143, 91 pp., Serv. Nac. Min. Geol., Buenos Aires.

DynaQlim – Upper Mantle Dynamics and Quaternary Climate in Cratonic Areas

Markku Poutanen, Doris Dransch, Søren Gregersen, Sören Haubrock, Erik R. Ivins, Volker Klemann, Elena Kozlovskaya, Ilmo Kukkonen, Björn Lund, Juha-Pekka Lunkka, Glenn Milne, Jürgen Müller, Christophe Pascal, Bjørn R. Pettersen, Hans-Georg Scherneck, Holger Steffen, Bert Vermeersen, and Detlef Wolf

Abstract The isostatic adjustment of the solid Earth to the glacial loading (GIA, Glacial Isostatic Adjustment) with its temporal signature offers a great opportunity to retrieve information of Earth's upper mantle to the changing mass of glaciers and ice sheets, which in turn is driven by variations in Quaternary climate. DynaQlim (Upper Mantle Dynamics and Quaternary Climate in Cratonic Areas) has its focus to study the relations between upper mantle dynamics, its composition and physical properties, temperature, rheology, and Quaternary climate. Its regional focus lies on the cratonic areas of northern Canada and Scandinavia.

Geodetic methods like repeated precise levelling, tide gauges, high-resolution observations of recent movements, gravity change and monitoring of postglacial faults have given information on the GIA process for more than 100 years. They are accompanied by more recent techniques like GPS observations and the GRACE and GOCE satellite missions which provide additional global and regional constraints on the gravity field. Combining geodetic observations with seismological investigations, studies of the postglacial faults and continuum mechanical modelling of GIA, DynaQlim offers new insights into properties of the lithosphere. Another step toward a better understanding of GIA has been the joint inversion of different types of observational data – preferentially connected with geological relative sea-level evidence of the Earth's rebound during the last 10,000 years.

Due to the changes in the lithospheric stress state large faults ruptured violently at the end of the last glaciation in large earthquakes, up to the magnitudes $M_W = 7–8$. Whether the rebound stress is still able to trigger a significant fraction of intraplate seismic events in these regions is not completely understood due to the complexity and spatial heterogeneity of the regional stress field. Understanding of this mechanism is of societal importance.

Glacial ice sheet dynamics are constrained by the coupled process of the deformation of the viscoelastic solid Earth, the ocean and climate variability. Exactly how the climate and oceans reorganize to sustain growth of ice sheets that ground to continents and shallow continental shelves is poorly understood. Incorporation of nonlinear feedback in modelling both ocean heat transport systems and atmospheric CO_2 is a major challenge. Climate-related loading cycles and episodes are expected to be important, hence also more short-term features of palaeoclimate should be explicitly treated.

Keywords GIA · Crustal deformation · Mantle dynamics · Quaternary climate

Introduction

The process of GIA with its characteristic temporal signatures is one of the great opportunities in geosciences to retrieve information about the Earth. It contains information about recent climate forcing, being dependent on the geologically recent on- and offloading of ice sheets. It gives a unique chance to study the dynamics and rheology of the lithosphere and asthenosphere, and it is of fundamental importance in geodesy, since Earth rotation, polar motion and crustal

M. Poutanen (✉)
Finnish Geodetic Institute, Geodeetinrinne 2, 02430 Masala, Finland
e-mail: markku.poutanen@fgi.fi

S. Cloetingh, J. Negendank (eds.), *New Frontiers in Integrated Solid Earth Sciences*, International Year of Planet Earth, DOI 10.1007/978-90-481-2737-5_10, © Springer Science+Business Media B.V. 2010

deformation, and therefore the global reference frames are influenced by it.

Despite the existence of long and accurate time series and extensive data sets on GIA, there still exist many open questions related to upper mantle dynamics and composition, rebound mechanisms and uplift models, including the role of tectonic forces as well as ice thickness during the late Quaternary. DynaQlim aims to integrate existing data and models on GIA processes, including both geological and geodetic observations. The themes of DynaQlim include Quaternary climate and glaciation history, postglacial uplift and contemporary movements, ice-sheets dynamics and glaciology, postglacial faulting, rock rheology, mantle xenoliths, past and present thermal regime of the lithosphere, seismic structure of the lithosphere, and gravity field modelling.

DynaQlim will probably lead to a more comprehensive understanding of the Earth's response to glaciations, improved modelling of crustal and upper mantle dynamics as rheology structure. An important aspect is to construct and improve coupled models of glaciation and land-uplift history and their connection to the climate evolution on the time scale of glacial cycles.

Observational Basis

During the Pleistocene, quasi-periodic variations between glacial and interglacial intervals prevailed, with dominant periods closely related to those present in the Earth-Sun orbit and 25.8 kyr rotational precession of the Earth (Berger, 1984). These Milankovitch variations have played a key role in shaping the landscape and driving the geodynamic evolution of cratonic regions such as northern Eurasia and North America during the Quaternary.

Extensive and diverse sets of observations can be applied to study and understand the key processes involved, including geodetic land uplift measurements, geological observations of past sea-level changes, late-glacial faults, terminal moraines and other glacial deposits as well as various palaeoclimatological proxies. These observations have played a vital role in a number of recent studies that have improved our understanding of the structure and dynamics of cratonic regions and the influence of ice sheet variations.

Abundant data have been collected in various cratonic regions, including Antarctica, Laurentia and Fennoscandia. Laurentia and Fennoscandia have a similar glaciation history during the Quaternary, though their tectonic evolutions are different. In Antarctica the glaciation history is distinctly different. DynaQlim will collect and compile observational evidence predominantly from geodetic and geophysical methods.

Geodetic Observations

Geodetic methods provide accurate measurements of contemporary deformation and gravity change. There are systematic postglacial uplift observations for the last 100 years based on repeated precise levelling, tide gauges, geodetic high-resolution observations of recent movements, gravity change and monitoring of postglacial faults. Until recently, horizontal motions could not be observed accurately. However, current GNSS (Global Navigation Satellite Systems, including GPS) observations are accurate enough to observe even minor horizontal motions over distances of several hundreds of kilometres.

Maps of vertical motion have traditionally been based on long time series of tide gauges and repeated precise levellings over several decades. Tide gauge time series reflect both vertical motions of the land and variations of the surface of the sea. Maps of relative sea level change for Fennoscandia were published by e.g. Ekman (1996), Kakkuri (1997), Mäkinen and Saaranen (1998) and Saaranen and Mäkinen (2002). The latest uplift models, based on repeated precise levelling, tide gauge time series and geophysical modelling have been published by Vestøl (2006), and Ågren and Svensson (2007), Fig. 1.

In North America repeated levellings of the rebound area are confined to regions near Hudson Bay (Sella et al., 2007) or other coastal areas. Overall, levelling data are much more scattered than in Fennoscandia.

Space geodetic techniques, such as GNSS, allow the construction of 3-D motions from relatively short (less than 10 years) time series. The project BIFROST (Baseline Inferences for Fennoscandian Rebound Observations, Sea Level, and Tectonics) was initiated in 1993 taking advantage of tens of permanent GPS stations separated by a few hundreds of km both in Finland and Sweden. Results are discussed e.g. in Milne et al. (2001), Johansson et al. (2002), and

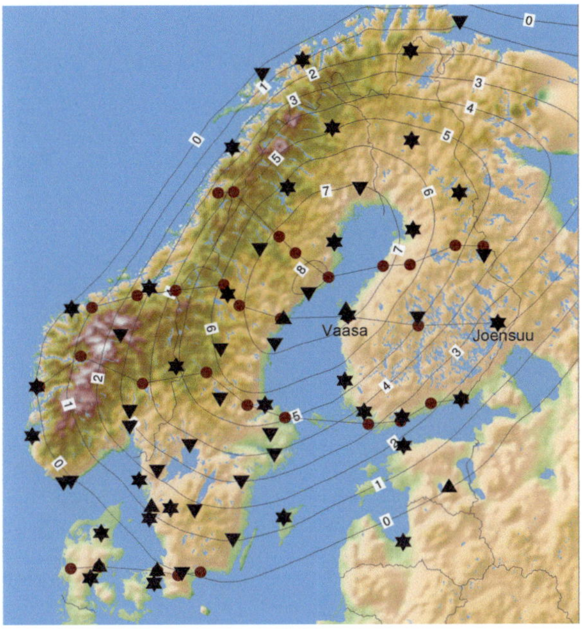

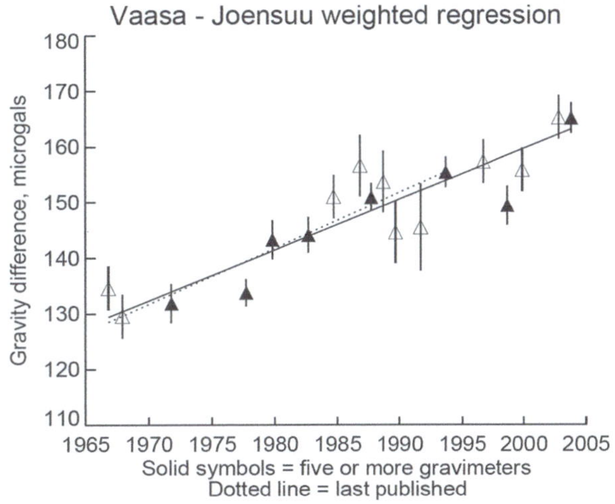

Fig. 1 GIA in Fennoscandia. *Left*: The upside-down triangles on the map are permanent GNSS stations, triangles stations where regular absolute gravity is measured as a part of the NGOS project, and dots with joining lines are the land uplift gravity lines, measured since the mid-1960s. Contour lines show the apparent land uplift relative to the Baltic mean sea level 1892–1991, based on Nordic uplift model NKG2005LU (Vestøl, 2006; Ågren and Svensson, 2007) *Right*: Diagram of the observed relative gravity change between Vaasa and Joensuu in Finland during 40 years of measurement on the land uplift gravity lines (Mäkinen et al., 2005)

Scherneck et al. (2002) (Fig. 3). Maps based on GPS time series were published e.g. by Mäkinen et al. (2003), Milne et al. (2001), Lidberg (2007), and Lidberg et al. (2007).

In North America several hundreds of continuous GPS stations have been used to compute contemporary velocities (e.g., Calais et al., 2006; Wolf et al., 2006; Sella et al., 2007). In Greenland a campaign with repeated GPS has been carried out over a period of close to 10 years (Dietrich et al., 2005) with uplift values of the order mm/year close to the ice cap.

The gravitational uplift signal can be detected by absolute and relative gravimetry (e.g., Ekman and Mäkinen, 1996; Mäkinen et al., 2005) or by the GRACE satellite mission (e.g. Wahr and Velicogna, 2003; Peltier, 2004; Tamisiea et al., 2007). The gravity satellites GRACE and GOCE are providing, or will provide, additional global and regional constraints on the gravity field (Pagiatakis and Salib, 2003; Müller et al., 2006). Recent studies have demonstrated that the GRACE data clearly show temporal gravity variations both in Fennoscandia and North America (Tamisiea et al., 2007; Ivins and Wolf, 2008; Steffen et al., 2008). The temporal trends and the uplift pattern retrieved from these data are in good agreement with previous studies and independent terrestrial data (Fig. 2).

The gravity change due to the postglacial rebound is about −2 μgal/cm of uplift relative to the Earth's centre of mass, or about −2 μgal/yr at the centre of the uplift area in Fennoscandia (Ekman and Mäkinen, 1996). Based on this, the peak geoid change rate is estimated to be 0.6 mm/yr. The results are based on land-uplift gravity lines in Fennoscandia (Fig. 1), observed regularly since the mid-1960s (Mäkinen et al., 2005). Currently, an increasing number of continuous GNSS sites are also monitored using repeated absolute gravity measurements.

Crustal deformation and sea level variation studies are based on stable reference frames. If effects at the 1 mm/yr level are to be studied, a stability of about 0.1 mm/yr in the reference frames is needed over several decades. Such stability is not yet achieved. Geodesy's response to this requirement is the Global Geodetic Observing System (GGOS), a new integral part of the International Association of Geodesy, (GGOS, 2008). There are several ongoing plans

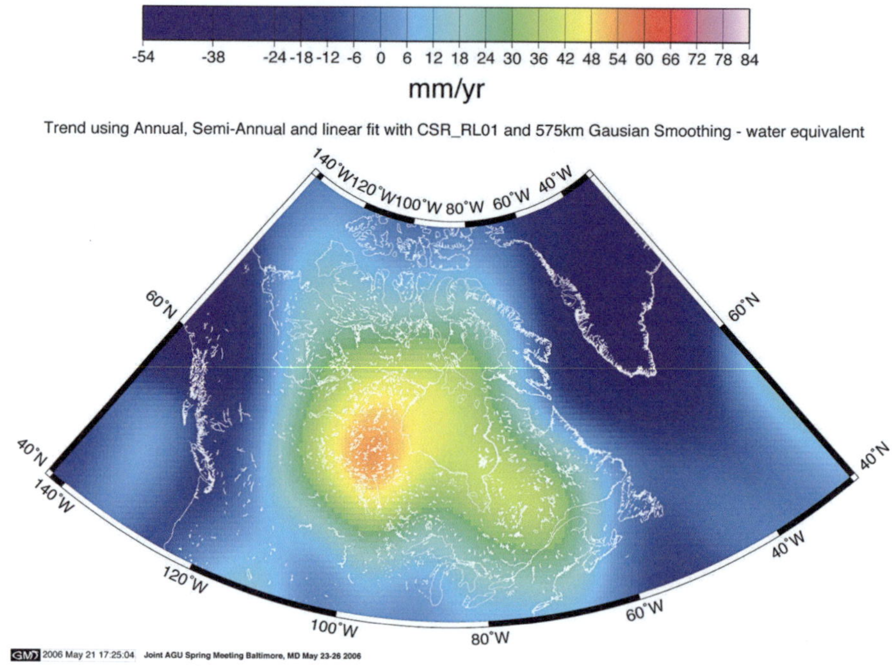

Fig. 2 GIA in North America shown as a GRACE-derived water-equivalent mass change. The GRACE signal is unfiltered by hydrological modeling. The GRACE Level 2 product employed is from Release 01 of the Center for Space Research (CSR) from the University of Texas at Austin, which uses the months January 2003 to December 2006, excluding July 2003. The harmonics are truncated at degree and order 60 and a Gauss filter of 575-km radius is applied. (Ivins and Wolf, 2008)

for regional implementation of GGOS, as an example the Nordic Geodetic Observing System, (NGOS, Poutanen et al., 2007). The NGOS plan includes also annual absolute gravity measurements at the permanent GNSS sites (Fig. 1).

Evidence from Geophysical Observations of Lithosphere Structure

Our present knowledge of the rheology and structure of the lithosphere is based on a combination of rock deformation experiments, petrophysical inference from seismology and heat flow (Blundell et al., 1992; Bürgmann and Dresen, 2008). Continuous GNSS observations of plate-wide strain, accompanied by seismological investigations, and followed by continuum mechanical modelling of GIA, studies of seismic source and wave propagation, and studies of the postglacial faults offer new insights into properties of the lithosphere. Observations and models of glacial and postglacial faulting can help to illuminate crustal stress fields and therefore crustal rheology issues.

Existing data on experimentally studied lower crustal and mantle composition and 3-D structure derived from xenolith data, lithospheric thermal models (Kukkonen et al., 2003; Hieronymus et al., 2007) and seismic studies (Bruneton et al., 2004; Sandoval et al., 2004; Yliniemi et al, 2004; Hjelt et al., 2006; Pedersen et al., 2006; Plomerova et al., 2006; Gregersen et al., 2006; Janik et al., 2007; Olsson et al., 2007) should be utilized for forward rheological modelling of the lithosphere and for testing of dynamic uplift models. The presence and volume of fluids in the upper mantle and the influence of fluids on the mantle rheology is an open question. As dissociated water may provide an effective mechanism for electrical conductivity in the upper mantle, important implications on mantle fluids and the lithosphere-asthenosphere system can be obtained from recent deep electromagnetic measurements (Korja et al., 2002; Hjelt et al., 2006; Korja 2007).

Inversion of deep temperature data in boreholes provides direct access to ground temperature histories during glaciation times (Kukkonen and Jõeleht, 2003). Kimberlite facies in crustal rocks contain mantle xenoliths and these provide a basis for extrapolating temperature and composition to larger depths using seismology (Stein et al., 1989; Kukkonen et al., 2003; Bruneton et al., 2004; Hjelt et al., 2006; Olsson et al., 2006, 2007; Pedersen et al., 2006). These results can be used to develop more realistic models of

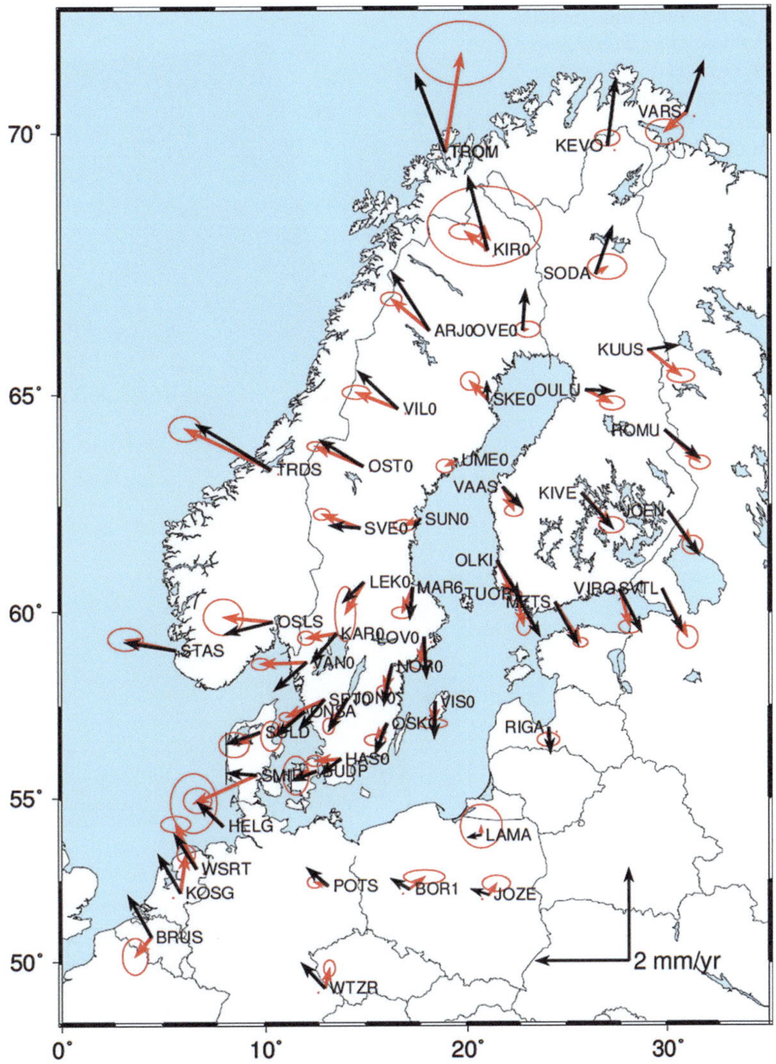

Fig. 3 Observed (*red*) and modelled (*black*) rates of horizontal displacement in Fennoscandia based on a model of Milne et al. (2001) and the GPS-derived velocity field of Lidberg et al., (2006), based on Nordic permanent GPS stations. (Lidberg et al., 2006)

mantle temperature and viscosity. These properties are key factors controlling the Earth's response to ice mass change.

Some of the largest fault scarps in northern Fennoscandia were formed at the end of the last glaciation (Kujansuu, 1964; Lagerbäck, 1979; Olesen, 1988), Fig. 4. These faults have lengths ranging from a few kilometers to 160 km and generally strike NNE, with maximum vertical offsets of 10–15 m. The faults generally dip to the east with downthrow to the west and they are almost exclusively reverse faults.

Quaternary deposits such as landslides and seismicity, trenching through the faults, dating using offset till sequences and radiocarbon dating of organic material, and geophysical investigations (e.g. Lagerbäck, 1979, 1990; Olesen, 1988, 1992; Bäckblom and Stanfors, 1989) have shown that the faults ruptured violently as large earthquakes. The magnitudes of these earthquakes is estimated to have reached MW 7–8, based on the distribution of triggered landslides, the distribution of current day seismicity and scaling relations for fault lengths (Lagerbäck, 1979; Arvidsson, 1996, Stewart et al., 2000).

As the faults are inferred to have ruptured just as the ice retreated from the respective area, these Glacially Induced Faults (GIFs) are frequently referred to as endglacial or postglacial, where the former is a more accurate description. The GIFs mostly ruptured through old zones of weakness (shear zones),

Fig. 4a Endglacial faults in Fennoscandia. *Blue squares* and *triangles* are Swedish permanent and temporary seismic stations, green triangles are Finnish seismic stations, and red triangles Norwegian seismic stations

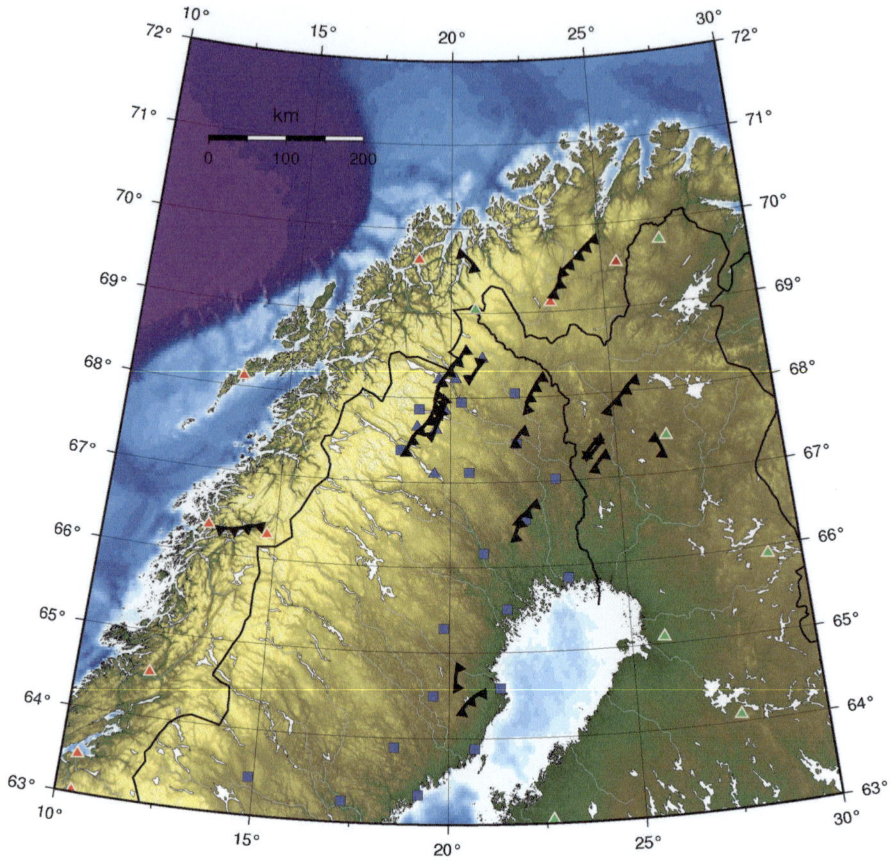

Fig. 4b Example of an endglacial fault in Fennoscandia: the Stuorragura reverse fault of northern Norway. View is due to the E and scarp height is c. 7 m (Olesen et al., 1992)

not necessarily following one zone but instead jumping to another to comply with the restraints set by the causative stress field.

As seen in the map in Fig. 4, there is no clear relationship between the orientation of the faults and the centre of rebound (Figs. 3 and 9), the GIFs consistently strike NNE-NE and mostly dip to the east irrespective of their location west of, north of or at the center of rebound. In addition to the Fennoscandian GIFs, end- or postglacial faults have been identified in most previously glaciated areas. A comprehensive review is presented in Munier and Fenton (2004), with examples from North America, the British Isles and Russia.

Although much effort has been spent on investigating the faults, key questions concerning the formation and current status of the GIFs are still unresolved. These include fault geometry at depth, fault strength, the interplay between the glacially induced stress field and the tectonic stress at the time of rupture, the influence of pore pressure and current deformation rates. Especially intriguing is the fact that, excluding the Berill Fault located in southern Norway (Anda et al., 2002), large GIFs have been identified exclusively in northernmost Fennoscandia (Munier and Fenton, 2004; Lagerbäck and Sundh, 2008), an observation which poses a difficult challenge to models of GIF formation.

The reconstruction of the former sea level is important for both the quantification of GIA and the reconstruction of palaeo-environments. Geologically, the uplift is documented in ancient shorelines (e.g. Lambeck et al., 1998, Tikkanen and Oksanen, 2002), but the accuracy of the timing of the shorelines is a limiting factor. The time of formation of shorelines and other indications of former sea level spans over the last glacial cycle. The evidence is based on so called sea level indicators (SLI), such as fossil samples of shells or morphological features like ancient shorelines and isolation basins.

Over the last decades many attempts were made to reconstruct palaeo-shorelines worldwide. This is because of the implications of sea level changes for the densely populated areas near the coasts. This in turn gave rise to a number of international campaigns for the compilation and interpretation of these SLIs, e.g. IGCP 61, 200 (van de Plassche, 1986), 247 (van de Plassche et al., 1995) and 367 (Shennan et al., 1998). For GIA, SLIs are a unique data source, because they allow the isolation of sea level change produced by crustal deformations and unaffected by presently relevant processes, such as sea level rise due to global warming. However, due to their usually indirect indication of former sea level, they have to be used with care in constraining GIA models (Klemann and Wolf, 2007). In the periglacial regions the palaeo-sea level is dominated by regional lithospheric flexure due to glacial loadings. Therefore, a link between GIA and reconstructions of palaeoclimate especially in the coastal regions is of interest.

Figure 5 shows the reconstruction of the topography in northern Europe with respect to sea level near the end of the last glaciation. A residual ice sheet is still present. The extent of the sea shows parts of the English Channel being above sea level and the Baltic Sea flooding large parts of Finland and northern Sweden. Also visible is the isolation of the Baltic from the ocean at this stage which defines the region to be at the Baltic Ice-Lake stage. This fact complicates the use of a standard GIA model for the reconstruction of this lake stage. The lake level exceeded the mean sea level by up to 26 m before its drainage (Påsse, 1996).

A physical model for the reconstruction of the lake needs palaeoclimatic information on precipitation and evaporation. Coupling with a dynamic ice model allows quantification of the inflow of water from the melting ice sheet and therefore an assessment of the relative salinity of the environment and the height of this lake above eustatic sea level. This additional height acts as an additional load which deforms the earth. On the other hand, this additional amount of water does not contribute to the eustatic sea level during the lake stages of the Baltic Sea.

Realistic regional modelling will need considerable improvement in the reconstruction of the palaeotopography. In addition, a spherically heterogeneous earth model may become necessary. Such models are designed to simulate the present time sea-level and geoid variations on a global scale as recorded by the recent space missions CHAMP and GRACE.

For GIA two further aspects are important: First, the surface heat flow will constrain the dynamics of the ice sheet (Näslund et al., 2005) and thus the deglaciation history. Second, the viscoelastic response of the solid earth is influenced by the rheological behaviour of the lithosphere. Again, lateral variations play a crucial role in the inference of the regional palaeotopography.

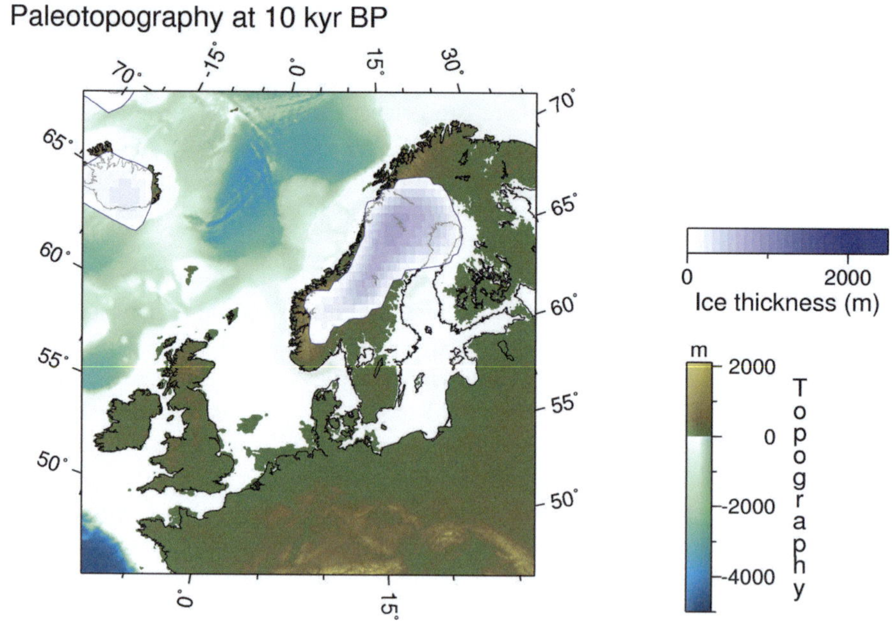

Fig. 5 Prediction of palaeo sea level near the end of the last glaciation phase in northern Europe based on a specific ice- and earth model

Seismicity and Stress-Field

The present-day seismicity in Fennoscandia as a whole is in general low to moderate in magnitude. Tectonic stress rates are small because this is an intraplate and cratonic region. This situation differs drastically from that at late-glacial times (i.e. 11–9 ky B.P.), when powerful earthquakes created impressive ground surface ruptures (e.g. Lagerbäck, 1979; Olesen, 1988). Presumably, the ice cap inhibited seismicity and strain-release during the Pleistocene glaciations. This caused earthquakes with magnitudes up to 8 when sudden global warming and ice-retreat occurred at the Pleistocene-Holocene transition (Johnston, 1989; Wu et al., 1999). A second mechanism, based on glacially induced stresses, was suggested by Wu and Hasegawa (1996)

The present-day seismicity in Fennoscandia is of intraplate type. The epicentres of these intraplate seismic events tend to be concentrated along ancient tectonic deformation zones. Due to the low seismicity level and relatively small number of permanent seismic stations in the past, studies of the sources of seismic events in Fennoscandia are relatively rare. The existing studies suggest that the sources are in areas of weakness in the crust which are favourably orientated with respect to the regional stress field and therefore can be reactivated (e.g. Slunga, 1991; Arvidsson and Kulhanek, 1994; Arvidsson, 1996; Uski et al., 2003). Thus studying both aspects (e.g. regional and local variations of stress field and distribution of zones weakness in the lithosphere capable to accumulate stresses) is important for the understanding of local seismicity and seismic hazard in Fennoscandia.

The stress state responsible for the observed seismicity appears to be the result of various stress-generating mechanisms (e.g. Fejerskov and Lindholm, 2000; Uski et al., 2003). In-situ stress measurements argue for relatively high magnitudes at shallow depths below the ground surface (Stephansson et al., 1986) The recent discovery of impressive stress-relief structures in different regions of Norway (Roberts, 2000; Roberts and Myrvang, 2004; Pascal et al., 2005) adds support to this conclusion. Observations to 6.5 km depth in the Siljan boreholes, central Sweden, suggest a strike-slip stress state at 5 km depth with maximum horizontal stress in the general direction NW-SE (Lund and Zoback, 1999). Although stress deviations are locally observed in Fennoscandia, maximum principal stress axes are in general horizontal and strike NW-SE (Slunga, 1991; Heidbach et al., 2008), suggesting ridge-push as the dominant mechanism (Fig. 6). A similar conclusion is reached for North America where the large majority of stress indicators show NE-SW compression in agreement with ridge-push (Adams and Basham, 1989).

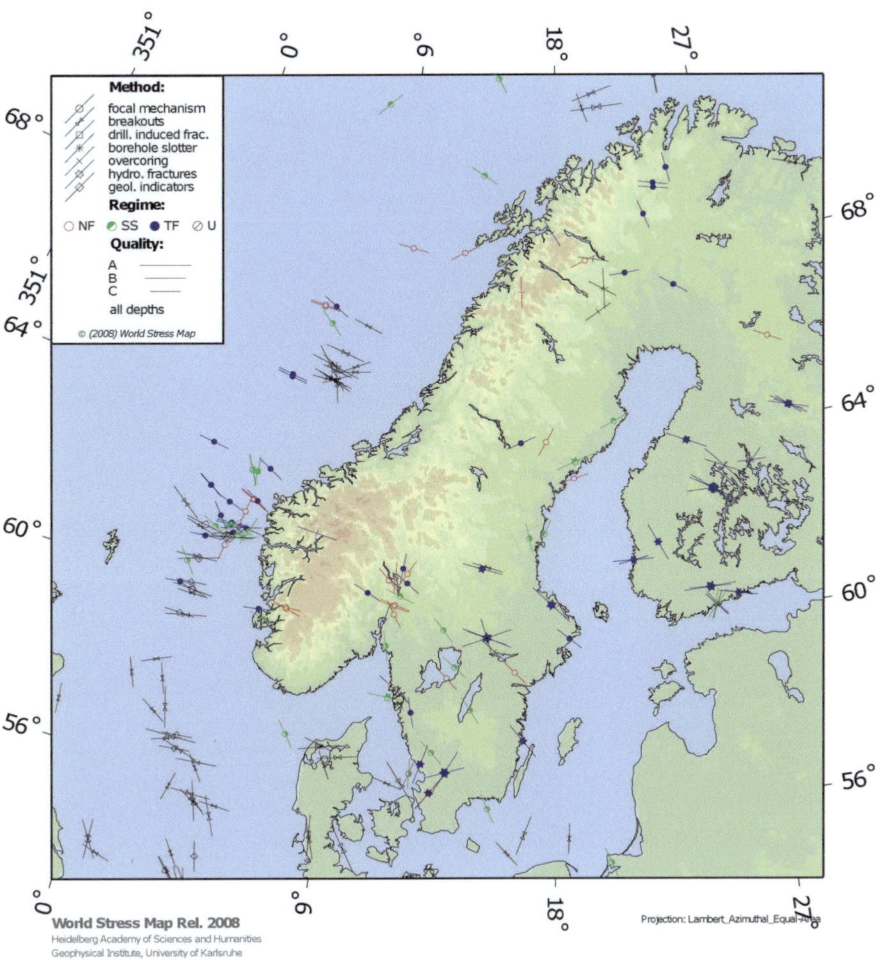

Fig. 6 Stress orientations in Fennoscandia and adjacent regions (World Stress Map website; Heidbach et al., 2008). Note the dominance of NW-SE maximum stress axes in agreement with ridge push force orientations in northern Europe

Postglacial rebound has often been advanced as a secondary source of stress modifying the tectonic stress. However, no clear radial pattern can be observed in the present-day stress compilations (Gregersen, 1992; Gregersen and Voss, 2009). This suggests that, in contrast with the situation that prevailed just after deglaciation (Wu, 1998), rebound stresses are nowadays in Fennoscandia, surpassed by plate motion forces or local stress sources.

A number of regions along the margins of the rebound domes in both Fennoscandia and Laurentia reveal complex and heterogeneous stress-fields. A similar situation exists in Antarctica and was possibly responsible for the $M_W \approx 8.1$ March 25, 1998 Balleny Is. earthquake in the Antarctic plate (Tsuboi et al., 2000; Ivins et al., 2003). This complexity is very unusual in the context of the homogeneity typically found in plate-tectonic stress fields. An additional stress source can be found in the anomalous elevation of southern and probably also northern Norway. It has been shown recently that the southern Scandes are prone to generate significant gravitational stresses acting on the adjacent regions (Pascal and Cloetingh 2009).

Due to the complexity and spatial heterogeneity of the regional stress field in Fennoscandia, it is not completely understood if the rebound stress (Fig. 7) is still able to trigger seismicity today. In addition, predictions of the onset time and mode of failure are very sensitive to the proper selection of models for ice sheet and mantle rheology (Wu et al., 1999; Klemann and Wolf, 1999; Lund and Näslund, 2008). However, the rebound processes seem to play a certain role in triggering seismicity in intraplate areas of northern America (see James and Bent, 1994; Wu and Johnston, 2000; Grollimund and Zoback, 2001). The same is claimed for Greenland by Chung and Gao (1997) and by Chung (2002).

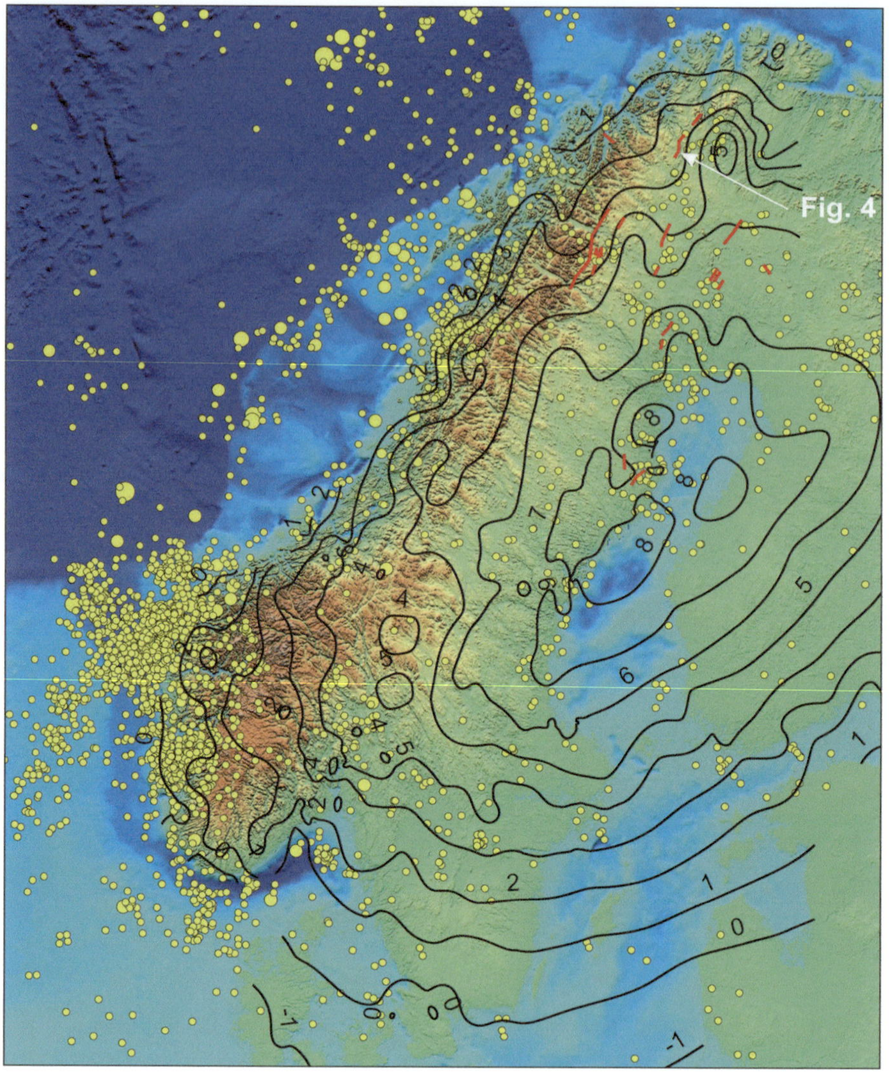

Fig. 7 Distribution of earthquakes in Fennoscandia superimposed on isolines of uplift in mm/yr (redrawn from Dehls et al., 2000)

The triggered earthquakes are concentrated either along zones of weakness or in regions of local stress concentrations. Recently van Lanen and Mooney (2007) have suggested that the structure of the lower crust, in particular the existence of deep faults rather than lateral variations in temperature, rheology or high pore pressure is a major factor controlling the spatial distribution of the intraplate seismicity in eastern North America. The faults are associated with ancient intracontinental rifts, palaeorifted margins or major terrain boundaries in the crust. These fossil structures can be easily reactivated by the rebound stresses. The importance of lateral variations in lithosphere structure for postglacial seismotectonics was also demonstrated by Wu and Mazzotti (2007), who showed that a narrow ductile zone that cuts vertically through the lower crust and lithospheric mantle generally has a larger effect on crustal motion and strain rates due to GIA than a horizontally uniform ductile layer.

Cryosphere and Palaeoclimate

Past and present changes in the mass balance of the Earth's ice sheets, ice caps and glaciers induce present-day deformation of the solid earth on spatial scales ranging from local to global. The Earth's deformational response to cryospheric change is complex due to a number of factors including complexities in the viscoelastic structure of the earth, the spatial and temporal variability of the ice mass changes and the interaction between the cryosphere and the ocean, which lead to a redistribution of cryospheric mass in a highly dynamic system. Both short and long term physical changes give important boundary conditions for the ice cap variations and their ablation. An example is provided by Greenland, where the ice sheet retreated after the last glacial maximum without disappearing. The associated sea level changes (e.g. Weidick 1995) supply important geophysical constraints together with those provided by GPS and gravity studies (e.g. Dietrich et al., 2005).

Recent advances in studies of the glacial history of northern Europe and Eurasia have significantly improved our understanding of the glaciation and deglaciation histories during the Weichselian and Holocene epochs over the past 100 kyr (Svendsen et al. 2004). In addition, the development of numerical modelling of glaciations has shed new light on the processes involved as well as their mutual couplings (Forsström, 2005; Zweck and Huybrechts, 2005; Näslund et al., 2005; SKB, 2005). As a result, the latest generation of ice sheet models is significantly better constrained and more realistic than previously.

Current Models and Problems to be Solved

The observations mentioned above can be used to constrain geophysical models of GIA using the observational evidence or to provide, e.g., GIA-induced uplift velocities for prespecified rheology (e.g. Lambeck et al., 1998). Milne et al. (2001) solved for Earth rheology using 3-D motion observations from the BIFROST continuous GPS network (Johansson et al., 2002) and the ice history of Lambeck. Current GIA models, however, suffer from regional deviations up to 0.2–0.5 mm/yr compared to the observed values. This will degrade the resolution of the past behaviour of the uplift, interpretation of the rheology, and therefore also the ice history.

The determination of the palaeoclimate and glaciation history of the Pleistocene are complex global problems and should therefore be solved in a joint effort. In view of the past observational and modelling activities, the two major formerly glaciated regions, Fennoscandia and Laurentia, are expected to be most promising. One main focus will be the global modelling of the climate forcing and the modifications due to the mechanical responses of the lithosphere and the asthenosphere (Bintanja et al., 2005; van den Berg et al., 2006).

Current GIA models are mostly based on radially stratified (3D) Earth models with linear rheology (e.g., Sabadini and Vermeersen, 2004). During the last few years progress has been made in the development of global, 3D-stratified earth modelling (Martinec, 2000; Wu and van der Wal, 2003; Whitehouse et al., 2006). However, due to computational restrictions, the latter models still have relatively low resolution.

High-resolution gravity data with spatial scales of 100 km or better and with nearly uniform quality will hopefully become available from the GOCE satellite mission scheduled for launch in 2009. The data are expected to reveal detailed 3D information on earth rheology and Late-Pleistocene ice sheet evolution (Vermeersen and Schotman, 2008). Collaborative projects have therefore been initiated on coupling interactively a regional (flat) 3D high-resolution finite-element model to a 3D thermomechanical ice-sheet model that includes dynamically consistent ice shelves. These models will especially be applied to northern Europe, where also information is available on mantle composition from xenoliths in the Caledonian lithosphere along the western margin of Northwest Europe. Simulation studies (Schotman et al., 2009) predict the appearance of distinct geoid anomaly patches expected to be visible in GOCE mission data for northwestern Europe, which in turn should provide valuable information on regional rheology and shallow Earth structure (Fig. 8).

Another task is to couple existing GIA uplift data, uplift models and the most recent geological and palaeoclimatological data on glaciation history. Northern Europe and Russia provide a study area with several recent contributions. Using land uplift models, the sensitivity of uplift data on variations in ice thickness and duration should be quantified, at least for the period of the last deglaciation, i.e. from the Last Glacial Maximum at about 22 kyr B.P. to the present time. Inverse modelling of the glaciation history may be a promising new approach previously not applied. The consideration of palaeotide models which are coupled to the land uplift history (e.g. Thomas and

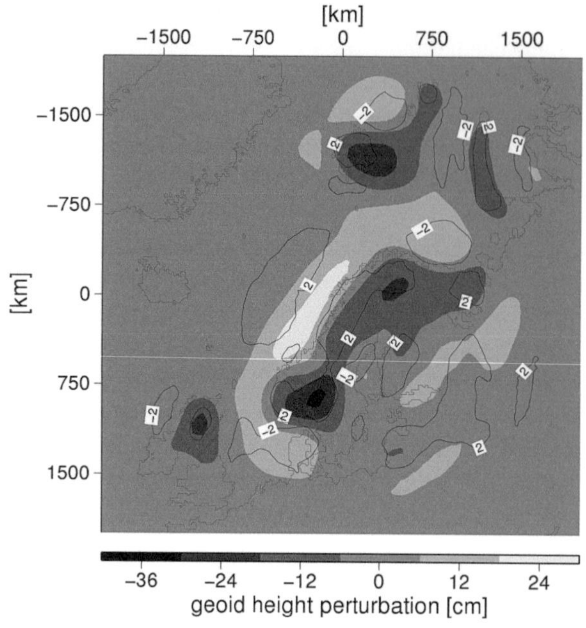

Fig. 8 Simulated geoid height perturbations induced by a crustal low-viscosity zone. For more details on models and parameters used, see Schotman et al. (2008). Such geoid anomalies should become detectable by the GOCE satellite mission in view of its expected resolution of the geoid down to characteristic spatial scales of 100 km and magnitudes of 1 cm

Sündermann, 1999) is expected to improve the reconstruction of palaeoenvironments and the interpretation of sea-level indicators (e.g. Klemann and Wolf, 2007). Also the influence of the Tornquist-Teysseire Zone (see Tesauro's map, this issue) on the uplift pattern during the last glaciation is of some evidence when focusing on sea-level variations in Denmark or southern Sweden.

An important step toward a better understanding of GIA in terms of the viscoelastic structure of the earth's lithosphere and mantle has been the joint inversion of different types of geodetic and gravimetric data related to GIA – preferentially connected with geological relative sea-level evidence of the earth' rebound during the last 10,000 years. An example of this is the study by Wolf et al. (2006), who reviewed and analysed the geological, tide-gauge, GPS and gravimetric evidence of GIA in the Churchill region of Hudson Bay, Canada. As a result, they were able to show that the different types of observable are consistent and lead to values of about 3.2×10^{20} Pa s and 1.6×10^{22} Pa s for the upper- and lower-mantle viscosities, respectively.

A similar situation exists in the region of Ny-Alesund, Svalbard, where, in addition to the data types mentioned above, also VLBI observations have been carried out (Hagedoorn and Wolf, 2003; Sato et al., 2006). However, in contrast to Hudson Bay, the situation on Svalbard is more complicated. The main reason is that, besides the contribution due to the Earth's GIA following the last Pleistocene deglaciation, the influence of the recent melting of the Arctic ice caps and glaciers cannot be neglected. Although some progress has been achieved in determining the recent glacial changes on Svalbard (e.g. Hagen et al., 2003; Pälli et al., 2003), uncertainties remain, which may explain the discrepancies existing between the different types of observation.

Whereas the spatial scale of Svalbard is too small for the detection of the recent glacial changes on the archipelago by the gravity satellite mission GRACE, the GIA signal over Canada recorded by GRACE during more than 5 years of operation is very prominent (Rangelova and Sideris, 2008), see Fig. 2. On the other hand, the density of the North American continuous GPS network is less than half of that of the Fennoscandian network, and the individual observation periods are usually less than half of those of the Fennoscandian stations (e.g. Sella et al., 2007), see also Fig. 9. Nevertheless, the joint inversion of Canada-wide geodetic and gravimetric data – supplemented by the vast geological evidence of postglacial relative sea-level change available – in terms of the Earth's viscoelastic structure is at the verge of becoming possible. At the same time, the need to consider lateral changes of this structure, such as variations of lithosphere thickness (e.g. Audet and Mareschal, 2004), or tectonic features becomes of ever increasing importance (e.g. Whitehouse et al., 2006; Klemann et al., 2008).

Models of the process of glacially induced faulting usually proceed along one of two lines. One is the strain accumulation mechanism suggested by Johnston (1989), whereby the glacial load inhibits earthquake strain release in the elastic part of the lithosphere and the accumulated strain is subsequently released instantaneously at deglaciation. A second mechanism, based on glacially induced stresses in the down-warped elastic lithosphere, was suggested by e.g. Wu and Hasegawa (1996).

Recent models of glacially induced faulting are based on the second mechanism, using increasingly

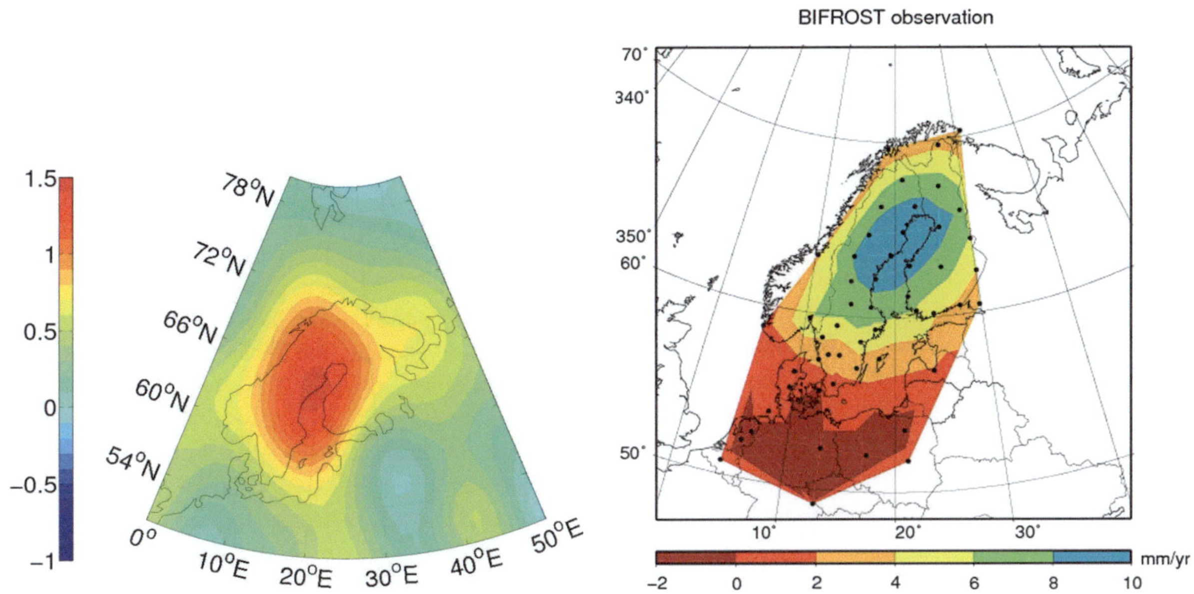

Fig. 9 Gravity trend from the GFZ GRACE monthly solution (01/2003–01/2008) and the GPS uplift velocity from Lidberg et al. (2007)

complex numerical models of the glacial history, earth rheology, rock mechanics and pore pressure (e.g. Wu et al., 1999; Hetzel and Hampel, 2005; Lund, 2005; Lund and Näslund, 2008). As alluded to above, these models are still in their infancy as they fail to predict many of the features of postglacial faulting observed.

One of the objectives of the DynaQlim initiative is to produce better models both of the glacial history and of the earth response in order to enhance our understanding of the phenomenon of glacially induced faulting and to better predict where and when it occurs. Not a purely academic exercise, as glacially induced faults are of utmost importance in the safety assessment for future nuclear waste repositories at northerly latitudes (e.g., Vidstrand et al., 2008).

Another important societal issue of the DynaQlim initiative is to investigate how much postglacial rebound stresses contribute to the triggering of seismicity that would eventually threaten human infrastructures. The largest historical earthquakes in Fennoscandia and the interior of North America occurred in the Rana region of northern Norway in 1819 with a magnitude of $M = 5.8$ (Bungum and Olesen, 2005), at mid-Mississippi River valley (New Madrid sequence) in 1811–1812 with M_W up to 8, and at Charleston in south-eastern USA in 1886 with $M_W \approx 7.7$ (Talwani, 1989). Because of a much denser infrastructure than in the 19th century, similar present-day earthquakes would be far more destructive.

As stated previously, the stress field responsible for the seismicity of formerly glaciated regions is dominated by ridge-push forces. It is however not known to what extent tectonic stresses interfere with postglacial rebound and other local stress sources to generate episodic destructive seismicity. Current models (e.g. Muir-Wood 2000) fail to account for the observations in a convincing way, probably because they do not incorporate reliable rheologies and do not involve the influence of other factors like topography (Pascal and Cloetingh, 2009) or glacial deep erosion and sediment loading (Stein et al., 1989). In brief, better illumination of this specific problem requires both more accurate rebound models and a more complete understanding of the pre-faulting stress regimes. Further aspects in the interference between GIA and surface processes are variations in river drainage during the glacial cycle which results in variations of sediment transport, trapping of sediments due to glacial variations and the resulting effects on palaeo-environments.

Climate

Ten Million Year Time Scale

The $\delta^{18}O$ isotopes deposited in ocean sediments record a long-term cooling trend sustained since the early Pliocene ($\sim$5.3 Ma to the present) (e.g., Raymo, 1994). There is great consensus that the ice sheet covering Antarctica evolved in concert with the opening of the Drake Passage 20–25 Ma that allowed the Antarctica ice sheet to become climatically buffered by a globally zonal current, the Antarctic Circumpolar Current, or ACC, the largest among all of the world ocean currents (Barker and Thomas, 2006). By direct inference, this is clearly an intersection of mantle convective and climate change time scales, and causality: plate tectonic reorganization ushering in a new stable state for the Earth's climate. What role the plate reorganization of about 4.5 Ma (Menard and Atwater, 1968) played in this context is poorly understood, although the physical isolation of the Pacific and Atlantic ocean water at the Panama isthmus likely occurred at this time (Haug and Tiedemann, 1998; Zachos, et al., 2001; Hay et al., 2002). Plate motions have been nearly constant since about 3–5 Ma (e.g., DeMets and Wilson, 2008).

Late-Pleistocene Ice Ages

Three orbital/rotational periodicities; eccentricity (400 and 100-kyr), obliquity (41-kyr) and rotational precession (23-kyr) that control summer insolation variability in the northern hemisphere, determine the time over which sustained growth and collapse of North American and Eurasian ice sheets can occur. Late summer insolation dominates the period at which glacial snow can melt during late summer. The shorter and less intense this period, the greater the chance that major ice sheets, ice caps and other large ice complexes may sustain secular growth. The relative importance of the three astronomical variables that form the "Milankovitch cycle", however, is poorly understood, in spite of the fact that the northern insolation variability may be calculated theoretically very accurately (e.g., Berger and Pestiaux, 1984). This is because the climate system has delicate and quite nonlinear feedback mechanisms at work, as well as late summer insolation variability. After about 4.5 Ma insolation minima cause prolonged ice ages.

While the waxing and waning of Northern hemispheric ice sheets are dominated by a 100-kyr cycle, benthic cyclicity in the $\delta^{18}O$ isotopic record is dominated by the 41-kyr-obliquity variability. Melt pulses of $\delta^{18}O$ from Antarctica, which are exactly out-of-phase with the Northern hemisphere at the 23-kyr precessional period, are likely the cause of this 41-kyr dominance in the benthic record, since the latter (obliquity variability) is always in-phase, and the former act to cancel one another. This fact is important because it means that both Northern and Southern hemispheric continental summer insolation is responsible for ice sheet growth and decay (Tziperman et al., 2006; Raymo et al., 2006) and that mean sea-level fluctuation from the last interglacial can be sourced to a continental origin with ever greater confidence. A substantial ramification may be the inference of 4–6 m higher last interglacial sea-level stand at Marine Isotope Stage 5e ($\sim$122 kyr BP). Rohling et al. (2008) suggests that the high-stand may be sourced to a substantial and rapid reduction of the volume of the Greenland ice sheet, well below its present day volume. At this time global mean surface temperatures were 2°C higher that today.

Jouzel et al. (2007) extracted high-resolution deuterium data (δD) from deep Antarctic ice cores (Dome C, EDC, and Dronning Maud Land hole, EDML) that are highly sensitive to the temperature of the ice that accumulated at the surface over the past 0.8 Ma. They show that the timing of warm and cold periods are tightly correlated with the global benthic $\delta^{18}O$ record determined by Lisiecke and Raymo (2005).

The ice core data also correlate with the more rapidly varying Dansgaar-Oschger (DO) events, short periods during which catastrophic marginal ice shelf breakup occurred, followed by massive expulsion of glacially eroded sediments into the ocean basins. The interval between these events is roughly $\sim$1500 yr (Hemming, 2004), and the coincident warming at discrete time intervals (<100 years) is also recorded in Greenland ice core $\delta^{18}O$ stratigraphy.

It is to be noted that the physical mechanism for the linkage of DO events with warm-cold oscillations is thought to be complex, involving internal ice sheet dynamics (e.g., Alley et al., 2001). The EDC and

EDML δD records also show a dominance of the in-phase obliquity cycle and support the bipolar seesaw hypothesis, in which thermohaline circulation cause warming and cooling in Greenland and Antarctica to be out-of-phase by roughly 0.5 kyr (Morgan et al., 2002). The Antarctic ice sheet is capable of providing the melt source for some of the millennial-scale benthic DO record (Blunier and Brook, 2001), a warming and melt pulsing that has probably been occurring since at least 2.2 Ma (Cowan et al., 2008).

A global ocean sediment record of the last 5.3 Ma studied by Lisiecki and Raymo (2005) reveals clear evidence of rhythmical deposition of ice rafted debris after 2.7 Ma. However, there is a time-dependence to the spectral trend in these global data, revealing stronger precessional and eccentricity oscillation later in the record, whereas prior to 1.6 Ma, the spectral power is dominated by the obliquity (41-kyr) variations. Raymo et al. (2006) used a coupled ice sheet-climate model that featured a dynamical Antarctic simulation that help explain the transition to relatively muted power in obliquity band since 1.6 Ma. A key model feature is that waxing and waning of ice grounded to the East Antarctic Ice Sheet (EIAS) margins provide huge mass contributions to ocean change during the northern hemispheric cooling associated with precessional (23-kyr) insolation changes, acting to cancel northern contributions.

The model EAIS volume changes would involve 30 m of eustatic sea-level change, about a factor of three larger than known for the last glaciation (Ivins and James, 2005). The ocean-modeled geographical distribution of marine sediment $\delta^{18}O$ observations supports these volumetrically equivalent contributions from north and south during 75–20 kyr (Rohling et al., 2004). The scenario proposed by Raymo et al. (2006) would explain the lack of 23-kyr benthic signature due to near complete cancellation between competing hemispheric responses to insolation forcing at the latter period. After 1.0 Ma, a longer-term cooling trend that is predicted by astronomical forcing (Pälike et al., 2001) drops solar insolation maxima at the coastal margins of the EAIS below a critical threshold, and the proximal ice sheet there experiences little terrestrial collapse during the warmest phases of post-1.0 Ma Milankovitch climate forcing. If the Raymo et al. (2006) scenario is correct, it has important implications for EAIS stability in a warming climate.

Last Ice Age, Postglacial Transition

Documentation of the Last Würm-Wisconsonian ice age has progressed rapidly during past 25 years. The greatest advance has come from Arctic Russia and Siberia (e.g., Serebryanny, 1985; Svendsen et al., 2004). One of the main upshots of the new glaciomoraine and other dated geomorphological indicators is the absence of large ice sheet complexes east of Severnaya Zemlya Is. and the Central Siberian Plateau at about 110°E longitude since about 55 kyr (e.g., Stauch et al., 2007). Additional constraints have become available from bathymetric and marine sediment core data and incorporated into comprehensive models that reconstruct the ice sheet that covered the Barents and Kara Seas at LGM (e.g., Elverhøi et al., 1993; Dowdeswell and Seigert, 1999). The Eurasian LGM features are now largely incorporated into the global ice models used for computation of GIA models for predicting solid earth deformation and sea-level variability (Lambeck et al., 2000; Peltier, 2004).

Exactly how the climate and oceans reorganize to sustain growth of ice sheets that ground to continents and shallow continental shelves continues to be poorly understood (Marshall et al., 2002; Zweck and Huybrechts, 2005) and incorporation of nonlinear feedback in modeling both ocean heat transport systems (e.g., Webb et al., 1997; Stastna and Peltier, 2007) and atmospheric CO_2 is a major challenge (e.g., Weaver et al., 1998). The entire stratigraphic column of $\delta^{18}O$, δD and bubble-captured atmospheric greenhouse gases, CO_2, NO_2 and methane in the deep ice cores of Antarctica show that anomalies in these greenhouse gases are present during all major atmospheric temperature excursions, although their lags, leads, and, indeed, their causality is still in debate (Brook et al., 2000; Blunier and Brook, 2001; Wolff, 2005). Lacking the "data-training" of modern global climate simulation for contemporary weather analyses, numerical simulation of LGM climate is frustrated by the need to properly introduce the relatively sparse regional proxy climate data (pollen records, etc.) into such models, having grid resolution of roughly 250 km by 250 km (e.g., Bush, 2004).

During the last global glacial transition-time (LGT), a period corresponding to the collapse phase of continental ice sheets from LGM, roughly between 23 and 11 kyr, palaeosealevel, marine and ice core records, consistently show warming and rapid sea level rise during the initial phases of northern hemispheric summer insolation maxima. This initial phase, termed "Termination I" for the last glaciation, occurs over the entire northern hemisphere and is believed to be well underway by 21 kyr, accelerating in pace until about 19 kyr (e.g. Clark et al., 2002; MaCabe et al., 2007) when a large rise (Meltwater Pulse $1A_O$, or mwp$1A_O$) characterizes global sea level records (e.g., Lambeck et al., 2000; Peltier, 2004).

Trapped gas composition ratios (O_2/N_2) from the Dome Fuji and Vostok ice cores of Antarctica have been used to calibrate levels of local summer insolation by Kawamura et al. (2007). The data unequivocally show that the northern hemispheric insolation maxima lead Antarctic climate change, and that, therefore, northern hemispheric termination phases over the 3 global glacial cycles promote Antarctic ice sheet loss. Figure10 shows the power spectra that are recovered from the Dome Fuji local ΔT_{site} and $\delta^{18}O$ stratigraphy, as it clearly records all of the Milankovitch periodicity that correspond to northern hemispheric insolation curve. The cresendo of Antarctica's reaction

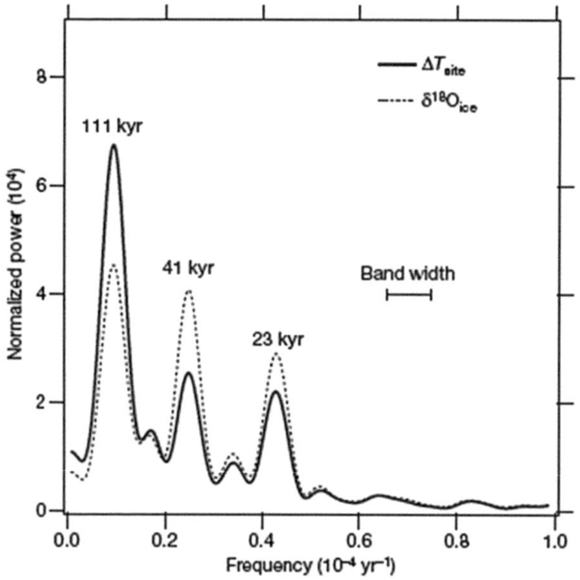

Fig. 10 Power spectra for temperature and oxygen isotopes anomalies at Dome Fuju ice core. Note that the amplitudes of the local 23-kyr and 41-kyr ΔT_{site} are quite similar. The $\delta^{18}O$ phase-lags are 1.0 ± 0.5 kyr and 2.1 ± 0.7 kyr, respectively, relative to precessional (23-kyr) and obliquity (41-kyr) Northern Hemispheric astronomical summer insolation bands. (From Kawamura et al., 2007)

to the Northern Hemisphere may have contributed in some measure to mwp$1A_o$ (Kawamura et al., 2007).

After mwp$1A_o$, transition climate and sea level rise are characterized by a bi-polar seesaw, with southern warming/cooling episodes leading those of the northern hemisphere. One of the most interesting facts about transition-time and early Holocene climate that has been revealed in the Greenland ice core record is how rapid the mean atmospheric rise may occur (e.g., Steffensen et al., 2008). The largest freshwater pulses during glacial transition times affected global thermohaline circulation (THC) in the ocean, such that deepwater formation may be interrupted (e.g., Alley et al., 2005).

Overall, deepwater formation moderates climate swings and keeps the surface relatively warm. During times when deepwater formation is interrupted global cooling is a possible outcome (Knorr and Lohmann, 2007). This is likely what drove the Antarctic Cold Reversal (ACR) at 14.6–12.5 kyr as fresh water entered the oceans from northern hemispheric deglaciation (mwp1a), although a southern fresh source is not entirely ruled out (e.g., Alley et al., 2005). The precise relationships during transition times are not clear, although recent ice core evidence from Dronning Maud Land strongly points to the bi-polar climate seesaw operating during glacial times in a mode in which Antarctic warming leads each of the longer-interval (~1500 yr) DO events of Northern origin (EPICA Community, 2006).

An apparent overlap in the time period of "southern freeze" (ACR) with the period when northern latitudes were experiencing relative warmth: the Bolling–Allerod period (14.5–12.9 kyr), is also poorly understood and is likely to involve time-lags in ocean circulation phenomena of 500 years or more. A recent breakthrough by Muscheler et al. (2008) in improving the calibration of the radiocarbon ^{14}C time-scale using correlation of tree-ring carbon, and ^{10}Be in Greenland ice cores, reveals that a short-lived, but dramatic reduction in atmospheric carbon may have occurred near the beginning of the Younger-Dryas (YD) event. The latter is a dramatic millennial time-scale return to glacial (stadial) conditions between 10.8 and 10.0 kyr. A reduction in atmospheric carbon may have played a role in promoting the YD event, possibly acting in phase with a large fresh water pulse that curtailed the THC. Clearly, other greenhouse gases, such as CH_4 and NO_2 play an important role in feedbacks as is evidenced in deep ice cores of the past 650,000 yr (Mayewski et al., 2009).

Holocene and Neoglacial Change

Modern climate begins, essentially, at the end of the YD, and with the return of warm climate by nearly 6–7°C and substantial increases in eustatic sea-level. Prior to the YD event there is the large pulse of fresh water supply from the continents: mwp1b at about 11.3 kyr (Fleming et al., 1998; Peltier, 2004).

The chronology of the decay of the Laurentide ice sheet in North America is relatively well-understood during glacial transition time, but less so during the early Holocene, due to the relative abundance of ^{14}C dates on glacial moraine materials and within lake sediments (Dyke, 2004). The remaining Holocene Laurentide ice sheet consisted of the Keewatin, Baffin Is./Fox, Eastern Nunavut and Labrador ice domes (Dyke, 2004; Carlson et al., 2008). Cosmogenic exposure dating of the rocks proximal to the Labrador ice dome across a comprehensive transect normal to the eastern and western retreating fronts reveal rapid disintegration of the dome between 7.4 and 6.8 kyr (Carlson et al., 2007). One-σ errors on these ^{10}Be-calibrated dates are generally about ±0.6 kyr, while the spatial resolution is 50 km along-transect.

Using a relatively primitive ice model of the dome and its shrinkage, Carlson et al. (2007) estimated the Labrador ice loss at this time accounted for about 3 m of eustatic sea level rise in a period in just 860 $\pm$ 170 years, and, thus, provides a plausible explanation for the source of the final meltwater pulse (mwp1c) that has recently been dated to 7.6 $\pm$ 0.4 kyr in southeastern Sweden by Yu et al. (2007). Loss of continental ice at this time is not too surprising due, in part, to the rise of northern summer insolation to their maximum during the Holocene "Optimum". Should the model of Carslon et al. (2007; 2008) be correct, there is an important cautionary footnote, for the average rate of collapse of the Labrador and remaining Baffin Is. would have been large enough to cause eustatic sea-level rise at a rate or 3.5–5.25 mm/yr, corresponding to ice loss rates of 1250–1880 Gt/yr, or about 7–10 times the current rate of loss of ice from Greenland (Rignot and Kanagaratnam, 2006), which now experiences rapidly warming climate at a similar latitude in the climatic environment of the western North Atlantic (e.g., Sidall and Kaplan, 2008).

Connections to Upper Mantle Dynamics

What does this increasingly fine-tuned knowledge base of Quaternary climate have to do with mantle dynamics? Mantle time scale responses to loading have a wide viscoelastic relaxation spectra, and for phenomenon with low spherical degree order, and for polar-wander responses, loading cycles are quite important, hence these more refined features of palaeoclimate should be explicitly treated in future mantle response models. An optimum response time scale for the upper mantle is bracketed by 500–5,000 years for craton-like Earth structure, hence the differences in a load model that assume the ice sheet configuration differences between the left and right panels of Fig. 11 should be quite substantial.

There are also many examples of plate boundary responses to Little Ice Age climate changes in which low viscosity parameters are demonstrated to be sensitive to the last 400 years of loading-unloading sequences (e.g., Larsen et al., 2005). Apparently the coupling between atmospheric CO_2 and glacier/ice cap growth and diminution also occurs during the waxing and waning of the global Little Ice Age (Cox and Jones, 2008), with temperature leading CO_2 by 50 years. The ramifications for the mantle are that ocean and atmospheric temperature changes that now accelerate global ice loss will be accelerating for 50–100 years, regardless of how robust are the mitigating factors brought to bear against anthropogenic forcing.

Challenges with DynaQlim

DynaQlim (Upper Mantle Dynamics and Quaternary Climate in Cratonic Areas) is a regional co-ordination committee of the International Lithosphere Program (ILP) scheduled to last from late 2007 to 2012. The kick-off meeting was arranged in Copenhagen in February, 2008, followed by a session during the European Geosciences Union General Assembly in Vienna, April 2008 (DynaQlim, 2008). A key aim

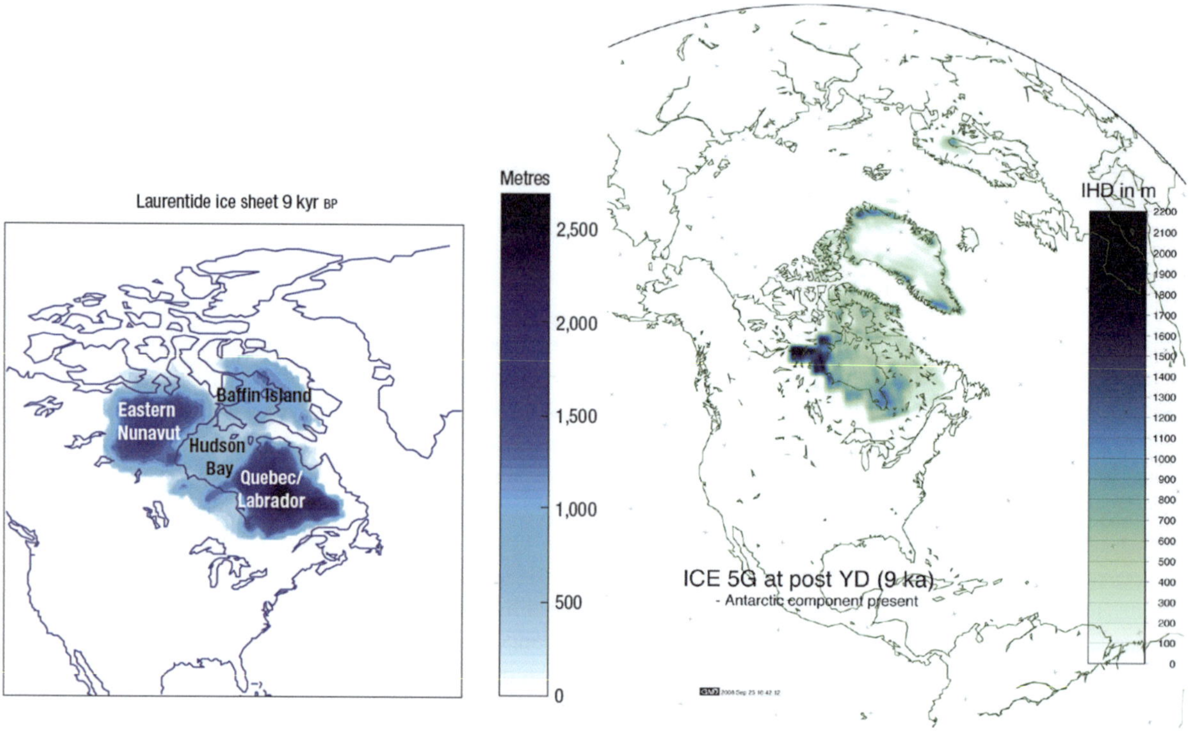

Fig. 11 Ice height differentials (IHD) at 9 kyr BP in a recent Laurentide $^{10}Be - ^{14}C$ based-reconstruction (left panel) and global ICE5G (Peltier, 2004) r.s.l. – ice dynamic based-reconstruction (right panel). Note that while both models portray the maximum Eastern Nunavut ice dome height amplitudes at the 1,500–2,000 m level, the Quebec/Labrador sectors differ by as much a 2 km in IHD. Left panel is from Carlson et al. (2008) right panel from a 256 degree-order spherical harmonic representation as discussed in Ivins and Wolf (2008)

is to facilitate the development of various models in order to generate more accurate predictions of Earth dynamics and ice sheet evolution during the Quaternary, and thus to understand the past and contemporaneous evolution of topography in previously glaciated terrains.

DynaQlim also provides a unique chance to study the dynamics and rheology of the lithosphere and asthenosphere in more detail. The results obtained during the running time of the project will have a number of applications, including present-day global change as well as future changes in response to a warming climate.

The tasks in DynaQlim have been divided into the following broad categories comprising the fields of

1. Geodesy, geodynamics, ocean dynamics;
2. Postglacial uplift, contemporary movements and gravity;
3. Dynamic ice sheets, glaciology;
4. Quaternary palaeoenvironments and climate;
5. Neotectonics and seismotectonics;
6. Dynamics, structure, properties and composition of the lithosphere;
7. IT, data management and outreach.

Access to relevant data is critical for successful scientific research. Concepts and technologies for an intelligent exchange of raw data (OGC, 2005), their desciription (OGC, 2007) as well as their representation in form of dynamic maps (OGC, 2006) exist.

For multidisciplinary projects conceptual challenges need to be tackled in order to make the data seamlessly accessible and portable for all users. Different disciplines (domains) describe the same datasets in their specific vocabulary, use their own data formats and organize them according to their requirements. In order to make a certain dataset originating from one domain usable to people from another discipline, the data need to be transformed and mapped between different profiles.

In the scope of the DynaQlim project, the exchange of data and information products among project partners needs to be realized based on sophisticated technologies. The installation of a web platform for the search and exchange of datasets and their descriptions is therefore intended.

Acknowledgments The research of Markku Poutanen is partly funded by the Academy of Finland, grant 120212. The research of Erik Ivins is funded by NASA's Earth Science Program, Solid Earth and Surface Processes Focus Area at the Jet Propulsion Laboratory, California Institution of Technology. The research of Jürgen Müller and Holger Steffen is funded by the Deutsche Forschungsgemeinschaft (DFG, German Research Foundation) through research grant MU1141/8-1 (SPP 1257) and that of Volker Klemann through the DFG research grant MA3432/2-2 (SPP1257).

References

Adams, J., Basham, P.W., 1989. Seismicity and seismotectonics of Canada's eastern margin and craton, in Gregersen, S., Basham, P.W. (eds.), *Earthquakes at North-Atlantic Passive Margins: Neotectonics and Postglacial Rebound*, 355–370. Kluwer Academic Publishers, Dordrecht, The Netherlands.

Ågren, J., Svensson, R., 2007. Postglacial Land Uplift Model and System Definition for the New Swedish Height System RH 2000. *Reports in Geodesy and Geographical Information Systems Rapportserie*, LMV-Rapport 2007:4, Lantmäteriet, Gävle.

Alley, R.B., Anandakrishnan, S., Jung, P., 2001. Stochastic resonance in the North Atlantic. *Paleoceanography*, 16, 190–198.

Alley, R.B., Clark, P.U., Huybrechts, P., Joughin, I., 2005. Ice-sheet and sea-level changes. *Science*, 310, 456–460 doi:10.1126/science.1114613.

Anda, E., Blikra, L.H., Braathen, A., 2002. The Berill fault – first evidence of neotectonic faulting in southern Norway. *Norsk Geologisk Tidsskrift*, 82, 175–182.

Arvidsson, R., 1996. Fennoscandian earthquakes: Whole crust rupturing related to postglacial rebound. *Science*, 274, 744–746.

Arvidsson, R., Kulhanek, O., 1994. Seismodynamics of Sweden deduced from earthquake focal mechanisms. *Geophys. J. Int.*, 116, 377–392.

Audet, P., Mareschal, J.-C., 2004. Variations in elastic thickness in the Canadian Shield. *Earth Planet. Sci. Lett.*, 226, 17–31, doi:10.1016/j.epsl.2004.07.035.

Barker, P., Thomas, E., 2006. Potential of the Scotia Sea Region for determining the onset and development of the Antarctic Circumpolar Current, in Futterer, D.K., D. Damaske, G. Kleinschmidt, H. Miller, D. Tessensohn (eds.), *Antarctica: Contributions to Global Earth Sciences*, 433–440. Springer-Verlag, Berlin Heidelberg New York.

Berg, J. van den, van de Wal, R.S.W., Oerlemans, J., 2006. Recovering lateral variations in lithospheric strength from bedrock motion data using a coupled ice sheet-lithosphere model. *J. Geophys. Res.*, 111, B05409, doi:10.1029/2005JB003790.

Berger, A., 1984. Accuracy and frequency stability of the Earth's orbital elements during the Quaternary, in Berger, A.L. et al. (eds.), *Milankovitch and Climate, Part 1*, 3–39. Reidel Pub. Co., Dordrecht, Netherlands.

Berger, A., Pestiaux, P., 1984. Accuracy and stability of the Quaternary terrestrial insolation, in Berger, A., Imbrie, J., Hays, J., Kukla, G., Saltzman, B. (eds.), *Milankovitch and Climate, Part 1*, 83–111. D. Reidel Pub., Dordrecht, Netherlands.

Bintanja, R., van de Wal, R.S.W., Oerlemans, J., 2005. Modelled atmospheric temperatures and global sea levels over the past million years. *Nature*, 437; 1 September 2005; doi:10.1038/nature03975.

Blundell, D., Mueller, S., Mengel, K., 1992. *A continent revealed; the European Geotraverse*, Cambridge University Press, Cambridge.

Blunier, T.,Brook, E.J., 2001. Timing of millennial-mcale climate change in Antarctica and Greenland during the last glacial period. *Science*, 291, 109–112.

Brook, E.J., Harder, S., Severinghaus, J., Steig, E.J., Sucher, C.M., 2000. On the origin and timing of rapid changes in atmospheric methane during the last glacial period. *Global Biogeochem. Cycles*, 14, 559–572, doi:10.1029/1999GB001182.

Bruneton, M., and 35 others, 2004. Complex lithospheric structure under the central Baltic Shield from surface wave tomography. *J. Geophys. Res.-Solid Earth*, 109(B10), B10303, doi:10.1029/2003JB002947.

Bungum, H., Olesen, O., 2005. The 31st of August 1819 Lurøy earthquake revisited. *Norwegian J. Geol.* 85, 245–252.

Bürgmann, R., Dresen, G., 2008. Rheology of the lower crust and upper mantle: Evidence from rockmechanics, geodesy, and field observations. *Annu. Rev. Earth Planet. Sci.*, 36, 531–567, doi:10.1146/annurev.earth.36.031207.124326.

Bush, A.B.G., 2004. Modelling of the late Qauternary climate over Asia: A synthesis. *Boreas*, 33, 155–163, doi:10.1111/j.1502-3885.2004.tb01137.x.

Bäckblom, G., Stanfors, R., 1989. Interdisciplinary study of post-glacial faulting in the Lansjärv area northern Sweden. Technical Report TR-89-31, Svensk Kärnbränslehantering AB, Stockholm.

Calais E., Han, J.Y., DeMets, C., Nocquet, J.M., 2006. Deformation of the North American plate interior from a decade of continuous GPS measurements. *J. Geophys. Res.*, 111, B06402, doi:10.1029/2005JB004253.

Carlson, A.E., Raisbeck, G.M., Clark, P.U., Brook, E.J., 2007. Rapid Holocene deglaciation of the Laurentide ice sheet. *J. Climate*, 20, 5126-5132, doi:10.1175/JCLI4273.1.

Carlson, A.E., Legrande, A.N., Oppo, D.W., Came, R.E., Schmidt, G.A., Gavin, A., Anslow, F.S., Licciardi, J.M., Obbink, E.A., 2008. Rapid early Holocene deglaciation of the Laurentide ice sheet. *Nat. Geosci.*, 1, 620–624, doi:10.1038/ngeo285.

Clark P.U., McCabe, A.M., Mix, A.C., Weaver, A.J., 2002. Rapid rise of sea level 19,000 years ago and its global implications. *Science*, 304, 1141–1144.

Cowan E.A., Hillenbrand, C.D., Hassler, L.E., Ake, M.T., 2008. Coarse-grained terrigenous sediment deposition on continental rise drifts: A record of Plio-Pleistocene glaciation on the Antarctic Peninsula. *Palaeogeography, Palaeo-*

climatology, Palaeoecology, 265, 275-291, doi:10.1016/j.palaeo.2008.03.010.

Cox, P., Jones, C., 2008. Illuminating the modern dance of climate and CO_2. *Science*, 321, 1642–1644, doi:10.1126/science.1158907.

Chung, W.-Y., 2002. Earthquakes along the passive margin of Greenland: Evidence for postglacial rebound control. *Pure Appl. Geophys.*, 159, 2567–2584.

Chung, W.-Y., Gao, H., 1997. The Greenland earthquake of July 11 1987 and postglacial fault reactivation along a passive margin. *Bull. Seism. Soc. Am.*, 87, 1058–1068.

Dehls, J.F., Olesen, O., Bungum, H., Hicks, E., Lindholm, C.D. and Riis, F., 2000. Neotectonic map, Norway and adjacent areas 1:3 mill. *Geological Survey of Norway*, Trondheim.

DeMets, C., Wilson, D.S., 2008. Toward a minimum change model for recent plate motions: Calibrating seafloor spreading rates for outward displacement. *Geophys. J. Int.*, 174, 825–841, doi:10.1111/j.1365-246X.2008.03836.x.

Dietrich, R., Rülke, A., Scheinert, M., 2005. Present-day vertical crustal deformations in West Greenland from repeated GPS observations. *Geophys. J. Int.*, 163, 865-874, 10.1111/j.1365-246X.2005.02766.x.

Dowdeswell, J.A., Siegert, M.J.,1999. Ice-sheet numerical modeling and marine geophysical measurements of glacier-derived sedimentation on the Eurasian Arctic continental margins, *Bull. Geol. Soc. Am.*, 111, 1080–1097.

Dyke, A.S., 2004. An outline of North American deglaciation with emphasis on central and northern Canada, in Ehlers, J., Gibbard, P.L. (eds.), *Quaternary Glaciations: Extent and Chronology 2: Part II North America*, 373–424. Elsevier, Amsterdam.

DynaQlim, 2008. Upper Mantle Dynamics and Quaternary Climate in Cratonic Areas. http://dynaqlim.fgi.fi .

Ekman M., 1996. A consistent map of the postglacial uplift of Fennoscandia. *Terra Nova* 8, 158–165.

Ekman M., Mäkinen J., 1996. Recent postglacial rebound, gravity change and mantle flow in Fennoscandia. *Geophys. J. Int.*, 126, 229–234.

Elverhøi, A., Fjeldskaar, W., Solheim, A., Nyland-Berg, M. Russwurm, L., 1993. The Barents Sea Ice Sheet – a model of its growth and decay during the Last Glacial Maximum, *Quaternary Sci. Rev.*, 12, 863–873.

EPICA Community Members, 2006. One-to-one coupling of glacial climate variability in Greenland and Antarctica, *Nature*, 444, 195–198, doi:10.1038/nature05301.

Fleming K., Johnston, P., Zwartz, D., Yokoyama, Y., Lambeck, K., Chappell, J., 1998. Refining the eustatic sea-level curve since the Last Glacial Maximum using far- and intermediate-field sites, *Earth Planet. Sci. Lett.*, 163, 327–342.

Fejerskov, M., Lindholm, C.D., 2000. Crustal stress in and around Norway; an evaluation of stress-generating mechanisms, in Nøttvedt, A. (ed.), *Dynamics of the Norwegian Margin*. Geological Society Special Publications, 167, 451–467, Geological Society of London, London, UK.

Forsström, P.-L., 2005. Through a glacial cycle: Simulation of the Eurasian ice sheet dynamics during the last glaciation. *Doctoral thesis*. University of Helsinki. http://urn.fi/URN:ISBN:952-10-2624-3.

GGOS, 2008. Global Geodetic Observing System. http://www.ggos.org.

Gregersen, S., 1992. Crustal stress regime in Fennoscandia from focal mechanisms. *J. Geophys. Res.*, 97, 11821–11827.

Gregersen, S., Voss, P., Shomali, Z.H., Grad, M., Roberts, R.G., Tor Working Group, 2006. Physical differences in the deep lithosphere of northern and central Europe, in Gee, D.G., Stephenson, R.A. (eds.), European Lithosphere Dynamics. Geological Society of London, Memoir 32, 313–322.

Gregersen, S., Voss, P., 2009. Stress change over short geological time: Case of Scandinavia over 9,000 years since the Ice Age, in Reicherter, K., Michetti, A.M., Silva Barroso, P.G. (eds.), Historical and Pre-Historical Records of Earthquake Ground Effects for Seismic Hazard Assessment. Geological Society of London Memoir special publications, 316, 173–178. doi:10.1144/SP316.10.

Grollimund, B., Zoback, M.D., 2001. Did deglaciation trigger intraplate seismicity in the New Madrid seismic zone? *Geology*, 29, 175–178.

Hagedoorn, J.M., Wolf, D., 2003. Pleistocene and recent deglaciation in Svalbard: Implications for tide-gauge, GPS and VLBI measurements. *J. Geodyn.*, 35, 415–423.

Hagen, J.O., Melvold, K., Pinglot, F., Dowdeswell, J.A., 2003. On the net mass balance of the glaciers and ice caps in Svalbard. *Arct. Antarct. Alp. Res.*, 35, 264–270.

Hay, W.M., E. Soeding, R.M. DeConto and C.N. Wold, 2002. The Late Cenozoic uplift – climate change paradox. *Int. J. Earth Sci. (Geol Rundsch.)* 91, 746–774, doi 10.1007/s00531-002-0263.

Hemming, S.R., 2004. Heinrich events: Massive late Pleistocene detritus layers of the North Atlantic and their global climate imprint. *Rev. Geophys.*, 42, RG1005, doi:10.1029/2003RG000128.

Hetzel, R., Hampel, A., 2005. Slip rate variations on normal faults during glacial-interglacial changes in surface loads, *Nature*, 435, 81–84, doi:10.1038/nature03562.

Haug, G.H., Tiedemann, R., 1998. Effect of the formation of the Isthmus of Panama on Atlantic Ocean thermohaline circulation. *Nature*, 393, 673–676.

Heidbach, O., Tingay, M., Barth, A., Reinecker, J., Kurfess, D., Müller, B., 2008. The 2008 release of the World Stress Map (available online at www.world-stress-map.org).

Hieronymus, C.F., Shomali, Z.H., Pedersen, L.B., 2007. A dynamic model for generating sharp seismic velocity contrasts underneath continents: Application to the Sorgenfrei-Tornquist Zone. *Earth. Planet. Sci. Lett.*, 262, 77–91, doi:10.1016/j.epsl.2007.07.043.

Hjelt, S.-E. Korja, T. Kozlovskaya, E. Lahti, I. Yliniemi, J. Bear and Svekalapko Seismic Tomography Working Groups, 2006. Electrical conductivity and seismic velocity structures of the lithosphere beneath the Fennoscandian Shield. Memoirs – Geological Society of London. 32, 541–560.

Ivins, E.R., James, T.S., 2005. Antarctic glacial isostatic adjustment: A new assessment. *Antarctic Sci.*, 17, 537–549, doi:10.1017/S0954102005002968.

Ivins, E.R., Klemann, V., James, T.S., 2003. Stress shadowing by the Antarctic ice sheet, *J. Geophys. Res.*, 108(12), doi:10.1029/2002JB002182.

Ivins, E.R., Wolf, D., 2008. Glacial isostatic adjustment: New developments from advanced observing systems and modeling. *J. Geodyn.*, 46, 69–77, doi:10.1016/j.jog.2008.06.002.

Janik, T., Kozlovskaya, E., Yliniemi, J., 2007. Crust-mantle boundary in the central Fennoscandian shield: Constraints from wide-angle P and S wave velocity models and new results of reflection profiling in Finland, *J. Geophys. Res.*, 112, B04302, doi:10.1029/2006JB004681.

Johansson, J.M., Davis, J.L., Scherneck, H-G., Milne, G.A., Vermeer, M., Mitrovica, J.X., Bennett, R.A., Jonsson, B., Elgered, G., Elósegui, P., Koivula, H., Poutanen, M., Rönnäng, B.O. and Shapiro, I.I., 2002. Continuous GPS measurements of postglacial adjustment in Fennoscandia 1. Geodetic results, *J. Geophys. Res.*, 107, doi:10.1029/2001JB000400.

Johnston, A.C., 1989. The effects of large ice-sheets on earthquake genesis, in Gregersen, S., Basham, P.W. (eds.), *Earthquakes at North-Atlantic passive margins: Neotectonics and postglacial rebound*, 141–173. Kluwer Academic Publishers, Dordrecht, The Netherlands.

Jouzel, J. 31 others, 2007. Orbital and millennial Antarctic climate variability over the past 800,000 years. *Science*, 317, 793–796, doi:10.1126/science.1141038.

Kawamura, K., 17 others. 2007. Northern Hemisphere forcing of climatic cycles in Antarctica over the past 360,000 years. *Nature*, 448, 912–916, doi:10.1038/nature06015.

Kawamura, K., Matsushima, H., Aoki, S., Nakazawa, T., 2007. Phasing of orbital forcing and Antarctic climate over the past 470,000 years from an extended Dome Fuji O2/N2 chronology. American Geophysical Union, Fall Meeting 2007, abstract# PP33A–1005.

Kakkuri J., 1997. Postglacial deformation of the Fennoscandian crust. *Geophysica* 33, 99–109.

Klemann, V., Martinec, Z., Ivins, E.R., 2008. Glacial isostasy and plate motion. *J. Geodyn.* 46, 95–103, doi:10.1016/j.jog.2008.04.005.

Klemann, V., Wolf, D., 1999. Implications of a ductile crustal layer for the deformation caused by the Fennoscandian ice sheet. *Geophys. J. Int.*, 139, 216–226.

Klemann, V., Wolf, D., 2007. Using fuzzy logic for the analysis of sea-level indicators with respect to glacial-isostatic adjustment: An application to the Richmond-Gulf region, Hudson Bay. *Pure Appl. Geophys.*, 164, 683–696, doi:10.1007/s00024-007-0191–x.

Knorr, G., Lohmann, G., 2007. Rapid transitions in the Atlantic thermohaline circulation triggered by global warming and meltwater during the last deglaciation. *Geochem. Geophys. Geosys.*, 8, Q12006, doi:10.1029/2007GC001604.

Korja T., Engels M., Zhamaletdinov A.A., Kovtun A.A., Palshin N.A., Smirnov M.Yu., Tokarev A., Asming V.E., Vanyan L.L., Vardaniants I.L., the BEAR Working Group, 2002. Crustal conductivity in Fennoscandia - a compilation of a database on crustal conductance in the Fennoscandian Shield. *Earth Planets Space*, 54, 535–558.

Korja T., 2007. How is the European lithosphere imaged by magnetotellurics? *Surveys Geophys.*, 28, (2–3), 239–272. doi:10.1007/S10712–007-9024-9.

Kujansuu, R., 1964. Nuorista siirroksista Lappissa. Summary: Recent faults in Lapland. *Geologi*, 16, 30–36.

Kukkonen, I.T., Jõeleht, A., 2003. Weichselian temperatures from geothermal heat flow data. *J. Geophys. Res.*, 108(B3), ETG-9, doi:10.1029/2001JB001579.

Kukkonen, I.T., Kinnunen, K., Peltonen, P., 2003. Mantle xenoliths and thick lithosphere in the Fennoscandian Shield. *Phys. Chem. Earth*, 28, 349–360.

Lagerbäck, R., 1979. Neotectonic structures in northern Sweden, *Geologiska Föreningens i Stockholm Förhandlingar*, 100(1978), 271–278.

Lagerbäck, R., 1990. Late Quarternary faulting and paleoseismology in northern Fennoscandia, with particular reference to the Lansjärv area, northern Sweden. *Geologiska Föreningens i Stockholm Förhandlingar*, 112, 333–354.

Lagerbäck, R. and Sundh, M. 2008. Early Holocene faulting and paleoseismicity in northern Sweden. SGU Research Paper C836, 80 pp.

Lambeck, K., Smither C., Johnston, P., 1998. Sea-level change, glacial rebound and mantle viscosity for northern Europe. *Geophys. J. Int.*, 134, 102–144.

Lambeck, K., Yokoyama, Y., Johnston, P., Purcell, A., 2000. Global ice volumes at the Last Glacial Maximum and early Late glacial, *Earth Planet. Sci. Lett.*, 181, 513–527.

van Lanen, X., Mooney, W.D., 2007. Integrated geologic and geophysical studies of North American continental intraplate seismicity, in Stein, S., and Mazzotti, S., (eds.), Continental Intraplate Earthquakes: Science, Hazard and Policy Issues: *Geological Society of America Special Paper* 425, 113–128, doi:10. 1130/2007.2425(08).

Larsen, C.F., Motyka, R.J., Freymueller, J.T., Echelmeyer, K.A., Ivins, E.R., 2005. Rapid viscoelastic uplift in southern Alaska caused by post-Little Ice Age retreat. *Earth Planetary Sci. Lett.*, 237, 548–560, doi:10.1016/j.epsl.2005.06.032.

Lidberg M., 2007. Geodetic Reference Frames in Presence of Crustal Deformations. *Doctoral thesis*. Department of Radio and Space Science, Chalmers University of Technology. Ny serie Nr 2705.

Lidberg M., Johansson, J.M., Scherneck, H.-G., 2006. Geodetic reference frames in the presence of crustal deformation – with focus on Nordic conditions. Symposium of the IAG sub commission for Europe (EUREF), June 14–17, Riga, 2006.

Lidberg M., Johansson, J.M., Scherneck, H.-G., Davis, J.L., 2007. An improved and extended GPS-derived 3D velocity field of the glacial isostatic adjustment (GIA) in Fennoscandia. *J. Geodesy*, 81(3), 213–230, doi:10.1007/s00190-006-0102-4.

Lisiecki, L.E. Raymo, M.E., 2005. A Pliocene-Pleistocene stack of 57 globally distributed benthic $\delta^{18}O$ records. *Paleoceanography*, 20, PA1003, doi:10.1029/2004PA001071.

Lund, B., Zoback, M.D., 1999. Orientation and magnitude of in situ stress to 6.5 km depth in the Baltic Shield. *Int. J. Rock Mech. Min. Sci.*, 36, 169–190.

Lund, B., 2005. The effects of deglaciation on the crustal stress field and implications for endglacial faulting: A parametric study of simple Earth and ice models. Technical Report TR-05-04, Swedish Nuclear Fuel and Waste Management Co., Stockholm, Sweden.

Lund, B., Näslund, J-O., 2008. Glacial isostatic adjustment: Implications for glacially induced faulting and nuclear waste repositories, in Connor, C.B., Chapman, N.A., Connor, L.J. (eds.), Volcanic and tectonic hazard assessment for nuclear facilities, 160–174. Cambridge University Press, Cambridge, UK.

McCabe, A., Cooper, J.A.G., Kelley, J.T. 2007. Relative sea-level changes from NE Ireland during the last glacial termination, *J. Geol. Soc. Lond.*, 164, 1059–1063, doi:10.1144/0016-76492006-164.

Mäkinen J., Engfeldt A., Harsson B.G., Ruotsalainen H., Strykowski G., Oja T., Wolf D., 2005. The Fennoscandian Land Uplift Gravity Lines 1966–2003, in C. Jekeli, L. Bastos, J. Fernandes (eds.), *Gravity, Geoid and Space Missions*. Springer, IAG Symposia 129, 299–303.

Mäkinen J., Koivula H., Poutanen M., Saaranen V., 2003. Vertical velocities in Finland from permanent GPS networks and from repeated precise levelling. *J. Geodyn.* 38, 443–456.

Mäkinen J., Saaranen, V., 1998. Determination of postglacial land uplift from the three precise levelings in Finland. *J. Geod.*, 72, 516–529

Marshall S.J., James, T.S., Clarke, G.K.C., 2002. North American Ice Sheet reconstructions at the Last Glacial Maximum, *Quat. Sci. Rev.*, 21, 175–192.

Martinec, Z., 2000. Spectral-finite element approach to three-dimensional viscoelastic relaxation in a spherical earth. *Geophys. J. Int.*, 142, 117–141.

Mayewski, P.A., et al. , 2009. State of the Antarctic and Southern Ocean climate system, *Rev. Geophys.*, 47, RG1003, doi:10.1029/2007RG000231.

Menard, H.W., Atwater, T., 1968. Changes in direction of sea floor spreading. *Nature*, 219, 463–467.

Milne G.A., Davis, J.L., Mitrovica, J.X., Scherneck, H.-G., Johansson, J.M., Vermeer, M., Koivula, H., 2001. Space-geodetic constraints on glacial isostatic adjustment in Fennoscandia. *Science*, 291, 2381–2385.

Morgan, V., Delmotte, M., van Ommen, T., Jouzel, J., Chappellaz, J., Woon, S., Masson-Delmotte, V., Raynaud, D., 2002. Relative timing of deglacial climate events in Antarctica and Greenland. *Science*, 297, 1862–1864, doi:10.1126/science.1074257.

Munier, R., Fenton, C., 2004. Appendix 3: Review of postglacial faulting. In: Munier, R. and H. Hökmark, Respect distances. Rationale and Means of Computation, Tech. Report, R-04-17, Swedish Nuclear Fuel and Waste Management Company, Stockholm, Sweden.

Muir Wood, R., 2000. Deglaciation Seismotectonics: A principal influence on intraplate seismogenesis at high latitudes. *Quaternary Sci. Rev.*, 19, 1399–1411.

Müller, J., Neumann-Redlin, M., Jarecki, F., Denker, H., Gitlein, O., 2006. Gravity Changes in Northern Europe as Observed by GRACE, in Tregoning, P., Rizos, C. (eds.), *Dynamic Planet.*, IAG Symposia 130, 523–527, Springer.

Muscheler, R., Kromer, B., Bjorck, S., Svensson, A., Friedrich, M., Kaiser, K.F., Southon, J., 2008. Tree rings and ice cores reveal C-14 calibration uncertainties during the Younger Dryas. *Nature Geosci.*, 1, 263–267, doi:10.1038/ngeo128.

Näslund, J.-O., Jansson, P., Fastook, J.L., Johnson, J., Andersson, L., 2005. Detailed spatially distributed geothermal heat flow data for modeling of basal temperatures and meltwater production beneath the Fennoscandian ice sheet. *Ann. Glaciol.*, 40, 95–101, doi:10.3189/172756405781813582.

OGC, 2005. OpenGIS Web Feature Service (WFS) Implementation Specification, Version 1.1.0, URL: http://www.opengeospatial.org/standards/wfs

OGC, 2006. OpenGIS Web Map Server Interface Implementation Specification, Version 1.3.0, URL: http://www.opengeospatial.org/standards/wms

OGC, 2007. OpenGIS Catalogue Service Implementation Specification, Version 2.0.2, URL: http://www.opengeospatial.org/standards/cat

Olesen, O., 1988. The Stuoragurra Fault, evidence of neotectonics in the Precambrian of Finnmark, northern Norway. *Norsk Geologisk Tidsskrift*, 68, 107–118.

Olesen, O., Henkel, H., Lile, O.B., Mauring, E., Rönning, J.S., 1992. Geophysical investigations of the Stuoragurra postglacial fault, Finnmark, northern Norway. *J. Appl. Geophys.*, 29, 95–118.

Olsson, S., Roberts, R.G., Böðvarsson, R., 2006. Analysis of waves converted from S to P in the upper mantle beneath the Baltic Shield. *Earth Planet. Sci. Lett.*, 257(1–2), 37–46. doi:10.1016/j.epsl.2007.02.017.

Pagiatakis, S.D., Salib, P., 2003. Historical relative gravity observations and the time rate of change of gravity due to postglacial rebound and other tectonic movements in Canada. *J. Geophys. Res. (Solid Earth)*, 108, 2406, doi:10.1029/2001JB001676.

Pälike, H., Shackleton, N.J., Rohl, U., 2001. Astronomical forcing in Late Eocene marine sediments. *Earth Planet. Sci. Lett.*, 193, 589–602.

Pälli, A., Moore, J.C., Jania, J., Glowacki, P., 2003. Glacier changes in southern Spisbergen, Svalbard, 1901-2000. *Ann. Glaciol.*, 37, 219–225.

Pascal, C., Cloetingh, S.A.P.L., 2009. Gravitational potential stresses on passive continental margins: Application to the Mid-Norwegian Margin. *Earth Planet. Sci. Lett.* 277(3–4), 464–473, doi:10.1016/j.epsl.2008.11.014.

Pascal, C., Roberts, D., Gabrielsen, R.H., 2005. Quantification of neotectonic stress orientations and magnitudes from field observations in Finnmark, northern Norway. *J. Structural Geol.*, 27, 859–870, doi:10.1016/j.jsg.2005.01.011.

Påsse, T., 1996. A mathematical model of the shore level displacement in Fennoscandia. *Technical Report* TR 96 24, Svensk Kärnbränslehantering AB, Stockholm.

Pedersen H.A., Bruneton, M., Maupin, V., 2006. Lithospheric and sublithospheric anisotropy beneath the Baltic shield from surface-wave analysis. *Earth Planet. Sci. Lett.*, 244, 590-605, doi:10.1016/j.epsl.2006.02.009.

Peltier, W.R., 2004. GLOBAL glacial isostasy and the surface of the iceage earth: The ice-5G (VM2) model and GRACE. *Annu. Rev. Earth Planet. Sci.* 32, 111–149, doi:10.1146/annurev.earth.32. 082503.144359.

Plomerová, J., Babuška, V., Vecsey, L., Kozlovskaya, E., Raita, T., SSTWG, 2006. Proterozoic–Archean boundary in the mantle lithosphere of eastern Fennoscandia as seen by seismic anisotropy. *J.Geodynam.*, 41(4), 400–410. doi:10.1016/j.jog.2005.10.008.

Poutanen, M., Knudsen, P., Lilje, M., Nørbech, T., Plag, H.-P. Scherneck, H.-G., 2007. The Nordic Geodetic Observing System (NGOS). *Proceedings of the IAG Dynamic Planet Symposium*, Cairns 2005, IAG symposium, 130, 749–756. Springer Verlag.

Rangelova, E., Sideris, M.G. 2008. Contributions of terrestrial and GRACE data to the study of the secular geoid changes in North America. *J. Geodynamics*, 46(3–5), 131–143, doi:10.1016/j.jog.2008.03.006

Raymo, M.E. 1994. The initiation of Northern Hemisphere glaciation. *Ann. Rev. Earth Planetary Sci.*, 22, 353–383.

Raymo, M.E., Lisiecki, L.E., Nisancioglu, K.H., 2006. Plio-Pleistocene ice volume, Antarctic climate and the global $\delta^{18}O$ record. *Science*, 313, 492–495, doi:10.1126/science.1123296.

Rignot, E., Kanagaratnam, P., 2006. Changes in the velocity structure of the Greenland ice sheet. *Science*, 311, 986–990, doi:10.1126/science.1121381.

Rohling, E.J., Marsh, R., Wells, N.C., Siddall, M., Edwards, N.R., 2004. Similar contributions to sea-level from Antarctic and northern ice sheets. *Nature*, 430, 1016–1021, doi:10.1038/nature02859.

Rohling, E.J., Grant, K., Hemleben, C., Siddall, M., Hoogakker, B.A.A., Bolshaw, M.,Kucera, M., 2008. High rates of sea-level rise during the last interglacial period. *Nature Geosci.*, 1, 38–42, doi:10.1038/ngeo.2007.28.

Roberts, D., 2000. Reverse-slip offsets and axial fractures in road-cut boreholes from the Caledonides in Finnmark, northern Norway: Neotectonic stress orientation indicators, *Quat. Sci. Rev.*, 19, 1437–1445.

Roberts, D., and Myrvang, A., 2004. Contemporary stress orientation features and horizontal stress in bedrock, Trøndelag, central Norway. *NGU Bull.*, 442, 53–63.

Saaranen V., Mäkinen J., 2002. Determination of post-glacial rebound from the three precise levellings in Finland: Status in 2002, in Poutanen M. , Suurmäki H. (eds.), *Proceedings of the 14th General Meeting of the Nordic Geodetic Commission, Espoo, Finland, October 1–5, 2002*. Finnish Geodetic Institute, 171–174.

Sabadini, R., Vermeersen, L.L.A., 2004. *Global Dynamics of the Earth: Applications of Normal Mode Relaxation Theory to Solid-Earth Geophysics, Modern Approaches in Geophysics Series*, 20, Kluwer Academic Publ., Dordrecht, The Netherlands, 328 pp.

Sandoval, S., Kissling, E., Ansorge, J., 2004. High-resolution body wave tomography beneath the SVEKALAPKO array – II. Anomalous upper mantle structure beneath the central Baltic Shield. *Geophys. J. Int.*, 157(1), 200–214. doi: 10.1111/j.1365-246X.2004.02131.x

Sato, T., Okuno, J., Hinderer, J., MacMillan, D.S., Plag, H.-P. Francis, O., Falk, R., Fukuda, Y., 2006. A geophysical interpretation of the secular displacement and gravity rates observed at Ny-Alesund, Svalbard in the Arctic – effects of post-glacial rebound and present-day ice melting. *Geophys. J. Int.*, 165, 729–743, doi:10.1111/j.1365-246X.2006.02992.x.

Scherneck, H.-G., Johansson, J.M., Elgered, G., Davis, J.L., Jonsson, B., Hedling, G., Koivula, H., Ollikainen, M., Poutanen, M., Vermeer, M., Mitrovica, J.X., Milne, G.A., 2002. BIFROST: Observing the three-dimensional deformation of Fennoscandia, in Mitrovica, J.X., Vermeersen, B.L.A. (eds.), *Ice Sheets, Sea Level and the Dynamic Earth*. American Geophysical Union, Geodynamics Series, 29, Washington, DC, 69–93.

Schotman, H.H.A., Vermeersen, L.L.A., Wu, P., Drury, M.R., de Bresser, J.H.P., 2009. Constraints of Future GOCE Data on Thermomechanical Models of the Shallow Earth: A Sensitivity Study for Northern Europe, *Geophys. J. Int.*, 178(1): 65–84. doi:10.1111/j.1365-246X.2009.04160.x.

Schotman, H.H.A., Wu, P., Vermeersen, L.L.A., 2008. Regional Perturbations in a Global Background Model of Glacial Isostasy, *Phys. Earth Planet. Inter.*, doi:10.1016/j.pepi.2008.02.010.

Sella, G.F., Stein, S., Dixon, T.H., Craymer, M., James, T.S., Mazzotti, S., Dokka, R.K., 2007. Observation of glacial isostatic adjustment in "stable" North America with GPS. *Geophys. Res. Lett.*, 34, L02306, doi:10.1029/2006GL027081.

Serebryanny, L.R., 1985. Mountain glaciation in the USSR in the Late Pleistocene and Holocene, in Velichko, A.A. (ed.), *Late Quaternary Environments of the Soviet Union*, University of Minnesota Press, 45–54.

Shennan, I., Long, A., Metcalfe, S., 1998. IGCP Project 367 'Late Quaternary coastal records of rapid change: Application to present and future conditions' and 25 years progress in research. *Holocene*, 8, 125–128.

Sidall, M., Kaplan, M.R., 2008. A tale of two ice sheets. *Nat. Geosci.*, 1, 570–571, doi:10.1038/ngeo286.

SKB. 2006. Climate and climate-related issues for the safety assessment SR-Can. Technical Report TR-06-23, Svensk Kärnbränslehantering AB, Stockholm.

Slunga, R., 1991. The Baltic Shield earthquakes. *Tectonophysics*, 189, 323–331.

Stastna, M., Peltier, W.R., 2007. On box models of the North Atlantic thermohaline circulation: Intrinsic and extrinsic millennial timescale variability in response to deterministic and stochastic forcing. *J. Geophys. Res. Oceans*, 112, C10023, doi:10.1029/2006JC003938.

Stauch, G., Lehkuhl, F., Frechen, M., 2007. Luminescence chronology from the Verhoyansk Mountains (North-Eastern Siberia). *Quaternary Geochronology*, 2, 255–259, doi:10.1016/j.quageo.2006.05.013.

Steffen, H., Denker, H., Müller, J., 2008. Glacial isostatic adjustment in Fennoscandia from GRACE data and comparison with geodynamic models. *J. Geodyn.*, 46(3–5), 155–164, doi:10.1016/j.jog.2008.03.002.

Steffensen, J.P., 19 others, 2008. High-resolution Greenland ice core data show abrupt climate change happens in few years. *Science*, 321, 680–684, doi:10.1126/science.1157707.

Stein, S., Cloetingh, S., Sleep, N.H., Wortel, R., 1989. Passive margin earthquakes, stresses and rheology, in Gregersen, S., Basham, P.W. (eds.), Earthquakes at North-Atlantic passive margins; neotectonics and postglacial rebound. *NATO ASI Series, Series C: Mathematical and Physical Sciences*, 266, 231–259, D. Reidel Publishing Company, Dordrecht-Boston, International.

Stephansson, O., Särkkä, P., Myrvang, A., 1986. State of stress in Fennoscandia, in *Proceedings of the International Symposium on Rock Stress and rock stress measurements*, Stockholm, 1–3 September 1986, Stephansson, O. (eds), Lulea, Sweden, 21–32.

Stewart, I.S., Sauber, K. and Rose, J.. 2000. Glacio-seismotectonics: Ice sheets, crustal deformation and seismicity. *Quat. Sci. Rev.*, 19, 1367–1389.

Svendsen, J.I., 29 others, 2004. Late Quaternary ice sheet history of northern Eurasia. *Quat. Sci. Rev.*, 23, 1229–1271, doi:10.1016/j.quascirev.2003.12.008.

Talwani, P., 1989. Seismotectonics in the southeastern United States, in Gregersen, S., Basham, P.W. (eds.) *Earthquakes at North-Atlantic passive margins: Neotectonics and postglacial rebound*, 371–392. Kluwer Academic Publishers, Dordrecht, The Netherlands.

Tamisiea, M.E., Mitrovica, J.X., Davis, J.L., 2007. GRACE Gravity Data Constrain Ancient Ice Geometries and Continental Dynamics over Laurentia. *Science*, 316, 881, doi:10.1126/science. 1137157.

Thomas, M., Sündermann, J., 1999. Tides and tidal torques of the world ocean since the last glacial maximum. *J. Geophys. Res.*, 104(C2), 3159–3183.

Tikkanen, M., Oksanen, J. 2002. Late Weichselian and Holocene shore displacement history of the Baltic Sea in Finland. *Fennia – Int. J. Geography* 180(1–2), 9–20.

Tsuboi, S., Kikuchi, M., Yamanaka, Y., Kanao, M., 2000. The March 25, 1998 Antarctic earthquake: Great earthquake caused by postglacial rebound. *Earth Planets Space* 52, 133–136.

Tziperman, E., Raymo, M.E., Huybers, P., Wunsch, C., 2006. Consequences of pacing the Pleistocene 100 kyr ice ages by nonlinear phase locking to Milankovitch forcing. *Paleoceanography*, 21, doi:10.1029/2005PA001241.

Uski, M., Hyvönen, T., Korja, A., Airo, M.-L., 2003. Focal mechanisms of three earthquakes in Finland and their relation to surface faults. *Tectonophysics*, 363, 141–157.

van de Plassche, O. (ed.),1986. *Sea-Level Research: A Manual for the Collection and Evaluation of Data.* Geo Books, Norwich.

Van de Plassche, O., Chrzastowski, M.J., Orford, J.D., Hinton, A.C., and Long, A.J., 1995. Coastal evolution in the Quaternary: IGCP Project 274. *Mar. Geol.*, 124, ix–xii.

Vermeersen, L.L.A., Schotman, H.H.A., 2008. High-harmonic geoid signatures related to glacial isostatic adjustment and their detectability by GOCE, *J. Geod.*, doi:10.1016/j.jog.2008.04.003.

Vestøl O., 2006. Determination of postglacial land uplift in Fennoscandia from leveling, tide-gauges and continuous GPS stations using least squares collocation. *J. Geodesy*, 80, 248–258, doi:10.1007/s00190-006-0063-7.

Vidstrand P., Wallroth, T., Ericsson, L.O., 2008. Coupled HM effects in a crystalline rock mass due to glaciation: Indicative results from groundwater flow regimes and stresses from an FEM study. *Bull. Eng. Geol. Environ.*, 67, 187–197.

Wahr J., Velicogna I., 2003. What might GRACE contribute to studies of postglacial rebound? *Space Sciences Series* 18, 319–330.

Weaver, A.J., Eby, M., Fanning, A.F., Wiebe, E.C., 1998. Simulated influence of carbon dioxide, orbital forcing and ice sheets on the climate of the Last Glacial Maximum, *Nature*, 394, 847–853.

Webb, R.S., Rind, D.H., Lehman, S.J., Healy, R.J., Sigman, D., 1997. Influence of ocean heat transports on climate of the Last Glacial Maximum, *Nature*, 385, 695–699.

Weidick, A. 1995. Land uplift and subsidence in Greenland since the Ice Age (in Danish), in Gregersen, S. (eds.), *The Physical Nature of Greenland.* Rhodos, Copenhagen.

Whitehouse, P.L., Latychev, K., Milne, G.A., Mitrovica, J.X., Kendall, R., 2006. The impact of 3_D Earth structure on Fennoscandian glacial isostatic adjustment: Implications for space-geodetic estimates of present-day crustal deformations. *Geophys. Res. Lett.*, 33, L13502, doi:10.1029/2006GL026568.

Wolff, E.W., 2005. Understanding the past-climate history from Antarctica. *Antarctic Sci.*, 17, 487–495.

Wolf, D., Klemann, V., Wünsch, J., Zhang, F.-P., 2006. A reanalysis and reinterpretation of geodetic and geomorphologic evidence of glacial-isostatic uplift in the churchill region, Hudson Bay. *Surv. Geophys.*, 27, 19–61, doi:10.1007/s10712-005-0641-x.

Wu, P., 1998. Intraplate earthquakes and Postglacial Rebound in Eastern Canada and Northern Europe, in Wu, P. (ed.), *Dynamics of the Ice Age Earth: A Modern Perspective*, 603–628. Trans Tech Publ., Switzerland.

Wu, P., Hasegawa, H.S., 1996. Induced stresses and fault potential in eastern Canada due to a disc load: A preliminary analysis. *Geoph. J. Int.*, 125, 415–430.

Wu, P., Johnston, P., Lambeck, K., 1999. Postglacial rebound and fault instability in Fennoscandia. *Geoph. J. Int.*, 139, 657–670.

Wu, P., Johnston, P., 2000. Can deglaciation trigger earthquakes in northern America? *Geoph. Res. Lett.* 27, 1323–1326, doi:10.1029/1999GL011070.

Wu, P., Mazzotti, S., 2007. Effects of a lithospheric weak zone on postglacial seismotectonics in eastern Canada and the northern United states, in Stein, S., Mazzotti, S. (eds.), Continental Intraplate Earthquakes: Science, Hazard and Policy Issues: Geological Society of America Special Paper 425, 113–128, doi:10.1130/2007.2425(09).

Wu, P., van der Wal, W., 2003. Postglacial sea levels on a spherical, self-gravitating viscoelastic Earth: Effects of lateral viscosity variations in the upper mantle on the inference of viscosity contrasts in the lower mantle, *Earth Planet. Sci. Lett.*, 211, 57–68, doi:10.1016/S0012-821X(03)00199-7.

Yliniemi, J., Kozlovskaya, E., Hjelt, S.-E., Komminaho, K., Ushakov, A., 2004. Structure of the crust and uppermost mantle beneath southern Finland revealed by analysis of local events registered by the SVEKALAPKO seismic array. *Tectonophysics*, 394, (1–2), 41–67. doi:10.1016/j.tecto.2004.07.056.

Yu, S.Y., Berglund, B.E., Sandgren, P., Lambeck, K., 2007. Evidence for a rapid sea-level rise 7600 yr ago. *Geology*, 35(10), 891–894, doi:10.1130/G23859A.1.

Zachos, J., Pagani, M., Sloan, L., Thomas, E., Billups, K., 2001. Trends, rhythms, and aberrations in global climate 65 Ma to present. *Science*, 292, 686–693.

Zweck, C., Huybrechts, P., 2005. Modeling of the northern hemisphere ice sheets during the last glacial cycle and glaciological sensitivity. *J. Geophys. Res.* D, D07103, doi:10.129/2004JD005489.

Ultradeep Rocks and Diamonds in the Light of Advanced Scientific Technologies

Larissa F. Dobrzhinetskaya and Richard Wirth

Abstract This is a review paper which summarizes recent achievements in studies of superdeep mantle rocks and diamonds from kimberlite and ultrahigh-pressure metamorphic (UHPM) terranes using advanced analytical techniques and instrumentations such as focused ion beam (FIB)-assisted transmission electron microscopy (TEM) and synchrotron-assisted infrared spectroscopy. In combination, they allow characterization of geological materials formed at varying pressures, temperatures, and stresses in different chemical environments, which has enabled us to make amazing advances in understanding large-scale processes operating in the Earth through plate tectonics. Mineralogical characterisations of the ultradeep earth materials using novel techniques with high spatial and energy resolution are resulting in unexpected discoveries of new phases, thereby providing better constraints on deep mantle processes. One of such results is that the nanometric fluid inclusions in diamonds from kimberlite and UHPM terranes contain similar elements such as Cl, K, P, and S. Such similarity reflects probably the high solubility of these elements in a diamond-forming C–O–H supercritical fluid at high pressures and temperatures. The paper emphasizes the necessity of further studies of diamonds occurred within geological setting (oceanic islands, foearcs and mantle sections of ophiolites) previously unrecognized as suitable places for high pressure minerals formation.

Keywords Mantle · Ultrahigh-pressure metamorphism · Diamond · Synchrotron · Focused ion beam · Electron microscopy

Introduction

Knowledge about the composition and structure of the Earth's interior is gained from observations on the deepest xenoliths/xenocrysts from volcanic sources, fragments of garnet peridotites, inclusions in diamond from various depths, eclogites and meta-sedimentary ultrahigh-pressure rocks occurring within collisional orogens, as well as from experimental mineralogy and gravitational and seismic observations. According to geophysical data, the Earth's upper mantle extends from 50 km to 660 km, followed by the lower mantle, the largest shell inside the Earth, at depths of 600–2,890 km. The mass of this region is $\sim$100 times larger than that of the continental crust. A layer D$''$ ($\sim$200 km thick) that exhibits anomalous seismic properties separates the mantle from the core. In general, the complex pictures of the lateral anisotropies and seismic discontinuities are often addressed to mineral phase transformations, formation of ultrahigh-pressure partial melt, and fluids, which all together might cause dramatic changes in the chemical and physical properties of Earth's interior.

Although the mineralogy of the Earth's mantle is well addressed by Ringwood's pyrolite model (Fig. 1) (Ringwood, 1991) the elucidation of the connections between the geophysics of the deep Earth and its

L.F. Dobrzhinetskaya (✉)
Institute of Geophysics and Planetary Physics, Department of Earth Sciences, University of California, Riverside, CA 92521, USA
e-mail: larissa@ucr.edu

S. Cloetingh, J. Negendank (eds.), *New Frontiers in Integrated Solid Earth Sciences*, International Year of Planet Earth, DOI 10.1007/978-90-481-2737-5_11, © Springer Science+Business Media B.V. 2010

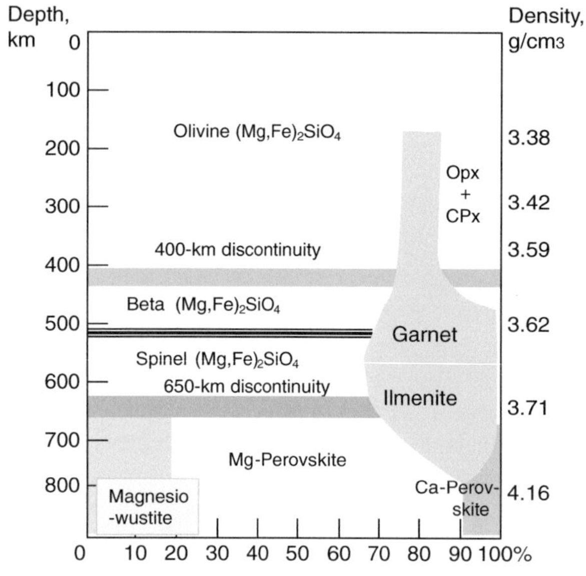

Fig. 1 Ringwood's diagram of phases distributions in the Earth's mantle based on pyrolite compositional model (Ringwood, 1991)

rocks/minerals/fluids has only recently begun because of the development of new scientific instrumentation and technologies. A trigger for many interesting discoveries has been a link of observations on natural ultrahigh-pressure rocks, mantle peridotites, eclogites, and diamonds flowing from ultrahigh-pressure experiments, advanced high-resolution electron microscopy, focused ion beam (FIB) techniques, and synchrotron assisted microX-ray, infrared and other spectroscopic techniques. This remarkable "blossom" of synergy between 21st-century science and technology will be reflected in the history of the geosciences because the current observations in minerals on a nanoscale level have significantly changed many of our assumptions about plate tectonics and have opened up many unexplored directions in geophysics and ultrahigh-pressure mineralogy. Within this context, the current paper reviews some existing advanced publications and summarizes our own data on the deepest mantle peridotites and diamonds that contain micro-structural and/or compositional records of their mantle origin.

Until now, the deepest fragments of natural rocks/minerals ever studied in scientific laboratories came from depths of 150 to 1,700 km (e.g. Hayman et al., 2005), and therefore they can only be used to characterize the upper mantle and the upper part of the lower mantle. Thus, the deeper horizons of the lower mantle and outer-inner core regions can only be explored with experiments on mineral syntheses and phase transformations in diamond anvil cell devices connected to synchrotron microX-ray in situ measurements at ~30 to 360 GPa.

Although mantle xenoliths and diamonds have been studied for many decades, some aspects of their history that are recorded in nanometric-scale inclusions became recognized only recently due to the introduction of new analytical and experimental instrumentation and scientific technologies. One of these branches – nanoscale mineralogy – has recently been recognized as a powerful tool for studying natural and synthetic minerals crystallysized at very high pressures and temperatures corresponding to the Earth's mantle-core conditions. Nanoscale mineralogy brings together geochemical and geophysical disciplines, providing clues for understanding the major seismic discontinuities, lateral anisotropy, and structure of the Earth's interior.

Methods

Focused Ion Beam-Assisted Transmission Electron Microscopy

Transmission electron microscopy (TEM) is a suitable tool for the investigation of geomaterials from the micrometre to the nanometre scale. Compared with other analytical instruments, the great advantage of TEM is that it provides simultaneously both chemical and structural information on mineral phases. We have performed our studies with a Technai F20 X-Twin transmission electron microscope with a field emission electron source coupled with energy dispersive X-ray analyses (EDS), an electron energy-loss spectrometer (EELS), and a high-angle, annular dark field (HAADF) detector. Chemical information from the sample is derived from the excited X-rays (EDS analysis), from EELS analysis, and from the HAADF, which collects elastically and inelastically scattered electrons for imaging in the scanning transmission mode (STEM). Elastic scattering of the incident electrons by target atoms depends upon the square of the atomic number Z of the scattering atom (e.g., Williams and Carter, 1996). Consequently, the image contrast contains information about the chemical composition of the mineral of interest. Bright field (BF) and dark field (DF) imaging, selected area electron diffraction (SAED), convergent beam electron diffraction (CBED), and high-resolution imaging (HREM) provide structural information on the sample. Fast Fourier transformation (FFT) of HREM images results in the corresponding diffraction pattern. The resolution of modern TEM is approximately 0.08 nm (O'Keefe et al., 2001).

Until recently, intensive use of TEM in the geosciences was hampered by difficulties in the sample preparation. The primary goal for TEM observations is to prepare a site-specific specimen, as thin as ~100 nm, in order to study tiny inclusions, mineral structures, or/and characteristics of grain boundaries starting from a petrographic thin section. Conventional TEM specimen preparation techniques, however, such as crushing the sample or precise argon ion milling, are not site-specific techniques and result in non-uniform sample thicknesses. This fundamental problem was overcome with the use of the focused ion beam (FIB) technique. Focused gallium ion beam devices were simultaneously developed at the University of Chicago and at the Oregon Graduate Institute in the mid-1970s for the computer manufacturing industry. Micrometre-size structures were milled from integrated circuits by applying a high-current-density focused ion beam from a liquid metal (usually gallium) ion source (LIMIS), (Puretz et al., 1984). In the late 1980s and 1990s, the era of TEM foil preparation with FIB began. The great success of that technique for materials science and industrial needs was based on the unique capability of FIB to prepare site-specific TEM foils (Young et al., 1990; Basile et al., 1992; Overwijk et al., 1993). For summaries of the FIB technique, see Orloff et al. (2003) and Giannuzzi et al. (2005).

FIB applications in the geosciences began about a decade ago, and since that time more and more applications related to TEM studies of geomaterials have been reported (e.g. Phaneuf, 1999; Vicenzi and Heaney, 1999; Wirth, 2000, 2001, 2003–2007, 2009;2007, 2008; Dobrzhinetskaya et al., 2001–2003; Heaney et al., 2001; Lee et al., 2003; Graham et al., 2004; Dobrzhinetskaya et al., 2005–2007; Obst et al., 2005; Smith et al., 2007; Loginova et al., 2006, 2008a, b).

Electron transparent membranes suitable for further TEM studies can be easily prepared with the FIB technique (Fig. 2) using standard polished rock/mineral slides and thin sections that geoscientists traditionally use for both optic microscopy and electron microprobe analyses. The FIB foils can also be prepared from any fine-grained minerals, or their powder can be embedded in epoxy, metals, alloys, glass, or ceramic materials. In this case, the surface roughness of such a specimen should be less than a μm because otherwise it might be difficult to find the sputtering site under the optical microscope for ex-situ lift-out of the foil. The shape of the specimen, however, is not crucial. TEM foils can even be prepared from sub-millimetre-sized spheres. Semiconductors or insulators need a thin coating of conducting material (e.g., carbon, gold) prior to FIB milling. A standard TEM foil has the dimensions of 15–20 μm × 10 μm × 0.150 μm and can be milled in a fully automated process in approximately four hours. The final thickness can be less than 50 nm using a dual-beam device and the in-situ lift-out technique. Further carbon coating of the TEM-ready foil is not required. More details of FIB technique applied to geomaterials are given elsewhere (e.g. Wirth, 2004, 2009).

FIB sample preparation opened up a new field of investigating sub-μm-sized inclusions in diamond (e.g. Dobrzhinetskaya et al., 2001–2003; Wirth and Rocholl, 2003; Wirth, 2004; Dobrzhinetskaya et al., 2005–2007; Klein-Ben David et al., 2006, 2007; Logvinova et al., 1996, 1998a, b). Argon ion milling of diamond resulted almost always in the removal or destruction of the lower-density phases because the sputtering depends on the hardness of the material. This problem is overcome with FIB. Because diamond milling with FIB is a very slow process, it can be significantly enhanced by using the selected carbon mill (SCM). The principle of SCM is that water vapour is discharged into the vacuum chamber close to the Ga ion beam and adsorbed by the sample surface. The Ga ion beam decomposes the water molecules, and the diamond is oxidized, thus enhancing the milling rate. Details of TEM foil preparation with FIB and the application of SCM are given in Wirth (2004).

In general, the FIB technique adds a new dimension to TEM specimen preparation, which includes (1) large, flat areas with uniform thickness; (2) successful preparation of extremely heterogeneous samples containing "difficult" material; and (3) precise selection of the foil position with respect to the minerals of the interest. In addition, the scar of the FIB cutting remains visible on the sample surface, thereby facilitating repeated FIB milling and additional examinations

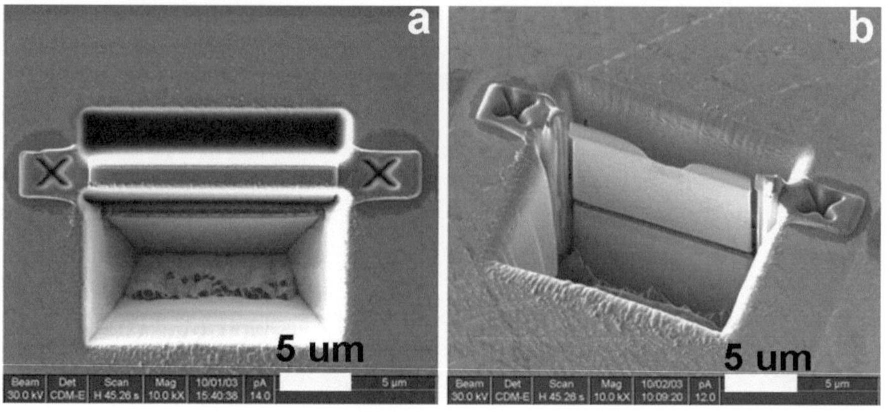

Fig. 2 Sample preparations with the FIB: (**a**) two crosses represent references for beam shift and sample drift correction during milling; two "trenches" are milled out at the both sides of the chosen location of foil preparation. (**b**) a membrane of 15 × 5 × 0.05 μm size is cut off the sample from the left, the right and the bottom sides; sample is ready to be "lift out"

at a precise area on the specimen in cases in which this is necessary to resolve controversial or confusing data.

Synchrotron Infrared (IR) Microspectroscopy

Synchrotron radiation, a bright source of infrared photons, was first exploited in micro-spectroscopy to probe organic and non-organic materials with sizes close to the diffraction limit (e.g., Holman et al., 2002; Liu et al., 2002). This provides IR spectra with high signal-to-noise ratios at spatial resolutions as fine as 3–10 μm. Mid-IR photons are too low in energy (0.05–0.5 eV) to either break the chemical bonds or cause ionization, and they also produce only a minimal heating of samples. Synchrotron micro-spectroscopy is a well-established method that for over a decade has been successfully applied to studies of biological tissues and cells, semiconductors, geological materials, interplanetary dust particles, forensics, the corrosion of metals, structural composites, and polymer laminates.

The intense synchrotron source provides IR spectra with high signal-to-noise ratios at spatial resolutions as fine as 3–10 μm. To measure light elements such as H, O, and N and their bonding or aggregation with/around C atoms of microdiamonds, we used the U2A beamline of the National Synchrotron Light Source (NSLS) at the Brookhaven National Laboratory (BNL) in Brookhaven, NY, USA. The optical layout of the beamline facility was described in detail elsewhere (Carr et al., 1995; Reffner et al., 1995). The diamond IR spectra were collected with a Bruker IFS 66 v/S vacuum Fourier transform interferometer interfaced with the synchrotron source and a modified Bruker IRscope-II microscope equipped with an HgCdTe type-A detector. The top aperture/field stop was set to 10 × 10 mm^2, and the spectra were acquired in the range of 600–4,000 cm^{-1} with a KBr beam splitter. The optical bench was evacuated, and the microscope was purged with dry nitrogen gas during the measurements to reduce or eliminate water vapour absorption. The details of application of this technique to studies of microdiamonds from ultrahigh-pressure terranes are given elsewhere (e.g., Dobrzhinetskaya et al., 2006b).

Samples from Earth's Interior: From What Depth Do They Originate?

Ultradeep Xenoliths from Kimberlitic Sources

The discovery (Haggerty and Sautter, 1990; Sautter et al., 1991) of ultradeep mantle xenolith from diamondiferous kimberlite in Africa containing exsolution lamella of clinopyroxene in pyropic garnet showed that such xenoliths represent rocks of the lowermost upper mantle and the mantle transition zone (~300–400 km). Based on the results of experiments (Akaogi and Akimoto, 1977; Irifune, 1997), the authors argued that conditions for Si in octahedral sites in the precursor majoritic garnet indicate pressure $P > 13$ GPa. Haggerty and Sautter (1990) suggested that the xenoliths represent the rocks that occurred at the 410-km seismic discontinuity zone, where, according to their interpretation, a transition from peridotite to eclogite took place. Their pioneering work has triggered further discoveries of ultradeep xenoliths containing majoritic garnet from potassic ultramafic magmas at the Ontong Java Plateau of Malaita, southwest Pacific (Collerson et al., 2000). Within one of these xenoliths, Collerson et al. (2000) have described majoritic garnet in association with Ca- and Mg-perovskite, Al-silicate phase with undetermined structure, and microdiamond. The conventional geobarometry based on the chemistry of majoritic garnet suggested pressure of ~22 GPa, whereas the Al-silicate phase was assumed to be crystallized at 27 GPa according to experiments. Taking in account the calculations and assumption above, Collerson et al. (2000) have suggested the depth of the Malaita xenolith formation at ~600–670 km. However such a deep origin was later questioned by Neal et al. (2001).

Because evaluation of depths from which mantle peridotites originate is always a subject of strong discussions, the majoritic garnet – or its product of decompression presented by pyrope with exsolution lamella of pyroxenes – remains one of the best indicators of the very high-pressure environments. It was verified by many experiments conducted in different laboratories that the majoritc garnet is stable at $P > 5$ GPa (e.g., Akaogi and Akimoto, 1977; Irifune, 1987). The composition of majorite is represented by the complex solid solution:

$$M_3{}^{VIII}\,(Al_{2-2n}M_nSi_n)^{VI}\,Si_3O_{12},$$

where $M = Mg^{2+}$, Fe^{2+}, and Ca^{2+}, $0 \leq n \leq 1$, and superscripts indicate cations oxygen coordination. When pressure rises above 5 GPa, the garnet-precursor is transformed into majorite (supercilisic garnet) with Si ($Si^{IV} + Si^{VI}$) > 3 cations per formula unit; the silica content also increases because the Al^{3+} and Cr^{3+} are replaced by M and Si^{4+} cations (e.g., Smith and Mason, 1970). Therefore, because the Si content in the octahedral site of majoritic garnet increases with increasing pressures (Akaogi and Akimoto, 1977; Irifune, 1997), and because the volume of the majoritic component dissolved in garnet is calculated from experimental data (Gasparik, 2003), the pyroxene exsolution lamellae in garnet can be used as the "pressure indicator."

Experiments on decompression of the majoritic garnet simulating the exhumation path of mantle peridotites shows that at high-T (1,400°C) decompression from 14 to 12GPa, exsolutions of interstitial blebs of diopside and Mg_2SiO_4- wadsleyite lamellae from a parental majoritic garnet take place (Dobrzhinetskaya et al., 2004, 2005a). These extend our interpretation of natural rocks, and allow reconstruction of the former majoritic garnet in peridotites based on the presence of the blebs of pyroxenes clustered around the decompressed garnet containing exsolution lamellae of olivine (former wadsleyite). Similar clusters of clinopyroxenes around pyropic garnet containing clinopyroxene lamellae exsolutions from the >300-deep African xenolith were reported by Haggerty and Sautter (1990), and Spengler et al. (2006) from the >600-km-deep garnet peridotite from the Western Gneiss region of Norway, an ultrahigh-pressure terrane.

Diamonds from Kimberlitic Source

Diamond is the oldest (~4,200 Ma) geological material (Menneken et al., 2007) although in general, diamond-beraing kimberlite/lamproite falling in range from Earlier Archean to Eocene contain diamonds which age is different than age of magmas formation (e.g. Heaman et al., 2004). Diamond due to its chemical inertness plays the role of a specific "container" delivering solid and fluid inclusions unchanged from Earth's interior to its surface. Due to strong covalent bonding of *sp3* between carbon atoms, its structure is stable at a wide range of pressures (4–>100 GPa) and temperatures (1,000–3,500°C) (Bundy, 1989). Because diamond is stable through geologic time in different geological environments (unless it is not oxidized and is transformed back to graphite) it remains an important material providing direct information about pressure, temperature, and chemical conditions that allow reconstruction of mantle mineralogy.

Diamonds from kimberlitic sources are the best natural samples for evaluating the composition of the mantle because they contain comparatively large (from several hundred nm- to mm-size) inclusions of different minerals. These inclusions are traditionally used for establishing P & T conditions and the depth of the diamond location during its growth. With progress in high-resolution TEM and FIB technologies the research on large diamonds has revealed new information that there is a continuum in inclusions size from those that are resolvable with electron microprobe (EPM) down to those that are sub-μm in size, including those, for instance, composed of only a few water molecules that have dimensions measured in angstroms. With advancement in nanobeam technologies and synchrotron-assisted spectroscopic applications, the existing gap in knowledge related to nanoscale inclusions in kimberlitic diamonds has recently began to be resolved (see Section "Submicrometre- and Nanoscale-Size Inclusions in Kimberlitic Diamonds").

The range of estimated depth from which kimberlitic diamonds originate is as wide as ~80 to >1,700 km (e.g., Stachel et al., 2005). Here, we limit our discussion to diamonds with an exceptional suite of mineral inclusions that suggest an origin from the deep upper mantle transition zone, at a depth of ~300–660 km (very deep diamonds) and to those that are believed to originate from a lower mantle depth of >660 km (superdeep diamonds).

The first diamonds containing ferripericlase (iron-magnesium oxide) inclusions indicative of their very-deep origin were found in Orroroo, South Australia, and Koffeifontain, South Africa (Scott-Smith et al., 1984). The authors have suggested the uppermost lower mantle origin of these diamonds because ferripericlase requires a minimum pressure of ~25.5 GPa. Such a pressure is expected below the 660 km seismic discontinuity. Later, ferripericlase was found in many other kimberlitic diamonds in Western and Eastern Africa, North and South America, and Siberia (e.g., Kaminsky et al., 2001). Numerous other lower-mantle minerals, such as Mg- and Ca-perovskites (high-pressure polymorphs of $MgSiO_3$ and $CaSiO_3$, respectively) and tetragonal polymorph of pyropic garnet (TAPP) existing at pressures >22 GPa (e.g., Harris et al., 1997; Harte et al., 1999) are known to exist in kimberlitic diamonds.

The alluvial diamond deposits of the São Luiz River, Juina Province, Brazil are known in the literature as a unique placer containing both very-deep and superdeep diamond groups (e.g., Harte and Harris, 1994; Harte et al., 1999; Kaminsky et al., 2001; Hayman et al., 2005; Harte and Cayzer, 2007). Numerous studies indicate that these diamonds are products of erosion of the Cretaceous kimberlitic pipes (e.g., Costa et al., 2003). Very-deep diamonds from this area contain abundant inclusions of garnet and clynopyroxene with a wide variety of textural geometries, which provides evidence that such diamonds must have come from a depth of ~450 km – and probably deeper (Harte and Crayzer, 2007).

Harte and Cayzer's (2007) paper presents a case showing that, even though clinopyroxene and garnet inclusions in Juina diamonds do not exhibit typical "exsolution lamellae" geometry, the clynopyroxene grains scattered inside the garnet and at its outer zone are, indeed the result of the decompression of the former majoritic garnet. Their electron back-scattered diffraction (EBSD) studies show that each inclusion of garnet is characterized by a constant crystallographic orientation, whereas the clynopyroxene grains have orientations consistent with the {110} and <111> directions of the garnet. The EBSD studies, along with calculations of the integrated bulk chemistry of a garnet precursor, therefore confirm that that Cpx and Grt inclusions were originally single-phase majoritic garnets and that they preserve various states of decompression during transport of the host diamonds from depths of ≥450 km to the Earth's surface (Harte and Cayzer, 2007). Many of the inclusions of garnet and clynopyroxene in the Juina diamonds studied by Harte and Cayzer (2007) now show compositions that reflect their re-equilibration at lower pressures corresponding to depths of ~180–210 km. Because the compositions of these re-equilibrated garnets and clynopyroxenes are similar to those from eclogite xenoliths that occurred in kimberlitic pipes, Harte and Cayser hypothesize that such eclogitic xenoliths might have originated from much greater depth within the mantle.

The superdeep diamonds originating from the lower mantle depth of 600~1,700 km occur in the Juina area in Mato Grosso, Brazil (e.g., Kaminsky et al., 2001; Hayman et al., 2005). The Mato Grosso diamonds contain inclusions of Fe-rich periclase (ferropericlase) with Mg# = 0.65 (Hayman et al., 2005). Although ferropericlase is the dominant inclusion in ultradeep diamonds, its presence alone does not signify their lower-mantle origin because ferropericlase is also stable in upper-mantle conditions. The coexistence of ferropericlase with $CaSiO_3$-perovskite or $MgSiO_3$-perovskite should be taken as strong confirmation of the lower-mantle conditions (e.g., Stachel et al., 2005). So far, in addition to ferripericlase, superdeep diamonds from Juina contain $CaSiO_3$-perovskite, Cr-Ti spinel, decompressed "olivine," Mn-Ilmenite, tetragonal almandine pyrope phase (TAPP), and native Ni. Many similar ultrahigh-pressure minerals or their assemblages are found as inclusions in other superdeep kimberlitic diamonds (e.g., Davies et al., 1999; Satchel et al., 2000; Huttchison et al.,2001; Kaminsky et al., 2001; Brenker et al., 2007). Corundum inclusions were also found also in superdeep diamonds in association with MgSiO3 perovskite and ferropericlase, suggesting that a free Al phase can exist in the lower mantle (Huttchison et al., 2001). By now, the superdeep diamonds are found at more than a dozen geographic localities on eight cratons (e.g., McCammon, 2001).

Submicrometre- and Nanoscale-Size Inclusions in Kimberlitic Diamonds

Using FIB-TEM techniques, R. Wirth has initiated studies of sub-μm-size inclusions in alluvial diamonds from Minas Gerais, Brazil, and from Ukraine and Siberia, as well as diamonds from kimberlites of Slave Craton in Northern Canada, South Africa, and the Siberian pipes Udachnaya, Mir, Internationalnaya, and Yubileynaya (Wirth, 2005; Kvasnytsya et al., 2005; Klein-BenDavid et al., 2006; Kvasnytsya et al., 2006; Logvinova et al., 2006; Wirth, 2006; Wirth et al., 2007; Wirth, 2007; Klein-BenDavid et al., 2007). Until recently, such tiny inclusions were not in the scope of the researchers because only large inclusions (up to several mm) were used for conventional EMP analyses in the field of diamond studies, and methods (e.g., like FIB) for studies of nanoscale inclusions was not available. The results show that sub-μm-size inclusions in diamonds from different locations in the world exhibit a surprising similarity. Usually, they consist of a fluid or melt associated with solid phases that are represented by silicates (e.g., phlogopite), carbonates, phosphate (apatite with fluorine and/or chlorine), chlorides (NaCl, KCl), sulphides (occasionally), ilmenite, and rare magnetite. Carbonates are usually represented by calcite with low strontium concentration, dolomite, and/or pure $BaCO_3$. Klein-BenDavid et al. (2006) studied diamonds from Slave Craton (Diavik Mine) and Siberia (Yubileinaya), and they have suggested that micro-inclusions consisting of a dense supercritical fluid were trapped by diamonds during their crystallisation in a media containing fluid phase. Later, during cooling, secondary mineral phases grew up from the trapped fluid.

Micrometre-sized olivine inclusions were found in an alluvial diamond from the Macaubas River (Minas Gerais) (de Souza Martins, 2006). The diamond crystal was mechanically polished in such a way that the olivine inclusions remained approximately 5 μm below the surface to keep them closed. A TEM foil was cut across the diamond-olivine interface using the FIB technique. The high-resolution TEM imaging showed that the interface consists of an approximately 50-nm-thick amorphous layer; EELS analyses revealed the presence of fluorine in this layer. Any contamination with fluorine can be excluded because the diamond was never exposed to HF and the olivine inclusion was "sealed" inside of the diamond prior to the FIB milling. The fluorine has been detected frequently with similar techniques in sub-μm inclusions in diamonds from other localities – the Koffiefontein mine, South Africa (Klein-BenDavid et al., 2007) and Siberia (Kvasnytsya et al., 2006; Logvinova et al., 2006). In addition to olivine, the KCl solid nano-inclusions co-existing with a fluid phase were observed in the same diamond; in this case, the KCl inclusion is surrounded by a corona of a brine.

Several FIB foils were prepared from superdeep diamonds from the Juina area of Brazil where we expected to find also some sub-micrometric fluid-solid inclusion pockets. To our great surprise, no fluid-rich nano-inclusions similar to those described above in diamonds from Minas Gerais, Brazil, or any other diamonds from the upper mantle, were observed in the prepared foils. Only carbonate nano-inclusions were found to be common for the Juina superdeep diamonds. Such contrasting observations may suggest that the superdeep diamonds have been grown in an anhydrous mantle environment that is different from the one that exists in the shallower levels of the Earth's mantle above the ~660-km seismic discontinuity zone. Understanding such differences in the presence/absence of nano-inclusions may lead to the conclusion that superdeep diamond environments represent a "dry" media, which could result in a different mechanism of diamond nucleation and growth. At the current stage of our ongoing research, more observations need to be collected to support or deny this working hypothesis.

One other interesting research avenue is emerging from the presence of carbonate nano-inclusions in superdeep diamonds. The FIB foil of one of the superdeep Juina diamonds containing nanometric carbonate inclusions in association with ilmenite was used for studies of carbon isotopes with nano-secondary ion mass spectrometry (nanoSIMS). Carbon isotopic data vary systematically within a single foil in the range of $\delta^{13}C$: –13.9 ± 1.9‰ and –25.1 ± 1.8‰. The range of these $\delta^{13}C$ values suggests that this diamond grew from partially biological carbon (Wirth et al., 2007), a finding that supports the existence of the deep carbonates and might indicate that the Earth's global CO_2 cycle has an ultradeep extension, as was proposed by Brenker et al. (2007).

We conclude that the nanometric–sub-μm solid inclusions co-existing together with the fluid phase represent the former fluid pockets entrapped by diamonds, and that their bulk chemistries reflect a composition of the fluid media from which the diamond grew. The large crystalline inclusions of eclogitic or peridotitic provenience, investigated in detail for many years with EMP and spectroscopic methods, represent an anhydrous environment in the lower Earth's mantle where the diamond has grown.

Ultrahigh-Pressure Metamorphic Rocks from Collisional Orogens

During continental collisions, rocks from passive margins, island arcs, and micro-continents are carried down to mantle depth through the subduction channels in which their partical melting and metamorphism (including UHP metamorphism) occur. The subsequent decoupling of the slices of UHP metamorphic rocks from the descending plate and their exhumation are a part of the mountain-growth process marking ancient continental collisional zones. The UHPM rocks were recognized and established as extended terranes in the 1990s due to discovery of well-preserved coesite and/or microdiamonds within garnet, pyroxene, zircon and other minerals. This was followed by international efforts that established the presence of UHP metamorphic rocks on all continents: Euro-Asia (Germany, Greece, Italy, Norway, China, India, Kazakhstan, and Kirgistan), Africa (Mali), Australia (New Zealand), South America (Brazil), Antarctica, and Greenland.

The great importance of the discoveries of UHPM rocks is that the paradigm that continental rocks are too buoyant and cannot be subducted very deeply was broken. With the discovery of coesite, diamond, relics of former majoritic garnet, olivine with oriented exsolutions of ilmenite and chromite, or magnetite, as well as titanite and omphacite containing exsolved coesite rods and plates, it became clear that the rocks were subjected to recrystallization at high pressures and high temperatures requiring depths >100–250 km (e.g. Bozilov et al.,1999; Chopin, 1984; Dobrzhinetskaya et al., 1995, 1996; Massonne, 1999; Mposkos and Kostopoulos, 2001; Ogasawara et al., 2002; Chopin, 2003; Perraki et al., 2006; Smith, 1984; Sobolev and Shatsky, 1990; van Roermund and Drury, 1998; van Roermund et al., 2002; Xu et al., 1992; Yang et al., 2003; Ye et al., 2000). Furthermore, evidence has been reported of former stishovite in metamorphosed sediments in the UHPM terrane of Altyn Tahg, western China, having a depth of subduction up to >350 km (Li et al., 2007).

The occurrences of UHPM rocks within continental collision zones corroborate the premise that light material from the continental crust was overcome by buoyancy and was subducted at least to the upper mantle – or even deeper, to the mantle transition zone – and that some portion of it has been uplifted to the surface from those depths (e.g., Dobrzhinetskaya and Green, 2007b; Ernst et al., 1997; Ernst, 2006; Ernst and Liou, 2008; Gerya et al., 2002, 2008; Liou et al., 2007; Stöckhert and Gerya, 2005). Though many interesting things may be learned from UHPM rocks, only the mantle garnet peridotites and diamonds are the focus of our consideration here. However, the studies mentioned above have helped not only to establish the depth to which continental rocks can possibly be subducted, but also to prove that some bodies of the garnet peridotite from UHPM terranes represent fragments of the Earth's lowermost upper or uppermost low mantle (e.g., Dobrzhinetskaya et al., 1996; Spengler et al., 2006; Liou et al., 2007).

Garnet Peridotites from UHPM Terranes

Garnet peridotites are usually reperesented by mantle xenoliths brough from Earth mantle to the surfice by kimberlitic magmas (e.g., Nixon, 1987), or small massifes or pods occurred within UHPM terranes incorporated in collisional orogenic belts (e.g., Ernst, 2006; Ernst and Liou, 2008; Liou et al., 2007). Two groups of grt-peridotites – mantle-derived and those that have protolith of the peridotite of former crustal source, metamorphosed under ultra-high pressure conditions together with surrounding sialic host rocks – are recognized within UHPM terranes (e.g., Zhang et al., 1994; Liou et al., 2007). Zhang et al. (1994) were the first to raise the question of whether both mantle-derived and crust-hosted garnet peridotites should be recognized. We focus here on mantle derived peridotites, and a more extended review related to both crustal and mantle source peridotite may be found in Liou et al. (2007).

An origin from depths >300 km for orogenic-belt garnet peridotite was proposed by Dobrzhinetskaya et al. (1996) based on their discovery of earlier unknown μm-size exsolution lamellae of ilmenite and chromite in olivine from the Alpe Arami garnet peridotite of the Central Alps. Based on the reconstructed abundance of ilmenite exolution lamellae (up to ~1 vol.%) incorporated in olivine, the morphology and crystallographic relationships of the ilmenite and chromite precipitates with the host olivine the authors have proposed that the precursor olivine was subjected to pressure ~10–15 GPa. Independent TEM observations of exsolved Ca-poor pyroxene displaying antiphase domains in diopside clustered around pyropic garnet provided additional confirmation of a deep (>300 km) origin for the Alpe Arami garnet peridotite (Bozhilov et al., 1999). Later experiments conducted in different laboratories (Dobrzhinetskaya et al., 2000, Tinker and Lesher, 2001) have revealed that the solubility of Ti in olivine is a function of pressure, and it was also pointed out that this process requires a certain availability of Ti in the starting material. The experiments confirmed progressively increasing Ti content in olivine from 0.3–0.4 vol.% (at 6–7GPa) to ~1 vol.% (at 12–14 GPa). The attempts to present these discoveries as a breakdown reaction of titanoclinohumite producing olivine with rods of ilmenite (e.g., Risold et al., 2003) were not convincing because no titanoclinohumite was ever found in the samples collected from the Alpe Arami garnet peridotite outcrops.

A unique fragment of Archean garnet peridotite (3.3 Ga) with relics of former majorite was discovered in the Otroy area of the Western Gneiss Region of Norway within Caledonian UHPM rocks (~400 Ma) containing diamond and coesite (van Roermund and Drury, 1998; van Roermund et al., 2002; Spengler et al., 2006). The peridotites contain coarse polycrystalline pyropic garnets, with pyroxenes represented by small inter-crystalline grains and tiny needles that were interpreted as products of the decompression of majoritic garnet formed at a depth of ~180 km (van Roermund and Drury, 1998).

Still later, a new quantification of pyroxene microstructures of the Otroy garnet peridotite yielded, in one polycrystalline garnet sample, >20.6 vol.% of pyroxene. Such a high content of exsolved pyroxene according to experimental data (e.g., Akaogi and Akimoto, 1977) corresponds to an unexsolved majoritic precursor that is stable at a minimum depth of 350 km (Spengler et al., 2006). Rare earth element (REE) studies in both exsolved pyroxene and host garnet from the Otroy garnet peridotite suggest that clinopyroxene needles and clinopyroxene blebs were in initial equilibration with garnet at high temperatures (≥1,300°C). The effect of deep melting (>30 vol.%) of the garnet-peridotite prior the decompression of majorite was confirmed by exceptionally poor concentrations of REE (Dy/Yb <0.07). The extremely light, REE-depleted nature of both needles and blebs of pyroxenes (Ce/Sm <0.08) is interpreted that they are a product of decompression, ruling out any possibility of their formation from the melt (Spengler et al., 2006). Experimental simulation of the exhumation path recorded in garnet lherzolite from Otroy confirmed that the interstitial blebs of diopside form as a product of exsolution reaction of parental majoritic garnet during its decompression from 14 to 12 GPa at T = 1,400°C (Dobrzhinetskaya et al., 2004, 2005).

Spengler et al. (2006) suggested that the Otroy garnet peridotite was melted at temperatures ≥1,800°C and depths of ≥250 km and that the melting was resumed when the ascending peridotites have reached the lower horizons (~150 km) of the Archean continent. The residue remaining after melting occurred at such extreme conditions has not been reported before. Spengler et al. (2006) have concluded that the lithospheric mantle fragment was incorporated into subducting sialic crust during the Caledonian continent-continent collision at ~400 Ma. This routeless fragment was "sealed" within eclogite-bearing metasediments and all together, they were subjected to ultrahigh-pressure metamorphism corresponding to the diamond stability field. The Otroy garnet peridotite represents one of the deepest kilometre-sized body of the rocks representing the lowermost upper mantle – a mantle transition zone fragment exposed now at the Earth's surface.

Uncounted numbers of small bodies of mantle-derived garnet peridotites are described in the Dabie-Sulu region of the Chinese Orogenic Belt (COB), a series of mountain chains that were created during subduction of the northern edge of the Yangtze craton to more than 150–200 km depth beneath the Sino-Korean craton (e.g., Zhang et al., 2000, Liou et al., 2007). Although the mantle origin of many peridotites has been accepted (e.g., Zhang, 2000; Yang and Jahn, 2000; Yang et al. 2007b), there is still debate about whether these peridotites were subjected to UHP metamorphism and about their travel path from mantle depth to the surface. One of the best examples of mantle-derived peridotite is the Rizhao garnet clinopyroxenite body from the Sulu region of COB. Liou et al. (2007) have suggested that the Rizhao clinopyroxene containing ~25 vol.% of exsolved garnet and ~4 vol.% of exsolved ilmenite originated from the former majoritic garnet. They assume that the following cations substitution has occurred during such majoritic breakdowns: $Ca^{2+}Ti^{4+} \rightarrow 2Al^{3+}$, $Mg^{2+}Si^{4+} \rightarrow 2Al^{3+}$, and $Na^{1+}Ti^{4+} \rightarrow Ca^{2+}Al^{3+}$ in octahedral sites (e.g., Zhang and Liou, 2003). These authors also hypothesized that a pure clinopyroxene could be a precursor for such a "lamellar" clinopyroxene-garnet assembly.

The origin depth for this and other mantle-derived peridotites from the Dabie-SuLu region of China is ~ 150–200 km (e.g., Liou et al., 2007). Other garnet peridotites of Chinese Central Orogenic Belt have been classified as originating from ~200–300 km based on the presence of clinoenstatite lamellae in orthopyroxene (Zhang et al., 2002). The authors assumed that the clinoenstatite lamellae could be formed either by inversion from orthoenstatite or by a displacive transformation of primary high-pressure clinoenstatitite due to decompression. According to Ye et al. (2000) the Yankoy (Sulu region) eclogites enclosed within garnet peridotite originate from a depth of >200 km. Ye et al. (2000) have proposed a possible subduction of continental material (eclogite) to the lowermost upper mantle. Although a depth of >200 km was proposed for many garnet peridotites of COB, in order to constrain the origin of mantle peridotite from UHPM terranes, more experimental work is necessary to achieve an understanding of how decompression microstructures are formed. Massonne and Bautsch (2002) have reported that a boudined garnet pyroxenite layer embedded in serpentinized garnet peridotite from the Granulitgebirge in Germany contains evidence of majoritic garnet as a precursor phase due to the presence of pyroxene exsolution lamellae in garnet. They hypothesized that uplift of such peridotites from a mantle transition zone (depth ~400 km) might has been triggered by the melt that occurred within the mantle plume. Electron microprobe analyses of minerals and geothermobarometry showed that the exsolution process in garnet took place at P-T conditions of ~25 kbar and 1,000°C (Massonne and Bautch, 2002).

Relics of ultrahigh-pressure minerals in combination with geochronology, REE, and isotope geochemistry have demonstrated that mantle-derived garnet peridotite from orogenic belts may have formed at great depths of the mantle regions long before their insertion into down-going continental plates. When they are involved in subduction channels by "being in the right place at the right time," they become "welded" together with the shallower mantle wedge fragments and compositionally diverse metasedimentary crustal rocks. Such a "welding" includes the garnet/pyroxene decompression reactions and/or ultrahigh-pressure metamorphism superimposed on the pre-existing mineral associations and their later retrograded recrystallisations during exhumation.

Diamonds from Ultrahigh-Pressure Terranes

Microdiamonds hosted by metamorphosed felsic and carbonate sediments were first discovered about 25 years ago in the Kokchetav massif of Kazakhstan but were not made known to the Western literature until 1990. Currently, five well-confirmed UHPM diamond-bearing terranes have been established: the Kokchetav massif of Kazakhstan (Sobolev and Shatsky, 1990), the Dabie and Qinlin regions of China (Xu et al., 1992; Yang et al., 2003), the Western Gneiss Region of Norway (Dobrzhinetskaya et al., 1995; van Roermund et al., 2002), the Erzgebirge massif of Germany (Massonne, 1999), and the Kimi complex of the Greek Rhodope (Mposkos and Kostopoulos, 2001; Perraki et al., 2006). In these localities, the diamonds are characterized by small (1–80 μm) crystals of skeletal, cuboidal, sub-rounded, and other imperfect morphologies. The nitrogen impurities in diamonds from Kazakhstan, Norway, and Germany suggest that all of them belong to the type 1b-1aA (e.g., Cartigny et al.,

2001; Dobrzhinetskaya et al., 1995, 2006a,b) implying a short residence time at high temperature (~900–1,100°C) of ~5 Ma (Stöckhert and Gerya, 2005). This distinguishes them from other nitrogen-bearing diamonds of kimberlitic sources (type 1aAB) that have much longer residence time in the Earth's interior (e.g., Jones et al., 1992; Cartigny et al., 2001). All known diamond-bearing metasedimentary rocks are formed at convergent plate boundaries in Paleozoic-Mesozoic time (~480–180 Ma). The occurrence of diamond within rocks of continental affinity suggests that these rocks, despite their intrinsic buoyancy, were subducted into the upper mantle to a minimum depth of 150 km and were subsequently exhumed to the Earth's surface.

Nanoscale Fluid and Solid Inclusions in Metamorphic Diamonds

Metamorphic diamond is an ideal material to provide insight into the conditions occurring during UHP metamorphism in subduction zones because it is highly resistant to graphitization during exhumation. The study of diamonds *in situ* in garnet and zircon shows that the diamond inclusions in rocks (from both Kokchetav and Erzgebirge massifs) are very frequently accompanied by hydrous phases, phosphates, and oxides (e.g., Dobrzhinetskaya et al., 2001–2003a, b). Chlorite, quartz, albite, and apatite are often intergrown with the Kokchetav microdiamonds (e.g., Dobrzhinetskaya et al., 2001, 2003a, 2005), while phlogopite, phengite, apatite, rutil, and quartz form intergrowth assemblages with Erzgebirge microdiamonds (Stöckhert et al., 2001; Dobrzhinetskaya et al., 2003b, 2007). Whether these additional phases were trapped simultaneously with diamond or are later alteration products was not always clear.

Because diamond is chemically stable (if not converted to graphite) any inclusions within diamond are pristine witnesses of the condition of their crystallization and of the composition of the medium from which they crystallized. Molecular water and carbonates also have been detected in some ~100-μm-size diamonds from the Kokchetav massif by conventional IR spectroscopy and by synchrotron-assisted IR spectroscopy for diamonds from the Erzgebirge massif. The results suggest diamond formation from a COH-rich fluid (e.g., De Corte et al., 1998; Dobrzhinetskaya et al., 2003a, b, 2005, 2007; Ogasawara, 2005). Using TEM to study diamonds from Kokchetav, we discovered nanometric inclusions of oxides composed of elements such as Ti, Si, Fe, Cr, and Th, as well as $MgCO_3$, $BaSO_4$, and $ZrSiO_4$, and the presence of empty cavities are indicative of the former presence of fluid (Dobrzhinetskaya et al., 2001, 2003b). Moreover, the COH-rich fluid inclusions containing Cl and S were directly observed in FIB-prepared TEM foils of diamonds from the Kokchetav massif (Dobrzhinetskaya et al., 2005, 2007).

We have recently studied several 40- to 50-μm-size diamonds from the Erzgebirge massif of Germany using advanced FIB, TEM, and synchrotron-assisted micro-infrared spectroscopy. The dozens of diamond crystals (10–50 ηm size) were separated from the felsic gneisses using our special microwave digesting procedure described in Dobrzhinetskaya et al. (2006b). The larger (40- to 50-ηm) crystals (Fig. 3) were used for TEM research followed by synchrotron IR studies. We provide below a description of one such inclusion as an example, and we refer the readers to our previously published data on similar microdiamonds (Dobrzhinetskaya et al., 2003–2007).

TEM studies of the foil #1042 (prepared from the diamond crystal shown on Fig. 3a) revealed the presence of several nanometer-sized, multi-component inclusions containing both the crystalline and fluid phases. Their multi-component character is recognized because a diffraction contrast has confirmed the presence of crystalline phases, whereas a "movement" of the fluid was recognized due to low density contrast caused by electron beam heating at the middle part of the inclusuion (Fig. 4a). Element maps performed for evaluation of the spatial distribution of different chemical components within the inclusion #1042-3 is presented in Fig. 4(b–j). The series of images in Fig. 4(b–j) represent maps of the K lines of Al, Ca, Fe, K, Mg, Si, O, P, and C, respectively. The bright contrast of high-Ca area partly correlates with high-moderate-Fe area, low-Mg and O probably representing (Ca,Mg,Fe)O phase. At least, they yielded Bragg diffraction spots indicating that they are crystalline matter (e.g., Wirth, 2004). The high-Al region very tightly defined at the upper right part of the inclusion pocket (Fig. 4b) correlates well with the Si (Fig. 4g) and O (Fig. 4h), indicating the presence of an Al_2SiO_5 polymorph. A "pulse-like" movement inside the inclusion caused by electron-beam heating overlaps with the

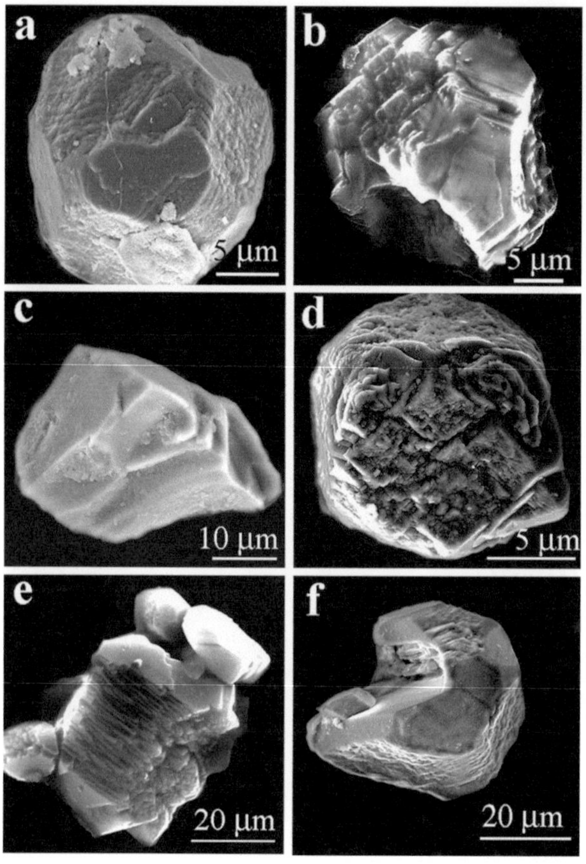

Fig. 3 Microdiamonds from the Erzgebirge quartzfeldspathic gneisses (ultrahigh pressure eclogite-bearing terrane, the eastern shore of Saidenbach Reservoir, ~1.5 km NW of the village of Forchheim, Erzgebirge, Germany). Microdiamonds were separated from the host rocks using method of microwave thermochemical digesting described elsewhere (Dobrzhinetskaya et al., 2006b). Diamonds exhibit imperfect morphologies. Their shapes are characterized by a combination of cube and octahedral faces (**a–d**) which are often complicated with sharp truncated corners (**b** and **c**) and presence of hillocks and triangle pits of various scale (**b, d, e**). Such diamond morphologies suggest that they were formed in a medium oversaturated with impurities, and that the rate of the carbon atoms deposition at the corners and on the faces of diamond nuclei was different, providing faster growth of the crystal edges (Dobrzhinetskaya et al., 2001)

large K-field (Fig. 4e), which weakly correlates with C, O and P; we assume that this confirms presence of fluid containing K, P, C and O. A fluid of similar composition was previously described from a Kokchetav microdiamond in which Cl and S coexist with K, C, and O instead of P; a K–P–C–O fluid was detected earlier in some Erzgebirge diamonds (e.g., Dobrzhinetskaya et al., 2003, 2005, 2007).

Our TEM onservations of the multi-component fluid-solid inclusion #1042-3 is supported by the results obtained with the synchrotron-assisted IR spectroscopy. The synchrotron IR spectrum (Fig. 5), which was recorded from the same diamond (Fig. 3a) that was used for the TEM studies, confirms the presence of carbon-rich fluid inside of the diamond. The IR spectrum, in addition to nitrogen, exhibits a clear absorption band at 3107 cm^{-1} corresponding to C–H bonds in the diamond matrix. The sample also contains a detectable amount of water presented as absorption bands at 1,630 cm^{-1}, which reflects bending motions of the H_2O molecule. The absorption bands at 3,420 cm^{-1} are identified as O–H stretches of the H_2O molecule. There is also a well-pronounced absorption band at 1,430 cm^{-1}, corresponding to the carbonate radical CO_3^{-2}; the latter might be incorporated in diamonds as carbonate microinclusions. The absorption bands between 800 and 600 cm^{-1} are interpreted as silicate inclusions.

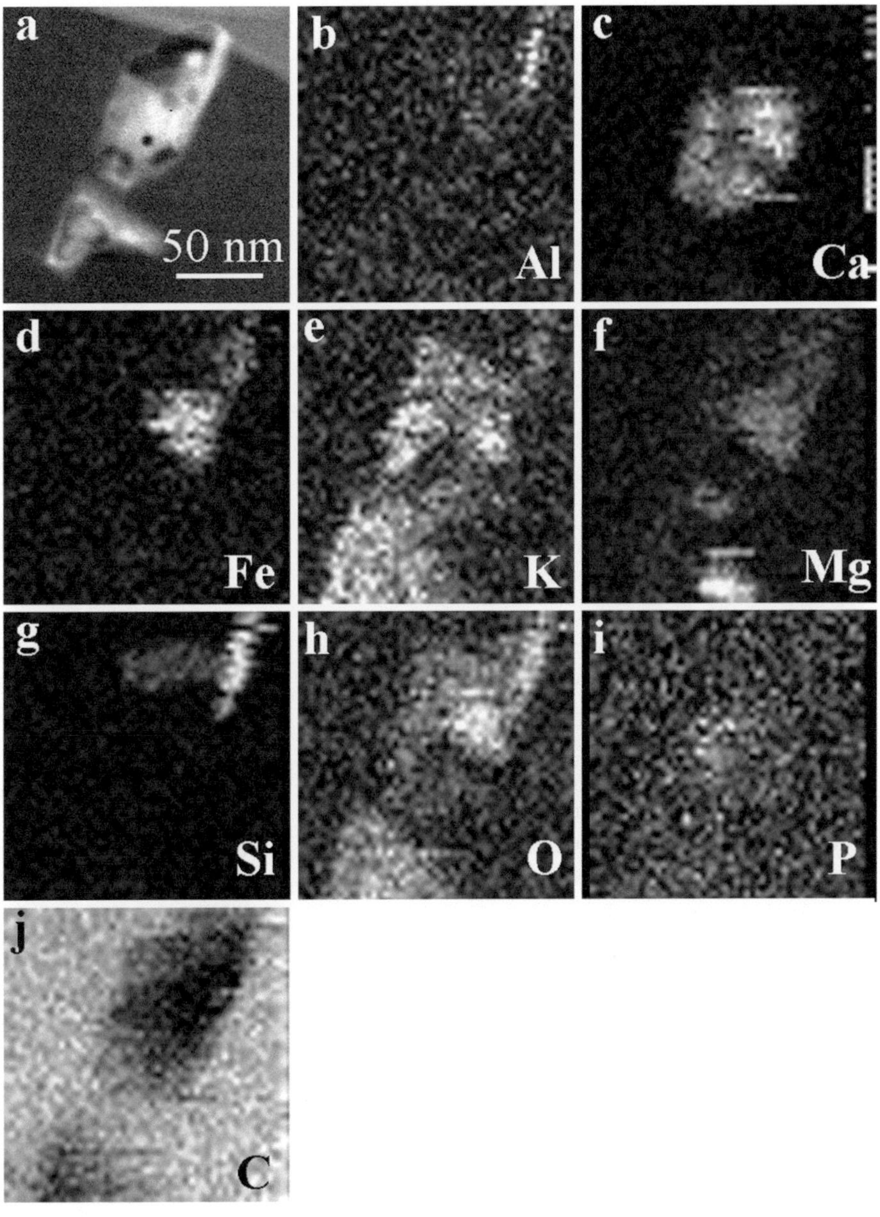

Fig. 4 TEM elements mapping of a multiphase inclusion (#1042-3) in diamond foil containing fluid and crystalline material. Bright-field HR TEM images of multiphase inclusion (**a**) are taken prior to analysis; density fluctuations causing contrast changes in the fluid phase due to electron beam interaction with the fluid was observed in the middle part of the inclusion. Individual EDX maps (**b–c**) of the K-lines of the following elements: Al, Ca, Fe, K, Mg, Si, O, P and C. See the text for further explanation

We have obtained similar synchrotron IR and TEM/FIB data, with a good correlation between them, from dozens of diamonds collected from both the Kokchetav and the Erzgebirge terranes, where diamond occurs in a high (>25 carat/tonne) concentration (de Corte et al., 1998; Dobrzhinetskaya et al., 2006, 2007). All data emphasize the role of carbon-oxygen-hydrogen (COH) fluid during the course of ultrahigh-pressure metamorphism as a trigger for microdiamond formation. Contrary to our observations, Hwang et al., 2001 have suggested that microdiamonds are crystallized from a silicate melt. Their conclusions are based on the TEM observations of the samples prepared by the precise ion polishing technique (PIPS), which has many disadvantages in comparison with the advanced FIB foil-milling technique (e.g., Dobrzhinetskaya et al., 2003b, Wirth, 2004, Obst et al., 2005). The main disadvantage of the PIPS technique is that the fluid inclusions are usually transformed to amorphous matter because of being damaged by the Argon

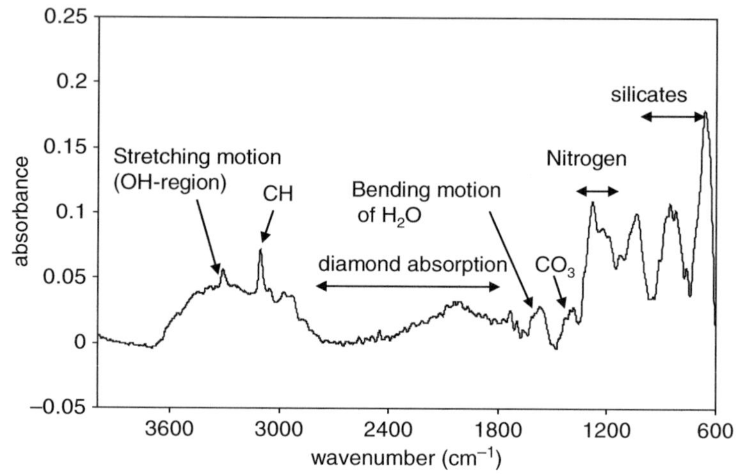

Fig. 5 Synchrotron assisted InfraRed spectrum (beam line U2A, Brookhaven National Laboratory, U.S.) obtained from the Erzgebirge diamond shown on Fig. 3a

beam. Therefore, such a technique of sample preparation practically excludes preservation of intake fluid inclusions in the TEM foil.

An alternative technique for obtaining records of H_2O in very small diamond crystals is synchrotron IR spectroscopy because the synchrotron source provides a low-diameter but high-energy beam. All microdiamonds from UHPM terranes studied with synchrotron and conventional IR techniques contain abundant fluid inclusions. Moreover, direct observation and documentation of fluid bubbles with FIB-assisted TEM studies serve as convincing evidence that C–O–H fluid is a media for the crystallization of these diamonds (e.g., Dobrzhinetskaya et al., 2005b).

Although the diversity of nanometric solid inclusions in the diamonds from UHPM terranes reflects compositional diversity of their host rocks (e.g., Dobrzhinetskaya et al., 2003a), the composition of their fluid inclusions have a surprising similarity. For example, diamonds from both felsic gneisses and marbles from the Kokchetav massif in Kazakhstan contain fluid inclusions rich in COH, K, Cl, P, and S, accompanied by nanometric crystals of carbonates. The diamonds from the Erzgebirge massif in Germany, which are hosted by felsic rocks with no carbonate lithologies around, contain fluid inclusions of similar composition (e.g., Dobrzhinetskaya et al., 2007). Interestingly enough, fluid inclusions of similar composition have now been established with FIB-TEM techniques in kimberlitic diamonds (Klein-BenDavid et al., 2007). Taking into account such uniform similarity, we may suggest that the most realistic explanation is a high solubility of Cl, K, P, and S in a supercritical COH fluid at high pressures and temperatures. Experimental confirmation of this concept is still awaited.

Some Notes Related to Microdiamond Morphologies

One of the interesting phenomena of the microdiamonds from the UHPM terranes is their diverse morphologies, which range from cube and cube-octahedral to skeletal and "uncertain" shapes. Our TEM studies of microdiamonds from the Kokchetav massif have shown that the morphology of the diamond inclusions in garnets extracted from one thick, polished slide (50–70 μm) of felsic gneiss ranges from skeletal forms composed of thin {111} plates through cuboidal and octahedral forms (Dobrzhinetskaya et al., 2001). In contrast, Hwang et al. (2006) have proposed that the morphology of microdiamonds depends upon melt composition. However, from our point of view, Hwang et al.'s (2006) statements remain unclear because no mechanism affecting the rate of nucleation or crystal face formation in a certain order or disorder with respect to the internal atomic arrangements of diamonds was proposed.

According to Chernov (1974) irregular skeletal forms develop if atoms are added to the edges and corners of a growing crystal more rapidly than to the centers of crystal faces. During rapid edge growth, internal cavities may develop on crystal faces, with the possibility of entrapment and preservation of the fluid from which the crystal grows. Our systematic observations of skeletal crystals with myriads of partially formed cavities bounded by {111} planes, of a single fully enclosed, faceted, cavity, and of diamonds with {111} bounded cavities open to cuboid surfaces, suggest that development of such cavities and their entrapment may have been common in growth of Kokchetav diamonds without any connection to the composition of the fluid (Dobrzhinetskaya et al., 2001).

Diamond growth can be divided into three broad categories based upon the ratio of driving force for growth to reaction kinetics. Sunagawa (1990, 1997) identified three kinetic-morphologic fields, including a region at low driving force and/or rapid kinetics where highly perfect crystals grow, a region at high driving force and/or slow kinetics where highly imperfect growth is expected to occur, and a region in between, where hopper and/or skeletal crystals predominate. Sunagawa considered the kinetic effects to be due to temperature alone, but Kanda (1985) and Pal'yanov et al. (1999) showed that the fluid from which diamonds grow (H_2O and carbonate melt, respectively) can have an important additional effect on kinetics and hence on morphology (H_2O increases skeletal morphology and carbonate melt yields perfect octahedra). Thus, increasing the driving force for growth and decreasing the kinetics of growth have similar effects on diamond morphology, but the kinetics cannot be simply described in terms of temperature.

We have proposed (Dobrzhinetskaka et al., 2001) that all of these morphologies are a consequence of growth dominated by {111} platelets in which the morphology that develops in an individual crystal depends on the ratio between the rate at which new plates nucleate at the margins of existing plates (N) and the rate plates grow in their own plane (G). Variation of N/G has the following consequences.

(i) If the ratio of new plate nucleation rate to in-plane growth rate is very small (N/G << 1), the result will be open skeletal forms consisting of large plates surrounding a wide variety of cavity sizes, some of significant size compared to the crystal as a whole. Progressively larger ratios of N/G will result in progressively smaller plate and cavity sizes.

(ii) If the nucleation of new octahedral plates along the edges of old plates is statistically random in space and time, and if N/G is moderate and is constant with time, symmetry constraints require that the macroscopic form of the crystal must show statistically cubic symmetry. Combining this constraint with the fact that {100} planes are furthest away in terms of orientation from the {111} planes that accomplish crystal growth, it follows that the hypothetical {100} faces that are approximated by cuboid surfaces are effectively the slowest growing and therefore dominate macroscopic form. The result, therefore, is development of cuboid forms with octahedral facets on their corners and along their edges, and with {111} bounded cavities in their irregular {100} surfaces. The cavity boundaries will intersect therefore in <110> directions. As N/G becomes larger, cavity sizes decrease and the morphology grades into fully dense cuboid crystals in which the irregular cuboid surfaces do not exhibit cavities, but nevertheless retain their <110> lineations defined by large numbers of intersecting small {111} facets.

(iii) In the limit as nucleation of new {111} plates becomes greatly dominant over plate growth (N/G >> 1), the barrier to nucleation of new layers on existing plates is removed and the concept of plate growth becomes meaningless; normal crystal growth then proceeds in which crystal habit is controlled by the inherent differences in growth rate of different forms. As a consequence, {111} faces will dominate at all times, internal cavities will not form, and octahedral habit will be observed.

The spectrum of morphologies and their abundant tiny inclusions can all be explained by a simple model based on the ratio of the rate at which {111} plates grow and the rate of random nucleation of new plates at their edges. Therefore, no differences in melt/fluid composition are required to explain the wide ranges of morphologies of microdiamonds.

Diamonds from Oceanic Island, Ophiolite, and Forearc Settings

Generally, oceanic islands, mid-oceanic ridge ophiolites, and forearc geodynamic structures are not suitable places for the formation of diamonds and their transportation to the Earth's surface. This is because magmas associated with the mentioned geodynamic environments originate at shallow depth, where pressure is too low, and oxygen pressure (fugacity) is too high for diamond stabilisation (e.g., Parkinson & Arculus, 1999; Wood et al., 1990). In shallow depths, at high oxygen fugacity and low confining pressure, carbon exists in the form of carbonate minerals or as a gas (CO_2), what is known from oceanic island basalts (OIB) and basalts from mid-oceanic ridges, and igneous and metamorphic rocks from modern convergent margins (e.g., Dixon and Clague, 2001; Heide et al., 2008; Wyllie, 1981). However, several recent discoveries of microdiamonds occurring in rocks located within previously "forbidden" for diamond geological settings, indicate that a new understanding is needed of the interactions among mantle convection beneath oceanic islands, forearcs, and the mid-oceanic floor, as well as the depths of their magma formation.

The first finding of nm-sized diamond in melt inclusions in mantle–derived garnet pyroxenite xenolith from Salt Lake Crater (Oahu, Hawaii) was only possible using FIB cut foils for TEM investigations (Wirth and Rocholl, 2003). Healed cracks (usually less than 10 μm in diameter) in orthopyroxene and clinopyroxene include small inclusions of former melt containing nanodiamonds (Fig. 6). The "melt" inclusions consist of Si-rich glass of basaltic composition. Diamonds occur within the Si-rich glass, forming small domains filled with dozens of ~20-nm-size crystals. Nanocrystalline native Fe and Cu, FeS, FeS_2, ZnS, AgS, and several Ti-, Nb-, Zr-, Ir-, In-, and Pd-rich phases of unknown stoichiometry and structure are associated with the diamonds. Carbon-bearing phases coexisting with diamond are $CaCO_3$, graphite, and amorphous carbon, which are scattered within the Si-rich glass matrix. Corroded corundum crystals scattered within the glass matrix have also been observed in several TEM foils (Fig. 7). They are indicative of the presence of a fluid that has partially dissolved the corundum rim zone. A defect-rich orthopyroxene crystal was also detected within one of the foils (Fig. 8); this might suggest a clinoenstatite to orthoenstatite phase transformation that has taken place during deformation.

The presence of nanocrystalline diamonds in glass inclusions in pyroxene crystals from garnet pyroxenite xenoliths of Hawaii has recently been confirmed independently by Raman spectroscopy by Frezzotti and Peccerillo (2007). The authors have received their sample from a collection of the Smithsonian Institution (Washington, D.C., USA). The confirmation of the presence of microdiamonds within dense fluid/melt inclusions by such an independent way shows that

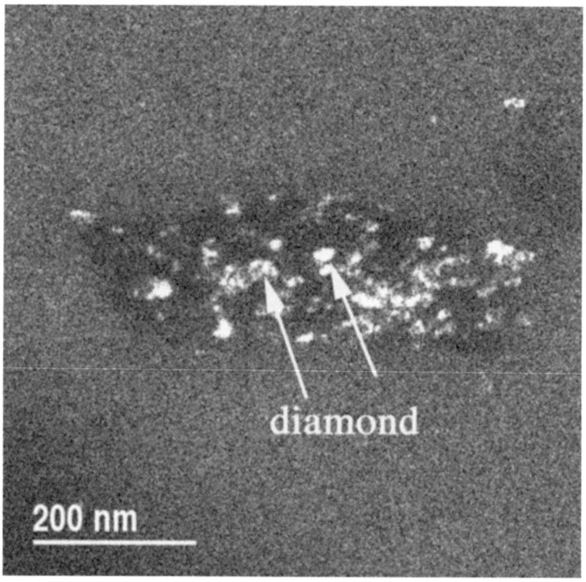

Fig. 6 Microdiamonds in FIB prepared TEM foil: diamond crystals (~20 nm size) are scattered within Si-rich glass of basaltic composition – a former melt inclusion in garnet pyroxenite xenolith from Oahu, Hawaii (Wirth and Rocholl, 2003)

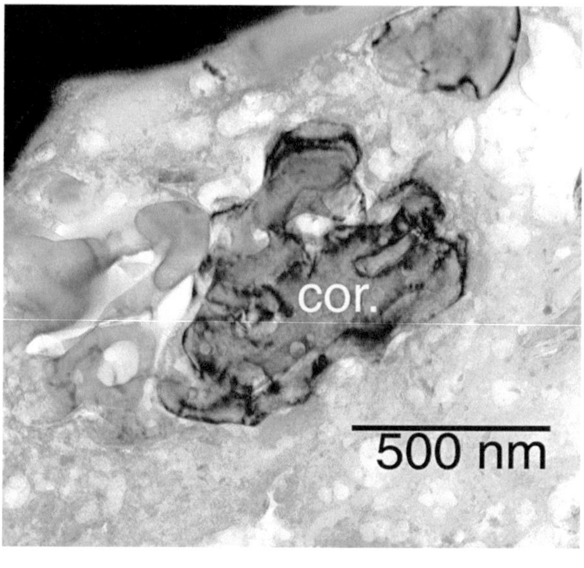

Fig. 7 Corroded corundum microcrystal coexisting with microdiamonds within Si-rich glass matrix

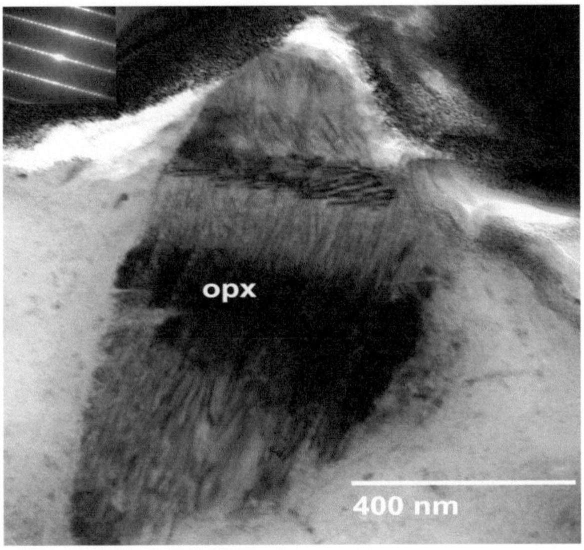

Fig. 8 Orthopyroxene microcrystal with defects in Si-rich glass matrix. Lamellar arrangement of the defects suggests a transformation of clinoenstatite to orthoenstatite

these diamonds are a typical rather a unique phenomenon. The observations suggest the presence of free diamond-bearing silica- and carbonate-rich fluids/melts formed in the diamond stability field, probably within the astenospheric mantle beneath Hawaii (Frezzotti and Peccerillo, 2007). The fluids or melts have been trapped inside healed cracks developed in pyroxenes, which are the major constituents of the garnet pyroxenite layer. The Salt Lake Crater garnet pyroxenites xenoliths are traditionally interpreted as basaltic cumulates trapped and adiabatically cooled within the Hawaiian lithosphere at T = 1,000–1,150°C and P = 1.6–2.5 GPa (Frey, 1980; Sen, 1988). However, the depth for this cumulate formation (50–80 km), if we consider the suggested P and T conditions, is not consistent with new observations because the minimum pressure required for diamond stabilization is ~4–5 GPa, which corresponds to a depth ~120–150 km. Following the discovery of the microdiamonds in the garnet pyroxenite xenolith in the Salt Lake Crater basalts (Wirth & Rocholl, 2003), it was recently suggested that this garnet pyroxenite represents high-pressure cumulates that were probably formed at pressures of ~5 GPa (Sen et al., 2005).

Another unusual microdiamond was discovered inside an OsIr inclusion from a chromite, a dominant mineral of chromitite pods enveloped in dunite-harzburgite that represents a mantle section of the Luobasa ophiolite formation of Tibet (Yang et al., 2007a). The Luobusa ophiolite, emplaced during closure of the Neo-Tethyan ocean in the Early Tertiary (~65 Ma), lies within the Yarlung-Zangbo suture zone – the geological boundary between Asia and India (Aitchison et al., 2002), and it consists of harzburgite, dunite-containing chromite pods, and sparse lower-crustal cumulates accompanied by minor basaltic pillow lavas and cherts. The ophiolite has been interpreted as a fragment of oceanic lithosphere formed at a mid-oceanic ridge and later modified in a mantle wedge above a subduction zone (Malpas et al., 2003).

Previous reports of diamonds and other ultrahigh-pressure minerals from the Luobusa chromitites (e.g., Bai et al., 2000; Robinson et al., 2004) were controversial because they were not found *in situ*, and because ophiolites at shallow levels above oceanic spreading centers are not an appropriate place for diamond formations. In a recent publication, Yang et al. (2007) have provided convincing evidence of in-situ diamond formation as part of a typical mineralogical assemblage of chromitite ore rich in OsIr and other platinum group elements (PGE). In addition to diamond, elongated crystals of coesite in association with FeTi alloy and a kyanite-bearing silicate assemblage was also reported. Although the minimum pressure required for diamond crystallization is 4 GPa, the coesite occurrence as prismatic domains, each of which is polycrystalline, strongly suggests that they are pseudomorphic after stishovite, thus implying a pressure > 9 GPa (Yang et al., 2007).

The new finds provide unequivocal evidence of UHP phases in the Luobusa chromitites and shed new light on the origin of these enigmatic deposits. To explain the superposition of ophiolitic structures on chromitites containing UHP minerals (Yang et al., 2007; Yamamoto et al., 2008), we envision that the chromitites were components of upwelling mantle beneath a spreading center and that at least parts of them did not melt, preserving their high-pressure "records" in what otherwise appears to be a normal ophiolite. This concept is supported by recent study of $^{187}Os/^{188}Os$ isotopes (Shi et al., 2007) on the Ru-Os-Ir alloys from massive chromitites in the Luobusa ophiolite. The Re-depleted model age is determined as 234 ± 3 Ma, which is older than the magmatic age (ca. 177 Ma) of the Luobasa ophiolite (Zhou et al., 1996). This suggests that the podiform chromitite contains older materials than products of a melt-extraction process under a mid-ocean ridge. Therefore, preservation of older fragments of massive chromite ores within the harzburgite section of ophiolites may not be an unusual phenomenon. Although the mantle peridotite is finally subjected to extensive partial melting under the ridge, chromite is a highly refractory mineral, so some petrological UHP evidence and a geochemical signature within the podiform chromites could have survived the shallow magmatic modification.

One more startling discovery of microdiamond associated with CO_2-rich fluid inclusion in clinopyroxene from a Cenozoic lamprophyre dike, which occurred in a forearc setting on Shikoku Island, Japan, was recently reported by Mizukami et al. (2008). The Raman spectra revealed that the multi-component solid-fluid inclusion pocket includes microdiamond, the gaseous phase of CO_2, Ca-Mg carbonate, probably dolomite, and other hydrous minerals which remain undetermined yet. Studies of associated minerals provide a pressure constraint of 5.5 GPa, which suggests that the newly recognized diamond-bearing rocks rose from depths of around 160 km and that their cooling started from $\sim$1,500°C. The authors argue that the placement of the lamprophyric rocks containing microdiamond into the forearc region close to the subducting plate implies that mantle flow in convergent plate boundaries occurs on a larger scale than previously recognized.

Although the geological situations of all three geodynamic settings that were not previously accepted as suitable places for diamond formation are slightly different from one another, they nevertheless brought to our attention that deep-seated, large-scale mantle flow/convection can reach the shallower levels of the mantle wedge and "deliver" fragments of UHP rocks/minerals into dominating shallower magmatic sources.

Summary Statements

Earth's mantle, including both its upper and lower regions, forms a huge shell starting from $\sim$50–70 km and extended to 2,700 km depth. Deep garnet peridotite xenoliths and superdeep diamonds from kimberlites, as well as fragments of mantle-derived garnet peridotites and eclogites within collisional orogenic belts are very small fragments that provide snapshots of mantle mineralogy, melt, and fluids at depths of >300 to 1,700 km. However, the lack of any geological relationships among such fragments, as well as their small sizes, makes it difficult to evaluate modal heterogeneities of the mantle, although more that 3,500 mantle xenolith localities are now known [see reviews by Nixon (1987)]. Geochemical and mineralogical characteristics collected from the larger-sized (mm-range) host minerals and inclusions in diamonds and xenoliths of garnet peridotites served as a foundation for characterization of the deepest mantle.

The additional valuable information was recently recognized due to studies of nanometric to submicrometric inclusions in kimberlitic diamonds, with the TEM assisted by focused ion beam and synchrotron IR spectroscopy. The studies show that the nanoscale–sub-μm fluid-solid inclusions reflect a composition of the fluid media from which the diamond has grown, whereas the larger crystalline inclusions represent the location in the Earth's mantle where the diamond has grown; the latter provide the pressure and temperature conditions for diamond crystallization. Most of fluid inclusions in kimberlitic diamonds beside those of C, H, and O contain Cl, K, P, and S (Klein-Ben David et al., 2007). Fluid inclusions of similar composition are found with FIB-TEM techniques in microdiamonds from ultrahigh-pressure terranes (e.g., Dobrzhinetskaya et al., 2005–2007). Given that diamonds from kimberlitic and UHPM sources are characterized by very contrasting bulk chemistry of their host rocks (mafic rocks for kimberlitic diamonds and felsic or carbonate rocks for UHPM diamonds), we suggest that the similarity of their fluid inclusions is due to high solubility of Cl, K, P, and S in a C–O–H supercritical fluid at high pressures and temperatures. The presence of Cl, K, P, and S in the deep mantle fluid may be inherited from the intermixture of the deeply subducted fragments of continental crust into the surrounding mantle and subsequent redistribution and dissolution of these components during mantle convection.

Further studies of nanoscale–sub-μm inclusions in kimberlitic diamonds will bring new data for establishing diversity and the compositional network for mantle fluids responsible for diamond formation. Moreover, we strongly believe that the further search for ultradeep minerals in garnet peridotites from UHPM terranes and microdiamonds, and for nanoscale inclusions in superdeep diamonds should be continued with advanced analytical techniques such as TEM-FIB, synchrotron-assisted X-ray, and Raman and IR spectroscopies. One of the more challenging studies with these techniques is an exploration of diamonds from unusual geological sources, such as rootless chromitite deposits within mantle peridotite of the ophiolite series. This research may open a new chapter for plate tectonics similar to that which happened about two decades ago with the discovery of microdiamonds within metasedimentary rocks in continent-continent collisional settings.

Acknowledgement We thank student S. Augustin for preparing TEM foils with FIB at GFZ Potsdam. LFD is thankful to GFZ for her travel grant, and for a friendly and highly intellectual environment during her long-lasting collaborations with the Department of Chemistry of the Earth – (division of Experimental Geochemistry and Mineral Physics). Our thanks to Z. Liu for assistance in synchrotron related studies. IR measurements were performed at the U2A beam line at NSLS of Brookhaven National Laboratory, the U.S. Department of Energy (DOE), Contract DEAC02-98CH10886. The U2A beam line is supported by COMPRES, the Consortium for Materials Properties Research in Earth Sciences under the U.S. NSF Cooperative Agreement Grant (EAR 01-35554) and DOE (CDAC, Contract No. DE-FC03-03N00144). This work is completed under auspices of the International Lithosphere Program, Task Force IV: "Ultradeep subduction of continental crust". Part of the research is supported (to LFD) by the U.S. NSF grants: EAR 0521896.

References

Aitchison, J.C., Abrajevitch, A., Ali, J.R., Badengzhu, D.A.M., Luo, H., Liu, J.B., McDermid, I.R.C., Ziabrev, S., 2002. New insight into the evolution of the Yarlung-Tsangpo suture zone, Xizang (Tibet), China. Episodes 25, 90–94.

Akaogi, M., Akimoto, S., 1977. Pyroxene-garnet solid solution equilibria in the system $Mg_4Si_4O_{12}$–$Mg_3Al_2Si_3O_{12}$ and $Fe_4Si_4O_{12}$–$Fe_3Al_2Si_3O_{12}$ at high pressures and temperatures. Physics of the Earth and Planetary Interiors 15, 90–106.

Bai, W.-J., Robinson, P.T., Fang, Q.-S., Yang, J.-S., Yan, B.-G., Zhang, Z.-M., Hu, X.-F., Zhou, M.-F., Malpas, J., 2000. The PGE and base metal alloys in the podiform chromitites of the Luobusa ophiolite, Southern Tibet. Canadian Mineralogist 38, 585–598.

Basile, D.P., Boylan, R., Baker, B., Hayes, K., Soza, D., 1992. FIBXTEM-Focused ion beam milling for TEM sample preparation. Proceedings of the Materials Research Society: Specimen Preparation for Transmission Electron Microscopy of Materials III, Anderson, R., Tracy, B., Bravman, J. (Eds.), 254: 23–41, Pittsburgh, PA: Materials Research Society.

Bozilov, K.N., Green, H.W., Dobrzhinetskaya, L.F., 1999. Clinoenstatite in the Alpe Arami peridotite: Additional evidence of very high pressure. Science 284, 128–132.

Brenker, F.E., Vollmer, C., Vincze, L., Vekemans, B., Szymanski, A., Janssens, K., Szaloki, I., Nasdala, L., Joswig, W., Kaminsky, F., 2007. Carbonates from the lower part of transition zone or even the lower mantle. Earth and Planetary Science Letters 260, 1–9.

Bundy, F.P., 1989. Pressure temperature phase diagram of elemental carbon. Physica A: Statistical Mechanics and its Applications 158, 169–178.

Carr, G.L., Hanfland, M., Williams, G.P., 1995. The mid-infrared beamline at the NSLS port U2B. Reviews of Scientific Instruments 66, 1643–1648.

Cartigny P., Harris J.W., Javoy M. 2001. Diamond genesis, mantle fractionations and mantle nitrogen content: A study of d13C-N concentrations in diamonds. Earth and Planetary Science Letters 185, 85–98.

Chernov, A.A., 1974. Stability of faceted shapes. Journal of Crystal Growth, 24/25, 11–31.

Chopin, C., 1984. Coesite and pure pyrope in high-grade blueschists of the Western Alps: A first record and some consequences. Contributions to Mineralogy and Petrology, 86, 107–118.

Chopin, C., 2003. Ultrahigh-pressure metamorphism: Tracing continental crust into the mantle. Earth and Planetary Science Letters 212, 1–14.

Collerson, K.D., Hapugoda, S., Kamber, B.S., Williams, Q., 2000. Rocks from the mantle transition zone: Majorite-bearing xenoliths from Malaita, Southwest Pacific. Science 288, 1215 – 1223.

Costa, V.S., Gaspar, J.C., Pimentel, M.M., 2003. Peridotite and eclogite xenolith from the Juina Kimberlite province, Brazil. Eighth International Eclogitic Conference, long abstract no: 0336.

de Corte, K., Cartigny, P., Shatsky, V.S., Sobolev, N.V., Javoy, M., 1998. First evidence of fluid inclusions in metamorphic microdiamonds from the Kokchetav massif, Northern Kazakhstan, Geochimica and Cosmochimica Acta 62, 3765–3777.

Davis, R.M., Griffin, W.L., Pearson, N.J., Andrew, A.S., Doyle, B.J., O'Reilly, S.Y., 1999. Diamonds from the deep: Pipe DO-27, Slave craton, Canada. In: Gurney, J.J., Gurney, J.L., Pascoe, M.D., Richardson, S.H. (Eds.), Proceeding of the VIIth International Kimberlitic Conference, Red Rood Design, Cape Town, pp. 148–155.

de Souza Martins, M., 2006. Geologia dos diamantes e carbonados aluvionares da Bacia do Rio Macaubas (MG). Ph. D. dissertation. Tese de Doutoramento, Universidade Federal de Minas Gerais, Instituto de Geociencias, Belo Horizonte, Brazil (in Spanish), 150 pp.

Dixon, J.E., Clague, D.A., 2001. Volatiles in basaltic glasses from Loihi Seamount, Hawaii: Evidence for a relatively dry plume component. Journal of Petrology, 42, 627–654.

Dobrzhinetskaya, L.F., Eide, E.A., Larsen, R.B., Sturt, B.A., Tronnes, R.G., Smith, D.C., Taylor, W.R., Posukhova, T.V., 1995. Microdiamond in high-grade metamorphic rocks of the Western Gneiss Region, Norway. Geology 23, 597–600.

Dobrzhinetskaya, L.F., Green, H.W., II, Wang, S., 1996. Alpe-Arami: A peridotite massif from depth of more than 300 kilometers. Science 271, 1841–1845.

Dobrzhinetskaya, L.F., Bozhilov, K.N., Green, H.W., 2000. The solubility of TiO_2 in olivine: Implication to the mantle wedge environment. Chemical Geology 163, 325–338.

Dobrzhinetskaya, L.F., Green, H.W., Mitchell, T.E., Dickerson, R.M., 2001. Metamorphic diamonds: Mechanism of growth and inclusion of oxides. Geology 29, 263–266.

Dobrzhinetskaya, L.F., Green, H.W., Weshler, M., Darus, M., 2002. Focused ion beam technique- a new avenue for transmission electron microscope studies of micro/nanoscale geological materials. Abstract, International Workshop, Geophysics, Structure and Geology of UHPM terranes. Beijin, China, Institute of Geology and Geophysics, Chinese Academy of Sciences, pp. 88–89.

Dobrzhinetskaya, L.F., Green, H.W., Bozhilov, K.N., Mitchell, T.E., Dickerson R.M., 2003a. Crystallization environment of Kazakhstan microdiamonds: Evidence from their nanometric inclusions and mineral associations. Journal of Metamorphic Geology 21, 425–437.

Dobrzhinetskaya, L.F., Green, H.W., Weschler, M., Darus, M., Wang, Y.-C., Massonne, H.-J., Stöckhert, B., 2003b. Focused ion beam technique and transmission electron microscope studies of microdiamonds from the Saxonian Erzgebirge, Germany. Earth and Planetary Science Letters 210, 399–410.

Dobrzhinetskaya, L.F., Green II, H.W., Renfro, A.P., Bozhilov, K.N., Spengler, D., van Roermund, H.L.M., 2004. Precipitation of pyroxenes and olivine from majoritic garnet: Simulation of peridotite exhumation from great depth. Terra Nova 16, 325–330.

Dobrzhinetskaya, L.F., Green II, H.W., Renfro, A.P., Bozhilov, K.N., 2005a. Decompression of majoritic garnet: Experimental investigation of the mantle peridotite exhumation. In: Chen, J., Wang, Y., Duffy, T.S., Shen, G., Dobrzhinetskaya, L.F. (Eds.), Advances in High-Pressure Technology for Geophysical Applications. Elsevier, pp. 265–287.

Dobrzhinetskaya, L.F., Wirth, R. Green, H.W. II., 2005b. Direct observation and analysis of a trapped COH fluid growth medium in metamorphic diamond. Terra Nova 17, 472–477.

Dobrzhinetskaya, L.F., Wirth, R., Green, H.W. II., 2006a. Nanometric inclusions of carbonates in Kokchetav diamonds from Kazakhstan: A new constraint for the depth of metamorphic diamond crystallization. Earth and Planetary Science Letters 243, 85–93.

Dobrzhinetskaya, L.F. Liu, Z, Cartigny, P., Zhang, J., Tchkhetia, N.N., Green H.W. II, Hemley R.J., 2006b. Synchrotron infrared and Raman spectroscopy of microdiamonds from Erzgebirge, Germany. Earth and Planetary Science Letters 248, 325–334.

Dobrzhinetskaya, L.F., Green, H.W., 2007a. Diamond synthesis from graphite in presence of water and SiO_2: Implications for diamond formation in quartzites from Kazakhstan. Internationa. Geology Review 49, 389–400.

Dobrzhinetskaya, L.F., Green, H.W., 2007b. Experimental studies of mineralogical assemblages of metasedimentary rocks at Earth's mantle transition zone conditions. Journal of Metamorphic Geology 25, 83–96.

Dobrzhinetskaya, L.F., Wirth, R., Green, H.W., 2007. A look inside of diamond-forming media in deep subduction zones. Proceedings of National Academy Sciences of the United States of America 104, 9128–9132.

Ernst, W.G., Maruyama, S., Wallis, S., 1997. Buoyancy-driven, rapid exhumation of ultrahigh-pressure metamorphosed continental? crust. Proceedings of National Academies of Science 94, 9532–9537.

Ernst, W.G., 2006. Preservation/exhumation of ultrahigh-pressure subduction complexes. Lithos 92, 321–335.

Ernst, G., Liou, J.G, 2008. High- and ultrahigh-pressure metamorphism: Past results and future prospects. American Mineralogist 93, 1771–1786.

Frezotti, M.-L., Peccerillo, A., 2007. Diamond-bearing COHS fluids in the mantle beneath Hawaii. Earth and Planetary Science Letters 262, 273–283.

Frey, F., 1980. The origin of pyroxenites and garnet pyroxenites from Salt Lake Crater, Ohahu, Hawaii. American Journal of Science 280, 427–449.

Gasparik, T., 2003. Phase Diagrams for Geoscientists. An Atlas of the Earth's Interior. Berlin, Heidelberg, New York: Springer-Verlag. 462 pp.

Gerya, T.V., Stöckhert, B., Perchuk, A.L., 2002. Exhumation of high-pressure metamorphic rocks in a subduction channel – a numerical simulation. Tectonics 21, 6–1 - 6–19.

Gerya T.V., Perchuk L.L., Burg J.-P., 2008. Transient hot channels: Perpetrating and regurgitating ultrahigh-pressure, high-temperature crust–mantle associations in collision belts. Lithos 103, 236–256.

Giannuzzi, L.A., Prenitzer, B.I., Kempshall, B.W., 2005. Ion-solid interactions. In Giannuzzi L.A., Stevie F.A. (Eds.), Introduction to Focused Ion Beams: Instrumentation, Theory, Techniques and Practice. Springer, pp. 13–52.

Graham, G.A., Grant, P.G., Chater, R.J., Westphal, A.J., Kearsley, A.T., Snead, C., Domínguez, G., Butterworth, A.L., MCphail, D.S., Bench, G., Bradley, J.P., 2004. Investigation of ion beam techniques for the analysis and exposure of particles encapsulated by silica aerogel: Applicability for stardust. Meteoritics and Planetary Science 39, 1461–1473.

Haggerty, S.E., Sautter, V., 1990. Ultradeep (greater than 300 km) ultramaphic upper mantle xenoliths. Science 248, 993–996.

Harris, J.W., Hutchison, M.T., Hursthouse, M., Light, M., Harte, B., 1997. A new tetragonal silicate mineral occurring as inclusions in lower mantle diamonds. Nature 387, 486–488.

Harte, B., Harris, J.W., 1994. Lower mantle mineral assemblages preserved in diamonds. Mineralogical Magazine 58A, 384–385.

Harte, B., Harris, J.W., Hutchison, M.T., Watt, J.R., Willding, M.S., 1999. Lower mantle mineral associations in diamonds from Sao Luiz, Brazil. In: Fei, Y., Bertka, C.M., Mysen B.O (Eds.), Mantle petrology: Field observations and high pressure experimentation: a tribute to Fransis R. (Joe) Boyd. Geochemical Society, Special Publication 16. pp. 125–163.

Harte, B., Cayzer, N., 2007. Decompression and unmixing of crystals included in diamonds from the mantle transition zone. Physics and Chemistry of Minerals 34, 647–656.

Hayman, P.C., Kopylova, M.G., Kaminsky, F.V., 2005. Lower mantle diamonds from Rio Soriso (Juina area, Mato Grosso, Brazil). Contribution to Mineralogy and Petrology 149, 430–445.

Heaman, L.M., Bruce A. Kjarsgaard, B.A., Creaser, R.A., 2004. The temporal evolution of North American kimberlites. Lithos 76, 377–379.

Heaney, P.J., Vicenzi, E.P., Giannuzzi, L.A., Livi, K.J.T., 2001. Focused ion beam milling: A method of site-specific sample extraction for microanalysis of Earth and planetary materials. American Mineralogist 86, 1094–1099.

Heide K., Woermann E., Ulmer, G., 2008. Volatiles in pillows of the Mid-Ocean-Ridge Basalt (MORB) and vitreous basaltic rims. Chemie der Erde-Geochemistry 68, 3537–368.

Holman, H.-Y.N., Bjornstad, A., McNamara, P., Martin, C., McKinney, R., Blakely, A., 2002. Synchrotron infrared spectromicroscopy as a novel bioanalytical microprobe for individual living cells: Cytotoxicity considerations. Journal of Biomedical Optics 7, 417–424.

Huttchison, M.T., Hursthouse, M.B., Light, M.E., 2001. Mineral inclusions in diamonds: Associations and chemical distinctions around the 670 discontinuity. Contribution to Mineralogy and Petrology 142, 372–382.

Hwang, S.L., Shen, P., Chu, H.T., Yui, T.F., Lin, C.C., 2001. Genesis of microdiamonds from melt and associated multiphase inclusions in garnet of ultrahigh-pressure gneiss from Erzgebirge, Germany. Earth Planetary Science Letter 188, 915–919.

Irifune, T., 1997. An experimental investigation of the pyroxene-garnet transformation in a pyrolite composition and its bearing on the constitution of the mantle. Physics of the Earth and Planetary Interiors 45, 324–336.

Jones, R., Briddon, P.R. & Oberg, S., 1992. First-principles theory of nitrogen aggregates in diamond. Philosophical Magazine Letters 66, 67–74.

Kaminsky, F.V., Zakharchenko O.D., Davies R., Griffin W.L., Shiryaev A.A., 2001. Superdeep diamonds from the Juina area, Mato Grosso State, Brazil, Contribution to Mineralogy and Petrology140, 734–753.

Kanda, H., 1985. Effect of metal and water on growth and dissolution morphologies of diamond crystals [thesis]. Sendai, Japan, Tohoku University, 137 p.

Klein-BenDavid, O., Wirth, R., Navon, O., 2006. TEM imaging and analysis of microinclusions in diamonds; a close look at diamond-growing fluids. American Mineralogist 91, 353–365.

Klein-BenDavid, O., Wirth, R., Navon, O., 2007. Micrometer-scale cavities in fibrous and cloudy diamonds - a glance into diamond dissolution events. Earth and Planetary Science Letters 264, 89–103.

Kvasnytsya, V.M., Taran, M.M, Wirth, R., Wiedenbeck, M., Thomas, R., Lupashko, T.M., Il`chenko, K.O., 2005. New data about Ukrainian diamonds. Mineralogical Journal (Ukraine) 27, 41–47.

Kvasnytsya, V., Wirth, R., Thomas, R., Taran, M., 2006. Ukrainian Geologist, 2, 11–16.

Lee, M.R., Bland, P.A., Graham, G., 2003. Preparation of TEM samples by focused ion beam (FIB) techniques: Applications to the study of clays and phyllosilicates in meteorites. Mineralogical Magazine 67, 581–592.

Liou, J.G., Zhang, R.Y., Ernst, G.W., 2007. Very high-pressure orogenic garnet peridotites. Proceeding of National Academy of Sciences of the United States of America 104, 9116–9121.

Liu, Z., Hu, J., Yang H., Mao, H.K., Hemley, R.J., 2002. High-pressure synchrotron X-ray diffraction and infrared microspectroscopy: Applications to dense hydrous phases. Journal of Physics of Condensed Matters 14, 10641–10646.

Liu, L., Zhang, J., Green, H.W., Jin, Z-M, Bozhilov, K.N., 2007. Evidence of former stishovite in metamorphosed sediments, implying subduction to >350 km. Earth and Planetary Science Letters 263, 180–191.

Logvinova, A.M., Wirth, R., Sobolev, N.V., 2006. Nanometre-sized mineral and fluid inclusions in cloudy Siberian diamonds: New insights on diamond formation. Abstract IMA, Kobe, Japan, p. 134.

Logvinova, A.M., Wirth, R., Sobolev, N.V., Seryotkin, Y.V., Yefimova, E.S., Floss, C.,Taylor, L.A., 2008. Eskolaite associated with diamond from the Udachnaya kimberlite pipe, Yakutia, Russia. American Mineralogist, 93, 685–690.

Logvinova, A.M., Wirth, R., Fedorova, E.N., Sobolev, N.V., 2008a. Nanometre-sized mineral and fluid inclusions in cloudy Siberian diamonds: new insights on diamond formation. European Journal of Mineralogy 20, 317–331.

Malpas, J., Zhou, M.-F., Robinson, P.T., Reynolds, P.H., 2003, Geochemical and geochronological constraints on the origin and emplacement of the Yarlung-Zangbo ophiolites, Southern Tibet. In: Dilek, Y., Robinson, P.T. (Eds.), Ophiolites in Earth History, Geological Society of London Special Publication 218, 191–206.

Massonne, H.-J., 1999. A new occurrence of microdiamonds in quartzofeldspathic rocks of the Saxonian Erzgebirge, Germany, and the metamorphic evolution. In: Gurney, J.J., Gurney, J.L., Pascoe, M.D., Richardson, S.H. (Eds.), The P.H. Nixon Volume. Proceddings of 7th International Kimberlitic Conference, Red Roof Design CC, Capetown, pp. 533–539.

Massonne, H.-J., Bautsch, H.-J., 2002. An Unusual Garnet Pyroxenite from the Granulitgebirge, Germany: Origin in the Transition Zone (>400 km depths) or in a Shallower Upper Mantle Region? International Geology Review 44, 779–796.

McCammon, C., 2001. Geophysics – deep diamond mysteries. Sciences 293, 813–814.

Mposkos, E.D., Kostopoulos, D.K., 2001. Diamond, former coesite and supersilicic garnet in metasedimentary rocks from the Greek Rhodope: A new ultrahigh-pressure metamorphic province established. Earth and Planetary Science Letters 192, 497–506.

Menneken, M., Nemchin, A.A., Geisler, T., Pidgeon, R.T., Wilde, S.A., 2007. Hadean diamonds in zircon from Jack Hills, Western Australia. Nature 448, 917–920.

Mizukami, T., Wallis, S., Enami, M., Kagi, H., 2008. Forearc diamond from Japan. Geology 36, 219–222.

Neal, C.R., Haggerty, S.E., Sutter, V., 2001. "Majorite" and "silicate perovskite" mineral compositions in xenoliths from Malaita. Science 292, 1015.

Nixon, P.H., 1987. Mantle xenoliths. Chichester: John Wiley, 844 pp.

Obst, M., Gasser, P., Mavrocordatos, M., Dittrich, D., 2005. TEM-specimen preparation of cell/mineral interfaces by Focused Ion Beam milling. American Mineralogist 90, 1270–1277.

O'Keefe, M.A., Nelson, E.C., Wang, Y.C., Thust, A., 2001. Sub-Ångstrom resolution of atomistic structures below 0.8 Å. Philosophical Magazine B 81(11), 1861–1878.

Ogasawara, Y., 2005. Microdiamonds in ultrahigh-pressure metamorphic rocks. Element 1, 1–6.

Ogasawara, Y., Fukasawa, K., Maruyama, S., 2002. Coesite exsolution from supersilicic titanite in UHP marble from the Kokchetav Massif, northern Kazakhstan. American Mineralogist 87, 454–461.

Orloff, J., Utlaut, M. Swanson, L., 2003. High resolution Focused Ion Beams: FIB and its Applications. NewYork: Kluwer Academic/Plenum Publishers, 300 pp.

Overwijk, M.H.E., van den Heuvel, F.C., Bulle-Lieuwma, C.W.T., 1993. Novel scheme for the preparation of transmission electron microscopy specimens with a focused ion beam. Journal of Vacuum Science and Technology 11, 2021–2024.

Phaneuf, M.W., 1999. Applications of focused ion beam microscopy to materials science specimens. Micron 30, 277–288.

Pal'yanov, Y., Sokol, A., Borzdov, Y., Khokhryakov, A., Sobolev, N., 1999. Diamond formation from mantle carbonate fluids. Nature 400, 417–418.

Parkinson I.J., Arculus R.J., 1999. The redox state of subduction zones: Insight from arc peridotites. Chemical Geology 160, 409–423.

Pearson, D.G., Canil, D., Shirey, S.B., 2005. Mantle samples included in volcanic rocks: Xenoliths and diamonds. In: Carlson, R.W. (Ed.), The Mantle and Core. Elsewier, pp. 171–275.

Perraki, M., Proyer, A., Mposkos, E., Kaindl, R., Hoinkes, G., 2006. Raman micro-spectroscopy on diamond, graphite and other carbon polymorphs from the ultrahigh-pressure metamorphic Kimi Complex of the Rhodope Metamorphic Province, NE Greece. Earth and Planetary Science Letters 241, 672–685.

Puretz, J., Orloff, J., Swanson, L., 1984. Application of focused ion beams to electron beam testing of integrated circuits. Proceedings of the International Society for Optical Engineering 471, 38–46.

Reffner, J.A., Martoglio, P.A., Williams, G.P., 1995. Fourier transform infrared microscopical analysis with synchrotron radiation - the microscope optics and system performance. Reviews of Scientific Instruments 66, 1298–1302.

Ringwood, A.E., 1991. Phase transformations and their bearing on the constitution and dynamics of the mantle. Geochimica and Cosmochimica Acta 55, 2083–2110.

Risold, A.Ch., Trommsdorff, W., Grobety, B., 2003. Morphology of oriented ilmenite inclusions in olivine from garnet peridotites (Central Alps, Switzerland). European Journal of Mineralogy 15, 289-294.

Robinson, P.T., Bai, W.-J., Malpas, J., Yang, J.-S., Zhou, M.-F., Fang, Q.-S., Hu, X-F., Cameron, S., 2004, Ultrahigh pressure minerals in the Luobusa ophiolite, Tibet and their tectonic implications. In: Tectonics of China, Geological Society of London Special Publication 226, 247–271.

Sautter, V., Haggerty, S.E., Field S., 1991. Ultradeep (>300kilometers) ultramaphic xenoliths: Petrological evidence from the mantle transition zone. Science 252, 827–830.

Scott-Smith, B.H., Danchin, T.V., Harris, J.W., Stracke K.J., 1984. Kimberlites near Orroroo, South Australia. In: Kornprobst, J. (Ed.), Kimberlites: The Mantle and Crust-Mantle Relationships. Elsevier, Amsterdam, pp 121–142.

Sen, G., 1988. Petrogenesis of spinel lherzolites and pyroxenite suite xenoliths from the Koolau shield, Oahu, Hawaii: Implications for petrology of post-eruptive lithosphere beneath Oahu. Contribution to Mineralogy and Petrology 100, 61–91.

Sen, G., Keshav, S., Bizimis, M., 2005. Hawaiian mantle xenoliths and magmas: Composition and thermal character of the lithosphere. American Mineralogist 90, 871 – 887.

Shi, R., Alard, O., Zhi, X., O'Reilly, S.Y., Pearson, N.J., Griffin, W.L., Zhang, M., Chen, X., 2007. Multiple events in the Neo-Tethyan oceanic upper mantle: Evidence from Ru–Os–Ir alloys in the Luobusa and Dongqiao ophiolitic podiform chromitites, Tibet. Earth and Planetary Science Letters 261, 33–48.

Smith, D., 1984. Coesite in clinopyroxene in the Caledonides and its implications for geodynamics. Nature 310, 641–644.

Smith, N.S., Kinion, D.E., Tesch, P.P., Boswell R.W., 2007. A high brightness plasma source for focused ion beam applications. Microscopy and Microanalysis 13, 180–181.

Smith, J.V., Mason, B., 1970. Pyroxene-garnet transformation in Coorara meteorite. Science 168, 832–833.

Sobolev, N.V., Shatsky V.S., 1990. Diamond inclusions in garnets from metamorphic rocks: A new environment of diamond formation. Nature 343, 742–746.

Spengler, D., van Roermund, H.L.M., Drury, M., Ottolini, L., Maso, P.R.D., Davies, G., 2006. Deep origin and hot melting of an Archaean orogenic peridotite massif in Norway. Nature 440, 913–917.

Stachel, T., Brey, G.P., Harris, J.W., 2005. Inclusions in sublithospheric diamonds: Glimpses of deep Earth. Elements 1, 73–78.

Stöckhert, B., Duyster, J., Trepmann, C., Massonne, H.-J., 2001. Microdiamond daughter crystals precipitated from supercritical COH+silicate fluids included in garnet, Erzgebirge, Germany. Geology 29, 391–394.

Stöckhert, B., Gerya, T., 2005. Pre-collisional high pressure metamorphism and nappe tectonics at active continental margins: A numerical simulation. Terra Nova 17, 102–110.

Sunagawa, I., 1990. Growth and morphology of diamond crystals under stable and metastable conditions. Journal of Crystals Growth 99, 1156–1161.

Sunagawa, I., 1997. Natural crystallization. Journal of Crystal Growth 42, 214–223.

Tinkler D., Lesher, C.E., 2001. Solubility of TiO_2 in olivine from 1 to 8 GPa. American Geophysical Union, Fall Meeting 2001, San Francisco, abstract #V51B–1001.

van Roermund, H.L.M., Drury, M.R., 1998. Ultra-high pressure ($P > 6$ GPa) garnet peridotites in Western Norway: Exhumation of mantle rocks from >185 km depth. Terra Nova 10, 295–301.

van Roermund, H.L.M., Carswell, D.A., Drury, M.R., Heijboer, T.C., 2002. Microdiamonds in megacrystic garnet websterite pod from Bardane on the island of Fjortoft, western Norway: Evidence for diamond formation in mantle rocks during deep continental subduction. Geology 30, 959–962.

Vicenzi, E.P., Heaney, P.J., 1999. Examining Martian alteration products using in situ TEM sectioning: A novel application of the Focused Ion Beam (FIB) for the study of extraterrestrial materials. 30th Annual Lunar and Planetary Science Conference, March 15–29, 1999, Houston, TX, abstract no. 2005.

Williams, D.B., Carter, C.B., 1996. Transmission Electron Microscopy for Materials Science. New York and London: Plenum Press, 729 pp.

Wirth, R., Rocholl, A., 2003. Nanocrystalline diamond from the Earth's mantle underneath Hawaii. Earth and Planetary Science Letters 211(3–4), 357–369.

Wirth, R., 2004. Focused Ion Beam (FIB>): A novel technology for advanced application of micro- and nanoanalysis in geosciences and applied mineralogy. European Journal of Mineralogy 16, 863–876.

Wirth, R., 2005. Micro- and nanometre sized inclusions in diamonds and microdiamonds: A new source of information about diamond genesis and fluid composition. ANAIS IV Simposio Brasiliero de Geologia do Diamante, Sociedade Brasileira de Geologia Nucleo Minas Gerais, Boletim N°14, Diamantina, 4–7. Sept. 2005, p. 184.

Wirth, R., 2006. On the origin of carbonado, a black diamond variety. Abstract, IMC 16 Sapporo, Proceedings 3, p. 1877.

Wirth, R., 2007. Nanometre-sized inclusions in diamond: FIB/TEM investigations provide new insights into diamond genesis and fluid composition. Abstract, Geological Association of Canada and Mineralogical Association of Canada Yellowknife, Canada, 2007. p. 87.

Wirth, R., Langer, K., Platonov, A.N., 2001. A TEM study of a chromium-bearing kyanite from a mantle xenolith: indications for an aluminium-rich exsolution precursor phase? European Journal of Mineralogy 13, 311–318.

Wirth, R., 2009. Focused Ion Beam (FIB) combined with SEM and TEM: Advanced analytical tools for studies of chemical composition, microstructure and crystal structure in geomaterials on a nanometre scale. Chemical Geology 261, 217–229.

Wirth, R., Pinti, D.L., Sano, Y., Takahata, N., Kaminsky, F., 2007. FIB/TEM and NanoSIMS investigations of micro- and nanoinclusions in diamond: New insights into the Earth's mantle and diamond formation. Abstract, AGU Fall Meeting, San Francisco. EOS, Transactions, American Geophysical Union, Supplements, 88 (52) 10–14 December, 2007. Abstract MR23D–10.

Wirth, R., Wunder, B., 2000. Characterization of OH-containing phases by TEM using electron; energy-loss spectroscopy (EELS): Clinohumite-OH, chondrodite-OH, phase A, and the (F OH)-solid solution series of topaz. Journal of Trace and Microprobe Techniques 18, 35–49.

Wood, B.J., Bryndzia, I.T., Johnson, K.E., 1990. Mantle oxidation state and its relationship to techtonic environment and fluid speciation. Science 248, 337–345.

Wyllie, P.J., 1981. Plate tectonics and magma genesis. Geology Rundschau 70, 128–153.

Xu, S.T., Okay, A.I., Ji, S.Y., Sengor, A.M.C., Wen, S., Liu, Y.C., Jiang, L.L., 1992. Diamond from the Dabie-Shan metamorphic rocks and its implication for tectonic setting. Science 256, 80–82.

Yamamoto, S. Komiya, T, Hirose, K., Maruyama, S., 2008. Coesite and clinopyroxene exsolution lamellae in chromites: In-situ ultrahigh-pressure evidence from podiform chromitites in the Luobusa ophiolite, southern Tibet. Lithos. (in press).

Yang, J-J., Jahn, B-M., 2000. Deep subduction of mantle-derived garnet peridotites from the Su-Lu UHPM terrane in China. Journal of Metamorphic Geology 18, 167–180.

Yang, J., Xu, Z., Dobrzhinetskaya, L.F., Green H.W. II, Pei, X., Shi, R., Wu, C., Wooden, J.L., Zhang, J., Wan, Y., Li, H., 2003. Discovery of metamorphic diamonds in Central China: An indication of a >4000 km-long-zone of deep subduction resulting from multiple continental collisions. Terra Nova 15, 370–379.

Yang, J., Dobrzhinetskaya, L.F., Bai, W.-J., Fang, Q-S., Robinson, P., Zhang, J-F., Green, H.W., 2007a. Diamond- and coesite-beraing chromitites from the Luobasa ophiolite, Tibet. Geology 35, 875–878.

Yang, J.S., Zhang, R.Y., Li,, T.F., Zhang, Z., Liou, J.G., 2007b. Petrogenesis of the garnet peridotite and garnet-free peridotite of the Zhimafang ultramafic body in the Sulu ultrahigh pressure metamorphic belt, eastern China. Journal of Metamorphic Geology 25, 187–206.

Ye, K., Cong, B., De, E., 2000. The possible subduction of continental material to depth greater than 200 km. Nature 407, 734–736.

Young, R.J., Kirk, E.C., Williams, D.A., Ahmed, H., 1990. Fabrication of planar cross sectional TEM specimens using a focused ion beam. In: Anderson, R. (Ed.), Proceedings of the Materials Research Society: Specimen Preparation for Transmission Electron Microscopy of Materials II 199, 205-216, Pittsburgh, PA: Materials Research Society.

Zhang, R.Y., Liou, J.G., Cong, B.L., 1994. Petrogenesis of garnet-bearing ultramafic rocks and associated eclogites in the Su-Lu ultrahigh-P metamorphic terrane, eastern China. Journal of Metamorphic Geology 12, 169–186.

Zhang, R.Y., Liou, J.G, Yang, J.S., Yui, T.F., 2000. Petrochemical constraints for dual origin of garnet peridotites from the Dabie-Sulu UHP terrane, eastern-central China. Journal of Metamorphic Geology 18, 149–166.

Zhang, R.Y., Shau Y.H., Liou, J.G., Lo, C.H., 2002, Discovery of clinoenstatite in garnet pyroxenites from the Dabie-Sulu UHP terrane, east-central China. American Mineralogist 87, 867–874.

Zhang, R.Y., Liou, J.G., 2003. Clinopyroxenite from the Sulu ultrahigh-pressure terrane, eastern China: Origin and evolution of garnet exsolution in clinopyroxene. American Mineralogist 21, 1591–1600.

Zhou, M., Robinson, P.T., Malpas, J., Li, Z. 1996. Podiform chromitites from the Luobusa ophiolites (southern Tibet): Implications for melt rock interaction and chromite segregation. Journal of Petrology 37, 3–21.

New Views of the Earth's Inner Core from Computational Mineral Physics

Lidunka Vočadlo

Abstract Although one third of the mass of our planet resides in its metallic core (divided into a molten outer part and a solid inner part), fundamental properties such as its chemical composition and internal structure remain poorly known. Although it is well established that the inner core consists of iron with some alloying lighter element(s), the crystal structure of the iron and the nature and concentrations of the light element(s) involved remain controversial. Seismologists, by studying the propagation characteristics of primary earthquake waves (P-waves), have shown that the inner core is anisotropic and layered, but the origins of these properties are not understood. Seismically observed shear waves (S-waves) add to the complexity as they show unexpectedly low propagation velocities through the inner core.

Interpretation of these seismic observations is hampered by our lack of knowledge of the physical properties of core phases at core conditions. In addition, the accuracy of derived inner core seismic properties is limited by the need to de-convolve inner core observations from seismic structure elsewhere in the Earth. This is particularly relevant in the case of shear waves where detection is far from straightforward. A combination of well-constrained seismological data and accurate high-pressure, high-temperature elastic properties of candidate core materials would allow for a full determination of the structure and composition of the inner core - an essential prerequisite to understanding Earth's differentiation and evolution.

Unfortunately, the extreme conditions of pressure (up to 360 GPa or 3.6 million times atmospheric pressure) and temperature (up to 6000 K) required make results from laboratory experiments unavoidably inconclusive at present. An alternative and complementary approach, that has only recently become available, is computational mineral physics, which uses computer simulations of materials at inner core conditions. Ab initio molecular dynamics simulations have been used to determine the stable phase(s) of iron in the Earth's core and to calculate the elasticity of iron and iron alloys at core conditions. Calculated S-wave velocities are significantly higher than those inferred from seismology. If the seismological observations are robust, a possible explanation for this discrepancy is that the inner core contains a significant amount of melt (possibly >10%). The observed anisotropy can only be explained by almost total alignment of inner core crystals.

Keywords Earth's inner core · Mineral physics · Ab initio calculations · Iron and iron alloys · Elastic properties

Introduction

The Earth's deep interior is inaccessible beyond samples from < 670 km depth that are brought to the surface in diamond inclusions (Collerson et al., 2000). Other than constraints on the density provided by measurements of the moment of inertia, the only direct observations that we have of the properties of the Earth's deep interior are from seismology.

L. Vočadlo (✉)
Department of Earth Sciences, UCL, London, WC1E 6BT, UK
e-mail: l.vocadlo@ucl.ac.uk

S. Cloetingh, J. Negendank (eds.), *New Frontiers in Integrated Solid Earth Sciences*, International Year of Planet Earth, DOI 10.1007/978-90-481-2737-5_12, © Springer Science+Business Media B.V. 2010

The velocity of seismic waves (both primary or P waves which are longitudinal, and shear or S waves that are transverse) depends on the physical properties (elasticity, density, temperature, pressure) and chemical composition of the minerals through which the waves propagate. Seismic wave velocities therefore give some indication as to the mineralogy, chemical composition and large-scale layered structure of the interior of our planet (see Fig. 1), but these data may only be interpreted with information from mineral physics. Experimentally or theoretically determined elastic properties of candidate materials lead to predicted wave velocities which can then be compared with those observed from seismology. Discontinuities in the observed seismic wave velocities with depth indicate major changes in mineralogy and/or chemistry, for example, the isochemical silicate transitions of olivine to wadsleyite at the upper mantle/transition zone boundary (410 km depth) (Ringwood, 1975), and the spinel to perovskite+ferropericlase transition at the transition zone/lower mantle boundary (660 km depth) (Liu, 1979). At the core-mantle boundary (approximately 2900 km depth), there is a dramatic increase in density as we move from the silicate mantle (density approximately 5500 kg m^{-3}) to the iron-rich core (11000 kg m^{-3}) (Oldham, 1906). Advances in seismological techniques allow the deep Earth to be sampled in increasing detail. Lateral variations in observed seismic wave velocities at a constant depth suggest that chemical and/or thermal heterogeneities are present in the deep Earth. Not surprisingly, improved techniques have resulted in more unanswered questions: for example, to what extent are elements unevenly distributed between the different mineral phases thought to be present in the Earth's deep mantle? What is the temperature profile through the deep Earth? What is the composition and structure of the core? While these questions have attracted considerable attention in recent years, it is the latter problem concerning the Earth's core that this paper seeks to address, through a review of recent ab initio calculations on iron and iron alloys at the pressure and temperature conditions of the Earth's core.

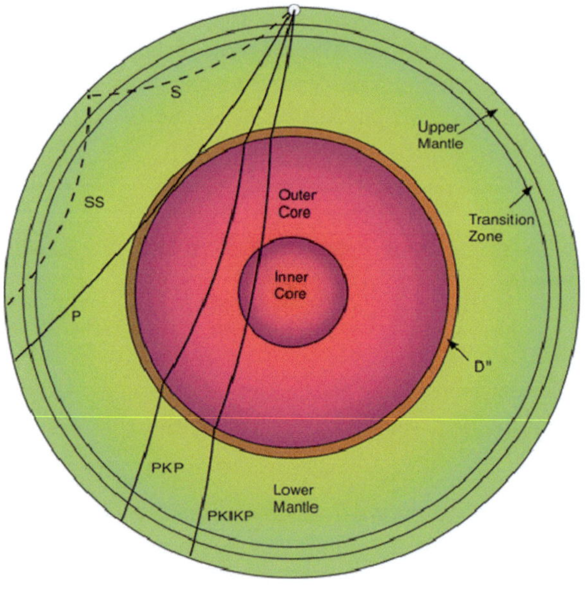

Fig. 1 Schematic illustration of the interior of the Earth showing the major internal boundaries and layers. Also shown are representative seismic waves that sample different parts of the planet. Taken from Brodholt and Vočadlo (2006)

The chemical composition of the Earth's core is not very well known. Cosmochemical and geochemical arguments (e.g., McDonough and Sun, 1995) suggest that the core is primarily an iron alloy with as much as 5 – 10 wt% Ni. From a compilation of density/sound-wave velocity systematics for a range of metallic compounds, Birch (1964) concluded that the iron in the core must be alloyed with a small fraction (up to 10 wt%) of lighter elements (such as S, O, Si, H and C). These light elements are those that have been suggested as being present in the Earth's core in relatively significant (>1%) amounts; in addition, very small amounts (fractions of a wt% to parts per million) of other iron-loving or siderophile elements (Cr, Mn and Co) and elements such as P and K (that are traditionally thought to prefer silicates over metals) could also be present due to increased siderophile behaviour with increasing pressure and temperature (McDonough and Sun, 1995; Poirier, 1994; Rama Murthy et al., 2003).

From seismology, it is known that the density jump across the boundary between the liquid outer core and solid inner core is between ~4.5 – 6.7% (Shearer and Masters, 1990; Masters and Gubbins, 2003). As a result, the density difference between pure iron and the Earth's inner core is much smaller than is the case for the outer core. This has led to the suggestion that the concentration of light elements in the inner core is significantly lower (maximum 3 wt%) than in the outer core (up to 10 wt%). Our present understanding is that the Earth's solid inner core is crystallising from the outer core as the Earth slowly cools, with nickel and the light elements continuously partitioning between the solid and liquid. More light element partitions into the outer core, which then becomes chemically buoyant giving rise to compositionally driven convection; this is the circulation that helps to drive the geodynamo, and hence the magnetic field. Such is the importance of the light element contribution to core evolution that several decades of work have addressed the nature of the light component(s), attempting to constrain core composition by matching seismic data with results from experiments and simulations. It is worth noting that, in contrast, relatively few studies have addressed the effect of nickel on the properties of iron at Earth's core conditions.

Observations tell us that the inner core is seismically anisotropic, with P wave velocities ~3% faster along the polar axis than in the equatorial plane (Creager, 1992; Song and Helmberger, 1993; Song, 1997; Song and Xu, 2002; Sun and Song, 2002; Beghein and Trampert, 2003; Oreshin and Vinnik, 2004). This would suggest that some of the crystals that make up the inner core have aligned in such a way that, overall, the seismically fast direction through the crystal aggregate is oriented along the polar axis of the Earth. Recent seismic observations suggest that it is also layered. The data require the presence of a seismically isotropic or weakly anisotropic outer layer, with lateral variations in thickness of between ~100 – 400 km, overlaying an irregular, non-spherical transition region to an anisotropic inner layer (Song and Helmberger, 1998; Ouzounis and Creager, 2001; Song and Xu, 2002; Ishii and Dziewonski, 2003; Sun and Song, 2008). The existence of an isotropic outer layer implies that the magnitude of the seismic anisotropy in the lower inner core should be significantly greater than previously thought, possibly as much as 5 – 10% (Ouzounis and Creager, 2001). The observed layering also implies that the upper and lower inner core could be compositionally or structurally different.

Correct interpretation of these seismic observations requires knowledge of the physical properties of core materials. These can be obtained through high pressure/high temperature laboratory measurements or computer simulation. Experimentalists have made great strides in recent years, attaining ever higher pressures and temperatures in the laboratory in order to recreate the conditions in the Earth's deep interior. Reaching the conditions of the Earth's core (135 GPa $< P <$ 360 GPa, 4000 $< T <$ 6000K) remains experimentally challenging. Compression devices such as diamond-anvil-cells (in which samples are squashed between the tiny tips of two opposing gem-quality diamonds) can reach ~300 GPa and, when combined with laser heating, they can reproduce conditions in the molten outer core, although not without considerable difficulty (Kuwayama et al., 2008). Shock apparatus (where simultaneous high pressure and temperature are achieved by sending a shock wave through a sample) can achieve even higher pressures and temperatures, although with greater uncertainties in the derivation of experimental conditions (Nguyen and Holmes, 2004).

An alternative (and complementary) approach to laboratory experiments is computational mineral physics. The advantage of computer simulations is that the properties of materials can be calculated at any desired pressure and temperature, and therefore the entire Earth is accessible at the touch of a button. Studies where results from simulations are combined with those from experiments (even if the latter were performed only at lower pressures and temperatures) are extremely useful as they are often used to confirm the validity of the simulation methodology (e.g., Vočadlo, 2007, Vočadlo et al., 1997; Vočadlo and Alfe 2002; 2003b).

The purpose of this paper is to show how ab initio simulation techniques can be used to improve our understanding of the Earth's inner core. After briefly outlining ab initio simulation methods, including the techniques to simulate high pressures and temperatures, results are presented for the ab initio simulation of both solid and liquid iron to illustrate how these simulations can provide much needed constraints on important geophysical quantities.

Computational Methods

Ab Initio *Techniques*

The aim of ab initio simulations is to calculate the total energy of a system (where a system is defined as a material at a particular temperature and pressure). From the total energy, and its variation as a function of pressure and temperature, other relevant physical properties such as seismic wave velocifies can be determined. Materials are described as a collection of positively charged ions and negatively charged electrons. Electrons are represented by a wavefunction and both the interaction between electrons, and between electrons and ions, is governed by Schrödinger's equation. For systems containing more than one electron (i.e., all systems except the hydrogen atom) the solution to Schrödinger's equation is not tractable. The effect on any one electron on all the other electrons therefore needs to be approximated. For the calculations described in this paper, the theoretical framework within which this approximation is formulated is called Density Functional Theory (DFT) (Hohenberg and Kohn, 1964), implemented in the code VASP (Kresse and Furthmüller, 1996). In this case, the system is described in terms of electron density, and electron-electron interactions are wrapped up into a term called the exchange correlation energy. The energetics of a system are calculated by numerically solving Schrödinger's equation for the system, and minimisation techniques are used to determine the ground-state (lowest) energy. The key feature of this type of calculation is that there is no material-dependent parameterisation: the only input values are the initial positions of the atoms, the atomic masses, the charge on the electron and the plane-wave representation for the core interactions.

Fig. 2 A schematic simulation box containing a collection of atoms which can be in any desired configuration; in the calculation, the atomic coordinates and cell dimensions are varied in order to produce the minimum energy (most stable) structure

Simulation of Pressure

In order to successfully model materials in the Earth's core, we need to simulate conditions of extreme pressure and temperature. Modelling pressure is fairly straightforward: a simulation box is set up containing a collection of atoms in the desired configuration (e.g. Fig. 2). Atomic coordinates and cell dimensions are varied in order to produce minimum energy E (Fig. 3), and a minimisation is considered successful when there are no residual forces on the atoms ($dE/dx = 0$ for all atoms, where x is a positional parameter) and the cell is under hydrostatic stress (i.e., the principal components of the stress tensor are equal). Pressure is derived from the forces between the atoms, which are obtained from the gradient of the potential function. These "static" minimisation calculations are performed at a temperature of zero Kelvin, i.e., there is no thermal energy in the system. Once the energy E has been calculated as a function of volume, E(V) can be fitted to an equation-of-state (EOS) which links energy and volume to pressure. The resulting variation of volume with pressure can then be compared with experimental measurements (Fig. 4).

Simulation of Temperature

Simulating temperature is computationally more demanding than modelling pressure, requiring more elaborate techniques. There are two approaches: (i) lattice dynamics, where the system is described in terms of atomic lattice vibrations (phonons) each with a certain wavelength and frequency, and (ii) molecular dynamics, where atoms are given initial positions and velocities which are allowed to evolve over a period of time *via* solutions to Newton's equations of motion.

The lattice dynamics method describes a system in terms of a simulation box containing vibrating particles whose frequencies vary with volume. The complex motions of the individual particles can be decomposed into a series of lattice vibrations or phonons, each with a frequency and a wavelength (Figs. 5 and 6). In ab initio lattice dynamics, the vibrational frequencies may be calculated by standard methods such as the small-displacement method. In this method, one set

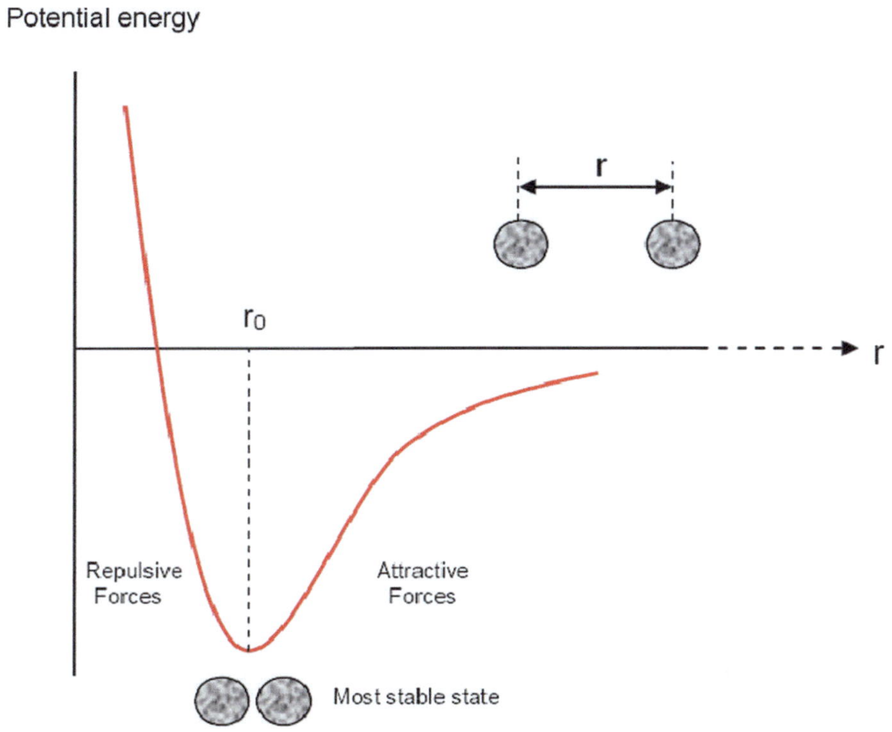

Fig. 3 Potential energy as a function of atomic separation, r; the equilibrium bond length, r_0, corresponds to an energy minimum

or multiple sets of fixed atoms are moved to positions slightly away from their equilibrium sites, causing a local strain field. From analysis of the forces between the atoms (similar to spring constants in an harmonic oscillator), the vibrational frequencies can be determined. Many thermodynamic properties, such as free energies, heat capacities, etc., may then be calculated at any desired temperature using standard statistical mechanical relations, which are direct functions only of these vibrational frequencies.

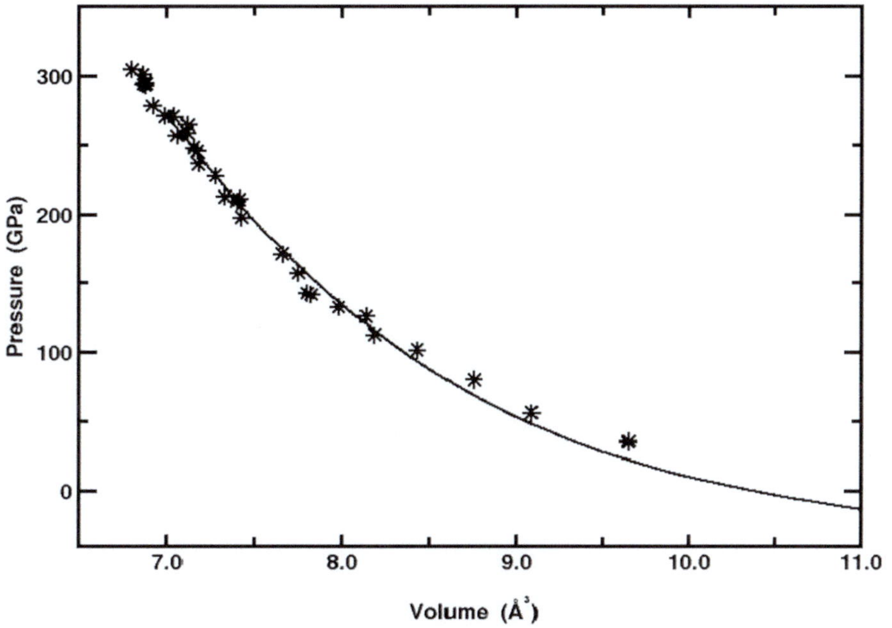

Fig. 4 Calculated pressure-volume curve at zero Kelvin for hcp iron (Vočadlo et al., 1997) compared with experiments (Mao et al., 1990)

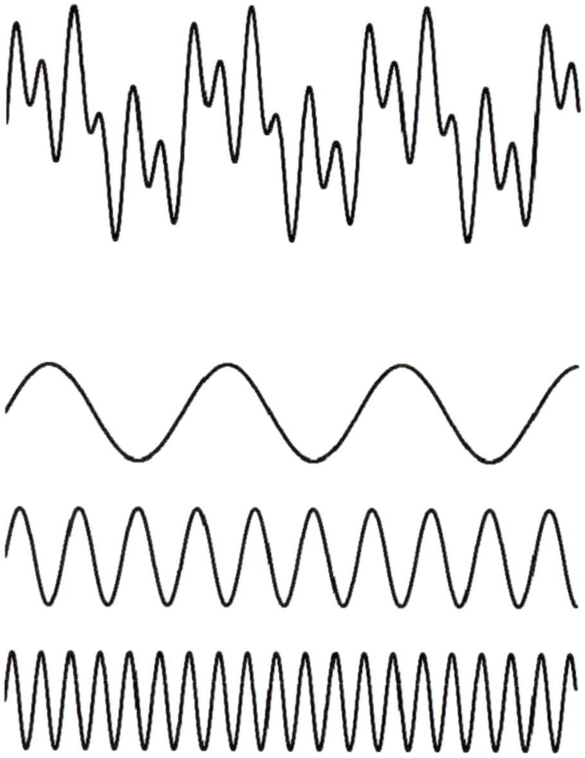

Fig. 5 Any complex motion, such as the trace at the *top*, can be decomposed into a set of simple sine waves (such as the three curves below), each with a frequency and wavelength

A limitation of the lattice dynamics method is that it treats the atoms as harmonic oscillators (i.e., the atoms vibrate symmetrically about their equilibrium position), whereas, in reality, at very high temperatures, the oscillations become anharmonic as their displacements from equilibrium become asymmetric (as seen in Fig. 3, where the slope in the potential energy when moving to longer bond lengths is different from the slope in the potential energy when moving to shorter bond lengths). This anharmonicity can accounted for when using the molecular dynamics method.

Molecular dynamics is routinely used for medium- to high-temperature simulations of minerals where atomic vibrations become more anharmonic as the atoms vibrate further away from their equilibrium positions. In first principles molecular dynamics, the ions are treated as classical particles and, for each set of evolving atomic positions, the electronic energy is solved quantum mechanically including the thermal excitations of the electrons. Newton's equations of motion are solved for all particles within a simulation box, to generate time-dependent trajectories and their associated positions and velocities (Fig. 7). The kinetic energy, which is directly related to temperature, is obtained from the velocities of the individual particles for each time step (e.g. Fig. 8). With this explicit particle motion, the anharmonicity is implicitly accounted for at high temperatures. Molecular dynamics has been used very successfully to calculate material properties at very high temperatures. Molecular dynamics is also an ideal tool for determining the properties of liquids, and is therefore the preferred method when simulating liquid iron alloys at the conditions of the outer core. Furthermore, by generating a simulation box containing both a solid and a liquid in coexistence (Fig. 9), or by calculating the relative free energies of the solid and liquid, melting temperatures can be determined as a function of pressure (Fig. 10).

Armed with these tools, computer simulations on deep Earth materials at any desired pressure and temperature can be achieved to predict material properties

Fig. 6 In lattice dynamics, the complex motion of atoms in a crystal above zero Kelvin (*left*) is represented as a series of vibrational waves or phonons (*right*) each with a frequency and wavelength

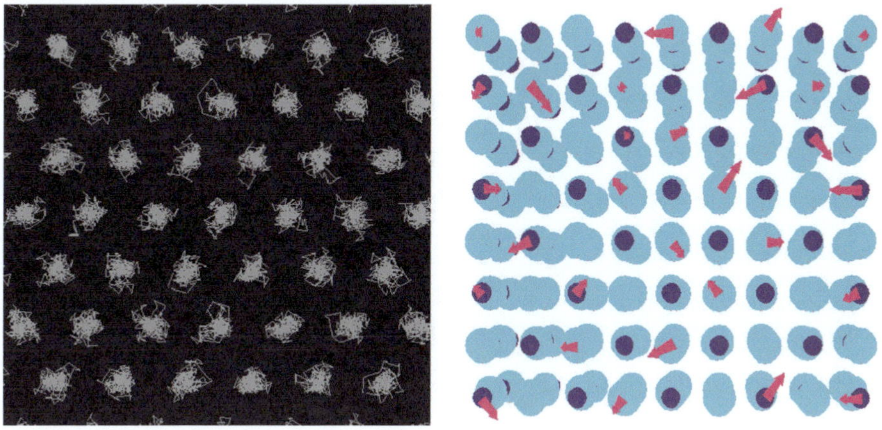

Fig. 7 In molecular dynamics, the complex motion of atoms in a crystal above zero Kelvin (*left*) is represented as a collection of moving atoms each with a trajectory that evolves over time

at core conditions. In the following section, examples are given of the successful application of these techniques to Earth's inner core materials.

The Ab Initio Simulation of Iron and Iron Alloys in the Earth's Inner Core

Constraints on the Structure of Iron in the Inner Core

Unfortunately for deep Earth scientists, the phase diagram of pure iron, particularly at core conditions, remains unknown (Fig. 11). At ambient conditions, iron crystallises in the body-centred-cubic (bcc) structure, with face-centred-cubic (fcc) iron stable at temperatures above 1183 K (at 1 atmosphere) and hexagonal-close-packed (hcp) iron stable above ~15 GPa (at room temperature). At these relatively low pressures and temperatures the phases are "well established", although even here there is doubt over the exact magnetic ordering of the hcp and fcc phases (Vočadlo et al., 2006; Wang and Wang, 2006). The iron phase diagram at very high pressures and temperatures, however is still hotly disputed. The three most likely candidates to exist in the inner core are the bcc, fcc and hcp phases (Vočadlo et al., 2003a, 2008; Nguyen

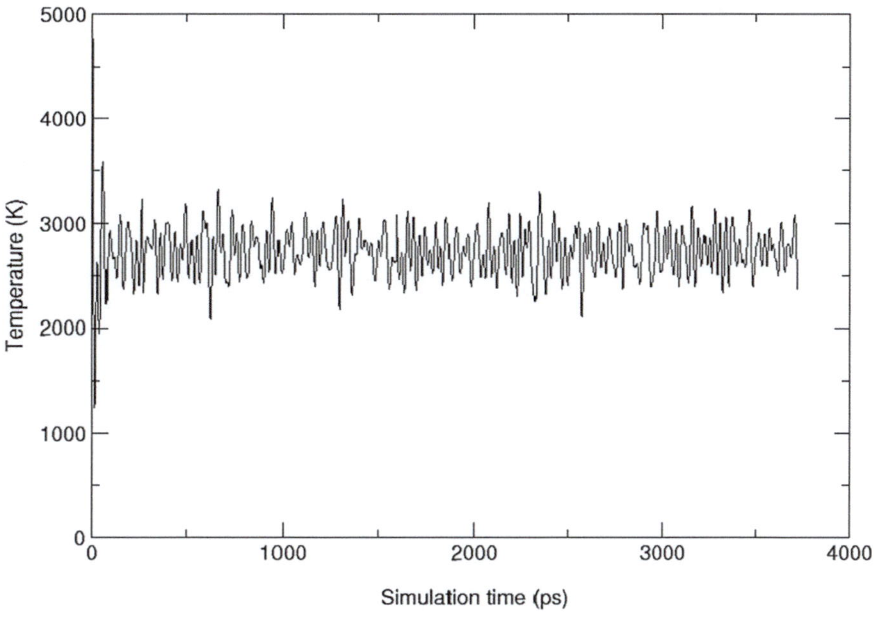

Fig. 8 Ab initio molecular dynamics simulations of iron showing the time evolution of the temperature of the system calculated from the particle velocities. A similar trace is seen for many time-evolving properties (stress, energy...)

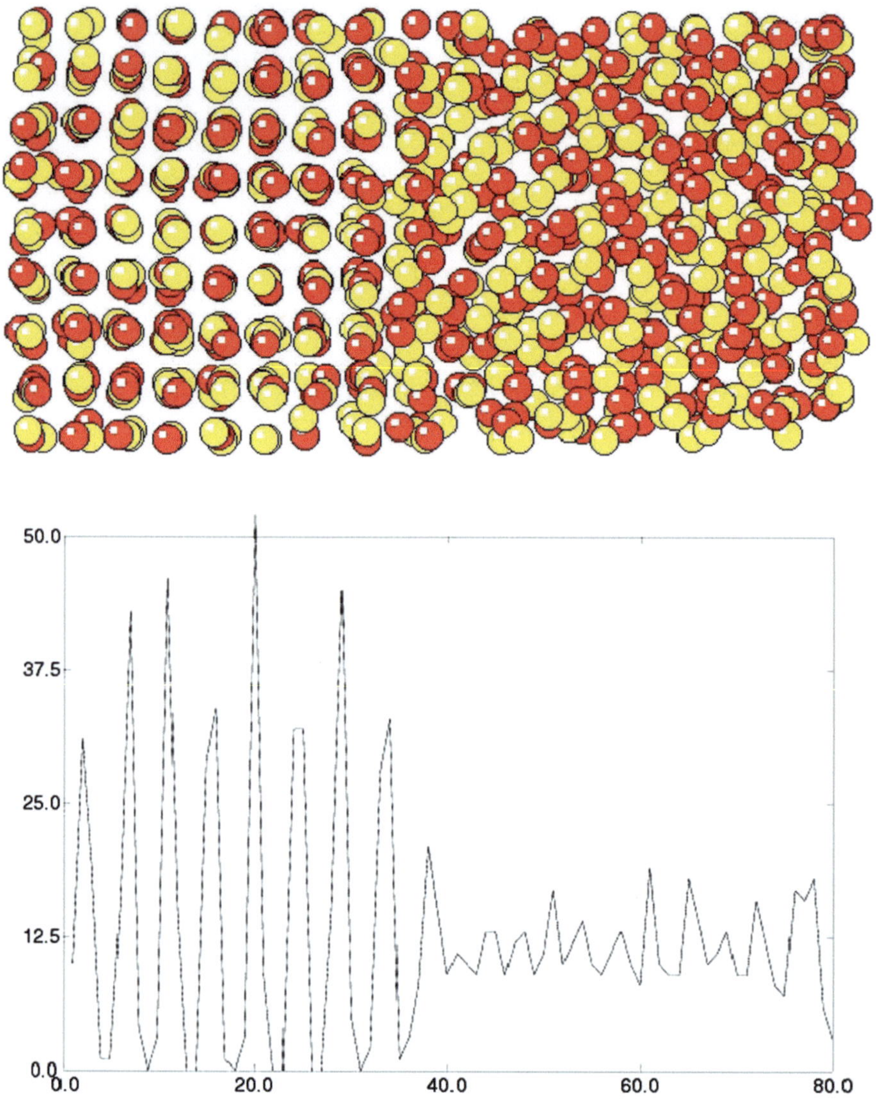

Fig. 9 Ab initio simulation of melting showing solid and liquid MgO in coexistence (*top*) together with the density profile (*bottom*) – the presence of the *solid* is identified by periodic oscillations of the density, while the density of the liquid phase has the form of random fluctuations with a much smaller amplitude (Alfè, 2005)

and Holmes, 2004; Ma et al., 2004; Mikhaylushkin et al., 2007) (Fig. 11), although at least three other more complex structures (dhcp or double-hexagonal close packed, distorted hcp, and bct which is a distorted bcc structure) have been suggested in the past (Vočadlo et al., 2000). As we will see in the next section, to complicate matters further, the presence of nickel and/or light elements in iron is known to affect the crystal structure that the iron alloy adopts in the core.

The observed complex structure of the inner core described in the Introduction begs the question: which mechanisms can lead to anisotropy and layering? The outer isotropic layer could be due either to the presence of randomly oriented crystals (containing, for example, bcc, fcc and/or hcp iron) or to the presence of a material with a low intrinsic anisotropy. The anisotropic lower layer in the inner core is thought to be due to the preferred orientation of aligned crystals, possibly, although not necessarily, of a different iron phase to the upper layer. The inner core could therefore, for example, be comprised entirely of bcc iron, entirely of hcp iron, entirely of fcc iron, or some combination of all three. There could also be a difference in chemical composition between the layers, or even different deformation histories and mechanisms. Clearly knowing the stable phase of iron at core conditions is essential if we are to interpret the seismological observations.

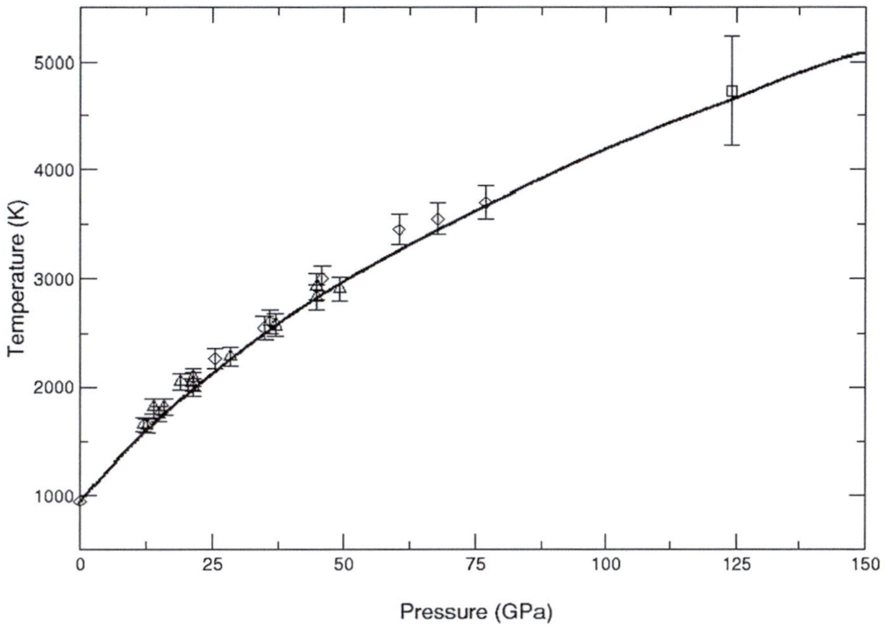

Fig. 10 Comparison of the calculated melting curve of aluminium with previous experimental results. *Solid curve*: Vočadlo and Alfè (2002). Diamonds and triangles: high-pressure measurements of Boehler and Ross (1997), Hanstrom and Lazor (2000) and Shaner et al. (1984)

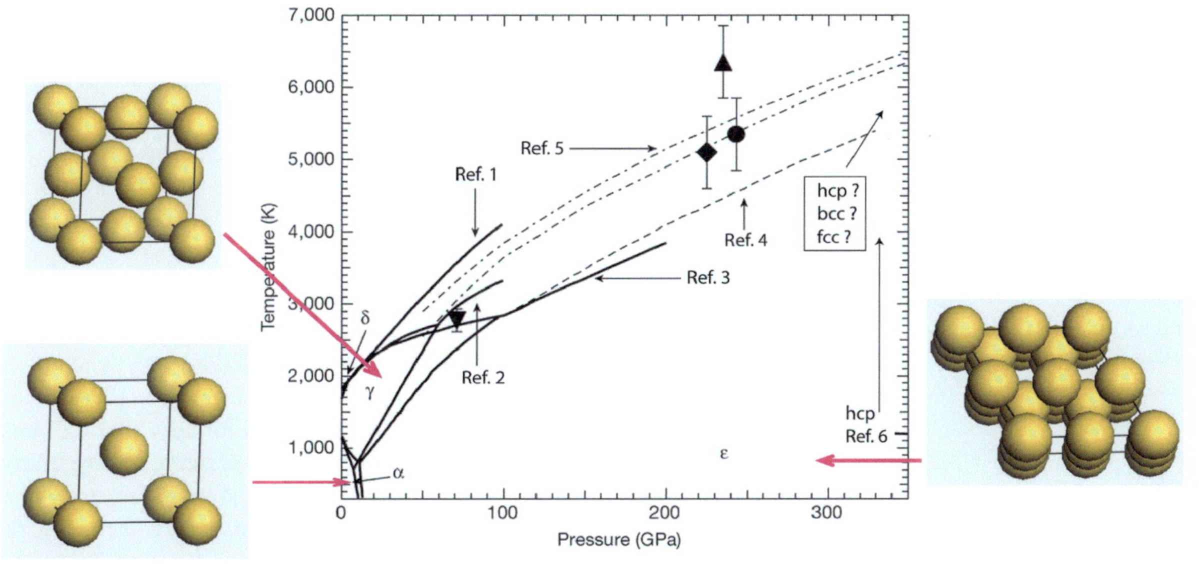

Fig. 11 Phase diagram of pure iron. The inner core boundary is at 330 GPa and ~5500 K. Solid lines represent phase boundaries and melt lines from DAC experiments; symbols with error bars are points on the melting curve from shock experiments; *broken lines* are melting curves from first-principles calculations. DAC data: Ref 1: Williams et al. (1987); Ref 2: Shen et al. (1998); Ref 3: Boehler (1993). Shock data: *triangle*: Yoo et al. (1993); *circles*: Brown and McQueen (1986); *reverse triangle*: Ahrens et al. (2002); *diamond*: Nguyen and Holmes (2004). First principles calculations; Ref 4: Laio et al. (2000); Ref 5: Alfè et al. (2002); Ref 6: Vočadlo et al. (2000, 2003a). Adapted from Nguyen and Holmes (2004)

In an attempt to resolve this issue, Vočadlo et al. (2000) calculated the free energies of the six suggested phases using lattice dynamics. They found that the bcc phase was vibrationally unstable at high pressures in agreement with previous computational work (Stixrude and Cohen, 1995; Söderlind et al., 1996). Only three phases remained vibrationally stable to core pressures – the hcp, fcc and dhcp phases. Free energies were calculated from the vibrational frequencies and showed that the hcp phase was the thermodynamically most stable phase at core conditions. However, it must be remembered that these lattice dynamics calculations are performed at zero Kelvin, with temperature incorporated explicitly via statistical thermodynamics relations. More recently, ab initio molecular dynamics calculations performed at the simultaneous high pressures and temperatures of the Earth's core (Vočadlo et al., 2003a) revealed that the bcc phase becomes mechanically stable at high temperature. Using thermodynamic integration methods, the free energies of the bcc, hcp and fcc phases were calculated (Vočadlo et al., 2003a, 2008); the results showed that, although the free energy of hcp phase was the lowest, the free energy differences are very small: at a density of 13,155 kg m^{-3} and a temperature of 5500 K, the Helmholtz free energy $F_{hcp} = -10.668$ eV, $F_{bcc} = -10.633$ eV and $F_{fcc} = -10.654$; thus, $\Delta F_{bcc-hcp} = 35$ meV and $\Delta F_{fcc-hcp} = 14$ meV (the uncertainty in the calculations is < 5 meV).

In summary, therefore, based on ab initio free energy calculations (Vočadlo et al., 2003a), the stable phase of pure iron at core conditions is hcp. However, the free energies of hcp-, bcc- and fcc-Fe are almost identical (within ~35 meV/atom), only marginally favouring hcp-Fe. As a result of these small free energy differences, the relative phase stability may be greatly affected by the presence of alloying elements. It should be noted that the calculation of the Helmholtz free energy, F, at constant volume, rather than Gibbs free energy, G, at constant pressure, is sufficient because the free energy difference, dF, at constant volume, is the same as the free energy difference, dG, at constant pressure, between different structures.

The Effect of Light Elements on the Stable Phase of Iron in the Earth's Inner Core

Both experimental and computational studies of the effect of S, Si, O and C on the high P/T phase diagram of iron have shown that the inclusion of light elements into solid iron generally enhances the stability of the bcc structure (Lin et al., 2002; Vočadlo et al., 1999; Côté et al., 2008a, b; Sata et al., 2008) (Figs. 12 and 13). This might be expected since the high-pressure polymorphs of several binary alloys of iron take the CsCl structure, which is topologically equivalent to bcc. Vočadlo et al. (1999) predicted a transition to the CsCl structure of FeSi at ~13 GPa,

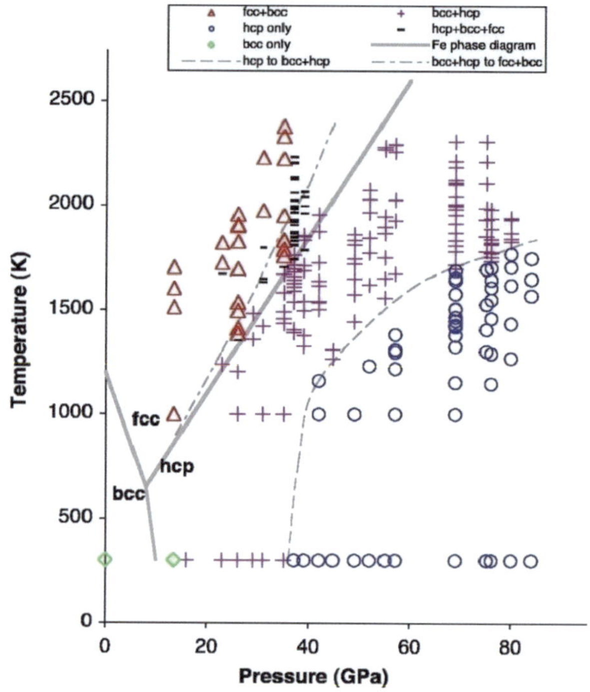

Fig. 12 The experimentally observed stability field of different crystal structures of iron containing 7.9 wt% of silicon (*symbols and grey dotted lines*) compared with the stability fields for pure iron structures (*grey lines*) The presence of silicon stabilises the bcc phase with respect to the hcp phase by up to ~40 GPa at high temperatures. Taken from Lin et al. (2002)

which was subsequently found experimentally (Dobson et al., 2002). Sata et al. (2008) found a new CsCl phase of FeS above 180 GPa using synchrotron X-ray diffraction techniques. Lin et al. (2002) used laser-heated DAC experiments to show that the presence of 7.9 wt% silicon stabilised the bcc phase of iron by up to 40 GPa (Fig. 12); this was subsequently confirmed by ab initio simulations at zero Kelvin (Côté et al., 2008a). Similar calculations at zero Kelvin (Côté et al., 2008b) also showed that small amounts (6.25 and 12.5 atom %) of C,S and O also stabilised the bcc structure over the hcp phase. Furthermore, even more recent lattice dynamics calculations (Côté et al., 2009) show that at the pressures *and temperatures* of the Earth's inner core, 7 wt% silicon stabilises the fcc phase over the hcp phase. Finally, we must not forget about nickel. The effect of nickel on phase stability is poorly understood. Recent experiments up to ~300 GPa and ~2000 K suggest that the presence of nickel significantly expands the stability field of the fcc structure (Kuwayama et al., 2008). Clearly light

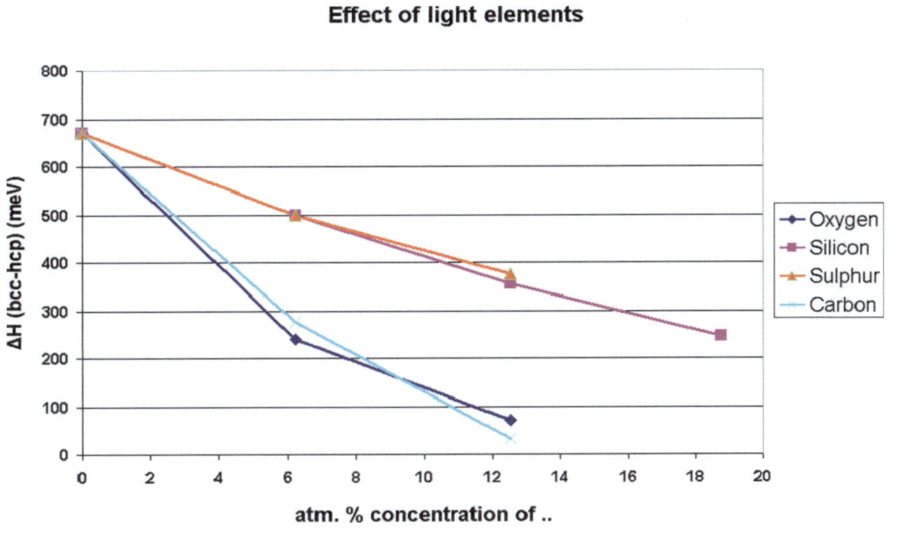

Fig. 13 The diagram shows the difference in enthalpy ΔH (free energy at zero Kelvin) between the bcc and hcp phases at 330 GPa from athermal ab initio calculations. The enthalpy difference drops significantly with the addition of lighter elements to a pure iron crystal, approaching zero at 12.5 atomic % of carbon or oxygen addition. Taken from Côté et al. (2008b)

elements (and nickel) have a significant, and relatively poorly quantified, effect on the stability of iron at core conditions.

Constraints on the Composition and Structure of the Earth's Inner Core from Calculated Seismic Wave Velocities

Theoretically determined seismic wave velocities have been obtained to try to further constrain the stable phase(s) of iron present in the Earth's inner core, and to further our understanding of the observed anisotropy and layering. Finite temperature ab initio molecular dynamics has been used to determine the elastic properties (and therefore seismic wave velocities) at core pressures and temperatures of three major candidates – hcp- bcc- and fcc-Fe – as well as iron alloyed with S, Si and Ni (Vočadlo, 2007; Vočadlo et al., 2008). Elastic constants were determined from the calculated time-averaged stresses associated with strains applied to thermally equilibrated structures. A deformation matrix was applied to the structures to enable distortions to be applied to each phase. Full details of these calculations may be found in Vočadlo (2007) and Vočadlo et al. (2008). The elastic constants are directly related to the shear modulus, G, and the isothermal incompressibility, K_T; the adiabatic incompressibility is then obtained from:

$$K_S = K_T(1 + \alpha\gamma T)$$

where the volumetric thermal expansion coefficient was taken to be $\alpha = 1\times10^{-5}\ \text{K}^{-1}$ and the Grüneisen parameter $\gamma = 1.5$ (Vočadlo et al., 2003b).

Standard relations may then be used to predict the P and S-wave velocities, V_P and V_S:

$$V_S = \sqrt{\frac{G}{\rho}} \quad V_P = \sqrt{\frac{K_S + \frac{4}{3}G}{\rho}}$$

To confirm the validity of this methodology, Fig. 14 shows the calculated P-wave velocities as a function

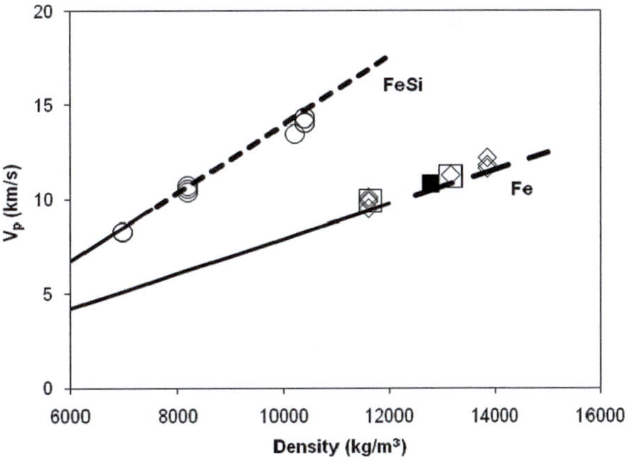

Fig. 14 Calculated P-wave velocity as a function of density (*symbols*) compared with the high-P ambient-T DAC experiments of Badro et al. (2007, *solid line*, extrapolations with *dashed lines*). Open circles: calculated FeSi V_P at different temperatures; open diamonds: calculated bcc-Fe V_P at different temperatures; open squares: calculated hcp-Fe at different temperatures; black filled square: shock datum of Brown and McQueen (1986). Taken from Vočadlo (2007)

of density for pure iron and FeSi together with the results from high-pressure, ambient temperature inelastic X-ray scattering experiments, extrapolated beyond the experimental range (Badro et al., 2007). The agreement is excellent, although it is interesting to note that, as a function of density, there appears to be little discrimination between the P-wave velocities of the hcp and bcc phases, and that the velocities are almost independent of temperature.

Table 1 gives an overview of the calculated values of V_P and V_S for pure iron and iron alloys, compared to values obtained by seismology. A surprising result of these calculations is that the calculated shear moduli, in all cases, are greater than those inferred from seismology by >10%. Table 1 also shows that the calculated values of V_S for pure iron phases are all >4.0 kms^{-1}, in agreement with inferences drawn from the extrapolation of lower pressure experimental data (Antonangeli et al., 2004) and also with the value of 4.04 inferred from shock experiments at a density of 12,770 kg m^{-3} (Brown and McQueen, 1986). The effect of light elements in general is to increase the shear wave velocities to over 5 km s^{-1}. Seismologially deduced values range between 3.5 km s^{-1} at the top of the inner core, and 3.67 km s^{-1} in the center of the Earth (Dziewonski and Anderson, 1981). If the uncertainties in the seismological values are well constrained, the difference between these observations and the results from mineral physics (both computational and experimental) suggest that a simple model for the inner core based on the commonly assumed phases is wrong.

It is important to assess whether this difference could be accounted for by anelasticity. The reduction in shear wave velocity due to shear wave attenuation is given by:

$$V(\omega,T) = V_0(T)\left(1 - \frac{1}{2}\cot\left(\frac{\pi\alpha}{2}\right)Q^{-1}(\omega,T)\right)$$

Where V(ω,T) and V_0(T) are the attenuated and unattenuated shear wave velocities respectively, Q is the quality factor and α is the frequency dependence of Q. For the inner core Q = 100 (Resovsky et al., 2005) and α of 0.2–0.4 (Jackson et al., 2000), which results in a decrease in the shear velocity of only 0.5–1.5%, nowhere near the >10% difference between the seismological observations and the calculated materials properties. It is important to note that the above analysis is necessarily approximate; anelasticity is a very complex issue that requires material data at the conditions of the Earth's inner core in order to draw irrefutable conclusions – clearly such data is unavailable at present.

Another possible explanation for this difference is that parts of the inner core are partially molten, with solute-rich liquid pockets trapped between solid grains. The amount of melt can be estimated by taking the Hashin–Shtrikman bound for the effective shear modulus of two-phase media (Hashin and Shtrikman, 1963):

$$\mu^{HS} = \mu_s + \frac{f_l}{(\mu_l - \mu_s)^{-1} + \dfrac{2f_s(K_s - 2\mu_s)}{5\mu_s\left(K_s + \frac{4}{3}\mu_s\right)}}$$

Where f is the fraction of solid or liquid and subscripts s and l denote solid and liquid respectively ($\mu_l = 0$). Using this equation, the minimum amount of melt in the inner core is estimated to be approximately 8% for the bcc and hcp phases, and as much as 25% for the fcc phase.

In order to be a viable candidate for the inner core phase, the iron phases must also be consistent with the observed P-wave anisotropy of ~3%. However, present theoretical and experimental estimates of this anisotropy in hcp-Fe are poorly constrained. Figure 15 shows the P-wave velocities as a function of propagation direction, calculated from the elastic constants (Table 1) of fcc-, bcc- and hcp-Fe. The equivalent plots for the fcc- and bcc-derived light element alloys are very similar to their pure iron counterparts, and are therefore not shown. For fcc-, bcc- and hcp-Fe the maximum P-wave anisotropy is 13.5, 3.3 and 6% respectively. However, this degree of anisotropy cannot be achieved through two mutually perpendicular directions. If it is assumed that the P-wave anisotropy in the Earth's inner core is due to crystal alignment then, by inspection of Fig. 15, we can examine the degree of anisotropy between polar and equatorial directions for different crystal orientations. Figure 15a indicates that, for fcc-Fe, the anisotropy will be < 1% for crystals aligned along < 110>, ~3% for crystals aligned along < 111> and as much as 6% for crystals aligned along < 100>; for < 110> and < 100> the anisotropy is such that the velocity along the alignment axis is lower than the average velocity in the plane perpendicular to it. Alignment

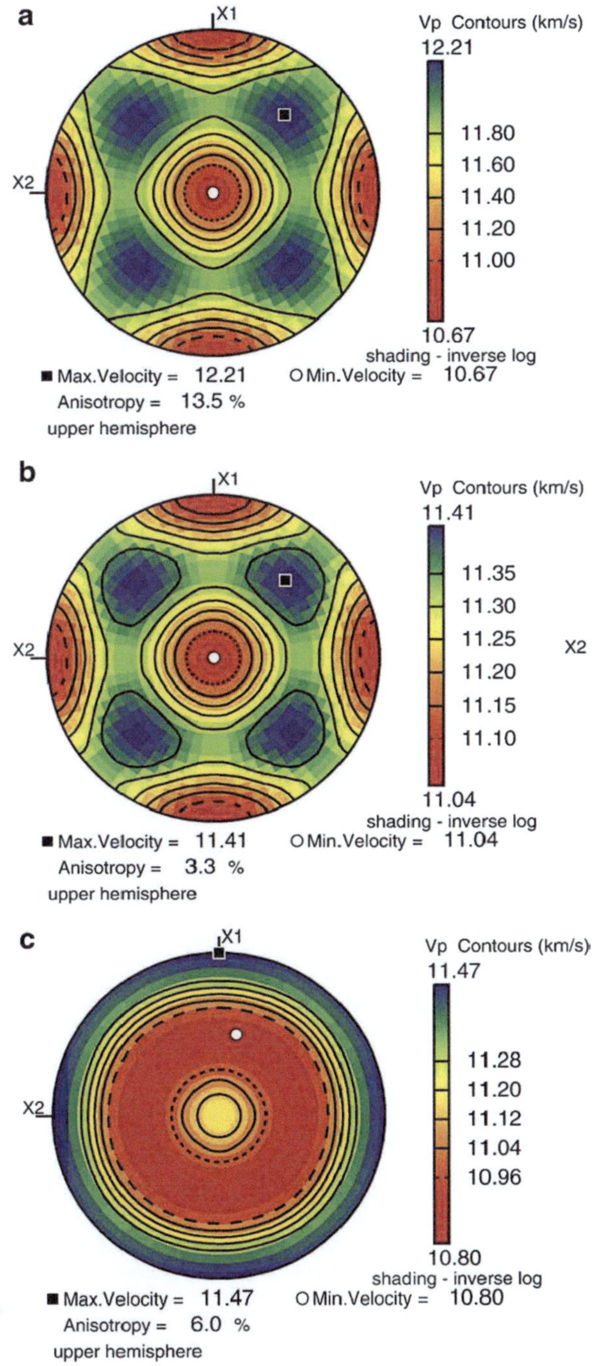

Fig. 15 Calculated single-crystal P-wave velocities for (**a**) fcc-Fe, (**b**) bcc-Fe and (**c**) hcp-Fe at inner core conditions as a function of propagation direction. Figure generated by Unicef Careware (Mainprice, 1990). After Vočadlo et al. (2008)

of fcc-Fe with <111> parallel to the polar axis can, in principle, give values comparable with the observed anisotropy, but would require full crystal alignment in the inner core; alternatively, the observations could be explained by crystals oriented with <100> in the equatorial plane, this arrangement requiring a lesser degree of alignment. For bcc-Fe (Fig. 15b) the P-wave anisotropy is similar in form to that for fcc but smaller in magnitude; the maximum anisotropy that can be achieved through mutually perpendicular directions is never more than ~2%. For the hcp-Fe phase (Fig. 15c) a ~3% anisotropy can be achieved through mutually perpendicular directions whereby, for the polar P-wave velocity to be faster, the c-axis of the crystals would lie in the Earth's equatorial plane. Thus, for the bcc phase of iron, the observed seismic anisotropy in the Earth's inner core cannot be achieved by alignment alone, while for the hcp phase it is just about possible to achieve the required degree of anisotropy by alignment but, as for the <111> orientation of fcc-Fe, the degree of alignment required would need to be very high.

Summary and Conclusions

Recent advances in ab inito simulation methods have enabled the calculation of high temperature properties of iron and iron alloys such as melting curves, high-PT equations of state, and elastic properties. Results from such simulations give excellent agreement with experiment where data exist, giving us confidence that these methods can be applied to the Earth's core. Using ab initio methods we have been able to show that the thermodynamically most stable phase of pure iron at core conditions is the hcp phase; however, the free energy differences between hcp-Fe and either bcc-Fe or fcc-Fe are very small (35 meV for bcc-Fe and 14 meV for fcc-Fe) – such small energy differences could easily be overcome by the presence of light elements (and also nickel). At the present time, however, it appears that no mineralogical model exists for which the predicted and observed seismic wave velocities and seismic wave anisotropy agree to within error. The high shear wave velocities predicted for all phases studied (including their light element alloy) are incompatible with seismology unless a significant amount of melt in the inner core is invoked (%: $>8_{bcc}$ $>12_{hcp}$ $>25_{fcc}$). The observed anisotropy can be accounted for by both fcc- and hcp-Fe only if there is a high degree of crystalline

Table 1 Shear moduli and sound velocities of fcc-Fe (Vočadlo et al., 2008), hcp-Fe and bcc-Fe (Vočadlo, 2007), together with those of Fe_3S, Fe_3Ni (Vočadlo et al., 2008), FeSi and FeS (Vočadlo, 2007); also shown are values taken from PREM at the core–mantle boundary and at the centre of the inner core (Dziewonski and Anderson,1981)

	ρ (kgm^{-3})	T (K)	G (GPa)	V_P (kms^{-1})	V_S (kms^{-1})
hcp	11628.1	4000	207	9.91	4.15
	13155	5500	209	11.14	4.01
bcc	11592.91	750	250	10.11	4.64
	11592.91	1500	228	9.98	4.44
	11592.91	2250	204	9.88	4.20
	13155	5500	223	11.29	4.11
	13842	2000	361	12.22	5.1
	13842	6000	249	11.83	4.24
fcc	13155	5500	285	11.64	4.64
FeSi	8199.34	1000	263	10.74	5.66
	8199.34	2000	244	10.58	5.45
	8199.34	3500	214	10.42	5.11
	10211.74	5500	401	13.53	6.26
	10402.15	1000	579	14.34	7.46
	10402.15	2000	526	14.08	7.11
	10402.15	3500	519	14.36	7.06
	10402.15	5000	468	14.12	6.71
FeS	10353	5500	203	12.02	4.43
	10894.13	1000	368	13.38	5.81
	10894.13	2000	356	13.43	5.72
	10894.13	3500	332	13.34	5.52
	10894.13	5000	311	13.41	5.34
fcc-Fe_3Ni	12097	5500	268	11.28	4.52
fcc-Fe_3S	13155	5500	250	11.97	4.55
PREM	12760		157	11.02	3.5
	13090		176	11.26	3.67

alignment; the anisotropy cannot be fully explained by alignment of bcc crystals.

Acknowledgements LV would like to that Alex Côté and Dario Alfè for Figs. 2 and 11 respectively; LV would also like to thank Ian Wood for helpful comments on the manuscript. Much of this research was funded by the Royal Society through their University Research Fellowship scheme.

References

Ahrens TJ, Holland KG, Chen GQ 2002 Phase diagram of iron, revised core temperatures. *Geophys. Res. Lett.*, **29**, #1150.

Alfè D 2005 Melting curve of MgO from first principles simulations. *Phys. Rev. Lett.*, **94**, #235701.

Alfè D, Gillan MJ, Vočadlo L, Brodholt JP, Price GD 2002 The ab initio simulation of the Earth's core. *Phil. Trans. Royal Soc. A*, **360**, 1227–1244.

Antonangeli D, Occelli F, Requardt H 2004 Elastic anisotropy in textured hcp-iron to 112 GPa from sound wave propagation measurements. *Earth Planet. Sci. Lett.*, **225**, 243–251.

Badro J, Fiquet G, Guyot F, Gregoryanz E, Occelli F, Antonangeli D, Requardt E, Mermet A, D'Astuto M, Krisch M 2007 Effect of light elements on the sounds velocities in solid iron: Implications for the composition of the Earth's core. *Earth Planet. Sci. Lett.*, **254**, 233.

Beghein C, Trampert J 2003 Robust normal mode constraints on inner core anisotropy from model space search. *Science*, **299**, 552–555.

Birch F 1964 Density and composition of the mantle and core. *J. Geophys. Res.*, **69**, 4377–4388.

Boehler R 1993 Temperatures in the Earth's core from melting point measurements of iron at high static pressures. *Nature*, **363**, 534–536.

Boehler R, Ross M 1997 Melting curve of aluminium in a diamond anvil cell to 0.8 Mbar: Implications for iron. *Earth Planet. Sci. Lett.*, **153**, 223–227.

Brockhouse BN, Abou-Helal HE, Hallman ED 1967 *Solid Stare Commun.* **5**, 211.

Brodholt J, Vočadlo L 2006 Applications of density functrional theory in the geosciences. *MRS Bull.*, **31**, 675–680.

Brown JM, McQueen RG 1986 Phase transitions, Grüneisen parameter and elasticity of shocked iron between 77 GPa abd 400 GPa. *J. Geophys. Res.*, **91**, 7485–7494.

Cohen RE, Stixrude L, Wasserman E 1997 Tight-binding computations of elastic anisotropy of Fe, Xe, and Si under

compression. *Phys. Rev. B* 56, 8575–8589. (errata, Phys.Rev. B 58 (1997) 5873).
Collerson KD, Hpaugoda S, Kamber BS, Williams Q 2000 Rocks from the mantle transition zone: Majorite-bearing xenoliths from Malaita, southwest Pacific. *Science*, **288**, 1215–1223.
Côté AS, Vočadlo L, Brodholt J 2008a The effect of silicon impurities on the phase diagram of iron and possible implications for the Earth's core structure. *J. Phys. Chem. Solids*, **69**, 9, 2177–2181.
Côté AS, Vočadlo L, Brodholt J 2008b Light elements in the core: Effects of impurities on the phase diagram of iron. *Geophys. Res. Lett.*, **35**, L05306.
Côté AS, Vočadlo L, Alfe D, Brodholt J 2009 Ab initio calculations on the combined effect of temperature and silicon on the stability of different iron phases in the Earth's inner core. *Submitted.*
Creager KC 1992 Anisotropy of the inner core from differential travel times of the phases PKP and PKIKP. *Nature*, **356**, 309–314.
Dobson DP, Vočadlo L, Wood IG 2002 A new high pressure phase of FeSi. *Am. Min.*, **87**, 784–787.
Dziewonski AM, Anderson DL 1981 Preliminary reference Earth model. *Phys. Earth Planet. Int.*, **25**, 297–356.
Hanstrom A, Lazor P 2000 High pressure melting and equation of state of aluminium. *J. Alloys Comp.*, **305**, 209–215.
Hashin Z, Shtrikman S 1963 A variational approach to the theory of the elastic behaviour of multiphase materials. *J. Mech. Phys. Solids*, **11**, 127.
Hohenberg P, Kohn W 1964 Inhomogeneous electron gas. *Phys. Rev.*, **136**, B864–B871.
Ishii M, Dziewonski AM 2003 Distinct seismic anisotropy at the centre of the Earth. *Phys. Earth Planet. Int.*, **140**, 203–217.
Jackson I, Fitz Gerald JD, Kokkonen H 2000 High temperature viscoelastic relaxation in iron and its implications for the shear modulus and attenuation of the Earth's inner core. *J. Geophys. Res.* **B10**, 23, 605–23634.
Kresse G and Furthmüller J 1996 Efficient iterative schemes for *ab initio* total-energy calculations using a plane-wave basis set. *Phys. Rev. B* **54**, 11169–11186.
Kuwayama Y, Hirose K, Nagayoshi S, Ohishi Y 2008 Phase relations of iron and iron-nickel alloys up to 300 GPa: Implications for the composition and structure of the inner core. *Earth Planet. Sci. Lett.*, **273**, 379–385.
Laio A, Bernard S, Chiarotti GL, Scandolo S, Tosatti E 2000 Physics of iron at Earth's core conditions. *Science*, **287**, 1027–1030.
Lin JF, Heinz DL, Campbell AJ, Devine JM, Shen GY 2002 Iron-silicon alloy in the Earth's core?. *Science*, **295**, 313–315.
Liu L 1979 The Earth: Its origin, structure and evolution. Ed. MW McElhinny, 177–202, Academic, New York.
Ma Y, Somayazulu M, Shen G, Mao HK, Shu J, Hemley RJ 2004 In situ X-ray diffraction studies of iron to Earth-core conditions. *Phys. Earth Planet Int.*, **143–144**, 455–467.
Mao HK, Wu Y, Chen LC, Shu JF, Jephcoat AP 1990 Static compression of iron to 300 GPa and Fe0.8Ni0.2 alloy to 260 GPa – implications for composition of the core. *J. Geophys. Res.*, 95, 21737–21742.
Mainprice D 1990 A Fortran program to calculate seismic anisotropy from the lattice preferred orientation of minerals. *Comput. Gesosci.* **16**, 385. (ftp://www.gm.univ-montp2.fr/mainprice//CareWare_Unicef_Programs/).
Masters G, Gubbins D 2003 On the resolution of density within the Earth. *Phys. Earth Planet. Int.*, **140**, 159–167.
McDonough WF, Sun S-S 1995 The composition of the Earth. *Chem. Geol.*, **120**, 223–253.
Mikhaylushkin AS, Simak SI, Dubrovinsky L, Dubrovinskaia N, Johansson B, Abrikosov IA 2007 Pure iron compressed and heated to extreme conditions. *Phys. Rev. Lett.*, **99**, 165505.
Nguyen JH, Holmes NC 2004 Melting of iron at the physical conditions of the Earth's core. *Nature*, **427**, 339–342.
Oganov AR, Brodholt JP, Price GD 2001 The elastic constants of MgSiO3 perovskite at pressures and temperatures of the Earth's mantle. *Nature*, **411**, 934–937.
Oreshin SI, Vinnik LP 2004 Heterogeneity and anisotropy of seismic attenuation in the inner core. *Geophys. Res. Lett.*, **31**, #L02613.
Ouzounis A, Creager KC 2001 Isotropy overlying anisotropy at the top of the inner core. *Geophys. Res. Lett.*, **28**, 4221–4334.
Poirier JP 1994 Light elements in the Earth's outer core: A critical review. *Phys. Earth Planet. Int.*, **85**, 319–337.
Rama Murthy V, van Westrenen W, Fei Y 2003 Experimental evidence that potassium is a substantial radioactive heat source in planetary cores. *Nature*, **423**, 163–165.
Resovsky J, Trampert J, Van der Hilst RD 2005 Error bars for the global seismic Q profile. *Earth Planet. Sci. Lett.*, **230**, 413–423.
Ringwood AE 1975 Composition and petrology of the Earth's mantle. McGraw-Hill, New York.
Sata N, Ohfuji H, Kobayashi H, Ohishi Y, Hirao N 2008 New high-pressure B2 phase of FeS above 180 GPa. *Am. Min.*, **93**, 492–494.
Shaner JW, Brown JM, McQueen RG 1984 in *High Pressure in Science and Technology*, edited by C Homan, RK Mac-Crone, and E Whalley, North Holland, Amsterdam, p. 137.
Shearer P, Masters G 1990 The density and shear velocity contrast at the inner core boundary. *Geophys. J. Int.*, **102**, 491–498.
Shen GY, Mao HK, Hemley RJ, Duffy TS, Rivers ML 1998 Melting and crystal structure of iron at high pressures and temperatures. *Geophys. Res. Lett.*, **25**, 373–376.
Söderlind P, Moriarty JA, Wills JM 1996 First principles theory of iron up to earth's core pressures: structural, vibrational and elastic properties. *Phys. Rev. B*, **53**, 14063–14072.
Song X 1997 Anisotropy of the Earth's inner core. *Rev. Geophys.*, **35**, 297–313.
Song XD, Helmberger DV 1993 Anisotropy of Earth's inner core. *Geophys. Res. Lett.*, **20**, 2591–2594.
Song XD, Helmberger DV 1998 Seismic evidence for an inner core transition zone. *Science*, **282**, 924–927.
Song XD, Xu XX 2002 Inner core transition zone and anomalous PKP(DF) waveforms from polar paths. *Geophys. Res. Lett.*, **29**, #1042.
Stixrude L, Cohen RE 1995 Constraints on the crystalline structure of the inner core – mechanical stability of bcc iron at high pressure. *Geophys. Res. Lett.*, **22**, 125–128.

Sun XL, Song XD 2002 PKP travel times at near antipodal distances: implications for inner core anisotropy and lowermost mantle structure. *Earth Planet Sci. Lett.*, **199**, 429–445.

Sun X, Song X 2008 The inner core of the Earth: texturing of iron crystals from three-dimensional seismic anisotropy. *Earth Planet. Sci. Lett.*, **269**, 56–65.

Vočadlo L 2007 Ab initio calculations of the elasticity of iron and iron alloys at inner core conditions: Evidence for a partially molten inner core. *Earth Planet. Sci. Lett.*, **254**, 227.

Vočadlo L, Dobson DP, Wood IG 2006 Ab initio study of nickel substitution into iron. *Earth Plant. Sci. Lett.*, **248**, 132–137.

Vočadlo L, Alfe D 2002 Ab initio melting curve of the fcc phase of aluminium. *Phys. Rev. B* **65**, 214105, 1–12.

Vočadlo L, Alfe D, Gillan MJ, Wood IG, Brodholt JP, Price GD 2003a Possible thermal and chemical stabilisation of body-centred-cubic iron in the Earth's core. *Nature*, **424**, 536–539.

Vočadlo L, Alfe D, Gillan MJ, Price GD 2003b The properties of iron under core conditions from first principles calculations. *Phys. Earth Planet. Int.*, **140**, 101–125.

Vočadlo L, Brodholt JP, Alfe D, Gillan MJ, Price GD 2000 Ab initio free energy calculations on the polymorphs of iron at core conditions. *Phys. Earth Planet. Int.*, **117**, 123–137.

Vočadlo L, Wood IG, Alfè D, Price GD 2008 Ab initio calculations on the free energy and high P-T elasticity of face-centred-cubic iron. *Earth Planet. Sci. Lett.*, **268**, 444–449.

Vočadlo L, Wood IG, Price GD 1999 Crystal structure, compressibility and possible phase transitions in epsilon-FeSi studied by first-principles pseudopotential calculations. *Acta Cryst. B*, **55**, 484–493.

Vočadlo L, de Wijs G, Kresse G, Gillan MJ, Price GD 1997 First principles calculations on crystalline and liquid iron at earth's core conditions. *Faraday Discuss* **106**, 'Solid-State Chemistry – New Opportunities from Computer Simulations', 205–217.

Wang J-T, Wang D-S 2006 Finite temperature magnetism of tetragonal iron. *Appl. Phys. Lett.*, **88**, 132513.

Williams Q, Jeanloz R, Bass J, Svendsen B, Ahrens TJ 1987 The melting curve of iron to 250 Gigapascals: a constraint on the temperature at the Earth's centre. *Science*, **236**, 181–182.

Yoo CS, Holmes NC, Ross M, Webb DJ, Pike C 1993 Shock temperature and melting of iron at Earth's core conditions. *Phys. Rev. Lett.*, **70**, 3931–3934.

Index

Note: The letters 'f' and 't' following the locators refers to figures and tables respectively.